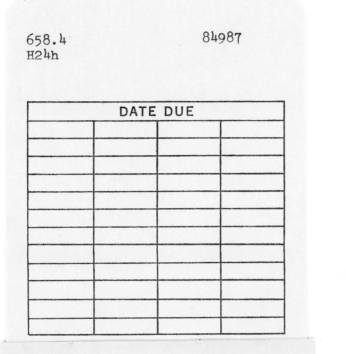

DATE DUE			

Handbook
of Industrial
Control Computers

HANDBOOK
OF INDUSTRIAL
CONTROL COMPUTERS

EDITED BY

THOMAS J. HARRISON

SENIOR ENGINEER
INTERNATIONAL BUSINESS MACHINES CORPORATION
BOCA RATON, FLORIDA

WILEY-INTERSCIENCE, a Division of John Wiley & Sons, Inc.
New York · London · Sydney · Toronto

658.4
H24h
84987
Sept 1973

Library of Congress Cataloging in Publication Data:

Harrison, Thomas J.
  Handbook of industrial control computers.

  Includes bibliographies.
  1. Electronic data processing—Process control.
I. Title.

TS156.8.H37   658.4   72-000068
ISBN 0-471-35560-7

Printed in the United States of America.

10 9 8 7 6 5 4 3 2 1

To Marilyn, Nancy, and Kristine

CONTRIBUTORS

ALEX. J. ARTHUR, Staff Programmer, Data Acquisition and Control Systems, IBM Corporation, Boca Raton, Florida.

J. PAUL HAMMER, Advisory Planner, Data Acquisition and Control Systems, IBM Corporation, Boca Raton, Florida.

THOMAS J. HARRISON, Senior Engineer, Data Acquisition and Control Systems, IBM Corporation, Boca Raton, Florida.

LYNN S. HOLMES, JR., Senior Field Systems Center Representative, IBM Corporation, Boca Raton, Florida.

FRED B. JONES, Senior Engineer, Storage Products Development, IBM Corporation, San Jose, California.

CHUNG C. LIU, Advisory Engineer, Machine Technology, IBM Corporation, San Jose, California.

RICHARD E. MACH, Development Engineer, Product Technology, IBM Corporation, San Jose, California.

GORDON W. MARKHAM, Program Manager, Sensor-Based Development, IBM Corporation, Palo Alto, California.

ROBERT C. MILLER, JR., Senior Planner, Advanced Systems Requirements, IBM Corporation, San Jose, California.

RICHARD N. POND, Development Engineer, Data Acquisition and Control Systems, IBM Corporation, Boca Raton, Florida.

SHANE L. SHANE, Installation Planning Engineer, SDD Laboratory, IBM Corporation, San Jose, California.

ROGER M. SIMONS, Senior Programmer, Engineering and Scientific Computation Laboratory, IBM Corporation, San Jose, California.

I. C. (JIM) TRAUTH, Senior Systems Engineer, TRW Controls (formerly Advisory Marketing Representative, Data Acquisition and Control Systems Center, IBM Corporation), Houston, Texas.

DONALD C. UNION, Software Systems Branch Manager, Gulf Electronic Systems, La Jolla, California (formerly Senior Industry Development Analyst, IBM Corporation, Houston, Texas).

H. WAYNE VAN NESS, Advisory Product Administrator, Sensor Based Marketing, IBM Corporation, White Plains, New York.

PREFACE

I do not remember when the idea of producing this book first came to me. My notebooks show an entry in May 1965, which was apparently the first time I recorded the thought. That entry was the result of some frustration in finding a single source of material which covered the practical aspects of analog circuit design for industrial control computers. The material existed, certainly, but it was scattered among hundreds of articles in dozens of journals and trade magazines. But the entry was only the start, and the original outlines that followed were concerned only with analog design, essentially the material in Chapters 3 through 5 of this book. As the outlines evolved over the years and as the idea was discussed with colleagues, it became obvious that the problem was more general; there was no single comprehensive reference on the practical aspects of the design and installation of industrial control computers. Thus the many outlines eventually evolved into the structure and scope which guided the production of this book.

During the outlining and writing of the book, the available literature was surveyed in detail. There are many sources of information concerning control theory and its application to industrial control. Industrial processes have been studied in terms of theoretical models, advanced techniques have been translated into practical control algorithms, and methods for characterizing processes have been developed and reported. The unique hardware and software problems related to the implementation of these results for processes controlled by a digital computer have not been considered in an organized manner, however. Some results, such as the common mode analyses of Chapters 3 and 4, were not readily available; neither were many of the practical installation and programming considerations. For this reason, the scope of the book was restricted to emphasize the computer aspects of control, rather than the control theory itself. This is why the title of the book contains the words "control computers" rather than "computer control." This is, perhaps, a subtle distinction but nevertheless a necessary one. The science and art of the application of digital computers to control and data acquisition are developing so rapidly that a single book cannot provide comprehensive coverage of all aspects of the subject. It is, in fact, difficult for the active workers in this field even to keep up with the new applications constantly being reported.

The material in the book is presented in terms of computers for use in continuous and batch industrial processes. This seemingly excludes other applications such as traffic control, manufacturing control, laboratory automation, and certain high-speed applications. This is not really true, however, since the computer hardware and software needed in these applications is essentially the same as that used in the process control applications. The system configurations are similar, if not identical, and the manner in which the system is used is very similar. Thus the material presented in this book should be useful in applications other than those which are included in the generally accepted definition of computer process control.

Every author, whether he is writing a novel or a highly technical treatise, visualizes the members of his audience. In the case of this book and this author, the readers were seen to be members of two groups. One group includes the members of the computer vendor's organization—the engineers who design the system hardware, the programmers who provide the programming systems, and the salesmen and technical consultants who sell the product. The other group encompasses the user personnel— the control engineer who is responsible for the installation of the control computer on an industrial process, the programmer who writes the application programs, and the members of management who make the basic go/no-go decisions concerning the use of computer control on a certain process. The backgrounds of the members of this readership are varied. The process engineer is probably not an electrical engineer and his experience with computers may be limited; the system design engineer and system programmer, on the other hand, probably have little knowledge of the practical control of an industrial process. My solution was to organize the book into parts which are aimed at each group of readers. In addition, the editing of the book has attempted to make the initial portion of each chapter somewhat tutorial so that a specialist in one aspect of computer control can use the material effectively as a starting point for understanding other aspects of industrial control computers.

The book consists of four main parts: the first chapter is a general introduction to industrial control and the control computer. This includes a definition of control, the evolution of control and control computers, the capabilities of the modern control computer, and a description of typical computer equipment. The material in this chapter is intended for all readers and forms a basis for an understanding of the remainder of the book.

The second part of the book, Chapters 2 through 9, deals with subjects of specific interest to computer system design engineers and programmers. The structure of a control computer is discussed in detail from an engineering point of view. This includes detailed descriptions of the hardware utilized to provide each of the functions required in the computer system, the advantages and/or disadvantages of alternate approaches, trade-offs between hardware and software, and details of the programming system design. This part of the book is written on a relatively high technical level and is intended as a useful reference for the system design engineer and programmer who are concerned with the problems of designing the system. It is also useful, however, to the technically inclined manager or process engineer in that it provides him with an appreciation of the hardware and software design considerations. This enables him better to match the capabilities of the computer to his control needs.

The next part of the book deals with those subjects which are normally of major concern to the user of the control computer. These chapters, Chapters 10 through 14, cover such topics as justification of a control computer in terms of direct and indirect benefits and costs, factors in the selection of a suitable computer, planning for the installation and use of the system, and the continuing use and maintenance of the control computer. Although this part of the book is intended primarily to provide the user with an understanding of his responsibilities with respect to the selection, installation, and use of a computer control system, it is also useful to the system design engineer and programmer in providing an understanding of the many factors which enter into the effective utilization of a control computer by the user.

The process or system engineer in the typical control computer situation is rarely an expert in all phases of computer technology. In fact, it is likely that his basic discipline is not computer engineering but is chemical, mechanical, metallurgical, or one of the many other engineering or technical disciplines. As such, it is likely that he requires a ready source of basic information concerning computer technology and use. Appendixes A and B are intended as a primer on computer hardware and software to provide such basic information. Appendix C provides a comprehensive bibliography on the use of computers in industry and research.

The production of the manuscripts for this book is itself an interesting application of computers. The entire book, more than 1500 manuscript pages, was produced with the aid of a document handling system called the Administrative Terminal System (ATS). The editor and several of the contributors had access to time-shared computer systems consisting of an IBM 360/40 computer located in the IBM Development Laboratory in San Jose, California, and an IBM 360/65 in the IBM Laboratory in Boca Raton, Florida. A number of IBM 2741 Communications Terminals are located in our office areas in the laboratories. The IBM 2741 is a remote terminal which is mechanically similar to an IBM Selectric* typewriter and is connected to the computer through telephone lines. Manuscripts are entered by typing the material in a free format. Once stored in the computer, corrections of single letters, words, phrases, or sentences are made using simple commands entered through the typewriter keyboard. For example, if a sentence contains the misspelled word "cmoputer," the correction is made by typing the command "123;cmoputer;computer." The number 123 is the reference number of the sentence assigned by the computer and the words following it are the original word and the word to be substituted for it. Upon receipt of the command, the computer corrects the document and types the corrected lines on the IBM 2741 terminal to provide verification to the operator. This is only one example of the many editing features that are available in the system. Other commands provide the capability of moving sentences or paragraphs from one point to another in the document, erasure of portions of text, and many other functions required in editing and correcting. The system also provides the ability to print the document on the terminal or on a high-speed peripheral printer. The document can be formated automatically to be printed with any line width or page depth. In addition, the system can provide right-hand justification of the copy, headings on the pages, and automatic page numbering.

The amount of time saved through the use of this system in producing this book was considerable. Despite extensive editing of almost every chapter in the book, each chapter was typed into the computer only once. Once entered, only the corrections and additions were retyped. When the manuscript was completed, the entire 1500 pages were printed on the high-speed printer in less than an hour! This is orders of magnitude faster than would have been possible with a manual system.

Every author wishes to acknowledge the contributions of others who have had a part in the production of his work. On the other hand, he

---

* Registered trademark, IBM Corporation.

faces the prospect of not including everyone in the formal acknowledgment. The problem is compounded in a book of this size which involves the work of 15 different authors. During the course of our work, more than 150 individuals have been directly involved in the review and the production of the manuscripts. This does not include the secretaries and wives who typed drafts, the librarians who tracked down references, the colleagues who listened to our problems and offered solutions, or the technical editors of IBM and Wiley who assisted in the production. As a result of not being able to list them all, the contributors and I can only say "thank you" to all of these people who know who they are and what they did. The assistance of IBM in providing the use of the ATS system described above is also acknowledged.

THOMAS J. HARRISON

Boca Raton, Florida
November 1971

CONTENTS

CONTENTS

CHAPTER 3 ANALOG-INPUT SUBSYSTEMS

THOMAS J. HARRISON

TRANSDUCERS AND ANALOG-INPUT FUNCTIONS

ANALOG-INPUT SUBSYSTEM CONFIGURATIONS

SYSTEM ERRORS

ANALOG-INPUT EQUIPMENT

CHAPTER 4 ANALOG-INPUT SUBSYSTEM ENGINEERING

THOMAS J. HARRISON

CHAPTER 5   ANALOG-OUTPUT SUBSYSTEMS

CHUNG C. LIU

CHAPTER 6   DIGITAL-INPUT/OUTPUT SUBSYSTEMS

J. PAUL HAMMER

CHAPTER 7   MAN-MACHINE INTERFACE

DONALD C. UNION

CHAPTER 8   SYSTEM DESIGN FACTORS

RICHARD E. MACH

CONTENTS

CHAPTER 9 PROGRAMMING SYSTEMS

ALEX J. ARTHUR

CHAPTER 10 PROJECT ORGANIZATION AND MANAGEMENT

H. WAYNE VAN NESS

CHAPTER 11 ECONOMIC JUSTIFICATION

LYNN S. HOLMES, JR.

CHAPTER 12   SYSTEM SELECTION

I. C. (JIM) TRAUTH, JR.

CHAPTER 13   INSTALLATION AND PHYSICAL PLANNING

RICHARD N. POND
SHANE L. SHANE

CHAPTER 14   APPLICATION PROGRAMMING

GORDON W. MARKHAM

CONTENTS

APPENDIX A   COMPUTER FUNDAMENTALS

J. PAUL HAMMER

APPENDIX B   PROGRAMMING FUNDAMENTALS

ROGER M. SIMONS

APPENDIX C   APPLICATION BIBLIOGRAPHY

THOMAS J. HARRISON

Handbook
of Industrial
Control Computers

Chapter 1

Process Control Computer Systems

ROBERT C. MILLER, JR.

The technological advances made by mankind during the past half-century have been phenomenal. To select the most important is difficult, but two at least are clearly outstanding in today's environment. One is the invention and construction of equipment that produces large quantities of specialized products from raw materials and energy. Examples include petroleum-refining complexes, paper-making plants, ammonia plants, and aluminum pot lines. These complex processes are large and variable enough to require some form of "control."

The other modern invention that must be placed high on the list is the digital computer, which has been mentioned frequently as having an impact equivalent to that of the Industrial Revolution. In slightly more than a decade, the digital computer has been widely adapted so that its arithmetic and logical power can be used to provide the "control" required to operate today's complex processes efficiently. Digital computers adapted for this purpose are known as "process control computer systems" (PCCS) and are the subject of this introductory chapter.

1.1   PROCESS CHARACTERISTICS

There are many processes. Some of the factors accounting for this variety include the raw materials used, the form and quantity of the energy required, the number of stages in the transformation of the raw materials, the response time of the various stages of the process, and, of course, the final product. A distinguishing characteristic of these processes is whether they are continuous, batch, or discrete in nature.

Continuous processes are those that produce a final product as long as the raw material, energy, catalysts, control, and other requirements are provided. For example, the various petroleum-refining processes are continuous. Batch processes, on the other hand, produce a limited quantity of the final product over a relatively short period of time, such as hours or days. Raw and partially refined materials are input in specific measures and sequences; mixing procedures and measures of energy are applied in a specified order. When the "recipe" is completed, a batch of the final product has been created. A well-known example of this is the blast-furnace process used to produce steel. Discrete processes, the third category, are associated primarily with the manufacture and testing of components, subassemblies, and final products. Here the process involves large volumes of separately distinguishable components that require control to ensure that the final assembled product has the proper mix of components or meets specified quality standards. An example of a discrete process is the automotive assembly line.

The process control computer system (PCCS) has application in all three categories but, to date, has enjoyed success predominantly in the continuous and batch types of processes. Certainly as automation techniques see greater application, the PCCS will be used increasingly in the discrete processes. The necessity for the continued use of the PCCS is created by the increasing demand for a greater variety and quantity of products as well as for improved quality.

1.2   THE CONTROL PROBLEM

A generalization of a process is shown in Fig. 1-1. The control function is represented as the interface between the inputs and the transformation, which together produce the final product. The general aspects of creating this interface and solving the control problem are briefly introduced here.

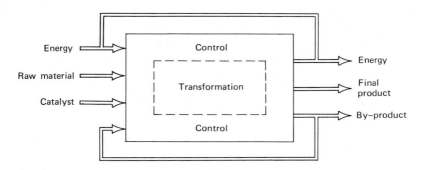

Fig. 1-1   Generalized process.

1.2.1   Need for Control

The characteristics of most processes are such that some minimum amount of control is mandatory, if only for the purpose of safety. The need for control, however, is usually a combination of several of the following elements:

- The transformation of raw material to a final product requires that energy, input mix, and flow of material be sustained at specific levels within safe tolerances.
- The variable composition of raw materials requires that adjustments be made to the other process factors to guarantee a consistent final product.
- The start-up and shut-down of some processes involve specific and accurately timed sequences that must be closely controlled.
- A change in market conditions or management objectives may demand that the product mix be changed quickly with minimum impact on equipment during the transition.
- Optimal use of equipment and materials is required to maximize the return on investment.

1.2.2   Elements of Control

In reality, control has been a part of the continuous and batch processes since their creation. In the first stages of development, man himself was the controller. He controlled the process by providing "measured" quantities of material and energy and by determining when the transformation was complete. These processes were simple from necessity. As complexity increased, however, the sophistication of control grew and the human element became a limiting factor. From the necessity for improvement came the inventions that allowed measurement of temperature, pressure, flow, and so on, as well as the creation of actuators that could control the critical elements of the process.

The sensors, which allow measurement, and the actuators, which provide the control, are two of the primary elements of control. Decision, or direction, is a third element that must be added if effective control is to result. Clearly, the temperature measurement at some point in the process cannot be used as a direct input to the movement of a control actuator. The measurement, however, may be used properly as the value of an independent variable for a computation from which the control action can be determined. From these computations, decisions are made and control maintained.

Before the advent of computing equipment, the experienced human operator provided the "computing" on an intuitive basis. The operators, through repeated experience, learned the process and its reactions to the control actions they applied. They learned, for example, how much to change a valve's position when a measured variable reached a point where action was necessary. In these early cases, the operator provided the control loop. He read the input, made decisions, provided intuitive computation, and carried out the control action. However, as the process became more complex, even the best operator was overwhelmed because of his physical limitations. The lack of consistent control action was another shortcoming of the human controller.

To solve these problems and extend the number of processes, the analog controller was developed. This device allows each control loop to be controlled individually in a process and at the same time applies reasonably consistent control action. Still, the controller has a major drawback. The many controllers installed within a process do not have the ability to communicate as a total set. Although several can be interconnected to produce cascade control when required, the degree of intercommunication is very limited. Thus the operator must still provide control decisions affecting interactions between many loops, even though the controllers provide control of the individual loops.

The great strength of the PCCS is that the process can be addressed in its entirety rather than as a set of independent parts. With this overview, the PCCS is capable of operating the process at its optimal point, which results in more revenue per dollar invested. Other advantages of total control include better utilization of equipment, quicker switchover when a change in product mix is required, proper start-up and shut-down procedures, sounding of alarms when required, and record keeping for accounting as well as for historical purposes.

1.2.3   Requirements for Control

The installation of a PCCS does not result automatically in the many benefits that are possible. The process characteristics first must be understood and translated into a control strategy. Frequently, the PCCS is used as the tool for studying the behavior of the process and its measured reaction to control actions. With the empirical and theoretical knowledge gained from these studies, a model of the process can be defined. Such a model specifies the process mathematically so that the PCCS can compute a reasonably complete picture of the total process. The creation of a new model for a process usually involves significant effort, but with the model, computations can be made to determine the necessary action for optimum operation. Without an overall understanding and description of the process, optimum control cannot be realized.

Control algorithms are the procedures that allow computation of the control action for the various segments of the process. Measurements from the transducers and inputs from status indicators are the inputs to the control algorithm. The values resulting from the algorithm computations specify the action needed to adjust the process through actuators that operate the control valves and other equipment in the process. The parametric values for the control algorithms are determined from the process model.

The process model, the control algorithms, the transducers measuring the required variables, and the control elements, when combined with the PCCS, provide the building blocks for modern computer-controlled process control systems.

1.3   THE DIGITAL COMPUTER

Not only is the invention of the digital computer one of mankind's outstanding technological accomplishments, as we indicated earlier; the rapid improvements in this computer have been equally significant.

The first commercially available digital computer was produced in 1950. It was capable of performing 1700 additions per second. Today, two decades later, existing large-scale computers are capable of performing in the range of 10–12 million instructions per second, and machines now in the planning stage are expected to perform over 300 million instructions per second. The speed of the giant computer overwhelms the mind, but the cost of the "small" computers, those capable of performing over 500,000 instructions per second, one can easily appreciate: they are available for less than $10,000.

The characteristics of the digital computer are discussed in this section. The small binary computer is the main point of focus because its capabilities are typical of those necessary to solve the control problem.

1.3.1   Digital Computer Characteristics

Let us briefly explore the characteristics of the digital computer system. As shown in Fig. 1-2, a digital computer consists of five fundamental parts:

1. *Storage,* frequently called memory, which is used as the residence for the data to be used in the computations and also for the instructions whose execution determines the behavior of the computer system.
2. *Control,* which interprets the instructions in a specific order and directs the system to make the logical decisions and to perform the arithmetic functions which the computer is capable of executing.

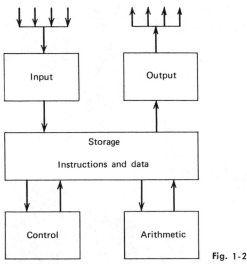

Fig. 1-2   Five parts of a digital computer.

3. *Arithmetic,* which gives the computer the ability to perform the normal operations of add, subtract, multiply, divide, and compare.
4. *Input,* which provides a path into storage for the program or the data (constants, parameters, raw measurements) upon which the system operates.
5. *Output,* which provides the path out of storage for the calculation results for direct control or meaningful instructions to an operator.

Each part of the computer is a subject unto itself. For example, input can be in many forms: the signal resulting from the operator's striking a typewriter key, the digital value resulting from conversion of a voltage signal of a thermocouple, and so on. These subjects are presented in detail in the following chapters; in particular, an introduction to the fundamentals of digital computers is presented in Appendix A. Detailed discussions of the system design are found in Chap. 2 through Chap. 8.

Typically, the binary number system is used in the digital computer. The binary number system has only two digital values: 0 and 1. By combining these two, any integer in the decimal number system can be represented; for example, 1110 in the binary system represents 14 in the decimal system. The importance of the binary representation is that only two states are required—ON and OFF, corresponding to 1 and 0. Therefore, the media used to store numbers in the binary computer can be simple and economic. The primary storage medium today is often an

Fig. 1-3   Magnetic cores. From IBM Corporation.

array of magnetic cores, such as those shown in Fig. 1-3, which can be electronically altered to be in either of two states, representing 1 or 0. Magnetic core developments are one of the reasons for the tremendous performance of the small binary computer. Storage units of the typical small binary computer have a read/write cycle of less than 1 or 2 μsec, which means that the computer can store or retrieve information in core storage at rates in excess of 1,000,000 operations per second. This core-cycle time, combined with operating speeds of the solid-state circuits used in the arithmetic and control sections, results in very fast internal performance.

The input/output (I/O) facility of the small binary computer is provided by paper-tape devices or punched-card readers and punches. Paper-tape I/O is the least expensive but also gives the least amount of flexibility. Punched-card I/O normally costs more but has the advantage of small unit-records (usually 80 characters/card). However, neither of these media results in a balanced system when combined with the very fast arithmetic, control, and storage units because the I/O devices are slow compared to these units. Some means of bringing the system into better

Fig. 1-4   IBM 1316 disk pack assembly. From IBM Corporation.

balance is required. The answer is to keep all of the programs in the system and thus avoid the necessity of entering programs every time they are to be executed. To do this, some form of bulk secondary storage is used.

Secondary storage exists in three common forms: (1) magnetic tape, (2) drum storage, and (3) disk storage. Each has its advantages and disadvantages. Magnetic tape gives almost unlimited storage and good recording speeds, but it utilizes sequential recording and is usually expensive. Its main drawback is the sequential characteristic, which means that the tape must be moved in one direction or the other to place the required record under the reading element. Reaching the desired information may require several seconds.

Drum storage utilizes a cylinder coated with a magnetic material on which data are recorded. Its outstanding attribute is that only a short time is required to reach an information record. Since the drum usually has a read/write head for each track, it requires only an average of half a revolution to be in the position to read or record data. However, the drum units available on the small systems are usually capable of storing only between 30,000 and 130,000 words of data, and they are expensive. In addition, the storage medium, the revolving cylinder-like device, is an integral part of the unit, so the amount of storage is not easily expandable.

The disk storage is a compromise between the other two media. The storage media are disks coated with a magnetic material on which information is recorded. The disk assembly, shown in Fig. 1-4, is usually removable so that practically unlimited storage exists. The average access time for the inexpensive units is below half a second and approaches the access time of the drums on the more costly units. The big advantages are the ability to access data randomly at reasonable speeds and to store large amounts of data at reasonable costs. Because of this, the modern small computer tends to have disk storage as the secondary storage. This is particularly true if the computer is to be used for many different applications that require new programs to be entered at frequent intervals.

The man-machine interface of the small computer is invariably minimal. It has the facilities to start and stop operations, to display some of the registers, and to determine the operating state of the system. But elegance and extensive functions are eliminated in order to achieve economies.

This is a snapshot of the small digital computer as it exists today and as it is likely to continue to exist for the next decade. It is capable of very fast computation once the data and program are in the main storage, but the input/output facilities are limited. Still, the range of applications that it is capable of satisfying is significant. It is on this foundation that the PCCS is built.

1.3.2   The Digital Computer as a Process Controller

Why is the digital computer so important to process control? First, it is economical. Today the PCCS is paying its way through higher process yield, fuel reductions, and shorter cycle times by maintaining the process near its optimum operating points. Indications for the future are that equivalent computer performance will be provided for fewer dollars as a result of gains in electronic technologies and that this will extend the use of the PCCS into new application areas.

Second, the program of the digital computer can be changed easily. Thus the program of the PCCS can be modified quickly to reflect changes in the equations that represent the model of the process under control or in the algorithms that represent the control strategy. This ability goes further by allowing the program to overlay itself if it becomes so large that it overflows the primary storage. That is, the program can be segmented and only the active portion need be resident in primary storage. This is very significant since few PCCSs available today can be economical and also have enough primary storage to contain the entire set of programs required for all phases of operation, such as start-up, normal control, emergency conditions, and execution of diagnostic tests to determine that the PCCS is in proper operating status.

Third, the logical capabilities combined with the high arithmetic speeds allow control computations and decisions to be made at rates far in excess of the frequency response requirements of most processes. This speed allows the process to be operated within close tolerances and frequently means that extra computing time is available. This "extra" time can be utilized for other computing chores such as engineering computations or for extension of computer control to other portions of the plant.

1.3.3   Small Binary Computer

The PCCS of the early 1970s probably will use the small binary computer for its central processing unit (CPU) in applications dedicated to control. The main reason is that the small binary computer provides sufficient computing power at a cost that allows the PCCS to be an economic tool for the user. In applications involving both control and general data processing, the general-purpose computer may become common. A detailed description of the small binary computer and its design is given in Chap. 2 as a guide to the development of the characteristics required in a control computer. However, we consider a few of its main attributes here.

At first, one might conclude that the small binary computer is capable of satisfying the requirements of only a few applications because it has a limited instruction set. Actually, the opposite is true since almost any function can be provided through the creation of subroutines. Subroutines are sequences of instructions that perform certain functions, for example, the computation of the sine function, the determination of square roots, or the interpretation of operation codes in a string of characters used in a programming language. Large machines may have specialized blocks of electronic components that implement these functions in hardware. Hence large computers have increased facility and faster execution time, but small computers have the same facility once the subroutines have been created by the programmer. Thus virtually unlimited capability exists so long as the primary storage can contain the program sequences or secondary storage is available to accept the overflow. The disadvantage of the small machine is that the execution time increases as a result of frequent accesses to the secondary storage. All this is not meant to imply that the small computer eliminates the need for the larger systems but rather to point out that the applications of the small machines form an extensive list that grows longer each year.

1.4   PROCESS CONTROL COMPUTER SYSTEM

The central building block of the PCCS is often the small binary computer discussed in the previous section. The reason for this choice is that the problems to be solved by the PCCS frequently are similar to those the small binary computer is designed to solve efficiently. Examples of these are mathematical operations such as equation evaluation, data smoothing, statistical computations, and optimization calculations. This is in contrast to the applications of the "commercial" data processor, which is specially designed for problems such as record keeping, financial

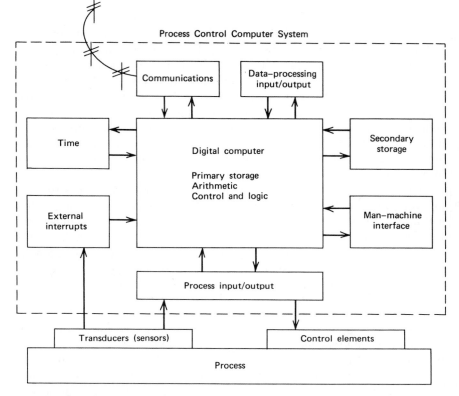

Fig. 1-5   Process control computer system.

computations, and inventory management. The "general-purpose" processor designed for both commercial and scientific applications also can be utilized in the PCCS if certain features are made available.

However, the standard small computer requires some additional functional elements before it can satisfy fully the process control application. These functions and their characteristics are introduced with the aid of Fig. 1-5, which shows the fundamental parts and their relationships within the PCCS. Brief reviews of this diagram during the following sections will serve to provide a clear understanding of the organization of a PCCS.

1.4.1   Central Processing Unit (CPU) and Storage

The classical CPU includes the controls necessary to make the computer function in a predictable manner, the primary storage to hold the program instructions and data, and the circuits that perform the arithmetic operations. These are presented in detail in Appendix A. The other functional capabilities required for a CPU in a PCCS are time, interrupt, storage protection, and secondary storage.

1.4.1.1 Time. Process control is a "real-time" problem. This description implies that "time" must be used in the control of the process. Actually, real time can be interpreted to mean that whatever is happening between the computer and the process is happening instant-by-instant, without any delay that would be detrimental to the control of the process. The last phrase in the previous sentence is important in that it implies that real time is relative; in one process or in response to a given process condition, a delay of 30 seconds may not be detrimental, whereas in another circumstance such a delay could have catastrophic results. Real-time control does not, however, require an explicit sense of time (much as a human body does not, in a strict sense, require a sense of time). On the other hand, process control applications have the requirement to schedule events within the process, produce a log of the day's events, or initiate control actions within the process at specific times after other events have occurred. Any one of these requires that time be available as a parametric value to the program being executed by the PCCS.

Two techniques are used to provide time to the PCCS program. First, a real-time clock can be attached as an I/O unit. The program reads the "real-time clock" (RTC) to determine whether the appropriate time has arrived for a specified action. Usually, the time value provided by the RTC is in hours, minutes, and seconds, or in hours and decimal-fraction hours with the smallest increment typically 0.001 hour.

The second technique is to provide an interval timer that functions as a counter operating at a specific rate. Given the interval between each count and the ability to set the counter to a desired numeric value, practically any interval of time can be measured. By appropriate programming, a real-time clock can be created and operated via a program in the PCCS. Although this requires more program complexity than does use of an external RTC, it provides enough flexibility for many applications and is often more economical.

1.4.1.2 Interrupt. The PCCS must be capable of responding to events out of the ordinary, or emergencies. This capability is provided by adding an interrupt function to the computer system. In a computer doing scientific calculations, the applications are usually such that the work can start at the beginning and proceed step by step in a known and unchanging pattern. Thus no mechanism is needed to deal with changing this pattern in response to outside events. But for the PCCS, this is no longer true. For example, an overheated motor may require immediate action. This means that the current computer activity must be redirected to a program subroutine that has been programmed to effect the required action. This technique of redirecting the computer operation is known as "program interrupt." From the programmer's viewpoint, an interrupt appears as follows: From any place in the program sequence, the computer shifts its interpretation of the instructions to a prespecified location in the primary storage. The storage location from which the next instruction was to be taken is stored in storage at another known location, and program execution begins at the new point. The PCCS program must be created such that the program routine stored at the location defined for this interrupt effects the necessary action for the particular situation. Once the corrective action is complete, the program returns to the storage location specified by the saved address and continues execution of the original program. Basically, this is how an interrupt system functions, but more is required.

Some events are more important than others and a means of establishing priority is needed. This is done by creating more than one level of interrupt and assigning a priority to each. Once priority is established, the PCCS operates in a manner such that if one interrupt is being handled and a more important event occurs in the process, the more important program is executed first. This is true because the higher priority interrupt is allowed to interrupt a lower priority task that is being handled by the PCCS. Once the higher interrupt has been completely serviced, the interrupted lower priority program is completed.

Adding a priority-interrupt facility to the computer increases its power in the real-time environment. It allows quick response to the emergency events associated with the process but it also has other good uses. Some of these are to indicate errors detected in the I/O system or during the read-out of storage, to signal completion of an I/O operation that has been progressing asynchronously to the computer program, or to indicate the end of an interval of time that the program has previously specified.

Programming a priority-interrupt system is, of course, more complex than is implied in this elementary discussion. However, the above discussion should clarify the concept of this function and its importance to the PCCS.

1.4.1.3 Storage Protection. It is very important that the PCCS stored program not be altered inadvertently if the process is to be controlled in a correct and consistent manner. Once a program has been checked and tested, presumably it always behaves in the same predictable manner. It should, therefore, cause no more problems within itself unless an error occurs or another program stored in the main storage alters it erroneously. Storage protection is a technique that shields the program from such random errors and the influence of outside programs.

Several storage protection techniques are used. A simple and straight-forward method is to use an indicator bit in each word of the primary storage. This bit is in addition to the bits used to designate the operation code, address, and so on. When the bit is ON, it indicates to the decoding hardware that nothing is to be stored in this word, and thus it prevents altering the contents of the storage location. Instructions that can change the storage protect bit are included in the machine. These instructions also must be controlled; that is, they are executable only under special conditions so that they cannot be used inadvertently and negate the desired protection.

Another method is to protect bands or blocks of the memory. The size of the band varies with the computer system but typically ranges from 256 to 2048 words. Each band is assigned a protection "key," which consists of a pattern of binary bits. Any program operating with that same protection key can read from, and store into, locations in the particular band. The protection keys can be altered by privileged instructions that can be executed only under controlled conditions.

Storage protection provides valuable protection even for single programs that have been thoroughly checked. The protection is primarily against a transient error that could change a critical instruction and thereby cause an improper control action. Significant design effort and hardware are used to eliminate such errors and storage protection is just one more level of defense.

As the cost of fast, powerful computers has decreased, the capabilities of the PCCS have increased. To utilize this power, multiprogramming operating systems have come into existence. These allow multiple programs, frequently with unrelated purposes, to reside in primary storage at the same time. Through this technique, the PCCS can be used to compile and debug new programs at the same time that it is controlling the process. The multiplicity and unproven nature of some of these programs demand that a means of storage protection exist in the PCCS. Hence it has become an accepted function of the modern PCCS.

1.4.1.4 Secondary Storage. As indicated previously, the small computer can use secondary storage very effectively. It is possible to solve very large computational problems by using an operator to enter sequential sections of a program that cannot be contained by the limited primary storage. This is not efficient or desirable, but for some applications it is possible and was the method actually used in the early computers that had limited storage. However, such an approach is not acceptable in process control applications. The program needed to represent the process and to produce optimum operation normally requires several times the amount of primary storage that can be obtained economically on the PCCS. In these cases, it is untenable for an operator to enter the programs as they are required; such a concept is entirely contrary to the real-time requirements of the application. Thus there is a need for another function to be added to the small binary computer in order to create the PCCS. Secondary storage answers this requirement.

The secondary storage required in the process control application is similar to that described for the small general-purpose or scientific computer, except for the need to have fast access to the data and instructions stored in this medium. The fast-access requirement can be solved by the magnetic-drum storage. A magnetic drum, however, usually has a small amount of storage compared to the disk storage units, which provide significant storage but slower access times. Thus a trade-off decision between the two media is required. The appropriate unit usually is determined by the application constraints.

1.4.2   Process Input/Output

In order for the PCCS to serve its purpose, it must have the capability to interface to the physical world. One of the major differences between the data-processing computer system and the PCCS is that the PCCS must be able to interpret, respond to, and control the physical environment in real time. The real world is primarily analog in character. However, man's efforts and impact on the world have made it possible, perhaps necessary, to consider it as a mixture of analog and digital signals. Thus the PCCS must have input and output devices that allow communication with the real-time, sensor-based world. Only a brief introduction of how this is accomplished is provided here since considerable detail is presented in subsequent chapters.

1.4.2.1   Analog Input. The voltage generated by a thermocouple varies as a function of the temperature of the environment in which the thermocouple exists. Knowing the function of temperature versus voltage allows the PCCS program to compute the temperature in desired units once the voltage signal has been converted to a digital representation that can be accepted by the PCCS. This conversion, and the conversion of other analog data, is done by the analog-input subsystem.

The analog-input subsystem consists of several sections. One is the "analog-to-digital converter" (ADC), which does the work of creating a digital value that corresponds to the voltage signal that has entered the PCCS from a sensor. The range of signals that can be converted depends upon the converter in use, but analog-input subsystems with variable gain can be controlled by the PCCS to cover a wide range of signal levels. Some subsystems have automatic gain control whereby the signal size is determined to be within one of several voltage ranges and the gain is set automatically to an optimum value. The ADC output to the computer storage contains an indication of the voltage range selected so that the digital value used in the program can be scaled properly. Typically, the basic analog-to-digital converter operates with a voltage input range of $0-\pm5$ V for what is known as high-level signals, and amplifiers or gain selection is provided for handling ranges of low-level signals between 0 and ±500 mV.

A second and equally important section of the analog-input subsystem is the multiplexing facility. If many sensor points are to be entered into the PCCS, it is prohibitively expensive to provide an analog-to-digital converter for each input point. To avoid this, a means of switching the signals from each of the input points into a single analog-to-digital converter is provided. This is called "multiplexing." The speed of switching that can be expected from a multiplexer, as well as the voltages that can be switched, depends upon the multiplexer design.

The third part of the analog-input system is the signal conditioning. From an electrical viewpoint, the world outside the PCCS is "noisy." Further, the signal levels do not always match the exact requirements of the converter or the multiplexer. To compensate for these realities, signal conditioning elements can be installed on each of the input lines. The signal conditioning elements may take different forms, depending upon the problem: attenuators to reduce the voltage level, amplifiers to increase the signal, or filters that eliminate electrical noise such as that created by electrical power lines.

The complete understanding of the analog-input subsystem is a career in itself. Details in significant depth are presented in Chap. 3 and Chap. 4.

1.4.2.2 Digital Input. Man's efforts to control and measure the physical world have resulted in discrete signals as well as in the analog signals just discussed. For example, it may be important to know whether a motor is ON or OFF. The indication of this can be made in a yes/no manner corresponding to the binary computer's unit of information, the binary digit (commonly called a bit). One input function that is used to enter this type of data is called "contact sense." It consists of a set of latch or storage circuits that operate in concert with the external relays or switches. In this way, the external conditions are always available to the PCCS. When the external relay is in an ON state, the corresponding bit of the contact-sense input register is ON, and vice versa, so that the external change is reflected immediately to the PCCS. The condition is not changed when the contact-sense register is read by the PCCS. Rather, the bits are altered only when a change occurs externally or when the PCCS actually sends a signal to cause a reset of the register.

This form of input has many uses. For example, it allows the PCCS to check its control action by matching an input point to an output point. When the output action has caused an external relay to be closed, the closure can be detected through the contact sense. It also provides a means of connecting special devices to the PCCS as long as the devices generate signals compatible with the PCCS facilities.

1.4.2.3 Process Interrupt.   This form of input is similar to the contact sense discussed above but has a few important differences. Some external events are of critical importance, and the PCCS must be made aware of them immediately. Process interrupt serves this function by causing a program interrupt and operates as follows: The signal is accepted and held in a latch similar to that used for the contact-sense function. The receipt of the signal immediately generates further action in the PCCS since this input is directly connected to the priority-interrupt circuit structure of the computer. The interrupt is accepted and diverts the program to a subprogram, as described previously. This subprogram determines the cause of the interrupt and takes the appropriate action. In determining the cause of the interrupt, the register that accepts and retains the signal indication is read into the computer storage. A subsequent instruction is executed to reset the register to zeros so that it is then ready to respond to the next interrupt that may occur in this set of latches.

A further difference from contact sense is that the interrupt signal probably is not available over a long time span. Therefore, its arrival is "remembered" until read and reset by the PCCS, whereas the contact-sense function may change state as the external state changes. Should the interrupt arrive when the computer is handing an interrupt of higher level, the lower priority interrupt is "remembered" until the computer can process it. Then, and only then, is the interrupt cleared. Furthermore, any other lower priority interrupts requiring attention during this time are accepted and held until the computer can finish its higher priority actions. Since delays are possible in the response to external interrupts, it is the responsibility of the process engineer to organize the PCCS so that those events requiring the least amount of delay in response are assigned to the higher priorities. In some PCCSs it is possible to assign several external interrupts to the same level, whereas in others each and every interrupt must be assigned a separate level. Each approach has both strengths and weaknesses.

1.4.2.4   Pulse Counters.   Some sensing devices generate electrical signals in pulse form as an indication of an event or a measure of quantity. An example is the turbine meter designed to measure liquid flow. The output of this device is a train of pulses, each pulse representing a known amount of liquid having passed the measuring point. The PCCS could count these pulses by using the interrupt system to indicate the event and a subprogram to cause a count to be added to a storage location. However, this approach adds burden to the computer and may interfere with the handling of other interrupts. The use of a separate pulse counter relieves this burden since its design provides for accepting the signal and increasing the count stored in a register external to the computer storage. The register normally is reset whenever it is read into the computer memory. Since the counter will overflow when the count exceeds its maximum value, some indication of this must be given. This can be done by connecting the overflow signal to an interrupt, which is then used to add the proper amount of flow to the accumulating register in the computer storage. This type of input is used commonly in discrete manufacturing applications as well as in continuous processes where pulse-generating sensors are employed.

1.4.2.5   Analog Output.   Control of the process is frequently accomplished through analog devices such as valve positioners and actuators. The computations to determine the proper control actions, however, produce digital results. The conversion from digital values to analog signals is provided by the analog-output subsystem. The digital-to-analog converter generates a voltage, or current, proportional to the digital value presented to the converter by the PCCS.

The analog-output facility provides the conversion interface between the digital computer and analog devices, such as valve actuators, recorders, and CRT displays. Frequently, the applications require only a few points of analog output. For this reason, some PCCSs do not provide multiplexing, which would cost more than several independent digital-to-analog converters.

1.4.2.6   Pulse Output.   Some control elements utilize stepping motors to move the control elements. These devices require a series of electrical pulses, rather than the analog voltage described above. The pulse-output facility of the PCCS electronically controls momentary switches, which typically are capable of switching external voltages up to 120 V with a duration of 3 to 10 msec. To get the desired control action, the PCCS program must determine the number of pulses required to accomplish the necessary movement. Some pulse-output units generate a specific number of pulses based on the contents of a register that has been set by the PCCS. This is, in essence, the reverse of the pulse counter. Other units give only a single pulse, and the number of pulses generated must be controlled through the computer program.

Another form of this type of output is known as "pulse duration." Here, a pulse is generated that lasts for a period of time specified by the computer program. This type of function is used to move control elements a specific distance where the movement is a direct function of the time the signal is applied. If this method of control action is not available in the PCCS, it can be simulated by use of the contact-operate feature described below. The simulation is accomplished by closing a contact and using an interval timer to measure the amount of time required to get the proper amount of movement.

1.4.2.7 Contact Operate. "Contact operate" is a feature used to close external relays that are used for ON/OFF controls such as those for starting or stopping an electrical motor. It is also used in conjunction with contact sense to connect special digital devices to the PCCS. In this latter case, contact operate is used for control signals and data output lines to the device.

Contact operate works as follows: A binary pattern is sent to latches that are part of this feature. These latches are used to switch voltages that are external to the PCCS and that are the actual signals causing the external action. Details of these forms of digital output are presented in Chap. 6.

1.4.3 Man-Machine Interface

Plant operations personnel must be able to communicate with the PCCS since they must obtain information on the state of the process, make parameter changes in the control procedures being implemented, and take emergency action if required. The PCCS normally has a man-machine interface, frequently called a "console." Such a console is used to communicate directly with the computer portion of the PCCS. The programmer may use the console during the program check-out phases, or the maintenance personnel may use it should the system require repair. Because the console can be used to change memory locations in the PCCS memory and thus alter the program that controls the process, this interface is an improper point for the plant operations personnel to take any action. Such activity could, obviously, have serious effects.

A second man-machine interface is provided that allows sufficient operator-machine communication but with proper security. This is usually known as the "process operator's console." There have been many designs for this type of console. The functions available on the consoles are usually very similar, but the approach and specific hardware used show great variety from one system to another. Since human factors are involved and the process being controlled varies, every PCCS user has specific requirements and different ideas about what the operating personnel find appropriate. Still, the following basic functions do not change:

- A visual display with quick response so that values requested are available for immediate interpretation.
- Functional keys that are clearly labeled and that, when activated, cause a specific function to be performed, for example, "display process variable."
- Numerical entry keys that allow new process parameters to be entered or a signed change to be made to an existing parametric value.

An important consideration is that this console have a program written to support it. The program resides in the PCCS storage with the other programs used in the control calculations. With this program, the functional keys depressed by the operator are interpreted and the required action is initiated through the appropriate elements of the PCCS. It is possible that all logic required by the console could be produced in hardware rather than via a PCCS program. However, the outstanding advantage of using a program is that it can be modified easily to permit changes in the functional operations of the console. Chapter 7 presents this facility and its considerations in detail.

1.4.4   Communications

As computers find greater acceptance in controlling the process equipment that man creates and at the same time play a greater role in operating the business of corporations, it becomes imperative to integrate the computers involved in the two areas. Communication facilities exist that allow one computer to "talk" to another. This interchange must be possible as these "integrated-system" concepts are implemented.

Several means of communication exist. Cost, speed of transmission, response requirements, and economic justification determine the method most satisfactory for the specific application. Communication can be effected by several means, some of which follow:

- Common carrier companies provide service at transmission rates from 600 up to 40,000 bits per second.

- Two CPU's can be connected over short distances through a multiple wire cable. This capability is usually provided by the system manufacturer.
- Microwave communications are used in special high-speed requirements.

Connecting two CPUs together requires a significant amount of sophistication in the programming support that allows the two asynchronous computers to communicate effectively. The timing of the operations that are being performed in the two must be coordinated. Normally, this means that one CPU must be prepared to interrupt the function that it is performing in order to respond to the other in a reasonably short time period. This interchange of data and programs allows greater control and more efficient operation of the total complex.

1.4.5   Programming Support

Before addressing programming support, a short introduction to programming is appropriate. The computer portion of the PCCS is based on the stored-program concept. In order for the computer to perform a specific task, a sequence of instructions designed for that task must exist in the primary storage. This sequence of instructions is called a program. The program defines the order in which the computer brings a specific unit of information from the storage, what operations (add, divide, test for zero, etc.) are to be performed, where the result is to be stored, and where the program should get the next instruction if certain conditions exist in the data or in the results of the calculations. By preparing the proper instruction sequence and entering it into main storage, the PCCS can be used to solve an "infinite" variety of problems. Preparing this sequence of instructions to obtain the desired results for a specific application is the function of the programmer. As such, the programmer must have two realms of knowledge. First, he must know the specific characteristics of the PCCS; second, he must understand the application for which a computer solution is desired. The programs he creates, along with the programs provided by the computer vendor, are called the system "software."

The actual language of the PCCS is based on the binary number system. Although the programmer might consider using the computer's true language to prepare a program, the effort required for a program of any length would be phenomenal because of the nature of the binary representation. To lighten this load significantly, special programs are prepared that use the computer itself to translate from higher level languages to the computer's own language. These programs are part of what is referred to as "programming support." The types of programs included in this conglomerate cover a wide range of sophistication. One of the main sources for them is the PCCS manufacturer, but a significant set of very valuable support comes from PCCS users. Frequently, the manufacturer's support is prepared to cover the general case and, as such, may not demonstrate as much apparent efficiency as the programs prepared by the user. The user can avoid generalized requirements and consider only his own use, which conforms to more specific objectives. The important function of all this support is to make it as easy as possible to prepare the desired program for the PCCS and to "debug" this program in as short a time as possible.

Typical programming support includes some of the following functions, many of which are discussed in detail in Chap. 9 and Appendix B.

ASSEMBLERS. These programs are the most primitive in the sense that they translate from symbolic instructions prepared by the programmer into the machine language of the PCCS, usually on a one-for-one basis. These are highly oriented toward a specific PCCS because of their close relationship to the machine's capability as reflected in its instruction set.

SUBROUTINES. These programs are a specialized sequence of instructions designed to compute selected functions or to execute common procedures. An example is the calculation required to compute the sine function of a variable. Another example is a program sequence that reads, translates to a binary image, and stores the data from a typewriter-like keyboard. Since these functions are required by a great number of users and are executed many times, considerable effort is expended to prepare a correct and efficient program. This relieves each user of the burden of preparing and debugging his own program for these functions.

COMPILERS. The purpose of these programs is to translate high-level languages into machine language. The distinction between an assembler and a compiler is that the compiler typically creates many machine instructions from a single statement. FORTRAN is an example of a compiler language. By means of these languages, a problem can be defined in a short time using statements that closely resemble algebraic and English phrases. The compiler checks the program statements for accuracy and translates them into another program in machine code. An important aspect of compiler languages such as FORTRAN is that the language is problem-oriented rather than machine-oriented. For example, the FORTRAN language is algebraic in character. As such, it allows the programmer to work with an efficient and familiar language when preparing a program rather than forcing him to learn the internal instruction set of the computer.

UTILITY PROGRAMS.   These programs are produced to serve as extra aids to the programmer. An example is a trace program that creates a map of the program execution as it operates with a specific set of data. Other utility programs include (1) loaders which enter programs into the computer storage from an input medium such as cards and (2) dump routines which list the contents of storage so that the programmer can determine what the program has done. Computer users frequently make many contributions in this area of support.

INTERPRETERS.   Application-oriented programs that interpret statements prepared in specific formats have been designed for use by personnel who understand the application but have little or no training in the use of computer languages such as FORTRAN. A recent development in this area is a program called PROSPRO, which allows the control engineer to define the process variables, such as input and the type of control desired, for each control loop. He can do this on a simple "fill-in-the-blanks" form which is then used to prepare the data as computer input. Through this language, the PCCS can be instructed by a process engineer or anyone else who understands the application but who may not have detailed knowledge of the PCCS. Language developments such as this have made the PCCS accessible to a greater number of people.

EXECUTIVES.   As the application program expands past the limits of primary storage, the need for executive programs becomes apparent. The executive programs handle functions such as the following:

- Determining the source of an external interrupt and transferring control to the proper program. If the program is located in the secondary storage, the executive loads it into the CPU so that the appropriate action can be taken.
- Maintaining a catalog of programs currently located in the CPU. Having this information increases the program efficiency when a subprogram can be located in primary storage.
- Maintaining the real-time clock and changing programs when specified times occur or when predefined intervals of time have elapsed.

The programmer creating the application program can certainly produce a program that takes care of all these functions. Where commonality exists, however, an executive program can be prepared once for all users.

This program can serve many users with excellent operating efficiency and a considerable savings in programmer effort.

Two other programming concepts that require introduction and that are dependent primarily on the executive design are "multiprogramming" and "time sharing." Multiprogramming seems the more important for the PCCS. Multiprogramming is a method of using the computer system in which several programs are in the primary storage simultaneously. There they can be executed whenever facilities and time are available. For example, if two programs are stored, the first may require the completion of I/O operation before it can proceed. Rather than remaining idle, the PCCS executes the second program until the I/O operation for the first is completed. Then it resumes the execution of the first program. In this manner, the total power of the PCCS can be realized. This concept plays a significant role in the PCCS since it allows the process to be controlled and also provides access to the PCCS for use in solving problems that need not be directly related to controlling the process.

In some ways, time sharing is similar to multiprogramming. Time sharing usually is defined as many users sharing the computer facilities at the same instant. The usual means of communicating with the system is via a terminal, the simplest of which is a typewriter-like unit. In order that each user be serviced, the time available on the computer is divided equally among the users; this is called "time slicing." Since the time slice is quite small, each user appears to be serviced immediately, which is in contrast to the situation wherein each user waits his turn to have the entire computer system available exclusively to himself. This means of using a computer system not only provides better service but also uses the system more fully. As noted in the Preface, the manuscript for this book was prepared on a time-shared computer.

Multiprogramming is far more applicable to the PCCS than time sharing, primarily because of the nature of the application. Control of the process is the most important function the PCCS must perform. When this is being done, the PCCS must be dedicated exclusively to that function. Only when a delay occurs for I/O, or when the control program is quiescent for some other reason, should the PCCS turn to other work. This is the basic concept of multiprogramming.

DIAGNOSTICS.   Another area of importance is diagnostic programming. The focus of this form of programming is on determining whether the PCCS is functioning correctly and, if it is not, locating the unit or component that is at fault. Once the fault has been repaired, the diagnostic programs are used to exercise the system to verify that the repair has indeed restored the PCCS to a completely operational system. Ideally, the diagnostic programs should be operable while the process is being controlled. Thus the diagnostics would be run during free time at predetermined intervals. To accomplish this, the diagnostics must be accessible to the supervisory or executive program that is directing the PCCS. Completion of the test without detecting an error can be recorded for historical purposes. If a portion of the PCCS is found to be at fault, repairs can proceed after the PCCS configuration has been modified logically by the executive program to exclude the malfunctioning element. Naturally, programming support involves a high degree of coordination, but the result is a PCCS that is able to control the process a greater percentage of the time than can a PCCS where diagnostic programming has been given only secondary consideration.

Programming is a world of flexibility. Programs produced by the vendor may take on a relatively rigid form but the facility to change exists. The application program frequently requires change as knowledge of the application and of effective control methods increases. The great flexibility of the PCCS does not exist in the hardware itself but in the fact that the action of the PCCS can be changed quickly by simply loading a new program. Thus programming is an area involving a great amount of creative effort. New control schemes, new languages for easier communication with the PCCS, and new optimizing techniques appear in the programming aspects of the PCCS. Since so much change occurs and so much more is possible in this area, only a very few aspects of programming have been introduced here.

1.5   CONTROL APPROACHES

The general components of a PCCS, including hardware and software, now have been explored. Due to the great versatility of the PCCS, the manner in which this tool is used can take many forms, some of which are discussed in the following material.

1.5.1   Data Collection

One of the simpler uses of the PCCS is to collect data. Here, the PCCS is connected to the process in the manner selected by the process engineer, as illustrated in Fig. 1-6. The process variables of interest are converted to digital form, accepted into the input system, and placed in storage. The values at this point are digital representations of the voltage generated by the sensors. These values are converted to engineering units by formula evaluation. For example, an equation of the form $AV^2 + BV$

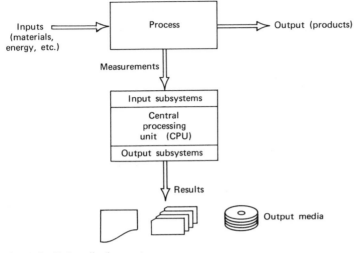

Fig. 1-6 Data-collection system.

$+ C$ can be used to calculate temperature based on the voltage V generated by a thermocouple. Depending upon the variable (pressure, temperature, flow, etc.), the appropriate formula is used. In addition, various mathematical analyses or other data manipulation may be performed by the computer.

The results of the calculations are recorded on a PCCS output device for subsequent use by the process engineer. The type of output selected depends upon the required frequency of measurement and the number of variables. Values that are strictly for historical purposes can be printed, whereas values to be used in further calculations usually are recorded on media that can be easily entered into a computer. These media include punched paper tape, punched cards, magnetic tape, and removable-disk storage.

The primary purpose of data collection is to assist in the study of the process under varied conditions. Through such activity the process engineer is able to define and/or refine the mathematical model of the process that is to be controlled by a PCCS. Data collection obviously has little, if any, direct affect on the process and, as such, represents a conservative approach to the introduction of process control via a computer system. Even when sophisticated modes of control are implemented using the PCCS, however, data collection for summary historical purposes and model refinement is almost always included as part of the computer's task.

1.5.2   Operator Guide Control

The next step in complexity is the application known as "operator guide." The computer is on-line but is used in an open-loop fashion, meaning that the output portion of the PCCS is not connected to the control elements of the process. The control action is actually performed by the process operator, who receives direction from the computer. This mode of control is illustrated in Fig. 1-7.

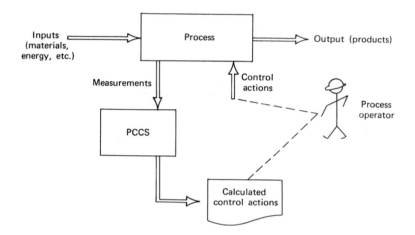

Fig. 1-7   Operator guide control system.

The PCCS functions as follows: At specific time intervals (depending upon the application requirements, but normally once every 5–10 min) the selected input parameters enter the PCCS as digital values through the analog-to-digital converter and the digital-input circuits. The values are transformed into engineering units if appropriate and are used in the control model for the process. From calculations using the model, the action required to adjust the process nearer the optimum is determined. These results are presented to the operator in a readable form that is usually printed by a typewriter. The actual modification to the control of the process is effected by the operator's changing the setpoints of controllers or performing other actions in accordance with instructions produced by the PCCS. The controllers are the means of keeping the process under optimum control, with the operator playing his role of monitor and director and making changes in accord with the raw material being used and the output product desired. The PCCS plays the role of guiding the operator unerringly and consistently in his attempts to optimize the process.

The number of input variables in an operator guide system is usually between 10 and 100, but, clearly, the PCCS can handle more if the economics are there. The number of control points calculated and new settings produced is relatively small since the operator must change the settings physically. Attempting to adjust 100 setpoints every 5 min might be possible, but it is certainly not advisable. The real limitation in operator guide control is this human element. Expecting an operator to read the new value of the setting, make the adjustment, and check the setting in less than 3 sec results in 20 adjustments every min. To assume that this pace can be sustained accurately for an entire work shift is not reasonable. Thus the human limitation is one of the serious disadvantages of this type of control.

Operator-guide control has advantages, however. It is a conservative approach for the new users of PCCS and was frequently employed as such by the PCCS pioneers. It is also a good approach for testing new process models with the process engineer functioning as the process operator. The operator, who usually has an excellent "feel" for the process, remains in the loop and is likely to detect an inappropriate combination of settings that may result from a PCCS program that is not thoroughly debugged. Alarm conditions can be monitored by the PCCS as a small addition to its other activities. This relieves the operator of that burden while he is adjusting the setpoints, an activity that may absorb a large portion of his time. Further, the PCCS can easily monitor a much larger number of alarm conditions than the operator.

Assuming that the program directing the PCCS has enough strategy built into it, the operator can be directed to adjust the process gradually to use a new feed stock or to produce a modified final product. This gradual adjustment is far more desirable from a safety viewpoint and can be done with ease via the PCCS.

1.5.3   Supervisory Control

The next use of the PCCS in degree of complexity is "supervisory control." In this approach the PCCS is used in a closed-loop mode, which means that the PCCS is connected to the process such that the setpoints of the controllers can be adjusted directly by the PCCS, as illustrated in Fig. 1-8. This is in contrast to operator guide, where the actual changes are implemented by the operator. The object and, of course, one of the major advantages is to keep the process near its optimum operating point through the quick action for which the PCCS is ideally suited.

The activity in a supervisory control system is not significantly different on the input side from that previously described for operator guide control. The calculations to determine the control action are also similar.

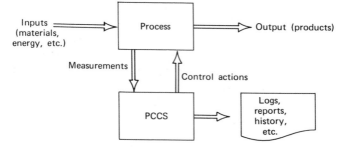

Fig. 1-8   Supervisory control system.

However, once the model calculations have been made, the action taken is very different. Where previously the settings had to be converted to a form suitable for an operator, they now must be converted to values that can be used to adjust the settings of the controllers. If the controllers accept a voltage, the computer values must be transformed to a binary value that, when sent to a digital-to-analog converter, results in a voltage of the proper level and sign. In this way the controllers are adjusted by small amounts to the corrected setpoint. For controllers that accept pulses, the calculated value must be converted to the number of pulses or the pulse duration required to move the setpoint the proper amount. Once the PCCS loop is closed, the operator's function becomes that of a monitor, acting during emergencies.

There is still a need for modifications to the control of the process based on changes in raw material or desired final product mix. This is handled by calculating new operating parameters for the equations representing the control loops. These calculations can be done either on another computer that is capable of solving large linear programming problems, or on the PCCS itself if enough computer time remains after the loop equations have been computed. If the latter alternative is followed, some means of program control is also required such that the computer portion of the PCCS can time share itself with the control of the process and the optimizing computations.

The optimization is done periodically, say, once a day or when material changes or product mix objectives are altered. The new parameters for the loop equations must be entered. This is done either by the operator via a keyboard or by the PCCS's reading punched cards that were output of the optimizing computer. Except for this type of change, the PCCS is capable of operating unattended for extensive periods of time. However, it is also capable of being changed quickly to represent a new control algorithm, to include more points of control, and, under proper programming, to be used to simulate changes in the process before the changes actually are implemented.

A major advantage of supervisory control is that the PCCS continually monitors and controls the process at the optimum point without the fluctuations that result from different operators who invariably have their own "special" way of adjusting the setpoints. Since the computations are done at electronic speeds, many more variables can be considered in the model of the process. These include variables that formerly could not be justified because of cost. An example of one such variable is the outside air temperature. This variable is obviously important in some processes, but to include it as an input to every controller would be prohibitive in cost. With the PCCS it is entered as one variable and can be used wherever appropriate.

1.5.4 Direct Digital Control (DDC)

In the applications discussed above, the fundamental differences were on the output side of the PCCS. In operator quide, there was no direct control of the process—the setting of the controls was done by an operator. In supervisory control, the control was still indirect. The setpoints of the controllers were adjusted but the actual signals for the control elements were supplied by the controllers, not by the PCCS. In "direct digital control" (DDC) the signals used to adjust the control elements come directly from the PCCS and the controllers are eliminated.

Controllers are essentially small analog computers that evaluate a single equation of the form:

$$\text{valve position} = K_0 + K_1 e + K_1 K_2 \int e \, dt + K_1 K_3 \frac{de}{dt} \qquad (1\text{-}1)$$

where K_0, K_1, K_2 and K_3 are adjustment parameters that allow the controller to be tuned to operate with many different processes, and the error parameter e is the difference

$$e = \text{measured variable} - \text{setpoint}$$

A nonzero error indicates that the control element requires adjustment to bring the process to the desired control point. Not all controllers compute all terms of Eq. (1-1). The simpler ones, known as "proportional" controllers, use only the first two items; if the third term is added it is called a "proportional-integral" controller; when all terms are used it is called "proportional-integral-derivative." For simplicity, they are referred to as 1-, 2- or 3-mode controllers. By changing the adjustment parameters, the setpoint controller can be adapted to a great variety of process-control problems. By interconnecting two controllers, more complex control techniques can be implemented.

The DDC concept calls for replacement of the setpoint controllers with a PCCS. Rather than compute the setpoints required for optimum operation, as in supervisory control, the PCCS computes the actual adjustment required and transmits the proper signal directly to the control element. This is done for each control loop. The number of loops may vary from 10 to 400, depending on the process and the capacity of the PCCS.

For a deeper understanding of the PCCS in a DDC system refer to Fig. 1-9, which illustrates a single control loop. The transducer signal for the measured variable enters the PCCS through the analog- or digital-

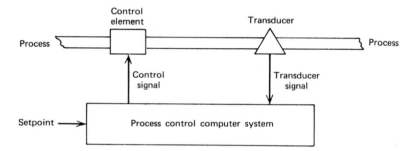

Fig. 1-9   Direct digital control loop.

input section. Once it has been converted to a digital form that can be used by the arithmetic section, the "error" (measured variable minus setpoint) is computed and used in the control algorithm for that loop. The control algorithm equation may be similar to Eq. (1-1), or an advanced control technique such as feed-forward control or adaptive tuning may be used. The results are then converted to a digital representation that is acceptable to the digital or analog-output section. This value is converted by the output devices to a signal that is acceptable to the control element. In this manner the control loop is monitored and adjusted *directly* by the PCCS. Each loop is handled in its turn or with a frequency commensurate with the process requirements. The setpoints for each loop are entered into the PCCS either by an operator or by another computer that does the optimization calculations.

With a DDC system, the operator must have the ability to change setpoints, monitor selected process variables, modify the limits acceptable for the measured variables, change adjustment parameters, and, in general, communicate with the control program. This requires a complete and functional man-machine interface (operator's console) such as those covered in Chap. 7.

One of the major advantages of using the PCCS in DDC is that the control algorithms for the loops can be changed, simply by modifying the stored program. This change, however, requires careful preparation in that the new program must be completely tested before being used for actual control. Although this is demanding, the flexibility is virtually unlimited. Some PCCSs are installed as a combination of supervisory control and DDC. Another advantage is that the data collected from the control loops can be used for other computations required in the operation of a process site. An example is material balance.

One obvious disadvantage to DDC occurs when the PCCS fails. The reliability of the system may be exceptionally high but a failure still will occur eventually. Plans for any DDC system must cover this eventuality.

1.5.5   Advanced Control

The control techniques considered advanced today probably will be commonplace and well understood in the next 5 years. Control theory is a complex subject that is beyond the scope of this chapter. However, it is appropriate to mention briefly two advanced control techniques at this point.

1.5.5.1   Feed-Forward Control.   To understand feed-forward control, let us first consider feedback control. Feedback control entails making a measurement downstream from a control element and then making a control adjustment as a function of that measurement, as shown in Fig. 1-10a. The control element is "back" upstream from the sensing point; hence the name "feedback."

Feed-forward, shown in Fig. 1-10b, is just as the name implies. The control adjustment is made downstream, or "forward" from the sensing point. This method of control allows corrections to be made at a future point in the process, thus keeping the process nearer the optimum than is possible with feedback. The feedback technique works on the basis of changing upstream conditions prior to their measurement at the sensing point. Feed-forward control attempts to anticipate changes and to adjust the stream at a point that still can influence the material that has been measured. Feed-forward control is particularly valuable in processes that have a long reaction time. Feedback control frequently cannot provide adequate response for those processes.

The use of feed-forward control requires comprehensive knowledge of the process characteristics. This is true because the PCCS must determine the value of the controlled variable that will compensate for changes in

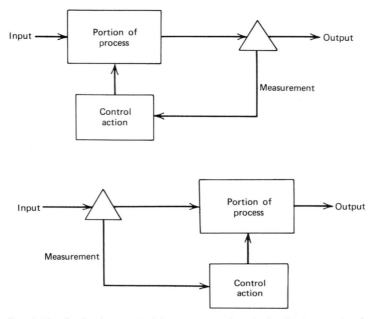

Fig. 1-10 Feedback and feed-forward control. (a) Feedback control; (b) feed-forward control.

the measured variable. It has the advantage, however, of detecting process disturbances before they reach the control point so that fluctuations can be kept to a minimum.

1.5.5.2 Adaptive Control. Adaptive control can be understood by considering its purpose. Any large process does not operate the same from day to day. Wide-ranging atmospheric changes, variable raw materials, catalyst contamination, and accumulating deposits that restrict flows contribute to changes in the process. These influences move the optimum operating point of the process. The goal of adaptive control is to compensate automatically for these changes so that the process operates at its optimum point all the time.

To accomplish this goal requires that a performance index be computed as a function of measured variables from the process. The maximum of this index, which represents the optimum operating point, usually is determined by sequential search methods, such as "peak seeking," "hill climbing," or "gradient" techniques. To locate the optimum, the parameters in the control algorithms are changed slightly and the performance index is recalculated. The result then is compared to the previous index value to determine whether the process is nearer the optimum. This process is repeated until the maximum is located. These new parameters "adapt" the control so that the process is kept at its best point of operation for existing conditions. The performance index is recalculated periodically to detect any change in optimal operating point.

Adaptive control has obvious advantages but implementation requires considerable sophistication. Determining the performance-index function for a particular process is not always simple. A method of locating the maximum value of the function must be selected and the implementation must ensure convergence. A reasonable degree of mathematical knowledge is required to understand the research papers that have been published on this subject.

1.5.6 Management Information Control System

The most complex and sophisticated application of the PCCS is as an element in a "management information control system" (MICS). In this configuration, the PCCS plays an important but somewhat secondary role. The management functions in large plants such as a refinery or chemical complex have increased significantly in the last several decades. These complexities are certain to increase in the future as the amount of available information grows, as the physical size increases, as the demand for more production continues, and as competition intensifies on a worldwide basis. Throughout all of this, the communication between the various departments in these complexes becomes more difficult. With the addition of new elements in the business, or the process, the problem becomes even more acute.

The natural approach to attacking these business complexities is to establish procedures designed to give more direct control over the total business. The computer for many years has played a substantial role in doing routine business calculations and preparing reports from the data that are collected. Thus the computer has become involved in establishing the procedures and actually has become the retainer of large data banks containing the details concerning such things as materials on hand, products produced, records of usage and performance. Through the coupling of this tremendous source of information and the fast, accurate arithmetic and logic capabilities of the computer, the data are processed and evaluated, and reports are prepared for dissemination to management. As better and more automatic methods of data collection are made available, the power of the computer to provide timely reports to management increases. The old bottleneck of passing all information through a central data-preparation facility is disappearing with the introduction of data-entry terminals. With this advance, data enter the management information system at the time the transaction actually occurs rather than hours or days later. This concept and its implementation allow a corporation to automate its procedures and bring the business more nearly under control.

The same problems, to one degree or another, exist in a process plant. The introduction of data-processing systems at these facilities began in the same manner as in many other businesses that did not have a process to control. Then the PCCS began to demonstrate its ability to control the process on an economic basis by operating the process at a more optimum level. When this was done, the process complex then had two computer systems installed: one doing the data processing and optimization calculations that produce the operation plans; the other, usually independently, controlling the process in real time. To attack the complexities of operating such a business, it is a natural step to connect these two systems together so that the *entire* operation is interconnected and operated in a coordinated manner. Such a computer complex, which normally includes several computer systems, is called a "management information control system" (MICS).

The PCCS in a MICS has basically the same functions to perform— collection of data from the process and the application of control in accordance with the established objectives and control algorithms. Added to this basic function in the MICS is the transmission of process data, in converted and edited form, to the central computer. From these data the central system can, for example, automate the material balance calculations for the entire plant. Other purposes served by these calculations include accurate and timely values for performance guides, forecasts and operation schedules, checking for unexpected loss of materials, and calculating flow in unmeasured streams. With timely and accurate data at the central computer, the operation of the entire complex can be coordinated. Effective implementation also allows information to flow from the central computer to the PCCS. This is as important as the PCCSs providing data to the central computer, since it allows a quick change to be implemented at the PCCS when information and calculations at the central computer dictate a change in the objectives that have been guiding the PCCS. Figure 1-11 shows the information flow that is possible within a MICS for a process plant.

The MICS requires some additional hardware. The central computer system must be installed and have communication facilities for transmission to and from the PCCS. The PCCS must have communication facilities as well; otherwise, there is little additional hardware required. The implementation of the MICS, however, requires a significant effort in programming. Programs must be prepared to coordinate the entire system. The programs for the PCCS must operate in real time, collect and edit data for the central computer, adjust themselves to reflect new objectives received from the central computer, and keep the process under control at all times.

1.6 RELIABILITY AND SERVICEABILITY

Reliability and serviceability are two very important considerations for the PCCS. They determine the system availability for control purposes. The need for high availability is clear since the major purpose of the PCCS is to control a process and this usually is a continuous phenomenon that must be under control 24 hours a day, seven days a week.

Reliability, simply defined, means how long the equipment will operate without failure. Failure, in this case, includes anything that keeps the system from performing its function. It may involve a simple element such as a single point of the analog multiplexer or a major element such as the power supply of the central processing unit. A statistical measure of reliability that is used frequently is "mean time between failure" (MTBF).

The reliability of a system is a function of two considerations—the elements used to build the system and the design techniques. The elements can take several forms; for example, the circuit technology used, a subsystem such as the digital-input section, or an I/O unit such as a printer. The reliability of the system may be inherently excellent but affected adversely by a poor I/O unit if the unit is required for effective control.

Electronic devices increase in reliability with each new technology. Their inherent reliability can be improved further through extensive testing, stringent specifications of their operational characteristics, and by use of wide tolerances in the circuit design. The space programs and military specifications have done a great deal to accent reliability requirements. Many of the improvements have been realized as a result of the seriousness of the demand and its resultant challenge, but the price paid has played an important role also. Since the entire success of a space mission may rest on a single transistor, the reliability of that element is vital, and cost becomes secondary. However, cost in the PCCS is important because its justification rests on an economic base entirely different from that of a space mission.

The second system-reliability consideration is design technique. In the case where a single electronic component plays a major role in the reliability of the system, the design can enhance reliability by provid-

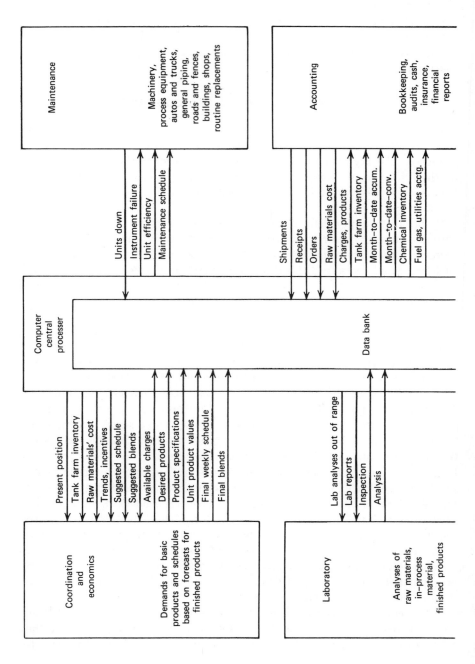

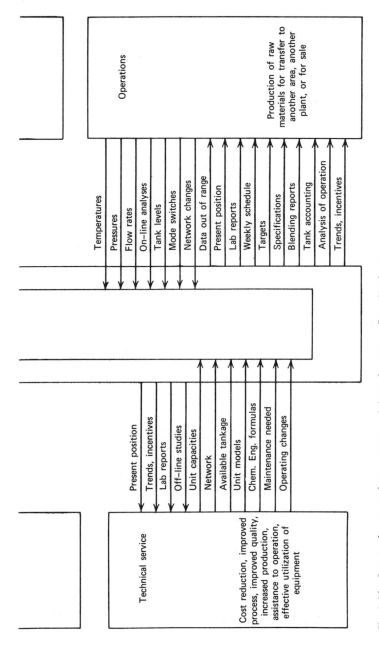

Fig. 1-11   Process plant operating management information system. From IBM Corporation.

The diagram contains the following labels:

Operations

Production of raw materials for transfer to another area, another plant, or for sale

Temperatures
Pressures
Flow rates
On-line analyses
Tank levels
Mode switches
Network changes
Data out of range
Present position
Lab reports
Weekly schedule
Targets
Specifications
Blending reports
Tank accounting
Analysis of operation
Trends, incentives

Technical service

Cost reduction, improved process, improved quality, increased production, assistance to operation, effective utilization of equipment

Present position
Trends, incentives
Lab reports
Off-line studies
Unit capacities
Network
Available tankage
Unit models
Chem. Eng. formulas
Maintenance needed
Operating changes

39

ing duality. Such duality allows the system to operate in spite of the failure of one element so long as the second element remains operable. To achieve even greater reliability, this principle can be extended, but the costs rise with the number of elements. Hence whereas duality solves part of the problem, it creates others. Circuit design based on minimum loading also increases reliability. Design for operation over environmental ranges and limits greater than are actually expected also improves reliability. Throughout the design, however, cost, physical size, and the increasing number of elements must be considered. The result is that any design created to solve a specific problem must be a reasonable compromise between cost and reliability.

Serviceability is also important in achieving high availability. Availability can be defined as follows:

$$\text{percentage availability} = \frac{\text{MTBF}}{\text{MTBF} + \text{down time}} \times 100$$

The down time includes the time to repair the system when a failure occurs and also the time between the failure and the beginning of the repair operation. Time to repair is dependent upon several things: the level of training of the serviceman, the complexity of the problem being repaired, and the design of the PCCS.

The design determines how a repair is made. The design should be such that the system can be easily repaired by replacing groups of electronic elements mounted on the same packaging unit, sometimes called a "card." Repair of the card then can be done after the PCCS is functioning, whereas making the repair directly on the PCCS would mean keeping the PCCS in a down state throughout the repairs.

Design of the I/O attachments also plays an important role. It should permit rapid disconnect, without the necessity of stopping the PCCS, to allow the I/O unit to be replaced quickly by an operable unit. This approach results in minimum down time and provides for off-line repair of those units for which spares are available.

The temporary removal of I/O units from the PCCS presents another consideration, however. The supervisory program controlling the PCCS should be capable of modifying itself so that the PCCS can continue to operate when certain units have been removed or are indicated to be out of service. This means that the program must be capable of by-passing those portions that are dependent upon that I/O unit instead of becoming "hung-up" while it waits for a response from a unit that is not capable of functioning in the system. This facility involves both hardware and software design. Thus availability is a function of the software design as well as of the hardware design.

Serviceability is also influenced by accessibility. Designing a very compact unit can result in difficult servicing simply because of the unit's physical size and the close proximity of the component parts.

There are other servicing considerations to be addressed, such as who maintains the PCCS. Several avenues are open to the user. If qualified personnel are available within the company, they can be trained to service the system. This training is usually available from the manufacturer for a fee. Another method is to obtain service from the manufacturer. Depending on the manufacturer, the cost of service and parts may be included in the monthly use charge if the PCCS is being leased, or it may be billed separately. If the system has been purchased, maintenance usually can be arranged for an additional fee if the user elects not to service the system with his own personnel.

The location of the service center relative to the PCCS installation is also significant. During the time a serviceman is traveling to make a repair, the PCCS may be completely inoperable and valuable production may be lost.

The methods used to service the PCCS should include diagnostic programs. These programs exercise and test the various parts of the PCCS to determine the location of a fault. An important aspect of the diagnostic programs is the ability to operate them during free time and under control of the PCCS supervisor or executive program. In this way, the fault can be located, repaired, and the repair tested without losing control of the process.

The diagnostic techniques developed by the manufacturer may appear inconsiderate of the user since the diagnostic programs require some of the primary storage and some of the process I/O points may be needed to test that portion of the PCCS. This requirement deserves adequate consideration, however, since it has direct bearing on the true availability of the system. The ability for the diagnostics to be integrated into the programming support should be understood. Although diagnostics are not used frequently, the times they are required can be critical.

Reasonable tests of the hardware also can be included in the application programs. For example, a known constant voltage can be applied to one input of the analog-input subsystem. Periodic checking of this input insures that portions of the analog-input subsystem are functioning properly. Elaborate checks of this kind can be developed, but the sophistication should be in balance with the cost and time required for the PCCS to execute the tests.

Maintenance and diagnostic programs used in interconnected systems such as the MICS present even more difficult problems which must be considered during the system design rather than after the problems occur. Major problems arise, for example, in determining which system in the complex is in error. Sophisticated techniques of maintaining such systems are being developed by vendors and users. In some cases, the central system may contain the diagnostics for a remote PCCS that is tested by calling in the diagnostics via the communication facilities. In this case, there are two computers, which may be of different designs, and a communication network. Any one of the three may contain the fault. Such complexities must be approached with a complete understanding of the situations that can develop as well as a plan for their solution.

One thing is clear. Eventually, there will be a failure during which the PCCS will not be able to control the process. One hundred percent availability is not economically practical. Duplexed systems give extreme availability, statistically, but not 100 percent. Therefore, the real cost of down time must be determined and, if necessary, some means of back-up must be provided. Today this is usually done by installing, or leaving, some setpoint controllers on the control loops that are critical and that, if maintained, allow the process to continue without creating a hazardous condition. Control elements that remain at their last setting should be used. With such precautions, the process remains under control until the PCCS can be put back on-line or until human intervention occurs, if it should be required. In any case, the process is still under control even though it may not be operating as optimally as is possible with the PCCS.

1.7   INSTALLATION CONSIDERATIONS

The installation of a PCCS is reasonably complicated. Normally, the manufacturer supplying the system provides consultation service and manuals that present a detailed guide for the solution of common problems. In some cases, the consultation is part of the normal service provided to customers, but in others an additional fee is charged. In any case, a first-pass understanding of the problems involved can be reached by studying the installation and physical planning manuals.

Two basic approaches can be used to accomplish the actual installation. The first is a cooperative effort between the PCCS user and the manufacturer's representatives. This places a requirement on the user and most probably means the assignment of a coordinator who can devote full time to the system during the last two or three months before installation.

The second approach is frequently called "system management." In this case, the manufacturer or a third party accepts, by contract, the responsibility for the complete installation. This could involve construction of new buildings, new or revised instrumentation and associated wiring, all coordination involving the PCCS manufacturer, programming, and so on. Once the installation is determined to be at an acceptable level of operation, the user takes control of the system. Obviously, such an arrangement involves a significant fee but it has the advantage of eliminating the need for unique talents within the PCCS user's company. It has the disadvantage, however, that once the system has been accepted, no one in the company has an experienced understanding of the installation. The choice of the approach that is to be taken deserves thorough and competent evaluation. To indicate briefly some of the complexity, the following elements are involved in the selection and preparation of the site for the PCCS:

LOCATION. The PCCS should be near the process to shorten instrumentation lead wires. The geographic area covered by the process, availability of utilities (primarily electrical power), and access by personnel also play important roles in the choice.

PHYSICAL SPACE. Adequate access to the equipment must exist for both operation and maintenance. Spare-part and test-equipment storage must be provided. The possibility of future expansion must not be overlooked.

FACILITIES. Adequate power for present and future demands, as well as a source of back-up power (if required), must be planned. Air-conditioning may be desirable even though the PCCS can operate over a wide temperature range. Proper lighting, ventilation, and other facilities for personnel are required.

ATMOSPHERIC ENVIRONMENT. Even though the PCCS has been designed to operate in an environment more severe than that anticipated for standard computer equipment, other factors require special consideration. Among these are the temperature and humidity range expected and the level of particle and gaseous contaminants in the surrounding air. Although not an atmospheric consideration, another element of the environment is the vibrations resulting from other equipment operating in the vicinity.

ELECTRICAL NOISE. The PCCS requires stable power and proper grounding. Instrument placement and the lead-wire connections to the PCCS must be designed such that the interference from electrical noise is minimal.

SAFETY. Standard safety regulations dictated by national and local codes must be met. Emergency and fire prevention equipment must be supplied for the protection of personnel, the PCCS, and valuable records located in the area.

The above discussion does not completely cover this vital subject. Chapter 13 addresses these topics in detail.

1.8   THE MANAGEMENT PROBLEM

Previous discussions have presented the PCCS functions and characteristics. There remain other elements of the problem that management must deal with when installing a PCCS to control a process. The subjects to be addressed are project management, system justification, system selection, and system use. These subjects are presented in detail in Chap. 10 through Chap. 14, but a brief introduction is pertinent at this point.

1.8.1   Project Management

A project manager should be assigned as soon as management decides to pursue the installation of a PCCS. Choosing the project manager as soon as possible provides the continuity that is always necessary in a complex undertaking. Be assured that the justification, selection, and installation of a PCCS are just that—complex. Even if the plan is to select a vendor to manage the entire installation, a project manager who acts as an interface to the vendor is required.

The PCCS project manager is a key element in the success of the project. He must have the capability to deal with operations, accounting, and plant management, to name only a few project areas. Further, he must have the capacity to understand and function within a wide range of technology. Although his training and experience probably will be in a specialty that is related to process control, he must be flexible in assimilating and working with the other areas of technical know-how. For example, he must be capable of understanding the role and problems of the programmers who adapt the PCCS to the selected process. Since the project manager plays such a significant role in the final success of the project, he should be involved in the selection of the process to be controlled, the PCCS, and the other members of the team. The importance of selecting this manager cannot be overemphasized.

There can be no set rule for the size of the team that has the responsibility for installing the PCCS. Furthermore, the various phases of the project have differing requirements and there may be a corporate or headquarters group that can provide many of the skills needed. However, if we assume that the location installing the PCCS has all the responsibility, a team of four to six competent people, in addition to the project manager, seems appropriate for typical projects. There should be two or three people from the operations department, including at least one process engineer who might come from the engineering department rather than from operations. It is imperative that operations personnel be involved since that department probably will assume responsibility for the system once successful installation is achieved. There should also be two or three programmers. To ensure continuity, there should not be less than two during preparations for the installation and during checkout. These people need not be professional programmers, but they must possess good programming aptitudes and skills. A common problem is whether one should use professional programmers who have studied the application or personnel who understand the application and have studied programming. The proper choice appears to be a person with application knowledge who can be trained to do programming. This choice is possible since the programming frequently can be done in a high-level language such as FORTRAN. At least one programmer should remain associated with the PCCS after it has been transferred to the operations department.

Selecting the members of the team is an important step in reaching a successful completion of the project. They should all be talented, aggressive people who have a high degree of self-motivation.

The skills represented by the team are not enough, however. Other proficiencies and services that must be available for the team include instrumentation, lab chemistry, facilities planning, mathematics or operations research, accounting, and electronic services.

A partial list of the PCCS team responsibilities includes the following:

- Create a schedule of events including management check points which will lead to economic operation of the PCCS.
- Execute a study program for the justification of the PCCS.
- Recommend the PCCS and vendor to be selected by management.
- Coordinate activities with other functions such as site preparation, instrumentation selection, and installation, routing of signal lines, and determination of control algorithm.
- Define and implement fallback procedures for use when PCCS or power failures occur.
- Prepare and debug programs for the PCCS.
- In cooperation with the operations department, check out the PCCS ability to control the process.
- Train process operators in the techniques of communicating with the PCCS.
- Update programs when modifications or improvements are required.

1.8.2   System Justification

In the early days of process control, few systems initially were justified on the basis of economic pay-off. These early systems were generally experimental. That stage, however, has passed. Today, most PCCSs must be economically justifiable, and they can be.

Chapter 11 presents a complete discussion of this subject. Some of the economic pay-off points addressed are improved control, consistent quality, improved manpower efficiencies, reduction of maintenance, increased process knowledge, and safety.

One of the major economic decisions that must be made is whether the PCCS is to be installed as a rental system or whether it is to be purchased. Some systems are available only by outright purchase. The purchase route can lead easily to a commitment of $200,000, and possibly more, for just the PCCS. The rental approach commits the company to $3000 to $6000/month, perhaps more, with a commitment for as few as three months. Although this method does not commit as much capital, it can be disadvantageous if the PCCS remains installed for many years. An offsetting factor is that the rental price may include some service, parts, and labor. In any case, it is a serious question which must have competent evaluation.

The true cost of a PCCS installation is obviously more than the price quoted by the vendor, whether the PCCS is rented or purchased. Many other items are involved. A partial list of the costs includes the following:

- The PCCS project team, which may be utilized for several installations within the company.
- Site preparation and upkeep.
- Purchase and installation of new instrumentation, including signal lines.
- Service training and salaries for company technicians if self-service of the PCCS is planned.
- Salaries or charges for programming modifications or updates during postinstallation period.

1.8.3 System Selection

Selection of a PCCS should be preceded by a careful definition of the requirements. This frequently takes the form of specifications against which several vendors are given the opportunity to bid. The preparation of this document should be the responsibility of the PCCS project team. Before these specifications can be prepared, however, a reasonable understanding of the process and its control requirements must be developed. For some companies, this step is straightforward because they have past experience from which to draw. For those inexperienced with PCCS concepts in process control, however, it can be a fairly lengthy process requiring several iterations. In such cases it is often desirable to seek outside assistance from a consultant or from an experienced vendor of PCCS. The important point is that a documented set of objectives and requirements should be prepared. Although this may not be simple, it provides indispensable guidelines throughout the duration of the project.

The following subjects should be considered during preparation of the requirements.

CONTROL CALCULATIONS. The form of control equations that must be evaluated or solved and the frequency of these calculations should be defined. Without this measure, it is impossible to determine the storage size, computing power, and system configuration for the PCCS. In some cases, the planned control techniques are considered proprietary, and this information cannot be made public. Benchmark programs can be used as a substitute. Benchmarks are programs designed to measure timed performance of selected elements of the system. Experienced PCCS personnel should be involved when benchmarks are used since the extrapolation from benchmark results to the real-life requirements is not simple.

NONCONTROL COMPUTATIONS. The computing power of today's PCCS usually exceeds the average control calculation requirements. This results in unused time that can be used for other work. If it is planned to use this time for secondary applications, this should be indicated because it usually makes a difference in the required programming support.

PROCESS INTERFACE. The number of PCCS inputs coming from the process, as well as the number of outputs for control, must be specified. This specification must include the reading or output frequency required for proper control as well as the signal level and other characteristics of each input and output.

Priority-interrupt requirements should be defined also. These should include the number of interrupts, their relative importance, the expected frequency, and the maximum time delay that can be allowed to handle them.

PROCESS OPERATOR INTERFACE. The methods and hardware used by the process operator have obvious importance. This aspect of the PCCS has had many innovations in recent years. Until reasonably standard approaches are developed, however, this problem should be addressed carefully in the requirements. Chapter 7 presents details on this important subject.

NONPROCESS INPUT/OUTPUT. The requirements for inputs and outputs that are associated with the process, but that are not directly connected to it, should be detailed. These may include data on raw material characteristics, the results of optimization calculations that are performed on other computers, changes in production requirements, special reports, and data that are transmitted to remote locations. The type of reports and their frequency are important since they can result in the necessity for additional output devices that in some cases must be remote from the PCCS site.

SYSTEM CONFIGURATIONS. The projected system configurations should be defined. The ease with which this can be done and the validity of the results are largely dependent upon the experience of the team. If their experience is inadequate, the consultant or vendor personnel can provide informed assistance. However, information on most of the above subjects must be provided to them.

When future expansions of the PCCS are planned, these details should be considered in the initial requirements. Installing a system that cannot be expanded easily can have far-reaching consequences.

PROGRAMMING SUPPORT. The type of programming support that is required to install and operate the PCCS should be determined. This affects the amount of primary storage that is required as well as the I/O and the secondary storage devices that must be included in the system configuration. This area is particularly difficult for the inexperienced, and competent assistance must be acquired. Programming systems were introduced in this chapter and are also presented in detail in Chap. 9.

RELIABILITY. Any special requirements for ultra-reliability must be addressed early in the program. Everyone desires absolute reliability, but its cost is still prohibitive. If any down time on the PCCS is projected to be disastrous, dual processors can be a partial answer. This technique or other alternatives must be investigated in the early planning stages. If power failures or wide voltage fluctuations are common, then alternate approaches such as motor generators or back-up battery power should be investigated. Any of these create additional costs that must be justified.

FACILITIES. If an existing site is to be used for the PCCS installation or if there is only limited space available, these points should be covered in the requirements. The physical environment (temperature, relative humidity, contaminants, etc.) within which the PCCS must operate should also be specified. For extremely hostile environments, special facilities may be required for the equipment as well as for the human operators.

INSTALLATION SCHEDULE. The completed installation date must be specified. The vendor's meeting this date is critical in cases where a new process has been designed around computer control. Because of the specialized configurations of some systems, the vendor may require a long lead time.

Once the objectives and requirements have been prepared and agreed upon, bids can be solicited from several vendors participating in the PCCS market. Evaluation of the responses and the eventual selection of the vendor are greatly simplified by the documentation. However, consideration of the responses is not always a straightforward exercise. Each response should be examined carefully as to its degree of conformity to the bid specifications. Some may deviate so far that they can be eliminated immediately. Others will have deviations that may not be serious and, in fact, may be system configurations that are in fact improved solutions. The prices quoted by the vendors must be evaluated against the economic justification studies that have preceded this point. As in most business judgments, selection of a vendor solely on the basis of the lowest bid is not always sound practice. For example, a higher bidder may match the bid request and also be able to provide for anticipated expansions of the system. The integrity of the responses and the accuracy of the equipment descriptions should be verified. This is particularly true in the case of systems for which the vendor may not have existing installations.

Other points that should play a role in the system selection are the following:

- The reliability and accountability of the vendor.
- The probable quality of service for the equipment and the proximity of the service personnel.
- The demonstrable success of other installations now using the proposed equipment.
- Support that can be expected in the form of consultations and training, and the charges involved.

Attention to detail and critical evaluations at this stage does much to ensure a successful and satisfying installation in the following months.

1.8.4 System Use

By the time the PCCS has reached the objective of controlling the process, significant effort has been expended. Almost every facet of the process site has been involved to some degree in the installation of this powerful tool. But every tool requires attention if it is to serve its owner, and the PCCS is no exception to this rule. With proper attention, it should provide excellent service for many years, in some cases for more than a decade. Normal upkeep of the equipment is an obvious necessity that must be provided either by the vendor or by specially trained in-house technicians. A few additional points that may not be so obvious are outlined below.

The operations department usually is assigned the responsibility for operating the PCCS. This quickly reduces the influence of the project team, which had the major responsibility during the installation. Although this is standard procedure, the new management may not be receptive initially to their new tool. Assuming a proper installation, this nonreceptive attitude should disappear quickly, particularly if some operations personnel were members of the project team. Close attention by management ensures a smooth takeover and a continuity of excellence. The PCCS project manager plays a significant role in the success of this transition.

The availability of competent programming talent is a continuing requirement. The project team provides this until the system is operational; but once the system reaches that point, the requirement for full-time programming talent can decline quickly. The dilemma is that the talent is needed but not on a full-time basis. A potential solution is for a competent person in operations to assume this responsibility along with his other duties. Clearly, this person must have programming aptitudes and should become involved shortly before the PCCS installation, or earlier. Practically, more than one individual should have the capacity to perform in this crucial area.

The back-up of skilled personnel is made necessary by absences resulting from vacation and illness. The ever-present possibility of attrition adds a further dimension to this problem. Application knowledge and an understanding of the programs that allow the PCCS to control the process are particularly susceptible. The company must have a plan to deal with this eventuality. Clearly, the beginning of a good solution is clear and accurate documentation. After that, an acceptable solution becomes an individual company matter. The important point is to recognize that a solution to this problem is significant in the continuing success of the PCCS.

Emergencies occur with equipment as well as with personnel, and again the PCCS is no exception. Electronic and electromechanical equipment that does not fail has not been realized yet and will not be realized in the near future. Emergencies caused by failures of the PCCS components should be expected. Failures of some sections—say, an output printer—may have little effect, but failure of others, such as power supply, could have disastrous results if control of the process is dependent absolutely upon the PCCS and no fallback capability has been provided.

It is mandatory that emergencies and their potential results be understood and that reasonable back-up equipment and procedures be provided. The form that these back-up solutions take is an individual matter dependent on the particular application.

Once the PCCS is successfully controlling the process, it is probable that all the available computer time is not being utilized. Giving the computer more work is a natural step since it reinforces the economic justification. Techniques and the facilities provided by multiprogramming or time-sharing operating systems make it possible to add other computing tasks while still controlling the process with the former effectiveness. In some cases, additional hardware is required to make these programming systems operable. These possibilities should be considered in order to take full advantage of the PCCS.

1.9   CONCLUSION

The general characteristics and uses of the process control computer system have been introduced. The coverage of its many facets has been brief by necessity. The remaining chapters of this handbook present the subjects covered here in significant depth for those who have a particular interest in or a particular need for an in-depth presentation.

It should be recognized that the PCCS has the ability to satisfy a much greater spectrum of applications than just process control. The applications collectively known as "data acquisition," "laboratory automation," and "factory or plant automation" have the same basic requirements as process control. There are differences in the input speeds, the number of interface points to the physical world, and the type and amount of computations. In addition, the control aspects are often either non-existent or secondary to the acquisition of measurements. Because of this, many PCCSs in the marketplace are used for both kinds of work. The PCCS is not a tool with limited application; rather, its total utility and influence on the world are only beginning to be understood.

BIBLIOGRAPHY

Chorafas, D. N., *Control Systems Functions and Programming Approaches,* Vols. A and B, Academic Press, New York and London, 1966.

Lee, T. H., G. E. Adams, and W. M. Gaines, *Computer Process Control: Modeling and Optimization,* Wiley, New York, 1968.

Miller, W. E. (Ed.), *Digital Computer Applications to Process Control,* Plenum Press, New York, 1965.

Savas, E. S. *Computer Control of Industrial Processes,* McGraw-Hill, New York, 1965.

Chapter 2

Processors and Peripherals

FRED B. JONES

The central processor represents the heart of the control computer system and is, in a sense, the part that justifies the whole. It brings a new dimension to the basic decision-making and problem-solving cabability essential to any automatic control system.

The power and flexibility of the stored-program digital computer have found application in virtually every field of human endeavor that involves problem solving or information handling. The ever-increasing use of digital computers in so many areas has fostered extensive research and development in the technologies used in computer manufacture. The results of this effort have, in turn, led to an even wider economic use of computers in previously inaccessible applications. This cycle continues, fed by man's desire to solve bigger problems, his need to control the physical world, and his ability to generate ever-increasing amounts of information.

The diversity of applications, the economic desire to satisfy as many applications as possible with a general design, as well as almost daily changes in technology, historically have given rise to a great variety of computer designs. Control computers are no exception. Many designs have been implemented with specific processes in mind, but the predominant trend is toward more general-purpose designs that satisfy the requirements of a spectrum of control applications.

The computer designer has three goals: low cost, high reliability, and increased function. Every engineering design is a compromise among these objectives that attempts economically to meet the requirements of an application or spectrum of applications within the constraints of currently available technology. It is our purpose here to discuss some of these compromises in relation to the industrial control application.

This chapter is a general discussion of computer concepts and functions as they relate to computer control systems. Space does not permit us to include a comprehensive discussion of all facets of computer implementation and design techniques nor is such detailed analysis appropriate here. Rather we attempt to emphasize those features that are most important in computers designed for computer control applications. For the reader totally unfamiliar with computers, a discussion of the basic logical building blocks of computers and basic computer concepts is given in Appendix A. For more comprehensive discussions of detailed computer design, there are many excellent texts, several of which are listed in the bibliography (e.g., Richards, 1966; Hawkins, 1968).

2.1   CONCEPTS AND DEFINITIONS

Before proceeding with a detailed discussion of specific control computer characteristics and functions, it is useful to introduce some concepts and definitions. The first are basic to the understanding of computers and the others serve to describe the environment and to set the limits of the subsequent discussion.

2.1.1   Storage and the Stored Program Concept

There are two fundamental, and related, concepts at the heart of the digital computer. The first, that of storage or memory, implies the ability of the computer to retain and randomly recall large amounts of arbitrary data. The importance of this concept hardly can be overstated. It enables the performance of complex and lengthly tasks by providing the means of storing large volumes of original operands and the intermediate and final results of calculations.

The predecessors of the modern digital computer did not have storage except, perhaps, a few registers for the temporary storage of intermediate results. Data for calculations had to be provided in the proper order via an input medium such as punched cards. If the results of intermediate calculations required storage beyond a few operations, they were punched into cards that were reloaded as input for subsequent calculation. The requirement of sequential processing and the inability to store a substantial amount of data greatly restricted the flexibility of these early computers. The sequence of operations involved in a calculation had to be planned carefully to minimize the storage of the intermediate data rather than to optimize the calculation, often to the detriment of efficient use of the machine.

The introduction of storage capability greatly enhanced the power and flexibility of the computer. Data could be loaded into the machine in arbitrary order, often according to a natural sequence such as by serial number, instead of the order dictated by the calculation. Intermediate results could be stored indefinitely for use in subsequent operations, and this substantially reduced the manual manipulation of cards and other input/output media.

In the modern digital computer, various types of storage are provided that differ primarily in the cost per unit of stored information and in the time required to access the stored information. In general, the slower the access, the lower the cost. High speed, random-access storage is provided by magnetic-core or integrated-circuit storage units at relatively high cost. The time required to access any given piece of data, however, is typically less than 1 μsec and may be less than 100 nsec with integrated circuits. Bulk storage units utilizing drums, disks, or strips of material coated with a magnetic oxide provide storage at a lesser cost with access times in the tens of milliseconds range. Magnetic tape provides storage that is relatively inexpensive but that may have an access time of seconds if the desired information is located at the far end of a rewound tape. Special storage units having the capacity to store a trillion bits of binary information have been developed using photographic techniques (Kuehler and Kerby, 1966, 1967). Although the access time is slow compared with electronic calculation speeds, the unit cost of storage is very low.

The availability of this hierarchy of storage devices on the modern digital computer provides great flexibility and is a significant factor in reducing the cost of computing. By assigning data to one or another of the various storage devices on the basis of its frequency of use, the operation of the computer can be optimized for efficient operation and minimum storage cost.

The second fundamental concept involved in the modern digital computer is that of the stored program. The program is the detailed and ordered list of the operations that the computer must perform to produce the desired result. The elements of the program are called instructions. Each instruction causes some specific action to be performed, such as "add variables A and B" or "compare two numbers." The instructions are expressed as groups of numbers or characters and as such can be stored in the computer storage units and handled as any other data.

In early calculators the program was wired on a patch panel by physically routing signals using patchcords. Sequence signals were connected to particular logical elements to cause a sequential series of events to occur. This mechanical means of programming limited the size of the programs to 60 or 100 steps and, coupled with the lack of significant storage capability, severely limited program flexibility. With the introduction of the stored-program concept and the necessary storage units, programs of 10^5 or 10^6 instructions are not uncommon.

In addition to merely allowing programs of greater size, storage facilities and the stored-program concept provide other significant advantages. Since the instructions are stored in a form identical to that of any other piece of stored data, they can be considered as arguments in a calculation or data-manipulation operation. Thus the program can be written in such a way that it actually modifies itself. This capability allows the flexibility to alter the sequence of instructions executed and to vary the data operated upon dynamically as the calculations continue.

The stored-program concept also eliminates the physical modification of the computer whenever a new program is to be executed. In the wired-program machine, the wiring of the patch panel essentially resulted in changing the physical configuration of the machine. Whenever a new program was to be run, the panel wiring required physical modification. With storage capability, the new program is merely loaded into the computer storage, and execution is initiated at a given storage location. This facility also allows more than one program to be available immediately for execution. The various programs are stored in the computer storage units and the execution of any one of them is initiated by instructing the computer to begin execution at the starting address of the desired program. This allows the computer to switch between several different programs at electronic speeds.

Among the many factors that contribute to the capabilities of the present generation of digital computers and that lead to their ever-increasing use, these two concepts are basic. They provide the computer with the flexibility and ready access to data that allow the machine to be adapted to a huge spectrum of problems and to perform complex calculations in a highly efficient manner.

2.1.2   Performance and the Hardware/Software Trade-Off

Control computers, like others, are available in a range of prices and capabilities. The general performance of a computer can be defined as the ability to do the job and the speed with which it is done. The actual performance of a general-purpose computer, however, is difficult to measure because it has an absolute value only with reference to a particular job. The primary characteristics involved in performance are input/output (I/O) capability, calculating and decision-making speed, storage capacity, and the characteristics of the computer program. Suffice it to say here that it is the computer designer's goal to maximize the performance-to-cost ratio within the constraints of the system specification and the intended application.

Major differences in computing power are accomplished in two general ways: by using faster components and by providing additional hardware* functions. New electronic-component technologies are constantly being developed that provide higher speeds and/or lower costs. At any particular point in time, however, the cost of a computer is more a function of the number of components than it is of their speed. For example, both small and large computers use storage units with access times of the same order of magnitude, and their logic circuits have approximately the same switching speeds.

Compared with component speed, the implementation of additional functions in hardware is usually a more significant factor in increasing the power of a computer. The additional functions may be algorithmic or architectural in nature. Enhanced algorithmic functions are those that perform the same operations as smaller machines but make use of faster, more expensive algorithms or parallelism. For example, many different binary addition algorithms can be implemented. In a simple adder, a limiting factor on adding speed is the propagation of carries, since these are determined sequentially starting at the low-order bit position. A more sophisticated design anticipates the existence of a carry, thereby reducing propagation time delay and the total addition time. The necessary hardware and, therefore, the cost is greater in this design but it results in increased performance.

Architectural functions are those that affect the basic system design characteristics. Examples of architectural functions that enhance performance include a richer instruction set, a larger word size, a larger addressing capability and variety of addressing techniques, and I/O channel capability.

This chapter deals primarily with the architectural aspects of the performance problem. This is done for two reasons: First, the architectural aspects of the computer are those with which the user becomes best acquainted; second, and more important to the control computer designer, is the relative contribution to the cost/performance index of the two types of additional functions. Control computers are usually in the small to medium performance range and of an organizational type

---

* Hardware is a generic term that refers to the physical equipment and devices that comprise a computer system. The term "software" refers to the programs executed by the hardware.

in which the architectural changes have a more pronounced effect. Additional algorithmic functions become the predominant factor in enhancing the performance of very large machines with already rich architectures, but these computers are less often dedicated to control applications.

The fact that a particular function is not implemented in a computer does not mean that the computer cannot perform the function. Most of the additional functions implemented in more powerful machines can be provided in smaller machines by means of programming. For example, a large machine might provide a hardware implementation of extended precision arithmetic operations. These same operations can be provided in the smaller machine by using a program subroutine that operates separately on the more significant and less significant portions of the arguments. The advantage of hardware implementation is that it can perform the function in less time since only a single instruction is accessed from storage. In the programmed function, a number of storage accesses are necessary to fetch the multiple instructions and operand parts in the routine, and each instruction is executed separately.

The decision as to whether a function should be implemented in hardware or via a program (software) is a matter of compromise generally referred to as a hardware/software trade-off. The hardware implementation generally requires more hardware and, therefore, increases the cost along with the increase in computing power. The software approach reduces the hardware cost but with increased execution time for the function. The trade-off, then, is one of hardware cost versus improved computing speed or efficiency.

Another factor in addition to speed of execution is the amount of storage, which represents cost, occupied by the program. The relative importance of this factor is dependent on the likelihood of the function's reappearing many times in the program. If it is so complex it is best handled by subroutine, the storage space requirement is probably not significant regardless of its frequency of use.

In making the hardware/software trade-off, the designer must answer the basic question of whether the increase in computing power is worth the added expense. Many factors enter into this consideration. The hardware/software trade-off for each additional function must be evaluated in terms of the intended application of the machine and against alternative functions, both architectural and algorithmic. As noted previously, performance and, therefore, increases in performance, have absolute meaning only in relation to the particular application. This significantly complicates the decision in the design of a machine intended for a broad spectrum of applications. The factors that must be considered are the frequency of use of the function in the application, the ease with which it could be provided through software, the relative speeds of execution of the two implementations, and the computing speed requirements of the application.

2.1.3 Scientific and Control Computers

Many of the problems encountered in control and related applications are similar to those involved in general scientific computing. It follows that those characteristics and hardware/software trade-offs that make a good scientific computer generally are applicable to control computers as well. Indeed, because of the economies of mass production, many control computers are essentially the same as the machines marketed for general scientific applications except for the addition of facilities unique to the control application.

The scientific computer design emphasizes characteristics that facilitate the extensive computations used in numerical solutions of scientific problems. An analogy to the differences in the problems solved by bookkeepers and applied mathematicians is not too far-fetched. The mathematician deals with arrays of numbers of varying precisions and magnitudes as opposed to columns of integers. He performs the sequences of arithmetic operations in equation solution as opposed to simple summations and percentages. The designer of the scientific computer emphasizes computational efficiency at the expense of efficient data-handling capability for operations such as searching, sorting, collating, and editing. These types of operations characterize commercial record-keeping and information-retrieval applications, although they are found also in scientific applications to some degree. In most general-purpose designs today, an attempt is made to balance the capabilities so that both scientific and nonscientific applications can be handled equally well. Since virtually all applications require both computation and data handling, any trade-off between the two provides acceptable performance for a given cost. In computers designed specifically for control and scientific applications, however, the computational capability is particularly important and the computer design is often biased to provide large computing capability at the lowest cost.

Control computers are distinguished from other computers principally in that they are directly connected to the physical world. Through various transducers, sensors, data-conversion equipment, and control elements, the computer is in direct contact with physical variables and control points in the process or experiment. The computer therefore is required to operate and react in the same relative times and at the same relative rates as those associated with the physical variables and controls to which it is connected and with which it interacts. That is, the computer system must often operate in "real time." In addition, the action taken by the computer is often dictated by an event occurring in the process rather than as a result of an operator, program, or internal computer action. That is, the control computer is characteristically "event driven."

Two fundamental aspects of the control computer design are derived from this environment. First, an emphasis is placed on the ability of the computer to accommodate the specialized characteristics of input/output devices that interface to the physical world. These devices include those necessary to measure analog and digital quantities that determine the state of the process, equipment that provides control signals necessary to operate process devices, and devices that recognize conditions or events requiring immediate action. With these facilities, the computer can be programmed to recognize and respond to conditions and events external to the computer system according to predefined control algorithms.

The second aspect is the need to react in real time with corrective control actions involving extensive computations that must be done in a short time period and at random intervals as determined by external events. This demands a peak computing power substantially greater than that which would be required if the computing load could be averaged in time. In computer control applications the peak loads implied by real-time operation are often of the type ascribed to equivalent scientific applications.

2.1.4 Dedicated Versus Time-Shared Systems

As discussed above, most data acquisition and control applications present an uneven computing load to the control computer, varying in time with random changes and upsets in the external environment. Since the computer performance must be adequate to handle the peak computing demands that may occur only infrequently, the average utilization of the computer may be very low; less than 20 percent utilization is not uncommon in some applications. Significant computing power is unused under these circumstances when the computer is dedicated to a single application. Time sharing is one technique of utilizing this otherwise wasted power to do nonprocess related work, called background jobs, during those time intervals that are unused by the control job. As discussed in Chap. 11, time sharing may provide some of the justification for the computer installation.

The general system requirements for time sharing are the ability to switch rapidly between the background and control jobs and a storage capacity sufficient to handle both programs. Time sharing is largely a programming-system function, but certain hardware functions discussed later, such as the interrupt system and storage protection, enable it to be accomplished more efficiently and reliably. These functions are available in many control-computer systems and, therefore, time sharing is often a natural extension of their use.

The nature of the background jobs in the time-sharing applications can affect the design of the control computer. If the background jobs are primarily scientific in nature, the scientific base of the control computer design is adequate to guarantee efficient background operation. If, however, the background jobs are commercially oriented, functions that are usually associated with these kinds of applications may be added.

2.1.5 Control Computer Structure

Commercially available control computers have traditionally used the basic binary, single-address, fixed-word structure common to small scientific computers. The basic advantage of this organization is that a fairly complete and powerful machine can be built with a minimum number of logical elements and, therefore, a minimum cost. In addition, a large number of optional features can be fitted readily to the structure so that the same basic machine can be used to provide a broad range of performance capability starting with a very low basic price. However, other architectures may be selected for the control computer. They may be dictated by the economies of using a mass-produced, general-purpose machine, the desirability of having compatibility between the control computer and other computers in a total management information and control system, or the desire to increase the time-shared performance in a mixed commercial/control application. In either architecture, however, the design emphasis required to produce an efficient control computer remains basically the same. The discussion of this chapter is limited largely to the small binary machine, both because of its widespread use and because it illustrates simply the most important principles and trade-offs.

The control computer system can be divided into major components as shown in the block diagram of Fig. 2-1. This diagram differs from that of a general-purpose computer only in the inclusion of the blocks labeled "Process I/O" and "Process operator console." These functions provide the interfaces between the control computer system and the process instrumentation and the process operator, respectively. This equipment provides the various signal transformations and other interfaces that are the subject of detailed consideration in Chap. 3 through Chap. 8.

The other four components—storage, arithmetic and control, channel, and DP I/O and auxiliary storage—are functionally similar to conventional general-purpose computers. The storage unit provides for the storage of data and programs. The capability of retaining arbitrary data

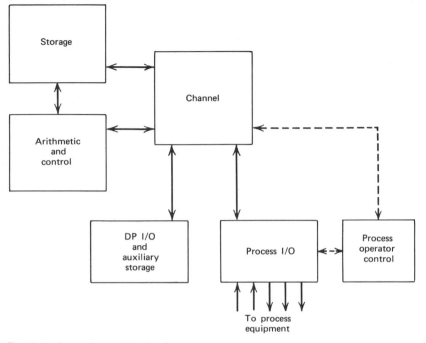

Fig. 2-1 General computer structure.

over a period of time is one of the most significant characteristics of the digital-computer control system for direct control that sets it apart from conventional analog controllers. The arithmetic and control section and the storage work in concert to perform the programmed instructions that direct the internal computer operations, such as data transfers with the input and output devices and arithmetic functions such as add, compare, and multiply. The channel provides the means of electrically connecting and controlling the operations of input and output devices. It provides the means of entering data and instructions into the storage unit and putting out the results of computations.

Although functionally similar to the components in a general-purpose computer, special considerations and functions are desirable for the control computer arithmetic and control section, storage, and channel. Among the primary considerations in the arithmetic and control section are the instruction set, word size, and addressing techniques. Storage protection is an important consideration in the design of the storage unit and its associated control logic. The characteristics of the channel are especially important in the control application. A wide variety of input/ output devices must be accommodated in an efficient manner and often with the additional requirement of high-speed data transmission. The channel and the arithmetic and control section also must be designed to respond to external events in the process. This usually takes the form of a hardware "interrupt" structure that can cause temporary discontinuence of a program in execution so that a special program can be executed to react to the external event. General-purpose computers also have this interrupt feature, but its primary use is in conjunction with conventional I/O devices such as printers and punches. In the control computer, a further refinement is often implemented in the interrupt hardware; classes of interrupts are established having a defined precedence of priority. In this case, an interrupt of lower priority cannot cause discontinuence of a program being executed on a higher priority level.

2.2   ARITHMETIC AND CONTROL UNIT

There are several important characteristics that vary widely in different digital-computer designs. These characteristics vary with different machine organizations, different performance levels, and the current state of technological development. Within the binary, fixed-word computer architecture at the lower end of the performance spectrum, the following characteristics may be considered the most critical: basic word size, instruction set, and addressing techniques. Within the framework of these major characteristics, other factors such as data and instruction formats, addressing range, and specific addressing techniques must be considered.

The following discussion is organized as a sequential consideration of the major characteristics listed above. In any real design situation, of course, they cannot be considered separately. The discussion is restricted in that no particular components or technologies are assumed. Only the ways in which the available technologies may influence the design parameters are mentioned here.

2.2.1   Word Size

The choice of the basic word size is the first important decision in the computer design. A word is defined in the class of machines under discussion as the amount of data that can be read or stored in storage in one operation. Word size forms a basic constraint on virtually all other system characteristics and has a direct bearing on cost in terms of the number of components in an optimally designed system. It also has an important effect on performance since the bandwidth (bits per second) of the storage unit is a fundamental constraint on performance.

The important hardware considerations affecting the choice of word size are the cost objectives as reflected in the number of logic circuits and the design point of available storage components. An optimum design almost invariably has one-word parallel data paths, and the number of logic circuits in data registers, gating, and arithmetic units increases almost directly with word size. Recognizing this fact, it can be seen that once the word size has been determined, a fundamental level of cost in terms of number of circuits has been established. The main storage cost is primarily a function of the number of bits of storage capacity, rather than of the word size. However, cost also varies with word size since many of the storage support circuits, such as sense amplifiers and inhibit drivers, are required in proportion to the number of bits in the word, whereas others, such as addressing logic, will decrease with increasing word size for a fixed total number of bits.

In addition to these hardware considerations, the most important characteristics that affect word-size selection are the formats of data and instructions, the two types of information that must be stored and manipulated in the computer. Data in the control computer come in the same four forms as in all computer applications. Fixed-point and floating-point numbers are the two numeric data forms especially suited to computation. Alphanumeric data consist of numeric digits, alphabetic characters, and special symbols such as punctuation marks and arithmetic operators. Data are required in this form for the bulk of the data-processing input/output equipment, such as printers and punched-card readers. The fourth type of data is bit-significant or logical data that represent the states of nonnumeric variables. That is, each bit in the word can represent an independent piece of information. Very often this data representation is used in control applications for binary states or events in the process such as on/off or out-of-limits/in-limits. Of the four types of data, consideration of the numeric data is the most important in word-size selection for the control computer.

Efficient utilization of the hardware is a constant goal. Ideally, all storage locations and other data-manipulating logic would be just large enough to hold a piece of information without wasting space. Information, however, comes in a variety of sizes or precisions, for example, Social Security numbers, batting averages, or the national debt. The desire to make the hardware efficiently fit data of variable precision has led some general-purpose computer designers to utilize variable word length designs in which words are made up of varying numbers of decimal digits or alphanumeric characters. Modern control and scientific machine designers, like those of the earliest machines, have found the fixed-word organization to be best. In control computers the bulk of numeric data with which the computer is concerned consists of physical variables such as temperatures and pressures. Although these types of data have a tremendous range of absolute magnitude, the precision with which they can be measured and controlled is generally equivalent to three, four, or at most five, significant decimal digits.

In addition to the basic required precision, there should be some spare bits for fixed-point computations in order not to make the programmer's "scaling" problem unmanageable. Scaling is the process of planning the location of the implied binary point when programming arithmetic operations with fixed-point arithmetic. The programmer must predict the variations in magnitude of the actual data during computation in order to avoid carry overflows and large losses of precision during arithmetic operations.

Another consideration in word-size selection is the representation of floating-point data. Floating-point computation requires a fairly complex set of algorithms, whether implemented as a programmed function or in hardware. There is a potential trade-off between speed of execution and storage efficiency based on the data format. If an entire word is devoted to the exponent and the fraction is in one or more separate words, the implementation is somewhat simplified since the exponent and fraction are handled separately in floating-point arithmetic operations. In addition, the fraction often has to be shifted, and this is performed most easily with separate words. This simplification and any resulting speedup of the operation, however, is usually at the expense of unused bits in the exponent word since the range of the exponent is typically limited to ± 64 (7 bits) or ± 128 (8 bits). If maintaining separate words for the fraction and the exponent requires that more than one word be used for the fraction even though there are unused bits in the exponent word, any benefit in the separate implementation is largely offset by the increased storage space required. For this reason, the decision is often made to use the extra bits in the exponent word to increase the precision of the fraction, even though this complicates the implementation of the floating-point arithmetic operations.

The discussion of the precision of variables often encountered in control applications in the fixed-point case also holds for the floating-point case, but with some modification. The implication of the floating-point representation is that the data are subject to extensive calculation, and some precision is lost in all practical computations. It is, therefore, desirable to provide extra bits of precision to minimize the rounding errors encountered in intermediate results.

Ten binary digits or bits are roughly equivalent to 3 decimal digits, 16 bits are over 4, and 20 bits equal 6. On this basis, therefore, one concludes that there is a word size somewhere between 10 and 20 bits with which most physical variables can be represented without too much waste. Fixed-word binary machines with as few as 8 bits and as many as 32 have been marketed successfully for control applications, the 16-bit machine being the most common. High-performance scientific machines use many more, from 32 to 64 bits being common.

The instructions that make up the machine programs are another class of information that must be stored and manipulated and, therefore, their format has a significant bearing on the word-size selection. There are two primary elements in most instructions: the operation code and the address. The operation code specifies the operation to be performed, such as "add to accumulator," "subtract," and "multiply." The address indicates the location in storage that contains the stored operands or is otherwise involved in the operation.* Some instructions do not involve a storage address and the address part of the instruction may be used for other control information or data. When the data are included in the instruction itself, they are often called "immediate data." In addition to these two primary parts, other modifier bits may be provided to represent further information about how the instruction is to be executed or to identify different instruction formats.

The number of bits required for an instruction depends on the size and capability of the computer as reflected in the number of different instructions to be implemented and the size of addressable storage. Because an instruction must be accessed from storage before any actual computation and this instruction readout takes time that is nonproductive, it is desirable to have the instruction formatted in a minimum number of words, preferably one. Single-word instructions also reduce the amount of storage required to store a program, thereby increasing the efficiency of storage utilization. In small machines with short words, however, two words may be required for some instructions to provide adequate addressing capability. Two formats, a short and a long, are often utilized in this case to increase storage and execution efficiency by requiring only the long format when necessary. In the case of two-word instructions, the

---

* The machine organization under discussion makes use of a register, usually called the accumulator, as the implied location when required by the operation for the second operand and result in arithmetic operations. Machines have been built with two and even three address instructions, with the intent of reducing the number of instructions required to perform functions, such as adding two operands that are in storage and put the result in storage. Any efficiency gained in the number of instructions, however, is offset by the inefficient use of bits when two addresses are not required and the relative frequency of immediately using the results of an operation for a subsequent operation. A popular variation on two or three address organizations has been to provide four to sixteen accumulators.

second is often used as the address. For this and other reasons, which are discussed in the following section on addressing techniques, it is convenient to have the word and the maximum address size coincide. If this is the case, the word size determines the amount of storage that can be addressed directly.

2.2.2   Addressing Techniques

One of the operations that must be performed by the computer logic in the execution of many instructions is accessing main storage to retrieve or store a word of data. Each instruction that requires access to main storage includes an address for this purpose. The number in the address portion of the instruction may be used without modification; this is called "direct" or "absolute" addressing. However, several algorithms have been devised for automatically modifying or otherwise interpreting the address portion of the instruction before it is actually used to access storage. The address that is used for actually accessing storage is called the effective address. In direct addressing, the address in the instruction is at once the effective address of the instruction. The other means of effective-address generation have been devised for one or more reasons of cost or performance, and each has its own particular uses, costs, and benefits. Listed somewhat in their order of importance, the basic addressing techniques are indexing and base addressing, indirect addressing, and relative addressing.

2.2.2.1   Indexing and Base Addressing.   One of the basic strengths of the digital computer is its ability to do repetitive calculations on large sets of data. Indexing is an address-modification technique that enhances this capability.

For the indexing function, the address portion of the instruction and another quantity are automatically added (or subtracted) to form the effective address. One or more registers are provided to hold the numbers that are to be added to the address portion of the instruction as they are executed. The number is called the index quantity and the register in which it is stored is the index register. Provision is made to modify quickly and easily the contents of index registers with one or more special instructions included in the instruction set, such as "increment index" and "decrement index."

The benefit of indexing is readily seen by considering the simple example where a series of computations must be repeated on each of m sets of n numbers. The sets of numbers are arranged in storage in contiguous groups of n locations. A program to perform the required computations is written using the addresses of the data set having the smallest addresses. The index register is loaded initially with $(m - 1)n$. When the program is first run, this quantity is added automatically to the addresses in the program to form the effective addresses. The program, therefore, operates on the mth set of numbers. At the completion of the computations on this set of numbers, one or more instructions are performed that subtract n from the index register and cause the program to be repeated, this time operating on the $(m - 1)$ set. The indexing instructions also test for the quantity zero in the index register before n is subtracted and terminate the "loop" if all the m sets have been operated upon.

The advantage of the indexing capability over the two alternatives is easily seen. Without this feature either the program would have to be rewritten with new addresses for each set of data or a subroutine would have to be added to the program actually to modify the address portion of each instruction. Both alternatives require more storage space, and the second also requires additional execution time.

In addition to simplifying the program, indexing has the advantage that, should either m or n need to be changed as a result of the addition of data, this is easily accomplished without any changes in the bulk of the program. Only the initial quantity loaded in the index register or the amount n subtracted need be altered. Thus indexing allows generalized programs to be written that are independent of actual data-array dimensions. Many other arrangements of data and programs can make use of the indexing facility to enhance the problem-solving capability of the computer by reducing program size, programming effort, and execution time.

Since not all the instruction addresses in a program need to be indexed, a bit in the instruction word is used to indicate whether the address should be added to the index. In machines that provide more than one index register, the one that is to be used, if any, must be indicated, and more than one bit in the instruction must be used for this purpose.

The implementation of hardware indexing is relatively expensive. To be most effective, several index registers must be provided and the time taken for the arithmetic operation of generating the effective address must be small. In order to minimize the cost, some small machines provide only one register; others use actual storage locations as index registers, preferring to sacrifice execution time for the flexibility of multiple index registers at low cost. Although the limitations of having only one index register rapidly become apparent, the optimum number of hardware index registers that should be provided is not so clear. However, the most popular compromise of cost and flexibility is three. In addition to the registers themselves, an adder to generate the effective address is a large part of the expense in providing the indexing capability. Virtually all small and medium-sized machines use the main data adder for indexing, even though it may increase the overall instruction execution time.

Base addressing is virtually the same as indexing, and most often the same hardware is used for both functions. The differences arise primarily from intent of use. The contents of a "base register" are added to addresses appearing in the program as the instructions are executed. Base addressing is used most frequently to solve two different problems: relocation and addressing range. In modern computers with complex programming systems, it is necessary that machine language programs be written without knowing the absolute storage location of the data and programs that will be used at execution time. In fact, the data and/or programs may be relocated during the course of the program execution. Given the foreknowledge that base addressing will be used and that the instruction addresses will be increased by the base amount, the program can be written using storage-location addresses starting at location zero. The subsequent addition of the base address to the addresses at the time of program execution allows the program and/or data to be relocated arbitrarily before or during execution by setting the contents of the base register to the actual address of the beginning of the program or data.

The base-addressing concept also can be used to overcome an addressing limitation of the instruction format. If the word length is too small to permit an address sufficient to address the total storage, the address portion of the instruction can be added to a base number that is large enough to provide the required address offset. To avoid the necessity of performing the arithmetic operation of adding two quantities, with the accompanying cost and execution time, "block or page addressing" may be used. The "block address register" provides the high-order bits of the effective address by concatenating them with those from the instruction. Storage therefore is divided into blocks or "pages," each as large as can be addressed by the address in the instruction. When the address portion of the instruction is short relative to the effective address, it is often called the "displacement."

2.2.2.2 Indirect Addressing. Indirect addressing is a technique in which the address portion of the instruction represents not the address of the data, but the address of the storage location containing the address of the data. Such an address is called an indirect address since it refers to the data indirectly. Indirect addressing lacks the power and flexibility of indexing but can be used to provide some of the same advantages. Normally it also costs execution time because an extra access to storage is required to obtain the effective address. It can be implemented much more economically than indexing, however, and it is provided in many low-cost computers.

With indirect addressing, as with indexing, a program can be written with indirect addresses that refer to a table of the actual addresses of the data. This table then can be modified prior to execution to relocate the data or during program execution to loop on different data. As can be seen, this is efficient only if the same data locations are referred to several times during the program. Such is often the case. Indirect addressing also may be justified if it is desired that the program itself not be modified during execution.

2.2.2.3 Relative Addressing. Relative addressing uses the same means of generating an effective address as indexing, except that the contents of the instruction counter is used as the index and is added to the address in the instruction. The instruction counter is the register in the computer whose contents is used to fetch the instructions themselves from storage as they are to be executed; it is incremented to the next sequential instruction automatically as each instruction is fetched. The effect of relative addressing is that the effective address is always related to the absolute location of the instruction itself by the amount in the address field of the instruction.

Relative addressing can be used to advantage for program relocation and also is a means of compensating for a small address field in the instruction format in the same manner as block addressing. It is used most often when a program must refer to itself or some parameter carried within itself. When used for general data, however, it has the undesirable characteristic of requiring that the data and program always be located with the same relationship. If their relative positions ever change, the instruction must be modified.

Given a basic indexing capability, the implementation of relative addressing is inexpensive. Some interesting programming techniques can be used when the instruction register is thought of as an index register.

The determination of which addressing technique or techniques are to be included in a particular design is, of course, done on the basis of cost and performance trade-offs. Specifically, is the cost of implementation justified by the increase in either performance or usability?

2.2.3 Instruction Set

The instruction set is the vocabulary of the machine-level language that directs the operation of the computer system, thereby enabling it to solve problems and perform predetermined functions. The size or richness of the computer vocabulary, as well as the speed with which the individual instructions can be executed, contributes to the overall power of the computer. The number and complexity of the instructions also have a direct bearing on cost since additional logic circuits are required to implement each additional instruction.

The size of instruction sets in commercially available computers varies considerably. Very small but usuable machines have been built with as few as 10 or 12 instructions, whereas large high-performance scientific machines may have 100 or more instructions. The instruction set of the small machine consists of the very basic arithmetic and logical operations, whereas that of a larger machine includes instructions that combine more complex sequences and combinations of these basic operations. Therefore, the small machine can provide essentially any function that the large machine can; it simply takes more program steps, which means more execution time and more storage. Usually, the first constraint encountered as bigger problems are put on small machines is storage capacity, so minimizing program steps is important. In addition, the real-time characteristic of control applications puts a high premium on speed. Therefore, both speed and storage space are primary considerations. This may justify the implementation of a relatively comprehensive instruction set on what generally is considered to be a small or medium-scale machine.

The instruction set generally provides the ability to perform four classes of functions: manipulate data, test and make binary decisions, control machine status, and effect input and output operations. The particular instructions implemented in these categories are highly dependent on the overall computer organization and hardware features provided. Some instructions may not fit nicely into a particular classification since the distinction between data, instructions, and status may become hazy under certain conditions. In addition, instructions often perform more than one of the basic functions.

However categorized, the instructions must be viewed as building blocks of more complex functions and therefore must complement each other. Also, since they are to be implemented in the same machine, the possibility of shared data paths and control functions is economically attractive. In addition, maintaining similar concepts and implementation algorithms may make them easier for the programmer to understand and use and, therefore, has a salutary effect on the efficiency of the programs he writes.

It is useful to discuss computer instruction sets by reference to particular machines. Since it is impractical to consider all the many different instruction sets available on current computers, we consider just two in the subsequent sections. The machines are similar in architecture but differ in the key characteristic of word size which, in turn, affects the instruction-format and addressing capability. The implementation of their instruction sets is distinctly different. They provide, therefore, an opportunity to contrast two approaches to the selection and implementation of an instruction set that is suitable for control computers.

The two computers selected for this discussion are the IBM 1800 Data Acquisition and Control Computer and the Programmed Data Processor-8/I (PDP-8/I)* manufactured by Digital Equipment Corporation (DEC). The PDP-8/I is a member of a family of small-scale general-purpose DEC computers that has found wide acceptance in data acquisition and control applications (*Small Computer Handbook,* 1968). The PDP-8/I is a 12-bit, single-address, fixed-word-length, general-purpose binary computer that is representative of the small or "mini" computer. The IBM 1800 is a similar machine in that it also is a single-address, fixed-word-length machine that can be utilized as a general-purpose computer (IBM 1800, 1966). It is a larger machine, however, and utilizes a 16-bit word. It is representative of the medium-scale computers used in control applications. It should be carefully noted that these machines were not selected on the basis of being equals and that the subsequent discussion is in no way intended as a comparison of machines of the same class. The prices and capabilities of the two machines differ significantly. Thus the selection, as well as the intention of the subsequent discussion, is based on two popular machines of similar architecture in which the instruction-set implementation is distinctly different. As the discussion illustrates, both machines have advantages and disadvantages that must be evaluated in terms of a particular application.

In the next section, the architecture, instruction format, and addressing algorithms of these two computers are discussed briefly. In the subsequent sections, eight categories of instructions are discussed and certain comparisons between the implementation in the IBM 1800 and the PDP-8/I are made.

2.2.3.1 Instruction Format and Addressing. IBM 1800 instructions appear in either a long or a short format. The short format consists of one 16-bit word and the long consists of two words stored in sequential storage locations. The formats are illustrated in Fig. 2-2. Generally, the format only affects the effective-address algorithm. In some cases, however, it may also affect the operation, and these differences are noted in the subsequent discussion.

---

* PDP is a registered trademark of Digital Equipment Corporation, Maynard, Massachusetts.

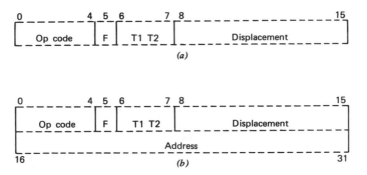

Fig. 2-2   IBM 1800 instruction formats. (a) Short format; (b) long format.

The one-word short format in the IBM 1800 includes four parts, as illustrated in Fig. 2-2a. The operation (Op) code is a 5-bit code specifying the particular instruction to be executed. There are, therefore, 32 distinct operation codes that are possible. As seen later, however, other bits in the instruction can be used as operation-code modifiers so that the number of distinct operations actually is greater than 32.

The format itself is identified by the next bit in the word, which is called the format or F bit; an F bit of 0 signifies that the instruction is short, whereas a 1 signifies the long format. Two bits, called the tag or T bits, specify which, if any, of the three index registers are involved in the operation. The remaining 8 bits are called the dispacement (Displ). The displacement has several uses but most commonly is used to modify the contents of the index register in effective-address generation. The long format, specified by F = 1, includes a second word called the address (Addr).

In instructions that access main storage, the effective address normally is generated by adding algebraically the contents of the index register specified by the T bits to the displacement in the short format or the address in the long format. All quantities are considered as two's complement numbers for this addition. Indirect addressing after indexing is provided in the long format under control of the first bit of the displacement, called the "IA bit." As there are only three hardware index registers provided, one state of the two T bits (T = 00) does not specify an index register. A tag of 00 in the long format indicates that the address is to be used without modification as either direct or indirect, depending on the IA bit. In the short format, a tag of 00 is used to specify the instruction counter (I) as an index register. This has the effect of providing a relative-addressing capability since the displacement is added to the instruction counter to derive the effective address. In all, therefore, there are 12 possible ways of generating the effective address for most instructions that access main storage. These are summarized in Fig. 2-3a.

	F = 0 Relative and indexed	F = 1, IA = 0 Direct and indexed	F = 1, IA = 1 Indirect and combination of indirect and indexed
T = 00	EA = I + Disp	EA = Addr	EA = C(Addr)
T = 01	EA = XR1 + Disp	EA = Addr + XR1	EA = C(Addr + XR1)
T = 10	EA = XR2 + Disp	EA = Addr + XR2	EA = C(Addr + XR2)
T = 11	EA = XR3 + Disp	EA = Addr + XR3	EA = C(Addr + XR3)

EA = effective address
I = contents of instruction counter
XRj = contents of index register j
$Addr$ = contents of address portion of 2-word instruction
$Disp$ = contents of displacement portion of 1 word instruction
$C(X)$ = contents of storage location X

Fig. 2-3a IBM 1800 address generation.

Address control bits	Effective address
00	EA = 0|Addr
01	EA = Current Page|Addr
10	EA = C(0|Addr)
11	EA = C(Current Page|Addr)

EA = effective address
$Addr$ = address portion of the instruction
$x|y$ = concatenation of x, a 5-bit page number, and y, the 7-bit address portion of the instruction

Fig. 2-3b PDP-8/I address generation.

Instructions in the PDP-8/I all appear in a one-word (12-bit) form. Two formats are used, one for instructions that reference locations in main store and one for those that do not. This latter group of instructions are called "augmented" and are divided further into two groups, I/O transfer and operate. The formats are illustrated in Fig. 2-4.

Three bits are provided for the operation code. Two of the eight possible codes are used to indicate the two types of augmented instructions. The six storage reference instructions provide 2 bits to control the addressing algorithm and 7 bits of actual address. The 2 address-control bits are called the indirect-address bit and the memory-page bit, respectively. Because of the limitation of only seven bits of address in the

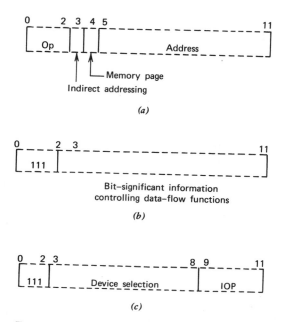

Fig. 2-4   DEC PDP-8/I instruction formats. (a) Memory reference instruction format; (b) operate instruction formats; (c) IOT instruction format.

instruction, a scheme of block addressing is used in the PDP-8/I similar to that described in the section on addressing. The 7-bit address portion of the instruction is concatenated with 5 high-order bits to provide an effective 12-bit address. The memory-page bit controls whether the 5 high-order bits are zero or equal to that of the current page. The indirect-address bit indicates whether the concatenated block address is to be used directly as the effective address or whether it specifies the storage location containing the 12-bit effective address. Figure 2-3b summarizes the four addressing algorithms possible. In order to facilitate operations in which indexing would be useful, a function called auto-indexing is provided. When certain locations of page zero are addressed indirectly (IA = 1, MP = 0), their contents are increased by one, as well as being used as the effective address.

The I/O-transfer form of augmented instructions uses the last 9 bits of the instruction to specify the particular I/O device and operation to be performed. Any data transfers to be performed take place between the I/O device and the accumulator.

The operate instructions utilize the last 9 bits to control operations within the arithmetic and control section of the machine without accessing storage. There are many of these functions, such as shifting the contents of the accumulator and incrementing the instruction counter, which are discussed later.

As will be seen, the shorter format of the PDP-8/I instruction not only affects the addressing but also greatly influences the selection of the instruction set. The limited number of storage-access instructions leads to the definition of multiple functions and a greater stress on the operate class.

2.2.3.2 Arithmetic Instructions. The arithmetic instructions are the most obvious members of the data-manipulating category. They provide the programmer with the basic calculating capability of the computer. Table 2-1 summarizes the arithmetic instructions available in the IBM 1800 and the PDP-8/I. The differences between the two computers are quite striking in this category of instructions. The only true arithmetic instruction provided in the PDP-8/I is the two's complement add (TAD), compared to the full complement of add, subtract, multiply, and divide in the IBM 1800. (An optional hardware feature on the PDP-8/I does provide these additional arithmetic operations.) Nevertheless, the PDP-8/I can be programmed to perform all the necessary arithmetic functions. For example, a subtraction can be performed by complementing the contents of the accumulator with the complement and increment accumulator instruction that forms the two's complement and then performing the TAD operation. This requires two instruction executions, however, so the relative execution time is greater.

Both machines utilize two's complement notation for representing negative numbers. This is common in modern computers in that arithmetic operations can be accomplished without specific reference to the signs of the operands. Appendix A discusses this notation and its advantages in more detail.

A consequence of addition and subtraction operations is the possibility

Table 2-1   Arithmetic Instructions

IBM 1800	DEC PDP-8/I
Add (A)	Two's Complement Add (TAD)
Double Add (AD)	Complement and Increment Accumulator (CIA)
Subtract (S)	Increment Accumulator (IAC)
Double Subtract (SD)	
Multiply (M)	
Divide (D)	

of generating a sum or difference with a magnitude too large to be represented in the accumulator. This is called an overflow. In two's complement notation the occurrence of such an event can be detected by an "exclusive or" of the carries, or borrows, out of the two high-order bit positions. A related situation arises when programming multiple precision addition and subtraction. The programmer must know of carries or borrows out of the high-order bit position when adding the lower order words of the operands.

In the IBM 1800, two testable 1-bit indicators, called the carry indicator (CI) and the overflow (OF), are used to enable the programmer to handle these situations. The CI is set or reset with each add and subtract instruction as appropriate. The OF indicator is set whenever an overflow occurs and remains set until tested by the programmer. This allows the programmer to test for overflows after sequences of arithmetic operations rather than after each add or subtract. In the PDP-8/I, the link (L) provides a function similar to the CI. In conjunction with the signs of the operands and results, the link can provide the programmer with carry and overflow information. Manipulation of the CI, OF, and L is provided by instructions such as the CML instruction listed in Table 2-3. Additional instructions involving the indicators are discussed in Sec. 2.2.3.7.

The complementing instructions in the PDP-8/I are required primarily to simplify the subroutines that provide the subtraction and division functions. The increment accumulator (IAC) instruction is provided since forming a two's complement requires the addition of one in the least significant bit position after complementing the register contents. The combined microinstruction CIA is a combination of the complementing and incrementing microinstructions. Although the equivalent of these operations can be performed using one or more instructions in the IBM 1800, they are not required since subtraction and division are provided as part of the basic instruction set.

The basic operations of add, subtract, multiply, and divide are more or less standard in machines of this type and size, although divide is often omitted. Divide is infrequently used relative to the others, is rather more complex and, therefore, is more expensive. It is usually implemented in hardware as a series of repeated subtractions, and consequently a programmed division is not very much slower in execution. With division, quotient overflow can occur if the divisor is small compared to the dividend.

In addition to these basic operations, the IBM 1800 includes double precision add and subtract because 32-bit operands are not uncommon and the implementation is not too costly as an accumulator extension Q has been provided for the multiply and divide functions. In the PDP-8/I, double precision functions are provided via subroutines.

Very rarely are more complex arithmetic operations such as square root or exponentiation implemented in hardware. Additional arithmetic instructions employed in larger general-purpose machines normally include the basic arithmetic operations with different data formats, such as decimal or floating point. Floating point is expecially useful in the general scientific computation associated with control applications and is included in some control computers. However, its implementation is relatively expensive and the program subroutine approach may be utilized instead.

2.2.3.3 Load and Store Instructions. The load and store functions are provided to transfer the contents of the accumulator to and from storage. Table 2-2 summarizes these instructions. In the IBM 1800 both load and store instructions are provided for both single and double words. The Load instruction causes the contents of the accumulator to be replaced by the contents of the addressed storage location. The Store instruction copies the accumulator contents into the addressed storage location without affecting the contents of the accumulator.

In the PDP-8/I, there is no specific load instruction. Rather, the addition operation (TAD) is utilized. In loading a new operand, this requires that the accumulator be cleared prior to execution of the TAD instruction either with a Clear Accumulator (CLA) or a Deposit and Clear Accumulator (DCA) instruction. The store function is provided by the DCA instruction that transfers the contents of the accumulator to storage and then clears the accumulator.

There is an interesting consequence of the slight difference between

Table 2-2   Load and Store Instructions

IBM 1800	DEC PDP-8/I
Load Accumulator (LD)	Deposit and Clear Accumulator (DCA)
Double Load (LDD)	Clear Accumulator (CLA)
Store Accumulator (STO)	Clear Accumulator and Link (CLA CLL)
Double Store (STD)	Set Accumulator (STA)
Load Status (LDS)	Clear and Increment Accumulator (CIA)
Store Status (STS)	Load Accumulator with Switch Register (LAS)
	OR with Switch Register (OSR)
	Clear Link (CLL)
	Set Link (STL)
	Get Link (GLK)

these two approaches to the load and store operations. In essence, the IBM 1800 clears the accumulator before loading, and the PDP-8/I clears following the store operation. A very common requirement in programming is to complete a calculation, store the result, and then load a new operand for the next sequence of instructions. The instruction sequence in this case would be:

IBM 1800	PDP-8/I	
. . .	. . .	
. . .	. . .	Final result of calculation
STO	DCA	Store final result
LD	TAD	Load new operand
. . .	. . .	Calculation on new operand

In both machines, the number of instructions executed in storing the final result and initializing the accumulator for the new calculation is the same. However, consider another common programming requirement —namely, the case where an intermediate result is to be stored, followed by continued calculation. In this case, the instruction sequence would be:

IBM 1800	PDP-8/I	
. . .	. . .	
. . .	. . .	Intermediate result
STO	DCA	Store result
	TAD	Reload result
. . .	. . .	Continue calculation

In this case, an additional instruction is required with the PDP-8/I instruction set when compared to that of the IBM 1800. This is an illustration of how slight differences in the instruction set can affect storage and execution efficiency. The advantage to the implementation in the PDP-8/I, of course, is that by using a CLA in conjunction with the TAD instruction, an additional load accumulator storage reference instruction is not required. With only a limited number of operation codes available, this is an important consideration.

In other machines, simple operations are sometimes combined with the load function. For example, instructions of the form Complement, Load Positive, and Load Negative are not uncommon. In these three instructions, the storage word is complemented, put in positive form if negative, and negative form if positive, respectively, prior to loading it into the accumulator.

The two IBM 1800 instructions Store Status and Load Status cause the status of the overflow and carry indicator bits to be placed in storage or set, respectively, from storage as a 2-bit quantity. The instructions usually are used in conjunction with the load and store accumulator instructions to retain and restore the state of the machine when an interrupt occurs. In the PDP-8/I, similar instructions provide for the manipulation in the link bit. The Store Status instruction is also used in the IBM 1800 to control the storage protect bits. Since this feature is not provided in the PDP-8/I, there is no equivalent instruction.

2.2.3.4 Logical Instructions. Logical instructions are similar in form to arithmetic instructions except that the functions performed are logical rather than arithmetic. The specified logical function is performed independently on each pair of corresponding bits in the two operands and is particularly important in control computers since many process data are bit-significant information. The logical instructions of the IBM 1800 and the PDP-8/I are shown in Table 2-3. Most machines of the IBM 1800 class provide the three most useful logical operators, AND, OR, and EXCLUSIVE OR. The implementation of these three operations can be economical by combining them with the binary adder since the AND and OR functions of the two operands may be used in implementing the adder and the EXCLUSIVE OR is the half-sum of the operands. Any other logical function can be implemented using not more than two of these three instructions.

The PDP-8/I implements two primary logical instructions, AND and the operate instruction Complement Accumulator. Any other logical function can be performed, although a few extra steps are required. Again, minimization of storage access instructions was important in determining which instructions should be implemented.

The uses of these logical instructions to manipulate data at the bit level are numerous and varied. To name a few, the AND is used to mask

Table 2-3   Logical Instructions

IBM 1800	DEC PDP-8/I
Logical And (AND)	Logical And (AND)
Logical Or (OR)	Complement Accumulator (CMA)
Logical Exclusive Or (EOR)	Complement Link (CML)

or extract bits; the OR is useful in combining elements of a word; the
EXCLUSIVE OR is useful in determining which, if any, bits in a word
have changed and for inverting bits; the Complement Accumulator, in
addition to logically inverting all the bits without storage reference, also
provides the arithmetic function of one's complement.

2.2.3.5 Shift Instructions. This is the fourth and last of the data-
manipulation instruction groups. The shift instructions provided in the
two computers being discussed are summarized in Table 2-4. These instructions cause the contents of the accumulator to be shifted to the left
or right within the accumulator. Since no storage access is required for
this operation, the portion of the instruction normally used for effective-
address generation in the IBM 1800 can be used to indicate the number
of bit positions shifted. The shift instructions require that the accumulator have the shift register function. This requirement is shared with
Multiply and Divide instructions whose implementation requires the
same capability.

The decision to include several different shift instructions is usually
made because of the need to handle both logical and arithmetic data
and to expedite the search for significant bits of a word. In the IBM
1800, logical data are handled by Shift Left Logical, Shift Right Logical, Shift Left Logical A&Q, and Rotate Right A&Q, the latter two shifts
involving both the accumulator and its extension Q. The Rotate Right
action differs from the others in that bits shifted out the right end of Q
are reinserted in the vacated positions at the left end of the accumulator.

The difference between logical and arithmetic shift operations is in

Table 2-4   Shift Instructions

IBM 1800	DEC PDP-8/I
Shift Left Logical A (SLA)	Rotate Accumulator Right (RAR)
Shift Left Logical A&Q (SLT)	Rotate Two Right (RTR)
Shift Left and Count A (SLCA)	Rotate Accumulator Left (RAL)
Shift Left and Count A&Q (SLC)	
Shift Right Logical (SRA)	Rotate Two Left (RTL)
Shift Right A&Q (SRT)	Clear Link and Rotate Accumulator Left (CLL RAL)
Rotate Right A&Q	Clear Link and Rotate Accumulator Right (CLL RAR)
	Clear Link and Rotate Two Left (CLL RTL)
	Clear Link and Rotate Two Right (CLL RTR)

dealing with the sign bit, the first bit in the word. In logical shifting, the first bit is shifted along with the other bits since it does not have a unique significance. In the arithmetic shift, the sign bit must be considered separately. In a shift right arithmetic operation, the vacated bits at the left end of the accumulator are loaded with the state of the sign bit, thus maintaining the algebraic sign of its contents.

The shift and count instructions in the IBM 1800 are used to search for significant bits in a data word. A maximum shift count is specified in the instruction. This count is decremented as the contents of the accumulator are shifted left. When a 1 is detected in the high-order position of the accumulator, shifting is terminated and the reduced count made available to the programmer in the index register. A similar instruction, called Normalize, is available on the PDP-8/I when the extended arithmetic element option is added.

There is less apparent similarity between the IBM 1800 and the PDP-8/I in this class of instructions. The PDP-8/I instructions emphasize the rotate instructions and, because of the lack of the ability to indicate a count in the instruction format, limit shifting to either one or two positions per instruction execution. The instructions are logical rotation instructions so that arithmetic shifts must be programmed by manipulating the link bit, which acts as a bit of the accumulator in the shifting operation with each shift. In the case of a single arithmetic shift, however, the rotate and clear link (CLL) instructions can be combined to provide the desired function in a single instruction execution time.

2.2.3.6 Branching and Testing Instructions. These instructions provide the basic decision-making capability of the program. Generally speaking, they test the value or state of a register or indicator and reflect the results of the test by a modification to the instruction counter. Thus the test may alter the normal instruction sequence of the program. Branching takes place when the instruction counter is loaded with an entire new value, and control is transferred to the new portion of the program. Skipping merely causes the instruction counter to be incremented by 1 so that the next sequential instruction is skipped. Because testing of conditions in the computer is done so often, it is the computer designers' objective to provide a fast and economical means to determine the state of specific indicators and to transfer program control accordingly.

As shown in Table 2-5, the PDP-8/I provides two jump or branch instructions that require an effective address. In addition, it provides a number of testing instructions that skip conditionally, depending on dif-

Table 2-5   Branch and Test Instructions

IBM 1800	DEC PDP-8/I
Branch or Skip on Condition (BSC)	Jump (JMP)
Branch and Store Instruction Counter (BSI)	Skip (SKP)
Compare (CMP)	Skip on Zero Link (SZL)
Double Compare (DCM)	Skip on Nonzero Link (SNL)
	Skip on Zero Accumulator (SZA)
	Skip on Nonzero Accumulator (SNA)
	Skip on Minus Accumulator (SMA)
	Skip on Positive Accumulator (SPA)
	Jump to Subroutine (JMS)

ferent conditions existing in the accumulator or link. These instructions are provided in the operate instruction format of the PDP-8/I, which does not allow storage access. Branching must be done by a jump instruction (JMP or JMS). Typically when testing a condition, the skipped-over instruction would be a Jump (JMP) so that transfer to another part of the program would occur if the skip did not.

In the IBM 1800, skip instructions equivalent to those provided in the PDP-8/I are provided by a single short-format instruction (BSC) in which the 6 lower bits of the displacement, called the condition bits, are used to indicate the condition or combination of conditions on which a skip is to be performed. In the long format, this same instruction provides branching to an effective address and, in effect, is equivalent to the sequence of Conditional Skip, Jump in the PDP-8/I.

The 6 condition bits provide for testing six conditions of the accumulator and its two indicators, carry out and overflow. In addition to the two indicators (carry and overflow), the high- and low-order bits of the accumulator and the condition of all zeroes in the accumulator may be tested. In conjunction with the Logical Shift instructions any bit of a word therefore may be tested. The sense of the condition bits is arranged such that any of six arithmetic sign conditions of the accumulator can be ascertained in one step (zero, not zero, positive and nonzero, negative and nonzero, positive or zero, and negative or zero). It also provides for an unconditional branch or skip since branch on both zero and nonzero can be specified.

The short format of the BSI instruction in the IBM 1800 and the JMS instruction of the PDP-8/I are equivalent. These instructions provide for a branch to EA + 1 following the storage of the contents of the instruction register at storage location EA, where EA is the effective address. These instructions provide a way to return to the mainline program following the execution of a subroutine since the address required to link back to the mainline program is stored at the beginning of the subroutine. The second use is in the handling of interrupts. In the IBM 1800, the occurrence of an interrupt actually forces the execution of the long form of the BSI instruction with indirect address. A functionally similar operation is performed in the PDP-8/I with the effective address of zero.

The long format of the BSI provides the same condition-testing capability of the BSC. Thus it is equivalent to a Conditional Skip, Jump to Subroutine sequence in the PDP-8/I.

The compare instructions in the IBM 1800 provide a skip of zero, one, or two instructions, depending upon a Less Than, Greater Than, or Equal To comparison of the accumulator and a word from storage. This instruction is actually implemented as a Subtract and test of the result without altering the contents of the accumulator. In the PDP-8/I, the equivalent function requires the execution of a series of instructions during which the contents of the accumulator are modified.

2.2.3.7 Machine Status Instructions. The machine status instructions are summarized in Table 2-6. The Wait and Halt instructions cause the execution of instructions to stop and the machine to enter the "wait" state. In this state data channel (indirect-program-controlled channel) operations may continue in the IBM 1800, but instruction execution does not resume until either the START button on the operator's console is manually depressed or an interrupt occurs. In the PDP-8/I, the computer remains in the wait state until manually restarted by depressing the START or CONTINUE console keys. These instructions are often inserted temporarily in a program during the debugging process to assist in following the progress of execution. The wait state also provides a preferable alternative in the IBM 1800 to an "idling" or "do nothing" routine.

The Interrupt Turn On and Interrupt Turn Off instructions enable

Table 2-6   Machine Status Instructions

IBM 1800	DEC PDP-8/I
Wait (WAIT)	Halt (HLT)
	Interrupt Turn On (ION)
	Interrupt Turn Off (IOF)
	No Operation (NOP)

and disable the program interrupt hardware in the PDP-8/I. These same functions are provided as I/O operations in the IBM 1800.

2.2.3.8 Indexing Instructions. This group of instructions, shown in Table 2-7, is a direct consequence of providing the indexing function in the effective-address generation discussed in Sec. 2.2.2. Since the IBM 1800 supplies unique index registers, means must be provided in the instruction set for controlling them. It is also desirable to provide an easy, fast means of performing the index, test, and branch function common in indexed program loops. In the PDP-8/I, the index registers are specific storage locations so that the regular storage reference instructions are used to manipulate the registers in conjunction with the auto-indexing feature.

Load Index in the IBM 1800 provides the capability of loading an index register with an initial value from storage. Store Index stores the contents of an index register into the specified storage location. These instructions are similar in function, and therefore in implementation, to Load and Store Accumulator.

Modify Index and Skip is provided basically to allow simultaneous modification, testing, and branching. It is used to facilitate the counting and register modification in repetitive program loops, as discussed in Sec. 2.2.2.1.

This instruction is also a good example of how additional functions can be derived from the implementation of a basic function. The basic function in this case is to add a quantity, the data, to the contents of an index register. The data are thought of as "immediate," which in the short format of the IBM 1800 implies the value represented by the displacement; that is, the displacement is added to the specified index register or to the instruction counter. If the instruction counter is the augend, the effect is that of a multiple skip or branch to a point relative to the current instruction address. In the long format the address portion of the instruction is added to the content of the index register. The indirect-addressing bit (IA) causes the contents of the word specified by the address to be added to the index register.

Table 2-7   Indexing Instructions

IBM 1800
Load Index (LDX)
Store Index (STX)
Modify Index and Skip (MDX)

When no index register is specified (T bits zero) in the long format, a different function called Add to Storage is performed. The displacement is added to the contents of the storage location specified by the address and the sum replaces it. This additional function, like modifying the instruction counter, has little importance relative to the primary function of modifying index registers. The added cost is small, however, to provide the utility of modifying indirect addresses and other data directly in storage.

The counting and testing function desired in an indexed program loop is obtained by incrementing the instruction counter an additional time whenever the specified index register (or storage word) changes its algebraic sign or goes to zero as a result of the modification. This skip function provides the means of automatically dropping out of the program loop after the prescribed number of repetitions.

2.2.3.9 I/O Instructions. In both the IBM 1800 and the PDP-8/I, input/output operations are performed using a single-instruction format. In the PDP-8/I, this is the IOT format illustrated in Fig. 2-4c. This format provides for the specification of a device-selection address and the generation of control (IOP) pulses. By providing the ability to decode the device-selection field and to respond to the possible combinations of IOP pulses in the I/O device adapters, a wide variety of I/O devices can be controlled using the same instruction format. Although only a single instruction format is utilized, the number of distinct instructions is equal to the product of the number of devices and the number of control actions available for each device. There are, therefore, 512 distinct I/O instructions possible in the PDP-8/I.

In the IBM 1800, there is only one I/O instruction, called Execute I/O (XIO). It is used in both the long and the short format and, therefore, provides for effective address generation by means of all 12 possible algorithms summarized in Fig. 2-3a. Associated with each XIO instruction, and stored at EA and EA + 1, is a two-word input/output control command (IOCC) that provides the detailed information necessary to effect the I/O action.

The format of the IOCC is illustrated in Fig. 2-5. The first word

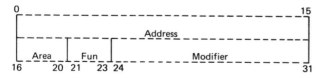

Fig. 2-5   IOCC format in the IBM 1800.

provides a 16-bit direct address, which is the source or destination of the I/O data in main storage for those devices and I/O actions requiring storage access. The second word of the IOCC provides data similar to those in the IOT instruction of the PDP-8/I. The area is a 5-bit device address field that may specify a single I/O device such as a card reader or a group of several devices. The 3-bit function (Fun) field provides for seven distinct control actions, such as read, write, and sense. The remaining 8-bit modifier field provides additional detail for either the function or the area code. Depending on the particular device, it may serve as a subaddress for some portion of a device, as an address of one of a group of devices specified by the area code, or as a refinement of the action specified in the function field.

In addition to controlling the data processing and process I/O subsystems in the IBM 1800, the XIO instruction and its associated IOCC are used to control other features of the system. Specifically, the internal timers, the console data switches (equivalent to the switch register in the PDP-8/I) and other console indicators and switches, the interrupt mask register, program-initiated interrupts, and the operations monitor ("watchdog timer") are all controlled utilizing the IOCC words.

The total number of devices and control actions that can be controlled by the XIO-IOCC combination is considerably greater than that possible with the PDP-8/I IOT instructions. On the other hand, the XIO and IOCC require three or four words of storage compared with the one storage location required for the IOT instruction. However, by combining the address of the affected data into the IOCC, the requirement for other storage instructions is avoided in programming I/O operations. That is, in the PDP-8/I, the I/O data must be retrieved or stored by means of preceding or succeeding TAD or DCA instructions, whereas this is not required in most of the IBM 1800 XIO instructions.

The final determination of which instructions should be included in a given design depends on several factors. The designer can ask himself several questions: Can the programmer implement the function performed by complex instructions by the use of several simpler instructions? Is the hardware implementation significantly faster than its programmed version? If so, will it be used often enough in production programs to provide a real overall performance increase, or should the cost be put into some other part of the machine, such as addressing techniques or additional memory?

2.3   STORAGE

As discussed in Sec. 2.1.1, the ability to store and subsequently to retrieve relatively large quantities of data and programs is fundamental to the operation of modern digital computers. Many techniques of performing this function have been developed so that a broad spectrum of storage devices offering speed, capacity, and cost trade-offs is available to the system designer. The design decisions made relative to storage are among the most important in the entire computer system. The overall performance of the system is very much dependent on both the speed and capacity of storage, and it represents a very significant part of the total hardware cost.

The storage elements of a system can be divided into two main categories: main store and auxiliary storage. Buffers can be thought of as another category of storage, but they are usually integral to some other primary function. For example, they may provide speed matching with relatively low speed electromechanical I/O devices or act in specialized functions such as CRT display refreshers. Main store appears as part of the central processor and is directly addressable by the program, whereas auxiliary storage is generally handled over the I/O channel as an I/O device.

2.3.1   Main Storage

The most important storage unit in the system is the main store. As discussed in the previous sections, the arithmetic and control unit accesses instructions from main store for execution, and the instructions obtain operands from it and store results in it.

The main store can be visualized as a sequential array of storage locations, each capable of holding one computer word and each having a unique address.* The addresses form a continuous set of integers from zero to $n - 1$, where n is the total number of locations in the storage unit.

The first popular stored-program computers used drum and delay-line techniques for main storage. The drums are coated with ferrous material and data are recorded magnetically on the surface. Read and write heads are mounted, one for each track, and the drum is continuously rotated on its cylindrical axis. By selecting individual heads, particular tracks can be read or written. Both drums and various delay lines are economically attractive, but they share the undesirable trait that the data are available only on a cyclic, nonrandom basis. They were supplanted as main storage units by magnetic-core memories. Drums continue to be used as auxiliary storage devices and delay lines find limited use in certain buffer applications. The advances into large-scale integrated circuitry,

---

* Very often words are subdivided for addressing purposes, usually to the character or byte and rarely to the bit, but for the purposes of this discussion, we shall consider the word as the smallest addressable unit. Each location or word in the main store is therefore directly accessible by means of its address.

although still more expensive than magnetic cores, are providing cost/
performance advantages with their high speed. They have the disadvantage that they are volatile; that is, the contents of storage are lost
with disruptions of power. Other techniques, such as thin films and
plated rods, are used for main stores, but they have not gained the wide
acceptance of magnetic cores or monolithic circuits.

The storage operation requires two basic functions—reading and
writing. Reading is the act of retrieving a previously stored word and
writing is the act of storing new data in storage. The basic timing of the
main storage unit is usually described as a "cycle." Depending on the
technology, a cycle may consist of a read and a write or it may consist
of only one or the other. The common term "access time" is the time of a
read operation. Most commonly, the cycle consists of a read followed by a
write to the same address since magnetic cores and some other memory
elements utilizing magnetic switching have a destructive readout characteristic. That is, the contents of a storage cell are destroyed by the act
of sensing its state. Therefore, every read operation must be followed
by a write to restore the contents of the location. The writing operation
in these technologies requires that the location be cleared prior to writing,
and this operation is the same mechanism as reading. With monolithic
electronic storage as well as with others that do not have these constraints, the designer has the freedom to define cycles which are simple
reads or writes. Other devices offer nondestructive readout but still require a clearing action before a write.

In either case, the speed (cycle time) of economically attractive main
storage units is usually both the limiting factor on operating speed and
the basis for the timing of other functions within the computer.

Two techniques of increasing the effective speed of main store often
are used in very large computers where the machine cycle is fast relative
to that of economical main storage units. One is to stagger in time the
cycles of two or more modules or sections of the storage unit so that reading and writing into different modules can be overlapped. Another is the
use of a very high-speed buffer of limited size, sometimes called a cache,
in conjunction with the larger, slower, less expensive main store. Taking
advantage of the fact that data and programs tend to be accessed
sequentially during high-activity periods that occur relatively infrequently, portions of main store are replicated automatically in the high-
speed buffer for use during the high-activity period. Changes to the data
in the buffer are then reflected back into main store as new high-activity
areas are transferred into the buffer. The net effect is to make all
storage appear nearly as fast as the buffer. Neither of these techniques
is common in current control computers.

2.3.2 Auxiliary Storage

Auxiliary storage devices fill the system's capability gap between the high cost of main storage and the very long access times of off-line storage. Data are said to be off-line when a manual operation is required to make them available to the system. Punched cards and magnetic tape are examples of off-line storage. Many billions of characters of data and programs are stored on magnetic tape. When it is desired to run these programs or to process the data, the manual operation of finding the proper reel of tape from the library and mounting it on the tape drive is required. This can take several minutes. Even after a tape reel is mounted, a significant amount of system time is required to search the tape for the requested information. The advantage of magnetic tape is, of course, the very low cost of data storage. When data or programs must be accessed more rapidly as a result of real-time requirements or too frequently to make magnetic tape economical because of system waiting time, on-line auxiliary storage devices are used.

Several types of bulk storage devices have been developed, providing a rather large spectrum of cost-per-bit versus access-time trade-offs. Most depend on magnetic-recording technology, not unlike the home tape recorder, as the storage means. The information is stored digitally as 1's and 0's rather than as an audio frequency signal. Magnetic drums, mentioned above as main memories, provide relatively fast access, but their relatively small capacities mean high cost. Drum capacities generally are in the order of hundreds of thousands of computer words. Access times are determined by the time, called rotational latency, for the desired location to come under the read/write heads. Since requests are usually random in time, the average access time of one-half revolution is usually quoted. The drums are turned at several thousand revolutions per minute and average access times of 3 to 20 msec are common.

At the opposite end of the cost/access spectrum of auxiliary storage devices in common use are devices utilizing magnetic tape strips. These devices have automatic electromechanical and pneumatic means of selecting and positioning strips of magnetic tape material under the read/write heads. The strips are kept in bins or cartridges and storage of hundreds of millions of words can be "on line." Access times vary but may be of the order of several seconds. Because of the relatively complex mechanisms involved in this kind of device, extreme reliability may be difficult to achieve; consequently this kind of device is seldom used in control applications.

By far the most widely used auxiliary storage devices are the disk storage units. Light aluminum disks between 1 and 2 ft in diameter are coated with magnetic material. From 1 to 20 of these disks are mounted on a spindle and rotated like phonograph records at several thousand revolutions per minute. Read/write heads are mounted on movable access arms, usually one per disk surface, which can position the heads on any of many "tracks." Usually the arms move together and may be actuated by either hydraulic or electromechanical means. A voice coil actuator is very common. The set of recording tracks, one per surface, that are under the heads at the same actuator position is called a cylinder. Each cylinder has the same characteristics as a magnetic drum. The effect is to provide an additional drum with each repositioning of the access arms. Up to 400 cylinders may be provided. The repositioning of the arm to a new cylinder is called a "seek." Arm movement requires from a few milliseconds to move to adjacent cylinders, to 50–100 msec to move from the innermost to the outermost cylinder. This seek time must be added to rotational latency to determine the total access time.

All auxiliary storage devices except magnetic tape have the characteristic of "random accessing." That is, any piece of data or location in the storage device can be accessed directly. This is in contrast to magnetic tape, where the information on the beginning part of the reel must be passed in order to gain access to that at the end. This random-access characteristic requires a method whereby the physical location of the information to be accessed can be specified by an address. Addressing in auxiliary storage devices nearly always specifies a "record" or "block" of data rather than a single word, as in main store. Addresses of this type then have components that refer to cylinder, track, and record number, respectively. The record number may be an actual angular position or "sector," or it simply may indicate the nth record from the beginning or index of the track. Variable length record formats are usually of the latter type.

Each record normally contains a copy of its own address so that the position of the head and disk can be verified by the hardware or software before any reading or writing of data is performed.

2.3.3 Storage Protection

Storage protection is the term used for any of several hardware means of preventing improper access to certain areas of storage. There are perhaps three basic uses of storage protection although, as with many computer functions, the ingenuity of programmers has led to other unforeseen uses of specific hardware functions.

The first use of storage protection is to prevent the erroneous alterations of programs or data that are essential to continued operation of the system or critical to the application. This is very common in control applications in which certain parts of the program concern portions of the process in which loss of control could be costly or even dangerous. Alteration of these programs by operator or system error must be prevented. Protection for these programs is especially important in time-shared applications where undebugged programs may be in the system at the same time as the control program, so that the possibility of inadvertent alteration is greatly increased.

A second and more complex use of storage protection has to do with space allocation in the main store. In many situations there is a supervisor program that allocates and schedules the storage space to various programs: problem programs, compliers, utilities, and so forth. There is a need to protect these programs and their data from each other.

The third use has to do with privacy. Many times the system contains information that the company or individuals do not wish to be available to the curious or unauthorized operator or programmer. This is a very difficult problem to solve in a general way. Unlike the first two cases, it applies primarily to auxiliary storage, as there is no practical way to prevent the unscrupulous operator from determining the contents of main storage. To make it impossible for him to do so would unreasonably constrain the maintenance procedures of hardware and programs.

These diverse protection objectives have given rise to several means of storage protection. The storage data may be "write protected"—that is, the information may be read but not altered; or it may be "read and write protected"—that is, data may not be accessed except under certain conditions. Read protection, where information may be written but not read, would seem to have value only in solutions to the privacy problem.

Storage protection schemes on auxiliary storage in current systems are intended primarily to guard against hardware failures, although supervisory programming systems may use them as a measure of protection against undebugged or faulty problem programs. As an example, in the IBM System/360, a mask character must be provided to the storage device. The character indicates whether writing and/or access-arm movement is permitted by subsequent commands.

In the IBM 1800 Data Acquisition and Control System the design objective was to provide the protection for programs and data essential to the operation of the system or process. An extra bit, called the storage protect bit, is provided with each word in main store. When the storage protect bit is set to 1, no alteration of that location is permitted; to 0, the write function is performed normally. The protection is simply provided by inspecting this bit in the storage buffer register whenever a word is read. If a 1 is detected during the read portion of the storage cycle, the contents are automatically restored in the write portion of the cycle and an error interrupt is generated if an operation which would alter the contents of the location was specified. A special instruction is used to set and reset the storage protect bit. The instruction is operative only when a console switch is in the ON position. At initial start-up time this instruction, in conjunction with the console switch, is used to set the storage protect bits in the locations containing the programs to be protected. Afterward, the switch is put in the OFF position to prevent their accidental reset. Because the objective is to protect vital programs and data, changing the protection bits is difficult and there is no way to override them.

The IBM System/360 provides a more comprehensive form of storage with read and write protection as well as write protect. Protection is on blocks or segments of main store instead of on a word-by-word basis, as in the IBM 1800. Each block of 2048 main storage locations has a 4-bit key register associated with it. These key registers can be set to any of 16 states by a special instruction. Whenever the main storage is accessed, a 4-bit key which has been assigned to the current program accompanies the storage address. This key must match the contents of the key register associated with the block of storage containing the address in order to effect the access. Each of 15 programs thus can be constrained to operate in certain blocks of main store. A storage-allocation program or supervisor assigns the keys to programs and storage blocks. The supervisor itself uses a key of zero, which allows it to access any location regardless of the associated key-register contents. The ability to override the protection system and the ease of changing the keys makes the system more appropriate to the second use discussed above. This was, of course, the primary intent of the designers.

2.4 PROGRAM INTERRUPT

The computer is basically a sequential machine performing the instruction steps of the stored program, one by one. Although it performs these steps very rapidly, it is solving but one problem or part of a problem at a time. In very small computers dedicated to a single task, the program proceeds from start to finish without interruption; that is, the computer executes the instructions starting at the beginning of a particular program and continuing until the solution is complete and/or until the program itself relinquishes control to the operator or another program. In the general case, however, this mode of operation results in inefficient use of the computer resources and, in many cases of computer control, may be unable to satisfy the application requirements. A means of interrupting the normal sequence of the program in progress in order to perform some other more important task and then return to the original program is called "program interrupt." This function is an essential characteristic of control computers for the detection and recognition of events in the process. The logic that provides this capability is called the "interrupt system."

2.4.1 General Concept

The two primary motivations for the interrupt concept are easily illustrated with examples. Consider a case in which the control computer has as one of its tasks the monitoring and controlling of a fuel-air ratio in a combustion chamber for reasons of safety and burning efficiency. Instrumentation that detects both a high and a low ratio has been installed and is connected to the computer data-acquisition subsystems. If either limit is violated, the computer must read a number of process parameters, determine new control settings, and effect a control action. All of this must be done within a short time interval, say, 5 sec, after the violation of limits occurs if the safety of the process is to be guaranteed. The calculations themselves take 1 sec and effecting the control action of moving the fuel valve may require as much as 3 sec.

The consequences of the inability of a control computer to cope with the requirements of this example are clear if the interrupt capability is not available. The computer may be initiating an optimization calculation that requires several minutes of computing at the instant the violation of limits occurs. Without the interrupt capability, it obviously could not respond to the process event in the required time.

This is only one of the many examples found in control applications that emphasize the need for an interrupt system in a control computer. The basic requirement in each case stems from the need for the control computer to be event-driven; that is, the control computer must respond to events in the process in a timely and efficient manner. In many control computers, a refinement of the interrupt concept is implemented also. Specifically, priority between events can be established so that the most important events are serviced first. This is called "priority interrupt" and is the subject of the next section.

The second primary motivation for the use of the interrupt function in computers is the efficient use of computer resources. In this case, it is primarily a matter of efficiently utilizing the computing capability of the machine while concurrently using the I/O resources with maximum efficiency. The optimization problem is based on the usual orders of magnitude difference in the operating speeds of the processor and I/O units. The processor can execute up to a million or more instructions per second compared to the relatively slow I/O devices; an example of such a slow device is a typewriter, which typically operates at only 10 to 30 operations per second. It is clearly inefficient to suspend processor operations whenever a character is to be printed since the processor can execute thousands of instructions in the time required to print a single character. However, it is also inefficient to operate the typewriter at less than its maximum speed.

The use of the interrupt system to control an I/O device is illustrated by the following example. The electric typewriter is attached to the computer such that the execution of the instruction Write causes a character to be transferred from computer storage to the printer, initiating the electromechanical operation of printing the character. This typically takes only a few microseconds, but the mechanical printing motion that ensues takes about 50 msec. The mechanical motion must be completed before another Write instruction can be executed to send another character from storage to the typewriter.

The completion of the mechanical motion causes a program interrupt that transfers program control to a particular location in storage. The instruction in this location is the first step in a short subprogram that operates the printing device and that consists of four basic steps: (1) the Write instruction mentioned above; (2) a means of selecting the next character to be printed, such as modifying the data address in the Write instruction; (3) a simple counter or other means of determining when all the characters to be printed have been sent; and (4) a return to the interrupted program. The execution time of this subprogram is so short that the typewriter can be operated easily at its maximum rate with very small effect on the program currently being executed in the processor, even though it is constantly being interrupted by the print routine.

The interrupt system is used in conventional computers to facilitate the operation of I/O devices concurrently with computing, as illustrated by the example. In addition, it is used to react quickly to detected error conditions or other unusual events within the computer system. In the control computer it is used for these purposes as well but, more importantly, it provides the basic means of responding to events in the process to which it is attached and for whose control it is responsible.

The interrupt function may be implemented in a number of different ways in a computer, and it provides a good illustration of hardware/software trade-offs. The implementation could consist entirely of software, or varying portions of it could be implemented in hardware. In the basic interrupt function, there are four actions that must be performed: (1) detect an interrupt condition and provide an indication of its source; (2) save the necessary information to resume the interrupted program subsequently; (3) transfer control to the interrupt service program, including the contents of all registers and indicators; and (4) return control to the interrupted program after the interrupt servicing program has been executed.

These actions all could be performed by the program. The computer program being executed would periodically sample the various inputs from the process and indicators internal to the computer at a rate determined by the required speed of response. In the examples given above, the typewriter status would be tested at 50-msec intervals and the air-fuel limit detectors would be monitored less often, perhaps every second. If an interrupt conditions were detected, the additional actions required by the interrupt function would be performed by the program.

Two significant reasons make this an unsatisfactory technique. First, all programs run on the machine would have to be aware of, and include, the monitoring function. This places an impossible burden on the programmer in all but the simplest system, since he must provide the necessary transfers to the monitoring function at periodic intervals. This requires that he accurately time the execution of groups of instructions in his program and include the necessary program steps to effect transfer of control. Furthermore, the procedure implies that he must store information concerning the state of his program every time that he transfers to the monitoring routine. In addition to this direct programming burden, the number of instructions executed is increased, thereby increasing the execution time and storage requirements.

Second, if the product of the number of conditions that can cause an interrupt and their respective sampling rates are summed, a total system capacity is derived that typically dictates a much higher level of computer performance—and therefore a much higher level of cost—than is necessary if hardware to detect interrupt conditions causes an interrupt automatically. With the appropriate hardware, the computer can be performing some other useful work, secure in the knowledge that the alarm condition will be recognized and responded to by the appropriate program through a program interrupt.

At the other extreme of the hardware/software trade-off, hardware that performs the total interrupt function can be provided. Generally, however, both control and general-purpose computers utilize a hardware/ software trade-off somewhere between the two extremes. Varying amounts of the status-switching and -saving functions are done by the hardware with the remainder left to the programmer. The minimum functions that must be performed by the hardware in a practical trade-off are saving the current contents of the instruction counter and loading it with the storage location of the interrupt service program to which control is to be transferred. The saved contents of the instruction counter is called the "link address" and is used to return or "link back" to the interrupted program after the interrupt program has been executed. The link address usually is stored at some fixed location or in a location related to the interrupt service routine. The new value of the instruction counter may be obtained from a fixed storage location. If these are the only functions performed by the hardware, the interrupt program must store all other information that is required to return control to the interrupted program. When control is returned, the machine status, including the contents of index registers and the accumulator, must be identical to that when the interruption took place. This means that the interrupt program must save this status and return it before linking back.

In some machines, this storing and restoring function may be performed automatically, usually with an increase in both performance and cost. An important alternative is to provide one or more additional sets of registers and indicators for use by the interrupt program. This is even more costly but the time taken for storing and restoring is eliminated, which improves both response and execution time.

In the IBM 1800, the interrupt function is identical to a Branch and Store Instruction Register (BSI) instruction with indirect address, which is discussed in Sec. 2.2.3.6. Indeed, the interrupt logic actually forces execution of this instruction with an address determined by the interruption source. The operation, therefore, is to get the effective address EA from a fixed storage location specified by the interruption source, store the link address at EA, and begin executing the program at EA + 1. An Interrupt Status indicator is also set, which is reset when control is returned later to the interrupted program.

2.4.2   Priority Interrupt

Interrupting conditions may arise from many sources. Many types of process signals as well as I/O devices and conditions internal to the computer itself may be connected to the interrupt system. Some means of establishing priority between these various interrupt sources must be provided since some obviously are more important than others. For example, an interrupt from the clock signaling that an hourly optimization program is to be run should be ignored if the computer is currently responding to a hazardous condition in the process.

The simplest priority-interrupt scheme is to permit only one interrupting condition to be acknowledged at a time. An interlock is provided to prevent any interrupt program from being interrupted itself. As one would suppose, this is the least costly method and works satisfactorily if the running time of the interrupt servicing program is less than the required response time for the most important interrupt condition. In the general control application, however, this is not the case and more sophisticated techniques are required.

A more general solution to the problem arising from interrupt sources of varied importance is called priority interrupt. In the usual implementation, the interrupt sources are assigned to "levels." Each level operates in basically the same manner as described in the previous section, except that the levels are assigned priorities. Interlocks are then provided that allow an event on a given level to interrupt a program due to a previous interrupt on a lower level but not one on the same or higher level. A series of events occurring in order of ascending priorities then can cause a situation analogous to "nesting" subroutines; each interrupt program in turn interrupts the previous lower-priority program, and, as each is completed, links back and completes the interrupted program in order of descending priority.

The term priority as related to the interrupt system is often misinterpreted. It refers to the execution precedence of the program associated with a particular interrupt condition. It does not relate necessarily to the importance of the interrupting condition in the process. A seemingly unimportant function such as a typewriter interrupt may be assigned to a very high-priority level even though this causes it to interrupt a service routine dealing with an important process condition. As shown in Chap. 14, this might be done since the typewriter interruptions have an insignificant effect on the execution time of the other interrupt program and better efficiency can be obtained by such an assignment. The assignment of particular devices and conditions to the available priority levels is done on the basis of both their importance to the control of the process and the total efficiency of the system.

As in the case of implementing the basic interrupt function itself, there are a variety of hardware/software trade-offs that may be considered in providing priority interrupt. A software implementation can be supplied using the single-level interrupt described in the previous section. When an interrupt occurs, control is transferred to a general interrupt routine that saves the necessary status information and identifies the source of the interrupt. Rather than proceeding with the routine that responds to the particular condition, however, the service program determines the priority of the interrupt and then decides if immediate action should be taken. If action is to be taken, the program does the necessary "housekeeping" to ensure that the information to link back to the interrupted program is retained and then transfers control to the subroutine that actually services that particular interrupt. If the interrupting condition has lower priority than the program currently in execution, however, control is immediately returned to the interrupted program. When the interrupted program has completed execution, the programming system then transfers control to the program that services the lower-priority interrupt.

Using the software implementation, as many interrupt levels as the programmer desires can be created. This method, however, reduces the speed of response to an interrupt because of the overhead of interrupt source and priority determination, and managing the delayed interrupts. Unless steps are taken to prevent it, it also means that even very high-priority programs may be interrupted continually for short periods of time by conditions of lower priority. For this reason, the programmer is often given the capability of inhibiting or masking the recognition of an interrupt. This facility must be used with care since masking could result in important process conditions' going uncorrected for a long period of time.

The more common implementation of priority interrupt involves the use of hardware to do the priority determination and to retain interrupt signals delayed by priority. Hardware is provided to remember the priority of the last interrupt taken and current interrupt conditions not yet serviced. It is useful to think of this information as contained in two registers. Each bit represents an interrupt level and its position indicates its priority. These registers will be referred to as the current level register and the interrupt request register. The occurrence of an interrupt condition turns on the bit of the request register corresponding to the priority level to which the signal is assigned. It is compared with the current level register to determine that no higher or equal priority program is in progress. If there is, the request remains in a pending state. If there are none, the interrupt is taken, the corresponding bit of the current level register is set, and the request register bit is reset. The program is then said to be operating on that level. The program can determine the level on which the interrupt occurred by testing the contents of the register. The current level bit is then reset when "linking back" after the routine associated with the interrupt is completed. Note that several bits of the current level register can be on simultaneously as a result of successive interruptions of higher priority. The highest priority bit that is on is the most recent interrupt.

A third register, called the "mask register," may be provided to permit the programmer to disable interrupts by individual levels. As with the other registers, each bit is associated with a particular level and controls the taking of interrupts on that level.

In the previous discussion, the burden of identifying the level and source of an interrupt has been left to the programmer. This causes programming overhead on each interruption and increases the response time. Additional hardware may be justified to simplify this function. For example, interruptions on each priority level may cause transfer of control to a different location in storage. The programs at each storage location than need not be concerned with interrupt sources not on the same level.

A further refinement is to have a register associated with each interrupt level. The various interrupt signals assigned to a given level are connected to the bits in the register with, possibly, more than one device connected to a single register bit. Each bit represents a priority sublevel. When any bit in the register changes state, and assuming that a program of lower priority is currently in execution, control is transferred to the storage location associated with the particular register. The routine associated with that storage location determines which bit or bits in the word caused the interrupt and takes the appropriate action. This technique is faster than that previously described because it minimizes the need to interrogate the connected devices to determine the exact cause of the interrupt. A hierarchy of many distinct levels of priority can be obtained using this technique. For example, 16 words of 16 bits provide for a minimum of 256 distinct interrupt conditions in a two-level hierarchy. Additional sublevels can be provided using the software techniques described previously.

The priority-interrupt structure is a unique characteristic of many control computers. Some small machines provide only a single level of interrupt and implement additional levels through software. Many control computers, however, provide some degree of hardware priority-interrupt capability. The assignment of devices to individual priority levels is generally quite flexible, although the hardware or programming-systems designer may require that some devices be associated with particular levels. The user typically can define the assignments of devices associated with the process. The assignment in the computer may be done either by software or by physical changes in the machine. The assignment capability provides a control computer that is extremely flexible and that can be used to satisfy a wide variety of process requirements.

2.4.3 Interrupt Implementation

The actual logic implementation of the interrupt function is closely interwoven with the architecture of the CPU. Thus the design of the interrupt structure for a given system must consider the characteristics of the CPU with which it must operate. In order to illustrate the type of logic structure typically required, however, the following paragraphs describe two sample implementations. The first is a single-level interrupt structure where all external interrupt signals have the same priority unless, of course, the software explicitly or implicitly provides priority. The second example shows how hardware priority can be built into the logic such that lower level interrupt signals are inhibited if any higher level interrupt is active.

Figure 2-6 illustrates the logic for the single-level interrupt implementation. The external interrupt signals are used to "set" or actuate bit positions in an interrupt register. The outputs from the bits in the inter-

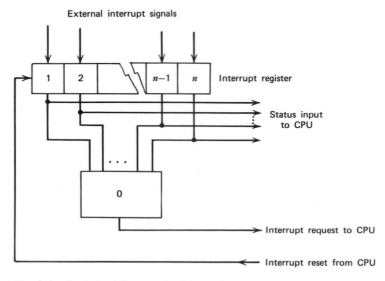

Fig. 2-6 Single-level interrupt implementation.

rupt register are ORed together to provide a single interrupt signal to the CPU. Whenever one or more bits in the interrupt register are set to the 1 or active state, the interrupt request signal is sent to the CPU. The CPU, either automatically or via software, can then read the status of the interrupt register and take appropriate action. After servicing the interrupt, the CPU issues a reset signal, which resets all the bits in the interrupt register. If all interrupt signals from the process have not been cleared by the service routine, a new interrupt request will be generated immediately following the register reset, and the procedure will be repeated.

In order to obtain higher efficiency and avoid other limitations inherent in the single-level interrupt approach, the hardware logic can be designed to provide priority or precedence in the processing of interrupt signals. An example of the logic required for this function is shown in Fig. 2-7. For clarity, the logic is shown in two levels of detail. Figure 2-7a shows

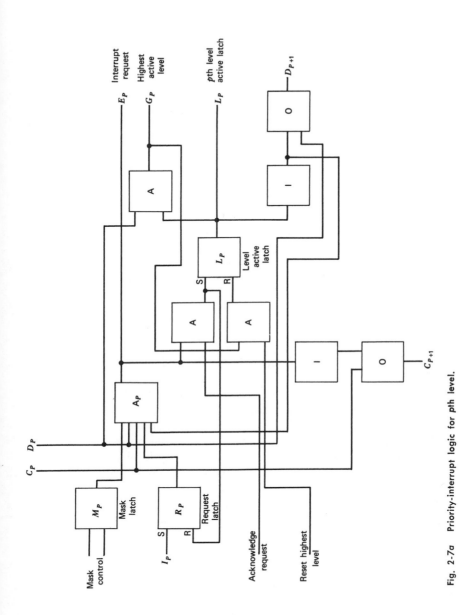

Fig. 2-7a Priority-interrupt logic for pth level.

103

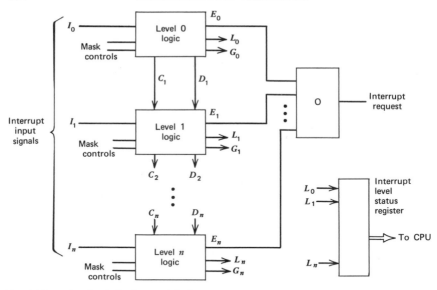

Fig. 2-7b *n*-Level priority-interrupt logic.

the detailed logic for each priority level; Fig. 2-7b shows how this level logic is combined to form an *n*-level priority-interrupt structure.

In the logic shown in Fig. 2.7a for the pth priority level, an external interrupt signal I_p sets the latch R_p indicating a request at this level. The actual signal E_p presented to the interrupting logic may be inhibited at AND gate A_p by several other signals. First is the mask latch M_p. The two signals C_p and D_p from the next higher priority level $(p-1)$ indicate that there is no higher unmasked request and no higher level active, respectively. The request is also inhibited if the pth level itself is active.

At the time the interrupting action takes place, the Acknowledge request signal and signal E_p set the Active latch L_p and reset the request. Signal G_p indicates the highest priority level active and, by definition, is the current level. It is also used when the interrupt task is completed to reset the Active latch and return to a lower level.

The interconnection of multiple levels is illustrated in Fig. 2.7b. The E, L, and G signal can be used by the CPU as logical registers as illustrated to determine the state of the interrupt system at any time. The L register will have bits on for all levels for which interrupting conditions have been acknowledged and G will have only one bit on, reflecting the current level of processing. The interrupt requests from each level are

ORed to provide the general interrupt request with the register of the bits E_p indicating the source level. Registers formed from bits M_p and R_p may also be made available to the CPU.

The automatic recognition and branching to a unique storage location that are commonly provided in priority-interrupt structures can be implemented as shown in Fig. 2-8. The status register described in the discussion of the previous figure provides inputs to address-coding logic. Since only one bit of the status register is in the 1 state at a given time, this address-coding logic is very simple. The output of the address coder is a unique address associated with each bit in the status register. This address is concatenated with fixed signals that define an operation code and other portions of an instruction such as those for the Branch and Store Instruction Register (BSI) instruction discussed previously. When an interrupt request R is pending, this artificial instruction is gated into the CPU instruction register when the execution of the current instruc-

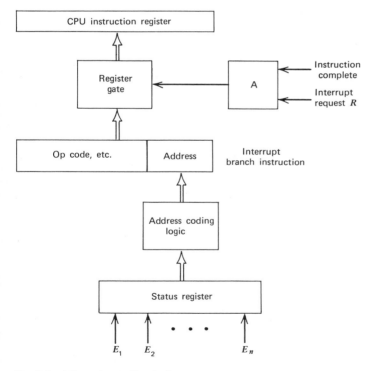

Fig. 2-8 Interrupt-execution logic.

tion has been completed. The next instruction to be executed, therefore, is the branch to the unique address created from the inputs from the interrupt status register. The processing then proceeds as described in Sec. 2.4.2.

2.5   INPUT/OUTPUT CHANNEL

The I/O channel provides the means of attaching I/O devices to the main store and the central processing unit. I/O devices, both those of the conventional data-processing type, such as punched-card readers, and those connected to the sensors and actuators of the process, have a wide variety of functional characteristics and data rates. The function of the channel is to provide the central processor with the ability to operate or control this diversity of devices and to control the transmission of data between them and the computer and/or storage.

The I/O devices differ in the rates at which they can send or receive data as well as in the amount of data usually transferred in a single operation. For the purposes of this discussion, the data transferred as the result of activating the channel are called a "record." Normally, a record consists of many characters or words. In most data-processing I/O devices, the differences are governed by the electromechanical characteristics of the device or the data medium. Punched cards, for example, are handled by an electromechanical transport that determines the maximum data rate, and the card itself is usually limited to 80 characters. Process I/O devices, those devices that are connected to the process, have speeds limited by such factors as analog-to-digital conversion speeds and relay operating speeds.

A simple I/O device is a relay that can be opened or closed by the computer. This device essentially requires that one bit of information be transferred to it at the relatively infrequent intervals when a motor should be started or a light turned on. This simple device is to be contrasted with large data-storage devices, such as disk files, which require several steps to perform the conceptually simple function of storing data for future reference. Such a device may require a "seek" or mechanical movement of read/write heads, a search for the proper record and, finally, the actual data tranfer, which can take place at a rate of several million bits per second. Between these two extremes of disk file and relay, there is a vast array of devices in a broad spectrum of data rates and degrees of complexity, many of which may be found at any particular installation. It is the function of the channel to accommodate this diversity of devices in a manner efficient in both implementation and usability by the programmer.

2.5.1 Overlapped Channels

In early computer systems the I/O operations were controlled by instructions that, like arithmetic, completed the specified operation before proceeding to the next sequential instruction. For example, an instruction might specify the printing of an entire line of characters on an electromechanical printer. When this instruction was encountered in a program, the next instruction (which might begin the assembling of the next line to be printed) would not be executed until the electromechanical motion of printing the entire line was completed. The arithmetic and logical circuitry and main storage essentially would be idle during the printing time.

The inefficiency of this mode of operation became more and more dramatic as the speeds of electronic circuitry and storage increased by orders of magnitude while the speeds of electromechanical devices remained relatively stable. This led directly to the concept of overlapped input/output operations where the I/O instruction simply initiated the input or output operation and proceeded to the next instruction in the program. The electromechanical operation was then performed concurrently with the continued execution of the program in the CPU. The CPU would thus be utilized for computational and other tasks much more of the time.

When I/O operations can be overlapped with CPU operation, the next step is, of course, to overlap them with the operation of other I/O devices as well. All but the most elemental of modern machines are designed to permit the concurrent operation of many, if not all, the attached devices and the CPU.*

Two general modes of overlapped I/O channel operation usually are available in modern control computers to accommodate the wide range of I/O devices required in most installations. The first, and simplest, will be called the "direct program controlled" mode. The second and more sophisticated mode is "indirect program controlled." (Various names are used by different vendors to describe this type of channel; these names include "cycle steal channel," "data channel," "direct memory access channel," and "data break channel.") Both control modes may, and

---

* Even with this capability, normal installations very often experience very low CPU utilization. Consequently, one of the most fertile areas for computer designers and architects is the attachment of I/O devices. The difficulty of the programmer in keeping track of multiple I/O operations and their associated tasks and the limitations on main store size make the simple addition of more and more I/O devices an unsatisfactory solution to low CPU utilization. Consequently, there is a continual search for better I/O devices and attachment means.

usually do, depend heavily on the interrupt system previously discussed for the efficient utilization of I/O devices.

In direct program controlled I/O operation, the input or output data-transfer operation is the execution of an I/O instruction. This is to be contrasted to the indirect mode of operation in which the data transfer takes place later during the execution of other instructions under control of the attached device. The differences in the implementation requirements of these two modes makes each appropriate for certain devices and design objectives. In general, the direct mode is more appropriate for low cost, slow speed devices such as typewriters, and the indirect mode is more appropriate for relatively expensive, high data rate devices such as magnetic tape and disk storage devices.

2.5.2   Direct Program Control

In early computers with unoverlapped channels, all I/O operations were performed under direct program control while other processor operations were suspended until the specified action was completed. For example, all activity in the central processor stopped while data from a card reader or magnetic tape were being loaded into storage or while a line of processed information was being transferred to, and printed on, a printing device. In order to overlap the operation of these devices with instruction execution, an electronic buffer large enough to hold all the data transferred by an instruction had to be provided. As buffers are expensive, this method of overlapping limited this means of attachment to devices whose data rates were slow enough to allow very short records to be transferred with each instruction without taxing the basic instruction execution rate of the CPU or to those that could absorb the cost of a large buffer. Certain punched-card equipment and line-printer attachments have been implemented this way.

Because of the cost of buffers and the provision of indirect mode channels for high data rate devices, direct program controlled operation in most modern machines is limited to the transfer of only one or two characters or computer words to or from the channel hardware for each separate I/O instruction. The execution of the I/O instruction is completed after the transfer and the processor proceeds to execute the next program instruction even though the I/O action may not be completed. Any mechanical or other relatively slow operation in the I/O device continues concurrently with execution of succeeding instructions in the computer. The interrupt system is used with this form of I/O control to signal the end of the relatively slow concurrent I/O operation so that the CPU can initiate the next I/O action. This allows the program to operate devices at their maximum rate while continuing to perform instructions in the main program at near maximum rate. If 10 characters of data are to be transferred to a printer, for example, ten separate I/O instructions must be executed. This form of attachment provides the lowest hardware cost, both in the CPU and channel and in the device, while requiring the program to transfer each character individually.

The details of the transfer of the data character or word vary in different computer designs. Basically, however, the following three fundamental pieces of information must be provided by the program instruction to effect a transfer: (1) the function to be performed such as read, write, or control, (2) the particular device to be used, and (3) the source or destination of the data involved in the operation. These three factors must always be present or implied although other information may be required for more complex operations or devices. The additional information is generally dictated by the nature of the device connected to the channel. For example, if the device is an analog-input subsystem, the address of the particular input point to be actuated must be provided. Because of word-size limitations, additional information such as this may be contained in the instuctions immediately following the channel instruction, or the channel instruction may specify the storage address of the additional information. Thus two or more instruction cycles may be required actually to initiate the I/O action.

Sometimes unique instructions for each device are defined, giving rise to instructions such as TYPE. This might be defined as causing a character in the storage to be transferred to the console printer for typing. In this case, the operation code alone specifies both the device to be used and the function. A storage address then must be provided if the source of the data in the CPU is in main storage. This kind of specialized I/O instruction is useful in very small computers or in special purpose machines where the variety of I/O devices is limited and the instruction format allows additional instructions.

The accumulator is often implied as the source or destination of the data so that no storage address is required in the I/O instruction. Using the accumulator with I/O devices in this manner also can lead to the use of conventional load accumulator and store accumulator instructions for I/O functions. In this usage, the address portion of the instruction may be used to specify the concerned device and/or other information.

Sometimes specific storage addresses are used to represent devices and sometimes functions within devices. Usually these storage addresses refer to locations that do not actually exist in the system. This instruction technique may be useful in very small or special-purpose machines where the formats allow addressing beyond the actual limits of real storage. The concept of thinking of I/O devices as specific storage locations may also provide programming advantages with some process I/O devices if instructions other than load and store may operate on them.

The decision of how to implement I/O instructions for direct program controlled operation is dependent on factors such as the number and variety of I/O devices, the instruction format, and the word size. In general, however, it depends largely on the means of implementing the second approach to I/O operations mentioned previously—indirect program control. It is generally desirable, from the viewpoint of both programming and implementation, to implement the two similarly.

2.5.3   Indirect Program Control

With indirect program control, the program initiates the I/O operation, but the actual data transfer between the device and storage is controlled by the channel itself, even though many words of data may be transferred. The operation of the channel proceeds concurrently with that of the CPU, interfering with normal program execution only as accesses to storage are required by the I/O device. The distinguishing characteristic between this and the direct program controlled mode is that a record consisting of many characters or words can be transferred by a single instruction without the requirement of an external buffer, as discussed above, and at a rate limited only by the main store itself.

The instructions that initiate the input or output operation provide the necessary information or parameters to the channel so that it can maintain the data transfer without further attention from the program. The minimum information required consists of the function to be performed, such as read or write, the particular device, the initial storage address to be used as the source or destination of the data, and the length of the record to be transferred. As in the direct case, additional information may be required, depending on the specific channel implementation and the characteristics of the I/O device. Given this information, the channel selects the I/O device, accepts or sends data to the device upon demand from the device, and accesses storage to store or fetch data as required, using the data address. Meanwhile, normal instruction processing proceeds in the CPU impeded only by the occasional storage cycles required by the channel. Following the transmission of the record, the channel can cause a program interrupt to advise the CPU that the input or output operation is complete, and the channel can be reinstructed.

The implementation of indirect mode channel has two significant aspects different from the direct mode. The first is multiplexing. In order to operate several I/O devices concurrently, the data paths and addressing structure of the main store must be designed to permit sharing by the various devices and channels. Their requests for storage cycles come asynchronously and to scattered storage locations. This sharing or multiplexing leads to increased complexity, which usually results in increased cost and something less than maximum theoretical bandwidths and response times. The second aspect is the need to provide registers, controls, and modification circuits to hold and update the storage address, word count, command, and device address discussed above. These factors mean a higher cost but also a higher performance attachment than the direct mode attachment.

A variation on this implementation uses actual main store locations with fixed addresses to contain the storage data address and word count. A minimum of three storage cycles is then required for each unit of data transferred—two to fetch and modify the address and count and one for the data. Thus, although three cycles are required and the interference with the CPU is increased, the cost of the registers is saved.

The advantages in performance and simplified programming with indirect mode are significant, and this type of channel is usually available on control computers. The programming simplification results from having to instruct the I/O device and channel only once per record rather than on every word. This, in turn, results in fewer interruptions and greatly reduced program overhead, permitting both higher data rate devices to be attached and more CPU time to be devoted to other tasks.

2.5.4 Chaining

Chaining is a functional extension to the indirect mode. It permits reinstruction of the channel and I/O device without program intervention. It is relatively expensive and is not usually found on control computers except in very elemental form.

At the completion of a record transfer, instead of causing a program interrupt, the channel simply accesses from main storage a new set of parameters (data address, word count, etc.) with which it can perform the next operation.

There are several variations possible. One is called data chaining, in which only a new address and count are fetched and the same operation with the same device is continued. Another is command chaining, in which a new operation is initiated, usually with the same device.

These commands and parameters may be fetched from a list in main store, or they may be associated with the data table. When arranged in a separate list, they have the form of a channel program; that is, a sequence of commands that are executed by the channel similar to the program or sequence of instructions executed by the CPU. This kind of concept requires another parameter called the command address to be supplied by the I/O instruction in the CPU. Additional circuitry in the channel is required to maintain and control this address. The concept can be extended to include logical decision-making capability in the channel program. However, the cost increases rapidly, and computers of the size considered here normally do not provide these functions, the designers preferring to allocate the decisions and computation associated with I/O operations to the CPU.

2.5.5   Channel Implementation

The implementation of a channel is highly dependent on a number of factors and, therefore, no single design can be described as standard. Significant factors that dictate the channel design are the characteristics of the processor, the nature of the instruction set, the word length, and the modes of channel operation. In addition to the two modes discussed in the previous sections, various hardware trade-offs are possible. One of these, that of multiplexed versus nonmultiplexed operation, is a question of sharing all, or portions of, the data paths between the CPU and channel. In a multiplexed channel, these data paths and some other common hardware are shared between multiple channels. This reduces the amount of hardware at the expense of some degradation in performance. Depending on the attached I/O devices and the application, however, this degradation may not be significant.

In the following sections, implementations of direct and indirect program controlled channels are described. Although the examples do not totally represent actual implementations, they illustrate the basic operation of a channel and some of the design considerations.

2.5.5.1   Direct Program Controlled Channel Example. As indicated above, the implementation of the channel is significantly affected by many aspects of the CPU and system architecture. Thus the registers involved in channel operations and the exact sequence of events varies from system to system. However, the configuration shown in Fig. 2-9 is provided to illustrate some of the general comments made in the previous section. In this very simple example, it is assumed that actual data transfers between I/O devices and the CPU originate or terminate in the accumulator (A-register) and that one word is transferred by each instruction. The operation of the direct program controlled channel illustrated in this figure is as follows.

The I/O instruction is accessed in storage and resides in the B-register as a result of the normal sequential operation of the CPU. The operation (OP) code decoder decodes the operation as an I/O operation and gates the remainder of the word in the B-register into the channel hardware.

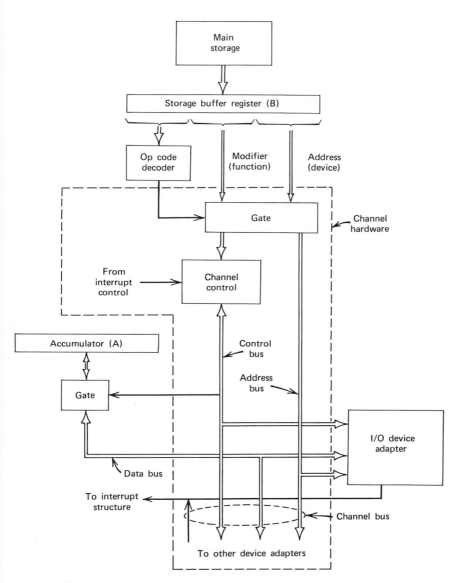

Fig. 2-9   Direct program-controlled channel example.

A portion of this data is used to specify the exact I/O function, such as read or write, and these data are gated into the channel control section. The channel control section generates the necessary timing and control signals for subsequent actions. The address portion of the word in the B-register, which in this case represents the address of the I/O device, is gated to all I/O device adapters attached to the channel, and the function to be performed is gated to the control bus.

All device adapters decode the device address, and the selected device responds by transmitting a signal back to the channel control section. The selected device also decodes the function to be performed on the control bus. If the function is a Read, the device adapter gates the data word onto the data bus, and the channel control gates the data bus to the A-register. The CPU then executes the next instruction which, in a system utilizing a channel similar to that described here, typically would be a store accumulator instruction to transfer the data from the accumulator to main storage.

The Write function is performed in a similar manner. Under control of the channel control section, the contents of the A-register are gated into the data bus and the addressed device gates the data bus into its buffer. The CPU then proceeds to the execution of the next instruction in the program.

This example is simplified and does not consider many of the other functions performed by the channel control section or various circumstances that can arise. For example, in a real channel the channel control typically must determine whether the device is busy with a previous request. Very often the address and control information shares the data bus with the data, and two cycles are required with appropriate control signals to distinguish the two. When transfers are with the main store rather than the accumulator, the roles of the A- and B-registers are reversed, or an additional register is provided to hold the address and function information. Error checking and other tests to validate the address and data also may be performed by the channel hardware. Conceptually, however, the direct program controlled channel operates in the manner suggested by this example. Its distinguishing characteristic is that each channel operation must be initiated by a separate instruction in the computer program.

2.5.5.2 Indirect Program Controlled Channel Example. As noted in the previous discussion, the indirect program controlled channel is characterized by its ability to transfer blocks of data to or from I/O devices without interference with the operation of the CPU except for periodic access to main storage. This type of channel control, known by various names—for example, cycle stealing and data break—is relatively common in control computers and can enhance greatly the performance of the system. Figure 2-10 illustrates an indirect program controlled channel that provides for the transfer of p words of data to or from storage as a result of a single I/O instruction. The figure and the following explanation are based on the data channel operation in the IBM 1800, although the example has been simplified (IBM, 1966).

In many respects, the operation of this channel is similar to that described in the previous section. One notable exception that can be discerned from the figure is that the accumulator in the CPU is not

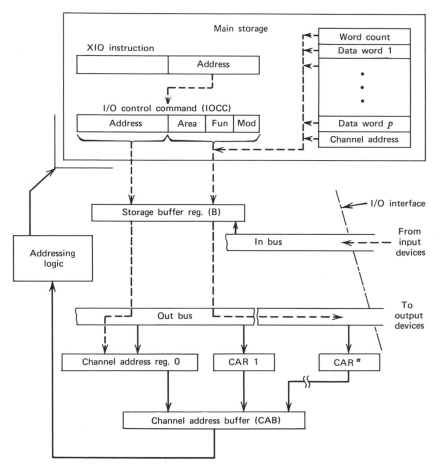

Fig. 2-10 Indirect program-controlled channel example.

involved in the data flow. Storage accesses are made directly through the storage buffer register (B-register). This register, as well as the addressing logic for main storage, is shared with the CPU so that CPU operation is momentarily interrupted whenever a storage access is required. In contrast to the normal interrupt operation, however, the status of the various CPU registers is not affected by these storage accesses so that program intervention to save the CPU status is not required. The CPU merely delays the execution of the next program instruction for the duration of the storage access cycle, which is "stolen" by the channel.

The figure indicates two double words and a table that are resident in main storage. The two-word execute I/O (XIO) instruction is the program-executable instruction that initiates the channel operation. Its address references additional data, called the I/O Control Command (IOCC), whose two words contain the data required in selecting the I/O device and controlling its operation. The table either contains the data to be transferred to an output device or provides the storage space for data transferred from an input device. The first word in the table, whose location is specified by the address in the IOCC, contains the number of data locations in the table which, in this case, is p.

When the XIO instruction is executed by the CPU, it causes the IOCC data located at the XIO address to be accessed from storage into the B-register. The area, function (Fun), and modifier (Mod) data from the IOCC are transferred via the output bus to the device adapter, whose address is specified by the data in the area portion of the word. The modifier may act as a subaddress that selects a secondary function in the selected device or distinguishes between similar devices connected to the same channel. The function portion of the IOCC specifies the operation to be performed, such as a read or a write. This information is used by the selected device adapter and the channel to control the specified operation.

After receiving an indication that the selected device has received the modifier and function data, the address portion of the IOCC that specifies the location of the first word in the table is transferred into a channel address register (CAR) via the B-register. There is a separate CAR for each data channel available on the system. The figure indicates that n channels are available for the purposes of this example. Each CAR is associated uniquely with a particular device connected to the channel. Provision is made to associate several devices with each CAR. In this case, however, only one of the devices may operate at a time. As a consequence, devices that are to be operated concurrently (i.e., multiplexed) must be assigned to different data channels.

When the selected device is ready, the adapter requests a "cycle steal" and the contents of the CAR are transferred into the channel address buffer (CAB). Thus CAB now contains the address of the first location in the data table to be utilized in the data transfer. The CAB accesses this address via the storage-addressing logic and the first word in the table is transferred to the B-register. Concurrently, the CAR is incremented by one count to prepare for the next access to storage. The word count in the B-register is transferred via the output bus to the selected device where it is stored in a word-count register. This ends the first cycle-steal operation.

When the selected device is ready to accept or transmit data, it issues a second cycle-steal request. Since the CAR had been incremented previously, it contains the address of Data Word 1 in the table. The contents of CAR are transferred to CAB and used to access that location. The CAR is again incremented as the first data word is transferred to or from the selected I/O device via the B-register and input or output bus. The word-count register in the selected device is decremented and this ends the first word of data transfer.

The action described in the preceding paragraph now repeats whenever the I/O device issues a cycle-steal request. Since the CAR is incremented during each cycle, each subsequent request accesses the next location in the data table. The main program execution continues in the CPU between the cycle-steal storage accesses. When the word-count register in the selected device is decremented to zero, the action specified by the original XIO instruction has been completed and the control unit requests a program interrupt to signal the end of the operation.

In the actual IBM 1800 channel implementation, an elemental form of data chaining is provided. The completion of the operation may cause an interrupt or it may reinitiate the same operation automatically using the same or a different (chained) data table. These options are specified by data contained with the word count in the first word of the data table. The control information is transferred to, and stored in, the selected I/O device adapter. At the completion of the operation described in the above example, an additional cycle-steal request is made with an indication to the channel that the contents of the addressed location should be gated to the CAR instead of to the output bus. This address will indicate the location of a data table, as did the address in the original IOCC. The operation then can be repeated.

In a system such as the one described in the example, the B-register, CAB, and certain other features are shared between the various channels attached to the system. Since these channels may be operating concurrently, there will be contention between the various active channels for this shared hardware. This can be controlled by assigning a priority similar to that used in the interrupt structure to each of the channels. When contention occurs, the channel request having the highest priority is serviced first.

2.5.6 Device Adapters

The I/O devices are connected to the channel by means of an adapter or control unit. The control unit has several basic functions: It must recognize and decode commands received from the channel and translate them into the specific control signals required by the device. In addition, it must translate signals from the device into signals that activate, control, and/or respond to the channel operation. The performance of these basic functions calls for a variety of specific operations to be performed that depend on the nature of the device and the structure of the channel. The control unit may be readily identifiable as a specific unit or may be considered part of, and packaged with, either the I/O device or the channel. Sometimes a single control unit may be used to control several devices in a multiplexed mode.

A partial list of operations that may be performed in the control unit includes data buffering, data-code translations, serializing and deserializing, control-signal sequencing, electrical transformation of signals, and data checking. In some systems, the control unit may also perform operations that in other systems are performed in the channel, such as providing storage addresses and record-length control. It may also test for specific control characters in the data and perform control operations as a result of detecting them. This is very often the case with magnetic tape and communications devices. The control unit generally reflects the status of the device to the computer and interfaces to the interrupt system to signal this status to the program.

The control unit for an analog-input subsystem illustrates some of the functions that might be required. The operation of this subsystem is typically asynchronous with respect to both the channel and the CPU, and the time required to read an analog input value and convert it to digital form may be variable. The minimum information required by such a subsystem is a "select" signal and an address of the analog-input point to be read. If the subsystem provides selectable gain under program control or other special features such as a digital comparator, the necessary information for these features must also be provided to the control unit.

The typical operation of the control unit for such a subsystem is as follows: When selected by the channel, the control unit acknowledges its selection and reads the analog-input point address and other data from the channel. It then sets a "busy" indicator to prevent the channel from selecting it again before the current operation has been completed. The control unit decodes the address to select the appropriate multiplexer switches. If several levels of multiplexing are provided (block switching), the second-level switches must also be selected. Similarly, if gain selection is required, the control unit selects the proper gain switches. Following this, the control unit provides a timed delay to allow the multiplexer and amplifier transients to settle. After this delay, it may be necessary to select a switch at the output of the amplifier if multiple amplifiers are included in the subsystem. This also requires a subsequent delay for transients to settle. After the proper delays have occurred, the control unit provides a "start convert" signal to the analog-to-digital converter (ADC). Following conversion, the control unit receives an "end of convert" signal from the ADC and generates a signal to indicate to the channel that it is ready to transfer data. Under control of the channel, the control unit transmits the data to the channel along with any other required information such as check bits or overload indications. The control unit then releases the selected multiplexer, gain, and amplifier switches and, perhaps after a short delay to ensure that the switches are opened, resets the "busy" indicator. Although this example assumes particular features and control sequences, it illustrates the decoding and control signal sequencing that commonly are provided by a control unit.

Note that this operation was described without reference to whether it was being operated in direct or indirect mode and could very well apply to either. In direct mode, the select signal that initiated the operation and the transfer of the analog input address and other control data would have resulted from the execution of an I/O instruction. The end of convert signal would cause the channel to interrupt the CPU program which, in turn, would execute another I/O instruction to cause the converted value to be transferred from the ADC to the CPU. The initiating instruction could then be executed again with a new analog-input address to repeat the cycle. With an indirect mode channel the transfers of analog input addressed from main store to the control unit and converted values from the control unit to main store would be performed by the channel without program interruption by stealing main-store cycles when requested by the end of convert signal.

This example is somewhat unusual in that data must be transferred in both directions during an operation. This situation is seldom found in conventional data-processing I/O devices and thus illustrates a unique requirement in control computers.

2.5.7 The Input/Output Interface

The interface between the I/O device and the channel is called the input/output interface. The channel may provide a unique logical and electronic interface for each device or a common interface that can be used to attach one or more of many different types of devices. Generally speaking, the smaller or more specialized computers provide interfaces optimized for each type of device. This is sometimes called "native" attachment of I/O devices. In larger general-purpose machines it is usually advantageous to provide a standard interface with enough flexibility to attach many different kinds of devices. This usually is more costly in terms of the number of circuits for small configurations of I/O devices. However, it provides greater flexibility in the number and diversity of devices that can be configured in a system from standard units.

In the case of native attachments, the channel and control unit functions are most often integrated into a single implementation and the distinctions between them become blurred. When a common or standard interface is used to enhance system configurability and/or reduce development costs when several systems and I/O devices are involved, the allocation of functions to the two sides of the I/O interface becomes especially significant. The allocation discussed previously is arbitrary and is based on the example of the IBM 1800. In most system-design situations, there is considerable latitude in locating the logical and physical interface, that is, in allocating functions to either the channel or the adapter.

Many factors enter into the decision process. Among the more important are the different modes of control provided, the physical distance over which the interface must operate, the number of concurrently operating devices and their individual and aggregate data rates, and the nature and variety of functional characteristics and associated control information that must be transferred.

Some of the considerations are the following. It would seem appropriate to move as much function to the channel side of the interface as possible, thus permitting it to be shared by the several attached devices. However, when this is done the implementation of the function tends to become more generalized and consequently more expensive. It also tends to fix the mode of operation, sometimes making it difficult to accommodate unusual devices or functions. Moving functions to the device or control-unit side of the interface seems appropriate, as it can be optimally designed for the operation of each individual device and often avoids duplicating functions on both sides of the interface. It also provides the most flexibility in the functional characteristics of devices that may be attached.

The word-count function in the IBM 1800 example can be used as an illustration. If this function had been allocated to the channel, each counter would have had to be the full size of the maximum possible record size, even though most devices transfer much shorter records.

Many devices, such as those with buffers to control or with a need to anticipate the end of a record, have to count the number of characters or words in either case. Putting the counter in the adapter also means the count specified by the initiating instruction can be interpreted uniquely for that device. Note also that provision must be made for transferring an initial count to the control unit and that a means must be provided for the programmer to determine the final count of data transferred. The counting function also must be implemented in every adapter or control unit so that the advantages of sharing data paths and modification circuits are lost. Similar discussions can be applied to the command controls, such as read and write, and the data address registers. Final decisions on these design questions can be made only after the alternatives are designed in enough detail to permit adequate cost and performance comparisons.

2.6 INPUT AND OUTPUT DEVICES

One of the unique features of the control computer is the wide variety of I/O devices that can be attached. These include virtually any I/O device found on a computer in a more general-purpose application as well as the various specialized subsystems discussed in Chap. 3 through Chap. 7. Generally speaking, the I/O devices can be categorized as either process I/O or data-processing (DP) I/O. Process I/O includes those subsystems and devices that are unique to the data acquisition and control application. The term "process I/O" is a carryover from the early days of computer use in the process industries and indicates that the I/O subsystem is directly connected to the process being monitored and controlled. Better terms that more nearly encompass the current use of the computer in industrial and laboratory applications are "data acquisition and control I/O (DAC I/O)" and "sensor based I/O." Neither of these terms is in common use, however, and each vendor generally chooses a term to his liking. For our purposes, the term process I/O will be retained.

Data-processing I/O includes those devices and subsystems that are common to the traditional data-processing applications. They include devices such as card readers and punches, paper-tape equipment, line printers, video display units, typewriters, and plotters. Some vendors and writers include bulk-storage devices such as disk and drum storage units in the category of data-processing I/O. For the purposes of this chapter, these are considered separately in Sec. 2.3.2.

In small and/or dedicated applications, the conventional data-processing I/O devices are often limited. They might include, for example, a console typewriter for operator communications and alarm messages, a line printer for logs and for the convenience of the programmer and engineer, and some type of input device, such as a paper-tape or card reader. Some installations have nothing more than a typewriter and paper-tape reader. At the other extreme is an application where the computer is used to control a very large process or for extensive computations of a data-processing nature. Here, the range of I/O devices may be similar to that found in a well-equipped scientific computation center with the addition, of course, of the specialized process I/O subsystems.

In this section, a few of these I/O devices are discussed briefly. Considerable details on the analog, digital, and operator I/O subsystems are included in the following chapters, so they are mentioned only briefly here. Although conventional I/O devices are not discussed in detail elsewhere in this book, for several reasons the material here provides only a review; First, the data-processing I/O devices are not unique to the control computer and their use is essentially identical to that found in ordinary computing applications. Second, there is already extensive literature describing DP I/O devices.

2.6.1   Process I/O Subsystems

The process I/O category generally can be subdivided on the basis of the input/output data form. The analog-input subsystems accept analog data from the various sensors and other sources in the process and convert this information into a digital form compatible with the internal characteristics of the digital computer. The analog-output subsystem converts digital data from the computer into analog signals for use by various control and graphical equipment.

Analog-input subsystems are characterized grossly by the rate at which analog information can be converted into digital data and by the number of analog inputs that can be accepted on a multiplexed basis. In addition, they are usually specified in terms of their analog parameters, such as accuracy, resolution, noise, and stability. The subsystems are available with a wide range of performance and, in addition to their primary function, may include special features such as automatic gain selection, digital comparison, and various filtering options. The detailed consideration of the design and performance of such subsystems is the subject of Chap. 3 and Chap. 4.

The operation of the analog-input subsystem is asynchronous with respect to the CPU and channel. The input data from the CPU and channel include, as a minimum, a start convert or initiate signal and the address of the particular analog-input point that is to be sampled. If programmable gain selection is provided, the gain information also must be transferred to the subsystem. Other special features, such as a digital comparator, may also require data transfer from the CPU or storage. Additional control signals may be required for features such as sample-and-hold amplifiers.

Since the conversion process is not instantaneous, a delay between initiating the operation and the availability of data is an inherent characteristic of this subsystem. After the conversion delay, the data available to the CPU consist of the converted value and, perhaps, status information, such as an overflow or error indication. When the system provides automatic gain selection, the value of the gain also must be transferred with the converted values from the subsystem to the CPU or storage. If a digital-comparator feature is provided, the only data output may be an indication of whether or not the analog input violated the preset limit values. Depending on the subsystem design, this information may be transferred to the CPU only when a violation occurs and then often in the form of an interrupt request.

The analog-output subsystem provides the control computer with the capability of controlling analog devices such as valves, strip-chart recorders, setpoints, and other adjustments on conventional analog controllers. The analog-output subsystem is generally specified in terms of its data rate, accuracy, resolution, noise, and stability. The subsystem may or may not include a multiplexing function; that is, a separate digital-to-analog converter (DAC) may be provided for each output, or the output of a single DAC may be multiplexed to circuits that have the capability of holding the analog value for a specified period of time. These circuits are similar to conventional sample-and-hold amplifiers except that the holding time is generally much longer.

The operation of the analog-output subsystem is typically asynchronous with respect to the channel and CPU. In contrast to the analog-input subsystem, however, there may be no inherent delay other than that due to the channel characteristics and the logic used in the control-unit implementation. If the system includes multiplexed outputs, there is a small delay to allow switch transients to settle and to charge the holding circuit on each output. If a DAC is provided for each analog output, the only delay is that required to read data into the DAC register. This is comparable to other logic delays in the system and, therefore, the DAC generally can accept data at the maximum channel rate. This does not necessarily imply, however, that the DAC output responds at the same rate. Because of bandwidth limitations, the rise time of the output may be much greater than the logic delays. The important point, however, is that the channel and analog-output control unit are available as soon as the DAC register is set, whether or not the output has actually responded to the new data.

The data required for operating the analog output subsystem include the data to be converted, the address of the DAC if more than one is available, and the address of the particular output point if a multiplexed output is provided. Control signals are minimal in most of these subsystems and the change in output is usually initiated automatically as a result of reading data into the DAC register. In some subsystems, a buffer register is provided such that data can be loaded into the buffers for a number of output channels without affecting the actual DAC output. On the receipt of a control signal, the data in the buffers are all transferred to their corresponding DAC registers simultaneously so that changes in the several outputs will occur simultaneously.

The second subdivision of process I/O subsystems, that of digital I/O, provides the control computer with the ability to sense and control devices that are essentially binary in nature; that is, the information from the input device is represented by one of two states, and the control of an output device is effected by outputting one of two states. For example, the state of a switch in the process that indicates that a motor or indicator is ON is a common type of input. Similarly, the computer can operate a relay to control a motor or indicator. Another type of digital input is a pulse train, such as that produced by a turbine flowmeter. The rate at which pulses are produced is proportional to the flow and the sum of the pulses is related to the total quantity of material that has passed through the meter. It is necessary here to be able to count pulses for some fixed period of time.

The significant feature of both digital-input and digital-output features is that they are concerned with "bit-significant" data. That is, each individual data bit may signify information independent of the state of any other input or output data bit. The bits in the data word need not be related as they are in the case of a binary word representing a numerical value. This does not imply that there might not be a defined relationship to other bits, but merely that it is not a requirement. A digital-input word could, for example, contain 3 bits that represent an address of an external device, 5 bits of data from that device, and 8 bits that represent independent data obtained from eight indicator switches. The associated software must provide the proper interpretation and manipulation of these data.

Even though the digital data may be bit-significant, the data generally are manipulated in groups that are related to the word size of the particular computer. This results in a more efficient implementation,

since data paths and functional operations are designed for word manipulation. Individual bits are manipulated by means of the logical instructions, and the software may provide special subroutines for this purpose.

In general, both the digital-input and digital-output features are
based either on registers or on counters whose length is typically equal
to the computer word size. The difference between the features that are
based on a register lies in the manner in which the individual bits are
controlled, in the case of digital input, and the manner in which the
register output is interfaced to the process, in the case of digital output.
The digital-input circuits may be designed to interface with commonly
available logic circuits. In this case, all the bits in the digital-input word
are usually assumed to be derived from a single source and the feature
often is called register input. This feature is usually used to connect
specialized digital equipment to the control computer. For example, a
high-speed analog-to-digital converter might utilize this type of interface
if the vendor does not provide such a device as a standard channel I/O
subsystem.

The analogous digital-output feature is register output. Here circuits
providing common digital logic signal levels are provided to interface
with external devices. As in the case of register input, the assumption
usually is that the attached device accepts the data on a word or partial
word basis, although this does not have to be the case. Register output
might be used, for example, to connect a special strip printer or a communication adapter.

The other digital-input feature that is based on a register is one that
provides sensing circuits for voltages or for switch contacts. A sensing
circuit is provided to control each bit in the register, independent of the
other bits. The contact-sense circuits essentially detect a low resistance
when the switch contacts to which it is connected are closed. The circuit
sets or resets the state of the register bit, depending on whether the
resistance is above or below a defined value. Voltage sensing circuits are
similar except that the register bit is set or reset according to whether
the externally applied voltage does or does not exceed a predefined level.
The sensing circuits are often identical to those used for interrupt-signal
detection; the only difference lies in the manner in which the circuits are
connected to the channel and CPU.

Analogous digital-output options are usually available. The output
may control a relay that, in turn, can be used to control equipment in
the process. The digital-output register also may control transistor
switches for high-speed switching.

Counters are the other major class of digital-input features. Input
pulses are counted in a binary counter until either the CPU reads the

count or the counter overflows. Usually, counter overflow results in an interrupt so that the CPU can either update the previous count stored in storage or take some other predefined action. The input circuit that drives the counter may be either a resistance- or voltage-sensing circuit. The logical implementation of the counter feature is relatively straightforward. The only major precaution that must be taken is to inhibit read-out operations whenever carries are propagating through the counter, since the ouput of the counter could be incorrect during this transient period.

The digital-output feature somewhat analogous to a counter input is the pulse output. This may provide a specific number of pulses in response to a single instruction, or the number of pulses may be specified in the instruction. An alternate form of output is pulse duration output, in which the duration of the pulses is varied according to either a preset selection or a data word.

The interface requirements for all digital-input features are, in general, quite similar. A request to read a particular input point requires the transfer of an address to the channel and the subsystem. The data in the addressed group are then returned to the CPU via the channel. Even though only a single bit may be of interest, the address refers to a group of inputs and the whole word is typically transferred to the CPU or storage. Software routines then extract the state of the particular input of interest.

The data requirements for digital output are similar. The CPU or channel must provide the address of the output group and the data word that specifies the desired combination of output states. Although some control computers provide the ability to change a single bit, most require that all bits in the word be specified. In the case of a pulse or pulse-duration output, the data consist of the number of pulses to be transmitted or a number representing the duration of the pulse.

2.6.2   Data-Processing I/O Devices

Although the primary data input/output transfers in a control computer are directly to and from sensors and actuators connected to the process, there is also a need for conventional data-processing I/O devices. As in general-purpose computer systems, means must be provided for program and data entry, program debug, data output, and system maintenance. There are also functions associated with the specific control application requiring on-line data entry, logging, and operator guidance.

In most cases, the same devices used in a general scientific computing system are used in control applications. The differences, if any, lie in emphasis on speeds. Since the bulk of the input and output activity is directly with the process, a smaller overall data-handling capability is required of conventional DP I/O devices in a dedicated control application.

The following sections review the characteristics of some of the devices commonly used in conjunction with control computers. Since information on specific devices is readily available from vendors and other sources, the discussion is quite brief and not all available types of equipment are considered.

2.6.2.1   Paper-Tape Equipment.   Paper-tape devices use rolls of paper or plastic tape 11/16–1 in. wide, depending on the number of bits in the character code. The standard roll is 8 in. in diameter and contains 800 ft of tape. Information is represented by holes punched in the tape; the binary values 1 and 0 are represented by a punched hole or no punch, respectively.

Paper-tape readers and punches are relatively simple and inexpensive devices compared with their main alternative, punched cards. While usually slower—10 to 300 characters/sec (although 1000 characters/sec readers are available)—they usually have a cost advantage and are simpler to attach to a computer because of their serial nature and the lack of complex controls. Inexpensive devices for manually preparing tapes off-line are readily available. The major disadvantage of paper tape relative to punched cards is the difficulty of making changes and the impossibility of sorting or collating without creating a new tape.

The two kinds of tape used are called chad and chadless. The chad is the small confetti-like bit of paper punched out of the tape. Chad tape is completely punched, whereas chadless tape is only partially punched, forming a tiny flap, as shown in Fig. 2-11. Chad tape is now commonly used so that photoelectric reading techniques can be utilized rather than the older and slower electromechanical means.

Characters are represented in a binary-coded decimal code along the tape with the bits of each character punched in a row perpendicular to the edge of the tape, as shown in Fig. 2-12. Each bit position is called a channel. Although 5-, 6-, 7-, and 8-channel tapes are all used, 8 is most common in computer application, the others having their origin in the communications industry. The number of channels is, of course, the number of bits available for the character code. With 5-channel tape, only 32 unique character representations are possible. In order to represent a larger character set, some bit combinations are used to represent two or more characters, such as a numeral and a letter. Other bit combinations are then defined as control characters and are used to define the particular representation that should be ascribed to the multiply-

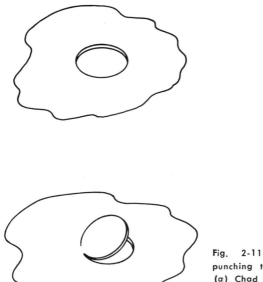

Fig. 2-11 Tape-punching techniques. (*a*) Chad tape; (*b*) Chadless tape.

defined bit combinations that follow. In the more common 8-channel tape, the first 4 channels are used for the BCD character code, channel 5 is a parity check bit, channels 6 and 7 are zone bits, and the last channel is an "end of line" indicator bit. Several examples of coding for alphanumeric and special symbols are illustrated in Fig. 2-12.

The drive mechanism of paper-tape devices consists of a toothed sprocket wheel whose teeth fit in a row of small holes punched between the third and fourth channels. The drive wheel pulls the tape from the supply reel across the reading or punching mechanism. From the sprocket wheel the tape goes to a takeup reel that maintains the necessity tension. Sometimes the supply reel is designed to permit the tape to be removed from the inside of the roll of tape so that rewinding is not required. Reading is performed by either photoelectric or electromechanical means, although the latter is less common in newer computer equipment.

The control unit for paper-tape devices can be very simple. When punching tape, data are transferred to the control unit on a character basis. The character remains in the control unit data register until the device signals that the punching operation is complete. The next character is then transferred to the control unit. In reading data, the data in the reader output register are transferred to the CPU or storage one

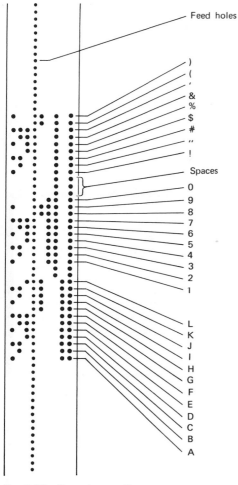

Fig. 2-12   Paper-tape coding.

character at a time. The tape movement is controlled on an incremental basis. For reading, the tape may be moved at a constant velocity, in which case it normally would be attached to an indirect programmed channel to guarantee no overruns.

2.6.2.2 Punched-Card Equipment. Punched cards have become the most widely used transfer medium for computer programs and data that are manually prepared. Everyone is familiar with the use of such cards for checks, statements, invoices, and other commercial applications. In a scientific computing environment, their chief advantage over paper tape is the ability to edit easily or to make changes and to assemble information from several sources. Punched-card equipment is generally complex and more expensive than that for paper tape but usually provides higher performance.

The use of punched-card equipment rather than paper tape in a control system is largely a matter of the extent to which programming changes and off-line, manually prepared data are a part of the overall system concept. In low cost, dedicated applications, paper tape or an on-line manual entry scheme are often satisfactory.

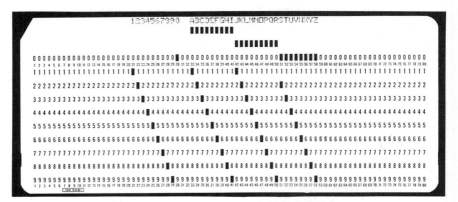

Fig. 2-13   Punched card illustrating Hollerith code. From IBM Corporation.

The IBM punched card, shown in Fig. 2-13, provides for 12 rows of 80 columns of punches. The 12-bit Hollerith code illustrated in Fig. 2-13 is commonly used for alphanumeric data with a single character punched in each column. They are often interpreted; that is, the character punched in each column is also printed along the upper margin to assist manual reading of the card. The card also may be used for pure binary information. The binary data may be arranged in either rows or columns. Early card readers read cards row by row, but new models read on a column basis. Data are transferred to main storage as 12-bit binary quantities.

Cards are read using either photoelectric or electromechanical means, although photoelectric techniques predominate in newer equipment. The cards are mechanically transported individually from an input hopper past the reading sensors at rates from 100 to 2000 cards/min. Program subroutines are used to convert the coded or binary card image data into computer words for further processing. After reading, the cards are stacked in an output hopper (stacker). Multiple stackers may be provided with automatic or programmed selection for sorting cards or indicating any that were read with a detected error.

Card punching is analogous to reading. The cards pass under the punch station and data are punched on either a row or column basis at a typical rate of 100 to 300 cards/min.

2.6.2.3 Typewriters. Typewriterlike devices play an extremely important role in control-system applications. They provide an economical answer to the system requirements of operator guidance and occasional manual entry of data. They are commonly included as a part of an operator console. They can often be at remote locations, attached to the computer by a single pair of wires or conventional telephone circuits, thus providing an economically feasible means of remote data entry and/or output.

Two types of these I/O devices are in common use in computer applications. One is an adaptation of the Selectric* typewriter, and the other is based on the teletypewriter in common use in the communications industry. Both devices have been modified to provide the interface necessary for use with a computer. The typewriter equipment may be used as both an input and output device or merely for ouput. In the latter case, the keyboard is often omitted.

The typewriter is a serial-by-character device and characters are transferred to the control unit sequentially. The control unit converts the character code into the necessary control signals to actuate the printing mechanism. It also generates the end-of-character interrupt signals, which can be used in conjunction with the control-computer interrupt system. Printing speed is typically between 10 and 30 characters/sec.

The operation as an input device is similar. The key depressions are coded into the corresponding character codes and transferred to the CPU or storage on a character basis. The control unit provides the necessary channel interface and interrupt signals to effect this transfer.

2.6.2.4 Line Printers. If a significant amount of data is to be printed, the slow printing speed of the typewriter output device makes it impractical. Output printers that handle data on a line basis, rather than on a character basis, are available with printing speeds from 300 to in excess of 3000 lines/min. Electromechanical printers are used at rates up to about 1000 lines/min, and electrostatic or xerography techniques are used for the higher speeds. Line printers are relatively complex, and expensive devices and many small and/or dedicated control applications

* Registered trademark of IBM Corporation.

do not use them. In larger systems or in systems on which both control and data-processing jobs are run, the line printer is the common output device.

A number of different types of printing elements are utilized in these high-speed printers. Some consist of individual print wheels or type bars for each print position, each of which contains a full set of characters. Others use a continuous chain to which the print elements are attached, or a single bar that traverses the line at high speed. An individual hammer is provided for each printing position, and the hammers are actuated when the proper character is in position. Electrostatic printers form characters by arc discharge between a matrix of electrodes and a conductive platen located behind special electrographic paper. Xerographic printers deposit charge on ordinary paper using a similar process to that used in xerographic copying machines.

The printer control unit contains a register with a character capacity equal to the number of characters in the line. Once the register is loaded from the CPU or storage, the control unit controls the actuation of the print hammers or the activation of the proper electrodes in the matrix to produce the desired character. Although the name "line printer" is commonly used, most of the electromechanical devices do not produce all characters in the line simultaneously. The characters are printed by the hammer striking the rapidly moving print element so that the characters are seemingly printed in a random order. The control unit times the firing of the print hammers such that the proper character on the moving chain or bar is struck at the time the appropriate character passes a particular print position. The action is so rapid, however, that the result is as though the total line of characters were printed simultaneously. To further increase the printing speed and to ease interrupt-response requirements, control units often contain a buffer register that is loaded with the next line while the line in the print register is being printed.

2.7 COMMUNICATIONS

Many times in configuring control systems, it is necessary or expedient to locate I/O devices—both conventional and process I/O—remote from the computer. It may also be desirable to connect local control computers to central supervisory or management information computer systems that are at different locations. These requirements lead to the use of communication equipment in conjunction with the computer and the I/O terminal equipment. Terminal and computer-to-computer communications in control applications are essentially the same as those found in general-purpose computer applications. However, the requirement to connect the control computer to a physically widespread process limits the economic trade-offs available to the control-system designer. This section provides a brief overview of computer communication equipment. More extensive discussions are available in the literature and standard reference works (e.g., Fink and Carroll, 1968).

Providing the multiwire cable normally required for attachment of I/O devices and direct computer-to-computer connection over any significant distance may be economically unfeasible or legally impossible. When going more than a few hundred feet or across public thoroughfares, a two- or four-wire system is much more desirable. When spanning several miles or more is required, dependence on the common carrier facilities is virtually a necessity.

Communication links connecting two or more stations appear in two general forms. The simplest is the *point-to-point* connection of two stations. With more than two stations, point-to-point connections may be established between each station and every other station, or all stations may be connected to a central control station. In this latter case, all communications between stations must be through the central control station. In a computer communication system, the central control is vested in one "master" computer and the connected terminals and/or computers are "slaves."

Multidrop or *multipoint* lines connect several stations on the same line, in a way very similar to the telephone party line. Technically, any station can send and receive to and from any other station. However, control of the line is usually vested in one station, called the *master station*. In a computer/terminal system, the computer is the master and the terminals are the slaves. In a multicomputer system, one computer may be assigned permanently as the master, or this authority may be shared on a "first come, first served" or a predefined priority basis.

The equipment used to interface the computer to the communications line is called a communications line adapter. The line adapter may connect directly to the communications line for low-speed DC signaling, as was used in early telegraph systems, or to a modem (modulator-demodulator). The modem is called a data set by the common carriers. When, as is often the case, several communication lines are to be attached and several line adapters are therefore required, a shared control unit is defined, called a communications multiplexor. As the data rates are relatively low, it is possible to share much of the control and data path circuitry that connect to the I/O channel between many line adapters. The design is somewhat complicated if several signaling rates and line control procedures are involved.

The functions performed in the line adapter vary depending on the communication technique and the characteristics of the modem, if one is used. Since the data transmission is usually serial-by-bit, serialization of the output data and deserialization of input data are common functions. In addition, the control unit provides the necessary control signals and signal level shifting for interfacing to the data set. Some error detection hardware is almost always included to detect transmission errors. Relatively sophisticated error detection and correction techniques may be included in the more complex communication adapters.

The basic function of the modem is to convert the serial-by-bit data to and from a modulated carrier signal. Frequency shift keying (FSK), phase modulation (PM), and pulse-coded modulation (PCM) are all used, depending on the quality and types of lines or other transmission media. Phase and frequency modulation are the techniques most often used on switched telephone networks. Frequency multiplexed modems are sometimes used for low data rate applications. In this technique, characters are coded as unique combinations of frequencies for transmission. The chief advantage of frequency multiplexing is that a very simple line adapter, not requiring serialized data with the accompanying timing and hardware, can be built for simple devices such as numeric keyboards. The push-button telephone instruments are an example of the use of frequency modulation in which combinations of frequencies are employed for each digit.

Communications lines vary in the direction that data can flow. Directional limitations are determined by the characteristics of the repeater amplifiers, modems, and other equipment in the line. Simplex is a term used to describe a line in which data can flow in only one direction. Half-duplex lines can transmit data in both directions but not concurrently, whereas duplex or full-duplex lines can transmit data in both directions simultaneously. Most modern common carrier facilities (lines and modems) are capable of full-duplex operation, but other system considerations, such as the nature of the I/O devices and line adapters, may dictate half-duplex operation. In addition, some cost saving may be realized by using half-duplex facilities. Some common carrier data sets have a capability called "reverse side channel" that also finds use in control applications. This is essentially a very low-speed simplex data channel superimposed on a medium or voice grade half-duplex service.

Data rates on communications facilities are, of course, a function of bandwidth and the modulation technique. Historically, data rate capability has been grouped into three general categories corresponding to less than 100, 100–600, and 600–5000 characters/sec. Low-speed lines are those that can accommodate data rates up to about 100 bauds* and are generally characterized as telegraph lines. They are also generally half-duplex. Medium-speed lines are those generally used for voice transmission. Good quality voice-grade lines can accommodate up to 5000 bauds. Over 5000 bauds is considered high-speed transmission. This kind of service is fairly expensive and does not find extensive use in control-system applications.

A wide variety of special- and general-purpose terminal equipment is commercially available for a multiplicity of applications. For remote man-machine communications, paper-tape and punched-card devices and teletypewriters are very common. CRT/keyboard combinations are becoming plentiful. Many low-cost "manual entry" devices such as numeric key entry and badge and card readers are used for factory data gathering and control. Special communication techniques to keep the cost low are often used with these devices. The whole range of process I/O functions are remotely attached by communication links.

The attachment of these I/O devices via communication links presents unique problems outside the scope of this chapter. However, the control-computer designer must be aware of the unique functional characteristics and problems associated with remote device attachment when configuring his system.

REFERENCES

Fink, D. G., and J. M. Carroll, *Standard Handbook for Electrical Engineers,* 10th ed., McGraw-Hill, New York, 1968.

Hawkins, J. K., *Circuit Design of Digital Computers,* Wiley, New York, 1966.

Kuehler, J. D., and H. R. Kerby, "A Photo-Digital Mass Storage System," *AFIPS Conf. Proc.,* 1966, pp. 735–742.

Richards, R. K., *Electronic Digital Systems,* Wiley, New York, 1966.

"IBM 1800 Functional Characteristics," Form A26-5918, International Business Corp., San Jose, Calif., 1966.

"Photo-Digital System Delivered to Lawrence," *Datamation,* December 1967, p. 83.

Small Computer Handbook—1968, Digital Equipment Corp., Maynard, Mass., 1967.

* The baud is defined as the reciprocal of the length of the shortest signal element measured in seconds. It is, therefore, a measure of the signaling rate, not the information rate. If the shortest signal element represents 1 bit of information, however, the baud rate is numerically equal to the information rate in bits per second. As an example, in 4-phase modulation, each of the four predetermined phase shifts represents 2 bits of information so that the information rate is numerically equal to twice the signaling rate in bauds.

BIBLIOGRAPHY

Aronson, R. L., "Data Links and Networks—Devices and Techniques," *Control Eng.*, February 1970, pp. 105–116.

Babiloni, L., E. de Agostino, and F. Fioroni, "On-Line Tradeoff: Hardware Versus Software," *Control Eng.*, October 1970, pp. 84–87.

Bailey, S. J., "Putting Memories to Work On-Line," *Control Eng.*, February 1970, pp. 126–134.

Buchholz, W., *Planning a Computer System,* McGraw-Hill, New York, 1962.

Borgers, E. R., "Characteristics of Priority Interrupts," *Datamation,* June 1965, pp. 31 ff.

Chu, Y., *Digital Computer Design Fundamentals,* McGraw-Hill, New York, 1962.

Cunneen, W. J., "Computer Time Allocation Eases On-Line Problem," *Control Eng.*, March 1970, pp. 71–75.

Dailey, J. R., and C. Kuntzleman, "Impact of Technology and Organization on Future Computer Systems," *Computer Design,* February 1970, pp. 49–54.

Desmonde, W. H., *Real-Time Data Processing Systems: Introductory Concepts,* Prentice-Hall, Englewood Cliffs, N.J., 1963.

Dinman, S. B., "The Direct Function Processor Concept for System Control." *Computer Design,* March 1970, pp. 55–60.

Flores, I., *The Logic of Computer Arithmetic,* Prentice-Hall, Englewood Cliffs, N.J., 1963.

———, *Computer Organization,* Prentice-Hall, Englewood Cliffs, N.J., 1969.

Foster, C. C., *Computer Architecture,* Van Nostrand Reinhold, New York, 1970.

French, M., "Rotating Disks and Drums Set Peripheral Memories Spinning," *Electronics,* May 26, 1969, pp. 96–101.

Hanson, M. L., "Input/Output Techniques for Computer Communication," *Computer Design,* June 1969, pp. 42–47.

Hellerman, H., *Digital Computer System Principles,* McGraw-Hill, New York, 1967.

House, D. L., and R. A. Henzel, "The Effect of Low Cost Logic on Minicomputer Organization," *Computer Decision,* January 1971, pp. 97–101.

Janson, P., "Common Bus Structure for Minicomputers Improves I-O Flexibility," *Control Eng.*, January 1971, pp. 50–53.

Klososky, R. A., "Computer Architecture for Process Control," *IEEE Trans. Industrial Electronics and Control Instrumentation,* June 1970, pp. 277–281.

Klososky, R. A., and P. M. Green, "Computer Architecture for Process Control," 1969 ISA Annual Conf., Paper 512, October 1969.

Kompass, E. J., "Modular Interfaces Mate Incompatibles," *Control Eng.*, February 1970, pp. 140–141.

———, "Peripheral Components—Incremental Tapes, Keyboards, CRT Display Units," *Control Eng.*, February 1970, pp. 135–140.

Lapidus, G., "High-Speed Input-Output Devices," *Control Eng.*, February 1970, pp. 117–125.

Louis, H. P., and W. L. Shevel, "Storage Systems—Present Status and Anticipated Development," *IEEE Trans. Magnetics,* September 1965, pp. 206 ff.

Ledley, R. S., *Digital Computers and Control Engineering,* McGraw-Hill, New York, 1960.

Maley, G., and J. Earle, *The Logical Design of Transistor Digital Computers,* Prentice-Hall, Englewood Cliffs, N.J., 1963.

Phister, M., *Logical Design of Digital Computers,* Wiley, New York, 1958.

Ragle, R. D., "Usable Flexibility with Process Control Computers," *Computer Design,* June 1969, pp. 36–39.

Rajchman, J. A., "Memories in Present and Future Generations of Computers," *IEEE Spectrum,* November 1965, pp. 90 ff.

Renwick, W., *Digital Storage Systems,* Wiley, New York, 1964.

Spivak, H. A., and L. E. Hawkins, Jr., "Reliability Considerations in Designing Industrial Control Computers," *Computer Design,* November 1968, pp. 34 ff.

Surgen, D. H., "Terminals: On-Line and Off, Conversational and Batch," *Control Eng.,* February 1970, pp. 96–104.

Thron, J., "How Computer Word Size Affects Performance," *Computer Design,* August 1967, pp. 14 ff.

Van Gelder, M. K., and A. W. England, "A Primer on Priority Interrupt Systems," *Control Eng.,* March 1969, pp. 101–105.

Chapter 3

Analog-Input Subsystems

THOMAS J. HARRISON

Digital computer evolution over the past two decades has been characterized by decreasing cost and increasing performance per dollar. As a consequence, the price of the actual central processing unit (CPU) is often a small fraction of the total process control computer system price, sometimes less than one-fifth. This would not be the case, of course, if the price of the specialized peripheral equipment required in control computers had similarly decreased. Although the price/performance characteristics of process input/output (I/O) equipment have improved, the improvement has been less than that in the CPU. This is due mainly to the precision nature of process I/O equipment, the need for custom features to satisfy special process requirements, and the inability of the designer of precision analog circuits to take full advantage of advances in microelectronics.

The previous chapter discussed the characteristics of the CPU and the types of data-processing input/output (DP I/O) available on modern PC computers. Although the CPU and DP I/O requirements differ from those of the scientific or general-purpose computer, the major difference lies in the process I/O equipment: The scientific and general-purpose computer communicates with the outside world in the form of numerical data, whereas the control computer also must be able to measure parameters of, and provide output signals to, the real world.

TRANSDUCERS AND ANALOG-INPUT FUNCTIONS

The following discussion considers conversion of physical data into computer-usable form. This conversion is performed by the analog-input subsystem or, as it is sometimes called, the *front end* of the process control computer system. Conversion of computer data into process-usable form is discussed in Chap. 5.

3.1 MEASUREMENT TRANSDUCERS

Number systems are the invention of man in his attempt to quantitatively describe and understand his universe. The world, however, is not discrete in a macrocosmic sense but rather is a continuum of values. Measurement involves converting a parameter, or measurand, of this continuum into a form to which a distinct number can be assigned. For example, when measuring temperature with a mercury-filled thermometer, volumetric expansion of the mercury is first converted to a linear displacement of the liquid level; a numerical value then is assigned to the temperature by comparing the level with a scale. The thermometer is a transducer and the comparison of the mercury level with a scale is a form of analog-to-digital conversion.

To be usable to the analog subsystem of most control computer systems, the measurand first must be converted to electrical form. This conversion is performed by a piece of equipment that generically is called a transducer. Examples of common transducers with electrical outputs are thermocouples and resistance elements for the measurement of temperature, strain gauges for the measurement of strain, pressure, or small displacement, and potentiometers for the measurement of position. The broad range of available transducers can be appreciated by scanning the *ISA Transducer Compendium* (Minnar, 1963; Harvey, 1968, 1970).

Since the transduction of basic physical quantities into electrical form usually is not considered to be a function of the analog front end of a computer system, this discussion is limited to the conversion of electrical signals into digital form. Several comprehensive treatments of instruments and their application have been published (Considine, 1957; Considine and Ross, 1964).

3.2 TRANSDUCER OUTPUT CHARACTERISTICS

The electrical output of most transducers is current or voltage. Some transducers, however, have change of resistance, capacitance, or inductance as the output quantity. Resistive outputs are converted to voltage or current quite simply by use of a constant current or voltage supply. These supplies usually are not included as part of the analog front end, although some manufacturers provide them as features. Equipment also exists for converting inductive and capacitive outputs into current or voltage form but, because of the low usage, this equipment rarely is provided as a standard feature in an analog subsystem.

The current or voltage output of most transducers is a DC signal, although some AC transducers are used. Again, however, AC transducers are relatively few in number and their outputs can be converted to DC by equipment external to the analog subsystem. Few manufacturers of control computers offer the ability to handle AC signals directly and the following discussion is limited to the consideration of DC signals.

The DC transducer signals range from full-scale values of millivolts to in excess of 100 V. Signals of less than 1 V will be classified as *low level* and signals in excess of 1 V will be called *high level*. The 1 V level was selected because the bulk of the transducers have outputs either in the millivolt range or in the range between 1–10 V. In addition, the precautions and methods required in designing circuits for millivolt-level signals are distinctly different from those required for higher level signals. The voltage level defined as the boundary between low- and high-level signals is the choice of the individual author since an accepted standard does not exist.

3.3   ANALOG-INPUT EQUIPMENT

The analog subsystems of commercially available control computers vary considerably in their constituent parts and arrangement. However, the equipment generally can be categorized into five distinct functions. Depending on the particular computer, the physical packaging and the interface between these functions may differ and, in some cases, one or more of them may not be offered. It is convenient to establish these categories and to discuss briefly the functions before proceeding with a detailed discussion of the hardware.

TERMINATION FACILITY.   The transducer signals are transmitted to the analog front end by single wires or wire pairs which may be shielded. In the case of single-wire transmission, the signal circuit is completed by means of a common ground path that serves several transducers. These signal wires must be terminated at the interface to the analog subsystem. The termination facility provided for this purpose may be merely a series of terminal strips, a cable connector, or special equipment designed specifically for each type of signal.

SIGNAL CONDITIONING.   This term describes any form of signal modification except amplification. (Some authors, however, include amplification.) Examples include filtering, attenuation, level shifting, linear or nonlinear compensation, and current-to-voltage conversion.

MULTIPLEXING.   The multiplexer consists of an electronic or electro-mechanical switch in series with each input line. The switches are controlled by the computer or by special logic circuits to route input signals to the analog-to-digital converter. In this way, a single analog-to-digital converter may be time shared among many input signals. Multiplexing is performed before and/or after amplification. Multiplexers are classified as low level or high level, high speed or low speed, and differential or single ended.

AMPLIFICATION.   Many transducer signals are low level and most analog-to-digital converters operate with either a 5-V or a 10-V full-scale input range. Therefore, amplification of low-level signals is necessary if the full resolution of the analog-to-digital converter is to be effectively utilized. The amplifier that provides this function is referred to as a *low-level* or *high-gain amplifier*. Typically, voltage gain ranges from 10–500 or 1000, with the gain being either fixed at the time of construction, selectable by means of a manual switch, selectable by the computer under program control, or automatically controlled by a special logic function. Depending on the type of multiplexer and the system configuration, the low-level amplifier may be differential or single-ended. An amplifier may be provided for each input signal or one or more amplifiers may be time-shared between many input signals.

Another type of amplifier employed in the analog-input subsystem is the high-level *buffer amplifier,* which provides a high input impedance, low output impedance, and a voltage gain of one or two. This impedance transformation is usually necessary since most analog-to-digital converters have a relatively low input impedance and significant errors would result in the measurement of high impedance sources. The buffer amplifier is often included as part of the analog-to-digital converter and, therefore, is not always recognized as a separate entity in system descriptions and specifications.

A third type of amplifier that is sometimes found in the analog front end is the *sample-and-hold* amplifier. The output of this amplifier is proportional to the input until a hold command is given. The amplifier output then remains constant for the remainder of the cycle, as shown in Fig. 3-1. There are several reasons for using sample-and-hold amplifiers:

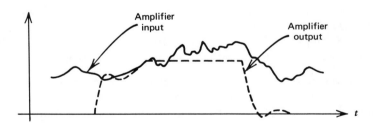

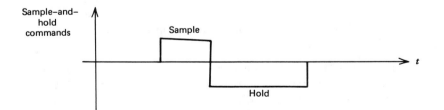

Fig. 3-1   Sample-and-hold amplifier operation.

If several amplifiers are provided in the system, the value of several inputs at a given instant of time can be determined by giving simultaneous hold signals to the amplifiers. In addition, the time uncertainty of the sample is decreased since the transition from the sample mode to the hold mode may be as little as a fraction of a microsecond.

ANALOG-TO-DIGITAL CONVERSION. The purpose of the analog-to-digital converter (ADC) is to provide a digital representation of an analog signal. Commercial ADCs are available with conversion rates ranging from a few conversions per second up to millions of conversions per second. In the process industries, conversion rates of less than 50,000 conversions/sec (cps) are most common. Most ADCs operate at an input level of 5 or 10 V full scale, although low-level ADCs have been used in process control computer systems (Funk *et al.*, 1963).

ANALOG-INPUT SUBSYSTEM CONFIGURATIONS

The five basic functions described above are combined to provide analog subsystems for a wide variety of applications. Each manufacturer of control computer systems offers a number of configurations to satisfy the needs of a broad market. It is possible, however, to describe a number of configurations that generally encompass the spectrum of available subsystems.

The system sampling rate and the signal level are the criteria for a convenient categorization of system configurations. The sampling speed, given in samples per second (sps), is the rate at which the total analog subsystem converts analog-input signals into digital values and transfers the values to the CPU or local storage; that is, it is the effective rate at which the system can generate numerical data. This rate is less than the ADC conversion speed since time must be allowed for multiplexer and amplifier transient settling time.

There are no standard definitions for *low, medium,* and *high* speed, and the terms have different meanings for low- and high-level systems. For example, a system using a low-level multiplexer and a time-shared amplifier that has a sampling rate of 10 ksps may be called a high-speed low-level system, whereas a high-level system operating at this rate is only a medium-speed system. In addition, the definitions change with time as new technologies and techniques are discovered and utilized. For our purposes, the classification shown in Table 3-1 is generally acceptable.

Table 3-1   Analog Subsystem Classification

Category	Low Level	High Level
Low speed	Less than 200 sps	Less than 5 ksps
Medium speed	200–10 ksps	5 ksps–50 ksps
High speed	Greater than 10 ksps	Greater than 50 ksps

3.4 HIGH-LEVEL SUBSYSTEMS

The two examples of high-level analog subsystems shown in Fig. 3-2 and Fig. 3-3 illustrate typical configurations that can be used in any of the speed categories. Many variations of the configurations are technically possible and many of them have distinct advantages for particular applications. The systems in the medium- and high-speed ranges represent relatively sophisticated machines. In general, high-speed high-level systems are found in applications in which fast transient phenomena are being monitored or where the number of input signals to be sampled is so large that a high-speed system is necessary to provide a reasonable per-channel sampling capability.

In Fig. 3-2 the high-level signals are brought into the system and

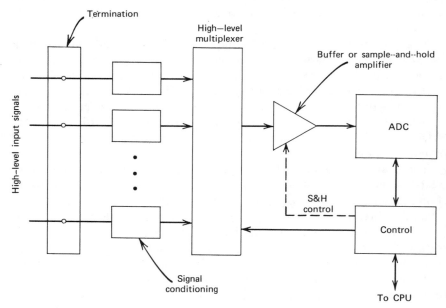

Fig. 3-2   High-level analog subsystem.

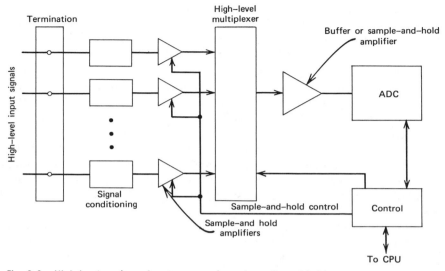

Fig. 3-3 High level analog subsystem—per-channel sample-and-hold.

terminated at the termination facility. Since this is a high-level system, termination considerations are largely mechanical rather than electrical; electrical effects such as thermal potentials and contact resistance are generally negligible. The signals then pass through signal conditioning circuits, if they are used, and are connected to the multiplexer.

In most systems, the high-level multiplexer utilizes solid-state switches and is single ended, although there are advantages to a differential system. Low-speed differential systems using relays are sometimes offered since this is a simple extension of some low-level configurations.

Under command of the control section, a multiplexer point is addressed (i.e., closed) and the input signal is applied to the buffer or sample-and-hold amplifier. In the case of a buffer amplifier, the main function is that of providing a low source impedance for the ADC. A sample-and-hold amplifier also provides this isolation function but, in addition, it may be utilized to reduce the aperture time of the system; that is, the actual time at which the sample is taken is known more precisely and a change in the actual input signal during the conversion period does not affect the converted value.

The output of the amplifier is connected to the ADC, which performs the actual conversion into digital form. The type of ADC used depends primarily on the speed and resolution requirements of the system.

Figure 3-3 shows another high-level analog subsystem used in some applications. It is similar to the previous configuration except that sample-and-hold amplifiers are provided for each input signal. The function of the other pieces of equipment is the same as before except that the control unit must provide the control signals for the multiple amplifiers. The control could include, for example, simultaneous sample-and-hold commands to a number of amplifiers or commands sequenced according to a preplanned or calculated schedule under control of the CPU. Because of the cost of the individual amplifiers, this configuration is expensive and is used only when the application requires the type of performance that can be obtained. A variation of this configuration is to provide a limited number of sample-and-hold amplifiers and a second multiplexer at the input to the amplifiers, thus providing a functional capability similar to that of the subsystem of Fig. 3-3 but at a reduced cost.

The main advantage of using per-point sample-and-hold amplifiers is that the value of a number of input signals can be determined at precisely the same instant of time. Using the time-shared system of Fig. 3-2, the value of two inputs can be determined only at points A and B' on the waveforms shown in Fig. 3-4. The values cannot be determined at the same instant, points A and A', because of the finite conversion time of the ADC. The per-channel sample-and-hold amplifiers allow instantaneous comparison since a hold command can be given to both ampli-

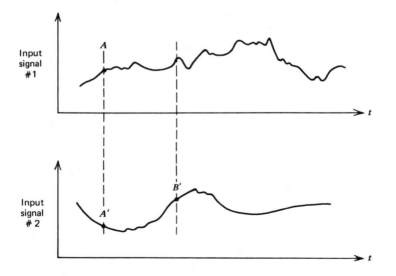

Fig. 3-4   Illustrating the time skew problem.

fiers simultaneously; the outputs of the two amplifiers can then be converted sequentially by the ADC.

3.5   LOW-LEVEL HIGH-SPEED SUBSYSTEM

The subsystem shown in Fig. 3-5 is typical of high-speed low-level configurations. It is similar to the high-level system of Fig. 3-2 except for the addition of differential amplifiers for each input signal.

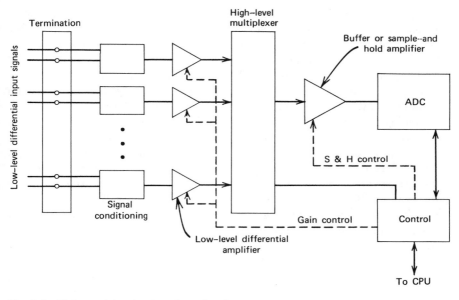

Fig. 3-5   High-speed low-level analog subsystem.

The differential input signals are terminated at the termination facility, as in the previous cases. The design of the termination facility includes all the mechanical factors involved in high-level termination plus additional electrical considerations that are important in a low-level situation. The most important of these are thermal and common mode potentials, which are discussed in Sec. 3.9 and Sec. 3.10. In addition, cold junction compensation of thermocouples may be provided either in the termination facility or in the signal conditioning equipment.

Signal conditioning may include all the functions mentioned in Sec. 3.3. Some signal conditioning may be combined with the amplification function; this would be the case, for example, when the amplifier is constructed with a prescribed filter characteristic.

The amplifiers are low-level amplifiers having a gain sufficient to amplify the full-scale low-level signals to the full-scale input level of the buffer amplifier. Typically, this means gains of 10–1000 corresponding to low-level full-scale ranges of 500–5 or 10 mV. Gains may be factory set, manually set by the operator, selected under program control by the computer, or selected automatically by logic in the control unit. These latter possibilities are indicated in the figure by the dotted control lines.

In addition to the gain requirement, these amplifiers must provide the basic common mode rejection capability of the system and they must accept the maximum common mode voltage to which the system is subjected. The amplifier output signal is typically a high-level single-ended signal and can be multiplexed by a high-level single-ended multiplexer. The multiplexer and the remainder of the system are essentially the same as the system discussed above and similar considerations apply.

The configuration of Fig. 3-5 represents the only low-level system capable of system sampling rates in excess of about 20 ksps with present technology. It is an expensive system because of the many amplifiers required. As a result, this configuration is used only in those applications that require a high bandwidth for the analysis of high-speed low-level data. This occurs infrequently in most continuous- and batch-process industries where time constants are long compared with the sampling interval of even the low-speed low-level subsystems. When integrated circuit amplifiers having the required performance are available, this system approach may become popular with manufacturers of control computer systems.

3.6   LOW-LEVEL MEDIUM- OR LOW-SPEED SUBSYSTEM

The configuration of Fig. 3-6 is probably the most popular subsystem for low-level signal processing in control computers. The main difference between the medium- and low-speed categories is not in the configuration but rather in the devices used for multiplexing and in the characteristics of the time-shared low-level amplifier. The system is identical to the previous one through the signal conditioning circuits, except that the signal conditioning is rarely combined with the amplification function.

The multiplexer is differential and must switch low-level signals without introducing an appreciable error. Thus the designer must be concerned with the effects of thermal voltages, noise, and common mode potentials to a much greater degree than in the case of a high-level multiplexer. Electromechanical relays and field effect transistors are the multiplexing devices commonly used.

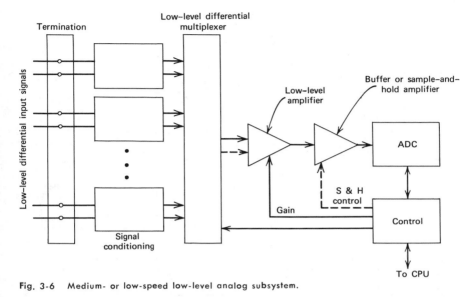

Fig. 3-6   Medium- or low-speed low-level analog subsystem.

The amplifier is similar to that required in the previous case except that, in general, it must have higher bandwidth. The gain selection options mentioned previously can be provided. If the multiplexer does not provide isolation from common mode voltages, the amplifier must provide common mode rejection. Since the common mode voltage is a step function that is applied when the multiplexer switch closes, the amplifier must have excellent common mode transient response. The increased cost of the amplifier may not affect the total system price appreciably because its cost is shared by many input channels.

If the multiplexer provides isolation from the common mode voltage, the low-level amplifier need not be differential. This is indicated in the figure by the dotted input line to the amplifier. The rest of the system is the same as that discussed previously.

3.7 COMBINATION SYSTEMS AND OTHER FEATURES

The above discussion treated each configuration as a separate entity. In reality, a single system might employ several types of multiplexers, or it might have several ADCs or any number of other combinations. In addition, some features are available that have not been discussed. An example of an actual subsystem is shown in Fig. 3-7 to illustrate a particular combination configuration and to show additional features.

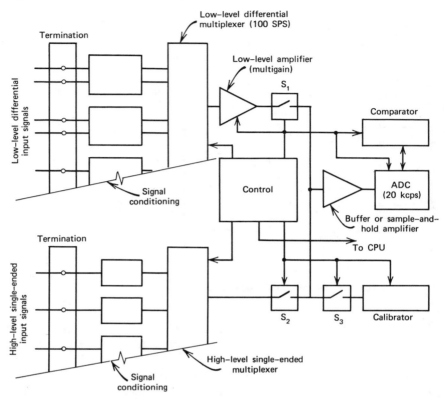

Fig. 3-7 Typical combination configuration with special features.

The configuration combines low- and high-level subsystems similar to those of Fig. 3-2 and Fig. 3-6. The operation of these parts of the combination configuration is the same as previously described.

3.7.1 High-Level–Low-Level Overlap Feature

Switches S_1 and S_2 in Fig. 3-7 permit high-level and low-level operation to be *overlapped* in an efficient manner. For the assumed system rates of 100 sps and 10 ksps, the time periods required for the entire conversion operation are 10 msec and 100 μsec, respectively. The actual conversion time of the 20 kcps ADC is 50 μsec, leaving a 50-μsec settling time for the high-level system and almost the full 10-msec period for settling of the low-level system. Assuming the use of electromechanical relays in the low-level system, almost all of the 10-msec interval may be required to allow for relay noise to decrease to an acceptable level, and during this time the ADC would be idle. However, if a low-level multiplexer switch is closed and switch S_1 is open, the transient output voltage of the low-level amplifier is not applied to the ADC input and most of the 10-msec settling time can be utilized for high-level conversions by closing S_2 and addressing high-level points. For the system rates selected, it is theoretically possible to make 99 high-level conversions while the low-level system is settling. In actual practice, the number is somewhat less because of the time required for addressing, transferring data, and other logical operations.

3.7.2 Programmable Calibration Verification

Programmable calibration verification consists of a precision voltage that can be applied to the ADC under program control to check conversion accuracy. In addition, the calibrator usually provides a short-circuited input channel for checking the zero drift of the converter. Based on these two readings, subsequent data can be corrected mathematically for changes in converter calibration. The calibration system illustrated is very simple, but it has the limitation that it checks only the operation of the buffer amplifier and the ADC. It is possible to provide for the insertion of calibration voltages either at the input to the low-level amplifier or at the inputs to the multiplexer. This provides a better calibration system at the expense of additional hardware.

3.7.3 Comparator Feature

This is a technique whereby input signals are checked to determine if they fall within an upper and lower bound. Conceptually, the simplest implementation is a special logic function that reads the two limit words from the CPU, initiates a conversion, and then mathematically compares the ADC value and the limit values. Based on the compare operation, an indication is returned to the CPU as to whether the ADC result is within limits. This method proceeds at the normal system sampling rate.

A second method that can be used with a successive approximation ADC and that requires an understanding of the operation of such an ADC is described in Sec. 3.15.3. This method permits limit checking at a rate limited primarily by the multiplexer speed.

SYSTEM ERRORS

Before proceeding with a detailed discussion of the analog-input equipment and the interconnection of this equipment, it is necessary to understand what system accuracy can be achieved in analog-input subsystems. Without this understanding, the techniques and procedures to be discussed cannot be appreciated. A detailed discussion of system accuracy and other criteria of system performance is found in Chap. 4, but the following sections provide an introduction to several of the errors of concern in the design of analog-input equipment.

3.8 DEFINITIONS AND VALUE

System accuracy is the exactness with which a system can measure a voltage. The difference between a perfect measurement and the measurement made by the system is the *system error* and is often expressed as a percentage of the full-scale range of the system.

$$\varepsilon = \text{(perfect measurement} - \text{system measurement/full-scale value)} \times 100 \quad (3\text{-}1)$$

The system error is a combination of individual errors in each of the pieces of equipment in a configuration and includes errors such as noise, drift, gain stability, and reference voltage inaccuracy.

As indicated by Eq. (3-1), system error is expressed in terms of the full-scale range. Since most process control analog subsystems have multiple ranges, a single error value does not describe the system. Typical values for commercially available systems are given in Table 3-2 where %F.S. and RTI are abbreviations for *percent full scale* and *referred to the input*, respectively. The *mean error*, which is the error in the average of a statistically significant number of readings from the same input signal, may be less than the errors listed in Table 3-2 by a factor of two or three, depending on the range.

The voltage errors listed in the table show that a very small potential can be a major source of error in a low-level system. Although errors in the various pieces of analog-input equipment cannot be added alge-

Table 3-2   Typical System Error

Full-Scale Range	Maximum-System Error	
(mV)	(%F.S.)	(μV RTI)
10	0.25	25
50	0.10	50
100	0.07	70
500	0.06	300
5000	0.05	2500

braically to obtain the system error because some errors of opposite polarity cancel, the error contribution of each piece of equipment in the analog-input subsystem must be considerably less than the total system accuracy.

Some errors are due to the tolerance of the electrical components used to build the system and, as such, are under partial control of the system designer. Other errors are dependent on the system configuration and the manner in which the system is physically constructed. These are discussed in detail in later sections. It is, however, necessary to have a basic understanding of two sources of error whose elimination often dictates the procedures used in the design of low-level systems. Specifically, these are the errors due to common mode voltages and thermal potentials. The following discussion considers common mode error in general terms; later discussions consider the problem in terms of specific circuit configurations. Thermal potentials are discussed in Sec. 3-10.

3.9 COMMON MODE ERROR

Single-ended instruments measure the voltage difference between a single input terminal and a ground reference, normally the instrument chassis. Such an instrument might be used as shown in Fig. 3-8, where G_1 is the reference ground. The source V_s is referenced to a ground point G_2 which may be at the same potential as G_1. The difference between the two ground points is represented by the voltage source V_g. If $V_g = 0$, the voltmeter measures the actual potential V_s; if G_1 and G_2 are not at the same potential, the voltmeter reading is in error by V_g volts. Although this is a simple and obvious case, potential differences between two ground reference points are not uncommon in a physical system. For example, if G_1 and G_2 are connected by a ground bus having a total resistance of 10 mΩ and there is a ground current of 100 mA, the difference in ground potential is 1 mV. For a voltmeter with a full-scale range of 10 mV, this is a 10%F.S. error.

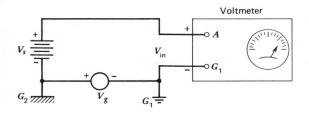

Fig. 3-8 Single-ended measurements.

The alternative, and the system most often used in low-level measurement applications, is the differential system in which the instrument measures the potential between two input terminals without regard to the instrument reference ground. This is illustrated in Fig. 3-9 where the input voltage is connected to the differential input terminals A and B. Theoretically, and within limits in practical situations, the voltmeter measures the voltage V_s without an error due to V_g.

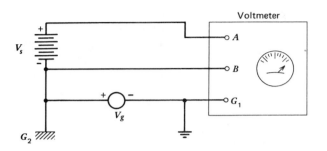

Fig. 3-9   Differential measurements.

3.9.1   Common Mode Voltage

In Fig. 3-9, the voltage between the instrument ground G_1 and input terminal B is equal to V_g, and the voltage between G_1 and the input terminal A is $V_g + V_s$. Thus V_g is common to both input terminals and is called the common mode voltage. That is, the *common mode voltage* is the voltage that is common to both terminals of a differential instrument when measured with respect to a reference ground. The voltage between the differential input terminals A and B is called the *normal mode* or *differential voltage*. Some authors define the common mode voltage as $(V_g + V_s)/2$; since V_s is generally much smaller than V_g whenever common mode error is of significance, the slight difference between the definitions is of no consequence.

3.9.2   DC Common Mode

If the measuring instrument is a perfect differential device, the common mode voltage V_g does not contribute an error to the measurement of the normal mode signal V_s. In any practical case, however, there is a finite error contributed by the common mode voltage.

Consider the DC model of the physical system as shown in Fig. 3-10. The R_L's represent the distributed resistances of the lines; the R_S's represent the distributed shunt resistance between the two input lines; and

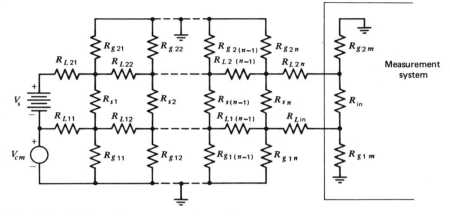

Fig. 3-10   Distributed DC common mode model.

the R_G's represent the distributed leakage resistances from the lines to the ground reference. The resistance R_{in} is the input resistance of the measurement system, and R_{g1m} and R_{g2m} are the internal system leakage resistances to ground. The common mode voltage V_{cm} is equivalent to the voltage V_g in the previous figures and the signal voltage is again denoted by V_s.

The voltmeter of the previous case has been replaced by a general measurement system because the analysis is valid for any differential input system. In a control computer system, the measurement system might include a multiplexer, differential amplifier, and analog-to-digital converter.

The model is quite complex and is of limited practical value because of the difficulty of determining the various resistances. The model can be simplified, as shown in Fig. 3-11, without destroying its usefulness in predicting the performance of most physical systems. The measurement system is replaced by a perfect measurement system having infinite input and leakage resistances. The resistances R_{in}, R_{g1m}, and R_{g2m} of Fig. 3-10 are combined with the resistances in the external circuit. The resistance R_s in Fig. 3-11 is the equivalent shunt leakage resistance of the lines in parallel with R_{in}, and R_{g1} and R_{g2} are the parallel combinations of the equivalent lumped leakage resistance to ground and the leakage resistance of the imperfect measurement system. Assuming that $V_s = 0$ does not affect the generality of the analysis since the model is linear and superposition applies. The input voltage V_{in} of the simplified model is

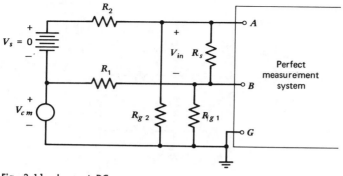

Fig. 3-11   Lumped DC common mode model.

$$V_{in} =$$

$$V_{cm} \frac{R_1 R_{g2} - R_2 R_{g1}}{(R_2 + R_{g2})(R_1 + R_{g1}) + (R_2 R_{g2}/R_s)(R_1 + R_{g1}) + (R_1 R_{g1}/R_s)(R_2 + R_{g2})} \quad (3\text{-}2)$$

For most systems, R_1 and R_2 are less than $1000\ \Omega$, R_s is on the order of megohms, and R_{g1} and R_{g2} are on the order of hundreds of megohms. Thus the second two terms can be neglected without an appreciable effect on the numerical result. This is equivalent to assuming that the parallel combination of the shunt line resistance and the input resistance of the measurement system is infinite. Applying this simplification to Eq. (3-2) results in

$$V_{in} = V_{cm} \frac{R_1 R_{g2} - R_2 R_{g1}}{(R_2 + R_{g2})(R_1 + R_{g1})} \quad (3\text{-}3)$$

If R_s had been assumed infinite initially, Eq. (3-3) could have been derived easily by noting that R_2 and R_{g2} form a voltage divider which determines the voltage at terminal A. Similarly, the voltage divider formed by R_1 and R_{g1} establishes the voltage at terminal B. The input voltage V_{in} is simply the difference between these terminal voltages.

As an example of common mode error, suppose that the signal lines are copper and constantan thermocouple wires and that the thermocouple junction is bonded to an electrode in an electrolytic plating tank. Values for the parameters of Eq. (3-3) might be

$$R_1 = 280\ \Omega \qquad\qquad R_2 = 10\ \Omega$$
$$R_{g1} = R_{g2} = 10^8\ \Omega \qquad V_{cm} = 50\ \text{V}$$

Substituting into Eq. (3-3) results in

$$V_{in} = 50 \left[\frac{280 \times 10^8 - 10 \times 10^8}{(280 + 10^8) + (10 + 10^8)} \right]$$

$$= 135 \, \mu V$$

If the measurement system has a full-scale range of 50 mV, this error represents 0.27%F.S. Since a typical analog-input subsystem has an error of 0.1%F.S. on a 50-mV range, the error due to the common mode voltage is almost three times as great as the basic system error.

The answer in the example is equal to the common mode voltage V_{cm} times the ratio of the difference of the line resistances to the value of the leakage rseistance. This is usually true since, in most cases, the leakage resistance paths are approximately equal and orders of magnitude larger than R_1 or R_2. Thus the input voltage V_{in} is given by the approximate expression

$$V_{in} \cong V_{cm} \frac{R_1 - R_2}{R_g} \tag{3-4}$$

where R_g is the value of either R_{g1} or R_{g2}. For the system designer or installer, this expression represents a rule of thumb for the common mode error due to line and leakage resistances.

3.9.3 Common Mode Rejection Ratio

According to Eqs. (3-3) and (3-4), the error is directly proportional to the common mode voltage when a linear model is assumed. It is common to consider the ratio of the common mode voltage to the error voltage as an index of system performance. The resulting ratio, called the *common mode rejection ratio* (CMRR), is defined as

$$\text{CMRR} = \left| \frac{\text{common mode voltage}}{\text{equivalent voltage at the system input due to the common mode voltage}} \right| \tag{3-5}$$

The CMRR for the resistive model of Fig. 3-11 is obtained directly from Eq. (3-3) since V_{in} is the voltage due to the common mode voltage when $V_s = 0$. Therefore, for this model

$$\text{CMRR} = \left| \frac{(R_2 + R_{g2})(R_1 + R_{g1})}{R_{g1}R_2 - R_{g2}R_1} \right| \tag{3-6}$$

Or, using the approximate expression of Eq. (3-4),

$$\text{CMRR} \cong \left| \frac{R_g}{R_1 - R_2} \right| \tag{3-7}$$

If the measurement system has gain, it is sometimes convenient to define CMRR in terms of the output voltage (or output value if the system includes an analog-to-digital converter).

$$\text{CMRR} = \left| \frac{\text{common mode voltage} \times \text{system gain}}{\text{output voltage due to the common mode voltage}} \right| \tag{3-8}$$

It is also common to express CMRR in terms of decibel (dB) units as

$$\text{CMRR(dB)} = 20 \log (\text{CMRR}) \tag{3-9}$$

where log refers to base 10 logarithms. Typical values of CMRR (dB) for the analog-input subsystem of modern systems range from 100 dB to 160 dB at DC with line unbalances of up to 1000 Ω.

3.9.4 AC Common Mode

The above derivation and examples are based on a DC model, but they are theoretically valid for AC common mode voltage. Usually, however, AC common mode rejection is governed by leakage capacitance from each line to ground rather than by leakage resistance. A simple lumped model for estimating the AC CMRR is shown in Fig. 3-12 where R_1 and R_2 represent the line resistances as in the previous case; the capacitances C_1 and C_2 are the parallel combination of the line to ground capacitance and the measurement system input terminal to ground capacitance for each line. The DC line-to-ground leakage resistance is generally negligible in comparison with the reactance of C_1 and C_2 at the frequencies of interest. As in the previous case, the line-to-line shunt capacitance and resistive leakage are neglected and the signal source V_s is assumed to be zero.

The resistance R_2 and the capacitance C_2 form a voltage divider, as

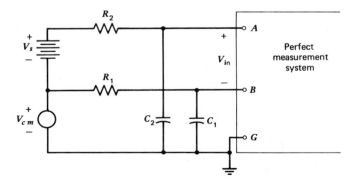

Fig. 3-12   Lumped AC common mode model.

do R_1 and C_1. The input voltage V_{in} is the difference between the voltages at terminals A and B. Therefore,

$$\frac{V_{in}}{V_{cm}} = \frac{s(R_1C_1 - R_2C_2)}{(1 + R_2C_2s)(1 + R_1C_1s)} \tag{3-10}$$

where s is the Laplace frequency variable. This is a complex expression and, in most cases, only its magnitude is of interest. After some manipulation, the absolute value is found to be

$$\left|\frac{V_{in}}{V_{cm}}\right| = \frac{\omega(R_1C_1 - R_2C_2)}{\{[1 + (R_2C_2\omega)^2][1 + (R_1C_1\omega)^2]\}^{1/2}} \tag{3-11}$$

where ω is the radial frequency and is equal to $2\pi f$, with f the frequency in Hertz.

The error can be reduced to zero by balancing the circuit so $R_1C_1 = R_2C_2$. In practice, this is difficult and is done only in critical system applications. It may be done, however, in equipment such as amplifiers where the circuit components can be controlled. This method of balancing is equivalent to null balancing a bridge circuit, since redrawing Fig. 3-12 shows that it is actually a bridge circuit.

Since most systems are reasonably symmetrical, the capacitances C_1 and C_2 usually are essentially equal and can be set equal to a capacitance C. Furthermore, at low frequencies with nominal line resistances, it is usually true that $R_2C_2\omega$ and $R_1C_1\omega$ are much less than unity. With these assumptions, Eq. (3-11) reduces to

$$\left|\frac{V_{in}}{V_{cm}}\right| \cong |(R_1 - R_2)\omega C| \tag{3-12}$$

Since V_{in} is the voltage due to the common mode voltage, Eq. (3-12) is the inverse of the common mode rejection ratio

$$\text{CMMR} \cong \left|\frac{1}{(R_1 + R_2)\omega C}\right| \tag{3-13}$$

As a numerical example, assume that the line unbalance is 150 Ω and that a CMRR of 1,000,000:1 at 60 Hz is desired. Solving Eq. (3-13) for C and substituting

$$C = \frac{1}{(R_1 - R_2)\omega\text{CMRR}} = \frac{1}{150 \times 2 \times 3.14 \times 60 \times 10^6} = 17.8 \text{ pF}$$

This is a very small value when considering a practical system. Special techniques, such as guard shielding, may be required to reduce the effective capacitance if reasonable values of CMRR are to be achieved. Typical values of AC CMRR for modern control computer systems are on the order of 100–120 dB at 60 Hz with up to 1000 Ω line unbalance.

3.10   THERMAL POTENTIALS

A thermal potential is produced at the junction of two dissimilar metals as a result of a temperature difference. This phenomenon is familiar to process engineers as the thermocouple effect. If a thermal potential exists in a low-level measurement system, it can produce a significant system error. It is quite easy, for example, to produce a thermal voltage of 10 μV merely by heating a junction with one's finger. This error at the input to a 10-mV system amounts to a 0.1%F.S. error.

3.10.1   Seebeck Effect

In 1821 T. J. Seebeck discovered that heating a bismuth-copper junction in a closed electrical circuit caused an electrical current. The same phenomenon can be demonstrated in any circuit consisting of dissimilar electrical conductors. If the two junctions in Fig. 3-13 are at different

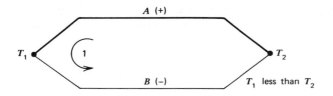

Fig. 3-13   Seebeck effect.

temperatures T_1 and T_2, a current I flows in the circuit as long as there is a temperature difference between the two junctions. The EMF causing this current is variously called the thermoelectric EMF, the total thermal EMF, or the Seebeck EMF. The value of the potential is dependent on the type of materials used in forming the circuit and the temperature difference between the junctions. If the current flows from conductor A to conductor B at the cooler junction, as shown in the figure, conductor A is said to be positive with respect to conductor B.

Another thermoelectric effect is the *Peltier effect*. This phenomenon is the heating of one junction and the cooling of the second junction when a current flows in the circuit. The effect is usually not a factor in low-level circuit design since the circuits are of high impedance and little current flows.

3.10.2   Laws of Thermoelectric Circuits

Although the theory of thermoelectric circuits can be deduced from thermodynamic principles and explained qualitatively by the electron theory of metals, the exact relationship between the composition or structure of the conductors and their thermoelectric properties is not established. However, three important laws concerning the thermoelectric EMF have resulted from experimental observations and theoretical studies.

Law of the Homogeneous Circuit. In a circuit of homogeneous conductors the thermoelectric EMF for a given pair of metals depends only on the temperatures at the junctions and is independent of the temperature distribution between the junctions.

In terms of the error sources in low-level systems, this law focuses the attention of the designer or installer on the junctions in a circuit rather than on the interconnecting conductors.

Law of Intermediate Temperature. If the EMF produced in a circuit of two dissimilar metals whose junctions are at temperatures T_1 and T_2 is E_1, and if the EMF is E_2 when the junction temperatures are T_2 and T_3, then the EMF is $E_1 + E_2$ when the junction temperatures are T_1 and T_3. This is illustrated in Fig. 3-14.

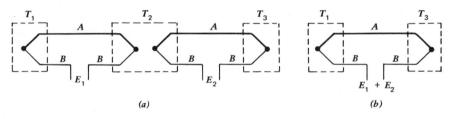

Fig. 3-14   Law of intermediate temperatures.

A practical consequence of this law is that if the thermoelectric EMF is known with respect to a reference temperature T_1, then the EMF with respect to another reference temperature T_2 differs by a constant equal to the EMF produced when the junctions are at temperatures T_1 and T_2.

Law of Intermediate Metals. If a thermoelectric circuit consisting of conductors A and B is broken and a third conductor C is added, the thermoelectric EMF is not changed if the two new junctions are maintained at the same temperature, even though this temperature may be different from the other junction temperatures. This is illustrated in Fig. 3-15.

The condition $T_1 = T_3$ in Fig. 3-15b is a situation of practical interest in terms of thermocouples and circuits; if a junction of dissimilar metals is formed by welding or soldering with a metal dissimilar to both con-

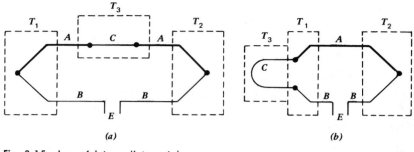

Fig. 3-15 Law of intermediate metals.

ductors, and if the junction is at thermal equilibrium, the EMF produced is not altered but depends only on the junction temperatures and the composition of the main conductors.

Another important consequence of the law is that if the EMF-temperature relationship is determined for conductors A and B, and for B and C, then the EMF for a junction formed by conductors A and C is the sum of the previously determined EMFs. Based on this relationship, a table of EMF-temperature values for various metals in conjunction with a common reference metal provides information for any other combination of metals.

3.10.3 Thermoelectric Power

The EMF-temperature relationships for particular pairs of metals are determined empirically by interpolation between known reference temperatures. It has been found that a cubic equation approximates these empirical relationships with fair accuracy over a range of 100°C or more. Generally, however, a quadratic equation of the form of Eq. (3-14) is sufficient to approximate the relationship. The constants a and b are determined empirically, T is the temperature of one junction, and the other junction is assumed to be at 0°C.

$$E = aT + \tfrac{1}{2}bT^2 \tag{3-14}$$

The gradient of the EMF-temperature curve is called the *thermoelectric power* and is approximated by the derivative of Eq. (3-14)

$$e = \frac{dE}{dT} = a + bT \tag{3-15}$$

The thermoelectric power of various metals with respect to a common reference metal is tabulated in many handbooks and provides the relative magnitude of the thermoelectric effect in various conductors. Table 3-3 gives thermoelectric data for metals that are commonly found in electronic

Table 3-3   Thermoelectric Power

	$\mu V/°C$	
Metal	a	b $\times 10^{-2}$
Silicon	−410.96	−48.18
Nickel	−21.83	−4.24
German silver	−13.62	−4.51
Mercury	−11.57	−4.55
Platinum	−5.80	−4.47
Aluminum, commercial	−3.14	−1.23
Lead	−2.76	−1.22
Tin	−2.53	−1.35
Brass (85.8 Cu, 14.2 Zn)	−2.05	−0.660
*Phosphor bronze	−1.50	−2.17
*Copper beryllium	−0.50	−1.67
Gold	−0.36	−1.03
Silver, annealed	−0.26	−0.07
Copper, hard drawn	0.00	0.00
Cadmium	0.09	2.67
Zinc	0.29	−2.21
†Solder, low thermal	0.59	−1.85
†Solder (60 Sn, 40 Pb)	2.50	0.97
Steel, piano wire	8.00	−2.80
Iron, transformer	13.89	−4.19
Germanium	299.74	71.28

The table gives the thermoelectric power in $\mu V/°C$ difference in temperature when the cold junction is at 32°F and the reference metal is hard drawn copper. The constants a and b are defined in Eq. (3-14). Entries denoted by an asterisk (*) are derived using Eq. (3-14) and data given in the *American Institute of Physics Handbook*, 2nd ed., 1963, pp. 4-7–4-10. Entries denoted by a dagger (†) were obtained experimentally by the author and should be regarded as approximate. All other entries are derived from the *Handbook of Physics and Chemistry*, 34th ed., 1952, pp. 2218 and 2219.

equipment. The values in this table are for a reference metal of hard drawn copper, although the original tables are based on reference metals of platinum or lead. Since the coefficient b is several orders of magnitude less than the coefficient a, it does not cause a substantial deviation from a linear characteristic for small temperature differences.

3.10.4   System Thermal Errors

Based on the thermoelectric laws and the data of Table 3-3, sources of thermal EMF errors in low-level systems can be identified. Typical electronic circuits use many dissimilar metals: copper or brass in terminals; nickel or cadmium plating on screws and other hardware; copper, aluminum, and steel in chassis material and cabinets; nickel-iron alloy in reed or mercury-wetted relay armatures; silver and gold in relay contacts; and tin-lead solder in connections. Thus most electronic circuits have junctions between dissimilar metals. This is one requirement for a thermoelectric EMF.

The second requirement is a temperature difference, and electronic circuits normally contain many possible sources of temperature differentials resulting from convection, conduction, and radiation of heat energy. Good electrical conductors have excellent thermal conductivity, and heat energy conducted through wiring or printed circuits can cause significant thermal differentials. High dissipation components, such as power transformers or transistors, can be the source of temperature gradients relatively far removed from the heat-producing component. High dissipation elements also can be sources of natural convection currents. In addition, most electronic equipment has forced air or, in some cases, liquid cooling because of the high power dissipation of modern microelectronic circuits.

Methods for minimizing thermoelectric errors in low-level circuits are based on the thermoelectric laws. Since the error potential is produced at a junction, the first basic principle is to minimize the number of junctions in the low-level circuit. This includes dry junctions such as connector and screw terminals as well as soldered connections that are in series with the input terminals of the measurement system. After the signal has been amplified, the thermal potentials are generally insignificant with respect to the signal level.

It is practically impossible to eliminate all connections between dissimilar metals in low-level circuits. However, those required should be made between metals which have low thermoelectric power with respect to each other. Referring to Table 3-3, gold, silver, copper, zinc, and cadmium have thermoelectric powers within 0.65 $\mu V/°C$. Of these five materials, gold, copper, and silver are preferable to zinc and cadmium because of better electrical and soldering characteristics.

Crimped connections between similar materials represents a low thermal method of connection when they can be used. Care must be taken that both conductors are free of oxide before crimping since the oxide has a high thermoelectric power. A value of more than 500 $\mu V/°F$ has been reported in the literature (Cromer, 1966).

According to the Law of Intermediate Metals, the presence of solder at a joint does not contribute an error if the joint is at thermal equilibrium. However, thermal equilibrium cannot always be guaranteed.

The use of *low thermal* solder composed of cadmium-tin alloy is recommended in critical applications. The thermoelectric power of this solder is appreciably less than that of the usual 60% lead–40% tin electronic solder.

Another useful technique in connecting a copper lead to a terminal of dissimilar metal is to wrap fine copper wire around the terminal several times and then sparingly apply normal electronic solder. The junction remains at the temperature of the terminal because of its thermal mass and gradients due to heat conduction through the copper wire are minimized.

It is not always possible to restrict junctions to those between the recommended metals. For example, the armature of a reed or mercury-wetted relay is nickel-iron and the leads are copper; a thermal error of 36 μV/°C is possible. According to the Law of Intermediate Metals, thermal potentials can be minimized by providing a second series junction maintained at the same temperature. For example, the junctions at the ends of the relay compensate for each other; the first connection is from copper to nickel-iron and the second is from nickel-iron to copper. Care must be taken to ensure that the two junctions remain at the same temperature since they may be several inches apart.

A compensating junction also can be easily provided in a differential system. If, for example, a junction is made between two metals in one side of a differential circuit, a second junction between the same two metals should be made in the other side of the circuit. If the two complementary junctions are at the same temperature, their thermal potentials tend to cancel. The use of the metals recommended above is also beneficial since the difference between two small potentials is likely to be less than the difference between two large potentials.

The second general means of reducing thermal potentials is to eliminate, or at least to minimize, temperature differences in low-level circuits. Temperature differences due to conduction, radiation, or convection can be minimized by physical separation of low-level circuits and heat-dissipating components. Baffles and heat shields are effective in minimizing radiation heating, but if it is not possible to use them, complemenatry junctions should be located equidistant from the radiation source to ensure reasonable cancellation of thermal potentials.

Convection currents can be avoided by careful placement of components or by use of an enclosure. Since most low-level circuits have very low dissipation, it is often possible to encapsulate them with an epoxy or silicone rubber material.

Heating of junctions due to conduction can be minimized by judicious placement of components. In addition, conductors having a large surface area or great length reach thermal equilibrium faster and, therefore, do not provide a conduction path to sensitive junctions.

Care also must be taken to prevent transient thermal effects or thermal shock due to sudden changes in ambient temperature. This can be done either by building the equipment in such a way that it responds quickly to a thermal shock and is, therefore, almost always at equilibrium, or by preventing the thermal transient from reaching sensitive connections. Most of the methods discussed above are effective in the thermal transient problem. Thus, for example, enclosures protect against thermal shock resulting from a sudden draft caused by opening a cabinet door; encapsulation or insulation of the enclosure tends to smooth out variations in ambient temperature; physically small connections minimize the chance of a transient thermal gradient across a connector or solder joint.

ANALOG-INPUT EQUIPMENT

With an introduction to general system configurations and two important sources of error, it is possible to discuss the main subsystem functions in terms of specific equipment and circuits. The following sections consider the problems the system designer faces and provide the process engineer or technically inclined manager with an understanding of the factors involved in the design of analog subsystems for control computers.

3.11   TERMINATION FACILITY

As described above, the termination facility provides a place and a means for terminating input signals at the interface between the process and the control computer system. The terminations generally consist of screw-down connections for the instrumentation wiring. The screw-down terminals are usually preferred over cable connectors in the process industries since they allow single wires to be added or removed without affecting other wiring. In addition, instrumentation personnel are familiar with this type of termination, and no special tools or training are required for their effective use.

3.11.1   General Considerations

Both electrical and mechanical considerations are of concern in the design of the termination facility, and many of these considerations are the same as those encountered in other portions of the system. Therefore, the discussion of Chap. 8 concerning physical access, terminal size, wiring ports, vibration, and contaminants is applicable. The discussion here is limited to those factors unique to the termination facility for high- and low-level signals.

These factors include noise, contact and leakage resistance, thermal potentials, common mode effects, and grounding methods. In addition, the termination of thermocouple signals requires special consideration because of the necessity for cold junction compensation. With the exception of thermocouple termination and thermal potentials, however, the design problems are similar in both low- and high-level subsystems. The emphasis differs between the two classes of subsystems depending on the source and magnitude of the various errors. For example, a CMRR of 80 dB may be acceptable in a high-level subsystem whereas it is unacceptable in many low-level applications.

3.11.2 Contact Resistance

The analog subsystem is usually a voltage-measuring device with a relatively high input impedance. This means that current flow from the signal source is quite small. Assuming a 10-MΩ input impedance and a total contact resistance of 400 mΩ for two terminals, a negligible error of $4 \times 10^{-6}\%$ results. A 400-mΩ contact resistance represents an end-of-life value of 200 mΩ/contact, which is an order of magnitude greater than one would normally expect in a dry contact operating in a low-level circuit. Thus contact resistance is rarely a problem in dealing with voltage signals.

The situation with signals from current transducers is somewhat different, and contact resistance can introduce a significant error. The most common current signals found in the process industries are 1–5, 4–20, and 10–50 mA signals derived from transmitters. The current signal is converted to a voltage by passing it through a resistor in the signal conditioning circuit. Physically, the resistor may be connected at the terminal points themselves or it may be mounted on a signal conditioning card. Assuming that a 50-mA signal is converted to a 5-V full-scale value, the potential error resulting from 100 mΩ of contact resistance is 5 mV or 0.1%F.S. Since typical system error might be less than 0.05%F.S. on this range, this is a considerable error.

Whether or not the error is significant depends on several factors: The 50 mΩ value for the contact resistance is quite conservative and the typical value of less than one-tenth of this results in only 0.01%F.S. error. This is a tolerable value considering the accuracy of typical process transmitters. In addition, the relationship of the contact resistance and the

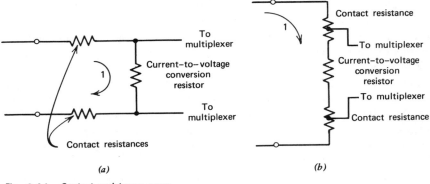

Fig. 3-16   Contact resistance error.

actual input wires is of importance. The voltage across the multiplexer input in Fig. 3-16a is not effected by the contact resistance. This figure is a model of the case when the current-to-voltage resistor is mounted on the signal conditioning card. In Fig. 3-16b, the signal current through the contact resistance produces a voltage drop across the input to the multiplexer and this causes a system error similar to the above example. This situation can occur when the current-to-voltage resistor is mounted directly across the input terminals of the system unless care is taken to ensure that contact resistances are negligible. Since this latter arrangement has the advantage that the current signal loop is not opened if the signal conditioning card is removed for some reason such as servicing, it is usually preferred by the user. It is important, therefore, to minimize contact resistance when the current-to-voltage resistor is installed at the input terminals.

Contact resistance is more dependent on the voltage across the contacts than it is on the current through the contacts. The voltage is rarely under control of the system designer or installer and, therefore, its control does not represent a good solution for ensuring low contact resistance. Other major factors affecting contact resistance are contact action, pressure, material, and degree of oxidation. These are under control of the designers and installers to a certain degree.

The contact material should be relatively corrosion resistant under the effects of air or vapor contaminants that might be present. Contact action should involve motion of the contacts when the card or plug is mated with its receptacle or when a screw fastener is tightened. This wiping action has the effect of breaking through any oxide film that might be present. Since contact pressure in mating connectors decreases with the number of insertions, servicing personnel should be cautioned against excessive or rough insertions.

In addition to an oxide film, a thin film of oil or other dirt can increase contact resistance. The perspiration film resulting from handling the contacts can cause difficulty and servicing personnel should be aware of this source of trouble.

3.11.3   Leakage Resistance

As opposed to low contact resistance, a very high insulation resistance is required for proper operation of an analog termination facility. The effects of low insulation resistance include errors due to signal or common mode crosstalk, loading errors due to reduced input impedance, and degraded common mode rejection in differential systems.

The insulation resistance between the input terminal and ground in a single-ended system or between the two input terminals in a differential system is in parallel with the system input impedance. This increases the loading error due to the voltage divider action of the signal source impedance and the input resistance. For example, a 10-MΩ input impedance system exhibits a 0.01% error when a signal having a 1000-Ω source impedance is measured. If the input impedance is reduced by a factor of 2 due to poor insulation resistance in parallel with the input terminals, this error is doubled. When compared to a typical high-level system accuracy of 0.05% F.S., the additional error may be appreciable.

A closely related source of error is signal or common mode crosstalk. Crosstalk is the condition when the signal applied to one terminal pair affects the signal value of another input. For example, if the leakage resistance between an input terminal for a 5-V signal and the input terminal for a 10-mV signal is 100 MΩ, a current of 50 pA flows between the terminals. Furthermore, if the 10-mV signal source has a source impedance of 1000 Ω, the current produces an error of 50 μV or 0.5% of the 10-mV signal. If the interfering signal is a common mode voltage of 50 or 100 V, the error is correspondingly greater.

The insulation resistance between the terminals and between the terminals and ground is primarily a function of the bulk resistance of the insulating material, the length of the leakage path, and the condition of the insulator surface. Most commercially available terminal strips have a sufficiently high bulk resistance to prevent any appreciable performance degradation under normal conditions. Some materials, however, are hygroscopic and the bulk resistance is appreciably lowered by long exposure to high humidity.

The length of the leakage path between insulated conductors or terminals is largely a matter of mechanical design. The path length can be increased by forming ridges or protrusions between the conductors or terminals. This is exemplified by the design of the ordinary barrier terminal strip.

The surface condition of the insulator is often the determining factor in the amount of leakage. Surface contamination such as perspiration, conducting oil films, or water vapor and dust can drastically reduce the leakage resistance between conductors. The insulation material should be thoroughly cleaned after assembly of the termination facility to remove any surface contamination, and care should be taken during installation to prevent excessive contact with the insulating surfaces. If access to conducting surfaces is not required after assembly, these surfaces and the adjacent insulation can be coated with a nonhygroscopic lacquer or varnish. A dust or vapor-tight enclosure is also helpful in maintaining clean surfaces. Periodic cleaning of the insulation with a nonconducting solvent that does not leave a residue may be required in stubborn cases. Under high humidity conditions, it may be necessary to provide an enclosure in which a desiccant can be effectively utilized.

It is difficult to separate the effects of bulk and surface resistivity in specifying a material to be used for insulation in a termination facility. For this reason, an all inclusive specification is normally used. A specification of 10,000 MΩ under normal humidity conditions between all conducting surfaces of the terminal strip is usually adequate. After 150 hr of soaking in a 95% relative humidity environment, a specification of 1000 MΩ can be achieved. Mica or glass-filled phenolic and nonporous ceramic materials have been successfully used in these applications.

3.11.4 Common Mode Errors

Assuming a terminal similar to those of Fig. 3-17 for illustrative purposes, resistive leakage paths exist through the insulation separating the stud from the mounting panel, through the wire insulation for both the input signal wires and the cables to subsequent equipment, and along any surface paths between conductors. According to the common mode analysis of Sec. 3.9, these leakages must be minimized if good common mode rejection is to be achieved. The insulation of the signal wiring is beyond the control of the designer and may be beyond the control of the system vendor if he does not have responsibility for the process wiring. However, the user should ensure that process wiring is nonhygroscopic and that it is of high quality. Both process and system wiring should be checked periodically for deterioration due to continued exposure to heat, chemical contaminants, and sunlight.

The insulation separating the terminal stud from the mounting panel

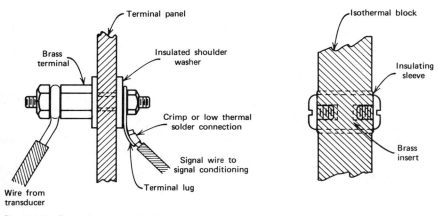

Fig. 3-17 Example termination designs.

must have characteristics similar to those of the wiring insulation. In addition, thermal insulation between the mounting panel and the terminal stud normally is not desired, particularly in the case of thermocouple junction compensation which is discussed in Sec. 3.11.8. A thin mica washer or an epoxy sleeve can provide a combination of suitable electrical and thermal properties. Mica has the disadvantage of poor mechanical characteristics, but this can be overcome with proper design.

Surface leakage is dependent on the characteristics of the surface, its cleanliness, and the environmental conditions in the enclosure. Assuming that initial cleanliness has been assured through careful cleaning, it is important to minimize contaminants that might enter the enclosure from the process environment. Gasketed dust covers may be sufficient for this purpose when coupled with periodic cleaning procedures in critical cases. Humidity in the enclosure can be controlled in severe cases by use of a desiccant. The desiccant requires renewal at periodic intervals as part of the system maintenance procedures. These measures are ineffective when forced ambient air cooling is used, and the termination enclosure should be designed to function properly without this type of cooling.

In Fig. 3-17, the brass stud and the panel form a parallel plate capacitor with the insulator acting as the dielectric between the plates. If good thermal coupling of the stud to the panel is desired, a thin insulator is required, and this tends to increase the capacitance with a corresponding decrease in the AC CMRR of the subsystem. A small stud reduces the capacitance but results in poor thermal transfer between the front and the back of the panel with an exposure of creating thermal potential errors. Thus the electrical and mechanical designers are faced with a compromise among several conflicting considerations. A single answer cannot be provided because the compromise also includes such factors as the desired performance, the overall system concept, and the cost of various approaches.

3.11.5 Grounding

If the terminal panel is electrically connected to the termination enclosure, normally it is grounded and the leakage capacitances and resistances are terminated at ground potential. As indicated by the common mode models, this affects the common mode performance of the system. On the other hand, isolation of the panel from ground means that the leakage resistance and capacitance coupling occurs between input terminals. This can result in crosstalk between both normal and common mode signals.

An additional factor is that noise picked up on the terminal panel is distributed to all input signals if the panel is ungrounded. Therefore, the decision relative to grounding the terminal panel depends on which of these phenomena dominate the performance of the system.

A theoretical answer to the grounding question is extremely difficult to obtain and the decision is generally made on the basis of empirical system test data. Additional information on system grounding practices is found in Chap. 4.

3.11.6 Noise

In low-level circuits, noise from other circuits or external sources can be a source of appreciable error. Noise control techniques in the termination facility are essentially the same as those that are required in other parts of the system, with the possible exception that the number of input signals concentrated in the termination facility may accentuate the problem. Noise control in the process wiring minimizes the amount of noise present in the terminal unit, but some noise is inevitable in any system. The enclosure for the terminals provides some shielding if it is metallic and properly grounded. In addition, twisted pair wiring or coaxial cables should be used for both signal wiring and for the wiring from the termination facility to subsequent circuits. Special shielding techniques, discussed in Chap. 4, are applicable to the terminal unit and may be used in critical applications.

3.11.7 Thermal Potential Errors

As discussed in Sec. 3-10, thermal potentials are generated at a junction of dissimilar metals when a temperature differential exists. The solution to the thermal potential problem is based on the use of metals that have similar thermoelectric powers and on the elimination of temperature differentials. The vendor usually must make a compromise when eliminating dissimilar metals in the termination facility. Thermocouples are the most common source of low-level signals and there are at least ten common metals normally used in thermocouples. The vendor cannot predict the mix of conductors that will exist in any given system, nor can he economically provide custom terminals that guarantee connections between similar metals. Considerations such as strength, ease of machining, conductivity, and corrosion resistance also restrict the choice of materials. Brass represents a good compromise between these factors and thus is often chosen. Since it is not the optimum metal, however, design emphasis must also be directed toward elimination of thermal differentials.

Radiation and air convection represent potential sources of temperature differentials in the termination facility. If the terminals are housed in a separate enclosure and there is a minimum of electronic equipment included in the enclosure, forced air ventilation may not be necessary and the problem of air convection is reduced considerably. In any case, however, the terminals should be protected from moving air and radiation sources. This can be done by placing them in a separate enclosure in the termination facility or by providing air baffles or shields. The enclosure or baffles should not place the front and the back of the terminal in separate closed volumes since this could result in a temperature difference between the front and the back of the panel on which the terminals are mounted. As long as there is no temperature difference, the presence of a dissimilar metal in the circuit, such as the brass terminal in Fig. 3-17, does not cause a thermal error.

Figure 3-18 illustrates a termination facility design which allows free air movement in front of and behind the terminal panel. The separator prevents air currents from other parts of the enclosure from causing local thermal unbalances due to air movement. Supporting the terminal panel by brackets also minimizes heat flow into the panel due to conduction from the enclosure itself. The brackets should be as small as possible to minimize heat conduction but must, of course, be capable of supporting the panel and withstanding any pressure applied during installation of signal wires. The brackets may provide an electrical connection between the terminal panel and enclosure, as discussed in Sec. 3.11.5.

Another source of conducted heat is through the incoming signal wires. Two procedures that minimize this potential source of thermal differentials are proper wire routing and stripping of insulation. Instead of wires terminating very close to the point where they enter the termina-

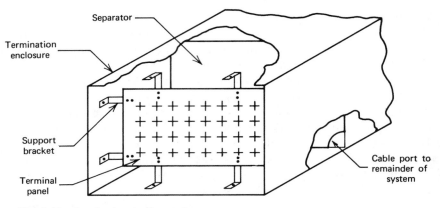

Fig. 3-18   Termination facility design.

tion facility, they are routed so that the end of the wire is at thermal equilibrium with the enclosure ambient before it reaches the terminal stud. A distance of 9–12 in. is normally satisfactory, except for heavy wires whose temperatures differ appreciably from the ambient temperature inside the enclosure. Allowance must be made in the mechanical design of the termination facility for this nonoptimum routing.

Materials that are good electrical insulators are generally also good thermal insulators. It is preferable, therefore, to remove all excess insulation so that heat conducted through the signal wires can be dissipated before it reaches the terminal stud. Care must be taken to retain sufficient insulation to withstand any normal or common mode voltages that might be present on the signal wires.

3.11.8   Thermocouple Compensation

Measurement of temperature with thermocouples requires that the temperature of the reference junction be known. In precision laboratory measurements, the reference junction is maintained at an ice point of 32°F and is composed of the same materials as the measurement junction. The use of a true ice point reference is difficult in a process even when provided by artificial cooling means. In addition, multiple reference junctions are required since different types of thermocouples are in use simultaneously. The basic thermoelectric phenomenon, however, does not require that one junction be held at the ice point; it only requires that the temperature of the reference junction be known. With a digital computer in the system, it is a simple matter to compensate numerically for a reference that is not at the ice point in addition to compensating for the usual thermocouple nonlinearities.

A common procedure is to connect all thermocouple input signals to terminals that are electrically insulated from, but thermally coupled to, a constant temperature terminal panel or *isothermal block*. The temperature of the isothermal block is measured with a resistance temperature detector (RTD). The RTD is used as one resistor in a bridge circuit, and the output of the bridge is proportional to the temperature. Since the bridge output is a function of the excitation voltage as well as the resistance of the RTD, the value of the excitation voltage must be known. Both the bridge output and the excitation voltage are measured as analog inputs to the subsystem, and the digitized values are used by the computer for numerically compensating the digitized thermocouple readings.

The RTD measures the temperature in a localized area, and since the isothermal block may be several square feet in area, it is necessary to ensure that all terminals are at the same temperature. A temperature variation of much less than a degree can be obtained with proper design of the isothermal block. All the considerations given above for avoiding localized temperature differentials in the termination facility apply to the design of the block. In addition, close thermal coupling of the terminal stud to the isothermal block is required.

The isothermal block should have good thermal conductivity and large thermal mass to ensure that localized temperature differences are minimized. For an isothermal block of several square feet installed in an enclosure, an aluminum block with a thickness of $\frac{1}{2}$–1 in. has been found to be satisfactory for all but the most critical applications.

In less critical applications, ordinary barrier terminal strips mounted in a closed enclosure may be satisfactory. A temperature differential of 1°F or less can be maintained if the practices of thermal isolation discussed in Sec. 3.11.7 are observed.

3.11.9 Safety

Although safety is an overall system consideration discussed in Chap. 8, its particular importance in the design of the termination facility must be emphasized. The termination facility is the interface between the process and the computer system. The service personnel responsible for the computer may not be intimately familiar with the process; similarly, process personnel may have limited familiarity with the computer system. Since both groups of personnel must work in the termination area, both the vendor and the user must ensure that adequate safeguards against hazardous conditions, particularly the presence of high common or normal mode voltages, are provided. Warning labels or tags should be used to alert personnel to potential danger. The design of the termination facility should minimize the chance of accidental contact with live circuits. Personnel should be thoroughly trained in both the system and in first aid and emergency procedures.

3.12 SIGNAL CONDITIONING

For our purposes, signal conditioning can be defined as any modification of the analog-input signal except amplification. This can encompass many different forms of modifications, some of which are unique to a given transducer or system. Common signal conditioning circuits, and the ones to which the present discussion is limited, include signal attenuation, current-to-voltage conversion, and normal mode filtering. One or more of these circuits are generally provided for each analog input.

3.12.1 General Considerations

Signal conditioning circuits generally are passive circuits composed of less than a half-dozen electrical components. They are physically small and, except in the case of a bridge or special purpose circuit, do not require electrical power. Thus the designer has some freedom in choosing where to mount the circuit. It may be located near the terminals in the termination facility if it is a separate enclosure, or it may be mounted in the enclosure that houses the multiplexer and other analog equipment.

Access to the signal conditioning circuits is required even after the system is installed since process wiring is always subject to change. If the user has special circuits or the responsibility for servicing the signal conditioning portion of the system, he requires access to the enclosure in which the circuits are mounted or the vendor must provide assistance whenever changes occur. Thus if a separate termination facility is available, this may be the logical location for the signal conditioning circuits. One disadvantage is that, if the termination facility and the multiplexer enclosure are separated, filter circuits provided as part of the signal conditioning are not effective against noise pickup between the termination facility and the multiplexer.

The signal conditioning required in a given process varies from input to input depending on the nature and source of the signal. Since the mix of required circuits cannot be predicted *a priori*, the vendor can either provide several standard circuits suitable for most signals, restrict the mix of signals on a given set of inputs, or provide a design that allows customization of each input. The first alternative is the one most often used and does not impose undue restrictions on the user if circuits are available in reasonably small groups. Probably 90% of the signal conditioning requirements of process instrumentation can be provided by five standard circuits.

The ability to customize requires a mechanical design that allows easy inclusion of several types of conditioning circuits on any input. One solution is to provide the circuit on a separate card that can be inserted in the analog-input circuit. This has the advantage that cards with an etched circuit but no components can be provided to the user for building special circuits.

The general principles and precautions concerning contact resistance, thermal potential sources, and leakage impedances discussed in previous sections apply equally well to the design of signal conditioning circuits. Providing a blank card to the user for the construction of custom signal conditioning circuits may provide additional problems not found in other parts of the system. The user should be instructed in the use of low thermal solder in low-level circuits to minimize the possibility of thermal errors in critical situations. The vendor should also instruct the user in the proper cleaning and coating of the cards to minimize leakage paths.

3.12.2   Current-to-Voltage Conversion

Many process signals are in the form of a DC current. Common current ranges for commercial process instrumentation are 4–20, 1–5, and 10–50 mA. Conversion of these signals to a voltage is accomplished by providing a resistor in parallel with the input circuit. The value of the resistor is determined by the maximum current value and the full-scale value of the range used in reading the resulting voltage. Generally, the highest available input voltage range should be used; this minimizes the effect of noise, thermal, and common mode errors since the signal is greater. The current-to-voltage conversion resistor may be mounted on the signal conditioning card if it is provided. One difficulty is that removal of the card for servicing or inadvertent removal by servicing personnel opens the current loop. If the current signal is also being used for control of process equipment, control is lost. Therefore, it is common to mount the resistor on the terminals in the termination facility. In this way, only physical removal of the resistor or process wiring can break the circuit. As discussed in Sec. 3.11.2, care must be taken in this case to avoid errors resulting from contact resistance.

The tolerance of the current-to-voltage resistor results in an error that adds to the total system error. The error cannot be calibrated out unless calibration ability is provided for each input, since the error varies from input to input. If the inputs are measured periodically, the computer can provide digital error compensation, but this requires storing tolerance values for all inputs and, in addition, the error varies with time and temperature.

Assuming an input range of 500 mV or more, typical system accuracy is within about 0.05%F.S.; the tolerance of the resistor should be at least this good when accuracy is of importance. Normally, however, the resistors used for this application need not have this good a tolerance since the current transducers from which the current is derived generally have an error considerably greater than the system error. Precision resistors are expensive, and the user should evaluate his requirements carefully before asking the vendor to provide special high-precision resistors.

3.12.3   Signal Attenuation

Some transducers found in control applications have outputs in excess of the maximum range available on most systems. For example, some instrumentation amplifiers have a 10-V full-scale output as compared with the common 5-V input range of many systems. Another application in which attenuation might be used is when the normal output of a transducer just exceeds a given input range of the system. Rather than utilize the next higher range and lose some of the available system resolution, the user might decide to attenuate the signal slightly and use the next lower range to obtain full resolution.

An attenuation circuit for a single-ended system is shown in Fig. 3-19. The attenuation ratio is

$$\frac{V_o}{V_{in}} = \frac{R_2}{R_1 + R_2} \equiv \alpha \qquad (3\text{-}16)$$

where V_o is the output voltage to the multiplexer and V_{in} is the input voltage applied to the input terminals. This expression is strictly valid only if the transducer has zero source impedance and the input impedance of the system is infinite. A more realistic situation is shown in Fig. 3-20 where finite source and input impedances are included. The effect of R_{in}

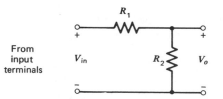

Fig. 3-19   Signal attenuation circuit.

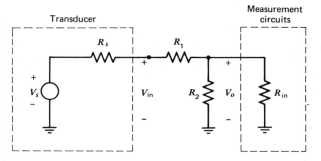

Fig. 3-20   Signal attenuation equivalent circuit.

is to increase the attenuation of the circuit since the effective value of R_2 is reduced. The series source resistance R_s tends to further increase the attenuation. The expression for the attenuation ratio of the circuit shown in Fig. 3-20 is

$$\frac{V_o}{V_{in}} = \frac{R_2}{R_1 + R_2 + R_s + (R_s R_2 + R_1 R_2)/R_{in}} \qquad (3\text{-}17)$$

The error due to R_{in} is predictable and can be compensated in the design of the attenuation network. Either R_2 can be made small enough that the parallel effect of R_{in} is negligible, or R_2 can be increased sufficiently to compensate for the shunting action of R_{in}. Assuming the second course of action, if R_2' is the value of the modified R_2, its value is given by

$$R_2' = \frac{R_2 R_{in}}{R_{in} - R_2} \qquad (3\text{-}18)$$

For $R_2 < 0.01\ R_{in}$, as is often the case, Eq. (3-18) is approximated by

$$R_2' \cong R_2 \left(1 + \frac{R_2}{R_{in}}\right) \qquad (3\text{-}19)$$

with an error of less than 0.01%.

Using the other approach of selecting R_2 small enough that the shunting action of R_{in} results in less than $\delta\%$ error, R_1 is first selected to satisfy

$$R_1 < \frac{R_{in}\delta}{100\alpha} \qquad (3\text{-}20)$$

where α is the desired attenuation ratio defined by Eq. (3-16). With this relationship satisfied, the appropriate value of R_2 is found from the definition of α.

Assuming that the error due to R_{in} is either compensated for or made negligible by the above methods, the error due to the nonzero source resistance R_s can be considered separately. The expression for a finite R_s, assuming no error due to R_{in}, is obtained from Eq. (3-17) by allowing R_{in} to approach infinity

$$\frac{V_o}{V_s} = \frac{R_2}{R_1 + R_2 + R_s} \tag{3-21}$$

The percentage error in this expression with respect to the desired ratio α is

$$\frac{100R_s}{R_1 + R_2} \% \tag{3-22}$$

This error can be compensated by reducing the calculated value of R_1 by the amount of the source resistance R_s. However, this cannot be done by the vendor except on special order since he has no way of predicting the particular transducers being used. If the transducer output is buffered by an internal amplifier, little error results even without compensation because the amplifier output impedance is normally small with respect to the value of R_1 and R_2.

The tolerances of the resistors R_1 and R_2 are an additional source of error. If the percentage errors in the resistors are $(100\ e_1)\%$ and $(100\ e_2)\%$, respectively, the percentage error in the attenuation α is

$$\frac{100(e_2 - e_1)}{1 + e_1 + (1 + e_2)[\alpha/(1 - \alpha)]} \% \tag{3-23}$$

In most situations, the exact value of the resistances is not as important as the attenuation α. Equation (3-23) shows that the error is more dependent on the match between the resistors than on their absolute value. Thus the designer can either specify high-precision resistors whose basic tolerances guarantee a low attenuation error or he can specify the resistors as a matched pair with nominal resistance tolerances and a tight ratio tolerance. This latter course of action is usually less expensive because it gives the manufacturer more freedom in building the resistors.

The resistances R_{in} and R_s, as well as R_1 and R_2, have a finite temperature coefficient. The temperature coefficient of R_1 and R_2 is easily controlled since vendors can supply matched resistors having a ratio temperature coefficient as low as several parts per million (ppm) per degree. The temperature coefficient of R_s and R_{in} is more difficult to control and it may be more of a factor than the ratio temperature coefficient of the attenuator network itself. The error analysis for any particular case uses the formulas derived previously.

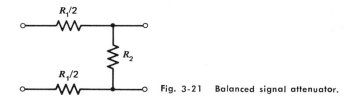

Fig. 3-21   Balanced signal attenuator.

The attenuator configuration of Fig. 3-20 is not suitable for use in differential systems when a high common mode rejection ratio is desired. The resistance R_1 constitutes an unbalance in the input circuit and results in degraded common mode performance. A differential attenuator is shown in Fig. 3-21; the resistance R_1 merely has been divided between the two input lines to form a balanced circuit. The only contribution to circuit unbalance is the tolerance of the two $R_1/2$ resistors. The error analysis presented in the single-ended case is valid and can be used for design.

3.12.4   Signal Filtering

The time constant of most process variations is quite long, ranging from seconds to hours. Therefore, frequency components of 1 Hz or less represent the information of interest in the analog-input signal. The typical transducer signal, however, has frequency components far in excess of the basic information frequencies. Typically, these consist of 60-Hz voltages induced in the instrumentation wiring or resulting from the conversion of common mode voltages to a normal mode signal, and high-frequency bursts of noise due to transients produced by equipment such as circuit breakers, relays, or motor controls.

The noise on the analog-input signal results in a lack of repeatability when an input is read several times. The noise is of more concern in a control computer system than in systems that use conventional analog controllers because of the wide effective bandwidth of the computer analog system. If the ADC has a conversion time of 50 μsec, for example, it easily converts the peak value of a 60-Hz noise signal. Because of the low bandwidth of most analog controllers, however, a 60-Hz signal does not affect the operation of an analog controller. By providing a filter for each input signal, the system bandwidth is reduced and the effect of noise is minimized.

In designing a suitable filter, the designer seemingly has many choices: passive or active; resistor-capacitor (RC), inductor-capacitor (LC), or resistor-inductor-capacitor (RLC). In a control computer application, however, the number of possibilities is limited by practical considerations.

The choice between active and passive filters is largely one of economics. An active filter contains an amplifier and, since filters are required on each channel if the system speed is not to be reduced drastically, the increase in system cost is rarely tolerable. Active filters may be used on a few inputs when the noise is unusually severe or when a special transfer function is required, but this is not the usual case.

Among passive filters, the LC and RLC designs have the highest performance in terms of obtainable attenuation and bandwidth characteristics. Theoretically, for example, LC filters can provide 80-dB attenuation at 60 Hz while maintaining a bandwidth of several Hertz. However, the required inductors are large, heavy, and relatively expensive. In addition, it is difficult to obtain the theoretical response because the many turns required to obtain sufficient inductance result in a high winding resistance if the physical size is to be reasonable. Thus the inductor has low Q and its use results in a filter that performs more like an RC filter than an LC filter. For these reasons, LC filters usually are not provided as standard signal conditioning circuits in analog subsystems designed for process control use.

The problems with the LC filter are the practical consequences of the low bandwidth requirement. There are a few notable exceptions to the low bandwidth requirements of process control signals such as signals from chromatographs or mass spectrometers. In these cases, LC or RLC filters may be used effectively. Although important, these applications represent the exception rather than the rule. A manufacturer usually can provide filters for these cases on a special order basis, or the user can design and build them. The design of LC filters is not discussed here, but convenient design techniques and tables of values are available in the references (Weinberg, 1962).

3.12.4.1 Single-Section RC Low-Pass Filters. The single-section RC low-pass filter is a simple design that is used in most analog subsystems. It is inexpensive, physically small, and has adequate performance for the large majority of process control applications.

The schematic diagram of the filter is shown in Fig. 3-22. The filter attenuation A, expressed in decibels (dB), is

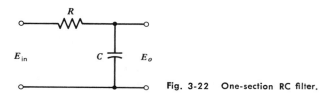

Fig. 3-22   One-section RC filter.

$$A = 20 \log \left| \frac{E_o}{E_{in}} \right| = -10 \log (1 + \omega RC)^2 \qquad (3\text{-}24)$$

where log is the base 10 logarithm. The relationship between the radial frequency ω in radians and the cyclical frequency f in Hertz is $\omega = 2\pi f$. A plot of the attenuation characteristic is shown in Fig. 3-23. From Eq. (3-24) and the figure, the attenuation is -3 dB when $\omega = 1/RC$. This is defined as the bandwidth of the filter and is denoted by ω_o.

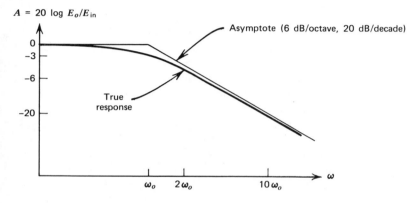

Fig. 3-23   Single-section RC filter response.

At frequencies much greater than ω_o, the unity term in Eq. (3-24) is negligible and the expression is approximated by

$$A \cong 20 \log \frac{\omega}{\omega_o} \qquad (3\text{-}25)$$

Comparing Eq. (3-24) and Eq. (3-25), the error in A due to the approximation is about 0.2% at $A = -20$ dB.

When plotted on semilogarithmic paper, Eq. (3-25) is a straight line originating at ω_o and having a slope of -6 dB/octave ($= -20$ dB/decade). As shown in Fig. 3-23, this line is asymptotic to the attenuation characteristic of Eq. (3-24).

The design of the filter can be based on the graphical approximation. For example, if 40-dB attenuation is required at 60 Hz, the -20 dB/decade asymptote intersects the 0-dB ordinate at 0.6 Hz or 1.2π radians. Therefore, $\omega_o = 1.2\pi$ and $RC = 1/1.2\pi = 0.265$. Suitable values for R and C are $R = 820 \, \Omega$ and $C = 330 \, \mu$F. These values are not exact but, considering the tolerance of available components, they are sufficiently close.

A nomogram for an exact design of the filter is given in Fig. 3-24. The use of the nomogram is illustrated in the figure using the above example. A line is drawn between the desired attenuation A and the frequency f at which A is desired. A second line joining any two points on the R and C scales is then passed through the intersection on the turn scale. Note that the R, C, and f scales may be multiplied by factors of 10 as long as the exponents sum to zero.

As the example demonstrates, the value of the capacitance for a low bandwidth can be quite large, and an electrolytic capacitor usually is required in a practical circuit. Since electrolytic capacitors are polarity sensitive, they theoretically limit the use of the filter to unipolar signals. In actual practice, however, most electrolytic capacitors withstand some reverse polarity voltage without an appreciable change in capacitance or increase in leakage current. The value of the reverse voltage that can be applied safely depends on the construction of the capacitor and its forward voltage rating. For most electrolytic capacitors a reverse potential of 0.5–1.0 V is a reasonable figure. Thus the use of electrolytic polarized capacitors is satisfactory in filters for low-level bipolar signals.

For high-level signals, the polarity of the input signal must be controlled or a nonpolarized electrolytic capacitor must be used. A nonpolarized electrolytic capacitor is generally bulkier and more expensive than the equivalent polarized unit. "Back-to-back" polarized capacitors can be used at the expense of using more capacitors in the filter.

Most electrolytics have an appreciable self inductance that results in a loss of attenuation at high frequencies. This loss of attenuation can be compensated by paralleling the electrolytic with a small mica or ceramic capacitor.

Another characteristic of electrolytic capacitors that alters the filter response is dielectric absorption. This phenomenon, present in all capacitors but particularly pronounced in electrolytic capacitors, is the result of bound charge in the dielectric material. Its effect is an increase in the effective capacitance of the component. This results in a filter time constant in excess of that predicted by theory. The effect is dependent on the magnitude and duration of the signal voltage.

The tolerances of the resistor and capacitor used in the filter circuit affect the bandwidth, as does a nonzero source impedance. Resistors with 5 or 1% tolerances are quite economical, but precision capacitors are relatively expensive. The standard tolerance on an electrolytic capacitor is 10 or 20%. However, since the filtering requirements are usually noncritical in most process control applications, the effects of this tolerance are rarely a problem.

The design of the RC filter is based on the use of a perfect capacitor,

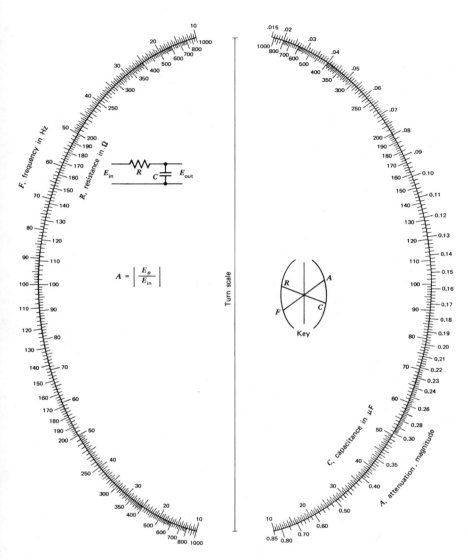

Fig. 3-24   Single-section RC filter nomogram. Prepared by M. Arthur, IBM Corporation, San Jose, Calif.

187

whereas a real capacitor has a finite DC leakage. The magnitude of the leakage is a nonlinear function that depends on the voltage, temperature, value, and type of capacitor. At a given operating point, a practical capacitor can be represented as a perfect capacitor in parallel with an effective leakage resistance.

At DC the reactance of the perfect capacitor is infinite, so the equivalent circuit of the RC low-pass filter reduces to that of the voltage attenuator of Fig. 3-19 with R_2 representing the equivalent leakage resistance. The error analysis developed for the voltage attenuator is valid and can be used in the design of the filter. The most important practical consequence of the equivalent circuit is that, given the value of R_2 and the maximum allowable DC attenuation error, the value of R in Fig. 3-22 is constrained. For a worst-case design, the effective leakage resistance at the maximum anticipated temperature and worst-case applied voltage is used in the error analysis.

Typical values of effective leakage resistance for a solid tantalum 50-μF capacitor with a working rating of 6 V when measured at a voltage of 5 V and a temperature of 60°C are less than 5×10^6 Ω. Special units have an effective leakage resistance of greater than 25×10^6 Ω under the same conditions. At 300 μF under these conditions, typical values are less than 2×10^6 Ω for a standard unit and greater than 5×10^6 Ω for a special capacitor.

The filter configuration shown in Fig. 3-22 is not suitable for use in differential systems because its unbalanced form causes a degradation of common mode performance. As in the case of the voltage attenuator, a balanced configuration is obtained by splitting the resistance value equally between the two input lines. Since the difference between the two $R/2$ resistors constitutes a common mode unbalance, 1% resistors may be desirable.

3.12.4.2 Two-Section RC Low-Pass Filter. If the attenuation and bandwidth requirements cannot be obtained with a single-section RC filter, the addition of a second section may provide the answer. The single-ended and balanced configurations of the two-section RC filter are shown in Fig. 3-25.

The additional section complicates the design because there is an interaction between the sections due to loading. In general, the single-section design procedure cannot be used. However, if R_1C_2 is much less than both R_1C_1 and R_2C_2, the two sections can be designed independently using the procedure discussed above, and the filter characteristic approximates that of two cascaded independent sections.

The approximate design of the filter is based on the graphical construc-

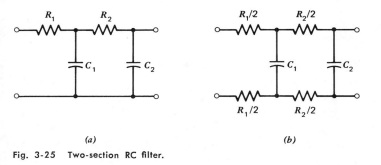

(a)   (b)

Fig. 3-25   Two-section RC filter.

tion of Fig. 3-26. The approximate response of each section is drawn using the asymptotic approximation introduced above. The breakpoints are given by $\omega_1 = 1/R_1C_1$ and $\omega_2 = 1/R_2C_2$ and the individual asymptotes have a slope of -6 dB/octave or -20 dB/decade. The response of the cascaded sections is the sum of the responses of the individual sections when measured in decibels. As the breakpoints approach each other in value, the attenuation at $\omega_1 \cong \omega_2$ approaches -6 dB and the 3-dB bandwidth approaches 0.64 ω_1. At frequencies much greater than ω_1 and ω_2, the slope of the attenuation characteristic approaches a -12 dB/octave or -40 dB/decade asymptote.

If the conditions for the approximate design cannot be satisfied, the interaction between the sections must be taken into account and the design becomes much more difficult. The attenuation function of the two-section filter is

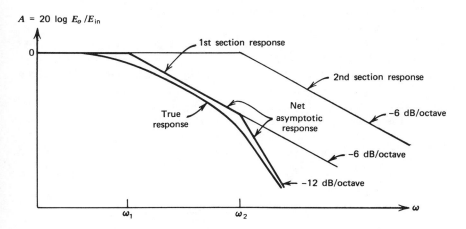

Fig. 3-26   Two-section RC filter response.

$$A = -10 \log \left\{ \left(\frac{1}{\omega_1 \omega_2 \omega_3} \right)^2 [\omega_3{}^2(\omega_1 \omega_2 - \omega^2)^2 + \omega^2(\omega_2 \omega_3 + \omega_1 \omega_3 + \omega_1 \omega_2)^2] \right\}$$

(3-26)

where $\omega_1 = 1/R_1 C_1$, $\omega_2 = 1/R_2 C_2$, and $\omega_3 = 1/R_1 C_2$. Given a particular set of values for the filter components, this expression is used to find the theoretical attenuation at any radial frequency ω.

In designing a filter, substitution of the desired attenuation A at the frequency ω into Eq. (3-26) gives the relationship between ω_1, ω_2, and ω_3. The component values are then determined from the definitions of these frequency parameters. This design calculation is tedious but has been reduced to a graphical construction using the graduated scales of Fig. 3-27 and the nomogram of Fig. 3-28 (Harrison and Bernacchi, 1967).

The graduated scales represent the relationship between the desired attenuation A, the position of the filter pole s_1 in the s-plane, and a normalized parameter Ω_o. The parameter Ω_o is defined as ω/ω_o where ω is the frequency at attenuation A and ω_o is the —3-db bandwidth. In the design procedure, Ω_o is calculated from the filter requirements and then s_1 is determined from Fig. 3-27. The pole position s_1 is used with the nomogram of Fig. 3-28 to determine the normalized circuit components.

The nomogram can be used to design a filter under any of four constraints: (1) $R_1 + R_2 = $ constant, (2) $R_2/R_1 = $ constant, (3) $C_1 + C_2 = $ constant, or (4) $C_1/C_2 = $ constant. The design procedure for each of the four cases is as follows:

$R_1 + R_2 = constant$
1. Define $R_o = R_1 + R_2$
2. Select a value for C_2 and calculate its normalized value C_2' according to $C_2' = C_2 R_o \omega_o$.
3. Draw a straight line through the values of C_2' and s_1 on the right-hand scales of Fig. 3-28 such that the line intersects the turn scale.
4. Draw another straight line from the intersection on the turn scale through the same value of C_2' on the left-hand C_2' scale.
5. Read the value of R_2/R_1 from the resistance scale.
6. If the ratio R_2/R_1 is acceptable, continue with Step 7. If not, return to Step 2 and select another value for C_2.
7. Calculate the component value from:

$$R_2 = R_2' R_o$$
$$R_1' = 1 - R_2'$$
$$R_1 = R_o - R_2$$
$$C_1 = 1/(s_1 s_2 C_2' R_1' R_2 \omega_o)$$

(3-27)

$R_2/R_1 = constant$

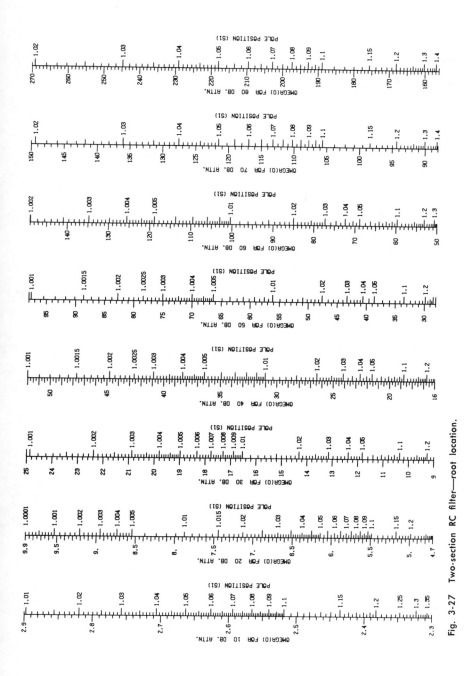

Fig. 3-27 Two-section RC filter—root location.

191

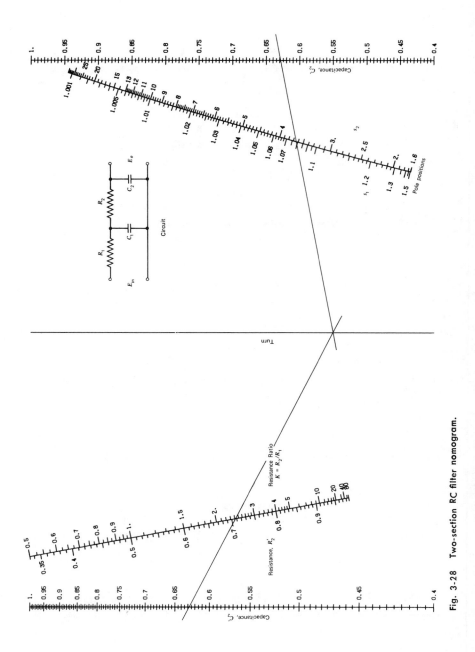

Fig. 3-28 Two-section RC filter nomogram.

In this case, a "trial and error" procedure is used. A straight line is passed through the pole position s_1 and some value of C_2'. A second straight line is passed through the desired ratio R_2/R_1, the intersection of the first line with the turn scale, and the same value of C_2' on the *left-hand* scale. Since C_2 cannot be picked arbitrarily, a solution may not be possible for some values of R_2/R_1. The sum $R_o = R_1 + R_2$ is picked arbitrarily and is used to determine component values using Eq. (3-27).

$C_1 + C_2 = constant$

Steps 1–6 of this procedure are identical to the case $R_1 + R_2 = $ constant, except that the following substitutions must be made in both the discussion and in Fig. 3-28:

$$\text{Replace } C_2' \text{ (left and right)} \quad \text{with} \quad R_1$$
$$\text{Replace } R_2' \quad \text{with} \quad C_1'$$
$$\text{Replace } R_2/R_1 \quad \text{with} \quad C_1/C_2$$
$$\text{Replace } R_o \quad \text{with} \quad C_o$$

The component values are calculated from:

$$\begin{aligned}
C_1 &= C_1'C_o \\
C_2' &= 1 - C_1' \\
C_2 &= C_o - C_1 \\
R_1 &= R_1'/(C_o\omega_o) \\
R_2 &= 1/(s_1s_2R_1'C_1C_2'\omega_o)
\end{aligned} \qquad (3\text{-}28)$$

$C_1/C_2 = constant$

This method is identical to the case of $R_2/R_1 = $ constant, except the above substitutions and the component calculations of Eq. (3-28) must be used.

As an example, suppose that a filter is desired with a bandwidth of 1.0 Hz, 50-dB attenuation at 35 Hz, and a total resistance of 1000 Ω. The bandwidth in radians is $\omega_o = 2\pi(1) = 6.28$ radians. The normalized frequency variable is $\Omega_o = 35(6.28)/6.28 = 35$. From Fig. 3-27 the pole position s_1 is 1.08 and from Fig. 3-28, s_2 is 3.6. The value of C_2 is arbitrarily selected as 100 μF, so $C_2' = (100 \times 10^{-6})\,1000(6.28) = 0.628$. The construction lines are shown in Fig. 3-28 and yield $K = 2.42$ and $R_2' = 707$. The component values are

$$R_2 = 0.707(1000) = 707 \ \Omega$$
$$R_1' = 1 - 0.707 = 0.293$$
$$R_1 = 1000 - 707 = 293 \ \Omega$$
$$C_1 = \frac{1}{(1.08)(3.6)(0.628)(0.293)(707)(6.28)} = 315 \ \mu\text{F}$$

Accuracy and component considerations for the two-section filter are identical to those discussed for the single-section filter. The DC attenuation error is affected by R_1, R_2, and the leakages of both capacitors so the error analysis for the single-section attenuator is not directly applicable. If the effective leakage resistances of C_1 and C_2 are denoted by R_{L1} and R_{L2}, respectively, the attenuation error in percent is given by

$$\varepsilon = 100 \frac{(R_1 + R_{L1})(R_2 + R_{L2}) + R_1 R_{L1} - R_{L1} R_{L2}}{(R_1 + R_{L1})(R_2 + R_{L2}) + R_1 R_{L1}} \% \qquad (3\text{-}29)$$

If R_{L1} and R_{L2} are approximately equal and several orders of magnitude greater than R_1 and R_2, as is usually the case, Eq. (3-29) is approximated as

$$\varepsilon \cong 100 \frac{2R_1 + R_2}{R_{L1}} \% \qquad (3\text{-}30)$$

For R_L's and R's of the same magnitude as in a single-section design, this error is about three times as great; this is the result of two sections of attenuation. In a practical sense, this implies that the capacitors must be three times as good for the two-section filter if comparable error performance is to be obtained.

3.12.4.3 Twin-T RC Filter. The single-ended and differential configurations of a third RC filter that is sometimes used in process control applications is shown in Fig. 3-29. The Twin-T filter is useful in cases when a single-frequency component of a signal is to be attenuated with minimum effect on other frequency components. The frequency response is a symmetrical "notch" rejection characteristic, as shown in Fig. 3-30. Theoretically, the attenuation at the notch frequency $\omega_n = 1/RC$ is infinite, but this cannot be attained in practice because of component limitations.

The practical aspects of building the Twin-T filter are similar to those of the previous filters. The leakage resistances of the filter capacitors cause a DC attenuation error, as do nonzero source and finite load impedances. In addition, the tolerance on the various components affects the null frequency and the null attenuation; the resistor $R/2$ and the shunt capacitor $2C$ are about 1.5 times more critical than the other components in determining the null attenuation and care must be exercised in their selection. Because of its sensitivity to component variations and the sharp attenuation characteristic, even a small change in component values can radically affect the filter attenuation at a frequency near the null frequency. This limits the use of the Twin-T in many practical applications unless the filter is cascaded with a low-pass filter.

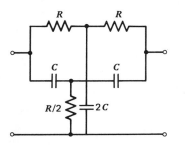

(a)

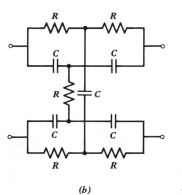

(b)

Fig. 3-29 Twin-T filter. (a) Single ended; (b) differential

3.13 MULTIPLEXING

Multiplexing is the process by which signal inputs are connected to time-shared amplification and/or analog-to-digital equipment in a sequential or random order. It may be performed at one or more of three points in an analog subsystem. Multiplexing after the ADC provides a system capable of converting each input signal at a rate equal to the conversion rate of the ADC. Multiplexing before the ADC allows it to be time shared between input channels at a sacrifice in system per-channel sampling rate. The use of a multiplexer before the amplifier allows both the amplifier and the ADC to be time shared.

In most process control applications, the primary multiplexing is performed before amplification even though this may result in a slower and less accurate system. The main reason for using this configuration is economic. Despite the increased cost of a multiplexer suitable for low-

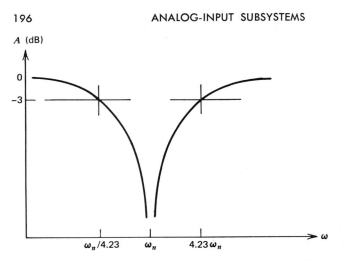

Fig. 3-30   Twin-T filter response.

level signals, the alternative of using per-channel amplifiers at a cost of $300 or more per channel results in a prohibitively expensive system. In addition, the accuracy and speed performance of the time-shared amplifier system is adequate for most control computer applications. In high-level systems, signal amplification is not required so high-level multiplexing is performed just prior to the ADC.

The electrical component of most concern in the design of a multiplexer is the actual switching device. Both electromechanical and solid-state switches are commonly used. In the following sections, the characteristics of the most popular switching devices are discussed.

3.13.1   Electromechanical Switches

The ultimate goal in the design of a multiplexer is to introduce no error into the measurement of an analog signal. This could be accomplished by using a perfect switch. In the OFF condition, the perfect switch has infinite impedance, and it has zero impedance in the ON condition. In general, electromechanical switches have electrical characteristics that are approximately ideal except for speed.

3.13.1.1 Reed Relays. The reed relay is an electromechanical, magnetically actuated switch. As shown in Fig. 3-31, the reed switch consists of overlapping cantilevered contacts hermetically sealed in a glass capsule containing an inert or reducing atmosphere under moderate pressure. Depending on the power level and the application for which the

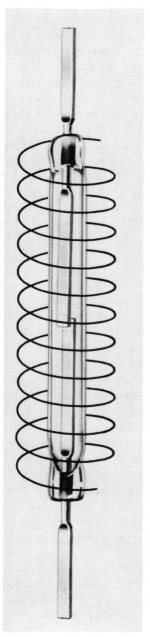

Fig. 3-31 Dry reed relay construction. From C. P. Clare & Co.

switch is designed, the capsule may be from 2.5 in. long to less than 1 in. with a diameter of 0.25–0.10 in. The contacts are a nickel-iron alloy with a reasonably high permeability and low retentivity. The contact portion of the reeds is usually plated with another metal to reduce contact resistance and provide a more reliable contact. In low-level multiplexing applications, this may be a gold alloy.

As indicated in Fig. 3-31, the capsule is enclosed in a coil. When current flows through the coil, the resultant magnetic field is predominantly concentrated in the reeds because of their high permeability. The flux in the gap between the reeds causes a force to be exerted on the reeds and they close when the force is sufficient to overcome the spring tension of the reeds. When the current is reduced below a certain level, the switch opens as a result of the restoring spring action of the reeds. In some switches the restoring action is produced by biasing the reed with a small permanent magnet mounted near the capsule. Magnetic biasing is also used to obtain a latching contact action.

Contact resistance of the closed relay is generally less than 200 mΩ and a specification of 50 mΩ is not uncommon. An OFF resistance of 10^{10} Ω in normal environments is quite common. The OFF resistance is determined mainly by the surface leakage of the capsule and the bulk resistivity of the encapsulation material. OFF resistances of 10^{13} Ω or more can be obtained by the use of special cleaning procedures and encapsulation material. The leakage resistance between the contacts and the actuating coil is determined by the physical configuration and the bulk and surface resistivity of the material between the contact leads and the coil conductors. A specification of 10^{9} Ω is obtainable in most commercially available units.

Stray capacitances are a function of the physical spacing between the various parts of the relay. A specification of 3 or 4 pF between contacts and coil and less than 1 pF between the open contacts is typical. The contact-to-coil capacitance can be reduced to a fraction of 1 pF by an electrostatic shield between the coil and the switch capsule at the expense of increased capacitance to the shield ground. As in the case of leakage resistances, these capacitances affect the common mode performance of the system.

The transition between ON and OFF requires a finite time; in addition, an electromechanical switch may bounce several times before the circuit is reliably closed. Pick times, including bounce, of less than 1 msec are typical for the reed relay. The transition from ON to OFF—the drop time—is less, perhaps only tens of microseconds. Both the pick and drop times are dependent on the characteristics and magnitude of the drive signal. The use of diode suppression on the drive coil results in longer drop time. Some form of suppression is essential, however, since over-shoots of up to 500 V are possible.

Despite the short pick and drop times, reed relays are rarely used for analog multiplexing above about 250 sps. The major limitation on system sampling rate is the settling time required for the relay noise to decrease to an acceptable level. The required settling time depends on the noise characteristics of the switch and on the accuracy, resolution, range, and bandwidth of the analog subsystem. As shown in Fig. 3-32, the noise immediately after closure may be as large as several millivolts peak and it may persist at a significant level for as long as 5 or 10 msec.

The noise is generated by electromagnetic induction and magneto-strictive effects as the reed vibrates in the magnetic field. Of the two effects, electromagnetic induction is essentially negligible. Since the compliance of the reed in a low ampere-turn relay is generally greater, a high ampere-turn relay is usually quieter. The noise is not appreciably affected by the impedance level of the external circuit. Neither is it affected by the applied ampere-turns if they are sufficient to saturate the reed material (Vitola and Breickner, 1964).

A thermal potential of about 36 μV/°C can be generated between the nickel-iron reed and the associated copper conductors. However, com-plementary junctions at each end of the relay reduce the net thermal potential. Thermal characteristics are shown in Fig. 3-33 for different ambient temperatures, rates of change of ambient, and coil duty cycle.

The operating life of a switching device is of prime concern since many applications demand high reliability. The life of any relay is dependent on the current being switched, the applied voltage, the reactive nature of the load, and the contact material. In most multiplexing applications, steady-state current is negligible and normal mode voltages are low. Thus, the typical 3–5 VA rating of a reed relay is not an obvious restriction. Manufacturers rate reed relay life at 3×10^6–20×10^6 operations at rated resistive load. Dry circuit (i.e., negligible current) life specifications range as high as several billion operations. To place this in perspective, a relay operating at 200 sps logs slightly more than 17×10^6 operations in 24 hr. However, since a particular relay in a multi-plexer is not operated at this rate for any appreciable period of time, 20×10^6 operations may represent many years.

One of the most important considerations in insuring long life is proper loading. In most multiplexing applications the assumption of a dry circuit is valid in the steady state. However, few actual loads are purely resistive and a particularly important case occurs in differential multiplexers when common mode voltages are applied. A typical amplifier input has shunt capacitance to ground due to the construction of the amplifier and multi-

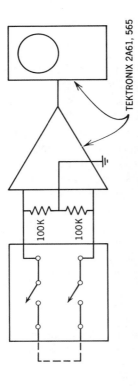

Test Circuit
Test Condition: Nominal coil voltage;
10 Hz; 50% duty cycle

TEKTRONIX 2A61, 565

100K

100K

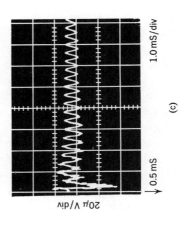

(c)

20μV/div

1.0 mS/div

0.5 mS

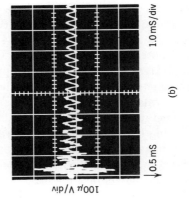

(b)

100μV/div

1.0 mS/div

0.5 mS

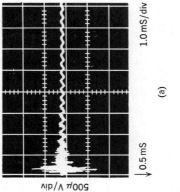

(a)

500μV/div

1.0 mS/div

0.5 mS

Fig. 3-32 Typical noise characteristics—dry reed relay. From C. P. Clare & Co.

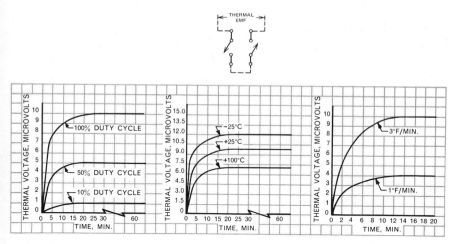

Fig. 3-33 Typical thermal characteristics—dry reed relay. From C. P. Clare & Co.

plexer. These shunt capacitances are charged by the common mode source through the multiplexer relay. Experience has shown that this capacitive charging current can reduce reed relay life drastically when high common mode voltages are present. A small resistor in series with the contacts limits the current and may extend the life of the relay.

The failure of a reed contact under the above conditions is similar to the erosion or pitting that occurs in other types of relays. One important difference is that the small particles broken off in the pitting operation may be nonconducting magnetic oxides. These are attracted into the gap between the contacts where they cause intermittent operation for short periods of time before they are thrown clear. Such imperfections in the surface of the contact can also result in high current densities and increased erosion and material transfer between the contacts. Eventually, the material build-up between the contacts results in complete welding.

Another phenomenon, known as cold welding, has been observed in some reeds that use softer contact materials such as gold. Continuous operation of the switch can cause the gold surface to cold work to an optically flat condition. The mechanism is not fully understood, but when this occurs a force of attraction may appear between the two closed contacts sufficient to overcome the restoring spring force. The problem may be elusive since the bond usually breaks if the relay is opened for inspection and the contact then appears to be in very good condition.

3.13.1.2 Mercury-Wetted Contact Relays. The mercury-wetted contact relay is similar to the reed relay in construction and operation.

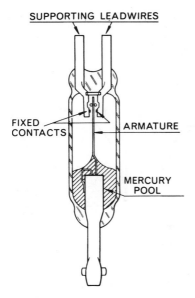

SUPPORTING LEADWIRES

FIXED CONTACTS

ARMATURE

MERCURY POOL

Fig. 3-34

It consists of nickel-iron poles and armature hermatically sealed in an inert or reducing gas that also contains mercury vapor. Figure 3-34 shows one form of the relay, although some capsules are used that are structurally identical to the dry reed capsule. The capsule is typically 1-2 in. long and about 0.25–0.5 in. in diameter. Because of the surface tension of the mercury and small capillary channels scribed on the surface of the armature, the mercury forms a continuous film over the surface of the armature and the active contact area of the poles. The action of the mercury during the break-before-make operation of an SPDT relay is shown in Fig. 3-35.

As opposed to the reed relay, the mercury-wetted relay must always be operated in a near-vertical position, typically within 30°. If the relay is inverted, up to 30 sec may be required for the mercury to drain from the contacts. This can cause a system error if a relay with a bridged contact is plugged into an operating system. Striking the relay to hasten clearing of the contacts should be avoided since this subjects the armature to extremely high shock forces that can alter relay operation permanently. The relay is very rugged, however, and operates at vibration levels well in excess of those normally encountered in analog subsystem applications.

The mercury-wetted relay is actuated by a magnetic field in a manner

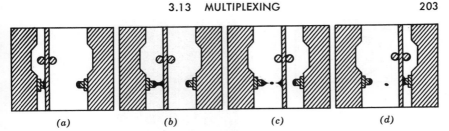

Fig. 3-35 Mercury-wetted relay contact action (Form C). From C. P. Clare & Co. (a) Mercury covers armature and contact faces. (b and c) As armature moves from open to closed position, mercury filament breaks before new contact is made. Rupturing mercury surfaces provides rapid breaking action. (d) Contact surfaces join. Mercury wetting dampens rebound, eliminates bounce.

analogous to that of a reed relay. Small permanent magnets usually provide the restoring force when the coil is deenergized. The bias magnets can also be adjusted to provide a latching action.

Electrically, the mercury-wetted relay resembles the reed relay. Contact resistance is generally lower, a maximum of 50 Ω being a conservative figure. The contact resistance over the life of the relay varies by only a few milliohms compared to the normal variation of 25% in a reed relay. Open contact resistance and leakage resistance effects tend to be lower than in reed relays as a result of Hg vapor in the capsule. A common specification is 10^9 Ω. Stray capacitances are somewhat higher, with values of 3–4 pF common in the larger capsules for open contact capacitance and up to 10 pF for contact-to-coil capacitance.

The pick-and-drop operate times may be longer than those of the reed relay. Depending on the type of relay and drive conditions, the pick time is from 0.75–6 msec, with drop times of up to 2 or 3 msec with no coil suppression. Coil suppression increases the drop time by as much as 50%. Because of the action of the mercury, there is no contact bounce when the contacts close.

Noise characteristics are at least as good as—and generally better than—those obtainable with reed relays of comparable size. The initial noise is typically less than 0.5 mV peak. Noise waveforms and data are shown in Fig. 3-36 for one of the larger capsules. Faster decay of noise may be obtained in some of the capsules with smaller internal structure and, in these cases, system speeds comparable to those obtainable with dry reed relays may be attained.

The life expectancy of the mercury-wetted relay is excellent compared to other electromechanical switching devices. Whereas the reed relay is occasionally specified at a billion operations under dry circuit conditions, the mercury-wetted relay is routinely specified in excess of a billion

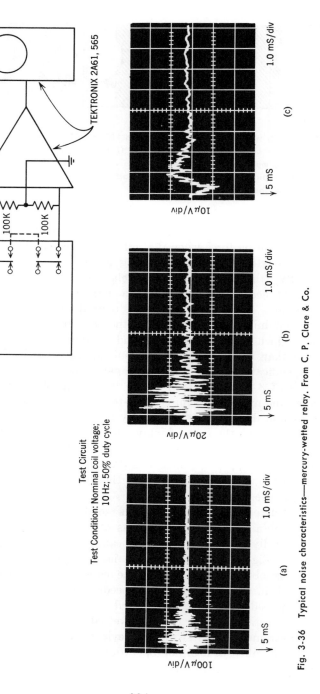

Test Circuit

Test Condition: Nominal coil voltage;
10 Hz; 50% duty cycle

TEKTRONIX 2A61, 565

100K 100K

20 μV/div

100 μV/div

10 μV/div

5 mS ↓ 1.0 mS/div

(a)

(b)

(c)

Fig. 3-36 Typical noise characteristics—mercury-wetted relay. From C. P. Clare & Co.

operations at rated loads with proper contact protection. Life expectancy under dry circuit conditions should exceed that specified for full load.

The mercury-wetted relay is electrically more rugged than the reed relay. Not only is its volt-ampere rating almost two orders of magnitude greater, but also it withstands momentary overloads with few detectable effects. Contact failure due to charging of stray capacitances by a common mode source has not been observed as it has in the case of the reed relay.

3.13.1.3 Chopper-Type Relays. Another dry contact type of relay utilized by several manufacturers of analog signal multiplexers is similar to the electromechanical chopper. The relay offers good electrical and life performance with excellent thermal characteristics. An example of this type of relay is manufactured by *James* Electronics under the name *Micro-Scan.**

The basic *Micro-Scan* mechanism is shown in Fig. 3-37. The fixed and transfer contacts at the base of the mechanism are constructed of a special gold alloy. The armature is driven by the magnetic field established

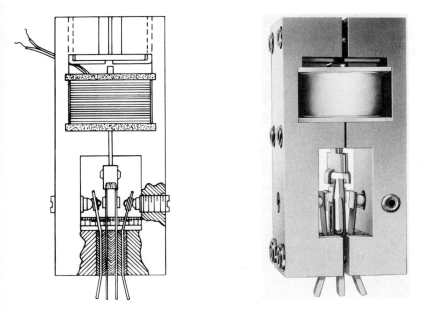

Fig. 3-37 *James Micro-Scan* mechanism. From James Electronics, Inc.

James and *Micro-Scan* are registered trademarks of *James* Electronics.

between coil-driven pole faces at the top of the relay. The armature is pivoted at its center and is electrically isolated from the transfer contact. The isolation of the signal and magnetic paths contributes to the low noise characteristics of the relay.

The relay has a maximum initial contact resistance of 25 mΩ. Operate and release times are less than 1 msec and there is no contact bounce when the relay is driven properly. The capacitance between the contacts is a fraction of a picofarad. Because of the physical isolation of the coil and the signal path, the coil-to-contact capacitance is very low, with a specification of 2×10^{-4} pF. The capacitance between the signal path and ground is higher than that encountered in a reed or mercury-wetted relay. It is specified at 25 pF but can be reduced with different shield arrangements at the sacrifice of increased open contact capacitance. Leakage resistances are specified at 10^9 Ω.

The life expectancy is 10^9 operations, better than the reed relay but not as good as the mercury-wetted relay. The device is intended only for low-power applications with a contact rating of 10 V at 1 mA maximum. A resistor in series with the contacts prevents degradation of the device's life due to capacitive common mode charging currents.

Thermal characteristics of the relay are excellent, as shown by the data in Fig. 3-38. Contact noise is on the order of 50 μV peak to peak initially and decreases to less than 5 μV within 100 msec after closure for a 200 KΩ load impedance and 6 KHz system bandwidth.

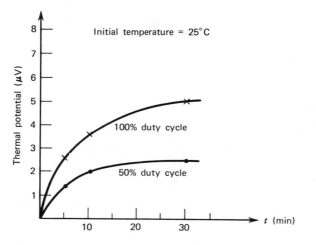

Fig. 3-38   *James Micro-Scan*—thermal characteristics. Data from James Electronics, Inc.

3.13.1.4 Other Electromechanical Switches. Although the reed, mercury-wetted, and chopper-type relays are the most common electromechanical switches used for multiplexing, other devices are available and may be applicable in particular situations. These include devices such as the stepping switch, crossbar matrix switch, and the air and mercury jet switches. All these devices have limited life or versatility, which accounts for their minimal use in multiplexers for process control analog subsystems.

3.13.2 Semiconductor Switches

The forte of a semiconductor switch in multiplexing applications is its speed. Whereas most electromechanical devices are limited to operating speeds of 250 sps or less, the semiconductor switch can be utilized at sampling speeds in excess of 100 ksps. Except for switching speed, however, the semiconductor switch does not approximate the ideal switch as well as the electromechanical switch does. Semiconductor switches have a number of limitations including low breakdown voltages, leakage currents, high ON resistances, and voltage offsets. Despite this, however, the semiconductor switch is an important device for analog multiplexing.

The semiconductor devices most often used for analog multiplexing are the bipolar transistor and the field effect transistor. The use of bipolar transistors is generally limited to signals of greater than 500 mV because of offset errors. The field effect transistor, however, has no offset and is used for both high- and low-level signals.

3.13.2.1 Bipolar Transistor Switches. The basic bipolar transistor switch is shown in Fig. 3-39. A pulse applied to the primary of the drive transformer provides base current that forward biases the base emitter junction and allows current flow between the collector and emitter

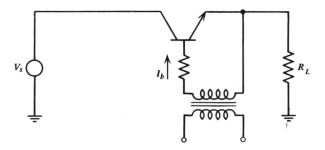

Fig. 3-39 Basic bipolar transistor switch.

terminals. In this condition, the transistor approximates the ON state of the ideal switch. Without a pulse applied to the transformer, the signal voltage V_s provides back bias for the transistor and the switch approximates the OFF condition of the ideal switch. The particular configuration shown here can only be used to switch positive signals since a negative signal forward biases the collector through the transformer secondary. For this reason, the single transistor switch is rarely used in multiplexers. It is convenient, however, to consider the characteristics of this switch before discussing the actual switch used in most multiplexers.

Figure 3-40 shows the V_c-I_c characteristics of an NPN transistor switch near the origin. The curves shown are the ones on which the operating points lie when the transistor is used as a switch. The curve

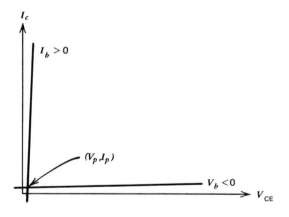

Fig. 3-40   Bipolar transistor switch characteristics.

in the saturation region for a positive base current I_b corresponds to the ON state of the switch. The curve related to the OFF condition is in the cutoff region. The curves do not intersect at the origin as they would if the transistor were a perfect switch. The point of intersection (V_p, I_p) represents a deviation from the characteristics of the ideal switch since a potential exists across the switch in the ON condition whether or not collector current is flowing. Similarly, a leakage current I_p flows in the OFF condition. The value of V_p may be as high as 10 or 20 mV, and I_p is generally much less than 1 μA for a silicon transistor at room temperature. In a germanium device, V_p is 1 or 2 mV, but I_p is on the order of 1 μA at room temperature.

R. L. Bright discovered experimentally that the values of V_p and I_p

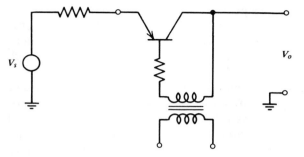

Fig. 3-41 Inverted transistor switch.

are reduced considerably when the transistor is driven between the base and collector, as shown in Fig. 3-41 (Bright, 1955). This connection, known as the *inverted* connection, is widely used in analog applications. Neglecting the voltage drop due to the intrinsic collector and emitter resistances and denoting the normal and inverted connections with the subscripts n and i, the voltage offset and current leakage are given by the approximate expressions

$$V_{pn} \cong -\frac{kT}{q} \ln \alpha_i$$

$$V_{pi} \cong -\frac{kT}{q} \ln \alpha_n$$

(3-31)

$$I_{pn} \cong I_{\mathrm{CBO}} \frac{(1 - \alpha_i)}{(1 - \alpha_i \alpha_n)}$$

$$I_{pi} \cong I_{\mathrm{EBO}} \frac{(1 - \alpha_n)}{(1 - \alpha_i \alpha_n)}$$

(3-32)

where α_n = normal common base current gain
$\quad \alpha_i$ = inverted common base current gain
$\quad I_b$ = base current
$\quad I_{\mathrm{CBO}}$ = collector to base leakage, emitter open
$\quad I_{\mathrm{EBO}}$ = emitter to base leakage, collector open
$\quad k$ = Boltzmann's constant = 1.38×10^{-23} J/°K
$\quad q$ = magnitude of electonic charge = 1.6×10^{-19} C
$\quad T$ = absolute temperature in °K = (°C + 273)

For most transistors, α_n is much closer to unity than α_i, with $\alpha_n = 0.98$–0.999 and $\alpha_i = 0.2$–0.6 being typical figures. Substitution of $\alpha_n = 0.99$ and $\alpha_i = 0.5$ into Eqs. (3-31) and (3-32) shows that $V_{pi}/V_{pn} \cong 0.015$ and, since $\alpha_n I_{\mathrm{EBO}} = \alpha_i I_{\mathrm{CBO}}$, $I_{pi}/I_{pn} \cong 0.01$. Thus both the leakage current and the voltage offset in the inverted connection are much less than those in the normal connection. The penalty for this improved switch is the requirement for a greater base drive as a result of the low gain of the inverted connection; this is rarely a problem in multiplexer applications since the signal current being switched is very small because of the high input impedance of the system.

The difference between the normal and inverted connection is a result of device geometry as well as of device physics. The collector in most transistors is physically larger than the emitter to ensure that almost all minority carriers are collected despite radial spreading. The difference in the relative areas of the emitter and collector accounts for the primary difference in the α's and the leakage currents I_{CBO} and I_{EBO}, although the relative doping of the emitter and collector regions is also a factor.

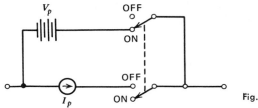

Fig. 3-42   Transistor switch equivalent.

Using the equivalent circuit of Fig. 3-42 to represent the imperfect switch in both the ON and OFF states, the effect of V_p and I_p on multiplexer operation can be qualitatively evaluated. As shown in Fig. 3-43, the offset voltage V_p is in series with the signal source V_s so the measurement system actually measures a potential $V_p + V_s$. Thus the offset voltage V_p is a direct contribution to system error. The second source of error is the leakage current of the OFF switches in the other channels. The leakage currents from all the OFF switches flow through the source impedance of the ON channel to produce an additional error of nI_pR_s where n is the number of OFF channels connected to the common multiplexer bus.

Equations (3-31) and 3-32) neglect the effect of intrinsic and extrinsic resistances of the transistor. Since the base current passes through these resistances, a term must be added to the expressions. Lumping the intrinsic and extrinsic resistance into a single collector resistance R_c, the total offset voltage V_o for the inverted connection is

$$V_o = I_bR_c - \frac{kT}{q} \ln \alpha_n \qquad (3\text{-}33)$$

The importance of the additional term may be considerable at high base

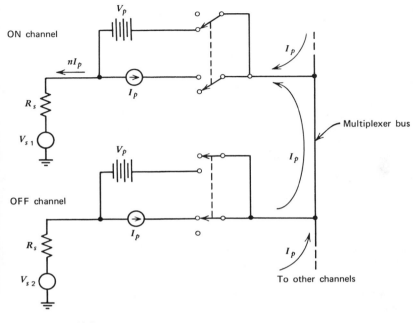

Fig. 3-43   Multiplexer error sources.

currents. For example, a resistance of 1 Ω at a base drive of 5 mA results in an additional 5 mV offset, which may be an order of magnitude greater than V_o.

Extrinsic resistances shunting the collector and emitter terminals also add a linear term to the expressions for I_p. However, careful mechanical design and the surface passivation of modern devices generally reduces this effect to a negligible value and $I_p \cong I_o$, the total leakage current.

Both V_o and I_o are temperature dependent. The major factor in the increase of I_o is the exponential increase in the leakage current of any semiconductor junction with increasing temperature. This is modified somewhat because the leakage current is dependent on α which is also temperature dependent. This effect is small, however, and assuming that the leakage doubles for every 10°C (=18°F), increase in temperature is sufficiently accurate for most calculations.

The expression for V_o given by Eq. (3-33) shows a linear increase in temperature due to the kT/q factor. This, however, is tempered by a simultaneous increase in α. The result is a net change on the order of a few microvolts per degree F, depending on the value of the base current. As shown in Fig. 3-44, it may be possible to choose a base current such

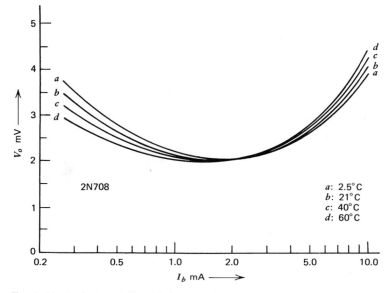

Fig. 3-44 Variation of V_o with I_b and temperature. From Verster, 1964, by permission of the publisher.

that the variation in V_o with temperature is minimized. The optimum I_b is affected by the value and temperature coefficient of the extrinsic resistance: If it is too high, the optimum I_b is too small to guarantee operation in the saturation region. If the extrinsic resistance is too low, the optimum I_b is large and the $I_b R$ offset may be excessive (Verster, 1964).

Figure 3-44 also shows that even at a constant temperature, the offset voltage is dependent on I_b because of the dependence of α on I_b. In some devices, the minimum is rather broad and a variation in I_b of as much as 5–1 may not affect V_o. In order to assure that the base drive does not vary significantly during the switching interval or under the influence of temperature, a relatively stiff current source is often employed as the switch driver. Although this reduces the switching speed of the transistor, the loss of speed is unimportant except in very high-speed systems.

The nominal value and variation in R_{on} is usually of minor importance in a multiplexer because the input impedance of the system is quite high. An expression for R_{on} of the inverted transistor when R_{on} is much greater than the extrinsic and intrinsic resistances in the base drive circuit is

$$R_{on} = \frac{kT}{qI_b}\left[\frac{1 - \alpha_i}{\alpha_i} + (1 - \alpha_n)\right] \tag{3-34}$$

The expression for the normal connection is the same as Eq. (3-34) except that the α_i and α_n are interchanged. Substitution of typical values shows that R_{on} is on the order of $(25 \times 10^{-3}/I_b)$ Ω in the inverted connection and $(1.25 \times 10^{-3}/I_b)$ Ω in the normal connection at room temperature where I_b is in milliamperes.

In addition to these static characteristics of transistor switches, aging and transient effects are of interest in low-level switching applications. The most significant aging effect for multiplexer applications is the change in V_o. A high-temperature test of germanium and silicon transistors showed that germanium devices continued to drift by tens of microvolts over a period of 30 days. Silicon devices also exhibited drift but it was an order of magnitude less. A subsequent test of the same devices at room temperature indicated that a permanent change in the value of V_o was possible but not necessarily certain. It was also found that the direction of the drift did not correlate with the material from which the device was fabricated. In fact, devices of the same type exhibited drift in opposite directions. Figure 3-45 shows the results of one of the tests at elevated temperature. The drift in the devices was traced to a hysteresis effect in the value of α_n (Verster, 1964).

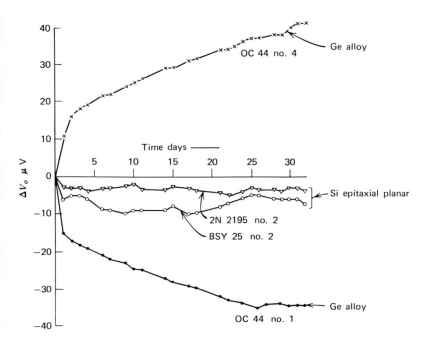

Fig. 3-45 Variation of V_o with time. From Verster, 1964, by permission of the publisher.

A source of transient error is the variation of V_o with base current as the switch is turned OFF and ON. The base current I_b does not change instantaneously and, since V_o is a function of I_b, V_o does not reach its final value instantaneously. If the risetime of I_b is short compared to the sampling time of the switch, this contribution of transient error is usually negligible.

The primary source of transient error in the transistor switch is current flow through the base emitter capacitance in the inverted connection as a result of the change in drive voltage. The magnitude of this current depends on the rate of change of the base voltage and the resistances and capacitances in the external circuit, particularly the interwinding capacitance of the drive transformer. The transient current passing through the source impedance of the signal circuit causes a spike of voltage to appear at the input of the amplifier. The exponential decay of this voltage spike is determined by the value of the interelectrode capacitance and the parallel combination of the source and load resistances. Generally, the time constant is quite short but, since almost ten time constants are required for an exponential transient to decay to 0.01% of its initial value, the transient may limit the practical switching speed in high-accuracy systems.

The transient due to capacitance feedthrough is present at both the turn-on and turn-off of the transistor. An additional contribution to the turn-off transient is the storage of minority carriers in the base region. High-frequency transistors are designed with small base regions, which reduces the stored charge and results in less of a transient. In most multiplexing applications, the turn-off transient is not of extreme importance unless the transient causes the amplifier to saturate with a resultant recovery time that limits system speed. Base circuits consisting of R's and C's can be designed to provide optimum drive waveforms that minimize the various transients (Verster and Boothroyd, 1963).

A summary of desirable characteristics of a transistor suitable for low-level switching when operated in the inverted mode include the following: (1) high α_i to minimize I_o and allow operation at low base current; (2) high α_n to minimize V_o; (3) low intrinsic and extrinsic collector resistance to minimize offset due to base current; (4) small base-to-collector capacitance to minimize transient current; (5) high cutoff frequency to minimize base minority carrier storage; and (6) small I_{EBO} to minimize offset due to I_o flow through the signal source resistance.

Even a transistor having the best of these properties available in practice may not satisfy the needs for multiplexer applications. A typical offset voltage for an inverted silicon transistor is on the order of 0.5–2 mV. This is an error of up to 2% of a 100-mV signal or 0.04% of a 5-V signal. In the article in which Bright described the characteristics of the inverted switch, he also introduced a circuit that reduces the offset voltage to a level acceptable for most high-level, and some low-level, applications.

The circuit, shown in Fig. 3-46, consists of two transistors whose collectors are connected together and driven in the inverted mode. The

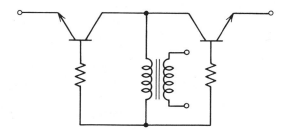

Fig. 3-46   The Bright transistor switch.

simple equivalent circuit of Fig. 3-42 shows that the net offset voltage is the difference between the individual offsets of the switches. Transistors can be selected such that offsets of less than 50 μV over moderate temperature ranges are realized using this connection. Since the offset is a function of I_b, the offset can be reduced further by providing independent adjustment of the two base currents. It is, however, not always possible to maintain the optimum currents over wide temperature ranges and, in addition, the task of adjusting each switch individually in a system having hundreds of multiplexer switches is considerable.

The two-transistor switch is also beneficial when the transistors are driven in the normal connection. The net offset is generally larger than in the inverted connection because the individual offsets may be several orders of magnitude larger. The normal connection is sometimes used, however, because the value of R_{on} is less than in the inverted connection and the switch breakdown voltage is greater. Low R_{on} is important when the input impedance of the amplifier following the multiplexer is on the order of hundreds of thousands of ohms rather than megohms.

Another advantage of the configuration introduced by Bright and its normal connection equivalent is that it multiplexes signals of either polarity. If the voltage V_o at the output of the single-transistor switch of Fig. 3-41 is more positive than the signal voltage V_s, as could be the case if a positive voltage is sampled on another channel, then the base emitter junction is forward biased and the transistor conducts. In the Bright switch, one or the other of the transistors has a back-biased base emitter junction and the switch remains in the OFF condition.

The breakdown voltage of the nondriven junction of the transistor switch limits the value of signal and common mode voltages that can be applied. Generally, the breakdown voltage of the emitter base junction is less than that of the collector base junction so the voltage the switch can withstand in the inverted connection is less than in the normal connection. Although this may be as high as 30 V, a more typical figure is 5–10 V. Since the sum of the common and normal mode voltages on any pair of channels may appear across the OFF switch, the system specifications for the sum of the normal and common mode voltage on each channel cannot exceed one-half the breakdown voltage of the switch.

A single device that operates on the principle of back-to-back transistors is known as the double emitter transistor or DET (Mitchell and Bell, 1962). A schematic diagram and pictorial representation of the DET switch are shown in Fig. 3-47. In the DET the base current and signal

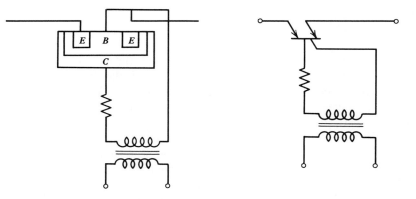

Fig. 3-47   Double emitter transistor switch.

current do not share a common path as they do in the usual transistor switch; the signal current flows from emitter to emitter and the base drive current flows from base to collector. These currents are perpendicular to each other, and this results in a significant reduction in the interaction between V_o and I_b. In addition, since only one base drive is used, variations in V_o due to unbalanced base currents are avoided. Since the DET is a single device, the thermal coupling between the emitters is excellent and thermal differentials that can affect the value of V_o have less effect than in the case of two separate transistors. The number of junctions between dissimilar materials is less in the DET than in the double-transistor switch. This is of some importance in low-level applications since the thermoelectric powers of germanium and silicon are very high compared to the gold lead wire.

Electrically, the characteristics of the DET are somewhat better than those obtained with matched transistors. A value of V_o of 20 μV or less can be obtained over a temperature range of 100°F. The saturation resistance R_{on} is on the order of 30 Ω with a base drive of 1 mA. The base-to-emitter capacitance that affects the transients is less than 5 pF. The emitter-to-emitter leakage current for an applied voltage of 10 V is less than 10 nA at room temperature in silicon devices.

3.13.2.2 Field Effect Transistors. The field effect transistor is a semiconductor device that overcomes some limitations of bipolar transistor switches. There are two types of field effect transistors, the *junction field effect transistor* (JFET) and the *insulated gate field effect transistor* (IGFET), also known as the *metal oxide semiconductor* (MOS) FET. Whereas the basic control phenomenon in the bipolar transistor is the control of the excess density of minority carriers in the base region, the FET is a majority device in which the load current is controlled by a field perpendicular to the direction of the load current. In a JFET the field is established by back biasing a *PN* junction, and in the MOSFET it is established between two conductors that act as the plates of a capacitor.

In its simplest form, the JFET takes the form of a bar of n- or p-type semiconductor into which two regions of opposite type semiconductor material have been diffused, as illustrated in Fig. 3-48. The two ends of the semiconductor bar are provided with ohmic contacts designated as the *source* and the *drain*. By convention, the source is the terminal at which electrons are injected into the device. As seen from the figure, however, the device is symmetrical and the designation is arbitrary for the geometry shown. The p-type regions are known as the gates and are connected together in the external circuit. The boundaries between the p-type and n-type materials form *PN* junctions. Since the majority

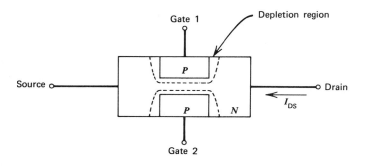

Fig. 3-48   Junction field effect transistor.

218   ANALOG-INPUT SUBSYSTEMS

carriers cannot flow through the depletion region surrounding the *PN*
junctions the current flow between source and drain is concentrated in
the *channel* between the depletion regions. Because there are no junctions
in the path between source and drain, the current is determined by the
applied source-to-drain voltage V_{DS} and the resistance of the channel
R_{DS}. This resistance is primarily determined by the dimensions of the
channel and the resistivity of the channel material.

The resistance is minimum when there is no reverse bias on the *PN*
junctions. In this condition, the JFET acts as an ideal ON switch in
series with R_{DS}. Since there are no *PN* junctions in series with the channel, R_{DS} is strictly passive and there is no junction offset voltage. The
absence of offset is the outstanding characteristic of the FET which permits its use in low-level multiplexing.

When the *PN* junctions are back biased by a voltage between the
gates and either the source or drain, the depletion regions surrounding the
PN junctions spread with a resultant narrowing of the channel. Application of sufficient reverse bias "pinches off" the channel and reduces I_{DS}
to a very small value. In the pinched off condition, the channel resistance
is thousands of megohms and therefore, the device approximates the OFF
state of the ideal switch.

The device geometry shown in Fig. 3-48 is difficult to fabricate because
of the diffusion on both sides of the channel material. The usual device,
therefore, utilizes a geometry such as that shown in Fig. 3-49. The *p*-type
substrate functions as one of the gates on which *n*-channel material is
epitaxially grown. Diffusion of the other *p*-type gate into the *n*-type
material completes the structure. The channel is the region between the
substrate and the diffused gate region.

For an *n*-channel MOSFET device, *n*-type material is diffused into a
p-type substrate to form source and drain regions. The surface of the
entire structure is covered with an insulating oxide, SiO_2 in the case of

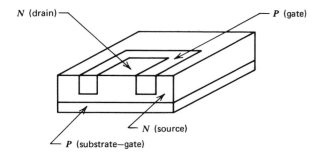

Fig. 3-49   Practical JFET structure.

most silicon devices. The oxide is etched through to the *n*-type source and drain regions and metallization is applied to provide ohmic contacts. Metallization also provides a gate region covering the area between the source and drain regions. The cross section of the completed device is shown in Fig. 3-50.

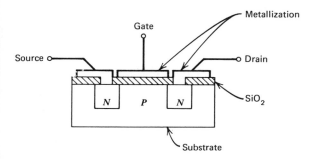

Fig. 3-50   MOSFET structure.

The source-to-drain current in the absence of a gate potential is extremely low since the source and drain are isolated by the substrate material and the *PN* junctions. The metallized gate and the substrate constitute a capacitor with the oxide layer acting as the dielectric material. When a positive potential is applied between gate and substrate, the positive charges at the metal-oxide interface induce negative charges in the semiconductor under the oxide and between the source and drain regions. If sufficient negative charges are induced, a layer of the *p*-type substrate immediately below the metal gate is converted to *n*-type material. This layer acts as a channel between the source and drain regions and a current flows when a voltage V_{DS} is applied. The dimensions of the induced channel and, therefore, its resistance are functions of the applied gate voltage. This type of operation is known as the *enhanced mode* since the current between source and drain is enhanced by the action of the gate potential. The gate voltage at which significant conduction occurs is called the *threshold voltage*.

Another possible MOSFET structure allows either enhancement mode operation or a depletion mode operation similar to that of the JFET, but this type of device is not generally used in multiplexers.

In comparing the MOSFET and the JFET for multiplexing applications, one of the major differences is the state of the switch with no gate signal applied. The JFET is a normally ON device, whereas the enhancement mode MOSFET is normally OFF. Care must be used in JFET multiplexing applications since an intentional or unintentional loss of system power results in all multiplexer switches being turned ON. This interconnects all process signal sources through the ON resistance of the FET in each channel. Whether this condition is hazardous or causes transducer damage depends on the particular system and process. In any case, however, either back-up power must be provided to keep the switches OFF or the consequences of having all switches ON must be carefully assessed.

The outstanding advantage of both the MOSFET and the JFET for multiplexing applications when compared with other semiconductor switches is the virtual lack of any offset voltage in the ON condition. The FET is not a perfect switch, however, and other sources of error must be considered. These are similar to those of the bipolar transistor; specifically, they include ON resistance, leakage currents, and transient effects.

The source-to-drain resistance R_{DS} is analogous to the R_{on} of the bipolar transistor. The resistance R_{DS} is determined by the dimensions of the channel and the resistivity of the channel material. The resistivity is a function of carrier mobility and, since the mobility of electrons is about three times that of holes, the n-channel device has a lower R_{DS} for a given geometry. The value of R_{DS} for an FET suitable for analog multiplexing is in the range of 20–2000 Ω. This is considerably higher than R_{on} for a bipolar transistor. As described earlier, the source impedance, switch resistance, and input impedance of the amplifier form a voltage divider that attenuates the input signal. Although this attenuation can be compensated partially by calibration, variations in R_{DS} between channels and with temperature and time contribute to system error. For temperature variations within 30°C of room temperature, the theoretical variation in R_{DS} for both n-channel and p-channel devices is about 1%/°C. Experimental results are about 0.6%/°C for n-channel silicon devices and 0.8%/°C for p-channel silicon devices.

In addition to source loading, the difference between the R_{DS} of two switches in a differential multiplexer channel contributes to the common mode unbalance of the system. If an FET with a high R_{DS} is used, the unbalance and its variation with temperature may be significant in terms of system performance. For this reason, devices with low R_{DS} are recommended for use in low-level differential systems.

Sources of error due to leakage currents are similar to those encountered in the bipolar transistor multiplexer. Leakage currents due to the *PN* junctions in the FET flow through the source resistance of the ON channel with a resultant voltage error. In the JFET, the junctions are those between the gate and the channel. The paths for the leakage currents are through the source impedance of the channel in which the switch is located and through the source impedance of the ON channel via the multiplexer bus. The error due to leakage through the source resistance of the OFF channel is of no concern since the channel is not being sampled. The leakage of all OFF switches through the ON channel, however, can be a significant source of error as described for the bipolar transistor multiplexer. Selecting devices with low leakage current, restricting the size of the source resistance, and the use of submultiplexing or block switching are effective solutions.

In the MOSFET the PN junctions are between the channel and the substrate. The leakage paths are basically the same as in the previous case and the error contribution is the same. An additional source of leakage current in the MOSFET is the resistance between the gate and channel. This resistance is determined by the resistivity and thickness of the oxide layer and by surface effects. Typically, the leakage resistance is greater than $10^{12}\,\Omega$ and is not a significant error source. In addition, the resistance is only slightly temperature dependent, so the leakage problem is not increased at high temperatures as it is in the case of PN junction leakage.

Although both the JFET and the MOSFET can be built as symmetrical devices, the source and drain areas with respect to the gate area are not the same in a typical FET. As a result, the source-to-gate and drain-to-gate leakage currents may differ by a factor of 2 or 3. This fact should be included in the evaluation of the effects of leakage currents since errors may be reduced by interchanging the source and drain in a given multiplexer design.

The input impedance of both the JFET and MOSFET between source and gate is very high. In the JFET this is the reversed biased PN junction and the impedance is on the order of thousands of megohms. The input impedance of the MOSFET is the impedance of the capacitor formed by the substrate and the metallic gate. The resistive component is determined by the resistivity of the oxide dielectric and is typically on the order of 10^{12}–$10^{14}\,\Omega$. The resistance is not particularly temperature dependent and remains essentially constant for temperatures of practical interest. Since the JFET looks like a reversed biased diode, its input impedance changes considerably with temperature according to the usual formula for diode leakage.

One consequence of the high input impedance is that a multiplexer switch may be driven from a direct coupled source. The transformer used in bipolar transistor circuits to provide isolation between the input signal and the drive is not required. A basic JFET multiplexing switch is shown in Fig. 3-51. The drive signal is normally negative to hold an

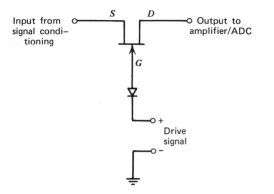

Fig. 3-51   Basic JFET multiplexing switch.

n-channel device in the OFF state. In this condition, the diode in the gate lead is forward biased and the source-to-gate junction is reverse biased. When the control signal is positive, the bias conditions on the diode and junction are interchanged and the switch is ON.

The minimum negative voltage required to hold the switch in the OFF condition is equal to the sum of the pinch-off voltage and the maximum negative voltage appearing on either the source or drain. The positive control signal must be greater than the most positive value of the input signal. In practice, larger voltages are used to guarantee that the switch is held in the desired state.

The basic multiplexer switch using the MOSFET is similar to that of the JFET switch. Whereas the JFET is a depletion mode device, the p-channel enhancement mode MOSFET is normally used for analog multiplexing. In a p-channel device the gate must remain more negative than the sum of the threshold voltage and the most negative signal if the switch is to remain ON. Since R_{DS} is dependent on the voltage beyond threshold, the ON drive is usually selected to exceed the minimum required voltage by enough to guarantee a low value of R_{DS}.

Care must be taken to insure that the drain and source-to-substrate PN junctions are never forward biased. This is sometimes done in a p-channel MOSFET by tying the substrate to a voltage source that is more positive than any input signal. This has the effect of increasing R_{DS} unless the gate drive in the ON condition is increased to compensate for the effect.

In both JFET and MOSFET circuits, the switch requirements are such that fairly large drive signals are required to guarantee proper operation for high-level signals. This can result in large variations in R_{DS} and leakage currents that are dependent on signal magnitude. A variety of driving circuits have been developed to minimize these effects. Many of them consist of either diodes or transistors that maintain the gate and input terminal at about the same potential independent of the signal amplitude. Transformer drive is also used to provide fixed voltage drive levels.

Breakdown voltages for the FET are of the same order of magnitude as in the bipolar transistor. The mechanism is basically the same in the JFET and the maximum allowable voltage is that which causes breakdown of either of the gate-to-channel junctions. When the gate is reverse biased, the source-to-gate voltage adds to the source-to-drain voltage. Thus the sum of these voltages must not exceed the breakdown voltage. Typically, this is on the order of 30–50 V, although the leakage currents may increase significantly before actual breakdown occurs.

In the MOSFET, the breakdown voltage of the gate-to-channel path is determined by the dielectric strength of the oxide layer. Although this can theoretically be increased by an increase in the thickness of the layer, such increase results in a higher threshold voltage. Typical devices have a gate-to-channel breakdown of 30 or 40 V for a threshold voltage of 3 or 4 V.

The drain-to-source breakdown mechanism in an MOSFET is slightly different. With $V_{GS} = 0$, the cutoff condition, and the substrate floating, the path between the source and drain is the back-to-back source-to-substrate and substrate-to-drain junctions. The breakdown between source and drain is determined by the breakdown of these junctions, which is typically on the order of 30 or 40 V. If the substrate is not floating, the breakdown voltage is determined by the individual substrate junctions.

Effects due to capacitance are approximately the same as in bipolar switches with the exception that there is no minority charge storage in the FET. The gate-to-output capacitance is the major contributor to the turn-on transient. The transient is a spike with an exponential decay whose time constant is dependent on the source impedance of the ON channel and the magnitude of the capacitances. The magnitude of the spike is dependent on the dV/dt of the drive waveform. The capacitances of the MOSFET are generally lower than those of the bipolar transistor or JFET with a resultant decrease in the transient size and duration.

The higher ON resistance of an FET compared with that of a bipolar transistor can be a limiting factor in the multiplexing speed. The signal source must charge the multiplexer bus capacitance through the switch and source resistance. The bus capacitance is the sum of the amplifier input capacitance and the gate-to-output capacitance of all the multiplexer channels. The high value of R_{DS} increases the risetime and may limit the multiplexing speed.

3.13.2.3 Photon-Coupled Devices. A desirable characteristic of any multiplexer switch is isolation between the signal and drive voltages. Neither the transistor nor the DET provide this isolation to the degree that is available in the FET and electromechanical switch. However, the reliability of the electromechanical relay and some of the FET characteristics are less desirable than those of other semiconductor switches. If the energy required to actuate a switch could be transmitted without resistive or capacitive coupling, a high degree of isolation could be achieved. Light or photon-coupled devices have this desired characteristic.

Photoconductive materials, such as cadmium sulfide, have a resistance that varies with applied illumination. Multiplexing switches composed of a photoconductive element and a source of light energy can be constructed, but their use is rather limited.

The main problems associated with the use of photoconductive devices are (1) source loading and common mode unbalance due to high ON resistance; (2) leakage current errors due to low OFF resistance; (3) system speed limitations caused by high ON resistance; (4) slow switching speed; and (5) lack of reliable light sources. These limitations have restricted the use of these devices to a few specialized applications in amplifiers and measurement instruments.

The development of solid-state light sources such as the light emitting gallium arsenide (GaAs) diode has renewed interest in photon-coupled switches (Bonin, 1965). The GaAs diode offers small physical size and the potential reliability of a solid-state device. Rather than using a photoconductive material for the switching element, a photoconductive diode or transistor is utilized. The characteristics of the switching device are the same as discussed previously except that the effects due to the drive current are eliminated. The photoconductive transistor is turned ON by the presence of excess carriers created in the base region by the photon excitation of the source.

One of the commercially available configurations is shown in Fig. 3-52. The light from the diode is coupled to the transistor through an optical coupling medium that serves to focus the light energy on the transistor junction. The physical separation of the diode and the transistor in the configuration shown is only a fraction of an inch, but this sustains a voltage of a 100 V or more without breakdown. In another experimental device, the light is coupled from the diode to the transistor by means of fiber optics, and the breakdown is even greater due to the larger separation. In addition, capacitive coupling is reduced to a fraction of a picofarad.

Fig. 3-52   Photon-coupled switch. From Texas Instruments, Inc.

Despite the seemingly favorable characteristics, the use of GaAs photon-coupled devices has been minimal because of high cost. Characteristics of the transistor or DET switching element render the switch unsuitable for use at low levels where the isolation would be most advantageous.

Another problem in the use of the device is the driving power required. Because of the nonlinear nature of the GaAs diode, a stiff current source is required. The light output is highly dependent on the drive and, because of the low quantum efficiency of the diode, the drive current is quite high. Currents up to 100 mA at a voltage of less than 2 V are required if levels of offset suitable for medium-level analog signal switching are to be obtained. Continued development and advances in manufacturing techniques may result in a device that will find extensive use in analog-multiplexing applications.

3.13.3   Multiplexer Configurations

All of the switching devices described in the previous sections can be used in a number of configurations for analog multiplexing. In certain configurations, one device might be preferred over another or the use of one device might result in a performance not attainable with another. Nevertheless, the following discussion of multiplexer configurations is generally applicable to any of the switching devices.

There are many possible ways in which a multiplexer can be built. In the following discussion, four distinctly different approaches are considered in detail. Many variations of these configurations, or even radically different approaches, are possible and may be technically valid. However, the discussion covers the configurations most often used in modern process control analog subsystems.

3.13.3.1 Single-Ended (SPST). The single-ended multiplexer utilizing a single-pole single-throw (SPST) switch is the simplest possible multiplexing configuration. It assumes the use of a single-ended signal source and consists of an SPST switch in series with each input signal. The output side of each switch is connected to a common amplifier or ADC input bus, as shown in Fig. 3-53.

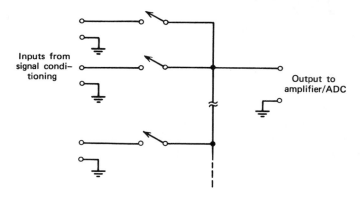

Fig. 3-53 Single-ended (SPST) multiplexer.

The switches are electronically controlled by the subsystem or computer logic. The control may be such that the switches are closed one at a time in a fixed sequence determined by the control logic. The usual case in this arrangement is to address the switches in numerical sequence since this requires the simplest logic. A common modification is to provide a starting address and an ending address or number of points to be scanned. The points between the two addresses or the points between the starting address and the address determined by the number of points to be scanned are then actuated in numerical sequence.

The second general method of control is random addressing in which the computer provides an address for each sampling operation. The addresses may be taken in a random sequence and the sequence can be modified by the computer program. This is obviously a much more flexible method than the fixed sequential scan. It does, however, require more hardware to implement and places a burden on available CPU time since at least one computer cycle is required to address each point sampled.

Combinations of the sequential and random addressing methods are also possible. Random addressing is provided on all the larger process control computer systems and many provide a programmable option for both types of addressing. A pure sequential method is normally restricted to smaller systems, remote terminals, or specialized data-logging equipment.

A major source of error in a single-ended multiplexer is the result of leakage currents flowing between OFF and ON channels. Suppose a switch is closed in a channel that has a full-scale input applied and, in the worst case, all other signals in the system are of opposite full-scale polarity. This is illustrated in Fig. 3-54 where V_{fs} is the full-scale input voltage, Z_{in} is the input impedance of the amplifier or ADC, R_{si} is the source impedance of the ith source and signal-conditioning circuit, and R_{Li} is an equivalent leakage resistance across the ith switch. In actual practice, R_{Li} is much greater than R_{si} which, in turn, is less than Z_{in} by several

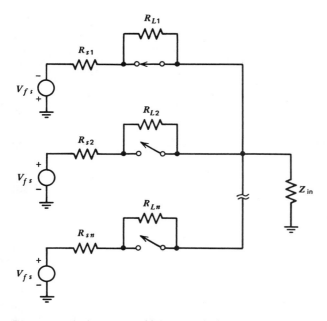

Fig. 3-54   Single-ended multiplexer equivalent.

orders of magnitude. The voltage across each open switch is equal to $2V_{fs}$ and a leakage current of $2V_{fs}/R_{Li}$ flows in each input circuit. Since Z_{in} is much greater than the source impedance of the selected channel R_{s1}, the majority of the leakage current flows through the resistance R_{s1}. If the system has n channels, the voltage generated across R_{s1} is

$$V \cong R_s \sum_i 2 \frac{V_{fs}}{R_{Li}} \quad i = 2, 3, \ldots, n \tag{3-35}$$

As an example, suppose all the R_L's are on the order of $10^9 \ \Omega$, $V_{fs} = 5$ V, $R_{s1} = 1000 \ \Omega$, and $n = 200$. The error potential is 1990 μV, or about 0.04% of the full-scale voltage. Since this is comparable to the accuracy of a high-level system, the error may be serious.

The error can be eliminated as a major factor by restricting n, restricting the magnitude of R_{si}, or using a technique that increases the effective value of the R_L's. Restricting the number of points n in the system tends to minimize the advantages which the time sharing provided by the multiplexer affords. Similarly, restricting the R_{si} decreases the versatility of the system and may limit its usage. Although both of these restrictions are normally necessary in any practical system, it is desirable that both n and the R_{si} be as large as possible. The third alternative offers this possibility and is commonly used in multiplexers.

The usual name for the technique is "block switching" or subcommutating. It consists of a second level of multiplexing, as shown in Fig. 3-55, in which the outputs of a small group of switches—say, m—are connected to the output bus through an additional series switch. Whenever a switch in a particular group is addressed, the associated block switch is also actuated. The effect of the block switch is to add an additional series resistance to each of the leakage paths from the unselected switches. If, for example, there are p groups of m channels each and the worst-case conditions used above are assumed, the leakage currents flowing through the source impedance R_s of the ON channel are

$$V = 2V_{fs}R_s \left[\frac{p-1}{R_L + R_L/m} + \frac{m-1}{R_L} \right] \tag{3-36}$$

When R_L/m is much smaller than R_L, this expression reduces to

$$V \cong 2V_{fs}(p + m - 2) \frac{R_s}{R_L} \tag{3-37}$$

The total number of switches in the system is pm, so Eq. (3-37) shows a considerably reduced error when compared to Eq. (3-35). For example, if there are 256 switches arranged in 16 groups of 16 each, the voltage without block switching is 511 $V_{fs}(R_s/R_L)$ as compared with 60

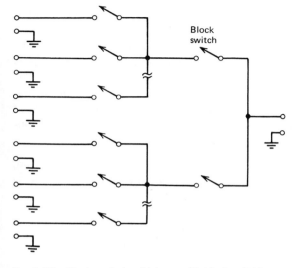

Fig. 3-55   Single-ended multiplexer with block switching.

$V_{fs}(R_s/R_L)$ with block switching. Thus the technique of block switching allows the system to be expanded without creating an excessive error due to leakage currents.

A second source of error due to the effects of many channels is source loading. If block switching is not used, the input impedance seen by the signal source is equal to Z_{in} in parallel with $R_L/(n-1)$ where R_L is the effective leakage resistance of each of the switches. If R_L is on the order of $10^9 \, \Omega$, $R_L/(n-1)$ is on the order of $10^7 \, \Omega$ for $n = 100$, and this is comparable with the typical value of Z_{in}. As a result, the effective value of the system input impedance is reduced and source-loading errors are increased. Block switching also provides a solution to this problem since, to a first approximation, the signal source sees the parallel combination of Z_{in}, the resistances of the other $(m-1)$ switches in the group, and the resistance of the other $(p-1)$ block switches. This is approximately equal to Z_{in} in parallel with $R_L/(p+m-2)$ and the improvement is the same as in the case of the leakage currents.

A similar situation exists in terms of shunt capacitance. The signal source sees the parallel combination of the amplifier or ADC input capacitance and the output capacitance of all the unselected switches and associated wiring when block switching is not used. Since the signal source must charge this capacitance through the source impedance, the risetime may be appreciable and can limit the system speed. Block switching reduces the value of the shunt capacitance in the same way that it increases the shunt resistance.

Other multiplexer errors are the direct result of imperfections in the switching device. These errors include transient spikes due to the inter-electrode capacitances of the switch and errors due to voltage offset. The magnitude of these effects depends on the type of switch and is not directly affected by the multiplexer configuration. Careful selection of the switch device and the considerations discussed in previous sections offer the best means of reducing these errors.

3.13.3.2 Differential (DPST). The simplest differential configuration is similar to the single-ended configuration except that a switch is provided in each input line, as shown in Fig. 3-56. The design consider-

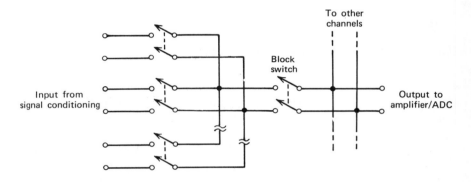

Fig. 3-56 Differential (DPST) multiplexer configuration.

ations are also similar to those previously discussed except that the effects of both switches must be considered. Block switching is typically employed in differential multiplexers for the same reasons and with the same benefits as in single-ended multiplexers.

The differences in the design of single-ended and differential multi-plexers stem from two main sources: First, differential multiplexers are typically used for low-level as well as for high-level signals and, therefore, sources of low-level error must be considered; second, the differential multiplexer has the additional performance requirement of common mode rejection.

The first consideration, that of dealing with low-level signals, has been discussed in previous sections and is only reviewed here. Low-level differential multiplexers normally employ either FET or electromechanical switches. The bipolar transistor switch and, to a lesser extent, the DET are limited in their use by the offset voltage error. They can be used, but elaborate matching and adjustment techniques are required for signals of less than 100 mV if the offset error is to be of negligible importance in high-resolution systems.

Thermal potentials can be another source of error in the low-level multiplexer. As discussed previously, they arise because of the thermo-couple action at the junction of two dissimilar metals. The solutions to the thermal problem include all the guidelines previously introduced in Sec. 3.10: reduction of the number of junctions to a minimum; use of complementary junctions; use of conductors that are thermoelectrically similar; and use of a mechanical design that minimizes temperature differentials. Sources of thermal error sometimes can be detected by heating the suspected junction with a heat gun or a low-temperature soldering iron whose tip has been covered with a heat-resistant electrical insulation.

The DPST differential multiplexer cannot eliminate the common mode potential; that is, a common mode potential applied to its input appears at the output when the multiplexer switches are actuated. Therefore, the amplifier or ADC that follows the multiplexer must have common mode rejection and must be capable of withstanding the applied common mode voltage. The differential multiplexer can, however, convert common mode voltages to normal mode error signals that are amplified and/or converted by the subsequent equipment. It is important, therefore, to consider the common mode effects in the multiplexer if the system common mode rejection is not to be degraded.

If there is no interaction between pieces of cascaded equipment, the reciprocal of the system CMRR is the sum of the reciprocal CMRRs of the cascaded equipment. That is, the common mode rejection figures combine in the same way as parallel resistance values. Thus, if a certain system common mode rejection is desired, the CMRR of the multiplexer must be even greater. In practice, it is difficult to measure the CMRR of the multiplexer independent of the rest of the system. This is because very few measuring instruments have a CMRR sufficiently high to avoid the interaction effect of cascaded equipment.

The DC and AC mode models introduced in Sec. 3.9 are directly applicable to the DPST differential multiplexer. The series resistance in these models is the series resistance of the switch, source, and signal conditioning circuit. Since the resistive difference or unbalance between the two lines is the major contributor to common mode error, the source resistance is more likely to be a problem than the other line resistances. With the possible exception of some FET devices, the ON resistance of the switches is low and the difference between them results in a negligible unbalance. Although the resistance of the signal conditioning circuit may compare with the source resistance, its contribution to common mode unbalance is small if balanced circuit configurations are used. Thus the elimination of common mode unbalance in the input circuits is not really a multiplexer problem but depends on the use of the system in the process.

The switches contribute to the value of the line-to-ground leakage resistance in the DC model. These resistances include card and package leakage from the signal line to system ground, leakage to ground through the multiplexer input and output wiring, and leakage from signal lines to system ground through the switch drive circuits. In the case of the electromechanical switch, this latter path is the contact-to-coil leakage. In semiconductor switches driven by a transformer, the secondary-to-primary leakage resistance is the major contributor. All of these resistances appear in parallel at the input or the output of the switch; when the switch is actuated, the input and output resistances are combined in parallel to form an equivalent resistance which is the leakage resistance used in the lumped model calculation. This means that the individual resistances must be very high if their parallel value is to be high enough to ensure adequate common mode performance. Block switching is beneficial since it isolates the shunt resistance of the switch groups by a high series resistance.

The importance of proper packaging in reducing leakage was considered in the discussion of the termination facility, and many of the same comments apply to the multiplexer design. Cards should be designed to provide physical and electrical isolation between signal lines and conductors leading to system ground. This implies a separation between the signal paths, switch drive circuits, and power supply conductors. In addition, insulation material for the components, including the card material, should have high resistivity and should be nonhygroscopic. Cards must be carefully cleaned and coated after manufacturing to remove traces of oils or conducting salts and to preserve this condition during their use. Even with these precautions, however, it may be necessary to provide a multiplexer enclosure that can be sealed and desiccated if high CMRR is to be maintained during long periods of high humidity.

In the AC common mode model, the pertinent parameters are the series resistance and the capacitance from the signal lines to system ground. The comments on the series resistance made above apply to the AC case. The capacitance is mainly stray capacitance due to coupling between signal conductors and ground, interwinding capacitance of drive transformers, or the coil-to-contact capacitances of electromechanical switches. These can be decreased by careful selection of the switching device and maintaining maximum physical separations. Block switching also acts to reduce the effective stray capacitance on any given channel. Shielding techniques are often employed to reduce the effective value of stray capacitance. The technique of guard shielding is discussed in Chap. 4 as a system consideration; suffice it to say here that it is effective in improving the AC CMRR at the expense of additional hardware complexity.

The above factors are the major sources of common to normal mode conversion in the DPST differential multiplexer. There are, however, several other sources that may be a problem in some systems. In the single-ended multiplexer, a transient due to the switch drive is coupled into the signal path. This same phenomenon occurs in the differential multiplexer except that it is now a common mode, as well as a normal mode, effect because there is a transient associated with each switch of the differential pair. If the transients were precisely the same, the effect would be only a common mode effect. They are rarely the same, however, and the difference results in a normal mode transient.

The common mode transient may be a problem because, according to the AC model, the CMRR of the system decreases with frequency and, therefore, a significant common to normal mode conversion of the high frequency components of the transients occurs. The resultant error is negligible if the transient decays to a small value before the conversion is initiated. This assumes that the normal mode error is not stored on a capacitor or other analog storage device prior to conversion.

The second possible problem is due to the characteristics of the amplifier that follows the multiplexer. Some types of differential amplifiers that have excellent steady-state CMRR cannot tolerate transient common mode signals. The transient causes the amplifier to saturate or become momentarily unstable. If the amplifier recovers before the conversion of the signal is attempted, no harm is done. If it does not, however, a conversion error of some degree results.

The system user should ensure that common mode transients are not introduced into the system on the process signal lines. Because of the basic operation of the DPST differential multiplexer, these transients feed directly through to the amplifier. Elimination of transient common mode sources and adherence to accepted wiring practices for control computer systems are effective preventatives.

3.13.3.3 Differential—Flying Capacitor. One of the characteristics of the differential DPST multiplexer is that the common mode potential appears at the output of the multiplexer when a channel is selected. This necessitates the use of a differential amplifier or ADC and, because of the manner in which the CMRR figures of cascaded equipment combine, it contributes to a degraded system CMRR. If the multiplexer eliminates the common mode voltage completely, a single-ended amplifier can be used and this is usually less expensive. A system in use by several manufacturers provides this capability. The method, known as the "flying capacitor" or "capacitive transfer" technique, is based on the principle that the subsystem can be electrically disconnected from the common mode source for the short conversion time if the signal is stored on a capacitor.

A schematic of the basic flying capacitor system is shown in Fig. 3-57. The multiplexer consists of a DPDT switch in each input channel and a capacitor connected between the armatures of the two sections of the

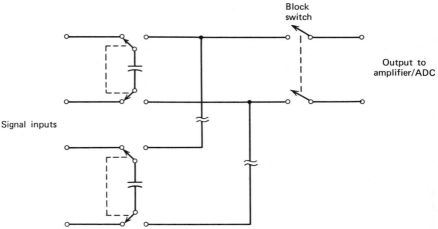

Fig. 3-57   Flying capacitor multiplexer.

switch. In its normal position, the capacitor is connected to the input signal and is, therefore, charged to the value of the signal. To read the value of the input voltage, the switch is actuated and the capacitor is transferred to the input circuit of the amplifier. Thus the stored voltage on the capacitor can be read without the amplifier being electrically connected to the input signal. This isolates the measurement system from the common and normal mode noise sources.

Despite the isolation, however, a common mode error is produced. The dominant common to normal mode conversion phenomenon is quite different from that in the models presented previously, and new models must be developed. The conversion predicted by the previous models is a secondary source of error in the flying capacitor multiplexer, and the

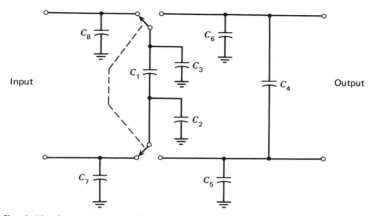

Fig. 3-58   Lumped model of flying capacitor multiplexer.

considerations discussed for the DPST multiplexer are applicable with particular emphasis on the input side of the multiplexer.

The error produced by the common mode potential in the flying capacitor differential multiplexer is the result of charge sharing. The circuit of Fig. 3-58 is a lumped representation of a single input channel in a flying capacitor multiplexer. The capacitor C_1 is the flying capacitor; C_5, C_6, C_7, and C_8 are stray capacitances from the signal lines to system ground, and C_4 is the shunt capacitance of the multiplexer output and amplifier input. The capacitances C_2 and C_3 are associated with the "armature" of the switch, be it electromechanical or solid state. When a common mode potential is applied to the input, C_2, C_3, C_7, and C_8 are charged to the value of this voltage. When C_1 is transferred from the input to the output of the multiplexer, charge is transferred between C_1, C_2, and C_3, and the capacitances on the multiplexer output. As a result, the charge on C_1 changes, as does the voltage across C_1. Thus the charge sharing results in an error that is partially attributable to the presence of the common mode voltage.

This qualitative explanation considers only the effects at the instant of switching and only the effects of the common mode voltage. In reality, the charge on the capacitors changes during the conversion period because of charge leakage through resistive paths as well as charging of the various capacitances by the common mode source through the impedance of the open circuit multiplexer switch. In addition, there is charge sharing between C_1 and the other capacitors whether or not a common mode potential is present. Finally, if residual charge is present on the capacitors C_4 through C_6, this affects the voltage on C_1 after the switching action.

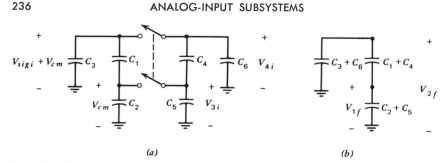

(a)   (b)

Fig. 3-59   Models for initial error calculation. (a) Before switching; (b) after switching.

The calculation of the voltages immediately after transfer is based on the principle of conservation of charge. Essentially, this principle states that the charge at a node in the network remains constant before and after the transfer of C_1. The analysis, therefore, consists of summing the charge at each node of the final network before and after the switching action. The capacitors C_7 and C_8 are not included since their charge remains constant because of the presence of the common mode source. Thus the equivalent circuits reduces to that of Fig. 3-59a and Fig. 3-59b. With an initial voltage assumed for each of the capacitances in the model, the charge equality equations for the two nodes of the final network are

$$(V_{sig} + V_{cm})C_3 + V_{sig}C_1 + (v_{4i} - v_{3i})C_4 + v_{4i}C_6$$
$$= (C_3 + C_6)v_{2f} + (C_1 + C_4)(v_{2f} - v_{1f}) \quad (3\text{-}38)$$

$$C_2 V_{cm} - C_1 V_{sigi} + C_5 v_{3i} - C_4(v_{4i} - v_{3i})$$
$$= (C_2 + C_5)v_{1f} - (C_1 + C_4)(v_{2f} - v_{1f})$$

The voltages of interest are v_{1f}, which is the common mode potential applied to subsequent equipment, and the difference $v_{2f} - v_{1f}$, which is the signal voltage after the transfer.

$$v_{1f} = \frac{\{(C_3 C_4 - C_1 C_6)(V_{sigi} - v_{4i} + v_{3i}) + [(C_1 + C_4)(C_3 + C_2) + C_2(C_3 + C_6)]V_{cm} + [(C_1 + C_4)(C_5 + C_6) + C_5(C_6 + C_3)]v_{3i}\}}{\text{DEN}}$$

$$(3\text{-}39)$$

$$v_{2f} - v_{1f} =$$
$$\frac{\{[C_1(C_2 + C_3 + C_5 + C_6) + C_3(C_2 + C_5)]V_{sigi} + [C_4(C_2 + C_3 + C_5 + C_6) + C_6(C_2 + C_5)](v_{4i} - v_{3i}) + (C_3 C_5 - C_2 C_6)(V_{cm} - v_{3i})\}}{\text{DEN}}$$

$$(3\text{-}40)$$

where

$$\text{DEN} = (C_1 + C_4)(C_2 + C_3 + C_5 + C_6) + (C_3 + C_6)(C_2 + C_5)$$

The first term of Eq. (3-40) is dependent on the signal voltage V_{sigi} and the coefficient is clearly less than unity. This shows that there is an error due to charge sharing between C_1 and the other capacitances even if the common mode and residual voltages are zero. If the capacitances are constant, this error is a proportionality factor and can be compensated by a change in the gain of the conversion equipment. However, the stray capacitances vary from channel to channel so the compensation cannot be entirely accurate. As shown later, C_1 is normally several orders of magnitude greater than the other capacitors. In this case, the signal attenuation ratio is approximately equal to $C_1/(C_1 + C_4)$. Thus C_4 should be minimized with respect to C_1 if excessive error is not to be introduced.

The second term of Eq. (3-40) depends on the residual voltage on C_4 at the time of transfer. The coefficient of this term is identical to that of the previous term except that C_4 and C_6 are substituted for C_1 and C_3, respectively. Since C_4 is generally orders of magnitude less than C_1, this error is small unless $(v_{4i} - v_{3i})$ is much larger than V_{sigi}. Since one source of the charge on C_4 is the previous signal which was sampled, the coefficient is a measure of the signal crosstalk between channels. This is of particular importance if the previous channel was a high-level signal and the subsequent channel is a low-level signal.

The third term in Eq. (3-40) is dependent on the common mode potential and the residual common mode voltage on C_5. The voltage on C_5 may be due to the common mode applied to the previous channel.

The form of Eq. (3-40) is such that the definition of CMRR is difficult to apply because the error is not solely due to V_{cm}. In order to conform to the definition, it is necessary to assume that V_{sigi}, v_{4i}, and v_{3i} are all zero. These conditions should be guaranteed when testing a system if a true measure of the CMRR is to be obtained. Under this assumption, the CMRR of the system is the reciprocal of the coefficient of the third term of Eq. (3-40). Since C_2–C_6 are in the picofarad range in a practical system, examination of the coefficient shows that a high CMRR requires that C_1 be orders of magnitude greater than the other capacitances. This is the usual case in practical systems and the CMRR is approximated by

$$\text{CMMR} \cong \left| \frac{C_1(C_2 + C_3 + C_5 + C_6)}{C_3C_5 - C_2C_6} \right| \tag{3-41}$$

The difference term in the denominator is the effect of the differential output of the multiplexer and it clearly shows an advantage in using a differential amplifier. However, the introductory paragraph indicated that the flying capacitor multiplexer has the advantage that a single-ended conversion device can be used. If this is done, C_5 is short circuited. This is equivalent to allowing C_5 to approach infinity in the above expressions. Thus in the single-ended case, Eq. (3-41) reduces to

$$\text{CMMR} \cong \frac{C_1}{C_3} \tag{3-42}$$

Although this is normally less than the CMRR obtained in the differential case, a CMRR of 10^6 can be obtained readily if C_1 is on the order of microfarads and C_3 is on the order of picofarads. It can be shown that although the differential scheme provides a higher DC CMRR, AC CMRR may be enhanced when the single-ended configuration is used.

Returning to the differential case, the value of v_{1f} given in Eq. (3-39) is the initial voltage applied to the conversion equipment which follows the multiplexer. Because of the relative sizes of the various capacitances, v_{1f} is typically quite small, but it is important to recognize that a common mode voltage is applied to subsequent equipment.

Since the period allowed for switch settling and conversion may not be negligible with respect to the time constants in the system, changes in the voltages calculated above must be considered. The analysis of these effects is based on the model of Fig. 3-60. Each of the three capacitances

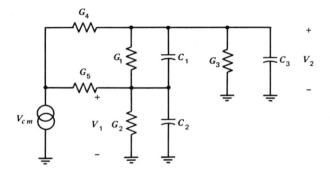

Fig. 3-60   Equivalent circuit during conversion time.

corresponds to the parallel combination of two capacitances in Fig. 3-59b. The voltages across C_2 and C_3 are initially equal to v_{1f} and v_{2f} as defined in the previous calculations. The conductances G_1, G_2, and G_3 are the resistive leakage paths associated with the capacitors, and G_4 and G_5 are the open circuit conductances of the input side of the DPDT multiplexer switches.

Two phenomena occur simultaneously after the initial switching action: The capacitors tend to discharge because of the capacitor conductances and, simultaneously, the capacitors are charged by the common mode source through the conductances G_4 and G_5. The analysis of the model is straightforward using the Laplace transform and, since the network contains only resistors and capacitors, the response is in the form of exponential functions. The expressions are simplified by recognizing that the time constants of the transients are very long compared to the period of interest even for a slow electromechanical multiplexer. Therefore, the exponential factors are approximated by the linear approximation $e^{-x} = 1 - x$. Solving for the signal voltage $V_2 - V_1$ and applying the approximation results in the expression

$$V_2 - V_1 \cong \frac{V_{cm}(G_4C_2 - G_5C_3)t}{C_2C_3 + C_1C_2 + C_1C_3}$$
$$+ v_{2f}\left\{1 - \left(\frac{G_1C_3 + C_2(G_1 + G_3 + G_4)}{C_2C_3 + C_1C_2 + C_1C_3}\right)t\right\}$$
$$- v_{1f}\left\{1 - \left(\frac{G_1C_2 + C_3(G_1 + G_2 + G_5)}{C_2C_3 + C_1C_2 + C_1C_3}\right)t\right\} \quad (3\text{-}43)$$

Because of the linear approximation, the response is a series of ramp functions. The first represents the charging action of the common mode voltage and the last two show the discharge of the initial voltages on the capacitors through the resistance network. The important point is that the signal voltage $V_2 - V_1$ can increase or decrease, depending on the magnitude of V_{cm} and the various components. The single-ended expression, derived by allowing C_2 to approach infinity, shows the same ramp action.

In both cases, the signal changes even if the common mode voltage V_{cm} is reduced to zero. This is the effect of the capacitor discharge through various leakage paths. In a practical system, the dominant effect is the discharge of C_1 through G_1, the system input conductance. Considering only this effect results in the equation

$$V_2 - V_1 \cong v_{2f}\left(1 - \frac{G_1}{C_1}t\right) \quad (3\text{-}44)$$

where v_{2f} now represents the initial signal voltage. Given G_1, the conversion period, and the allowable change in the signal during conversion, Eq. (3-44) provides a minimum value for C_1. In most systems, however, this bound is less than that established by Eq. (3-41) and Eq. (3-42) based on the CMRR requirement.

The results of the above analyses depend on the initial values of voltages across the capacitances associated with the multiplexer output bus. The initial voltage depends on whether the channel is sampled immediately following the sampling of another channel or if it is sampled after a long period of multiplexer inactivity. In the case of immediate sampling, the initial voltages on capacitors C_4, C_5, and C_6 of Fig. 3-59 are approximately the same as their value at the end of the previous sampling period. These can be estimated from Eq. (3-43) and a corresponding equation for the common mode voltage on C_2 in Fig. 3-60.

If the multiplexer has been at rest for a long period of time, voltages on C_4, C_5, and C_6 in Fig. 3-59 are determined by resistive and capacitive voltage dividers formed by the various leakage resistances and stray capacitances of the switch and multiplexer output bus. The magnitude of the initial voltages can be estimated from a model similar to Fig. 3-60.

An additional AC common mode error is a feedthrough effect during the conversion interval. The qualitative existence of this phenomenon is seen by replacing the open circuit switch conductances G_4 and G_5 in Fig. 3-60 with the interelectrode capacitance of the open switch. The resulting capacitive voltage dividers cause a small AC signal across the input to the amplifier. The magnitude of this signal is directly related to the value of the switch capacitance, and this emphasizes the necessity for careful switch selection.

This discussion demonstrates that the error mechanisms in the flying capacitor multiplexer are considerably more complex than was the case for previous multiplexers. Despite the relatively straightforward analysis that is possible using the models, the practical situation is vastly more complicated and the models serve only as a guide in design. It is practically impossible to determine the values of the components used in the models with any degree of precision. Furthermore, the models were developed for only a single channel whereas a complete analysis must account for the parallel effects of many channels. The models and equations can be used, however, for estimating system performance and, perhaps more important, they provide qualitative understanding of the source of errors in this multiplexer configuration.

This discusson has been concerned mainly with common mode and other error phenomena that are unique to the flying capacitor multiplexer. However, many of the error sources considered in the discussion of the DPST multiplexer also apply. For example, the precautions required in designing a multiplexer for low-level signals are basically the same. Similarly, switch transients and leakage currents affect the performance in similar fashion. A major difference is that leakage currents do not flow through the source impedance of the ON channel because of the open switches on the input side of the DPDT switch. Rather, the leakage current acts as a current source that charges the flying capacitor. Transients due to coupling of the switch drive voltage into the signal path result in charge being added to that already existing on the flying capacitor. This may be more significant than in the DPST multiplexer because of the long time constant of the flying capacitor. On the other hand, the size of the flying capacitor usually means that the voltage change due to a given transient is less than in the case of the DPST multiplexer.

Since the flying capacitor is normally in the range of microfarads to hundreds of microfarads, the considerations on capacitor selection discussed in the section on signal conditioning apply. An additional precaution must be observed: If a strict break-before-make sequence of switching is not obtained, the common mode potential may appear across the capacitor or the amplifier input. The problem can be avoided by appropriate specifications on the switch, by sequencing the break-and-make drive signals when separate switches are used, or by delaying the action of the block switch by an amount sufficient to guarantee that both contacts have transferred.

The outstanding characteristics of the flying capacitor multiplexing technique is that the system is essentially disconnected from the source during conversion time. Thus the amplifier and ADC common mode requirements are less. In fact, the amplifier may be single-ended and, therefore, of simpler design. A consequence of being disconnected from the source during conversion time is that noise introduced into the signal lines during conversion is not coupled into the conversion equipment. The major disadvantage of most practical flying capacitor designs is that the large value of capacitance usually required limits the bandwidth of the system. This is not, however, a severe limitation in many process applications.

3.13.3.4 Differential—Transformer Coupled. Another useful technique for isolating the conversion equipment from the common mode source is transformer coupling of the signal from the multiplexer input to the conversion equipment. There are many variations of the basic technique, but Fig. 3-61 shows a typical configuration that illustrates the main principles of operation. Each input signal is connected to the primary of a transformer through switches represented by S_1 and S_2. The second-

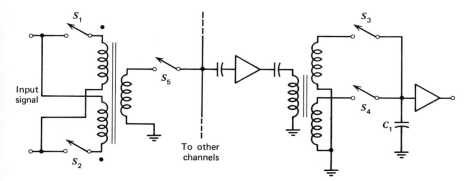

Fig. 3-61 Transformer-coupled multiplexer.

ary of each transformer is connected through a series switch to an AC or DC time-shared amplifier. The amplifier output is transformer coupled through switches S_3 and S_4 to a time-shared DC amplifier.

In operation switch S_5 is closed when a channel is addressed. Switches S_1 and S_3 are then closed, essentially simultaneously, for a short period of time. Assuming a positive signal, the closure of S_1 produces a positive pulse at the transformer output with a duration determined by the switch closure time, and an amplitude determined by the value of the input and the transformer turns ratio. The pulse is amplified by the amplifier and is applied to capacitor C_1 through switch S_3. Switch S_2 is then closed for an equal period of time and the negative pulse produced at the transformer secondary is amplified and applied to C_1 through switch S_4 whose operation is synchronized with that of S_2. In the configuration shown, this sequence of events is repeated many times during the system conversion time. Thus switches S_1 and S_2 function as a square wave modulator or chopper and S_3 and S_4 serve as a synchronous demodulator.

Another transformer-coupled multiplexer is essentially identical to the configuration shown except that the demodulator and C_1 are not used. The switches S_1 and S_2 operate in sequence to produce a single doublet pulse which is amplified by the amplifier. The conversion time of the ADC is less than the closure time of the switches, so the conversion is accomplished while the amplitude of the pulse is essentially constant, as shown in Fig. 3-62. An alternative is to use a sample-and-hold ampli-

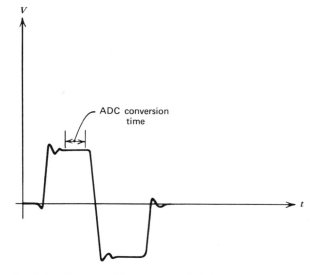

Fig. 3-62   Single doublet conversion technique.

fier to reduce ADC speed requirements and to minimize the effect of transformer droop. Even though this mode of operation requires only the positive pulse, the use of the doublet resets the core material. If this is not done, the core tends to saturate with resultant nonlinearities and a decrease in system input impedance.

The switch S_5 is similar to the block switch except that a switch is provided for each input. The isolation provided by the switch reduces the capacitance of the amplifier bus. It also reduces interaction between channels due to common mode voltages or ringing induced by the sampling pulse.

The design considerations in a transformer-coupled multiplexer include those discussed for the other multiplexers. Low-level error sources such as thermoelectric potentials and offset voltages affect the operation of the modulator and demodulator. Leakage currents flowing through the source and load impedances introduce similar errors. These factors, as well as those that are unique to the transformer-coupled multiplexer, must be considered in a thorough design.

A unique problem is that, since it is not possible to wind a transformer with a turns ratio tolerance sufficient to meet stringent gain accuracy requirements, a gain adjustment must be provided on each channel. This might simply be a trimming potentiometer on the transformer secondary, but it represents a per-channel adjustment for the manufacturer and for servicing personnel.

The transformer specifications are probably the most critical parameters in the design of this multiplexer. The important considerations can be developed from the equivalent circuit of Fig. 3-63. The resistance R_1 represents the source and primary resistance and C_1 is the primary winding capacitance. Similarly, R_2 and C_2 are the load resistance and capacitance referred to the primary of the transformer. The inductance L_1 is the open circuit, or self, inductance, and L_2 represents the leakage inductance.

The exact transient analysis is unduly complex for our purposes and a

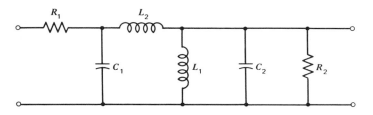

Fig. 3-63 Transformer equivalent circuit.

more heuristic approach provides the required information. When the multiplexer switches close, a step funtcion is applied to the primary of the transformer. Since the risetime of the output waveform should be small, it is reasonable to assume that the self inductance is infinite during the initial rise. In addition, source loading errors dictate that the load resistance R_2 be very large so that its effect can be neglected in an approximate analysis. It is also clear that the leakage inductance L_2 must be minimized if a fast risetime is to be achieved. With these assumptions, the equivalent circuit is reduced to a simple low-pass RC network with a time constant equal to $R_1(C_1 + C_2)$. Therefore, this product must be minimized in addition to minimizing L_2. Since R_1 includes the signal source impedance, the transformer risetime is a factor in establishing a maximum allowable source impedance for a given system speed.

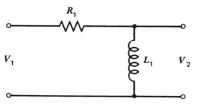

Fig. 3-64   Equivalent circuit at top of pulse.

Once the top of the pulse has been reached, the output voltage is essentially constant and the capacitances can be neglected. As before, R_2 can be neglected and, under the assumption that L_2 has been minimized, the equivalent circuit reduces to that of Fig. 3-64. The response is an exponential decay.

$$V_2 = V_1 e^{-R_1 t/L_1} \tag{3-45}$$

where V_2 is the secondary voltage and V_1 is essentially equal to the signal voltage. Since conversion takes place at the top of the pulse when the doublet technique is used, the variation of V_2 is to be minimized. This requires that R_1/L_1 be made as small as possible. As before, minimizing R_1 is beneficial but not always possible since it includes the source resistance. Using the maximum R_1 and knowing the allowable variation in V_2, the minimum value of L_1 can be determined from a linear approximation of Eq. (3-45).

$$V_2 \cong V_1 \left(1 - \frac{R_1}{L_1} t\right) \tag{3-46}$$

The self inductance and winding capacitances are also important in considering the system input impedance seen by the signal source. The linearized approximation of the current flow in the equivalent circuit of Fig. 3-64 during the conversion time is

$$I \cong \frac{V_1 t}{L_1} \tag{3-47}$$

Since the voltage at the transformer input is constant, the increasing current results in a decreasing input impedance during the conversion time. The average input impedance over a conversion interval of T seconds is

$$Z_{in(avg)} \cong \frac{2L_1}{T} \tag{3-48}$$

Therefore, maximizing the self inductance and minimizing the sample time result in the highest system input impedance.

The effect of the capacitances were neglected in this analysis but can be easily included in an approximate analysis. When the multiplexer switch closes, a charging current flows that has an initial value of V_1/R_1 and an exponential time constant approximately equal to $R_1(C_1 + C_2)$ if the leakage inductance is neglected. This response, the inductive current of Eq. (3-47), and the total current are shown in Fig. 3-65. It is clear that the average input impedance is decreased by the capacitive current. This emphasizes the importance of specifying low winding capacitances and the use of the series switch on the transformer secondary.

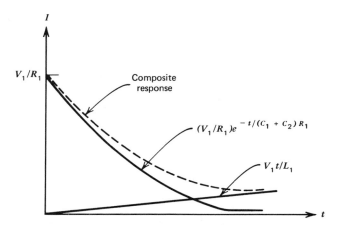

Fig. 3-65 Input current flow.

The above discussion is also applicable to the continuous or carrier method. When the chopped signal is demodulated, its average value depends on the shape of the output pulses. For example, excessive droop or a slow risetime decreases the average value of the demodulator output. If these effects are strictly proportional to the input signal, the degradation is merely a scale factor and can be compensated by a change in the channel gain. However, many of the effects are not linear and can result in a nonlinear system characteristic.

In carrier systems, the turns ratio matching between the two primary windings or the two halves of a center tapped primary is an important parameter. A mismatch results in a nonsymmetrical output wave which causes a ripple frequency in the demodulator output. Failure to attenuate this ripple in the demodulator filter results in a degradation of system repeatability.

System repeatability is also degraded by noise pickup in the input transformer. Since the transformer contains many turns of wire, it is quite susceptible to time-varying magnetic fields. Locating the transformer as far as possible from sources of magnetic fields such as power transformers and enclosing both the source and the input transformer in magnetic shielding enclosures are effective measures.

Another source of normal mode noise is a change in transformer parameters due to physical stress or movement. For example, the permeability of the transformer core changes under applied stress and, if a magnetic field is present in the core, a voltage is induced in the secondary winding. Similarly, a change in effective leakage inductance or core cross section as a result of movement of core laminations produces a voltage. Reducing external fields, specifying that the core be demagnetized after assembly and testing, and construction techniques that prevent movements within the transformer minimize the generation of these voltages. In addition, the transformer and vibration-inducing components should be physically separated and acoustically insulated from each other.

The DC common mode performance of transformer-coupled multiplexers is usually very good. This is due to the excellent DC isolation provided by a well-designed transformer. The common to normal mode conversion mechanism is predominantly a voltage divider action between the primary and secondary resistance and the input resistance of subsequent equipment.

The sources of AC common to normal mode conversion include the current in the primary due to the common mode source and the current in the load due to the primary to secondary capacitance. The common mode analysis depends on the state of the switches and the length of the sampling interval compared with the period of the common mode voltage.

Although all possible cases cannot be considered in this dicsussion, a case of interest is the steady-state crosstalk analysis in which the switches are open for an interval that is long compared with the period of the common mode voltage. An equivalent circuit is shown in Fig. 3-66 where the R_s's and C_s's are associated with the switches, the L's and R_w's

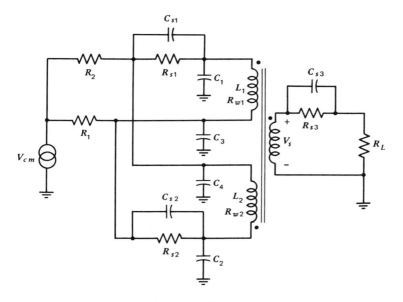

Fig. 3-66 Common mode equivalent circuit.

are the self inductances and winding resistances of the primary winding, C_1–C_4 are stray capacitances, and R_1 and R_2 are line and source resistances. Although reasonably complete except for interwinding capacitance, this equivalent circuit is so complex that the analysis is difficult to interpret. The basic principles are adequately demonstrated using the simplified circuit of Fig. 3-67. The simplification is justified by noting that the R_s's and C_s's present a high impedance in the common mode path compared with the path through the primary inductance. The capacitances C_3 and C_4 are neglected on the basis that the voltage drop across the source resistances is very small and, therefore, the voltages across C_3 and C_4 are approximately equal to V_{cm}.

High CMRR requires that V_s be minimized and this, in turn, requires that V_{p1} and V_{p2} be minimized. These primary voltages are determined by the relative reactances of the resistance, inductance, and capacitance

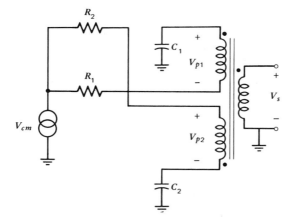

Fig. 3-67   Simplified common mode equivalent circuit.

of each primary path. Therefore, reducing the inductance and capacitance increases the CMRR. However, since previous discussion indicated that L_1 should be as large as possible, the stray capacitances C_1 and C_2 must be minimized.

This analysis neglects the effect of primary to secondary coupling capacity. Neglecting the primary current flow considered above, the equivalent circuit of Fig. 3-68 applies where L_3 is the secondary self inductance. The voltage V_o is directly proportional to the capacitance value and, therefore, C_5 and C_6 should be as small as possible. This is accomplished by the physical construction of the transformer and by the guard shielding techniques discussed in Chap. 4.

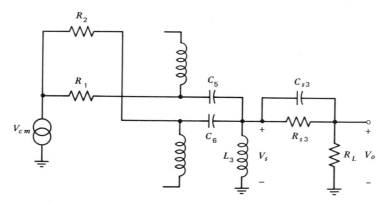

Fig. 3-68   Effect of interwinding capacitance.

The total secondary voltage is the result of both the capacitive current and the primary current discussed earlier. If the secondary switch S_3 is not used, V_s is applied to the amplifier input. If S_3 is included, only a small fraction of V_s appears across R_L when the switch OFF reactance is much larger than R_L. This emphasizes the importance of S_3 in large systems and the necessity for selecting a switch with high OFF reactance.

The above discussion concerns common mode crosstalk; a common mode effect also occurs as a result of applied common mode voltage on the sampled channel. The common to normal mode conversion phenomenon is due to the transient produced when the multiplexer switches close. In the doublet conversion method, the analysis need only consider the closing of one switch; in the carrier method, however, the repetitive action of both switches must be considered. The exact analysis of these effects is complicated by the complexity of the transformer transient response.

The equivalent circuit of Fig. 3-69 provides a qualitative example of the conversion mechanism and its complexity. Prior to closing S_1 and S_3, the common mode error at the transformer secondary is due to the steady-state phenomena discussed previously. When the switches close, the primary voltage across L_1 decreases with a resultant decrease in the induced secondary voltage. However, the switch closure results in a tran-

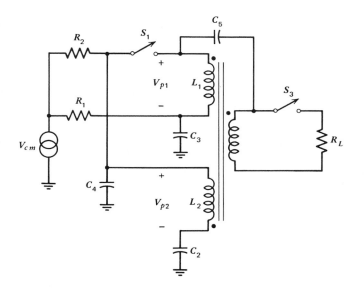

Fig. 3-69 Transient common mode equivalent circuit.

sient current through L_1 and the interwinding capacitance with a resultant voltage across R_L. The duration of the transient error depends on the characteristics of the transformer and the magnitude of R_L.

The common mode error due to steady-state current flow through the other primary winding is still present while S_1 is closed. Its effect on the system is increased, however, because the closure of S_3 eliminates the voltage divider effect and the total secondary voltage is applied to the measurement system.

3.14   AMPLIFIERS

The type of amplifiers employed in an analog subsystem depends on both the performance characteristics and the configuration of the subsystem. Most subsystems use at least one "buffer" amplifier to provide isolation between parts of the subsystem. For example, a buffer is usually used at the input of the ADC to provide a low source impedance for the comparator and a high load impedance for the multiplexer. In low-level subsystems, a high-gain amplifier is required to raise the signal value to a level compatible with the ADC range.

The amplifier configuration depends, in part, on the multiplexer configuration. If the multiplexer does not eliminate the common mode voltage, a differential amplifier is used to provide this important system characteristic. On the other hand, some multiplexers, such as the flying capacitor configuration, do provide isolation from the common mode source and a single-ended amplifier can be used. In general, a multiplexer that provides common mode isolation allows the use of a simpler amplifier at the expense of a more complex multiplexer.

Despite the wide variety of possible amplifier designs from which the engineer must choose, he has the assistance of copious design information in texts and the literature. Almost every advanced textbook on transistor electronics includes a chapter on the design of DC amplifiers; few discuss the design of multiplexers and ADCs. To balance the abundant supply of amplifier design information, however, is the fact that the design of a high-gain amplifier is probably the most difficult design task encountered in an analog subsystem.

Because of the many possible designs and the availability of many excellent articles and books, this discussion is limited to consideration of the major amplifier requirements, a few important design criteria, and a sample of possible amplifier configurations that may be used in analog subsystems.

The relative importance of a particular amplifier specification depends on the amplifier configuration and the manner in which the amplifier is used in the analog subsystem. Parameters that are generally important in most systems are gain, noise, bandwidth, input and output impedance, common mode rejection, and the variation of gain and zero offset with time and temperature. Other factors important in some systems are power supply sensitivity and overload recovery.

3.14.1 Gain and Sensitivity

Closed-loop gain requirements are easily defined since the desired input ranges are known, as is the full-scale input of the ADC. Most analog subsystems use ADCs with full-scale input ranges of either 5 or 10 V. System input ranges are not standardized but generally range from about 10 mV on the most sensitive range to 500 mV or 1 V in a low-level system, and up to 5 or 10 V in a high-level system. Thus the closed-loop gain requirements are from 500 to 1000 down to unity.

The closed-loop gain is established by the ratio of two resistors in the feedback loop of the amplifier. If a gain adjustment is provided in the amplifier, the ratio precision of the resistors need not be very great. However, variation of the ratio with temperature and time affects the long-term stability and the temperature characteristics of the amplifier. Therefore, the gain-determining resistors are usually specified as matched pairs or assemblies with a specified stability and temperature coefficient.

The relationship between the resistor ratio and the open- and closed-loop gains is given by the classical feedback amplifier equation.

$$G = \frac{A}{1 + A\beta} \qquad (3\text{-}49)$$

where G is the closed-loop gain, A is the open-loop gain, and β is a ratio determined by the gain resistors. For large open-loop gain $A \gg 1$, the closed-loop gain is approximately equal to the reciprocal of β. Gain variation with time and temperature are dependent on the characteristics of A as well as on those of β. A quantitative measure of the effect of changes of A on the closed-loop gain is obtained by differentiating Eq. (3-49).

$$\frac{dG}{dA} = \frac{1}{(1 + A\beta)^2} \qquad (3\text{-}50)$$

Or, in terms of a sensitivity factor dG/G,

$$\frac{dG}{G} = \frac{1}{1 + A\beta} \frac{dA}{A} \qquad (3\text{-}51)$$

Therefore the fractional change in G is related to the fractional change in A by the feedback factor $(1 + \beta A)$ which, for most practical cases, is equal to the loop gain $A\beta$.

In a worst-case design approach to gain stability, A is specified such that, if A varies from a minimum design value to infinity, the closed-loop gain varies less than a specified amount. Although conservative, this approach is not realistic and results in an open-loop gain that is much higher than needed in most cases. In addition, the excess open-loop gain increases the problem of stability compensation. A more realistic approach is to recognize that the amplifier is calibrated periodically and then to require that the gain variation over the calibration interval be less than a specified value. For example, an analysis of the amplifier may show that a 20% open-loop gain variation may occur over a 6-month calibration interval; if the closed-loop gain is 100 and a gain variation of less than 0.01% is specified, Eq. (3-51) requires an A of 200,000 as compared with 1,000,000 in a worst-case design.

An additional factor that must be considered in the design of the gain characteristics of the amplifier is the required gain adjustment. The maximum adjustment is determined by considering the worst-case tolerances on the gain network in conjunction with either the minimum or maximum open-loop gain, whichever results in a greater number. If the adjustment is small compared with the gain resistor value, the temperature coefficient due to the adjustment potentiometer is negligible.

The resolution of the adjustment is also of concern since it contributes to the calibration accuracy of the amplifier. The potentiometer should be selected for adequate setting stability to minimize the chance of discrete gain changes resulting from vibration or other mechanical disturbances.

3.14.2   Bandwidth and Settling Time

The required bandwidth of the amplifier is determined by the subsystem configuration with time-shared amplifiers representing the most stringent case. In the time-shared system, the bandwidth is dictated by the sampling rate of the system rather than by the bandwidth of the input signal. For an accurate conversion the output of the amplifier must settle to within the quantizing interval of the ADC before the conversion is initiated. If an integrating ADC is used, this requirement is relaxed because error due to insufficient settling is averaged over the conversion interval.

Although the analytical relationship between the amplifier bandwidth and the settling time is extremely complex for a practical amplifier, the transient response generally is dominated by one or two poles in the transfer function. If a single pole dominates, the response to a step function is a simple exponential function.

$$V_o = V_{in}(1 - e^{-\omega_o t}) \qquad (3\text{-}52)$$

where ω_o is the —3-dB bandwidth. If the settling time T is defined as the time required for the amplifier to reach $(100 - r)\%$ of the theoretical final value, the relationship between T and ω_o is

$$\omega_o T = \ln \frac{r}{100} \tag{3-53}$$

This relationship is tabulated in Table 3-4 for settling margins r corresponding to the quantizing intervals for 8-, 10-, 12-, and 14-bit systems.

Table 3-4   Bandwidth–Settling Time

Bits	Settling Margin (r)	$\omega_o T$
8	3.9×10^{-3}	5.54
10	9.8×10^{-4}	6.93
12	2.4×10^{-4}	8.32
14	6.1×10^{-5}	9.70

When the transient response of the amplifier is dominated by a pair of poles, the transfer function is approximated by

$$F = \frac{\omega_n{}^2}{s^2 + 2\zeta\omega_n s + \omega_n{}^2} \tag{3-54}$$

where ω_n is a frequency parameter and ζ is the damping factor for the amplifier. The 3-dB bandwidth ω_o is related to ω_n by the expression

$$\omega_o{}^2 = \omega_n{}^2\{1 - 2\zeta^2 + [(2\zeta^2 - 1)^2 + 1]^{1/2}\} \tag{3-55}$$

The damping factor ζ is experimentally determined by observing the percentage overshoot of the amplifier step response. The definition of overshoot and its relationship to ζ is shown in Fig. 3-70.

The response of a linear second-order system to a step input is

$$\frac{V_o}{V_{in}} = 1 - \frac{e^{-\zeta\omega_n t}}{\sqrt{1 - \zeta^2}} \sin\left(\omega_n \sqrt{1 - \zeta^2}\, t + \cos^{-1}\zeta\right) \tag{3-56}$$

This response is bounded by the exponential factor which, in turn, determines an upper bound on $\omega_n T$ for a given margin r.

The value of $\omega_n T$ is converted to a more useful parameter $\omega_o T$ by use of Eq. (3-55). The result of these calculations is plotted in Fig. 3-71 with ζ as a parameter.

The bounds on the settling time represented by the data of Table 3-4 and Fig. 3-71 are only estimates of actual amplifier performance because of the assumptions which were made. An actual amplifier is only approximated by the first- or second-order function and, in addition, the assump-

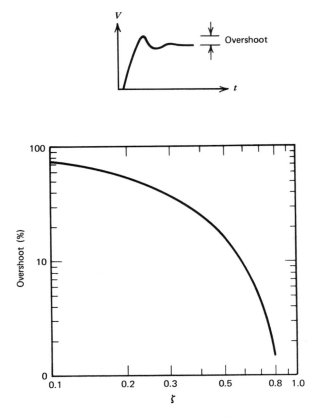

Fig. 3-70   Overshoot vs. dampening factor (ζ).

tion of linearity is often invalid. Many amplifiers have a nonlinear response to a step input, the largest deviation from linearity occurring for large input signals. This nonlinearity results from slew rate limiting of the output signal; that is, the output does not increase any faster than a fixed number of volts per unit time. Slew rate limiting is usually due to limited current drive capability in the amplifier output stage. Its effect increases the settling time of the amplifier for large signal excursions.

3.14.3   Temperature Coefficient and Drift

The effect of temperature and time on the amplifier characteristics is an important consideration in high-accuracy systems. One temperature effect is the variation of gain with temperature. Variation in h_{FE} with

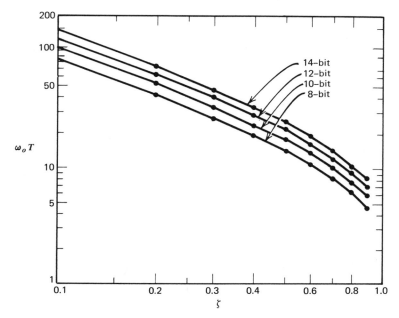

Fig. 3-71 Bandwidth—settling time vs. dampening factor (ζ).

temperature result in changes in the gain of each amplifier stage and, in turn, variations in the overall open-loop gain. For example, a variation of 10% in each of two cascaded stages results in an open-loop change of greater than 20%. The variation in the closed-loop gain, however, is considerably less because of the feedback action described by Eq. (3-49). The effect of temperature on open-loop gain must be a consideration when utilizing the gain sensitivity criteria of Eq. (3-51).

The second source of temperature-dependent gain variation is the characteristic of the gain-determining resistors. With sufficiently high open-loop gain, Eq. (3-49) shows that the closed-loop gain is primarily determined by the ratio of the feedback resistors. If this ratio varies with temperature, the amplifier exhibits a temperature-dependent gain. This is not a serious problem, however, since precision resistor assemblies are available that have ratio temperature coefficients of less than 10 ppm/°C.

A more serious temperature dependency is associated with the zero offset of the amplifier. Zero offset is the voltage appearing at the amplifier output when the input is zero. The zero offset is usually specified as an equivalent input voltage to make the specification independent of the amplifier gain. For example, a specification of 1 μV/°C RTI (referred to the input) results in an output zero offset of 0.5 mV/°C in an amplifier with a gain of 500.

In the high closed-loop gain amplifier typically used in low-level systems, the two primary sources of zero offset are the input currents of the first stage and the difference between the V_{BE}'s of the two input transistors. A secondary source is the base current of the second stage flowing through the collector resistors of the first stage. These sources are illustrated in Fig. 3-72, which shows the first stage of a differential amplifier with subsequent stages represented as a lumped amplifier.

The secondary temperature effect of the second-stage input current results in changes in the first-stage collector voltage. Assume, for example, that the amplifier is initially adjusted for zero output voltage. When the second-stage input current I_{in1} changes as a result of tempera-

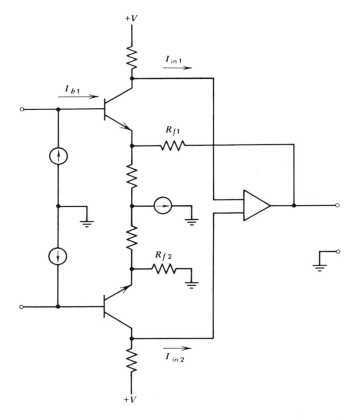

Fig. 3-72   Differential amplifier configuration.

ture, the first-stage collector current must change since the emitter current is fixed by the current source. The change in collector current causes a change in collector voltage. Coupled with the change in I_{in1}, however, is a simultaneous change in I_{in2}. If the changes in the two currents are exactly the same, the collector voltages change by the same amount and the differential input voltage to the second stage is unchanged. Therefore, only the difference $I_{in1} - I_{in2}$ results in an output offset voltage. The change in the first-stage collector voltage is dependent on the difference in the second-stage base currents and the magnitude of the collector resistors. Minimizing both the currents and the resistors results in less effect at the amplifier output. There are other considerations, such as a requirement for high first-stage gain, which often result in relatively high values of collector resistance, however, and emphasis usually is placed on the second-stage design. The temperature effect of the second-stage input current is a secondary factor in the overall amplifier performance since the voltage changes are not amplified by the first stage. As a result, its contribution to the amplifier output voltage is less than effects associated with the first stage.

The change in the first-stage input current can be a major contributor to zero offset temperature effects. As before, however, it is the difference between the two input currents that is important. The magnitude of the base current is determined by the emitter current and the leakage current through the back-biased transistor junction. The leakage current is quite small in silicon devices but changes rapidly with temperature. The magnitude of the DC current gain h_{FE} determines the emitter current-dependent portion of the base current. Since the emitter current is established by the emitter current source, variations in h_{FE} result in changes in input current. The base currents are typically established by current sources as shown in the figure so that excess base current must be either supplied or absorbed by the signal applied to the amplifier input. The unbalanced base current flowing through the source resistance results in a potential that is amplified by the closed-loop gain. The use of low emitter currents minimizes base currents and the possibility of significant differences between the two base currents. In addition, the use of transistors with matched h_{FE}'s reduces changes in the difference of the h_{FE}'s. Minimizing the source impedance also results in lower zero offset voltage due to zero offset current. This is not always practical, however, since it unduly restricts the allowable source impedance when some types of multiplexers are used.

In a silicon transistor, the voltage drop between the base and emitter is on the order of 700 mV; the temperature coefficient of this voltage is about -2 mV/$^\circ$C. A change in V_{BE} is equivalent to a change in the input signal voltage so that even a minimal change in temperature can result in a V_{BE} change which is significantly larger than some low-level signals. Again, however, the construction of the differential amplifier limits the effect of V_{BE} variations to the difference between the V_{BE}'s of the two input stage transistors. By selecting transistors with similar characteristics, both the initial V_{BE} difference and its temperature variation can be minimized. Commercial matched pairs of transistors are available with less than 5 mV initial offset and a temperature coefficient of less than 10 $\mu V/°C$. These devices usually also have h_{FE}'s matched to better than 10% to minimize unbalanced base currents. The best of these devices are monolithic structures diffused into the same chip to ensure thermal equilibrium between the transistors.

Even the low temperature coefficient of the matched pair is not sufficient for some amplifier applications. For example, a variation of 10 $\mu V/°C$ on a 10-mV full-scale range represents a 0.1%/°C error; typical subsystem requirements are for a temperature coefficient at least an order of magnitude less. This can be achieved in several ways: The temperature can be controlled, the variations can be compensated by other temperature dependent components, or a dynamic zero correction technique can be utilized. Although the second course of action has been most common, recent semiconductor developments have made the first approach practical. In one commercial device, the temperature of the matched pair transistor substrate is controlled at a temperature greater than the highest operating ambient temperature. The substrate is heated by the dissipation of other transistors diffused into the chip. The temperature is controlled by a regulator circuit which is also diffused into the chip. This technique results in a matched pair with low initial offset and an offset temperature coefficient of as low as 0.1 $\mu V/°C$ (Emmons and Spence, 1966).

The second general compensation method is to inject a temperature-dependent current or voltage into the input circuit in such a way that it counteracts the changes in the circuit parameters. For example, instead of connecting the balancing feedback resistor R_{f2} in Fig. 3-72 to ground, it can be connected to a temperature-dependent voltage generated by a diode-resistor network. If the characteristics of the diode junction approximate those of the transistor V_{BE}, the variations in V_{BE} can be reduced to less than 1 $\mu V/°C$. However, this requires a temperature coefficient adjustment for each amplifier since the polarity of the V_{BE} difference cannot be predicted. With more sophisticated compensation circuits in the emitter and collector paths, the effects of temperature can be further reduced (Hoffait and Thornton, 1964).

Another method of correcting for zero offset error in a time-shared amplifier is to measure the magnitude of the error between input samples and then to subtract the error from subsequent input signals. The basic concept is illustrated in Fig. 3-73 where the capacitor on the amplifier

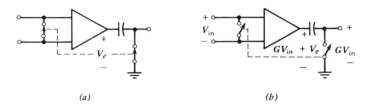

(a)   (b)

Fig. 3-73 Dynamic zero correction amplifier. (a) Correction interval; (b) data interval.

output provides storage for the correction signal. During the interval between sampling the input signal, the switches S_1 and S_2 are closed, as shown in Fig. 3-73a. Since S_1 shorts the amplifier input, the amplifier zero offset V_e is applied across the capacitor at the output. When a signal is sampled by closing the multiplexer switches, switches S_1 and S_2 are opened simultaneously. The signal at the output of the amplifier is the sum of the amplified signal GV_{in} and the zero offset voltage V_e. The signal at the input to the ADC, however, is GV_{in} since the stored voltage on the capacitor cancels the offset. In this example, the signal is AC coupled to the ADC; similar compensation techniques for direct coupled systems are possible. Using more sophisticated methods, errors due to both voltage and current offsets can be minimized. Since dynamic correction provides error cancellation based on the instantaneous value of zero offset, it is effective in reducing the temperature variation in offset as well (Ellis and Walter, 1960; Prinz, 1947).

3.14.4 Common Mode Rejection

If the multiplexer used in a differential subsystem does not isolate the amplifier from the common mode source, then the amplifier characteristics are the major factors in determining the common mode rejection of the subsystem. Even in systems in which the multiplexer provides common mode isolation, such as the flying capacitor, a differential amplifier still may be desirable if a very high CMRR is required.

Both steady-state and transient common mode situations must be considered. The transient common mode case is of particular concern in time-shared amplifiers used with direct coupled multiplexers since the common mode voltage is applied as a step function when the multiplexer switches close. An amplifier with high CMRR under steady-state con-

Fig. 3-74   Common mode transient protection.

ditions may even saturate or oscillate when subjected to a common mode step function. One technique utilized in some commercial amplifiers is illustrated in Fig. 3-74. The transformer connected in the input leads of the amplifier reduces unbalanced transient input currents due to common mode sources without affecting the normal mode operation.

Other principles involved in achieving a high CMRR are similar to those employed in the multiplexer design and are based on the concepts developed in conjunction with the common mode models of Sec. **3.9**. Specifically, the design techniques involve either passive or active balancing of an equivalent bridge circuit or increasing the effective impedance of the common mode leakage paths.

The design concepts of concern in the passive balancing method can be illustrated with the aid of Fig. 3-72. A common mode source connected to the two input terminals causes current flow in the input leads. There are two current paths in each base circuit: One path is through the base emitter junction to the amplifier ground via the emitter resistors and current source or via the feedback path; the other is through the base collector junction, the collector resistors, and the collector power supply. The current results in changes in the emitter and collector voltages of each input transistor. If the voltage changes are not the same for the differential pair, a net voltage is applied to the subsequent amplifier stages as a normal mode signal, indistinguishable from the true signal. The common to normal mode conversion, therefore, is a function of the difference between the common mode leakage paths.

The impedance of the common mode leakage paths is dependent on both the passive components and the transistor characteristics. The impedance of the path through the base collector junction depends on the value of the collector resistors and the output impedance $(1/h_{ob})$ of the transistor. For DC common mode sources, the resistive portion of $1/h_{ob}$ is orders of magnitude greater than the collector resistors and, therefore, is of primary concern. If the common mode source is AC, the important transistor parameter is C_{ob}, the base-to-collector capacitance. Since the impedance of C_{ob} at common mode frequencies of interest may be relatively low, the collector resistors are important in establishing the impedance of the common mode path.

As in the common mode models, the difference between the common mode leakage paths is of most concern. In most transistors, the magnitude of h_{ob} is such that the balance between the base collector paths is not a serious problem unless a CMRR of greater than about 100 dB is required at low frequencies. A higher CMRR requires that the h_{ob} of the two input transistors be matched or that compensating components be added to the input circuit to balance the leakage paths. Adjustment of the collector resistors is not effective since this results in a change in the amplifier zero offset.

Similar comments apply to the path through the base emitter junction. The impedance of this path is determined by the effective input impedance of the amplifier, the impedance of the current source, and the impedance of the feedback path. In Fig. 3-72, the resistance R_{f2} provides a path that balances the common mode path through R_{f1}. If a capacitance is used in parallel with R_{f1} for frequency compensation, maintenance of common mode balance requires a similar capacitance in parallel with R_{f2}.

The overall CMRR characteristic of the amplifier is the result of the composite action of the various leakage paths. Thus an unbalance in the collector circuit of the top transistor may be compensated by an unbalance in the emitter path of the bottom transistor. This allows compensation of unbalances at one or two points in the circuit rather than at the specific point where the unbalance actually occurs.

Passive balancing of the differential amplifier is the simplest method of enhancing CMRR performance. Although it can result theoretically in infinite CMRR, in practice it is effective only for CMRR requirements of less than 100 dB unless frequent adjustments are made to compensate for changes in the circuit due to time and temperature. These adjustments are tedious since they typically interact with other adjustments such as gain and zero offset. For this reason, passive balancing is combined with more sophisticated schemes when a high CMRR is desired.

One effect of negative feedback in amplifier design is an increase in the amplifier input impedance. Negative feedback of the common mode signal results in an analogous increase in the effective impedance of the common mode leakage paths. A simplified example of this technique is illustrated in Fig. 3-75 which shows the first two stages of a differential amplifier with an applied common mode source. Current sources and the closed loop signal feedback path are not included in this simplified schematic. The common mode feedback action can be qualitatively explained by considering a small perturbation of the common mode signal. If the common mode voltage increases, the voltage at the collectors of the first stage decreases as a result of the inversion of the first stage. This change in voltage causes the voltage at the summing point between the second-stage emitter resistors (1) to decrease. This decrease in voltage is inverted by the common mode amplifier with the result that

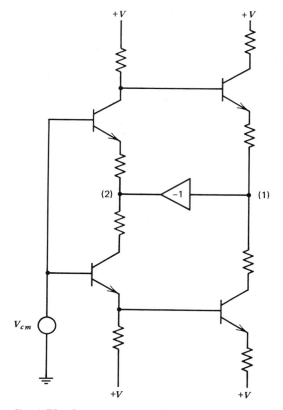

Fig. 3-75   Common mode feedback example.

the voltage at the first-stage emitter summing point (2) is increased. The increase in emitter voltage counteracts the decrease in the first-stage collector voltage and, therefore, the effect of the common mode source.

The example of the figure is only one of several ways in which common mode feedback can be utilized in the amplifier. The common mode voltage can be sensed in any of the differential stages of the amplifier, including the input bases, and the feedback signal applied to the collectors, emitters, or the emitter summing point of the first stage. Coupled with a reasonably balanced configuration, a CMRR of 120 dB is obtainable using this technique.

Another method of increasing the common mode performance of an amplifier involves a dynamic bridge of a number of separate amplifiers

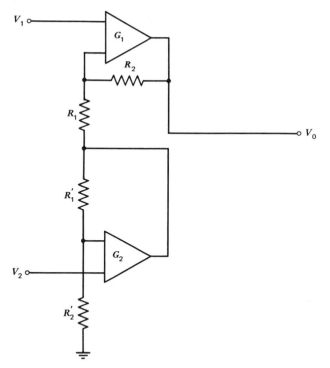

Fig. 3-76 Dynamic bridge amplifier.

(Brown, 1961; Meyer-Brötz and Kley, 1966). A configuration using two amplifiers is shown in Fig. 3-76. The amplifier output voltage is

$$V_o = \frac{(R_1 + R_2)G_1 V_1}{R_1 + R_2 + G_1 R_1} - \frac{G_1 G_2 R_2 (R_1' + R_2') V_2}{(R_1 + R_2 + G_1 R_1)(R_2' G_2 - R_1' - R_2')} \qquad (3\text{-}57)$$

With the usual assumption that the open loop gains G_1 and G_2 are very large, Eq. (3-57) reduces to

$$V_o = \frac{(R_1 + R_2) V_1}{R_1} - \frac{R_2}{R_2'} \frac{(R_1' + R_2') V_2}{R_1} \qquad (3\text{-}58)$$

The voltage V_2 is the applied common mode voltage V_{cm} and V_1 is the sum of the signal voltage V_{sig} and the common mode voltage. With these substitutions, Eq. (3-58) becomes

$$V_o = \frac{(R_1 + R_2) V_{sig}}{R_1} + \left(1 - \frac{R_2 R_1'}{R_1 R_2'}\right) V_{cm} \qquad (3\text{-}59)$$

The coefficient of V_{sig} is the closed loop gain of the overall amplifier; that is,

$$G = \frac{R_1 + R_2}{R_1} \tag{3-60}$$

The second term of Eq. (3-59) is the common mode error at the output of the amplifier. Its value is independent of gain and depends only on the match between the resistors in the network. For an infinite CMRR, the second term must be zero. This requires that

$$\frac{R_1'}{R_1} = \frac{R_2'}{R_2} \tag{3-61}$$

which is the condition for a balanced bridge.

The gain independence of the output error due to the common mode voltage results in a CMRR that is dependent on the gain. This is because the definition of CMRR is in terms of the common mode error referred to the amplifier input. Equation (3-62) for the CMRR shows the dependence on the amplifier gain.

$$\mathrm{CMRR} = \left| \frac{G}{1 - (R_2 R_1'/R_1 R_2')} \right| = \left| \frac{R_1 R_2' G}{R_1 R_2' - R_2 R_1'} \right| \tag{3-62}$$

The gain dependence of the CMRR in the dynamic bridge amplifier is unique since most amplifiers have a CMRR that is independent of gain.

The intrinsic isolation characteristics of the transformer exploited in the transfermer-coupled multiplexer are also applicable to amplifier designs. An example of this type of amplifier, often referred to as a floating amplifier, is shown in Fig. 3-77. The input signal is chopped or

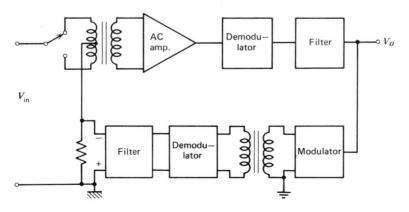

Fig. 3-77 Floating amplifier configuration.

modulated at the amplifier input and the resulting AC signal is amplified by an AC amplifier, demodulated, and filtered to provide a DC output. Overall feedback is provided by modulating the output signal, demodulating it after passing it through an isolation transformer, and subtracting it from the input signal.

The configuration shown is only one of several possible configurations for the floating amplifier. A common variation is to provide partial amplification of the signal before modulation. This has the advantage of relaxing the low-level requirements on the transformer and input modulator but, in return, it requires a floating power supply to power the preamplifier.

The sources of common to normal mode conversion are similar to those in the transformer-coupled multiplexer. They include the resistive leakage through the transformer and through the floating power supply in the case of the preamplifier approach. The AC common mode rejection is typically limited by the interwinding capacitance of the transformer. Because of the strong dependence on this capacitance, guard-shielding techniques are normally required in floating amplifiers when a CMRR of 120 dB is required.

Other error sources are also similar to the transformer-coupled multiplexer. These include transient feedthrough of spikes produced by the modulation-demodulation process, susceptibility to high-frequency common mode voltages, offset voltages in the modulator, and interference from the modulator-demodulator drive signal.

The floating amplifier can provide a CMRR in excess of 160 dB at DC and from 120–140 dB at low frequencies. In addition, common mode voltages limited only by the breakdown voltage of the transformers can be accepted. This is in contrast to the previously discussed amplifiers in which the maximum common mode voltage is typically limited to about 10 V by semiconductor breakdown considerations.

The requirement for shielded transformers, special circuit shielding, and the modulator-demodulator are disadvantages that result in relatively high cost. In addition, the floating amplifier usually has limited bandwidth as a result of the transformer characteristics and the modulation-demodulation requirements. Because of these limitations, the floating amplifier is not widely used in time-shared applications. It is, however, used extensively in per-channel applications when high common mode voltages are present.

3.14.5 Time-Shared Requirements

Time sharing of the amplifier in analog-input subsystems imposes design requirements in addition to those normally associated with instrumentation amplifiers. One of these, settling time and transient response, is discussed in Sec. 3.14.2. In this section, several other requirements stemming from the time-shared application are considered briefly.

In an analog subsystem, a multiplexing switch is usually located before and/or after the amplifier. When the amplifier is used on a per-channel basis, the multiplexer switch is a single-ended switch at the amplifier output. A time-shared amplifier is normally preceded by a multiplexer and may be followed by a high-level switch if the ADC is time shared between several amplifiers or multiplexers. These switches are a possible source of additional system errors owing to interaction between the amplifier and the switch.

As discussed in Sec. 3.13, a transient due to the multiplexer drive may be coupled into the signal path through the switch capacitances. Although of low energy and short duration, the magnitude of this spike may be several volts. The application of this pulse to the amplifier input or output may cause a transient reaction in the amplifier which results in a system error.

When multiplexing the output of the amplifier, the transient spike appears across the amplifier output impedance. Although this is a low impedance at DC because of the negative feedback employed in the amplifier, the impedance may be appreciable at the high frequencies contained in the spike. Since the voltage produced at the amplifier output by the spike is not in agreement with the amplifier input conditions, the feedback action of the amplifier attempts to restore the amplifier to its steady-state condition. Because of the delay in the feedback network, however, the reaction of the amplifier is delayed and the corrective action arrives at the amplifier output after the pulse has disappeared. At the very least, this action can result in an amplifier transient that must settle before useful data can be obtained. In the worst case, the amplifier may oscillate in a manner analogous to certain systems with dead time in classical control system problems.

The solution to the transient problem is to reduce the magnitude of the spike or to provide the amplifier with the necessary transient characteristics. The reduction of the spike is similar to the problem in multiplexers; basically it involves reducing the dV/dt of the drive signal, if possible, and minimizing the coupling capacitance between the drive source and the signal path.

In order to eliminate the unfavorable amplifier reaction, the output impedance must be reduced even at high frequencies, and delays in the feedback path must be eliminated. This latter requirement is satisfied by avoiding capacitances in the feedback network. The necessary frequency compensation for amplifier stability can be provided by networks between stages rather than in the feedback path. Low output impedance at high frequencies requires a high amplifier bandwidth. The required bandwidth may be greater than that necessary to provide satisfactory settling characteristics for a step input signal.

Multiplexing at the amplifier input poses similar problems. The spike due to the multiplexer drive may be sufficient to saturate a high-gain amplifier in a low-level system. The solution is similar to the previous case with increased bandwidth being the general requirement. An additional consideration in input multiplexing in differential systems is that, if one switch closes before the other, a normal mode signal equal to the applied common mode voltage is applied to the amplifier for a short period. Besides the possibility of saturating the amplifier, there is a chance that such a signal may cause destructive or nondestructive breakdown of the first-stage semiconductors. In severe cases, voltage clamping at the amplifier input may be necessary. The use of a transformer in the input leads as illustrated in Fig. 3-74 is also beneficial in minimizing effects due to switch skew.

The multiplexer switch at the input of a time-shared amplifier also places a requirement on the first-stage biasing. In a per-channel application, base bias can be provided through the source impedance of the channel. Such self-biasing circuits may not be acceptable in the time-shared application since the bias path is opened when the multiplexer switch is opened. If the amplifier saturates under this condition, the saturation recovery time must be included as part of the settling time for reading a channel. In addition, long periods of saturation cause internal heating of the amplifier owing to dissipation in the output stages with a resultant possibility of increased zero offset. For these reasons, self-biasing input stages are not commonly used in time-shared amplifiers. The input bias is provided by fixed current sources connected to the bases of the input transistors. This is usually sufficient to guarantee that the amplifier output remains at a voltage near zero when the multiplexer switches are open.

Overload recovery is another amplifier requirement associated with time-shared applications of amplifiers. In per-channel use, an amplifier overload is usually due to a malfunction of the transducer or an incorrect gain setting. The overload condition is detected by the computer and the operator corrects the situation manually. Only the data from the affected channel are lost. In the case of the time-shared amplifier, the system design must ensure that data on subsequent channels are not affected. This requires that the amplifier recover from saturation before the next channel is read. An overloaded amplifier is a nonlinear device, and the usual linear design criteria are not applicable; therefore, increased bandwidth alone is not sufficient. The nonlinear nature of the problem makes it difficult to provide specific design rules. In general, however, the amplifier should be designed with a minimum number of energy storage devices. Typically, this means minimizing the number of capacitors in the circuit. Providing clamping circuits on the individual amplifier stages is also effective in preventing deep saturation if the linearity of the amplifier can be maintained over the required range.

3.15   ANALOG-TO-DIGITAL CONVERTERS

The actual transition between the analog and digital domains in the analog subsystem occurs in the analog-to-digital converter. This device, usually referred to as an ADC or A/D converter, accepts an analog signal as an input and provides a digital representation as an output. In the case of a modern process control computer system, the digital signal is represented as a binary word or a sequence of pulses since these are compatible with the data format of the CPU.

There are many methods of converting an analog signal to a corresponding digital signal, and many of them have unique advantages for particular applications. Our interest is confined to the methods most often used in the time-shared analog subsystems of a modern process control computer system. If the many possible variations of these conversion methods are neglected, this limits the discussion to a few distinct methods.

3.15.1   Basic Conversion Principle

An analogy to an ADC that is familiar to every engineer is the common laboratory voltmeter. The application of a voltage to this instrument causes the deflection of an indicator or needle; the position of the indicator is compared to a calibrated scale to obtain a digital value. In this case, the conversion of the analog signal to its corresponding digital value is a two-step procedure; first, the voltage is converted to an angular position by an electromechanical phenomenon; second, the position of the indicator is converted to a digital value by the observer, who compares the indicator position to a calibrated scale.

The correspondence between this analogy and the methods used in many electronic ADCs is quite exact. The analog-input signal is first converted to an intermediate form, such as a frequency or pulse duration, that is more amenable to electronic measurement. The intermediate quantity is then compared to a calibrated reference which, corresponding to the form of the intermediate quantity, might be a frequency or time-duration standard. Although there are exceptions to the idea that the input signal is converted to an intermediate form, all ADCs incorporate some means of comparing the unknown signal with a standard or reference.

3.15.2   Multiple Comparator ADC

The multiple (or parallel) comparator ADC is the simplest electronic method of converting an analog signal into a digital value. Although it is rarely used in the analog subsystems of concern here because of its lack of resolution, its description serves to introduce basic ADC concepts.

The fundamental circuit in this ADC is the comparator. The comparator is a circuit in which the output voltage has one of two values or states depending on the polarity of the difference between two input voltages. If the two input voltages are V_1 and V_2, and if the two output states are denoted by the binary symbols **1** and **0**, the ideal comparator is described mathematically by

$$
\begin{array}{cc}
\text{output state} & \text{input condition} \\
\hline
\mathbf{1} & V_1 > V_2 \\
\mathbf{0} & V_1 < V_2
\end{array}
\tag{3-63}
$$

Assuming that the binary states **1** and **0** are equal to output voltage levels of 6 and 0 V, respectively, the transfer characteristic of an ideal comparator is shown as the solid line in Fig. **3-78**.

The transfer function of a practical comparator is not discontinuous at $V_1 - V_2 = 0$ but approximates the dotted curve in Fig. **3-78**. Thus, if $V_1 - V_2$ is slightly greater than zero, the output of a practical comparator is some voltage between the 6- and 0-V binary conditions. The input voltage that causes the output to reach the 6-V level is called the threshold voltage V_t; in practice, V_t may be as small as several hundred microvolts.

The block diagram of a simple multiple comparator ADC is shown

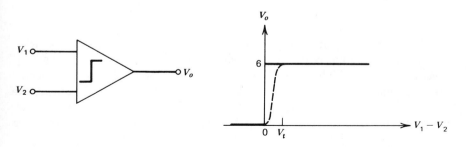

Fig. 3-78   Comparator transfer characteristics.

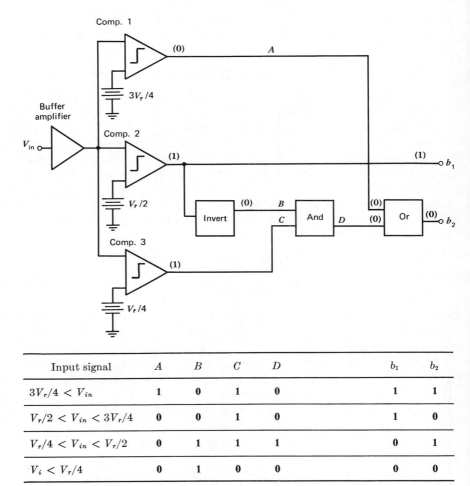

Input signal	A	B	C	D	b_1	b_2
$3V_r/4 < V_{in}$	1	0	1	0	1	1
$V_r/2 < V_{in} < 3V_r/4$	0	0	1	0	1	0
$V_r/4 < V_{in} < V_r/2$	0	1	1	1	0	1
$V_i < V_r/4$	0	1	0	0	0	0

Fig. 3-79 Multiple comparator ADC.

in Fig. 3-79. The output of the buffer amplifier, which provides isolation between the comparators and the signal source, is connected to each of the comparators. The second inputs of the comparators are connected to reference voltages $3V_r/4$, $V_r/2$, and $V_r/4$. The logic following the comparators transforms the output state of the three comparators into an appropriate digital code. The particular logic shown here provides a binary word whose value is proportional to the analog-input voltage.

As an example of the operation of this ADC, suppose that the value of the input voltage is $0.6V_r$. Assuming a unity-gain buffer amplifier, the input difference voltage applied to Comparator 1 is negative but that applied to Comparators 2 and 3 is positive. Thus the output voltage levels of Comparators 1, 2, and 3 correspond to the binary states **0, 1**, and **1**, respectively. Tracing through the logic, the output binary word $[b_1 \ b_2]$ is **[1 0]**. The state of the logic at the input and output of each logic block is indicated in the figure by the numeral in parentheses. The state of the logic and the output word for input voltages ranging from 0 V to voltages greater than the reference voltage V_r are shown in the table in the figure.

The time interval between the application of the input voltage and the appearance of the output word $[b_1 \ b_2]$ is determined by the settling time of the buffer amplifier and comparators and by the logic delays. With present technology, the logic delay is measured in nanoseconds and the settling time of the amplifier and comparators may be a few microseconds or less. The time required to generate the output word, known as the *conversion time,* is essentially unchanged if additional comparators and associated decode logic are added. Therefore, this type of converter is capable of conversion rates of hundreds of thousands of conversions per second.

In contrast to its high speed capability, the multiple comparator has serious limitations that render it unacceptable for the type of analog subsystems normally used in control computer systems. Specifically, the precision of the conversion is a function of the number of comparators. In the example, the output word does not change if the input voltage varies between the limits shown for each line of the table; the output word is **[1 0]** for any voltage between $V_r/2$ and $3V_r/4$, for example. This is equivalent to measuring a length with a yardstick that is marked only at 1-ft intervals.

The distinct intervals that can be distinguished by an ADC are known as the *quantizing intervals* or *quantizing steps.* In some fields, notably nuclear physics intrumentation, the term *channel* is used, but in this chapter a channel is the analog-input signal path in a multiplexed system. In most ADCs the quantizing intervals are equal, although this does not have to be true.

A term closely related to quantizing interval is *resolution.* Expressed in terms of the input voltage, the resolution is the smallest change in the input voltage that always results in a change in the digital output word. Thus, for example, in an ADC having a full-scale range of 10.24 V and 1024 equal quantizing intervals, the resolution is 0.01 V. Resolution is also referred to in terms of the number of possible output digital words. In the previous example, there are 1024 output words corresponding to the 1024 quantizing levels. The resolution then can be referred to as "one part in 1024." If the output word is in binary code, it is common to relate resolution to the number of bits in the word; thus the resolution of 1 part in 1024 is commonly called 10-bit resolution since the number of 10-bit binary words is $2^{10} = 1024$. Similarly, if the output word is in a decimal code, resolution may be stated in terms of the number of digits; for example, a resolution of 1 part in 10,000 is equivalent to 4-digit resolution. Although reference to resolution in terms of parts, bits, or digits is somewhat imprecise because of the attendant assumptions, this nomenclature is used universally and is convenient since it is independent of the full-scale voltage of the ADC.

The resolution of the parallel ADC can be increased by adding more comparactors. In general, n-bit resolution requires $2^n - 1$ comparators. For n greater than 3 or 4, the cost of the comparators results in an ADC cost that is excessive compared to other conversion methods. The low cost of integrated amplifiers suitable for use as comparators reduces the cost appreciably, but the problem of adjusting each comparator independently and the variations in quantizing steps due to zero offset drift still limit the use of the technique to applications where low resolution is satisfactory.

3.15.3   Successive Approximation ADC

The successive approximation ADC is probably the most popular ADC used in modern analog subsystems for computers. Conversion speeds in excess of 50,000 conversions/sec (cps) are possible at resolutions of 14 or 15 bits, and higher speeds can be obtained at lower resolution.

The successive approximation ADC involves the basic principle of comparison common to all analog-to-digital conversion techniques. It does not, however, require the conversion of the input signal to an intermediate quantity since the input voltage is compared directly to a precise fraction of the reference voltage. At each successive step in the conversion process, the approximation is refined until the fraction of the reference voltage equals the input voltage within the resolution of the converter.

In a sense, the operation of the successive approximation ADC is analogous to the game of "20 Questions." Suppose, for example, that one was asked to determine an unknown number to 2-digit precision by asking the question, Is the number greater than X? when the other person can only answer Yes or No. Knowing that the number lies between 0 and 10, one might proceed through the sequence:

question. Is the number greater than:	response
3	Yes
8	No
5	Yes
6	No
5.5	Yes
5.8	No
5.6	Yes
5.7	No

with the eventual result that the number is determined to be between 5.6 and 5.7.

This is essentially what a successive approximation ADC does. The input voltage is compared to a precision reference voltage and the output of a comparator indicates whether the reference voltage is greater than, or less than, the input signal. The reference is then adjusted up or down to provide a new approximation for the next comparison. This sequence continues until the desired precision is obtained. Assuming no *a priori* knowledge, the most efficient convergence is obtained when the interval between each successive guess is half the interval between the two successive approximations. Thus, in the above example, the values of the questioner would be 5.0, 7.5, 6.25, 5.625, and 5.3125. The unknown number is determined to within the last interval which, in this case, is equal to 3.125% of the maximum value of 10. Thus the quantizing interval is also equal to 1 part in 32 which, in turn, is a resolution of 1 part in 2^5. This is not a coincidence, since 5 comparisons were made and, in general, n comparisons are required for n-bit resolution.

The electronics required to perform a 4-bit successive approximation procedure are shown in Fig. 3-80. The circuits consist of a comparator, ladder network, voltage switches, precision reference source, and a number of logic blocks. The characteristics of the comparator are, ideally, the same as those discussed in the previous section.

The ladder network and voltage switches are nothing more than a digitally controlled attenuator network that determines the voltage applied to one side of the comparator. This attenuator is one form of digital-to-analog converter (DAC) whose design is considered in detail in Chap. 5; only the basic explanation of its operation is required to understand the successive approximation ADC. When the flip-flop corresponding to each voltage switch is in the 1 state, the voltage switch is ON. If the word represented by the state of the flip flops in the output register is denoted by $[b_1 \; b_2 \; b_3 \; b_4]$, then the output of the ladder is

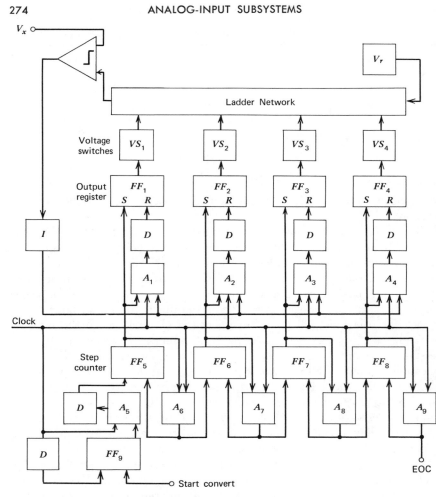

Fig. 3-80   Four-bit successive approximation ADC.

$$V_o = (b_1\, 2^{-1} + b_2\, 2^{-2} + b_3\, 2^{-3} + b_4\, 2^{-4})V_r \qquad (3\text{-}64)$$

where the b_i's can be either 0 or 1 corresponding to the binary states. For example, if VS_1 and VS_3 are ON, the output is $(\frac{1}{2} + \frac{1}{8})V_r = 0.625\ V_r$. For ADC operation, the DAC must be designed such that the switches add negligible error to the ladder output voltage. This, as discussed in Chap. 5, requires that the switches have basically the same characteristics as low-level multiplexing switches.

The operation of the ADC is easily understood by considering a specific

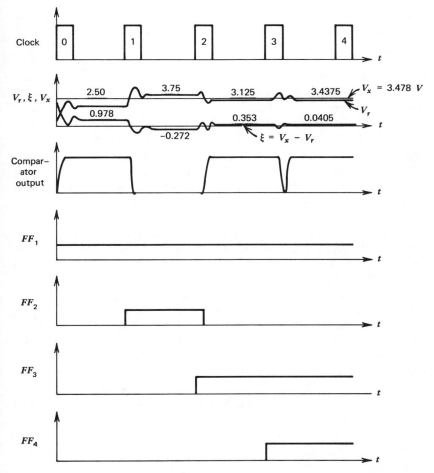

Fig. 3-81   Successive approximation ADC waveforms.

example. If the unknown voltage V_x in Fig. 3-80 is assumed to be 3.478 V, the waveforms at the various points in the circuit are as shown in Fig. 3-81. Initially, the flip flops in the output register and the step counter (FF_1, FF_2, ..., FF_8) are all in the **0** state. A pulse at the "start convert" input sets FF_9 such that at the next clock pulse (clock pulse 0 in Fig. 3-81), a **1** is forced into FF_5; FF_9 is reset by the delayed clock pulse. The delay on the output of A_5 prevents the initial clock pulse from being transferred into FF_6 via A_6 during the 0 clock pulse time. The **1** in FF_5 causes FF_1 to go to the **1** state and VS_1 is turned ON. Assuming that the

ADC has a full scale of 5 V, the output of the ladder increases and, after some settling time, is equal to 2.500 V. The error signal at the comparator input is of $3.478 - 2.500 = 0.978$ V, so the comparator output switches to the 1 state. Clock pulse 1 causes the 1 in FF_5 to be transferred to FF_6 with the result that FF_2 is set and VS_2 is turned ON. Switch VS_1 remains in the ON state since the output of the inverter is 0 and the input conditions on A_1 are not satisfied.

Voltage switches VS_1 and VS_2 are now ON and the ladder increases from its previous value of 2.500 V to 3.750 V. This results in a comparator input voltage of $3.478 - 3.750 = -0.272$ V, and the comparator switches to the 0 state. At the time of the third clock pulse (pulse 2), the conditions on A_2 are satisfied, FF_2 is reset, and this turns VS_2 OFF. Clock pulse 2 also shifts the 1 in FF_6 into FF_7 causing FF_3 to be set and VS_3 to be turned ON.

The conversion continues in this manner until the output register is set to [1 0 1 1] and clock pulse 4 shifts the 1 out of FF_8. This pulse can be used as an end-of-convert (EOC) indication. If the EOC pulse is coupled back to FF_9, the conversion is repeated. In most analog subsystems, the EOC pulse is used to interrupt the computer which then transfers the data in the output register (FF_1 through FF_4) to a designated storage location. An alternate readout procedure is to AND the output of the comparator inverter and the clock pulses to obtain a serial data format. This technique is commonly used in telemetry systems where data are transmitted as a train of pulses.

A diagram that illustrates all possible comparison sequences in a 4-bit ADC is shown in Fig. 3-82. The binary words represent the possible states of the output register at the time of each of the five clock pulses; the arrows between states indicate the possible transitions that can occur as a result of the comparison. In each case, the upper arrow of the pair is the transition that results when the input voltage is greater than the ladder output voltage. The output register states for the above example are circled. An important point illustrated by this diagram is that if an incorrect transition is made at any point in the cycle due, for example, to a noise pulse, it is impossible to arrive at the proper result.

It is obvious that the repetitive logic of this ADC can theoretically be extended to obtain a conversion to any number of bits resolution with an appropriate ladder network. Practical limitations on the precision of the ladder network, the stability of the reference voltage, and the switching threshold stability of the comparator limit the number of bits to 14 or 15.

There are many variations of the basic successive approximation technique illustrated by the above example as indicated by the fact that

Clock pulse 0   1   2   3   4

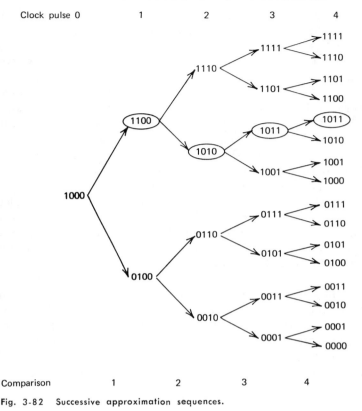

Comparison   1   2   3   4

Fig. 3-82   Successive approximation sequences.

hundreds of patents relating to the technique have been issued. Two important variations worth noting are bipolar operation and the use of output codes other than binary. Bipolar (i.e., positive and negative) signals can be handled in several ways. The basic concept is to determine which polarity signal is applied and then either to reverse the input leads or to modify the comparison voltage. Polarity can be determined by setting the ladder output to zero and observing the output state of the comparator.

The ADC output word can, obviously, be converted to any other code by appropriate logic decoding. It is possible, however, to build the converter to operate directly in codes such as decimal or BCD. All that is required is a counter that operates in the desired code and a ladder network design that provides for output voltages proportional to the number in the register. For example, in a 4-digit BCD converter, a 4-digit counter is provided and the ladder network supplies output ratios of 0.8,

0.4, 0.2, 0.1, 0.08, 0.04, . . . , 0.0002, 0.0001. The use of a code consistent with the computer code results in the most efficient hardware since additional code conversion logic is not required.

Another system feature that is easily implemented in the successive approximation ADC is limit comparison. As mentioned in Sec. 3.7, this feature allows the computer to compare the value of an input voltage with a predetermined number. In many process applications, for example, the exact value of a parameter is not particularly important, but if it exceeds a certain value, it might indicate an alarm condition. If the logic of the ADC is arranged such that a word from the computer can be entered into the output register, then the resultant output of the ladder network equals the predetermined limit value. A single comparison indicates whether the input voltage is greater than or less than the ladder output voltage. This decision is made in one or two clock cycles, which is considerably less than the n cycles required for a complete conversion An alternative is to do a complete conversion and then mathematically to compare the result with the limit value.

Despite many variations, the basic concept of the successive approximation ADC remains the same. This is also true of the detailed design requirements for the major components of the ADC. Each of the major components—ladder network, voltage switches, reference voltage source, comparator, and logic—deserves some discussion. The requirements for the first three of these items are basically the same as in the case of digital-to-analog converters and are thoroughly considered in Chap. 5; the design criteria are only summarized here. The present discussion is directed toward the design of a high-performance ADC, typically 14 or 15 bits resolution at conversion speeds of 20 kcps to 50 kcps and proportionally faster at lower resolution. At lower speeds and/or resolution, the design requirements may be correspondingly less severe.

For a 50 kcps ADC, the total conversion time is 20 μsec, or an average of about 1.5 μsec/bit for a 14-bit ADC. Allowing for comparator settling time and logic delays, the comparison voltage generated by the ladder voltage and the reference voltage source must settle to within the quantizing interval in something on the order of 700 nsec. Assuming a single-order transient respose and realizing that the quantizing interval in a 14-bit ADC is approximately 0.006%F.S., a settling time of about 10 time constants is required. Thus the 10–90% risetime of the ladder must be on the order of 50 nsec. Ladder networks fulfilling these stringent requirements are available in both wire-wound and metal-film technologies, but they represent a significant cost in an ADC.

A second important characteristic of the ladder network is the precision of the ratios between the input and the output. Failure to achieve the required precision results either in nolinear conversion or in a lack of monotonicity in the converter output. The characteristics of the voltage switches are also important in this regard since the switch ON resistance affects the transfer ratio of the ladder. Although nominal initial values of switch resistance can be accounted for in the ladder design, deviation from the design value and changes with time and temperature must be consistent with the stability performance expected of the converter. The OFF leakage current of the switch and the offset voltage, in the case of bipolar transistors, also causes errors at the ladder output.

The requirements on the voltage reference include accuracy and stability consistent with the specifications of the overall converter. Similarly, noise and other disturbances must be less than the converter quantizing interval if consistent readings are to be expected. The regulation requirements are quite stringent because the reference must settle to within the resolution limit after each comparison despite the severe load changes that occur during the conversion. Depending on the impedance level of the ladder and the value of the reference source, current requirements may be as high as several hundred milliamperes. A higher impedance decreases the current requirement but increases the ladder risetime.

The demands on the logic in a 50 kcps successive approximation converter are not particularly rigorous. Comparisons and logic action occur in synchronism with the clock pulses, so the logic switches approximately every microsecond. This, in addition to the fact that the logic is relatively simple with few cascaded functions, means that most of the presently available logic technologies are satisfactory. Since logic noise injected into the analog circuits may cause conversion errors, the logic power supplies should be adequately decoupled.

Major specifications of concern in the design of a suitable comparator are gain, zero offset, threshold stability, common mode rejection, and the input impedance in relation to the ladder output impedance. Since the comparator must switch when the input difference signal exceeds the voltage resolution of the converter, the minimum comparator gain is

$$G_{min} = \frac{2^n V_L}{V_{fs}} \qquad (3\text{-}65)$$

where n is the resolution in bits, V_L is the logic threshold level, and V_{fs} is the full-scale range of the ADC. For a typical resolution of 14-bits and a logic level of 3 V, the gain must be on the order of 10,000 for a 5-V converter.

Tight control of the gain—that is, good gain stability—is normally achieved in an amplifier by use of negative feedback. In a comparator, however, feedback has the undesirable characteristic of increasing the overload recovery time. Therefore, minimum feedback is used and the comparator is designed for a nominal gain several times higher than the required minimum to allow for relatively large variations in gain.

Overload recovery time is the single most important parameter in a comparator. In the successive approximation ADC, for example, consider an input voltage equal to $0.25\ V_r \pm V_r/2^n$—that is, an input voltage within one quantizing interval of $0.25\ V_r$. At the first comparison, the ladder output is $0.5\ V_r$, so the comparator input voltage is approximately $0.25\ V_r$. This results in severe comparator overloading since the comparator is designed to saturate at $V_r/2^n$. On the very next conversion, the ladder output is exactly $0.25\ V_r$, so the input to the comparator is $\pm V_r/2^n$. If the comparator does not recover, or if it overshoots its final value, the decision made by the comparator is in error. As noted previously, the error cannot be corrected and the remainder of the conversion cycle results in subsequent bits being either all turned ON or all turned OFF.

The overload recovery problem can affect the comparator decision on any comparison, but it is most severe on the first few comparisons since this is when the overload is the greatest. A design trade-off between speed and comparator design is possible; by allowing more than one clock cycle for the initial comparisons, the comparator is allowed extra time to settle to final value. Thus by sacrificing some speed, the comparator design requirements are eased. It is also possible to design the converter in such a way that an error made on one comparison is eventually corrected. This requires additional logic, however, and is normally employed only when the required conversion speed is such that the comparator requirements exceed the state of the art. This is not the usual case in ADCs designed for process control subsystems, and the extra settling time solution is more often used.

Overload recovery time can be decreased by designing the comparator with minimum gain and energy storage elements. This latter requirement is dictated by the fact that if discrete and transistor junction capacitances are charged by saturation of the comparator stages, the charge must be removed before the amplifier recovers. High-speed transistors are designed for minimum carrier storage and are recommended for this application. Discrete capacitances should be avoided except when necessary to stabilize the comparator. Minimum gain also improves the overload recovery time. The gain must, of course, be high enough to satisfy Eq. (3-65) under all operating conditions.

Another condition that affects the comparator performance is dissipation in the comparator itself. Self heating causes bias level changes that alter the switching threshold and thereby cause an offset in the output data. A source of excess dissipation is comparator saturation between conversions when the ADC input is open. In general, solutions are to use a conservative low dissipation design for the comparator or to prevent saturation between conversions. One approach in the latter case is to provide a shorting switch at the ADC input which is closed except when a conversion is in progress.

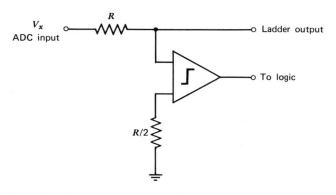

Fig. 3-83   Alternate comparator configuration.

A slightly different comparator arrangement, shown in Fig. 3-83, reduces the severe overloading to which the comparator is subjected by a factor of 2. The equivalent circuit is shown in Fig. 3-84 where V_c and the associated resistor R are the Thevinen equivalent of the ladder network and Z_{in} is the comparator input impedance. If Z_{in} is much greater than R, the normal mode input signal to the comparator is

$$V_e \cong \frac{V_{in} - V_c}{2} \tag{3-66}$$

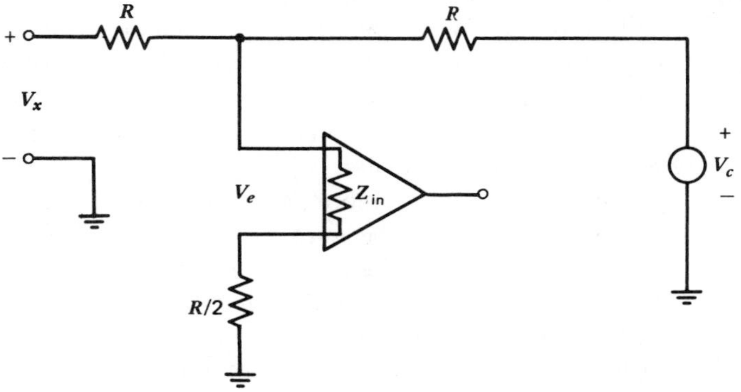

Fig. 3-84   Comparator equivalent circuit.

which is only half the voltage applied to the comparator in Fig. 3-80. Thus the degree of saturation is only half as great, and the overload recovery of the comparator is, in general, improved.

An additional advantage of the circuit of Fig. 3-83 is that the effect of comparator input current is decreased. The input current to each input of the comparator flows through an equivalent resistance $R/2$ and equal potentials are produced if the two input currents are the same. As the input currents change with time and temperature, they tend to track, and only the difference contributes to a comparator offset.

Another advantage is that the comparator is subjected to minimum common mode voltage. During the conversion process, the drop across $R/2$ resistor is small compared to the drop across the input impedance Z_{in}. Therefore, the negative terminal of the comparator is essentially at ground and the comparator does not require appreciable common mode rejection capability in this case.

The situation in Fig. 3-80 is quite different since a common mode signal equal to the ladder output voltage is applied to the comparator throughout the conversion cycle. If the common mode rejection of the comparator is poor, the applied common mode voltage causes an equivalent input voltage with a resultant shift in the effective comparator switching threshold. The shift is not constant during the conversion cycle since the common mode voltage changes after every comparison. The common mode rejection of the amplifier should be high enough to guarantee that the normal mode error due to the maximum common mode voltage is less than one quantizing interval. This requires a minimum rejection ratio of 2^n. Although not exceedingly high, this is an additional comparator requirement that can be avoided by using alternate comparator arrangements.

The effect of input current differences in producing an offset in the comparator output which is then reflected in the digital data has been mentioned. A similar effect is the voltage offset of the comparator. The initial offset can be compensated, but changes due to time and temperature result in an offset in the data. Although this is not a severe problem, it is a noticeable effect in a 14- or 15-bit converter where the quantizing interval may be only a few hundred microvolts.

The successive approximation ADC is more noise sensitive than an ADC that utilizes an integrating principle because each comparison takes place at the instant the comparator is strobed by the clock pulse. If, just prior to the clock strobe, a noise pulse causes the input level to change enough that the comparator switches to the wrong state, an erroneous comparison results. Although such an error can occur on any of the n comparisons in the cycle, it is most likely to occur when the comparison and input voltages are approximately equal since a small change in the input signal can change the state of the comparator. This type of error contributes to the "spread" or "scatter" in the converter output when the same voltage is read repeatedly. The noise that causes this type of error may be due to sources external to the analog subsystem, or it may be generated internally. Although external sources are not directly controllable by the designer, the effect of internal noise sources can be lessened by the use of proper shielding, isolation, and decoupling techniques. Adequate decoupling of logic circuits or power supply lines that are common to both logic and analog circuits is of particular importance in high-resolution systems.

3.15.4 Ramp ADC

In the successive approximation ADC the unknown voltage is compared to a sequence of known reference voltages. The value of the comparison voltage at any step is dependent on the result of previous comparisons. In the successive approximation ADC described above, the voltage sequence and the logical decisions are such that an n-bit digital representation of the input voltage is obtained after n comparisons. Other sequences of voltages can be used to obtain the desired result; for example, the comparison voltage can be increased one quantizing step at a time until it matches the input voltage as indicated by a change in comparator state.

A slight extension of the use of a monotonic staircase voltage is to compare the unknown input voltage with a constant slope ramp voltage. Given the slope of the ramp and the time interval between when the ramp voltage is zero and when it is equal to the unknown voltage, the value of the unknown voltage can be determined. The block diagram of an ADC that operates on this principle is shown in Fig. 3-85.

Initially the output of the ramp is slightly less than zero, as indicated in Fig. 3-86, the output of Comparator 1 is in the **0** state, and the output of Comparator 2 is in the **1** state. When the start-convert signal is given, the counter is reset to zero and the ramp generator is activated. When the output of the ramp generator passes through zero, Comparator 1 changes state and clock pulses are gated into the counter. When the output of the ramp generator is equal to the unknown voltage, Comparator 2 changes state and pulses are inhibited from entering the counter by the AND gate. The effect of this sequence of events is to convert the unknown voltage into a time duration that is measured by the combined action of the comparators, counter, and AND gate.

If the time interval between changes in state of the comparators is T

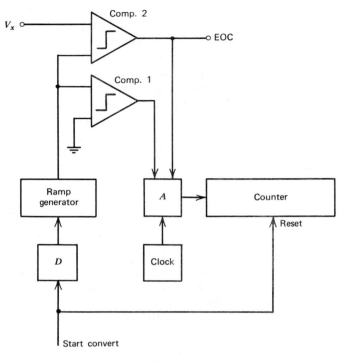

Fig. 3-85   Conventional ramp ADC.

and if the slope of the ramp voltage is K V/sec, the unknown voltage V_x is

$$V_x = KT \tag{3-67}$$

During the time interval T, Nf_c pulses enter the counter where N is the final number in the counter and f_c is the frequency of the clock. Combining this with the previous equation,

$$V_x = Kf_cN \tag{3-68}$$

Thus the unknown voltage is equal to a constant times the number in the counter.

Equation (3-68) provides the basis for understanding several of the possible error sources in the ramp ADC. Since V_x is directly proportional to f_c, variation in clock frequency during the conversion interval affects the accuracy of the result. Long-term variation in the frequency contributes to a difference between successive readings of the same V_x. The magnitude of these effects is directly proportional to the frequency variation. For a 12-bit converter, a frequency variation of approximately

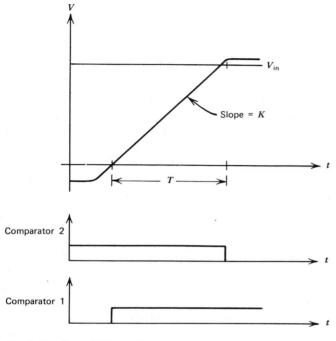

Fig. 3-86   Ramp ADC waveforms.

0.025% results in a 1-bit error. In practice, the clock of a modern computer may have a short-term frequency variation an order of magnitude greater than this; thus a special high-stability oscillator is often required in this type of ADC.

The slope of the ramp also is a potential source of error. If the linearity of the slope is 1%, the linearity of the converter can be no better than this figure. It is particularly difficult to build a ramp generator whose output begins at zero without some initial nonlinearity. This is the basic reason for starting the ramp voltage at a value less than zero and utilizing a comparator to indicate the zero crossing. A ramp linearity of 0.1% or somewhat better is typically achieved in practice. A change in the linearity with temperature contributes to the converter temperature coefficient.

Other errors affecting performance are not indicated by the basic equation. For example, comparator and logic delay can have significant effect. When the input to Comparator 1 changes sign, the output change is delayed because of circuit delay and the overload recovery time; the same is true for Comparator 2. The delay caused by Comparator 1 prevents pulses from entering the counter until the ramp has reached some finite value V_t. Similarly, the delays associated with Comparator 2 do not inhibit counter pulses until the ramp has exceeded the unknown voltage V_x by some voltage V_T; the difference $V_T - V_t$ represents an error in the conversion. Relating this to the delay times, only the difference in delays contributes to an error. The delays normally differ since the comparators are not operating under the same conditions; Comparator 1 is not usually in heavy saturation, whereas Comparator 2 is. In addition, Comparator 2 has applied common mode voltage, although this can be reduced by using the comparator arrangement described in the previous section.

The noise characteristics of the ramp ADC are similar to those of the successive approximation ADC since the comparison action is basically the same. A noise transient may cause Comparator 2 to change state prematurely or it may delay the action.

The logic requirements for the ramp converter are quite severe and are often the limiting factor in the conversion speed that can be obtained. If an n-bit conversion is to take T_o seconds, the oscillator frequency is $2^n/T_o$ and the counter must be capable of operating at this rate. As an example, a 12-bit converter with a 1-msec conversion time requires a 4-MHz clock and counter. With a successive approximation ADC, a 4-MHz clock could, theoretically, result in a conversion time of 1.3 μsec. Thus, for equal speed, the logic requirements in the ramp converter are orders of magnitude more severe.

The ramp ADC has an outstanding advantage over the successive approximation ADC; It is less expensive when nominal performance is required. The elimination of the ladder network and the voltage switches, and the somewhat simplified logic configuration are the primary sources of cost savings. The advantage disappears, however, if the ramp technique is used in a converter with resolution greater than 10 or 12 bits and speeds in excess of several thousand conversions per second.

In some application areas the ramp converter is preferred because of its differential linearity characteristics. Differential linearity is the variation in the quantizing intervals when compared with the theoretical quantizing interval. A ramp ADC can achieve a differential linearity of better than 1%, whereas the successive approximation technique may result in a figure as much as an order of magnitude greater unless special precautions are taken. The difference is due to the fact that in the ramp converter, the ramp is a continuous function and differences in the quantizing levels are attributable mainly to the comparator characteristics. In the successive approximation ADC, the quantizing intervals are determined by the ratio matching of the ladder network and the comparator characteristics. It is very difficult, for example, to guarantee equal quantizing levels between the ladder transitions [011111] and

[**100000**] and between [**000001**] and [**000000**] because of the different resistors involved in the two transitions. Differential linearity is important in applications where the converter output is used to construct statistical histograms; a variation in quantizing intervals results in unequal numbers of readings in different quantizing intervals even though the input distribution is uniform. The ramp converter is, for example, used extensively in nuclear physics instrumentation.

3.15.5 Integrating Multiramp Converter

The integrating multiramp converter principle is relatively new compared with the previously discussed techniques (Propster, 1963). In its simplest form, it retains the simplicity of the ramp converter but provides significantly better performance (Ammann, 1964). In slightly more sophisticated form, the accuracy and the resolution normally associated with the successive approximation ADC can be obtained.

The block diagram of a dual-ramp integrating converter is shown in Fig. 3-87. The types of circuits are similar to those used in the normal

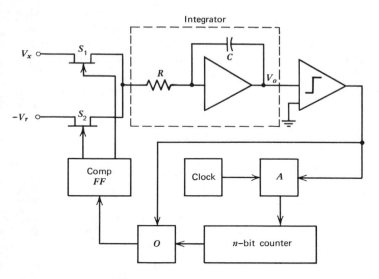

Fig. 3-87   Dual-ramp integrating ADC.

ramp converter except for the substitution of an integrator for the ramp generator and the addition of another analog switch. The integrator shown is a standard operational configuration consisting of a resistor R, a capacitor C, and an amplifier A.

The analog-to-digital conversion cycle consists of two distinct phases,

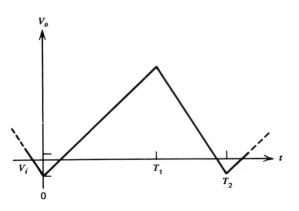

Fig. 3-88   Dual-ramp integrating ADC waveform.

as indicated by the up and down portions of the integrator output wave-form shown in Fig. 3-88. At time $t = 0$, the output of the integrator is equal to the effective comparator threshold voltage $-V_i$; the control logic resets the counter to zero and closes switch S_1 which connects the un-known voltage V_x to the integrator input. The voltage V_x is integrated for a fixed time interval $[0, T_1]$ which is established by counting a fixed number of clock pulses. Although not necessary, it is convenient to have this number of pulses equal to the full-scale count of the counter, 2^n in the case of an n-bit counter, so that the overflow pulse can be used to trigger subsequent operations.

At time T_1, the integrator output voltage is

$$V_o = \frac{1}{RC} \int_0^{T_1} V_x \, d\tau - V_i = \frac{1}{RC} \bar{V}_x T_1 - V_i \qquad (3\text{-}69)$$

where $\bar{V}_x$ is the average value of V_x over the interval. The overflow pulse from the counter at time T_1 turns switch S_1 OFF and turns S_2 ON, thereby connecting the reference voltage $-V_r$ to the integrator. The ouptut of the integrator during integration of $-V_r$ is

$$V_o = -V_i + \frac{1}{RC} \bar{V}_x T_1 - \frac{1}{RC} \int_{T_1}^{t} V_r \, d\tau \qquad (3\text{-}70)$$

The integration continues until the comparator changes state. At this time the integrator output voltage is equal to $-V_i$ so that Eq. (3-70) becomes

$$-V_i = -V_i + \frac{1}{RC} \bar{V}_x T_1 - \frac{1}{RC} V_r(T_2 - T_1) \qquad (3\text{-}71)$$

where the reference voltage V_r is assumed constant. The V_i's drop out and the integration constant $1/RC$ cancels in the remaining terms. Solving for the average of the input signal,

$$\bar{V}_x = V_r \frac{T_2 - T_1}{T_1} \qquad (3\text{-}72)$$

If N is the number on the counter at time T_2, then $T_2 - T_1 = N/f_c$, where f_c is the clock frequency. Since $T_1 = 2^n/f_c$, Eq. (3-72) reduces to

$$\bar{V}_x = V_r \frac{N}{2^n} \qquad (3\text{-}73)$$

Thus the average value of the input voltage is proportional to the number in the counter at time T_2.

The cancellations that take place during this derivation have considerable practical importance. The elimination of V_i in Eq. (3-71) shows that long-term offset drift has no effect on the accuracy of the converter. Similarly, the clock frequency f_c canceled out of the final expression, indicating that long-term frequency variations do not affect the result. The cancellation of the integration constant $(1/RC)$ shows that neither the value of the RC product nor its variation with time and temperature affects the operation of the converter. This is of considerable practical significance since low temperature coefficient capacitors and resistors are expensive components.

Another advantage of an integrating converter when it is compared with the conventional ramp and the successive approximation ADC is less noise sensitivity. The ramp and successive approximation ADCs are noise sensitive because the comparison takes place between the instantaneous value of the input signal and the reference voltage. In integrating converters, of which the dual ramp is an example, the comparison is between the reference and the time integral of the input voltage. If a noise perturbation has a small time integral compared to $\bar{V}_x T_1$, it has only slight effect on the comparison. In essence, the input signal is being heavily filtered as a result of the integration. In addition, the dual-ramp ADC is isolated from the signal source at the time of comparison since S_1 is OFF. This isolation further decreases the sensitivity of the ADC to noise on the input signal.

A quantitative measure of the noise rejection is derived by integrating a sine wave over the converter integration period. If the rejection ratio R_n is defined as the attenuation resulting from the integration, this calculation results in

$$R_n = 20 \log \frac{T_1}{RC} + 20 \log \frac{\sin^2 \pi T_1 f_n}{\pi T_1 f_n} \qquad (3\text{-}74)$$

where T_1 is the integration time and f_n is the frequency of the noise. The

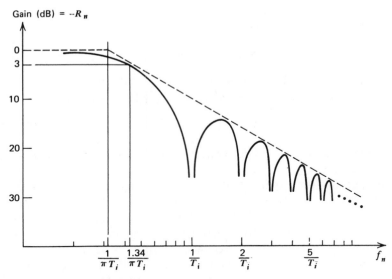

Fig. 3-89   Integrating ADC frequency response.

first term is a constant that can be compensated by adjustment of the ADC gain. The normalized attenuation characteristic is plotted in Fig. 3-89. As expected from physical reasoning, the rejection is infinite whenever the integration interval is an exact multiple of the period of the noise.

An extension of the dual-ramp integrating ADC that provides additional performance with only a slight increase in hardware is shown in Fig. 3-90. The integrator output waveform shown in Fig. 3-91 consists of three ramp voltages, giving rise to the name "triple-ramp integrating ADC" (Aasnaes and Harrison, 1968). The configuration is similar to the dual ramp except for the addition of the second reference voltage, a second comparator, and slightly different control logic.

The operation of the converter is also similar to that of the dual-ramp ADC. The counter is set to zero and switch S_1 is closed, thereby connecting the unknown voltage V_x to the integrator. The period of integration is determined by counting a fixed number of clock pulses. In the triple ramp, however, these pulses enter the counter at the $n/2$ stage so that each pulse adds $2^{n/2}$ to the value in the counter. If the clock frequency is f_c, the integration time T_1 is determined from $N_1 = 2^{n/2} f_c T_1$ where N_1 is the full-scale count of the n-bit counter. The overflow pulse from the counter terminates the integration by turning switch S_1 OFF; simultaneously, it turns switch S_2 ON to connect the first reference volt-

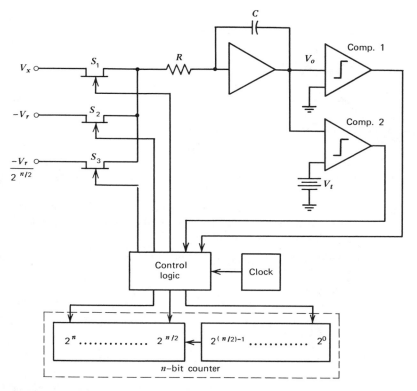

Fig. 3-90 Triple-ramp integrating ADC.

age $-V_r$ to the integrator. The integration of the negative reference voltage causes the output of the integrator to decrease, as shown in Fig. 3-91. During this interval, clock pulses are counted by the upper $n/2$ bits of the counter. The integration continues as the integrator output voltage passes through a threshold voltage V_t as determined by Comparator 2. The exact value of V_t is not critical but it must be greater than $V_r/2^{n/2}$. The first clock pulse after Comparator 2 changes state turns S_2 OFF and connects another reference voltage $-V_r/2^{n/2}$ to the integrator via switch S_3. At the same time, the control logic routes clock pulses into the lowest order position of the counter. Integration continues until a change in state of Comparator 1 indicates that the integrator output is approximately zero, thus completing the conversion cycle. During the last interval of integration, $2^{n/2}$ pulses may enter the lower stage of the counter. If the number of pulses exceeds $2^{n/2}$, a carry pulse from the lower portion of the counter adds to the number in the upper

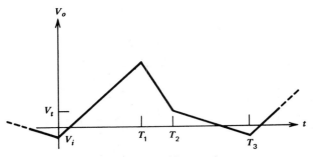

Fig. 3-91   Triple-ramp integrating ADC waveform.

portion of the counter to correct the approximate number obtained during the second integration period. In effect, the third integration acts as a vernier adjustment of the initial result.

The mathematical expression for the waveform of Fig. 3-91 is

$$-V_i + \frac{1}{RC} \int_0^{T_1} V_x \, dt \qquad \text{(first ramp)}$$

$$-\frac{1}{RC} \int_{T_1}^{T_2} V_r \, dt \qquad \text{(second ramp)} \qquad \text{(3-75)}$$

$$-\frac{1}{RC} \int_{T_2}^{T_3} \frac{V_r}{2^{n/2}} \, dt = -V_i \qquad \text{(third ramp)}$$

The reference V_r is constant and denoting the average value of V_x by $\bar{V}_x$, Eq. (3-75) becomes

$$\bar{V}_x T_1 - V_r(T_2 - T_1) - \frac{V_r(T_3 - T_2)}{2^{n/2}} = 0 \qquad \text{(3-76)}$$

Noting the different rates of counting resulting from pulses entering the counter at different bit positions, the effective number of pulses entering the counter during each time interval is

$$\begin{aligned} N_1 &= f_c T_1 2^{n/2} \\ N_2 &= f_c(T_2 - T_1) 2^{n/2} \\ N_3 &= f_c(T_3 - T_2) \end{aligned} \qquad \text{(3-77)}$$

Substituting into the previous expression and solving for $\bar{V}_x$

$$\bar{V}_x = V_r \frac{N_2 + N_3}{N_1} \qquad \text{(3-78)}$$

The number N_1 is the full count of the n-bit counter and $(N_2 + N_3)$ is the number in the counter at the end of the conversion cycle. The result is, therefore, similar to that obtained in the dual-ramp case in that the average value of the input voltage over the first integration interval is proportional to the final number in the counter. In addition, the cancellation of RC, V_i, and f_c shows that the triple-ramp ADC retains the advantages of the dual-ramp converter.

The main advantage of the triple ramp when compared to the dual ramp is one of speed of conversion. In the dual ramp, the total number of pulses that enter the counter for a full-scale input is $2(2^n)$; in the triple ramp, the maximum number is $3(2^{n/2})$, or a factor of $2(2^{n/2})/3$ less. For equal clock frequencies, the conversion time of the triple ramp is less by the same factor. For example, at 14-bits resolution, the triple-ramp ADC is about 85 times faster than a dual-ramp converter.

The above descriptions of the dual and triple ramp neglect second-order effects that are important in the design of a converter for high accuracy and resolution. One of these is the nonlinearity of the integrator. The output of a real integrator circuit is an exponential when a step function of voltage is applied. Because of a time constant that is long compared with the integration time, the output approximates an ideal ramp. An approximate, but more exact, expression for the output of the integrator is derived by expanding the exponential in a series and neglecting higher order terms: If this is done in the derivation of Eq. (3-73) for the dual-ramp integrating ADC, the value of $\bar{V}_x$ is

$$\bar{V}_x \cong V_r \frac{N - KN^2}{2^n - K2^{n+2}} \qquad (3\text{-}79)$$

where K is a constant, less than unity. The difference between this expression and Eq. (3-73) is the error due to nonlinearity. After some simplification, this results in

$$\frac{\Delta \bar{V}_x}{V_r} \cong \frac{KN(2^n - N)}{2^{n+2}} \qquad (3\text{-}80)$$

Therefore, the error increases and then decreases with increasing input voltage and is, in fact, zero for a full-scale voltage $(N = 2^n)$. A similar result applies to the triple-ramp integrating converter.

The initial analysis of the dual-ramp configuration showed that the conversion was independent of the effective comparator threshold. Since the effect of zero offset in the comparator is a shift in the effective threshold level, zero offset variations do not affect the converter operation if they are small during any conversion cycle.

The delay in the comparator is also a potential source of error which is canceled by the up-down integration action. The effect of the delay is to allow the integrator output voltage to overshoot the threshold level and to allow extra pulses to enter the counter. The effect is identical to that due to a change in threshold level. If the change in the threshold level is the same at the beginning and the end of the cycle, there is no effect on the accuracy of the conversion.

Another delay that can contribute an error is the delay in the switches S_1 and S_2. After the integration of the unknown voltage V_x, there is a slight delay between the overflow pulse from the counter and the time the reference is actually connected to the integrator. A similar delay occurs at the end of one cycle and the beginning of the next when the unknown is reconnected. It can be shown that the difference between these two delays contributes to an offset in the result of the conversion. The offset can be compensated by calibration of the converter, but changes in the delay with time and temperature are not canceled. Thus this source of error contributes to the temperature and time stability performance of the converter.

3.15.6 Voltage-to-Frequency Methods

Another conversion method that provides the effect of input signal integration is voltage-to-frequency conversion (VFC) (Skyles, 1967). The converter consists of a circuit that converts the input voltage into a signal or a series of pulses, the frequency of which is proportional to the applied voltage. By counting the pulses or the cycles of the output signal for a precise period of time, a number proportional to the input voltage is obtained.

A number of circuits are available for use as VFCs. These include voltage variable capacitors used as the frequency-determining element of an oscillator, and the use of an FET as a voltage-controlled resistor in a resistance-controlled oscillator. Another method that uses an integrator and two comparators is shown in Fig. 3-92. An input applied to the integrator produces a ramp voltage at the output as in the case of the dual- and triple-ramp converters. When the ramp voltage reaches the threshold level V_t of Comparator 1, the multivibrator MV fires to produce an output pulse. The change in state of Comparator 1 also sets a flip flop FF which gates a restoring voltage V_r to the summing point of the integrator. When the integrator output reaches zero, as indicated by the change in state of Comparator 2, the restoring voltage is removed and the integration cycle is repeated. By counting the pulses produced during a precise time interval, a number proportional to the input signal is obtained.

Critical design areas include the integrator linearity and offset, the stability of the comparator switching level, and the generation of a precise counting interval. With suitable care, the VFC technique can be used to obtain accuracy and resolution performance at least as good as

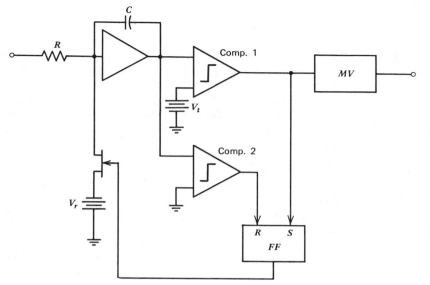

Fig. 3-92   Voltage-to-frequency converter.

that obtainable with the conventional ramp converter and at about the same cost. If the integration interval is made very long compared with the period of the pulses, accuracy comparable to a successive approximation ADC can be obtained. In addition, of course, the VFC provides the noise immunity that is lacking in the conventional ramp and successive approximation techniques.

REFERENCES

Aasnaes, H. B., and T. J. Harrison, "The Triple Integrating Ramp ADC," *Electron.,* April 29, 1968, pp. 69–72.

Ammann, S. K., "Noise-Proofing a Digital Voltmeter with Off-the-Shelf Microelectronics," *Electron.,* November 16, 1964, pp. 92–96.

Bonin, E. L., "Light-Coupled Semiconductor Switch for Low-Level Multiplexing," *Electron.,* February 8, 1965, pp. 54–59.

Bright, R. L., "Junction Transistors Used as Switches," *AIEE Trans. Commun. Electron.,* March 1955, pp. 111–121.

Brown, S. C., "Wideband Differential Amplifiers—Part II," *Electron. Des. News,* February 1961, pp. 84–87.

Considine, D. M. (Ed.), *Process Instruments and Controls Handbook,* McGraw-Hill, New York, 1957.

Considine, D. M., and S. D. Ross, (Eds.), *Handbook of Applied Instrumentation,* McGraw-Hill, New York, 1964.

Cromer, E. G., Jr., "Selecting and Using Nanovolt Instrumentation," *Electron. Instrum. Dig.,* July 1966, pp. 22–28.

Ellis, D. H., and J. M. Walter, "An Analog and Digital Airborne Data Acquisition System," *Proc. IRE,* April 1960, pp. 713–724.

Emmons, S. P., and H. W. Spence, "Very Low-Drift Complimentary Semiconductor Network dc Amplifiers," *IEEE J. Solid-State Circuits,* September 1966, pp. 13–18.

Funk, H. L., T. J. Harrison, and J. Jursik, "Converter Digitizes Low Level Signals for Control Computers," *Autom. Control,* March 1963, pp. 21–23.

Harrison, T. J., and J. R. Bernacchi, "Graphs Speed Two-Section Filter Design," *Electron. Des.,* October 25, 1967, pp. 104–107.

Harvey, G. F. (Ed.), *ISA Transducer Compendium,* 2nd ed., Plenum Publ. Corp., New York, 1968 (Part I), 1970 (Part II).

Hoffait, A. H., and R. D. Thornton, "Limitations of Transistor DC Amplifiers," *Proc. IEEE,* February 1964, pp. 179–184.

Meyer-Brötz, G., and A. Kley, "The Design of Differential D.C. Amplifiers with High Common Mode Rejection," *Electr. Eng.,* February 1966, pp. 77–81. (Also, in German, *Int. Elektron. Rundsch.,* January 1964, pp. 607–610.)

Minnar, E. J. (Ed.), *ISA Transducer Compendium,* Plenuum Press, New York, 1963.

Mitchell, B., and B. Bell, "The INCH—Discussion and Applications," *Solid-State Des.,* October 1962, pp. 34–42.

Prinz, D. G., "D.C. Amplifiers with Automatic Zero Adjustment and Input Current Compensation," *J. Sci. Instrum.,* December 1947, pp. 328–331.

Propster, C. H., Jr., "Analog to Digital Converter," *IBM Tech. Discl. Bull.,* January, 1963, pp. 51–52.

Skyles, R. W., "An Integrating DVM," *Instrum. Control Syst.,* January 1967, pp. 75–78.

Verster, T. C., "Silicon Planar Epitaxial Transistors as Fast and Reliable Low-Level Switches," *IEEE Trans. Electron Devices,* May 1964, pp. 228–237.

Verster, T. C., and A. R. Boothroyd, "Operation of the Junction Transistor as a Fast Switch," *Proc. IEE,* February 1963, pp. 353–368.

Vitola, J. J., and J. P. Breickner, "Dynamic Noise Generation in Reed Switch Contacts," *Electron. Ind.,* December 1964, pp. 66–69.

Weinberg, L., *Network Analysis and Synthesis,* McGraw-Hill, New York, 1962.

BIBLIOGRAPHY

Arlowe, H. D., and R. C. Dove, "Signal Conditioning Sensor Outputs," *Instrum. Tech.,* April 1967, pp. 39–45.

Baird, P. G., "Extraneous Noise and DC Voltage Measurements," *WESCON Conv. Rec.,* 1965.

Ballinger, D. O., "Applications and Characteristics of Dry Reed Relays," *Proc. 11th Natl. Conf. Electromagn. Relays,* 1963, pp. 161–165.

Barbour, C. W., "Simplified PCM Analog to Digital Converter Using Capacity Charge Transfer," *Proc. Natl. Telemetering Conf.,* May 1961, pp. 22–24.

Barbour, C., "Final Report: Development and Fabrication of a Prototype Airborne and Ground Encoding System," *U.S. Gov. Rep. AD 426999,* January 13, 1964.

Barr, P. (Ed.), "Voltage to Digital Converters and Digital Voltmeters," *Electromech. Des.,* January 1964, pp. 147–159.

Barton, K., "The Field Effect Transistor Used as a Low-Level Chopper," *Electron. Eng.,* February 1965, pp. 80–83.

Bannochie, J. G., and R. A. E. Fursey, "Sealed Contact Relays," *J. Br. IRE,* February 1961, pp. 193–199.

Bazilinski, M., "The D-C Chopper Amplifier," *Electron. Des. News,* March 1966, pp. 98–103.

Beneteau, P. J., and B. Murari, "D.C. Amplifiers Using Transistors," *Electron. Eng.,* April 1963, pp. 257–259.

Beneteau, P. J., G. M. Riva, B. Murari, and G. Quinzio, "Drift Compensation in dc Amplifiers," *Solid-State Des.,* May 1964, pp. 19–23.

Benedict, R. P., "Thermoelectric Effects," *Electr. Manuf.,* February 1960, pp. 103–118.

Block, D. E., "Digital Data Acquisition System," *Instrum. Control Syst.,* August 1964, pp. 97–102.

Boysel, L. L., and D. R. Hill, "Large-Scale Integrated Metal Oxide Semiconductor Applied to Data Acquisition," *NEREM Rec.,* 1966, pp. 74–75.

Breuer, D. R., "An Integrated DC Differential Instrumentation Amplifier," *Proc. Natl. Electron. Conf.,* 1965, pp. 946–951.

Brooker, R. D., and R. B. Stephens, "Low-Level High-Speed Data Scanning Systems," *J. Br. IRE,* October 1962, pp. 277–286.

Brown, G. A., "Converter Compendium," *U.S. Gov. Rep. AD 636189,* July 1, 1966.

Buchmann, A. S., "Noise Control in Low Level Data Systems," *Electromech. Des.,* September 1962, pp. 64–81.

Budde, C. A., "A/D Converter Goes Adaptive," *Electron. Des.,* February 15, 1966, pp. 77–80.

Burgess, T., "Specifying Transformers for Low-Level Systems," *Electron. Prod.,* February 1963, pp. 36–38.

Burns, W. L., and K. W. Foulke, "An Advanced Microelectronic PCM Multicoder," *IEEE Trans. Commun. Technol.,* April 1966, pp. 162–169.

Cerni, R. H., "Transducers in Digital Process Control," *Instrum. Control Syst.,* September 1964, pp. 123–126.

Cerni, R. H., and L. E. Foster, *Instrumentation for Engineering Measurement,* Wiley, New York, 1962.

Colin, D., "High-Speed Techniques in Analog-to-Digital Conversion," *Electron. Instrum. Dig.,* July 1966, pp. 49–54.

Crawford, R. H., *MOSFET in Circuit Design,* McGraw-Hill, New York, 1967.

Cronhjort, B. T., "A Time Coding Analog-to-Digital Converter," *Proc. IEEE,* November 1963, pp. 1541–1549.

Davis, W. R., "Exploiting the Mercury Jet Switch," *Control Eng.,* February 1957, pp. 69–72.

Delhom, L., "Choosing Low-Level Switches—Bipolar Transistors or FET's?" *Electro-Technol.,* December 1968, pp. 42–46.

De'Sa, O., and L. Molyneux, "A Transistor Voltage-to-Frequency Converter," *Electron. Eng.*, July 1962, pp. 468–469.

Effenberger, J. A., and R. Steele, "A Transistorized Integrator," *Electron. Eng.*, April 1965, pp. 230–234.

Ekiss, J. A., "Transistor Choppers," *Semicond. Prod.*, October 1960, pp. 23–27.

Fattal, L., "Field Effect Transistors as Choppers," *SCP and Solid-State Technol.*, April 1964, pp. 13–18.

Feulner, R. J., and J. B. Robertson, "A Unique Analog-to-Digital Converter for Industrial Telemetry," *Autom. Control,* January 1963, pp. 11–16.

Finlay, D. J., "The Analysis and Performance of Transistor Choppers," *Electron. Eng.,* September 1966, pp. 572–578.

Flynn, G. J., "Processing Digital Data," *Electron.,* May 18, 1964, pp. 64–69.

Flynn, G., "Analog-to-Digital Converters," *Electron. Prod.,* October 1967, pp. 18 ff.

Fuchs, H., and K. B. Rhodes, "A Transistorized Data Amplifier," *Electron. Eng.,* December 1961, pp. 778–782.

Fuller, D. W., "Low-Level Multiplexing for Digital Data Processing," *Autom. Control,* March 1962, pp. 40–41.

Gatti, E., "General Survey: Fast Analog to Digital Converters," *Proc. Conf. Autom. Acquisition and Reduction,* 1964, pp. 259–265.

Gibbons, J. F., "Super-Saturated Transistor Switches," *IEEE Trans. Electron Devices,* November 1961, pp. 443–452.

Gray, J. R., and S. C. Kitsopoulos, "A Precision Sample and Hold Circuit with Subnanosecond Switching," *WESCON Conv. Rec.,* Part 2, August, 1963.

Greenberg, R. I., "Propagation-Type A/D Converters Are Fastest," *Electron. Des.,* April 27, 1964, pp. 64–67, 69.

Griffin, C. E., J. P. Knight, and J. H. Searcy, "Low-Level Multiplexing for Digital Instrumentation," *Electron.,* October 7, 1960, pp. 64–66.

Grigson, C. W. B., "Some Precision Direct Coupled Transistor Amplifiers and Their Approximate Design," *Electron. Eng.,* July 1964, pp. 454–459.

Grove, A. S., E. H. Snow, and B. E. Deal, "Stable MOS Transistors," *Electro-Technol.,* December 1965, pp. 40–43.

Gruenberg, E. L. (Ed.), *Handbook of Telemetry and Remote Control,* McGraw-Hill, New York, 1967.

Hackel, I., and H. Hagemann, "Anwendung von Transistoren als Präzisionszerhacker," *Elektron. Rundsch.,* March 1963, pp. 122–132.

Hadlock, C. K., and D. Lebell, "Some Studies of Pulse Transformer Equivalent Circuits," *Proc. IRE,* January 1961, pp. 81–83.

Hakimoglu, A., and R. D. Kulvin, "Sampling Ten Million Words a Second," *Electron.,* February 7, 1964, pp. 52–54.

Hedin, R. A., "A Logical Approach to Comparator Design," *Electron. Des.,* October 11, 1965, pp. 50 ff.

Hendricks, L. V., "A Programmed Course on Field-Effect Transistors," *The Electron. Engineer,* September 1967, pp. 54–62.

Hilbiber, D. F., "Stable Differential Amplifier Designed Without Choppers," *Electron.,* January 25, 1965, pp. 73–75.

Hoehn, G. L., Jr., "Semiconductor Comparator Circuits," *Tech. Rep. No. 43*, Stanford Electronics Laboratories, July 28, 1958.

Hoeschele, D. F., Jr., *Analog-to-Digital/Digital-to-Analog Conversion Techniques*, Wiley, New York, 1968.

Hoffman, P. A., "A Precision Logarithmic A/D Converter," *Proc. Natl. Telemetering Conf.*, May 1963.

Howden, P. F., "A Review of Chopper Amplifiers," *Electro-Technol.*, June 1963, pp. 64–67.

Hutcheon, I. C., and D. Summers, "A Solid-State D.C. Amplifier with Floating Input and Low Drift," *Electron. Eng.*, September 1961, pp. 592–596.

Hudson, P. H., F. A. Lindholm, and D. J. Hamilton, "Transient Performance of Four Terminal Field-Effect Transistors," *IEEE Trans. Electron Devices*, July 1965, pp. 399–407.

Jensen, O. W., "The Junction FET as a Switch," *EEE*, April 1966, pp. 99–102.

Jasper, L., and H. K. Haill, "Improving A-D Converter Accuracy," *Electron.*, September 13, 1963, pp. 38–40.

Jones, J. D., "High-Speed Low-Level Data Acquisition," *Instrum. Control Syst.*, April 1965, pp. 96–101.

Jursik, J., "Rejecting Common Mode Noise in Process Data Systems," *Control Eng.*, August 1963, pp. 61–66.

Kane, J. F., and D. L. Woolesen, "Field-Effect Transistors in Theory and Practice," *Monitor*, pp. 13–24.

Klein, G., and J. J. Zaalberg van Zelst, "General Considerations on Difference Amplifiers," *Phillips Tech. Rev.*, July 1961, pp. 345–351.

Klein, G., and J. J. Zaalberg van Zelst, "Difference Amplifiers with a Rejection Factor Greater Than One Million," *Phillips Tech. Rev.*, August 1963, pp. 275–284.

Knight, J. P., L. R. Klinger, and D. C. Yoder, "Low-Level Data Multiplexing," *Instrum. Control Syst.*, August 1963, pp. 86–89.

Korn, G. A., "Performance of Operational Amplifiers with Electronic Mode Switching," *IEEE Trans. Electron. Comput.*, June 1963, pp. 310–312.

Kurkin, Y. L., and N. S. Kurkina, "Precision Transistor Integrator," *Autom. Remote Control*, July 1961, pp. 799–805.

Lee, R., *Electronic Transformers and Circuits*, 2nd ed., Wiley, New York, 1955.

Levin, M. H., "Advantages of Direct-Coupled Differential Data Amplifiers," *Hewlett-Packard J.*, July 1967, pp. 13–15.

Loeffler, R. F., "Thermocouples, Resistance Temperature Detectors, Thermistors—Which?" *Instrum. Control Syst.*, May 1967, pp. 89–93.

Luecke, G., "An Integrated Circuit Multiplex Switch," *Solid-State Electron.*, March 1964, pp. 205–214.

Lyden, A., "Single and Matched Pair Transistor Choppers," *Electron. Eng.*, March 1965, pp. 186–189.

McCullough, W. R., "A New and Unique Analog-to-Digital Conversion Technique," *IEEE Trans. Instrum. Meas.*, December 1966, pp. 276–286.

Manfredi, P. F., and A. Rimini, "Analysis of Error and of the Precisions Obtainable in the Operation of Analog to Digital Conversion," *Energ. Nucl.*, June 1963, pp. 301–314.

May, P. T., "An Analogue to Digital Converter," *U.S. Gov. Rep. AD 424982*, September 1963.

Martin, M. E., and W. M. Phillips, "Voltage-Transfer Low-Level Commutation System," *Proc. Natl. Telemetering Conf.*, 1959, pp. 161–167.

Meyer-Brötz, G., "Grundschaltungen von Gleichspannungsverstärkern in der Messwertverarbeitung," *Int. Elektron. Rundsch.*, July 1963, pp. 341–344.

Meyer-Brötz, G., and A. Kley, "The Common-Mode Rejection of Transistor Differential Amplifiers," *IEEE Trans. Circuit Theory*, June 1966, pp. 171–175.

Middlebrook, R. D., *Differential Amplifiers*, Wiley, New York, 1963.

Middlebrook, R. D., and A. D. Taylor, "Differential Amplifier with Regulator Achieves High Stability, Low Drift," *Electron.*, July 28, 1961, pp. 56–59.

Moss, R. Y., "Errors in Data Amplifier Systems," *Hewlett-Packard J.*, July 1967, pp. 17–20.

Munoz, R. M., "A Nonlinear Analog/Digital Converter for More Constant Accuracy," *EEE*, April 1965, pp. 57–59.

Muth, S., Jr., "Reference Junctions," *Instrum. Control Syst.*, May 1967, pp. 133–134.

Okada, R. H., "Stable Transistor Wide-Band D-C Amplifiers," *AIEE Trans. (Commun. and Electron.)*, March 1960, pp. 26–33.

Onstad, R. C., "Solid-State 30-Channel Multiplexer Designed for Minimum Components," *Electron.*, October 6, 1961, pp. 77–79.

O'Reilly, T. J., "The Transient Response of Insulated-Gate Field-Effect Transistors," *Solid-State Electron.*, August 1965, pp. 947–956.

Ostler, R. I., "Practical Method of Analogue-to-Digital Conversion," *Instrum. Pract.*, December 1960, pp. 1305–1312.

Palmer, J., "Spikes on Chopper Transistor Waveforms," *Electron. Eng.*, March 1962, pp. 163–167.

Pieczonka, W. A., T. H. Yeh, P. Polgar, and M. M. Roy, "GaAs-Si Photon-Activated Switch," *IEEE Trans. Electron Devices*, December 1965, pp. 643–649.

Rado, J. J., "Input Impedance of a Chopper-Modulated Amplifier," *Electro-Technol.*, June 1962, pp. 140 ff.

Rea, D. E., "Field Effect Transistors—Simpler Switches for Analog Signals," *Control Eng.*, July 1965, pp. 66–69.

Rensch, H., "Characteristics and Applications of Reed Contacts," *Electr. Commun.*, March 1965, pp. 385–397.

Richards, J. C. S., "A D.C. Differential Amplifier Using Field Effect Transistors," *Electron. Eng.*, September 1965, pp. 598–601.

Rose, J., "Straight Talk About Data Amplifiers," *EDN*, November 23, 1966, pp. 26–33.

Rose, J., "A 100-Db Dynamic-Range Amplifier!", *Electron. Des. News*, April 1967, pp. 66–69.

Roth, J. F., "Collection of Analogue Plant Data for On-Line Computer Processing," *Ind. Electron.*, March 1963, pp. 326–329.

Ruegg, H. W., "An Integrated FET Analog Switch," *Proc. IEEE*, December 1964, pp. 1572–1575.

Sandlin, W. C., "Case History: Designing a Microelectronic Encoder," *Electron. Des.*, October 12, 1964, pp. 82 ff.

Schantz, S. C., "Which Reed Relay Package?" *EEE,* June 1966, pp. 67–71.

Schindler, H. R., "Using the Latest Semiconductor Circuits in a UHF Digital Converter," *Electron.*, August 30, 1963, pp. 37–40.

Schmid, H., and B. Grindle, "Pulse-Width Modulator Offers Precision Performance," *Electron.*, October 11, 1963, pp. 29–31.

Schmid, H., "Digital Meters for Under $100," *Electron.*, November 28, 1966, pp. 88–94.

Schover, D. S., and M. Stein, "High Speed Threshold Logic A/D Converter with Error Correction," *IEEE Int. Conv. Rec.*, Part 1, 1964, pp. 22–29.

Sear, B. E., "Research for High Speed Analog-to-Digital Conversion Techniques," *U.S. Gov. Rep. AD 604421,* August 1964.

Sear, B. E., "Gigabit Circuit Design for an Encoder with a 200-Megacycle Sample Rate," *Proc. Natl. Electron. Conf.*, 1964, pp. 263–271.

Sevin, L. J., Jr., *Field Effect Transistors,* McGraw-Hill, New York, 1965.

Shea, R. F., *Amplifier Handbook,* McGraw-Hill, New York, 1966.

Shipley, M., "PHOTOFET Characteristics and Applications," *Solid-State Des.*, April 1964, pp. 28–30.

Shipley, M., Sr., "Analog Switching Circuits Use Field Effect Devices," *Electron.,* December 28, 1964, pp. 45–51.

Simpson, J. H., "Temperature Effects in Low-Level Transistor Choppers," *Solid-State Des.*, March 1964, pp. 22–28.

Smith, K. L., "Data Acquisition Systems for Control Applications," *J. Br. IRE,* December 1962, pp. 497–506.

Sommer, B. I., and G. W. Plice, "Specification and Testing of Shielded Transformers," *Electro-Technol.*, May 1963, pp. 102–105.

Sommer, B. I., "Stop Thermal Drift in Low-Level D-C Circuits," *ISA J.*, February 1965, pp. 85–86.

Steiger, W., "A Transistor Temperature Analysis and Its Application to Differential Amplifiers," *IRE Trans. Instrum.*, December 1959, pp. 82–91.

Stuttard, E. B., "Some General Features of Digital Data Acquisition Systems," *J. Br. IRE*, October 1962, pp. 263–267.

Stewart, R. D., "Solid-State Optoelectronic Commutator," *Electron.*, February 16, 1962, pp. 38–39.

Susskind, A. K. (Ed.), *Notes on Analog-Digital Conversion Techniques,* Technology Press, Cambridge, Mass., and Wiley, New York, 1957.

Taskett, D., "Common-Mode Considerations for Differential Operational Amplifiers," *Electro-Technol.*, May 1967, pp. 40–41.

Tavares, S. E., "A Comparison of Integration and Low-Pass Filtering," *IEEE Trans. Instrum. Meas.*, March–June 1966, pp. 33–38.

Timothy, K. O., "Singularities in Differential Amplifiers," *The Electron. Engineer,* September 1967, pp. 47–51.

Toney, P. A., "Design Considerations for Differential Low-Level Junction FET Commutators," *IEEE Trans. Aerosp. and Electron. Syst.*, November 1967, pp. 871–879.

Verster, T. C., "A Method to Increase the Accuracy of Fast Serial-Parallel Analog-to-Digital Converters," *IEEE Trans. Electron. Comput.*, August 1964, pp. 471–473.

Walker, A. D. "A Voltage Controlled Oscillator Using a Standard Integrated Circuit," *Electron. Eng.*, January 1966, pp. 21–23.

Walton, C. A., "An Analog Input and Output System for a Real-Time Process Control Computer System," *Proc. Joint Autom. Control Conf.*, 1962, Paper 13-4.

Weir, B., and T. Prosser, "Temperature Controlled Microcircuits," *WESCON Conv. Rec.*, August 1964.

Widler, R. J., "A Fast Intergrated Circuit Comparator and Five Ways to Use It," *Electron. Des. News,* May 1965, pp. 20–27.

Williams, A. J., Jr., J. U. Eynon, and N. E. Polster, "Some Advances in Transistor Modulators for Precise Measurements," *Proc. Natl. Electron. Conf.*, 1957, pp. 40–54.

Wunderman, I., "Optoelectronics at Work," *Electron.*, July 27, 1964, pp. 58–62.

Wunderman, I., "Applications for Optic Couplers," *Electron.*, July 27, 1964, pp. 62–67.

Yee, S., "Characteristics and Operation of a High-Speed Electronic Analog Switch," *IEEE Int. Conv. Rec.*, Part 4, 1963, pp. 25–36.

Yoder, D. C., "Random Access Low-Level Multiplexing," *Proc. Joint Autom. Control Conf.*, 1966, pp. 103–115.

Young, E. J., "A Low-Level, High-Speed Sampling Switch," *Proc. Natl. Telemetering Conf.*, 1959, pp. 176–180.

Young, F. M., "Factors Limiting A-D Conversion 'State of the Art,'" *Data Syst. Eng.*, May 1964, pp. 35–40.

Analog-Digital Conversion Handbook, Digital Equipment Corp., Maynard, Mass., 1964.

"Analog-to-Digital Conversion—How and How Fast," *Electron. Instrum. Dig.*, July 1966, pp. 45–48.

"Analog-to-Digital Conversion—Why and When?," *Electron. Instrum. Dig.*, July 1966, pp. 42–44.

"IBM 1800 Data Acquisition and Control System Installation Manual—Physical Planning," Form A26-5922, International Business Machines Corp., San Jose, Calif., 1966.

"IBM 1800 Data Acquisition and Control System—Process Input/Output Features," Form 227-5959, International Business Machines Corp., San Jose, Calif.

"IBM 1800 Functional Characteristics," Form A26-5918, International Business Machines Corp., San Jose, Calif., 1966.

Chapter 4

Analog-Input Subsystem Engineering

THOMAS J. HARRISON

The previous chapter discusses equipment that is normally included in an analog-input subsystem. The chapter is directed toward an identification of major design considerations. The analog-input subsystem, however, is more than a collection of independent circuits. In a sense, the subsystem is greater than the sum of its parts. Even though each individual circuit is well designed, the design engineer faces additional problems posed by interactions between the subsystem components.

This chapter is concerned with overall subsystem design considerations, including subsystem common mode performance, grounding and shielding for noise control, errors related to the use of the subsystem, the problem of specifying overall subsystem performance, and consideration of maintenance problems.

4.1   SUBSYSTEM COMMON MODE PERFORMANCE

The common mode performance of an analog-input subsystem is a good example of a performance parameter that is dependent on the overall subsystem design as well as on the design of the equipment that comprises the subsystem. Overall performance can be degraded or enhanced depending on the common mode rejection of cascaded circuits. In addition, the normal mode characteristics of the signal conditioning circuit can increase the subsystem common mode performance.

4.1.1   Cascade Common Mode Rejection

If an individual piece of equipment in an analog-input subsystem does not provide isolation from common mode potential, its common mode error is combined with the error generated in subsequent equipment. As a result, the total common mode error is the net effect of errors in the cascaded termination, signal conditioning, and multiplexing circuits. In addition, if the multiplexer does not provide common mode isolation, the common to normal mode conversion of the amplifier or ADC also must be considered.

As in the case of the common mode analyses in the previous chapter, an exact analysis is difficult because of the distributed nature of the leakages which produce the common mode error. Lumped models, however, provide guidance in understanding the qualitative relationship between the common mode rejection ratios of cascaded equipment and the overall performance.

A lumped model for a DC analysis is shown in Fig. 4-1. It consists of two cascaded sections of the form considered in Sec. 3.9. The first section consisting of R_1, R_1', R_2, and R_2' might represent the series and shunt resistances in the termination facility and signal conditioning circuits

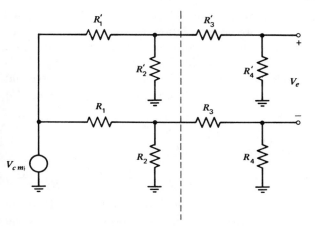

Fig. 4-1   Two-section cascade DC common mode model.

with the remaining resistors representing the model of a DPST differential multiplexer. Although the model of an actual system may consist of three or four cascaded sections, the analysis of the two-section model illustrates the basic concepts of cascaded common mode performance.

Converting the first section of the model in Fig. 4-1 to a Thevenin equivalent circuit as shown in Fig. 4-2 provides a circuit that can be analyzed by inspection. The modified common mode source $R_2 V_{cm}/(R_1$

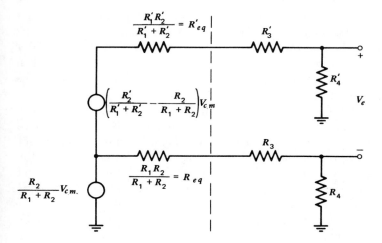

Fig. 4-2   Equivalent two-section cascade model.

$+ R_2)$ is the attenuated common mode voltage that would be applied to the second section if there were no interaction between the two sections. Similarly, the normal mode source is the error voltage due to the first section when the two sections are isolated. The total error voltage V_e is the sum of the voltage produced by the modified common mode source and the equivalent normal mode voltage. By superposition, the expression for V_e is

$$V_e = \left(\frac{R_2'}{R_1' + R_2'} - \frac{R_2}{R_1 + R_2}\right)\left(\frac{R_4'}{R_3' + R_4' + R_{eq}'}\right) V_{cm}$$
$$+ \left(\frac{R_4'}{R_3' + R_4' + R_{eq}'} - \frac{R_4}{R_3 + R_4 + R_{eq}}\right)\frac{R_2 V_{cm}}{R_1 + R_2} \quad (4\text{-}1)$$

Each term in this expression for V_e can be related to the CMRR of the single-section model derived in Sec. 3.9 if an assumption concerning the magnitude of the parameters is made. Typically, the leakage resistances R_2, R_2', R_4, and R_4' are orders of magnitude greater than the series resistances R_1, R_1', R_3, and R_3' so that the second factor in the first term is very close to unity; that is,

$$\frac{R_4'}{R_3' + R_4' + R_{eq}'} \cong 1 \quad (4\text{-}2)$$

This is equivalent to assuming that the second section does not attenuate the equivalent normal mode source in Fig. 4-2.

With the substitution of Eq. (4-2) into Eq. (4-1) and some minor rearrangement of form, the first factor is the reciprocal of the CMRR of the first section derived as Eq. (3-6) except for the omission of the absolute value signs. The absolute value in the definition of CMRR is a convention that restricts the CMRR to nonnegative numbers. Recognizing this convention, the first term of Eq. (4-1) is written as $V_{cm}/(\pm \text{CMRR}_1)$ where CMRR_1 is the common mode rejection ratio of the first section when considered as an isolated circuit and where it is understood that the positive sign is associated with the case $R_1 > R_1'$.

A similar reduction of the second term of Eq. (4-1) is possible. The second factor of this term is approximately unity because of the relative magnitudes of R_2 and R_1. Assuming that the factor is unity is equivalent to assuming that the common mode voltage is not attenuated by the resistor divider action of R_2 and R_1. The remaining factor in the second term is reduced using the previous assumption and the fact that $R_4 \cong R_4'$ in most practical subsystems.

The result of these manipulations is a reduced form of Eq. (4-1) which gives the approximate magnitude of the common mode error voltage V_e.

$$V_e \cong \left(\frac{1}{\pm \text{CMRR}_1} + \frac{R_3 - R_3'}{R_4} + \frac{R_1 - R_1'}{R_4} \right) \qquad (4\text{-}3)$$

The second term of Eq. (4-3) is the approximate expression for the CMRR of the second section, analogous to Eq. (3-7) in the previous analysis. Using the nomenclature $\pm \text{CMRR}_2$ to represent the CMRR of the second section with the understanding that the positive value is used when $R_3 > R_3'$, Eq. (4-3) is written as

$$V_e \cong V_{cm} \left(\frac{1}{\pm \text{CMRR}_1} + \frac{1}{\pm \text{CMRR}_2} + \frac{R_1 - R_1'}{R_4} \right) \qquad (4\text{-}4)$$

Therefore, the reciprocal of the common mode rejection of the cascaded circuit is

$$\frac{1}{\text{CMRR}} = \left| \frac{V_e}{V_{cm}} \right| \cong \left| \frac{1}{\pm \text{CMRR}_1} + \frac{1}{\pm \text{CMRR}_2} + \frac{R_1 - R_1'}{R_4} \right| \qquad (4\text{-}5)$$

This expression shows that the CMRR cannot be expressed as a simple function of the common mode rejection ratio of the individual sections. The interaction effect, represented by the last term, cannot be neglected since it is usually the same order of magnitude as the other terms. For example, if $R_2 \cong R_4$, the third term is approximately equal to $1/(\pm \text{CMRR}_1)$. The interaction means that estimating the overall common mode performance of an equal subsystem is difficult since, even though the actual CMRR of the cascaded sections may be known empirically, the unbalance $R_1 - R_1'$ and the leakage resistance R_4 can be only grossly estimated.

Despite this difficulty, however, Eq. (4-5) provides qualitative information concerning the common mode performance of cascaded subsystems. It shows that the common mode rejection of the cascaded subsystem may be significantly better than the performance of either section. Although the first and the third terms of Eq. (4-5) must have the same sign, the remaining term may have the opposite sign so that unbalances in the two sections compensate to form a system with less effective unbalance.

If the signs of all three terms in Eq. (4-5) are the same, the CMRR of the cascaded system is less than the CMRR of either section. A simple expression for the lower bound on the CMRR is difficult to obtain because of the interaction term. However, it is reasonable to assume that CMRR_1 and CMRR_2 are of the same order of magnitude and, in addition, the leakage resistances R_4 and R_2 are usually similar in magnitude. If this is the case, the CMRR of the cascaded system may be as small as one-third the CMRR of the individual sections. This result, although not strictly worst case, is a reasonable design guide for most analog-input subsystems.

The analysis of cascade AC common mode performance is similar to the DC case. The essential difference is that the reactances of the leakage capacitances to ground are substituted for the leakage resistances R_2, R'_2, R_4, and R'_4.

4.1.2   Common Mode Guard Shield

In the development of the AC common mode model in Sec. 3.9, it was shown that a capacitive unbalance of less than 20 pF is necessary to achieve a CMRR of 120 dB. Maintaining this degree of capacitance balance is difficult in a practical system and special techniques are sometimes required in high-performance subsystems. A common technique, borrowed from the field of high-precision laboratory instrumentation, is guard shielding.

The basic guard shield concept is shown in Fig. 4-3. The input lines to the subsystem, shown as a twisted pair, are enclosed in a shield that is connected to the common mode source. The capacitance between the shield and system ground is represented by C_3. The capacitances C_1 and C_2 are the conductor-to-shield capacitances. If the guard shield were not present, C_1 and C_2 would terminate at system ground. For a given installation, the conductor-to-shield capacitances C_1 and C_2 in the guarded configuration usually are greater than the conductor-to-ground capacitances in the nonguarded situation because of the proximity of the conductors to the shield.

It appears that the probability of unbalance is increased by the presence of the shield since C_1 and C_2 are usually greater. Theoretically, however, there is zero current flow through C_1 and C_2 because both the shield and the conductors are at the same common mode potential. Since no current flows through the line resistance of the conductors, there is no normal mode voltage produced. In reality, there is a small potential

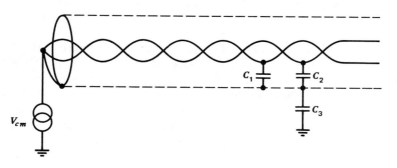

Fig. 4-3   Guard-shielding technique.

difference because of the voltage drop in the shield caused by current flow from the common mode source through C_3 to ground. The potential is small, however, since the resistance of the shield is low and the guard technique can result in a significant improvement in common mode performance.

The situation shown in Fig. 4-3 illustrates the use of a guard shield on the input lines to the subsystem. Guard shielding is also used internally in a subsystem to reduce common mode errors. Figure 4-4 illustrates the

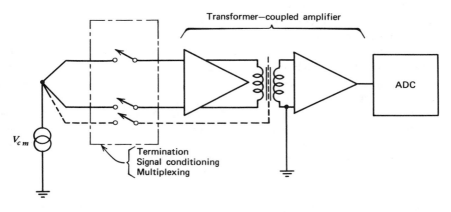

Fig. 4-4 Subsystem guard shield.

use of the technique both internal and external to a total subsystem. The shield encloses the input lines and is switched in the multiplexer with a third multiplexing switch. The switch is necessary since a shield common to two inputs can result in grounding a common mode source or in connecting two common mode sources of different potentials together through the low-resistance shield. The shield at the output of the multiplexer encloses a portion of the amplifier. The amplifier is transformer coupled and the guard shield includes the shield between the transformer windings. Since the differential signal is converted to a single-ended signal at the output of the transformer, the output section of the amplifier and the ADC are not included in the guard shield.

The figure illustrates a system that takes maximum advantage of the guarding technique since the shield encloses all the circuits in which common to normal mode conversion can take place. This is not the only way in which the guard can be handled in a system, however. The guard shield can be terminated at the termination facility and not multiplexed, or it can enclose the multiplexer switch and be terminated at the multiplexer output. From the discussion of cascade common mode performance in the previous section, it is obvious that the most effective general situation is when the guard encloses the entire differential portion of the system. This is also the most expensive design and adequate, even excellent, performance can be obtained with partial shielding.

Although guard shielding provides enhanced common mode rejection performance, it has disadvantages in some applications and subsystems. Shielding of the entire subsystem generally results in more expensive hardware. This is due not only to the cost of the shields, but also to the necessity for an additional multiplexer switch on each input. Furthermore, a guarded system requires that the user provide shielded inputs if the technique is to be even partially effective. Subsequent analysis shows that floating or grounding the shield usually results in a CMRR that is lower than the CMRR of the equivalent nonguarded system. Thus the use of shielded inputs is mandatory rather than elective if significant common mode voltages are present.

The guard technique also introduces design problems in the subsystem. System settling time is sometimes increased owing to the necessity of charging the additional shield capacitance in the guarded system. High-frequency noise can be coupled into low-level circuits through the shield because of its reduced effectiveness at high frequencies. In addition, the common mode performance of the guarded system is quite frequency sensitive. This is because the reactance of C_3 is small at high frequencies and significant currents can flow through the relatively large leakage capacitances C_1 and C_2 even though the potentials across these capacitances are low.

The models for the analysis of the guarded configuration and its comparison to the unguarded technique are shown in Fig. 4-5. In both models, the resistances R_1 and R_2 represent the total resistance of the input lines and the series resistance of the input circuit. The capacitances C_1 and C_2 are the conductor-to-ground leakage capacitance in the unguarded model, Fig. 4-5a, and the conductor-to-shield capacitances in the guarded model, Fig. 4-5b. The resistance R_3 in the guarded model is the sum of the shield resistance and the resistance between the shield and the common mode source. The capacitance C_3 is the shield-to-ground capacitance. Capacitance between the input conductors is neglected for simplicity. As shown in Chap. 3, this capacitance improves the common mode performance of the system so that neglecting it represents a conservative analysis.

The nonguarded model is equivalent to the model of Fig. 3-12 used in the derivation of Eq. (3-11). Utilizing the nomenclature $X_i = 1/\omega C_i$ for the reactance of the capacitors and $Y_i = 1/R_i$ for the admittance of the resistances, the CMRR for the nonguarded case is

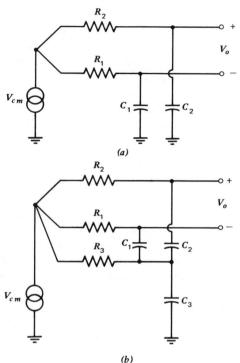

Fig. 4-5   Common mode models. (a) Nonguarded; (b) guarded.

$$\text{CMRR}_{\text{no guard}} = \left| \frac{V_{cm}}{V_o} \right| = \left| \frac{(1 + X_1Y_1)(1 + X_2Y_2)}{X_2Y_2 - X_1Y_1} \right| \qquad (4\text{-}6)$$

A similar analysis for the guarded model results in the expression

$$\text{CMRR}_{\text{guard}} = |(V_{cm}/V_o)| = \left| \frac{(1 + X_3Y_3)(1 + X_1Y_1)(1 + X_2Y_2)}{X_2Y_2 - X_1Y_1} \right.$$
$$\left. + \frac{X_3Y_1Y_2(X_1 + X_2) + X_3(Y_1 + Y_2)}{X_2Y_2 - X_1Y_1} \right| \qquad (4\text{-}7)$$

The effectiveness of the guard shield is examined by considering the ratio $R = \text{CMRR}_{\text{guard}}/\text{CMRR}_{\text{no guard}}$. Showing that this ratio is always greater than unity is equivalent to showing that the guard always results in improved common mode performance if the parameters in the two models are comparable. The respective line resistances in each model are comparable and, in fact, they are probably equal. The same assumption concerning the capacitances C_1 and C_2 requires some qualification. If the conductors in the nonguarded case are in close proximity to a ground, the leakage capacitances C_1 and C_2 in the two cases may be of the same order of magnitude. This could be the case, for example, if the input conductors to a subsystem are enclosed in a grounded conduit. Internal to the system, however, it is probable that C_1 and C_2 in the guarded case are appreciably greater than the equivalent capacitances in the unguarded case owing to the proximity of the shield and the conductors. Thus the assumption of comparable parameters in the two models must be made with caution in some situations.

Forming the ratio $R = \text{CMRR}_{\text{guard}}/\text{CMRR}_{\text{no guard}}$

$$R = \left| 1 + X_3 Y_3 + \frac{X_3(X_1 + X_2)Y_1 Y_2 + X_3(Y_1 + Y_2)}{(1 + X_1 Y_1)(1 + X_2 Y_2)} \right| \quad (4\text{-}8)$$

It is immediately noted that, if the shield resistance is negligible and if the guard is connected directly to the common mode source so that $R_3 = 0$ or $Y_3 = \infty$, the guarded case is infinitely better than the non-guarded case because of the second term in Eq. (4-8).

If $R_3 \neq 0$, however, a detailed examination of the expression is required. Even though the ratio consists of unity plus an apparently positive term, the expression is complex and its magnitude may be less than unity.

An easily proved sufficient condition that the magnitude of Eq. (4-8) is never less than unity is derived from a vector diagram of the equation. The vector diagram, shown in Fig. 4-6, consists of three vectors representing the three terms in the equation. The resultant vector **R** is the vector sum of the three components, and its magnitude is equal to Eq. (4-8). To satisfy the condition that R is never less than unity, the terminal point of **R** must never fall inside the unity circle shown dotted in the figure.

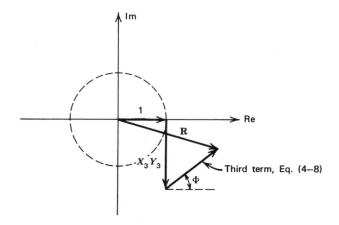

Fig. 4-6   Vector diagram illustrating Eq. (4-8).

It is obvious that a sufficient condition is that the angle Φ of the third term be bounded by $\pm\pi/2$. This, in turn, requires that the real part of the third term be nonnegative.

Considering only the third term of Eq. (4-8) and separating both the numerator and denominator into real and imaginary parts results in

$$\frac{X_3(X_1 + X_2)Y_1Y_2 + X_3(Y_1 + Y_2)}{(1 + X_1X_2Y_1Y_2) + (X_1Y_1 + X_2Y_2)} \tag{4-9}$$

where the first term in the numerator and denominator is the real part. Rationalizing the fraction and extracting the real part of the numerator provides the expression

$$X_1X_3Y_1{}^2(X_2{}^2Y_2{}^2 - 1) + X_2X_3Y_2{}^2(X_1{}^2Y_1{}^2 - 1) \tag{4-10}$$

The factors in parentheses are negative real and, since the first factor in each term is also negative real, the expression is nonnegative.

The conclusion drawn is that the use of a guard shield always results in enhanced common mode performance. The increase in performance depends on the magnitude of X_3 and Y_3 so that R_3 and C_3 should be minimized.

As noted in the introductory comments, a design that utilizes a guard shield normally makes the use of a shield mandatory. That is, the shield must be tied to the common mode source if it is to be effective. This can be shown by examining the performance of the system when the shield is floating as compared with a configuration with a properly terminated shield. The CMRR in the case of a floating shield is derived from Eq. (4-7) by allowing R_3 to approach infinity, the equivalent of Y_3 approaching zero. This results in the expression

$$\text{CMRR}_{\text{float guard}}$$
$$= \left| \frac{(1 + X_1Y_1)(1 + X_2Y_2) + X_3Y_1Y_2(X_1 + X_2) + X_3(Y_1 + Y_2)}{X_2Y_2 - X_1Y_1} \right| \tag{4-11}$$

The effect of floating the shield is evaluated by considering the ratio of Eq. (4-7) and Eq. (4-11). That is, showing that the expression of Eq. (4-12) is always greater than unity proves that floating the guard always results in a decrease in CMRR.

$$\frac{\text{CMRR}_{\text{guard}}}{\text{CMRR}_{\text{float guard}}}$$
$$= \left| 1 + \frac{X_3Y_3(1 + X_1Y_1)(1 + X_2Y_2)}{(1 + X_1Y_1)(1 + X_2Y_2) + X_3Y_1Y_2(X_1 + X_2) + X_3(Y_1 + Y_2)} \right| \tag{4-12}$$

As before, this is a complex expression and its interpretation can be based on the vector diagram of Fig. (4-7). Showing that the terminal

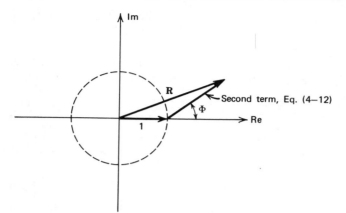

Fig. 4-7   Vector diagram illustrating Eq. (4-12).

point of **R** must lie outside the dotted unity circle in the diagram is equivalent to showing that its magnitude is always greater than unity. As before, this condition is satisfied if the magnitude of the angle Φ does not exceed $\pi/2$. This requires that the real part of the second term in Eq. (4-12) be nonnegative.

The real part of the second term in Eq. (4-12) is found by rationalizing the fraction and identifying real and imaginary terms. The real part is

$$X_3{}^2 Y_3[Y_1 Y_2(X_1{}^2 Y_1 + X_2{}^2 Y_2) - Y_1 - Y_2] \qquad (4\text{-}13)$$

This expression is nonnegative owing to the even orders of the reactance terms and the negative sign associated with the admittance terms. Thus the ratio R of Eq. (4-12) is never less than unity.

This result provides the important conclusion that it is always better to connect the shield than to leave it floating. Equation (4-12) shows, however, that the improvement in common mode performance is directly proportional to the admittance of the shield and the reactance of the leakage capacitance C_3. Thus both the resistance R_3 and the capacitance C_3 must be minimized if optimum common mode performance is to be achieved.

Since floating the shield is clearly not acceptable, both the designer and the user face a problem if it is not possible to terminate the shield at the common mode source. This might be the case, for example, if a guarded subsystem is installed in a preexisting plant in which shielded input lines have not been provided. Grounding the shield is probably not acceptable since this reduces the system to the nonguarded case except that the capacitances C_1 and C_2 may be significantly greater than in a nonguarded subsystem. Tying the shield to one or both of the input lines through resistors is unacceptable because of its deleterious effect on the DC CMRR and/or input impedance. A technique that is sometimes useful, however, is to couple the shield capacitively to the input lines as close to the common mode source as possible. This reduces the AC input impedance but does not seriously affect the DC CMRR if high-quality capacitors are used. Although the resultant AC CMRR is less than if the guard were terminated at the common mode source, it is probably greater than if the guard were left floating.

4.1.3   Gain-Dependent Common Mode Performance

In Sec. 3.14.4 it is shown that the common mode performance of some types of amplifiers is gain dependent. Another source of gain-dependent common mode error is capacitive leakage around gain-producing portions of the subsystem.

The basic problem is illustrated in Fig. 4-8, which shows an AC common mode source connected to an amplifier though a direct-coupled multiplexer. Ideally, the output voltage V_o of the amplifier is dependent only on the amplifier input voltage so that $V_o = GV_{in}$. In reality, however, the finite amplifier output impedance R_o and the voltage dividers formed by the stray capacitances C_1, C_2, and C_3 result in an output voltage that is dependent on the common mode voltage V_{cm}. This common mode error is independent of the gain and, because of the manner in which CMRR is defined, results in a gain-dependent CMRR.

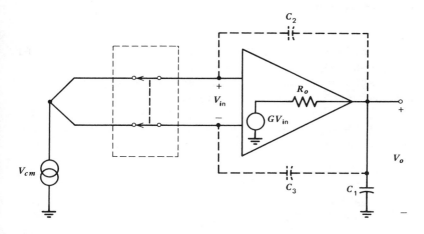

Fig. 4-8   Common mode feedthrough.

The phenomenon associated with capacitive leakage around the gain-producing element of the subsystem is not restricted to configurations that utilize direct-coupled multiplexers. Figure 4-9 shows the same situation for a flying capacitor multiplexer. The magnitude of the error in this case is not the same as in the direct-coupled case because the effective common mode voltage V_{cm} is less than the actual common mode voltage V'_{cm} owing to the divider action of C_5–C_8. In addition, the amplifier used in the flying capacitor configuration may be single ended, as shown by the dotted ground connection in Fig. 4-9.

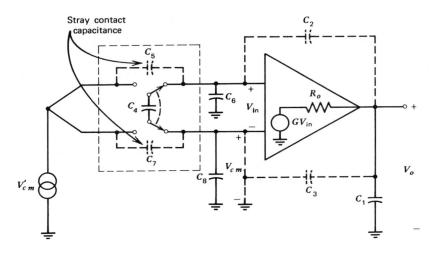

Fig. 4-9   Common mode feedthrough—flying capacitor multiplexer.

The error due to common mode feedthrough is negligible if the amplifier output impedance R_o is small compared with the reactances of C_2 and C_3. As a pessimistic assumption for the purposes of analysis, however, R_o is assumed to be large compared with these reactances. Neglecting the amplifier output voltage GV_{in} by superposition, the configurations of Figs. 4-8 and 4-9 are represented by the model of Fig. 4-10. The output voltage V_o is the error due to the common mode voltage and is given by

$$V_o = \frac{C_2 V_1 + C_3 V_2}{C_1 + C_2 + C_3} \tag{4-14}$$

For all practical purposes, $V_1 = V_2 = V_{cm}$ in the differential case so that

$$V_o = V_{cm} \frac{C_2 + C_3}{C_1 + C_2 + C_3} \tag{4-15}$$

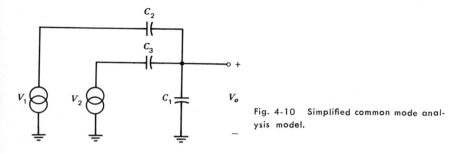

Fig. 4-10   Simplified common mode analysis model.

Utilizing the definition for the CMRR in terms of the output voltage as defined in Eq. (3-8), the CMRR of this differential model is

$$\text{CMRR} = G\frac{C_1 + C_2 + C_3}{C_2 + C_3} \qquad (4\text{-}16)$$

If the single-ended amplifier is utilized, as is possible in the flying capacitor multiplexer configuration, the common mode source V_2 in Fig. (4-10) is shorted to ground. The common mode error and the common mode rejection ratio of this single-ended configuration is

$$V_o = V_{cm}\frac{C_2}{C_1 + C_2 + C_3} \qquad (4\text{-}17)$$

$$\text{CMRR} = G\frac{C_1 + C_2 + C_3}{C_2} \qquad (4\text{-}18)$$

The equations for the CMRR of the two configurations show the dependence of the common mode rejection on the amplifier gain. As an example of the magnitude of the effect, suppose that $C_1 = 30$ pF and $C_2 = C_3 = 1$ pF. The CMRR in the differential case ranges from 16:1–16,000:1 for gains of 1–1000. In the single-ended case, the CMRR ranges from 32:1–32,000:1 for the same range of gains. These CMRR figures are very low compared with the common mode performance expected from most analog subsystems. Although this partially results from neglecting the effect of the amplifier output impedance, this example emphasizes the importance of reducing the feedthrough capacitances C_2 and C_3 to a fraction of a picofarad. Physical isolation of amplifier input and output circuits and avoidance of capacitive feedback paths from input to output are effective measures for reducing the magnitude of these capacitances.

From the example and Eqs. (4-15) and (4-17), the output error is reduced and, therefore, the CMRR is enhanced when the amplifier is single ended. The single-ended configuration is possible only when the multiplexer provides the required common mode isolation, such as in the flying capacitor multiplexer. This analysis substantiates the statement made in Sec. 3.13.3.3 that the AC CMRR of the flying capacitor system is increased when a single-ended amplifier is used. As shown there, however, the DC CMRR may be decreased.

4.1.4   Filters and Common Mode Performance

Low-pass filters used for signal conditioning in the analog-input subsystem can enhance both the DC and AC common mode performance of the subsystem. The improvement in DC CMRR is due to the transient characteristics of the filter and is related to the per-channel sampling rate of the subsystem. The increase in AC CMRR is related primarily to the attenuation characteristics of the filter, although transient characteristics may be a factor if the system sampling time is small compared with the period of the common mode voltage.

The control of leakage capacitances which result in common to normal mode conversion is particularly difficult in the termination facility and in the input portion of the multiplexer. This is because of the mass of the terminals and the large number of wires that are concentrated in these areas. In addition, the common mode performance of the installed system is highly dependent on the instrumentation wiring, which is not under the direct control of the system designer. As a result, there may be significant common to normal mode conversion before the input signals are multiplexed.

If the filters used for signal conditioning are located after the cause of the common to normal mode conversion, the filters attenuate the AC normal mode signal which is the common mode error. As a result, the error signal actually measured by the subsystem may be considerably less than the actual error signal. Since the error is less, the CMRR is greater and the performance of the subsystem is improved.

If the majority of the AC common to normal mode conversion takes place before the signal reaches the filter, the improvement in CMRR is approximately equal to the filter attenuation at the frequency of the common mode interference. For example, if the common mode source has a frequency of 60 Hz and the filter has 40-dB attenuation at this frequency, the CMRR of the subsystem is increased by approximately 40 dB at 60 Hz.

The filter does not eliminate the common mode potential, however, and subsequent equipment is subjected to the common mode voltage. If significant common to normal mode conversion takes place after the filter, the increase in CMRR is not as great.

The enhancement of DC common mode performance due to the transient properties of the filter are analyzed using the model of Fig. 4-11

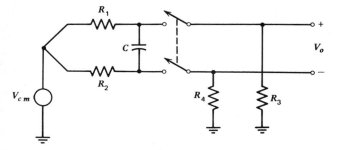

Fig. 4-11   Common mode filter model.

(Jursik, 1962, 1963). The resistances R_1 and R_2 are the sum of the line and filter resistances, R_3 and R_4 are the leakage resistances on the output side of the multiplexer, and C is the filter capacitor. The leakage resistances on the input side of the multlplexer also affect the CMRR but this effect can be estimated using the static model described in Sec. 3.9. The present analysis is simplified by neglecting these leakage resistances. The model of Fig. 4-11 assumes a single-section RC filter, although the effect of the filter and the analysis of the model is similar for other types of low-pass filters.

Qualitatively, the action of the filter in increasing the CMRR of the system is as follows. When the multiplexer switches close, the unbalanced line and/or leakage resistances produce a normal mode voltage. The rate of change of this voltage, however, is limited by the filter capacitor C. If the sampling time of the multiplexer is short compared with the time constant of the filter, the common mode error V_o does not reach the maximum value determined by the unbalanced line and/or leakage resistances. Therefore, the error voltage measured by the system is less than it would be if there were no filter and the common mode performance of the system is increased.

An equivalent circuit for the model of Fig. 4-11 is shown in Fig. 4-12 where $R = R_1 + R_2$ and it is assumed that $R_3, R_4 \gg R_1, R_2$. The equivalent input voltage to this circuit is a train of pulses whose duration h is equal to the closure time of the multiplexing switch and whose repetition period T is the time between samples on a particular input channel. The amplitude of the pulses is a function of R_1, R_2, R_3, and R_4 and is given by

$$V_m = V_{cm} \frac{R_2 R_3 - R_1 R_4}{(R_1 + R_3)(R_2 + R_4)} \tag{4-19}$$

As seen by comparison with Eq. (3-3), this amplitude is the value of the common mode error when the filter capacitor C is not present.

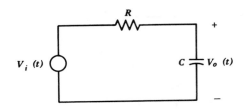

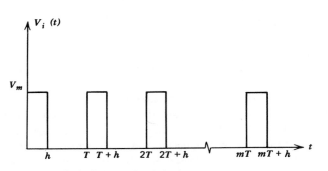

Fig. 4-12   Equivalent circuit and voltage source.

The output of the circuit of Fig. 4-12 in response to the equivalent input waveform is an exponential rise during the time intervals h, followed by an exponential decay between input pulses, as shown in Fig. 4-13. The mathematical expression for this output voltage is

$$
\begin{aligned}
V_o(t) = V_m\{(1 - e^{-t/RC}) - (1 - e^{-(t-h)/RC})U(t - h) \\
+ (1 - e^{-(t-T)/RC})U(t - T) - (1 - e^{-(t-T-h)/RC})U(t - T - h) \\
+ (1 - e^{-(t-2T)/RC})U(t - 2T) - \cdots\}
\end{aligned}
$$

(4-20)

where

$$
\begin{aligned}
U(t - \tau) &= 1 \qquad t > \tau \\
&= 0 \qquad t < \tau
\end{aligned}
$$

The first term in Eq. (4-20) represents the initial rise of the common mode error so that the maximum common mode error if the point is sampled only once is given by the value of this term at $t = h$. That is,

$$
V_o(h) = V_m(1 - e^{-h/RC})
$$

(4-21)

Since V_m is the common mode error with no filter, the second factor in Eq. (4-21) is the inverse of the improvement in the apparent CMRR of

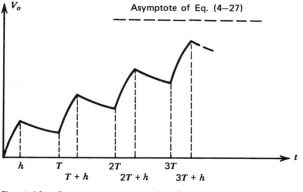

Fig. 4-13   Common mode error signal.

the model. As an example, a relay multiplexer might have $h = 10$ msec and a filter with $R = 1000\,\Omega$ and $C = 300\,\mu$F so that $RC = 0.3$. With these parameters, Eq. (4-21) becomes

$$V_o = V_m(1 - e^{-10^2/0.3}) \cong 0.03V_m \qquad (4\text{-}22)$$

The improvement in the apparent CMRR is a factor of $1.0/0.03 \cong 30$.

If the same input channel is sampled repeatedly, the common mode error continues to increase at a rate determined by the relative sizes of h and T. The interval of interest is the period during which the multiplexer is closed for the mth time; that is, $mT < t < (mT + h)$. By summing the terms in Eq. (4-20), the error voltage during this time interval is expressed as

$$V_o(t) = V_m - V_m e^{-t/RC}\left\{\frac{e^{(m+1)T/RC} - 1}{e^{T/RC} - 1} = e^{h/RC}\frac{e^{mT/RC} - 1}{e^{T/RC} - 1}\right\}$$
$$mT < T < mT + h \quad (4\text{-}23)$$

After a number of samples, $mT \gg RC$ and Eq. (4-23) reduces to

$$V_o(t) = V_m\left\{1 - e^{-(t-mT)/RC}\frac{1 - e^{-(T-h)/RC}}{1 - e^{-T/RC}}\right\} \qquad RC \ll mT < t < mT + h$$
$$(4\text{-}24)$$

The maximum common mode error occurs at the end of each charging interval, so let $t = mT + h$. Equation (4-24) reduces to

$$V_o(mT + h) = V_m\frac{1 - e^{-h/RC}}{1 - e^{-T/RC}} \qquad RC \ll mT \qquad (4\text{-}25)$$

This expression shows the relationship between the common mode error and the rate at which a point is sampled when many samples are taken on a single input channel. As the sampling rate is increased, that is, as T approaches h, the common mode error approaches V_m, which is the error when no filter is present. As the sampling rate decreases so that $T \gg h$, the common mode error approaches the value it had at the time of the first sample as given in Eq. (4-21).

Using the same values of R and C as in the previous example and a time interval of $T = 0.3$ sec, Eq. (4-25) has the value

$$V_o(mT + h) = V_m \frac{1 - e^{-10^{-2}/0.3}}{1 - e^{-0.3/0.3}} \cong V_m \frac{0.03}{0.6} = 0.05 V_m \qquad (4\text{-}26)$$

The common mode error is greater than in the previous example but it is significantly less than the error when no filter is used. The apparent increase in the CMRR is a factor of $1.0/0.05 = 20$.

This analysis indicates that a lower sample rate on filtered channels results in higher common mode performance, other factors being constant. In addition, Eq. (4-25) indicates that the CMRR is improved by increasing the filter time constant. A subsequent analysis of normal mode error owing to continued sampling of a filtered input shows that it is preferable to increase the time constant by increasing the value of C rather than R.

In some applications, it is of interest to know how long it takes the common mode error sketched in Fig. 4-13 to reach a steady state. This can be expressed in terms of a time constant for the output voltage V_o at the instants of time $t = mT$. Substituting these times into Eq. (4-23) results in the expression

$$V_o(mT) = V_m(1 - e^{-mT/RC}) \frac{1 - e^{-h/RC}}{1 - e^{-T/RC}} \qquad (4\text{-}27)$$

If all the parameters except m are fixed, this equation is a function of m only, and the effective time constant of the output voltage is given by RC/T. For $mT > 3\ RC/T$, the common mode error exceeds 95% of its final value. For the previous example, $3\ RC/T = 3$ so that the common mode transient has exceeded 95% of its final value after the first $m = 3/T = 10$ samples.

4.2 SAMPLING ERRORS

Although concerned primarily with the effects of common mode voltage, the analysis of the previous section illustrates that system errors can be related to system sampling rate. In addition to the common mode error, there are other errors associated with the sampling rate and the sampling time of the system. These include errors due to successive samplings on the same channel, errors due to sampling a low-level channel immediately following a high-level reading, and errors due to the aperture time of the system. In addition, errors can be introduced into data as a result of violation of the sampling theorem which relates the sampling frequency and the signal frequency.

4.2.1 Normal Mode Crosstalk

A channel scanning sequence in which a low-level channel is sampled immediately following a high-level channel can result in a normal mode error that is dependent on the value of the previously sampled signal. In reality, this error exists for any scanning sequence, but it is generally insignificant except in the case of a high-level to low-level sequence. The error results from the storage of charge on the stray capacitances at the output of the multiplexer and the input of the amplifier. These capacitances are charged to the potential of the high-level signal and, if a low-level point is then sampled, stored charge is transferred to or from the filter capacitor on the low-level channel. If the filter bandwidth is so low that the source cannot equalize this charge before the ADC is activated, an error results. Since the error depends, in part, on the inability of the low-level channel to equalize the excess charge, this source of error is found only in systems that utilize low-pass filters as signal conditioning elements.

The error is also dependent on the time that has elapsed between the end of the high-level sample and the beginning of the low-level reading. If the time between samples is long compared with the time constant of the capacitance and amplifier input resistance, the charge on the capacitance decays to a small value before the low-level channel is sampled and there is little crosstalk effect.

The error is also dependent on the multiplexer configuration since some multiplexers, such as the flying capacitor type, do not provide charging current from the signal source to neutralize the error during the conversion period. This is because the capacitors are physically disconnected from the source during the conversion time. Others, such as the DPST multiplexer, do allow equalization of charge during conversion.

The model for the analysis of the normal mode crosstalk phenomenon in the DPST multiplexer is shown in Fig. 4-14. Two channels, each having an individual low-pass RC filter, are connected to the common multiplexer bus through multiplexer switches S_1 and S_2. The capacitance of the multiplexer bus and the input circuit of subsequent equipment are represented by C_3 and the shunt resistance R_3.

The multiplexer output waveform of Fig. 4-15 is derived heuristically by consideration of the model. Multiplexer switch S_2 closes at some time

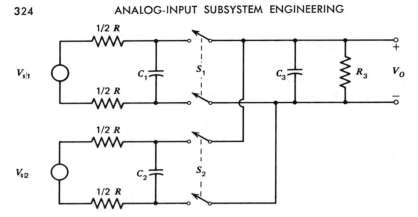

Fig. 4-14   Normal mode crosstalk model—DPST multiplexer.

$t = mT$ and remains closed for a period h. At time $mT + h$, S_2 opens and the voltage V_o decays owing to discharge of C_3 through R_3. At $t = (m + 1)\ T$, switch S_1 closes and the output voltage jumps to approximately V_{s1}, the signal voltage on channel 1. During the period that the switch is closed, $t = (m + 1)T$ to $(m + 1)T + h$, the capacitors C_1 and C_3 charge through an equivalent resistance determined by the filter resistance R and the input resistance R_3.

The waveform is a series of exponential functions with time constants

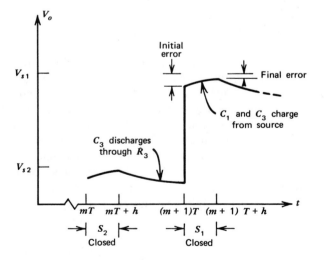

Fig. 4-15   Multiplexer bus output voltage.

determined by the various resistors and capacitors. During the interval between samples, the output voltage is a simple exponential decay whose value at $t = (m + 1)T$ just prior to the closure of S_1 is

$$V_o[(m + 1)T] = V_o(mT + h)e^{-(T-h)/R_3C_3} \qquad (4\text{-}28)$$

At the time S_1 closes, charge is shared between C_3 and C_1 according to the conservation of charge equation

$$C_3 V_o[(m + 1)T] + C_1 V_{s1} = (C_1 + C_3)V_o[(m + 1)T^+] \qquad (4\text{-}29)$$

where $V_o[(m + 1)T^+]$ is the output voltage immediately after the switch closes. This voltage is slightly less than V_{s1} because of the charge sharing.

During the time that S_1 is closed, the parallel combination of C_1 and C_3 is charged by an equivalent source $V_{s1}R_3/(R + R_3)$ through an equivalent source resistance of $RR_3/(R + R_3)$. The output voltage during this interval is

$$V_o(t) = \frac{R_3}{R + R_3} V_{s1} + \left(V_o[(m + 1)T^+] - \frac{R_3}{R + R_3} V_{s1} \right) e^{-(R+R_3)t/RR_3(C_1+C_3)}$$

$$(4\text{-}30)$$

This expression is simplified by recognizing that in a practical system $R_3 \gg R$ and $C_1 \gg C_3$ so that the coefficient of t in the exponential factor reduces to RC_1. In addition, the coefficient of V_{s1} is very close to unity. With these simplifications and substitution of $V_o[(m + 1)T^+]$ from Eq. (4–29), Eq. (4–30) becomes

$$V_o(t) = V_{s1} + \left\{ \frac{C_3}{C_1 + C_3} [V_o(mT + h)e^{-(T-h)/R_3C_3} - V_{s1}] \right\} e^{-t/RC_1} \qquad (4\text{-}31)$$

The second term of this expression is the error term since the true value of the signal on channel 1 is V_{s1}. The evaluation of the error term requires knowledge of $V_o(mT + h)$ which, in turn, depends on the signals sampled prior to the closing of switch s_2. As a practical matter, however, $V_o(mT + h) \cong V_{s2}$ if the time constant R_3C_3 is long compared with the time between samples $T - h$.

The error varies over the sampling interval and the actual error recorded by the system depends on the type of system since, for example, a sample-and-hold system can always have a maximum error if the hold command is given just before s_1 opens. The error is bounded, however, by its value at the beginning and end of the conversion interval. Defining the error as a percentage of the full-scale voltage V_{fs} by $\mathcal{E} = 100(V_{s1} - V_o)/V_{fs}$, the bounds are

$$\varepsilon_{\max} = \frac{100 C_3}{V_{fs}(C_1 + C_3)} [V_{s1} - V_{s2} e^{-(T-h)/R_3 C_3}] \tag{4-32}$$

$$\varepsilon_{\min} = \varepsilon_{\max} e^{-h/RC_1} \tag{4-33}$$

As an example of the magnitude of this error, a relay multiplexer system might have the following parameters:

$$
\begin{array}{ll}
h = 10 \text{ msec} & V_{fs} = V_{s1} = 10 \text{ mV} \\
C_1 = 300 \ \mu\text{F} & C_3 = 300 \text{ pF} \\
R = 1000 \ \Omega & R_3 = 10^7 \ \Omega \\
V_{s2} = 5 \text{ V}
\end{array}
\tag{4-34}
$$

With these parameters and the assumption that channel 1 is sampled immediately after channel 2 so that $T - h = 0$, the maximum and minimum errors are 0.05% and 0.0485%.

The analysis of the error for a flying capacitor multiplexer is similar. The error at the time switch s_1 closes is the same as in the previous case except that it is now the minimum error; the error increases during the conversion interval owing to discharge of C_1 and C_3 through R_3. With the same assumptions as in the previous analysis, the errors as a percentage of the full-scale voltage of channel 1 are

$$\varepsilon_{\min} = \frac{100 C_3}{V_{fs}(C_1 + C_3)} [V_{s1} - V_{s2} e^{-(T-h)/R_3 C_3}] \tag{4-35}$$

$$\varepsilon_{\max} = \frac{100}{V_{fs}} \left[V_{s1} \left(1 - \frac{C_1}{C_1 + C_3} \right) e^{-h/R_3 C_1} - \frac{C_3 V_{s2}}{C_1 + C_3} \ e^{-(T-h)/R_3 C_3} e^{-h/R_3 C_1} \right] \tag{4-36}$$

Using the parameters of Eq. (4-34) as an example, the maximum error is slightly greater than in the DPST multiplexer example.

Examination of the expressions of Eqs. (4-32), (4-33), (4-35), and (4-36) provides a number of conclusions. The error is clearly a cross-talk effect, as indicated by the presence of the V_{s2} factor. The equations also show that the error is dependent on the time between samples $T - h$ and on the length of the sample h. The effect is, therefore, affected by the system sampling rate. The equations show that increasing C_1 reduces the error. This substantiates the comment made in the previous section that increasing the filter time constant by increasing C_1 rather than R is preferable.

This analysis of normal mode crosstalk errors and the numerical examples show that the phenomenon can be a source of significant error in systems that sample both low- and high-level signals in a random sequence. Since the error is due to charge stored on the capacitance of the multiplexer output bus and the amplifier input circuit, it can be reduced by ensuring that this charge is negligible at the start of each conversion. This is easily accomplished by providing a switch that shorts out these capacitances between sampling intervals. Care must be taken in the design of the shorting switch to ensure that the capacitances are not recharged by the control signal used to turn the switch OFF. With a shorting switch, V_{s2} in the previous equations is zero and the error expressions for the DPST multiplexer reduce to

$$\varepsilon_{\max} = \frac{100 C_3 V_{s1}}{(C_1 + C_3) V_{fs}}$$

$$\varepsilon_{\min} = \varepsilon_{\max} e^{-h/RC_1} \qquad (4\text{-}37)$$

In the flying capacitor configuration, the errors are

$$\varepsilon_{\max} = 100 \left(1 - \frac{C_1}{C_1 + C_3} e^{-h/R_s C_1} \right) \frac{V_{s1}}{V_{fs}}$$

$$\varepsilon_{\min} = \frac{100 C_3 V_{s1}}{(C_1 + C_3) V_{fs}} \qquad (4\text{-}38)$$

4.2.2 Effect of Continued Sampling

The analysis of the previous section shows that charging the multiplexer and amplifier capacitance reduces the filter output voltage because of the charge-sharing action. This same phenomenon was noted in the flying capacitor common mode analysis of the previous chapter. If the charge removed from the filter capacitor is not replaced before another sample is taken, the error is further increased by the effect of successive samples.

The model for an analysis of the effect of continuous sampling on a filtered channel is shown in Fig. 4-16 along with a sketch of the waveform at two points in the circuit. When the multiplexer switch s_1 closes at $t = mT$, there is charge sharing owing to the difference in potential across the two capacitors. During the interval mT to $mT + h$ when the switch is closed, both capacitors are charged by the signal source through an equivalent resistance determined by R_3 and R. When the switch opens at $t = mT + h$, C_1 continues to charge through the resistance R but C_3 begins to discharge through R_3. At time $(m + 1)T$, the switch closes and the cycle repeats. The output voltage V_o never actually reaches the input voltage value V_{s1}, but it does reach a steady-state condition where the voltage at any point in the cycle is the same as it was in the previous cycle.

The exact analysis of this model results in nonlinear difference equations whose solutions are quite difficult. A less exact but manageable analysis is obtained by making several assumptions. First, the shorting

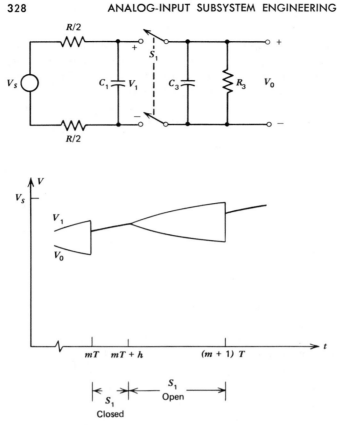

Fig. 4-16   Model for successive sampling error.

switch described in the previous section is used to fix the voltage across
C_3 at the time the switch closes. The effects of both R_3 and C_3 are
neglected during the charging interval since they have a small effect
on the time constant in a practical system. This means that the capacitance C_1 charges for the entire interval from mT to $(m + 1)T$ with a
time constant RC_1. The simplified circuit is shown in Fig. 4-17.

During the interval mT to $(m + 1)T$, the capacitor is charging
through a resistance R. The voltage across C_1 just before the switch
closes at $t = (m + 1)T$ is

$$V[(m + 1)T] = V_s + [V(mT^+) - V_s]e^{-T/RC_1} \qquad (4\text{-}39)$$

where $V(mT^+)$ is the voltage just after the switch closed at $t = mT$.
Using a conservation of charge equation similar to Eq. (4-29), this
voltage is

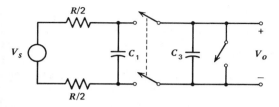

Fig. 4-17   Simplified model for successive sampling error.

$$V(mT^+) = V(mT) \frac{C_1}{C_1 + C_3} \equiv kV(mT) \qquad (4\text{-}40)$$

Substitution of this quantity into Eq. (4-39) results in

$$V[(m+1)T] = V_s(1 - e^{-T/\tau}) + kV(mT)e^{-T/\tau} \qquad (4\text{-}41)$$

where $\tau = RC_1$ has been defined.

The solution of this recursion equation for the output voltage at $t = (m+1)T$ is a series

$$V[(m+1)T] = V_s \left\{ k^{m+1}e^{-(m+1)T/\tau} + (1 - e^{T/\tau}) \sum_{i=0}^{m} (ke^{-T/\tau})^i \right\} \qquad (4\text{-}42)$$

This expression is simplified by some manipulation and the use of the formula for the sum of terms in a geometric progression with the result

$$V[mT] = V_s \left\{ \frac{1 - e^{-T/\tau} + k^m e^{-(m+1)T/\tau}(1 - k)}{1 - ke^{-T/\tau}} \right\} \qquad (4\text{-}43)$$

The error just after the switch closes at $t = mT$ as a percentage of the full-scale voltage V_{fs} is $\mathcal{E} = 100(V_s - kV[mT])/V_{fs}$. After substituting and using the definition of k in some factors, the error is

$$\mathcal{E} = \frac{100C_2V_s}{(C_1 + C_2)V_{fs}} \left\{ \frac{1 - k^{m+1}e^{-(m+1)T/\tau}}{1 - ke^{-T/\tau}} \right\} \qquad (4\text{-}44)$$

As the channel continues to be sampled, the exponential term in the numerator becomes small for $mT \gg \tau$. Thus, the error reaches a steady-state value of

$$\mathcal{E} = \frac{100C_2V_s}{(C_1 + C_2)V_{fs}} \left\{ \frac{1}{1 - ke^{-T/\tau}} \right\} \qquad (4\text{-}45)$$

This expression shows that the error is decreased as the size of the filter capacitor is increased and that increasing the sampling interval T also decreases the error.

The analysis for a flying capacitor multiplexer is similar except that the charging period is reduced to $T - h$. The resulting expressions are the same as those above except that T is replaced by $T - h$.

4.2.3 Aperture Time

The process of converting an analog quantity to a digital number requires a finite period of time. Effectively, this is the conversion time of the ADC in a system without sample-and-hold, or the time uncertainty in the output of a sample-and-hold amplifier owing to the finite switching time from sample to hold mode when this type of amplifier is used. This period of time is a "window" or "aperture" through which the analog subsystem sees the input signal. In applications in which the value of the input signal with respect to time is important, the finite width of this aperture results in errors if the input signal varies appreciably during the aperture period. These errors can be interpreted as time errors in the voltage value or as voltage errors at a given time, as shown in Fig. 4-18; either interpretation is valid and preference for one or the other depends on the particular application under consideration.

The term "aperture time" has several accepted definitions in the literature. One definition states that the aperature time is the effective time duration during which conversion takes place. Thus, for example, the aperture time for a ramp or successive approximation ADC is the total conversion time. In a system using an integrating ADC, the aperture time is the integration period. A second definition states that the aperture time is the uncertainty in the conversion interval. In a sample-and-hold system, for example, this definition considers the variation in

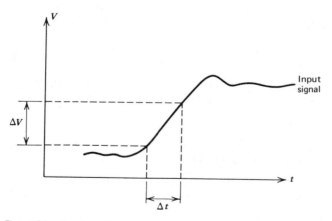

Fig. 4-18   Aperture time error.

the sample-to-hold mode transition as the aperture time. Either definition is acceptable and the development which follows can be applied to either case.

A measure of the effect of aperture time is the maximum rate of change of the input voltage which can be tolerated without introducing a fixed level of error into the measurement. A convenient level of error to consider is one-half the least significant bit (LSB), since this is the fundamental limitation on the accuracy of a quantizing system. Since the LSB corresponds to a quantization interval of $V_{fs}/(2^n - 1) \cong V_{fs}/2^n$ for an n-bit system with a full-scale voltage of V_{fs}, the maximum rate of change of the input voltage that can be tolerated with an error of less than $\frac{1}{2}$ LSB is

$$\left(\frac{dV}{dt}\right)_{max} = \frac{V_{fs}}{\Delta T 2^{n+1}} \qquad (4\text{-}46)$$

where ΔT is the effective aperture time of the system. As an example, the maximum rate of change for a 14-bit system with a 5-V full-scale range and an aperture time of 50 μsec is about 3 V/sec.

Another measure of aperture time effects is the maximum-frequency sine wave that can be sampled without introducing an error in excess of $\frac{1}{2}$ LSB. The rate of change of a sine wave is

$$\frac{dV}{dt} = \frac{d(A \sin \omega t)}{dt} = A\omega \cos \omega t \qquad (4\text{-}47)$$

where A is the amplitude of the sine wave. The maximum rate of change occurs at $\omega t = 0$ so that limiting the error to $\frac{1}{2}$ LSB requires

$$\left(\frac{dV}{dt}\right)_{max} = A\omega \leq \frac{V_{fs}}{\Delta T 2^{n+1}} \qquad (4\text{-}48)$$

The maximum frequency f_{max} in Hertz that can be sampled is

$$f_{max} = \frac{V_{fs}}{A\pi\Delta T 2^{n+2}} \qquad (4\text{-}49)$$

Because of the linear approximation, this equation is valid only if the aperture time is small compared with the period of the sine wave.

Using the same parameters as before for an example, the maximum frequency is on the order of 0.1 Hz for $A = V_{fs}$. This low value is the result of a relatively large amplitude for A and the high resolution of a 14-bit system.

The results of this discussion cannot be applied indiscriminately but must be interpreted in terms of the system under consideration and the application. It must be realized that the digital value produced by the

ADC accurately represents the input value at *some* time during the sampling or conversion interval. Thus the aperture error only has importance if it is necessary to know the value of the input voltage at a particular point in the conversion interval.

The basis for the derivation of the above equations was that the input signal not vary more than $\frac{1}{2}$ LSB during the aperture time. This requirement is quite severe for most process control applications since the exact relationship between the signal voltage and time is not usually of prime importance. Notable exceptions to this are analytical instruments in which a peak value must be determined in a time relationship to other output peaks. But even here, the conversion time in most systems is small compared with the signal variation, and the time aperture error is insignificant compared with the requirements of the application.

4.2.4 Sampling Theorem

The sampling process used in time-shared systems is a modulation process that produces sidebands in exactly the same way that modulation of an RF carrier with an audio signal produces upper and lower sidebands. The modulating signal in the case of a sampled data system is a periodic impulse function with a period T equal to the time between sampling instants. In the frequency domain, this signal is represented by a fundamental frequency component $\omega = 1/T$ and an infinite number of harmonics $2/T$, $3/T$, $4/T$, ... The result of modulating an input signal with these multiple carrier frequencies is to produce upper and lower sidebands around each of the modulating frequencies. Thus the original spectrum of the input signal is repeated as a modulated spectrum around each hormonic carrier frequency. Figure 4-19 illustrates an input signal, the impulse modulating signal, and the sampled signal in both the time and frequency domains.

This figure is an idealization in that the impulses are assumed to have zero width, whereas the multiplexing action in an actual system is equivalent to modulation with finite width pulses. The frequency domain representation of a train of finite width pulses is a series of $|(\sin X)/X|$ functions centered on the harmonic carrier frequencies. Modulating an input signal with these pulses results in additional sidebands not shown in the figure. The idealization of zero width pulses, however, is satisfactory for the purposes of this discussion.

The original signal can be recovered from the sampled spectrum by using a low-pass filter that passes the frequencies $0-W$ radians and rejects all higher frequencies. However, if the sampling frequency $\omega = 1/T$ is less than $2W$, the lower sideband of the first harmonic spectrum overlaps the original low-frequency spectrum. The energy passed by the reconstruction filter now includes energy from the modulated spectrum

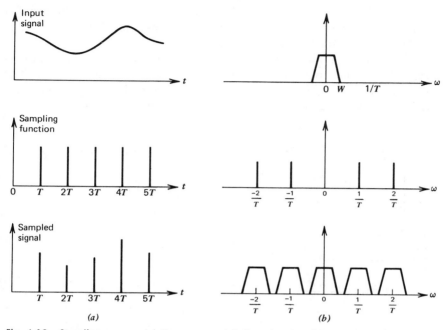

Fig. 4-19 Sampling as a modulation process. (a) Time domain; (b) frequency domain.

and the resulting signal is not the same as the original signal. The action of modulation that results in this overlap phenomenon is known as *aliasing* or *folding* and represents an error that is dependent on the relationship between the frequency components in the signal and the sampling frequency. An example of aliasing known to every movie fan is the low-frequency rotation of the stagecoach wheels in a western movie resulting from the stroboscopic sampling of the movie camera; the high-speed rotation is aliased into a low-speed rotation.

The discussion of the previous paragraph is a nonmathematical demonstration of a fundamental theorem of sampling which states that the signal must be sampled at a frequency at least twice as great as the highest significant signal frequency if the original signal is to be reconstructed accurately from the sampled signal. The mathematical derivation of this theorem, the effects of nonuniform sampling intervals and finite pulse width, and the resultant errors in a given application are beyond the bounds defined for this chapter. The purpose of this section is merely to introduce the aliasing concept and to provide means for estimating the magnitude of the error that is introduced into the reconstructed signal. The reader is referred to the literature for detailed mathematical treatment of the sampling theorem and the effect of its violation in particular control applications (Susskind, 1957; Gruenberg, 1967; Harman, 1963; Terao, 1967).

An estimate of the magnitude of the aliasing error is easily derived from the frequency domain representation of the sampled signal. The construction is illustrated in Fig. 4-20 for a signal having an effective bandwidth of ω_o and an attenuation characteristic of -6 dB/octave.

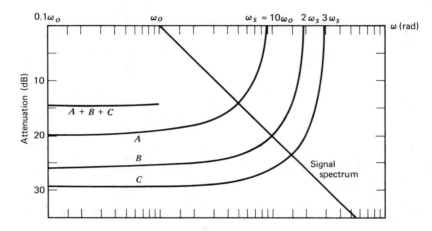

Fig. 4-20   Estimate of aliasing error. After Melsheimer, 1964.

A sampling frequency of $\omega_s = 10\omega_o$ is assumed so that modulated spectra are centered on this frequency and its harmonics. Only the lower side-bands of the first two harmonics are shown since the contribution of the upper sidebands and the additional lower sidebands is insignificant. The lower sideband curves are derived by folding the original spectrum about a frequency equal to one-half the frequency of the carrier due to sampling. For example, the original spectrum is down 3 dB at $1.4\omega_o$. The lower sideband of the first modulated spectrum is, therefore, down 3 dB at $(10.0 - 1.4)\omega_o = 8.6\omega_o$. Similarly, the lower sideband of the second modulated spectrum is down 3 dB at $(20.0 - 1.4)\omega_o = 18.6\omega_o$.

Using the above construction, the maximum aliasing error at any given frequency in the original spectrum is the algebraic sum of the contributions of the modulated spectra at that frequency. This is shown as curve $A + B + C$ in Fig. 4-20. In most cases, the major contributions to the error are from the first few modulated spectra. It is obvious from this construction and from the previous figure that the aliasing error decreases as the sampling frequency is increased. It also decreases as the slope of the spectrum characteristic is increased.

The aliasing error given by this construction is a maximum bound since frequency components are added algebraically. In reality, the summation is a vector sum and, therefore, the true result is usually less than the algebraic sum.

According to the sampling theorem, a sampling frequency in excess of two times the highest significant signal frequency allows the signal to be reconstructed from the sampled data. The example used for Fig. 4-20 demonstrates the importance of the word "significant" in this statement. Even though the sampling frequency was selected as ten times the bandwidth, the error at $0.1\omega_o$ is -15 dB or about 1.3%. This relatively large error is due to the -6 dB/octave characteristic since the signal is still significant, about 1%, at the sampling frequency. Because of the nonideal characteristics of the spectrum and the reconstruction filter, and to ease the problems of reconstruction, a sampling rate considerably greater than that required by the sampling theorem is usually used.

The error introduced by aliasing is not a result of hardware except, possibly, by a limitation in multiplexer speed. It is, rather, the result of the manner in which the system is used. Whether aliasing represents a significant error in the control of a process is highly dependent on the analysis of the sampled data. Generally speaking, aliasing errors pose the greatest problem if the analysis involves spectrum analysis or the identification of particular frequency components in the signal. This is not a common requirement in control situations, although it is not uncommon in some laboratory automation applications. A potential problem in control is the introduction of a significant aliased frequency component which results in oscillatory control action. Although unlikely because of the magnitude of the aliasing error, this problem can be identified through a programmed analysis of parameter trends.

4.3   GROUNDING AND SHIELDING FOR NOISE CONTROL

The term noise is used to describe any extraneous signal or disturbance in an analog subsystem. Noise signals may be either steady state or transient in nature and may contain frequency components from DC to very high frequencies. The nature and spectrum of the noise depends on the source and on the modification that occurs between the source and the point of entry into the subsystem. Examples of noise include the common mode potentials discussed in this chapter and in Chap. 3, spikes and transient disturbances due to electromechanical or electronic switches, and "pickup" from radiation sources such as oscillators or arc discharges.

The specific manifestations of noise in an analog subsystem depend on the characteristics of the noise and the subsystem. In extreme cases, noise results in a complete malfunction or inoperative condition in one or more parts of the subsystem. More typically, however, the effect is a lack of precision or repeatability in reading input signals and intermittent errors in the conversion process. Although the result of these effects depends on the application and the control method being used, noise generally causes an increased software burden and a loss of accurate and precise control. The software burden is the result of the increased digital filtering and data-checking procedures that are required. The lack of precision control results from the variability of measurements which masks actual process variations.

A survey of the process control literature of a decade ago reveals that, although the topics of grounding, shielding, and noise control were considered, there is markedly less emphasis on them than there is in the present-day literature. The primary reasons for the increased interest are related to both the hardware and the application.

One of the significant differences between conventional control equipment and the time-shared analog subsystem of the process control computer system is the difference in bandwidth. Conventional analog controllers have a bandwidth of less than 10 Hz and, in many cases, less than 1 Hz. This is satisfactory in most cases because process variations are relatively slow. In time-shared equipment, however, the bandwidth of the actual measurement equipment may be several orders of magnitude greater. The difference between the measurement equipment bandwidth and the channel bandwidth is particularly noteworthy. The channel bandwidth, normally established by the signal conditioning circuits, is often comparable to the bandwidth of conventional analog control equipment. The time-shared portion of the subsystem, however, must have a greater bandwidth in order to respond to the sampled signal.

As a specific example, the 60-Hz noise typically found in industrial processes does not affect the operation of conventional analog control equipment because the restricted bandwidth attenuates the noise signal. This same attenuation is usually provided by the low-pass signal conditioning circuits in the analog subsystem for a process control computer system. If, however, 60-Hz noise is injected into the signal path following the low-pass filter, the wide bandwidth of the time-shared amplifier and ADC does not provide significant attenuation. As a result, the ADC converts the variation of the signal due to noise into a variation in the digital data used in the control calculation.

4.3.1   Elements of Noise Control

Effective noise control requires knowledge of the possible noise sources, the means by which noise is coupled into the subsystem, and the methods of noise elimination and control.

The possible sources of noise in an analog subsystem are many and varied. They include sources internal to the analog subsystem, such as digital control logic, oscillators, and electromechanical switching devices. In addition, noise sources from other subsystems in the process control computer can be coupled into the analog subsystem. These include sources such as relay transients from digital I/O or data-processing I/O equipment, high-energy driver circuits associated with storage or communication circuits, and radiation sources such as oscillators in television display or communication equipment. Sources of noise external to the computer system are also of concern since the analog subsystem is connected to the process via transducer and control wiring. Although the methods of noise control are similar for sources both internal and external to the subsystem, external noise sources are considered as a separate subject in Chap. 13.

Despite the varied sources of noise, the means by which noise enters a subsystem are more easily categorized. Noise is coupled through conductive paths, by means of electrostatic or electromagnetic induction, or by reception of radiated energy. Conductive coupling is the result of noise currents flowing through impedances common to both the noise source and the noise-sensitive circuit. Electromagnetic induction is generation of a voltage owing to a time-varying magnetic field. Radiated energy is coupled into circuits in a manner analogous to the reception of radio frequency energy by a communications receiver.

Despite the easily categorized coupling phenomena and noise control methods, the implementation of effective noise control can be extremely difficult. There is, seemingly, an art as well as a science of noise control. This is the result of the distributed nature of the system. The coupling means and the impedances in the system are not easily described in mathematical terms that can be analyzed effectively. Thus the designer must be aware of the principles involved in noise control and do his best to incorporate them into a design. He must also use an empirical approach in arriving at a final design. Often merely moving a grounding point or changing the length of a wire results in dramatic improvement even though such a change cannot be justified precisely.

4.3.2   Sources of Noise

As noted above, the possible sources of noise in an analog subsystem are many and varied. The designer must examine his subsystem and the environment in which it must operate to identify possible sources. The information in this section provides guidance and gives examples of such sources.

4.3.2.1 Power Equipment. Because of the sensitivity of low-level analog subsystems, sources of electrical energy orders of magnitude greater than signals of interest are immediately suspect. A noise signal as much as 120 dB below the amplitude of a high-energy source can be significant in a low-level system. For example, a 120-μV signal resulting from the primary 120-V power source represents a potential noise error of 1.2% on a 10-mV full-scale range.

A process control computer system and its environment typically include a number of high-energy sources. Primary among these is the power supply and distribution system for the computer. Power transformers produce significant time-varying magnetic fields that can result in electromagnetically induced potentials in sensitive circuits. If the output of the supply contains an appreciable ripple frequency, the power distribution system acts almost as an antenna that can induce significant potentials in circuits located an appreciable distance from the supply.

Certain types of power supply regulation equipment are also a source of noise. Regulators that use switching control rather than analog control are of particular concern. The switching regulator produces high-frequency transients once or twice per cycle at the power line frequency. The transients may be capacitively coupled into sensitive circuits or they may be radiated by the power distribution wiring. This type of regulator typically uses silicon-controlled rectifiers (SCR), bipolar transistors, or magnetic amplifiers.

An inadequate regulator transient response may also result in noise in the analog subsystem. If the regulator cannot respond to periodic heavy loads resulting from such things as relays or electromagnetic punches, the voltage distributed to sensitive circuits varies and causes changes in offset, gain, or other critical circuit parameters.

Other power equipment in the computer control system includes motors, relays, and electromechanical actuators such as punches and file positioners. Relays are particularly common in the power distribution system and electromechanical equipment such as printers. In addition, control relays are normally provided in the system as a digital output feature to control equipment in the process. All of these devices can produce noise as a result of power supply transients, electromagnetic fields, and radiation due to contact arcing.

4.3.2.2 Radiation Sources. Sources of radiated energy exist both internal and external to a process control computer system. The digital

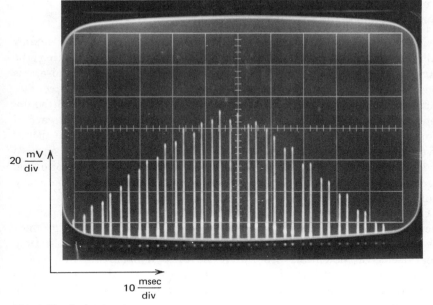

$20 \frac{mV}{div}$

$10 \frac{msec}{div}$

Fig. 4-21   Radar interference.

clocking pulses that control the logical operation of the computer itself are a prime example. These are usually generated by a crystal-controlled pulse oscillator operating in the megahertz range. Since clock pulses are required in all parts of the system, the signal is widely distributed. The chance of coupling between cables in the subsystem and sensitive circuits is appreciable, although the high frequency of the clock makes control of this noise source relatively easy.

Oscillators are also associated with other subsystems. Examples include oscillators for ADC timing, communication circuits, or graphic displays. These sources usually are confined to a particular portion of the system, which eases the noise control problem.

Contact arcing in motors and relays is another source of radiated energy found in most systems. Since arcing usually is not periodic, it represents an elusive source of noise. Proper methods of arc suppression and shielding are effective means of control.

Sources of radiation external to the system are also worthy of note. Powerful radio, television, or radar installations in the vicinity of sensitive circuits can be the source of intermittent noise. Even hand-held "walkie-talkies" or aircraft flying over an installation can cause errors. The photograph of Fig. 4-21 shows the response of a $G = 500$ amplifier to the radiated energy of a radar station located a number of miles from a system. The shielding inherent in the subsystem enclosure was sufficient to eliminate the noise in this case.

4.3.2.3 Miscellaneous Sources. One source of noise in the analog subsystem can be attributed to the flexure of cables connecting the various circuits. This noise, known as "triboelectric cable noise," is the result of the cable dielectric momentarily losing intimate contact with the cable conductors when the cable is mechanically distorted by bending or tapping. The potential produced can be as great as several hundred microvolts in a high-impedance circuit. However, the energy level is very low, so it rarely affects circuits other than the ones to which the cable is connected. Effective control consists of minimizing mechanical flexure of the cable by proper tie-down clamps and cable routing.

Other sources of noise often identified include ground loops and thermoelectric sources. Ground loops are the result of improper grounding practice and are considered in more detail in the next section. Thermoelectric potentials are the result of a temperature differential existing in a circuit containing a junction of dissimilar metals. This phenomenon and its significance in low-level circuits are discussed in Sec. 3.10.

4.3.3 Grounding Practice

A ground in an electric circuit is a conductor that acts as a reference point for other potentials in the circuit. A process control computer system has four grounding systems. The safety or frame ground ensures that the noncurrent-carrying conductors remain at ground potential even in the event of a fault between a potential and the ground system. This ground is connected to cabinets, panels, enclosures, trim strips, and other metallic parts of the system. Owing to possible noise problems in the analog subsystems, this ground should be a noncurrent-carrying conductor. This usually requires a separate grounding conductor connected to the main system ground, usually located at the building service entry. The basic requirements for the frame ground are defined in codes such as the National Electrical Code and other local codes. Its primary purpose is to minimize the risk of personnel injury in the case of a system fault.

The power supply ground is the return conductor for the DC power supply distribution system. It may or may not be common with the signal grounding system that provides the reference point for information signals in the system. The signal ground is often divided into a digital ground and an analog ground, each of which provides the voltage reference point for the respective types of circuits. In low-level systems,

it is almost mandatory that a separate analog ground system be established if satisfactory noise control is to be obtained.

The four grounding systems are not entirely independent since, at
some point in the system, they are common at a system grounding
point. This ground point is connected to the primary power distribution system ground, usually at the building service entrance. As an
alternate approach, a driven ground rod can be installed at the system
if the resistance between this ground and the primary power distribution ground is less than a fraction of an ohm. If this approach is taken,
the primary power distribution ground is disconnected from the system
ground and the driven ground is connected to the system ground point.
Although the fact that the systems are eventually common seems to
negate their independence, the key point is that the separate ground
structures prevent cross coupling of signals as a result of impedances
common to noise-producing and noise-sensitive circuits.

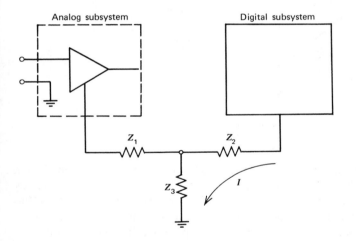

Fig. 4-22   Effect of common ground impedance.

The result of a shared ground system is illustrated in Fig. 4-22, which
shows a digital subsystem and an analog subsystem represented by an
amplifier. The impedances Z_1, Z_2, and Z_3 represent the impedances of
the ground system. A ground current I from the digital subsystem flowing through the common ground impedance Z_3 results in a potential IZ_3,
which is equivalent to an input signal for the analog subsystem. If the
current I is DC, the error introduced is an offset which, if it is constant,
can be compensated by calibration of the analog subsystem. It is more

likely, however, that the ground current varies as the digital circuits are actuated in the logical operation of the digital subsystem. In this case the error introduced into the analog subsystem also varies. This results in a variation in the output of the analog circuit which is eventually converted into variation in the digital data.

The alternative to this type of grounding system, and a fundamental principle in grounding low-level circuits, is a single-point ground structure, as shown in Fig. 4-23. A ground circuit is established for each of

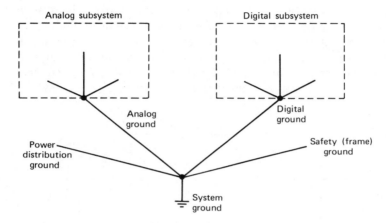

Fig. 4-23   Single-point grounding philosophy.

the required ground systems and the systems are then commoned at a single point. The same philosophy is applied to each subsystem by establishing a single subsystem grounding point. The connection between the subsystem ground points and the system ground is critical if the situation of Fig. 4-20 is not to be duplicated. The connection must be an extremely low-impedance connection; low resistance is not enough since appreciable inductance can result in potentials owing to high-frequency noise signals. As a test of the sufficiency of a ground connection, a second conductor can be paralleled temporarily with the ground circuit. Any variation in the system performance as the parallel ground is connected and disconnected is an indication of a poor ground system. In performing this test, care must be taken that ground loop errors are not inadvertently introduced since this can invalidate the results.

A common result of multiple ground points is what is popularly known as a "ground loop." A ground loop is a circuit in the grounding system through which circulating currents flow. These currents cross couple portions of the system through the impedance of the ground

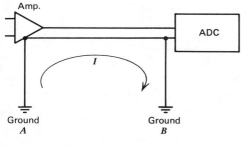

Fig. 4-24   Illustration of ground loop.

system. An example of a ground loop is shown in Fig. 4-24. Owing to a difference in ground potential, a current flows between the two grounds through the common connection. The circulating current results in transient and offset noise being coupled into the ADC. The similarity of this diagram and the illustrations used in connection with the common mode development of this chapter and Sec. **3.9** should be noted. The common mode problem is really just one aspect of the more general grounding and noise problem.

Shields represent a common source of ground loops in many systems. Designers often ground a shield at every convenient point with the idea that this improves the effectiveness of the shield. This is a valid philosophy in shielding RF circuits where the impedance of the shield may be significant. In low-frequency circuits, however, this is not a problem, and multipoint grounding of the shield more often than not causes ground loop problems.

Detecting ground loops and other grounding problems is not always a simple matter since the manifestations of poor grounding may be quite subtle. One useful method is to draw a pictorial diagram of the entire subsystem grounding structure. A schematic representation is not satisfactory because it does not show the physical relationship between portions of the ground system. An illustration of the difference between pictorial and schematic representation is shown in Fig. 4-25. Both representations in the figure show the electrical connections made between the amplifier and the ADC. The schematic representation, however, does not clearly show that the amplifier and signal grounds are commoned at the ADC. As discussed above, this physical relationship is really the most important factor in identifying and eliminating noise due to grounding problems.

Owing to the inevitability of change during the process of developing and testing a subsystem, the grounding diagram should be drawn first when the subsystem is initially designed. It should then be redrawn

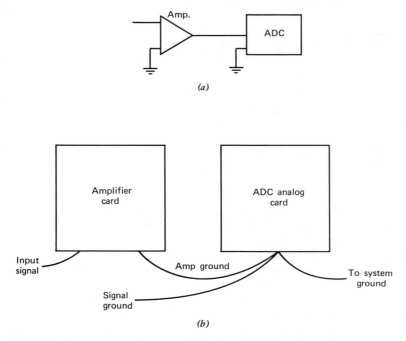

Fig. 4-25   Ground system diagrams. (a) Schematic; (b) Pictorial.

after the system has been completely debugged to verify that grounding problems have not been added to the system as a result of changes made during the debugging process.

The performance of the subsystem is often an indicator of ground problems. Differences in the performance on the bench and in the system environment often mean that either the bench or the system configuration contains a ground loop. Another clue is a change in subsystem performance when an apparently unrelated circuit is actuated or removed. For example, removing a card in another subsystem or removing an inactive multiplexer card should not affect subsystem performance. In the same category as changing the system configuration, physical disturbances can sometimes uncover noise and grounding problems. Opening doors, pounding on the frame, and flexing control and signal cables may lead to the discovery of intermittent grounds or noise sources. Standard laboratory instruments such as oscilloscopes and ohmmeters are also useful in tracking down problems. The scope is used to determine whether significant ground potentials exist or whether the ground structure is noisy with respect to the system single-point ground.

By disconnecting a ground tie and checking for continuity between the disconnected point and the system ground with an ohmmeter, multiple ground connections can be detected.

The "cut and try" philosophy of proper grounding cannot be over-stressed. The development of a suitable grounding system for an analog subsystem requires considerable patience and empirical engineering. The single-point grounding principle and other guidelines only provide guid-ance to the engineer.

4.3.4   Shielding

The effect of noise sources that are coupled into the subsystem by electrostatic or electromagnetic induction or by radiation can be greatly diminished by the use of shields. A shield consists of a metallic enclosure surrounding some portion of an electronic subsystem. A metallic structure that only partially encloses a circuit or subsystem provides some shield-ing effect, but its effectiveness is very dependent on the configuration of the shield and the character of the noise.

Electrostatic shields, sometimes called Faraday shields, are used to reduce the effect of electrostatic coupling. Electrostatic coupling is the phenomenon by which energy is transferred from one conductor to another through the capacitance between the conductors. The capaci-tance is a function of the conductor spacing and size. These two factors and the frequency of the noise source determine the noise current that flows.

The requirements for an effective shield are that it be a conductor of relatively low impedance at the frequency of the disturbance and that it be located between the noise source and the noise sensitive circuit. The shield must be grounded to provide a path to ground for the capacitively coupled noise energy. Figure 4-26 illustrates the shielding action of an electrostatic shield. The capacitance C_1 represents the capacitance be-

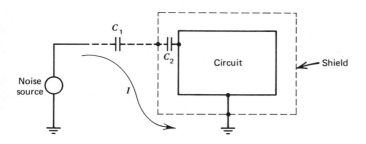

Fig. 4-26   Electrostatic shielding.

tween the noise source and the shield, and C_2 is the capacitance between the shield and the circuit. If the impedance of the shield is low compared with the path through C_2 and the circuit, the majority of the noise current flows through C_1 to ground. Since little current flows through C_2, the operation of the circuit is not affected by the noise source.

The mathematics and conditions for good shielding are developed in Sec. 4.1.2 for the common mode guard shield. The models considered in that analysis are conceptually valid in the general case, as shown by the comparison in Fig. 4-27. The difference between the models is the placement of the ground and noise source. The major conclusions in the previous analysis are that the shield must never be floating and that the effectiveness of the shield depends on its impedance. The guard shield analysis also showed that the shield was most effective when connected directly to the common mode source. The analogous conclusion in the general case is that the shield must be tied directly to ground. If the noise source can be identified, making the ground tie close to the noise source increases the effectiveness of the shield by reducing the impedance of the path to ground.

A shield enclosing the entire subsystem does not provide protection for sensitive circuits inside the enclosure if the noise sources are located inside the shield or if noise is conducted into the enclosure through ports or cables. For this reason, additional enclosures may be required around noise-sensitive circuits. In less severe cases, a partial shield may be created by running one or more grounded conductors around the periphery of the sensitive circuit or by mounting conducting planes adjacent to the circuit.

A second type of shield used in analog subsystems is a flexible shield consisting of woven, uninsulated wire. This material is used to shield cables, bundles of wire, or other circuits where flexibility is required. This type of shield is often covered with an insulating jacket to prevent grounding at other than predetermined points. This is necessary to avoid ground loops.

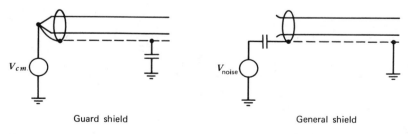

Guard shield   General shield

Fig. 4-27 Comparison of shielding application.

Two common types of shielded cable are the coaxial cable and the twisted shielded pair. The coaxial cable consists of a single insulated conductor covered by a flexible shield, either woven or foil tape. The twisted pair consists of a pair of insulated conductors twisted together and enclosed in a woven or foil-tape shield. The shield provides electrostatic protection and, as discussed later, twisting reduces noise owing to electromagnetic induction.

Both the coaxial cable and the twisted shielded pair are effective in reducing the effects of electrostatic induction. The coaxial cable has the disadvantage that the shield, as a signal conductor, has high capacitance to ground, and this can cause degradation of common mode performance. Another potential problem in the use of coaxial cable is that ground loops may occur if the shield is required as a signal conductor or if the signal return path is through the ground system. This can be prevented by the use of two cables, one for the signal and one for the signal return, both of which have shields grounded at a single point. Despite these disadvantages, however, both coaxial cable and twisted shielded pairs are equally effective if applied properly.

The second type of induced noise propagation, electromagnetic induction, is the result of a time-varying magnetic field in the vicinity of a conductor, or a moving conductor in the presence of a static magnetic field. The basic phenomenon is shown in Fig. 4-28. The magnetic field $\Phi(t)$, passing through a loop of wire containing N turns, induces a voltage V equal to

$$V = -N\frac{d\Phi}{dt} \qquad (4\text{-}50)$$

Since the flux passing through the loop is equal to βA, where β is the flux density and A is the area of the loop, a voltage is also induced if the area of the loop changes in the presence of a static magnetic field.

In the subsystem, the loop might consist of a pair of wires connecting

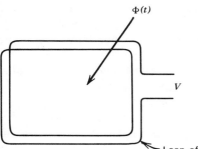

Fig. 4-28   Electromagnetic induction.

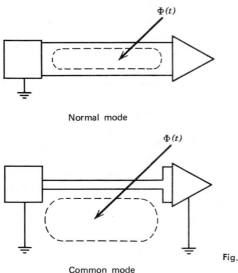

Normal mode

Common mode

Fig. 4-29 Possible induced noise sources.

two pieces of equipment, as shown in Fig. 4-29. In this case the induced potential is a normal mode source of noise. If the loop consists of a closely spaced pair of wires and a grounded conductor located some distance away, the induced potential is primarily common mode. The normal mode component is minimized by the close physical spacing of the pair of wires.

The field in the vicinity of a conductor or circuit can be eliminated by enclosing the conductor or circuit in a shield of high-permeability material. The flux concentrates in the shielding material and does not pass through the circuit. In contrast to the electrostatic case, the shield does not have to be grounded. If it is, however, it provides a measure of electrostatic protection as well. Most ferrous metals provide electromagnetic shielding and special high-permeability metals have been developed specifically for shielding purposes. The common magnetic shield found in the analog subsystem consists of a steel enclosure or conduit. Cables having a high-permeability foil shield are available where the use of conduit is not feasible.

Generally, the requirements for electromagnetic shielding in an analog subsystem are not severe if reasonable precautions are taken to avoid placing circuits and cables in strong magnetic fields. As a result, extensive use of magnetic shields is not common. A simpler method of providing effective reduction of electromagnetic induction is the use of twisted pair cables. The polarity of an induced potential is dependent

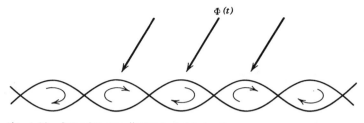

Fig. 4-30   Potential cancellation in twisted pairs.

on the relationship between the field and the loop. As shown in Fig. 4-30, the effect of twisting is to transpose this relationship in each loop of the twisted pair so that the potential induced in one loop is canceled by the potential in the next loop. This is indicated in the figure by the arrows which show the direction current would flow if each loop were independent.

The effectiveness of twisting and electromagnetic shielding is illustrated by the test data in Fig. 4-31. This test consisted of producing an electromagnetic field by passing a heavy current through a 25-ft conductor. The potential induced in a pair of wires located 1 in. from the current-carrying conductor was measured. The effect of twisting and/or shielding is clearly demonstrated in the results. In all cases, steel conduit provided the best shielding performance. Merely twisting the wires, however, provides comparable performance at considerably less expense.

Similar tests and experience with low-level circuits lead to the general principle that electrostatically shielded and twisted pairs should be used for all signal interconnections between low-level circuits in the analog subsystem. It is only in extreme cases that extensive magnetic shielding is required.

4.3.5   Isolation and/or Elimination

Grounding and shielding for noise control concentrates on protecting sensitive circuits at the location of the circuits. The noise problem does not exist, however, if the noise source is eliminated or if it is completely isolated from sensitive circuits. Although total elimination or isolation is not always possible, the degree of interference can often be reduced by attacking the source of the noise.

The effect of noise sources often can be minimized by shielding the source with electrostatic or electromagnetic shields. Other control measures include the redesign of a noise-producing power supply, additional decoupling circuits, and separate analog power supplies. The effect of pulse transients can be reduced by slowing the pulse risetime when this is possible. Sometimes changing the length of a conductor connected

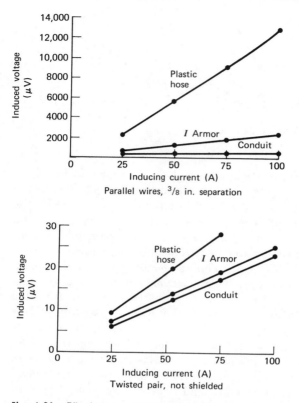

Fig. 4-31 Effectiveness of twisting and shielding. From Maher, 1965, by permission of the publisher.

to a radiation source reduces interference by changing the wavelength of the "antenna." Most of the techniques effective in dealing with noise-sensitive circuits are also beneficial when applied to the noise source.

If the noise source cannot be eliminated, its effect on sensitive circuits may be reduced by effectively isolating it from them. This isolation may be electrical or spatial. Many of the methods described above are forms of electrical isolation. Shielding, separate ground systems, decoupling, and by-passing are examples. In severe cases, it may be necessary to provide a completely separate subsystem to ensure adequate isolation. This requires techniques such as separate power supplies using isolation transformers and the use of pulse transformers for coupling digital signals to other subsystems.

Although a completely isolated subsystem is required in only the most critical applications, spatial isolation is a general principle that should be observed in any analog subsystem. Of the four coupling means associated with noise sources, all except conductive coupling are dependent on the physical separation between the noise source and the receiving circuit. That is, the interference effects of electrostatic induction, electromagnetic induction, and radiation decrease as the distance between the source and the sensitive circuit is increased. Thus considerable benefit can be obtained by spatial separation.

Sensitive circuits should not be located near noise-producing sources such as relays, power transformers, oscillators, or high-speed digital circuits. Analog signal cables connected to the analog subsystem must be isolated from digital cables to prevent electrostatic induction of digital pulses into the analog circuits. In addition, analog cables carrying signals of widely differing amplitudes should be separated from each other. Wire pairs carrying 5-V signals, for example, should not be located close to pairs carrying 10 mV since an error of 0.5 mV is induced in the 10-mV signal even if there is 80-dB isolation between the pairs.

The design implications of spatial separation include provision for separate raceways or cable channels for analog and digital signals. They may also require that analog input signals in the termination facility be segregated into groups to ensure that signals of various levels are spatially isolated. The design of circuit cards containing both analog and digital signals, such as the multiplexer cards, must be such that spatial separation is maintained. The card layout should also allow for spatial separation of backpanel wiring. When it is necessary to transpose the position of two conductors, one analog and one digital, the conductors should cross at right angles since this minimizes the possibility of induction between the two wires.

4.4 SUBSYSTEM MAINTENANCE

The application of a computer in a process control application requires a high degree of system reliability. The emphasis on reliability results from the profit or safety justification behind most process control installations. The use of the computer provides increased profits through greater quantity, better quality, and more systematic procedures. Safety of personnel and equipment is sometimes guaranteed by the use of the computer control system. In a few cases, the process may be entirely inoperative if the computer fails because manual or semiautomatic control may not provide adequate control.

These needs of the real-time control situation demand reliability appreciably greater than that usually required in data-processing or scientific computing applications. The result of machine failure in these latter applications is not to be minimized; realistically, however, the most common result of a failure is user inconvenience and irritation. In critical applications, a back-up computer, perhaps many miles away, can be used to complete the interrupted calculation. This course of action is rarely possible in the process control situation because of the required interface to the process.

The reliability of a computer is dependent on many factors including the intrinsic reliability of the components, the architecture of the machine, and the amount of error checking/correcting hardware and software. Techniques such as the use of majority logic, redundant analog circuits, and duplexed systems can provide extremely high levels of reliability. Reliability, however, costs money. Both the user and the vendor must analyze the trade-offs between additional hardware and software and the application cost of a computer failure.

Reliability is closely allied with the subject of maintenance since, obviously, a more reliable computer requires less frequent maintenance. The vendor designs his equipment to fail as rarely as possible within cost constraints. He must also provide means of diagnosing failures and effecting repairs with the minimum loss of service to the user.

4.4.1 Hardware and Software Features

The interface between the process control computer and the process instrumentation is a particularly vulnerable source of system failure. The possibilities of damage are many: A bulldozer can tear up instrumentation lines; an electrician can inadvertently connect a high-voltage line to a low-level input; or a lightning strike can subject the subsystem to high-voltage transients. The broad geographical limits of the typical process and the variety of applications require that the engineer design the system for maximum self-protection.

A general philosophy of self-protection is to restrict damage to a single input channel if at all possible. With proper installation and application planning, it is unlikely that the loss of a few input points will completely disable the entire process. The importance of this approach is emphasized by the time-shared characteristics of the typical analog subsystem. If the failure is not restricted to a single channel, time-shared equipment such as the amplifier or ADC can be damaged, with the result that no valid input data can be obtained from the process.

The most common cause of damage to the analog-input subsystem is the application of excessive voltage to an input channel. The method of protection depends on the subsystem configuration as well as on the components used in the subsystem. For example, mercury-wetted relay and transformer-coupled multiplexers withstand higher levels of overvoltage because of the intrinsic ruggedness of the devices. Similarly, the flying capacitor configuration provides some protection since an excessive normal mode voltage usually results in the capacitor's breaking down. This either causes the capacitor to open or at least minimizes the voltage drop across the capacitor.

Each configuration and component must be considered by the designer in determining possible failure modes. The effect of open and closed failures of multiplexer switches in various combinations must be analyzed, for example. The cost of providing clamping circuits on each channel versus the cost of losing a channel is another consideration. Analysis of the failure modes must also include the possibility of transducer damage due to a system malfunction.

Techniques that can be used to protect the system include clamping circuits which limit the input voltage, ensuring that a component fails in such a way as to protect the time-shared equipment, sensing the overvoltage and electrically disconnecting a portion of the subsystem, or limiting current flow with resistors. In some cases, fusing the input circuits with thermal or semiconductor fuses provides protection.

Protection circuits often affect other system performance parameters and these effects must be considered. For example, clamping circuits on inputs channels can affect the input impedance and the common mode performance. Performance degradation severely limits the designer since the normal system operation is of prime competitive importance.

Failure due to factors other than faults on the input channels of an analog subsystem are similar in nature to those that occur in any electronic system. These failures may be the result of failure of a component or of degradation of the performance of a component or circuit. This type of failure is minimized by conservative circuit design and an accurate evaluation of indications of imminent failure.

Since it is virtually inevitable that the system will fail at some time, it is important that the initial design of the subsystem include considerations of maintenance features and procedures. One of the biggest problems in the maintenance of a large complex system is isolation of the cause of failure.

Although it is easy to state that failure isolation is a requirement in effective maintenance, implementation of the statement may be difficult. Many of the failure manifestations are ambiguous and can be the result of several causes. For example, a gross error in the digital data obtained from the analog subsystem can be the result of defective channel hardware, a defective amplifier or ADC, failures in the channel addressing hardware, failure in the digital channel between the analog subsystem and the computer, a failure in the computer, or a programming bug.

Failure isolation can be enhanced by providing hardware maintenance features. These features can supply means for physically or electronically isolating portions of the system. For example, providing a physical means of disconnecting the process inputs between the termination facility and the multiplexer allows servicing personnel to determine clearly whether the problem is associated with the computer control system or the process instrumentation. Similarly, disconnect features at the block switch level, at the input to the amplifier, and at the input to the ADC allow the serviceman to isolate a problem to a well-defined functional portion of the subsystem.

Providing test points for the injection or observation of test signals is also a necessity. Test points must be carefully selected to ensure that they provide meaningful data for diagnosing failures. They must also be selected such that connecting test equipment does not alter the significant features of the waveform. For example, waveforms at a very low-level voltage test point can easily be altered by connecting the test instrument if the instrument introduces a ground loop, alters the distribution of stray capacitances, or causes an impedance mismatch. The analysis of the waveforms obtained at the test points under normal and abnormal conditions must be carefully documented. In addition to a general description of the waveform, documentation should include the waveform analysis for common failures. Since all failures cannot be anticipated, it is important that servicing personnel be taught enough of the theory of the system to appreciate abnormalities not specifically considered in the documentation.

The system designer must also consider the test equipment that is available to servicing personnel at the site since it is often limited in accuracy and function when compared with the standard instrumentation available in a laboratory. Usually an oscilloscope is the most commonly available equipment other than a basic volt-ohm-milliammeter. Because of the capital expense of providing costly test equipment for every installation, the versatility of the scope may be limited by a restriction on the number of available plug-in features. As a result, test procedures must be developed with an *a priori* knowledge of the test equipment capability.

Precision test equipment normally is not available to the serviceman because of its expense and the fact that is must be periodically calibrated to be of value. The analog-input subsystem itself is probably the most precise equipment available to the service personnel and, with proper design, it can be used for self-diagnosis of the analog subsystem. With the proper software support and a precision reference source, the subsystem can be used for calibration, isolation of input channel failures, and for testing other analog features such as analog output.

A precision reference source is a necessity in servicing a high-accuracy system. This source should provide test signals for each range of the analog input subsystem. Its precision should be greater than the subsystem precision to be effective. Means should be provided such that the precision signal can be inserted into the system at various test points as an aid to isolating a malfunctioning circuit. In addition, a dedicated input channel can be provided for on-line checking of the system. This on-line checking capability, if supported by the software, can also eliminate some service calls since it allows the user to determine whether the analog subsystem is operating correctly before calling the serviceman.

Another hardware feature that is useful in failure analysis is a single-cycle operation. In this mode, the system performs a single logical step under control of the serviceman. Using this feature, service personnel can observe the system statically at any point in its cycle of operation. For example, in a successive approximation ADC, the result of each comparison can be observed. The feature is limited, however, in that it does not allow analysis of errors caused by the dynamics of the system.

Ease of maintenance is increased and the cost of stocking spare parts is often reduced by a careful study of system modularity. The ability to interchange cards or to substitute spare cards is a valuable diagnostic tool. Ease of replacement in a modular system reduces repair time to a minimum. It also reduces the cost of repairing the defective module since it localizes the problem to a smaller portion of the subsystem. In many cases, the use of small modules allows "throw-away" replacement, in which the defective module is merely discarded. The increased packaging density made possible by microelectronics is increasing the cost of the individual modules, however, so the economics of throw-away versus repairable modules must be carefully evaluated. A high degree of modularity can also increase the cost of manufacturing and ordering procedures, so this factor must also be included in the analysis.

Experience has shown that many of the problems attributed to a failure in the analog subsystem are, in reality, due to problems in the process instrumentation. Also, problems in the analog subsystem may not manifest themselves except under actual operating conditions. These factors, and the need to minimize interruptions of computer control in a real-time situation, often require that servicing be performed on-line. This places special requirements on the design of circuits and the manner in which service personnel diagnose system errors. If circuits must be

replaced with power applied to the machine, personnel safety requirements are of paramount importance. Secondarily, the circuits must be
designed such that removal or replacement with power on does not result
in machine or process damage as a result of transients or voltage
sequencing.

On-line servicing also requires that servicing personnel have an intimate knowledge of both the computer system and the process. The
serviceman must recognize, for example, that removing the amplifier in
a time-shared system invalidates all the analog data. If the processor
were to use the invalid data for control, the result could be a runaway
process dangerous to both personnel and equipment. Designing for complete on-line servicing at a reasonable cost is usually impractical because
of the safeguards that are required. It should be possible, however, to
make noncritical card replacements, adjustments, and some simple repairs without taking the system completely off-line.

The design of adequate diagnostic software for the analog subsystem
is an extremely difficult job because of the many types of errors and
failures possible. On the other hand, effective diagnostic programs can
reduce the diagnosis and repair time materially. The cost of program
preparation and storage space versus increased availability is a trade-off
that must be asssessed carefully.

Diagnostic programs should provide more than mere exercise for the
analog subsystem. The programs should include algorithms to assist in
the isolation of the defective functional module. Many errors can be
located precisely by a deductive procedure that can be included as part
of the program. For example, if the data from all analog points are
invalid, the problem is occurring in the time-shared portion of the subsystem. If the data from all multiplexer channels that are common to a
single block switch are in error, the failure is probably associated with
the block switch or its associated addressing circuit. Such deductive procedures can be quite sophisticated, but their implementation requires an
analysis of their economic value.

Providing the customer with the ability to run diagnostic programs
can often shorten or eliminate service calls. This is particularly true in
the analog-input and -output subsystems where experience shows that the
problem is often the process instrumentation. The diagnostic programs
can be included as part of the programming system and can be called
when either the operator or another program detects an error.

4.4.2 Service Personnel

Servicing of the process control computer system may be provided by
the vendor under the terms of the lease or a separate service contract, or

it may be provided by the user. The servicing of a complex computer system is a difficult and exacting job that requires skilled servicing personnel. The availability of the system is strongly dependent on their ability to quickly locate, diagnose, and repair a failure.

The characteristics of the analog subsystem pose unique problems in the training of personnel and the maintenance of their skills. Although the analog-input subsystem may be quite extensive, the number of analog circuits is small compared to the total number of circuits in the total system. On the basis of component failure statistics alone, the number of failures is less in the analog circuits than in the digital hardware. As a result, personnel have only infrequent opportunity to gain experience in diagnosing analog problems. Although extensive training is necessary, it is not the total answer since the results of training dissipate with time without the experience and practice that guarantees quick and accurate diagnoses. The problem is often accentuated by the low population of systems in the area serviced by a particular branch of the vendor service organization.

These same problems are not as severe in servicing the digital portions of the system. Not only are the number of digital failures greater, but the service personnel have the opportunity to gain experience on digital systems other than process control computer systems. Therefore, their skill in diagnosing digital problems is being constantly increased and refined.

Even the best prepared and most experienced service personnel are hampered by a lack of adequate documentation. This must, of course, include the detailed logic diagrams for the location and identification of test points, individual circuits, and the like. Second-level diagrams showing the functional relationships between groups of circuits are also required. These, and similar types of documentation, are required for adequate servicing of either analog or digital subsystems.

A unique requirement of the digital and analog input/output subsystems is documentation showing how the system is interconnected with the process. One of the reasons for this type of documentation is semantics. The service personnel and the process operator often do not talk the same language and documentation is required for effective translation of the operator's trouble report into a machine trouble definition. For example, the process operator may say that the computer system is not reading the output of FRC 21. The operator knows that FRC 21 is the designation for a flow controller in a particular loop. It is unlikely that he knows that the output of FRC 21 is connected to address 0048 of the analog-input subsystem. Service personnel must have the documentation to translate the FRC 21 notation into the actual machine location if they are to isolate the cause of the problem efficiently. Experience has shown that service personnel who have this documentation and who understand the basic operation of the process are appreciably more effective in system maintenance.

4.4.3   Reliability via Application

Increased reliability and, therefore, reduced maintenance can be obtained by intelligent use of the system. With an adequate understanding of the system and the process, the installation engineer and the process engineer can often design a system configuration to minimize the chance of a failure that can cause downtime or loss of productivity. The manner in which this is done depends on both the system and the process. The following example, however, is indicative of the protective measures that can be designed into the application of the process control computer.

The installation engineer, knowing the details of the system design, and the process engineer, knowing the critical control parameters of the process, can assign analog inputs such that the chance of losing critical data is minimized. For example, critical measurements can be duplexed; that is, the same measurements can be obtained with two independent analog-input channels. The two analog input channels should be selected such that they are addressed using unique circuits. For example, in a 64-channel multiplexer using an 8×8 addressing matrix, points 00, 08, 16, 24, . . . are actuated by the same driver circuit. In addition, the 16 points between 00 and 15 many be connected to the amplifier via a common block switch. Therefore, the use of addresses that are neither common to the same driver nor to the same block switch reduces the chance of losing both measurements at the same time. In the example, addresses 00 and 17 satisfy this requirement.

This technique can be extended to protect against loss of control from many types of single failures. It does not prevent the failure nor does it eliminate the need for service. It does, however, allow the process to continue operating under computer control, and it reduces the pressure on the serviceman to make an instantaneous repair. With less pressure and more time to think, he can often provide a better diagnosis and effect repairs in a more efficient manner.

4.5   SUBSYSTEM PERFORMANCE CRITERIA

The assessment of subsystem performance consists of four steps. First, one must decide on what criteria performance is to be judged. Then, the performance criteria must be rigorously defined, including a definition of the conditions under which each criterion applies. Third, methods of testing must be defined and executed. Last, the criteria and the test results must be interpreted in terms of the application in which the machine is used. This last step, of course, is highly dependent on the application, and an adequate evaluation requires the knowledge and understanding of both the process engineer and the installation engineer. It would appear, however, that the first three steps would be governed by formal or informal agreement among system vendors and users. Although a number of organizations are considering standards relating to analog subsystems, no generally accepted standards exist at present.

In addition to other problems it has created, the lack of generally accepted standards has spawned a plethora of terminology relating to the performance and specifications for analog subsystems. The term "accuracy," for example, is modified by the adjectives total, mean, measured, average, typical, long-term, short-term, calibration, and initial, among others. The term "stability" is applied to system characteristics associated with time periods as short as a few minutes or as long as a year. To explore thoroughly and relate all the terminology now in use would require a discussion equivalent to this total chapter.

Since such an exploration is a practical impossibility in this book, the discussion of this section is centered around a set of terms that is sufficient to describe the performance of the analog subsystem macroscopically. The performance of the subsystem, as reflected in this set of terms, is the result of a combination of errors and other deficiencies in the subsystem. These factors are categorized and the effect of each category on overall performance is considered. The manner in which the factors in these categories can be combined to provide an estimate of overall system performance is also discussed. Lastly, the game of "specsmanship" is described.

The message for the reader in this section is summarized in two statements. First, when dealing with specifications and other descriptions of subsystem performance, define your terms. Second, know the definitions and limitations that are associated with each term.

4.5.1 Performance Criteria

An analysis of the analog-input subsystem shows there are many sources of errors and limitations that contribute to the rather intangible concept called "subsystem performance." Many of these factors are discussed in detail in this and the previous chapter. In this section, the net effect of these factors is reduced to three basic performance criteria: accuracy, repeatability, and stability. In addition, several criteria that relate to the manner in which the system is used are discussed. Although the terminology and the boundary conditions vary from author to author, these basic concepts are generally accepted as measures of subsystem performance.

If a fixed reference source is connected to the input of an analog-input subsystem and a large number of conversions are performed, the resulting digital values are not all identical. This is the result of small perturbations, generally referred to as noise, which occur during the conversion interval. The results of this large number of conversions are conveniently summarized by constructing a histogram showing the number of times a particular digital value is obtained. Such a histogram is shown in Fig. 4-32. Since the output data are quantized, the curve is discontinuous. It is common, however, to connect the data points as a smooth distribution curve, as shown by the dotted line in the figure.

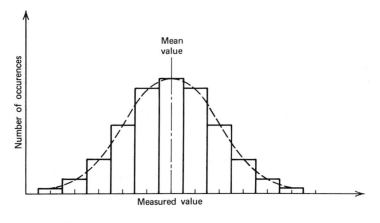

Fig. 4-32   Distribution of measurements.

The smooth curve that approximates the histogram of Fig. 4-32 approximates the normal distribution curve of theoretical statistics. Experience shows that the distribution of data from a properly adjusted analog-input subsystem is typically of this form. Marked deviations from the normal distribution are often an indication of a subsystem malfunction. For example, a double-humped curve may indicate that a significant noise source is altering the input signal for an appreciable period of time or that the subsystem is drifting the data collection time. Other indicators of malfunction are a skewed distribution in which the number of readings on one side of the mean is significantly different from the number of readings on the other side of the mean. A sharp peak in the distribution may indicate the presence of a noise source or a misadjustment of the ADC since the system is favoring a particular value. Experience with a particular subsystem often allows the design engineer to relate the characteristics of the distribution curve to specific subsystem problems.

Two important performance criteria, accuracy and repeatability, are defined in terms of the distribution of measured values. *Accuracy* is the measure of the degree to which a measurement approximates the "true" value of the measurand. The *error* is the difference between the measurand and the "true" value. Error is typically expressed as a percentage of the full-scale range or as a percentage of the reading according to the definitions

$$\varepsilon(\%\text{F.S.}) = \left(\frac{\text{"true" value} - \text{measured value}}{\text{full-scale value}}\right) \times 100 \qquad (4\text{-}51)$$

$$\varepsilon(\%\text{R}) = \left(\frac{\text{"true" value} - \text{measured value}}{\text{measured value}}\right) \times 100 \qquad (4\text{-}52)$$

The abbreviations %F.S. and %R represent "percent of full scale" and "percent of reading," respectively. In Eq. (4-52) the "true" value is sometimes substituted for the measurand value in the denominator. In most practical cases, the difference is negligible. There can, however, be a significant difference between the values obtained from Eq. (4-51) and Eq. (4-52). For readings that are much less than full value, the second definition requires a much smaller absolute error. For example, if a 10-mV signal is read on a 50-mV range, an error of 0.1%F.S. is 50 μV, whereas an error of 0.1%R is only 10 μV.

In the industry at this time it is common to use percentage of full scale for overall specification of analog subsystems. The preliminary standards being developed by some standards organizations, however, are tending toward the use of percentage of full scale *plus* a percentage of reading. This is a realistic approach since some errors, such as zero offset, are independent on the value of the reading.

The previous paragraphs started with a definition of accuracy but were primarily concerned with error. Obviously, the two concepts are related in an inverse relationship such that the smaller the error, the greater the accuracy. It is well to point out, however, that the accepted terminology of the industry uses the terms incorrectly. A typical statement, for example, is that a subsystem has an accuracy of 0.1%F.S., whereas proper use of the term requires the statement that the accuracy is 99.9%F.S. Despite the obvious difference, the misuse of the term is well entrenched and almost universally accepted.

The term accuracy must be modified with an adjective to describe adequately the performance of an analog subsystem since a number of different accuracy criteria can be defined based on the distribution of measured values. Two of these, mean accuracy and total accuracy, are illustrated in Fig. 4-33.

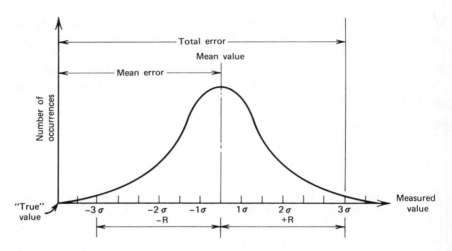

Fig. 4-33   Illustrating accuracy and repeatability.

The mean error, corresponding to the mean accuracy, is the difference between the "true" value and the arithmetic mean or average of a statistically significant number of readings under specified environmental conditions. In operational terms, it is indicative of the accuracy that can be achieved if data are obtained by averaging a number of input readings taken over a relatively short period of time. The number of readings required depends on the dispersion or repeatability of the individual readings. (Repeatability is discussed below.) If too few are used, or if the dispersion is significant, the mean may be significantly different from the true mean. If so many readings are taken that the input signal or the condition of the subsystem varies appreciably during the measurement period, the mean can again differ from its true value.

A statistically significant number of readings is a sample whose statistics closely approximate the true statistics of the subsystem and source characteristics. That is, increasing the number of readings or repeatedly performing the data collection procedure will not result in substantially different calculated statistical measures such as the mean and deviation. If the assumption is made that the analog subsystem will produce statistically normal samples of readings, a guideline as to the number of readings that should be collected can be derived by considering the standard deviation of the mean value. This provides the result that the number of readings in the sample should satisfy:

$$N \geq \left(\frac{R}{\delta E}\right)^2 \tag{4-53}$$

where R = repeatability specification in %F.S.

E = mean error specification in %F.S.

δ = allowable fractional error in the calculated mean

This guideline assumes that the subsystem is operating nominally within specification. It provides a probability in excess of 99% that the calculated value of the mean error is within $\pm \delta E$%F.S. of the true mean error if the distribution of readings is statistically normal. The use of the guideline must, of course, be tempered by experience. A value of $\delta = 0.1$ has been found appropriate through experience with a number of recent subsystems.

For the user who does not, or cannot, average a large number of readings, the mean accuracy specification is of minimal value in evaluating the performance of the subsystem. What is required is a measure of the error in any single reading. This error is highly dependent on the dispersion or spread of the distribution. This dispersion is typically specified as the *repeatability* of the subsystem. It is a measure of the difference between the mean value of the distribution and some maximum expected value for a particular reading. Because of the statistical nature of the noise that causes the spread in the distribution, a confidence level must be assigned to the repeatability specification. The confidence level is generally stated in terms of the standard deviation σ of the distribution. The confidence levels $\pm 1\sigma$, $\pm 2\sigma$, and $\pm 3\sigma$ correspond to confidences of 68.2%, 95.5%, and 99.7%, respectively, when the sample of readings is statistically normal. A repeatability specification of ± 0.1% with a 3σ confidence level is generally interpreted to mean that 99.7% of the readings in the sample will be within ± 0.1% of the mean value. Although not universally accepted, the most common specification for analog subsystems is based on this 99.7% criterion. It should be noted that the calculated value of the $\pm 3\sigma$ limits may differ from the theoretical value so that more or less than 99.7% of the reading may be included within the calculated limits. The difference, however, is usually small, and current usage generally equates the $\pm 3\sigma$ and 99.7% criteria.

As shown in Fig. 4-33, the repeatability is defined as the maximum deviation between the mean value and the limits that encompass 99.7% of all values in the distribution. If the curve is skewed, the plus repeatability R^+ and the negative repeatability R^- may differ. The normal interpretation of a specification such as "repeatability $= \pm 0.1\%$" is that the maximum of the plus and minus repeatability is less than this value. The repeatability can also be specified as the sum of R^+ and R^-, in which case it is sometimes called spread. This interpretation is both less stringent and less indicative of system performance.

The sum of the mean error and the repeatability is a limit on the error for 99.7% of the readings. This measure, called *total error* in Fig. 4-33, provides the user with the maximum deviation between the "true" value and any single measured value with a 99.7% probability. It is, therefore, the most useful performance parameter for the user who requires high accuracy but cannot average readings.

Another type of accuracy, which is sometimes called calibrated or corrected accuracy, can be meaningful if on-line calibration capability exists in the system. By reading the calibration source, most of the mean error can be compensated by correcting the data mathematically. This assumes that the calibration source is accurate and stable and that calibration readings are averaged to reduce the effects of repeatability. In high-accuracy applications, this method provides the best accuracy obtainable with the subsystem at the expense of programming and calculation time burden.

The use of "true" value in the preceding discussion of accuracy and repeatability deserves consideration. What is the "true" value of a measurement? Strictly speaking, the "true" value is the value of the input when compared with an accepted primary standard such as those maintained by the National Bureau of Standards. Although it is impractical to maintain a primary standard in the system, a secondary standard whose value is traceable to a primary standard is possible. This normally consists of a stabilized zener reference source in the analog subsystem. With proper calibration techniques, the measurements taken by the subsystem are actually relative to this calibration source. In this case, the "true" value is the value that would be obtained by measuring the unknown voltage with a perfect measurement instrument that uses the calibration source as its reference. This is an interpretation often used in the specification of an analog subsystem; that is, the accuracy of the system is specified relative to a defined secondary reference included in the subsystem.

The third basic performance criterion specified for most systems is stability. *Stability* is the change in the mean accuracy of a system over a specified interval of time. Since factors other than time, such as environment, affect the mean accuracy, measurement conditions also must be stated. Most specifications do not require that the conditions be constant between the times when stability data are collected; they merely require that the conditions at the time of data collection always be the same.

The specification on stability usually is provided as a single value rather than as a plus-and-minus value. The number is interpreted as the maximum deviation in the mean value during the stated interval of time. This interpretation is necessary since there is no way of ensuring that the system is at its nominal mean accuracy at any point in the test.

The statement of a time period in the stability specification is important. Experience shows that the drift in the mean value is often non-monotonic; the value merely meanders between limits and the difference between the limits increases with time. In addition, time periods must be specified to clearly differentiate between stability and repeatability. Typically, repeatability is specified over a period on the order of 10 min and stability over 8 hr, 24 hr, 6 months, or 1 yr. Stability is often specified over several of these intervals since its importance over a particular time period depends on the application.

In addition to the basic performance criteria of accuracy, repeatability, and stability, other parameters that are dependent on the manner in which the subsystem is used are also specified. Included in this category are specifications on the temperature coefficient and common mode rejection. The importance of these specifications depends on the application of the subsystem since, for example, the temperature coefficient effects are small if the system is always in an air-conditioned environment.

The *temperature coefficient,* as a percentage error per degree, is the measure of the change in the mean error owing to temperature variations. The primary effect of temperature in most analog subsystems is a change in the zero offset voltage. Secondarily, temperature variations result in a change in subsystem gain. Since error due to gain is dependent on the value of the reading, the temperature coefficient may be stated as the sum of a percentage of full scale and a percentage of reading. If the subsystem accuracy is stated only in terms of the full-scale range, the temperature coefficient specification is generally stated in similar terms.

In differential subsystems, the common mode performance is also specified. Since the CMRR is particularly sensitive to operating conditions, a complete specification includes a statement of these conditions. Factors that should be included are the per-channel sampling rate, the environmental conditions, the source unbalance, and the frequency of the common mode source.

In addition to these specifications which relate to the normal performance of an analog subsystem, a specification listing usually includes parameters limiting the extent of other specifications or defining the operating limits on the subsystem. For example, maximum and minimum operating temperatures and maximum common and normal mode voltages are typically included.

4.5.2 Types of Errors

The number of different errors that contribute to the overall system specification is quite large if the detailed operation of each piece of equipment in the subsystem is considered. These many errors, however, fall into four categories based on their effect in the subsystem. The first category is that of fixed errors. Fixed errors are independent of time, temperature, and the manner in which the system is used. Examples of fixed errors are gain inaccuracy and the resolution of the gain adjustment, initial zero offset, and any variation of gain as a function of input signal value. Fixed errors contribute to the mean accuracy specification of the subsystem.

Variable errors are errors that are a function of time and temperature. Examples include the zero offset variation with temperature, zero offset drift with time, and changes in the gain with time and temperature. These errors contribute to the stability and temperature coefficient specifications of the subsystem.

Random noise sources introduce random errors into the subsystem. Neither the exact magnitude nor the polarity of the error on any particular sample can be predicted. The random nature of the error may be a result of the characteristics of the source, as in the case of component noise, or it may be a result of lack of synchronism between the source and the sampling interval. This latter case is exemplified by induced 60-Hz noise voltages.

Another category of noise is systematic noise. The presence of this noise at a particular point in the sampling process is known *a priori*. Examples of systematic noise include interference from the clock oscillator, modulation-demodulation ripple, and multiplexer switch transients.

The last category of errors, operational errors, are those errors resulting from the manner in which the subsystem is used. Examples include common mode error, source loading error, and errors due to sampling rate. Operational errors usually are not included as part of the accuracy and repeatability specification but are listed as separate performance parameters.

Each piece of equipment in the analog-input subsystem contributes to each category of error. A list of error sources for a typical subsystem and their categorization is shown in Table 4-1. Since the configuration sometimes affects the categorization of the error, some entries are shown as contributions in several categories.

Table 4-1   Error Contributions

Source	Fixed	Variable	Noise Systematic	Noise Random	Operational
Termination/signal conditioning					
Thermals		X			
Contact resistance					X
Source loading					X
Multiplexer					
Thermals		X			X
Switch transient			X		
Leakage current	X	X			
Switch offset	X	X			
Amplifier					
Thermals		X			
Noise				X	
Overload					X
Source loading	X				
Chopper noise			X		
Gain error	X	X			
Zero offset	X	X			
Linearity	X				
ADC					
Quantization	X				
Noise				X	
Gain error	X	X			
Zero offset	X	X			
Linearity	X				
Subsystem					
Common mode					X
Scan sequence					X
Scan rate					X
Induced noise			X	X	

4.5.3   Estimating Subsystem Performance

In order to estimate the overall performance of the subsystem, the errors listed in Table 4-1 must be combined to form overall accuracy, repeatability, and stability specifications. Although the methods used by various vendors differ, the following paragraphs describe a common and generally acceptable method.

The fixed errors are added arithmetically as a "worst-case" contribution to the mean error. To this is added the arithmetic sum of the variable errors which are applicable under the definition of mean error. That is, if the mean error includes a temperature variation of $\pm 10°F$ from some reference condition and 24-hr stability, the appropriate contributions of time and temperature are added to the fixed errors. A third contribution to the mean error is the arithmetic sum of those systematic errors which are known to affect the subsystem as a constant offset from one sample to the next. For example, the normal mode charge-sharing error always results in a negative error when a shorting switch is used. The sum of these fixed, variable, and systematic errors is an estimate of the subsystem mean error.

The variable errors that are dependent on time and that are not included in the mean error are arithmetically summed to obtain an estimate of the worst-case subsystem stability. The same summation is performed on the variable errors which are temperature dependent to obtain an estimate of the subsystem temperature coefficient.

The estimate of the repeatability is the sum of two components. The first is the arithmetic sum of the systematic noise errors that are not included in the mean error calculation. The second component owing to random noise is determined as the square root of the sum of the squares of the random errors. This method of combination is based on the statistical theorem which states that the sum of two or more independent random functions is the square root of the sum of the functions.

In performing all of these calculations, a compatible system of units and definitions must be used. Thus, for example, the confidence level of all the random errors must be the same. Also, the difference between errors that are a percentage of reading must be distinguished from those that are independent of reading if the overall specification is stated in these terms. Errors referred to the input must be multiplied by the gain of the system before converted to a percentage error to be used in the calculation.

The result of the calculations is merely an estimate of the subsystem performance and must be verified by extensive testing. The estimate assumes many errors combine in a "worst-case" manner, whereas errors often cancel because of opposite polarity. If extensive testing verifies that certain errors cancel, then the estimate can be refined as a result of the empirical data.

4.5.4 "Specsmanship"

Specsmanship is the art of determining the specification for a subsystem and/or interpreting written specifications. Specsmanship poses a problem for the vendor and user alike. The vendor faces the problem of providing specifications that are both competitive and understandable to a wide variety of users. The user has the problem of interpreting competitive specifications from many different vendors and relating them to his particular application.

The vendor's problem has many facets. Primary among these is the problem of competitive evaluation. The lack of accepted definitions and standards often results in misinterpretation of performance specifications. The possibility of misinterpretation is significant, partially because many users are unsophisticated and inexperienced in matters relating to precision analog measurement equipment. The sophisticated user recognizes that no single subsystem can be best in all respects because of the design compromises required. He selects his subsystem by matching its characteristics with the needs of his control problem. The unsophisticated user tends merely to compare numbers. Since the computer process control field is relatively young, the inexperienced users represent the majority at present. The result is a reluctance on the part of all vendors to provide detailed performance specifications.

A second major problem for the vendor is that his systems are applied to a wide range of applications. The performance of the subsystem in a particular situation is highly dependent on the process. For example, the noise in the process affects the variability of the data collected by the subsystem. If the vendor essentially guarantees the system performance, he must be very conservative in his specifications to account for a reasonable amount of process interdependency. A related problem is the matter of subsystem size. Even though many subsystems can accommodate up to several thousand points, the "typical" subsystem in many applications is less than several hundred points. Since the performance of a smaller subsystem is generally better, the vendor might be justified in specifying the subsystem on a "typical" basis. This, however, can lead to a problem when the performance of a large system must be guaranteed.

A partial solution to the vendor's problem is the growing number of experienced users who recognize the trade-offs involved in the system design and application. As the experienced group grows, the vendor will provide more complete data with assurance that they will be understood and interpreted wisely.

For the user, the game of specsmanship involves understanding and comparing the performance of competitive systems. The plethora of definitions and specification formats makes this an extremely difficult task. One vendor may provide pages of detailed specifications, such as those in Table 4-1, and expect the user to combine them in a manner appropriate for his particular application. Another vendor, however, may provide overall system specifications and very little information to assist the user in evaluating a parameter of particular concern in his application. No one can say that either vendor is correct or that one's method of specification is better than the other's. Nor can anyone blame the user for his irritation and confusion. The vendor must recognize the user's problem and assist him with supplementary information and engineering assistance to estimate the system performance in a particular application. The user must understand the vendor's problem and educate himself in the subject of analog-input subsystems so that he can communicate effectively with the vendor.

The rules of the game of specsmanship are not easily defined. In the following paragraphs, however, an attempt is made to assist the user in understanding the interpretation of specifications as well as to caution him against certain practices common in specsmanship.

Perhaps the most important factor the user must understand is that specifications are not binary in nature. A common misinterpretation is that exceeding a limit results in instantaneous malfunction. This is analogous to the fifteenth-century idea that sailing beyond the horizon meant falling into the endless space of the universe. Most analog specifications are such that exceeding a limit results in gradual degradation of subsystem performance. For example, if the specifications state that the CMRR is 120 dB at a per-channel sample rate of 3 sps, it is unlikely that the CMRR is 80 dB if the sample rate is increased to 4 sps. The vendor knows that the sample rate affects the common mode performance, so he places conditions on the parameter, but the limits are not absolute. If the application calls for a slightly higher sampling rate, the chances are that the system will operate within specification.

Some limits must be respected. For example, if the normal mode overvoltage is specified at 50 V, the chances are the system will not be catastrophically damaged if 51 V or even 55 V is applied. At some point however, a very slight increase in voltage results in damage.

Deciding whether or not the limitation on a specification can be exceeded requires knowledge of how the system operates. This emphasizes the necessity of the user's knowing the basic operation of analog-input subsystems and of working closely with the vendor in evaluating the estimated performance.

The importance of understanding the definitions of terms used in a specification is stressed in previous sections. However, the relationship between specsmanship and definitions is so crucial that it deserves further comment. Without adequate definitions, what is unstated in the specification may be as important as what is stated. For example, a specification that lists a common mode rejection figure without stating the maximum source unbalance is misleading. It cannot be compared with a specification in which the unbalance is stated.

The following list of questions and comments, although not exhaustive, provides a guide to the specsmanship of definitions and limitations. Not all of these questions can be adequately addressed in a simple summary specification. However, they should be considered prior to a comparison of competitive systems.

1. Is the stated accuracy a mean accuracy, a total accuracy, a calibrated accuracy, or "typical" accuracy? The difference is significant and the definition of "typical," in particular, requires careful investigation.

2. What are the environmental conditions under which accuracy specifications are valid? Specifying accuracy at a single temperature, for example, is unrealistic since even air-conditioned rooms are subject to some temperature and humidity variation from day to night and from season to season.

3. What is the effect of continuously sampling a single channel at a given rate?

4. What effect does sampling rate have on common mode performance?

5. What is the channel bandwidth?

6. What is the time period covered by the stability specification?

7. In low resolution systems, the $\frac{1}{2}$ LSB quantizing error is significant and must be included in an accuracy evaluation.

8. What is the value and the nature of the system input impedance? Does an input current flow when the input is shorted?

9. Is the common mode performance specified at a frequency other than DC? Most process noise is AC so the AC CMRR is an important parameter in determining system performance.

10. What is the effect of environment, particularly high humidity, on the common mode performance?

11. Is a guard shield required on input signal conductors to meet specified performance? If so, what are the specifications without the guard shield?

12. What is the recovery time after an input channel is subjected to a large overvoltage? If certain types of capacitors are used in signal conditioning circuits, dielectric absorption can extend recovery time far beyond that predicted by the usual RC decay equations.

13. Are there restrictions on the placement of different types or levels of signals in the termination facility?

14. In high-bandwidth systems, the repeatability of the subsystem may be dependent on the source resistance.

15. What is the specification on normal mode and common mode crosstalk?

16. Do the specifications apply for the maximum size system? Depending on the philosophy of the vendor, specifications may be based on the maximum system or on a "typical" system. In general, specifications that are dependent on interactions within the system, such as common mode performance and repeatability, are improved in small systems.

17. Are percentage specifications based on full-scale value, the value of the reading, or a combination?

18. In comparing competitive specifications, are temperature-dependent specifications stated in the same units? Specifications stated in terms of "per degrees Fahrenheit" look better than those stated in degrees Centigrade since $1°C = 1.8°F$.

From the vendors point of view, there are equally important questions that should be answered by the user. The most common complaints in attempting to answer a bid request from a user are (1) the user does not understand time-shared analog-input subsystems; (2) the user has not defined his terms; (3) the user has asked for an impossible combination of specifications as the result of searching through a pile of catalogs and picking the best specification from different pieces of equipment; (4) the bid request is iron-clad and does not allow for alternate, and perfectly valid, solutions; (5) the user has not assigned a priority to his requirements, that is, which requirements must be met and which are subject to negotiation; and (6) the user has asked for elaborate, detailed specifications on individual circuits even though it is doubtful whether he or the vendor could precisely relate them to overall system performance.

The user should answer the following questions in preparing a request for quotation or in evaluating competitive equipment.

1. Do you understand the performance requirements of the application? For example, is the requirement one of high accuracy or high stability and low repeatability?

2. Is a sufficient explanation of the application included in the bid request so that the vendor can propose meaningful alternative solutions,

or are the specifications written in terms of functional requirements rather than a predefined configuration?

3. Is the importance of particular specifications indicated?

4. Are special environmental or operating conditions clearly stated in the request? Examples include the presence of contaminants, corrosive gases, and sanitation requirements. (In one application, the entire computer system, including the electromechanical I/O equipment, was subjected to cleaning with live steam!)

5. Have the acceptable price limits been determined by merely adding up component prices from a variety of catalogs? This is unrealistic since it does not recognize the cost of system engineering and testing which is required to integrate the component parts into a working system.

The rules for the game of specsmanship are extensive and ever-changing, but these questions are indicative of the considerations that must be included in the evaluation of a subsystem.

If the whole subject of dealing with specsmanship has to be reduced to a single comment, it is probably the following. The evaluation of analog specifications requires extensive knowledge and experience in analog technology. A thorough knowledge of the operation of an analog subsystem and error phenomenon such as presented in these chapters allows the evaluator to assess the significance of specifications. It also ensures that unstated specifications and dependencies are not overlooked. If the user does not have personnel with the necessary knowledge, he must make every attempt to obtain them. If this is not possible, the vendor's experience, reputation, and capability for providing technical assistance must be prime considerations in selecting a computer system.

REFERENCES

Gruenberg, E. L. (Ed.), *Handbook of Telemetry and Remote Control,* McGraw-Hill, New York, 1967.

Harman, W. W., *Principles of the Statistical Theory of Communication,* McGraw-Hill, New York, 1963.

Jursik, J., "Rejecting Common Mode Noise in Process Data Systems," *Control Eng.,* August 1963, pp. 61–66.

Jursik, J., "An Introduction to Process Measurements for Computer Control Systems," unpub. ms., 1962.

Maher, J. R., "Cable Design and Practice for Computer Installations," *TAPPI,* May 1965, pp. 90A–93A.

Melsheimer, R. S., "Bessel or Butterworth? Choosing Filters for Modern Instrumentation Systems," *Bulletin 41103,* Astrodata, Inc., Anaheim, Calif., November, 1964.

Susskind, A. K. (Ed.), *Notes on Analog-Digital Conversion Techniques,* Technology Press, Cambridge, Mass., and Wiley, New York, 1957.

Terao, M., "Quantization and Sampling Selection for Efficient DDC," *Instrum. Technol.,* August 1967, pp. 49–55.

BIBLIOGRAPHY

Barr, P., "Influence of Aperture Time and Conversion Rate on the Sampling Accuracy of A-D Converters," *Data Syst. Eng.,* May 1964, pp. 30–34.

Buchman, A. S., "Noise Control in Low Level Data Systems," *Electromech. Des.,* September 1962, pp. 64–81.

Cerni, R., "Sources of Instrument Error," *ISA J.,* June 1962, pp. 29–32.

Cerni, R. H., and L. E. Foster, *Instrumentation for Engineering Measurements,* Wiley, New York, 1962.

Clifton, H. W., "Design Your Grounding System," *IEEE Trans. Aerosp.,* April 1964, pp. 589–596.

Dunbridge, B., "Analysis of Analog-to-Digital Converter Time Aperature Error," *Telemet. J.,* August/September 1966, pp. 32–36.

EID Staff, "Grounding Low-Level Instrumentation Systems—A Coherant Philosophy," *Electron. Instrum. Dig.,* January/February 1966, pp. 16–18.

Engle, C. E., "Techniques to Analyze and Optimize Noise Rejection Ratio of Low Level Differential Data System," *Elteknik,* March 1966, pp. 35–42.

Friauf, W. S., "Dynamic Characteristics of Analog-Digital Converters," *Instrum. Control Syst.,* January 1965, pp. 111–114.

Gaines, W. M., and P. P. Fischer, "Tentative Recommendations for the Definition and Testing of the Dynamic Performance of Analog-to-Digital Converters," *AIEE Trans. (Commun. and Electron.),* September 1961, pp. 387–394.

Gaines, W. M., and P. P. Fischer, "Terminology for Functional Characteristics of Analog-to-Digital Converters," *Control Eng.,* February 1961, pp. 97–98.

Gaines, W. M., and W. N. Patterson, "Accuracy in Process Control Instrumentation Systems," *AIEE Trans. (Appl. and Ind.),* March 1962, pp. 5–10.

Gardenhire, L. W., "Selecting Sample Rates," *ISA J.,* April 1964, pp. 59–64.

Goff, K. W., "Dynamic Considerations in Applying DDC," *ISA Paper 6.4-1-66,* October 1966.

Gordon, B. M., and B. K. Smith, "How to Specify Analog-to-Digital Converters," *Electron. Des.,* May 10, 1961.

Hoffart, H. M., "Instrument Grounding Keeps Space Signals Clean," *Instrumen. Technol.,* August 1967, pp. 41–44.

Linebarger, R. N., and G. K. L. Chien, "Word-Length Sample-Rate Tradeoffs in Computer Control Systems," *JACC Prepr.,* 1964, pp. 124–130.

McAdam, W., and D. Vandeventer, "Solving Pickup Problems in Electronic Instrumentation, *ISA J.,* April 1960, pp. 48–52.

Marimon, R. L., "Ground Rules for Low-Frequency, High-Gain Amplifiers (Part I)," *Electron. Des.,* April 12, 1963, pp. 56–64.

Morrison, R., *Shielding and Grounding in Instrumentation,* Wiley, New York, 1967.

Moon, W. D., "Measurement Accuracy," *IEEE Trans. Aerosp.*, October 1964, pp. 1161–1165.

Nalle, D., "Elimination of Noise in Low-Level Circuits," *ISA J.*, August 1965, pp. 59–68.

Robinson, T. A., "Definitions of Grounding Terms for Power, Electronics, and Aerospace," *IEEE Trans. Aerosp.*, August 1963, pp. 979–981.

Robinson, T. A., "The Role of Grounding in Eliminating Interference," *IEEE Spectrum*, July 1965, pp. 85–89.

Schneider, D. B., "Designing for Low Level Inputs," *Electron. Ind.*, January 1961, pp. 81–85.

Slaughter, J. B., "Quantization Errors in Digital Control Systems," *IRE Trans. Autom. Control*, January 1964, pp. 70–74.

Smith, B. K., "Practical Information Theory Aspects of High-Speed Data Handling," *IRE Trans. Instrum.*, March 1958, pp. 95–98.

Stewart, E. L., "Grounds, Grounds, and More Grounds," *Simulation*, August 1965, pp. 121–128.

Stewart, E. L., "Noise Reduction on Interconnect Lines," *Simulation*, September 1965, pp. 149–155.

Terao, M., "The Selection of the Quantization and the Sampling for Efficient DDC," *2nd IFAC/IFIP Conf.*, Menton, France, June, 1967.

van der Grinten, P. M. E. M., "Control Effects of Instrument Accuracy and Measuring Speed," *ISA J.*, December 1965, pp. 48–50.

Van Doren, A., "Solving Error Problems in Digital Conversion Systems," *Electromech. Des.*, April 1966, pp. 44–46.

Vogelman, J. H., "A Theoretical Analysis of Grounding," *IEEE Trans. Aerosp.*, April 1964, pp. 109–111.

Walton, J., "Noise Reduction by Isolation," *Instrum. Control Syst.*, February 1966, pp. 109–111.

"Bibliography on Interference in Low Level Circuitry," *IEEE Trans. Ind. Electron. and Control Instrum.*, September 1964, pp. 50–54.

"Control Data 1700 Computer System—Analog/Digital Systems General Information Manual," *Publication 60187600*, Control Data Corp., La Jolla, Calif., n.d.

"External Noise Levels," *Instrum. Control Syst.*, October 1965, pp. 134–137.

Grounding and Noise Reduction Practices for Instrumentation System, SDS Data Systems, Pomona, Calif., n.d.

"IBM 1800 Data Acquisition and Control System Installation Manual—Physical Planning," *Form A26-5922*, International Business Machines Corp., San Jose, Calif., 1966.

"System Performance Characteristics of Multiplexer Instrumentation," *Redcor Data Handl. Notes*, Redcor Corp., Canoga Park, Calif., May 1967.

"Throughput Rate (Definitions, Discussion, and Limitations)," *Redcor Data Handl. Notes*, Redcor Corp., Canoga Park, Calif., November 1966.

Chapter 5

Analog-Output Subsystems

CHUNG C. LIU

The analog-input subsystem, described in the two preceding chapters, provides the means whereby the process control computer can measure and "observe" the physical world. This is a necessary function since the state of the process must be measured if proper control action is to be formulated. This ability to collect data under control of the digital computer is particularly important in real-time situations where manual measurement or manual entry of the data into the computer would result in intolerable delays. In a few applications, the functions provided by the analog-input subsystem are sufficient, but this is not the usual case. Since a process is not totally analog in nature, it must also be possible to sense digital information—for example, whether a motor is ON or OFF. This input capability is provided by the digital-input subsystem which is discussed in the next chapter. With this function, the process control computer can accurately determine the analog and digital status of the process on a real-time basis.

Before considering the digital-input capability, however, the control capability of the system deserves consideration. Just as "observing" the process requires analog- and digital-inputs, controlling the process requires both analog- and digital-output signals. Analog signals are used to control devices such as valve positioners and, in some cases, setpoint controllers. In addition, continuous analog signals are required as input to display devices such as strip chart recorders and oscilloscope displays.

Digital-output signals provide the capability of controlling switches and relays. This function is used extensively in consoles and panelboard displays.

The analog-output subsystem closely parallels the analog-input subsystem. The analog-input subsystem is the interface between the process and the computer which provides the measurement capability; the analog-output subsystem is the interface between the digital process control computer and process load devices. The need for this interface is evident because many load devices in process control applications require analog signals as input and the world within the computer is purely binary, a form not readily usable by many load devices. The circuit that performs this necessary conversion from digital to analog form is generally called the digital-to-analog converter, or DAC.

5.1 SUBSYSTEM FUNCTIONS

The role of the analog-output subsystem in a process control computer is twofold: (1) It provides the capability for the process control computer to exercise control over analog process load devices; and (2) it provides a suitable means for the computer to communicate with the process operator through graphic displays. These functions are provided by one or more DACs in the analog-output subsystem in conjunction with appropriate logic controls. In addition, however, other functions such as analog storage and back-up analog controllers must be provided in some applications. In this section, the required functions are reviewed briefly. Subsequent sections consider the detailed aspects of the subsystem design.

5.1.1 Voltage Output

The output from a DAC is usually an analog voltage. Depending on the application, the DAC may include an analog driver amplifier which acts as a buffer to isolate the load from the high impedance of the DAC and to increase the output voltage and current driving capability. The voltage output is commonly used in applications such as display, setpoint control, automated testing, hybrid computation, and position control for servomotors and electronic beams. This type of output is generally characterized by medium to high speed and moderate to high accuracy.

Typically, the maximum output voltage is either ±10 or ±100 V DC and has a resolution of 10–13 bits. In process control applications, the ±10-V output is most common. The output settling time varies from 1–30 μsec and the accuracy is in the range of ±0.01–$\pm0.1\%$. The output current capability may range from ±10 mA–±1 A, as indicated by the application; the 10-mA range is the most common in process control equipment.

5.1.2 Current Output

Current output signals are commonly used in process control applications. Most domestic electronic instruments for process control applications operate with one of the three current ranges: 1–5 mA, 4–20 mA, or 10–50 mA DC. The nonzero minimum output value is called a "live zero" and is produced by an offset in the analog-output circuit. In other words, a zero digital-output signal from the computer corresponds to the minimum current of the range and the current never goes to zero. The signals used outside the United States, however, may be zero based: 0–5 mA, 0–20 mA, or 0–50 mA DC. Current output signals are typified by a 1–10 msec response time and a ±0.2–±0.5% accuracy. A resolution of 10 bits is sufficient for most applications.

Because the conventional output from a DAC is a voltage, voltage-to-current conversion is usually required to obtain the current output signal. The design of the current output station is discussed in Sec. 5.5.

5.1.3 Analog Memory Devices

Frequently, the analog-output subsystem is required to store its output values for long periods of time. There are two storage techniques: digital storage and analog storage. The digital storage utilizes an individual DAC for each output. Thus the data are stored in the digital holding register which controls the DAC switches. The main advantage of the digital storage lies in its ability to retain the output value indefinitely if no power interruption occurs.

The alternative to providing a DAC per channel is to time share a single DAC among a number of analog storage or memory devices. Under control of the computer and/or the subsystem control logic, the DAC output is multiplexed to the input of the analog memory device. The output of the analog memory provides the continuous signal required for the control of the process or display equipment.

This section provides a description of a number of analog memory devices suitable for process control analog storage applications. The performance parameters typically desired for these applications are the following: (1) Setting accuracy of ±0.5%; (2) resolution of ±0.1% (10 bits); (3) repeatability of ±0.1%; and (4) setting stability or droop rate of less than 1%/hr. There are three devices that satisfy these criteria: a motor driven potentiometer, a multiaperture magnetic core, and a combination of an FET and capacitor.

Motor driven potentiometers are electromechanical analog memory devices. The motor is usually a single-phase, high-torque, synchronous motor designed for pulse inputs. The input to the motor is generally a pulse or series of pulses. When a pulse of suitable form arrives at the input, the magnetic force between the rotor and stator steps the shaft by a discrete amount. A potentiometer that is mechanically coupled to the motor can be used to control a voltage. Therefore, discrete values of output voltage can be obtained by stepping the shaft through the range of the potentiometer. This type of arrangement is usually quite expensive owing to the cost of the stepping motor. It does provide, however, an inherent fail-safe and nonvolatile analog memory.

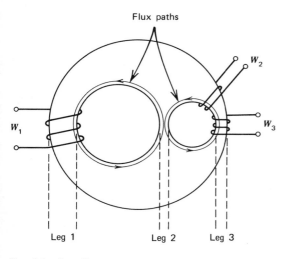

Fig. 5-1   Transfluxor core.

The transfluxor is an AC memory element that features nondestructive readout (Rajchman and Lo, 1956). It is a ferrite core that has square loop properties and multiple flux paths. Figure 5-1 shows the transfluxor core with its major and minor aperture. Two magnetic paths may be traced on the core, one formed by leg 1 and leg 2 around the major aperture and another formed by legs 2 and 3 around the minor aperture. Three windings are used: The control winding w_1 is used to control the magnetic state around the major aperture. The excitation winding w_2 is used to excite the output winding w_3 by modulating the magnetic path around the minor aperture. The output voltage is the voltage induced in winding w_3.

When pulses are applied to winding w_1, the magnetic state of the material at the perimeter of the major aperture is affected first since this represents the shortest magnetic path. The magnetization of this path effectively reduces the cross section of leg 2 and, therefore, controls the reluctance of the path around the perimeter of the minor aperture. By modulating the flux in leg 2 with a signal applied to winding w_2, an output is obtained at winding w_3 which is a function of the magnetic state existing around the perimeter of the major aperature and, specifically, the flux established in leg 2 by the pulses applied at winding w_1.

For analog memory use, the transfluxor is incorporated in a closed loop circuit, as shown in Fig. 5-2. The DAC, error amplifier, and compensation network can be shared by a number of transfluxor channels by using multiplexing switches at the points indicated by the X's. The performance of such a system has been reported as follows (Walton, 1968): The setting accuracy of the system is typically $\pm 0.1\%$ and, at worst, $\pm 0.25\%$; the temperature coefficient varies from $\pm 0.08\%/°C$ at midscale to $\pm 0.5\%/°C$ at both minimum and maximum scale; the settling time varies from 50 to 100 msec for 20 and 100% input steps, respectively.

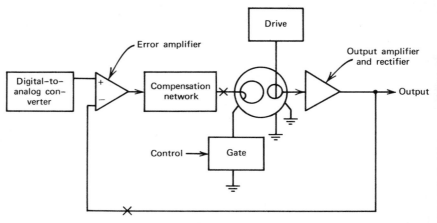

X = Multiplexing points for additional channels

Fig. 5-2   Transfluxor closed-loop system.

A third type of analog memory device utilizes a charge storage principle with field effect transistors (Maxfield and Steele, 1967). Charge is stored on a capacitor to hold a defined voltage level and can be retained for a long period of time when a field effect transistor is used as a source follower between the capacitor and the output load. Figure 5-3 shows such

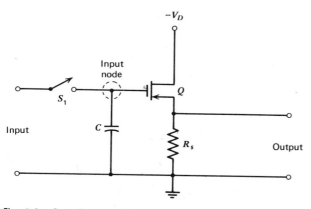

Fig. 5-3 Capacitor and FET analog memory.

a memory circuit. It consists of an input switch S_1, a charge storage capacitor C, and a metal oxide semiconductor field effect transistor (MOSFET). The input voltage derived from a DAC is stored on the capacitor when the switch S_1 is closed. The stored voltage on the capacitor remains essentially constant, as does the output voltage, owing to the extremely large leakage resistance at the input node.

This type of analog memory device is very attractive because of its simplicity, infinite resolution, and low droop rate. Since the MOSFET input gate leakage resistance is on the order of 10^{12} to 10^{15} Ω, the droop rate of the circuit is essentially determined by the leakage resistances of the input switch when it is turned OFF, the leakage resistance of the charge storage capacitor, and the leakages associated with the physical environment of the input node. With a closed-loop design, a setting accuracy of ± 0.1–0.2% is easily achieved. Typical repeatability is better than $\pm 0.1\%$, and a droop rate of less than 0.1%/hr is easily obtained under laboratory conditions. The droop rate is degraded somewhat under high-humidity conditions but is less than the typical requirement of 1%/hr when proper packaging techniques are used. Junction field effect transistors (JFET) may also be used for those applications in which the droop rate requirements are less stringent. The practical design of this type of analog memory with a current output is discussed in Sec 5.5.

There are other types of analog memory components such as the Solion* electrochemical device, transpolarizers (the electrostatic analog of the transfluxor), magnetostrictive integrators, the Memistor† electrochemically variable resistor and second harmonic integrators (magnetic

* Trademark of Texas Research Corp.
† Registered trademark of Memistor Corp.

core storage devices) (Nagy, 1963). These devices are not commonly used in the analog-output subsystems of process control computers because of performance limitations, however.

5.1.4 Back-up Devices and Techniques

In the introduction of the digital computer to process control applications, the final acceptance of the computer may rest on the confidence instilled in the user by attainment of basic computer reliability and the provision for fail-safe techniques. At present, a failure due to either a programming problem or a hardware failure is still a possibility. The concern is what happens if the computer fails. For supervisory control where the computer is used to adjust the setpoints of the analog controllers, a computer failure leaves the analog controller with the last computer control setting and, therefore, the process is still under conventional control. In this case, the controller must contain a setpoint memory circuit.

For direct digital control (DDC) applications, the requirement for back-up depends on the dynamic characteristics of the process, the expected failure rate of the digital computer, its repair time, and the effect of failure on the plant. As in the case of supervisory control, back-up must be employed on a per-channel basis. The back-up devices and techniques can be divided into four classes: (1) Analog memory (hold last value) and manual back-up; (2) analog controller back-up; (3) digital back-up; and (4) dual digital computer back-up.

Two necessary prerequisites of any back-up technique are computer fault detection and the isolation of the process load from the computer in the event of computer failure. Fault detection is usually achieved by suitable computer programs which may take a number of forms. When a fault is detected, a "computer down" signal arrives at the analog-output subsystem, and this initiates transfer from computer mode to a fallback or back-up condition.

Among the four classes of back-up devices, the hold last value analog memory with manual back-up is the most widely used. When the computer fails, the analog output retains the last computer output value and, therefore, the process control element holds its last position. This technique is acceptable for most noncritical continuous process loops because once the process is operating, maintaining the control element position provides the necessary control if the computer downtime is short. Should the computer be off-line for an extended period, the operator can adjust the outputs manually. This is the simplest form of back-up and can be provided at a minimum system cost.

The back-up analog controller technique is reserved for critical process loops where the additional cost can be justified. In the petroleum and chemical industry, for example, approximately 10–30% of the plant control loops are typically considered critical. In this back-up technique, the back-up controller normally acts as a simple current output station whose output is controlled by the computer. The two- or three-mode control response associated with an analog controller can be provided through the digital computations in the control algorithm. In the case of a computer failure, however, the output station automatically is converted into a two- or three-mode analog controller. The advantage of this technique, as compared with the conventional supervisory setpoint control using standard analog controllers, is that the advanced control concepts associated with DDC can be implemented. But in case of a computer failure, the system reverts to conventional control.

Various options for the setpoint control in the fallback mode are possible: (1) Fallback to last process variable; (2) fallback to computer updated setpoint (setpoint tracking); (3) fallback to last process variable as initial setpoint and then decay to a predetermined safe value; and (4) fallback to a preset setpoint value. The selection of the setpoint fallback mode is dependent on the customer's philosophy of control, the plant operating requirements, the dynamic characteristics of the loop, and other factors.

The digital back-up and dual computer back-up techniques generally are more sophisticated and more expensive (Morley and Cundall, 1965; Cundall, 1965; Lombardo, 1967). The programming effort in the dual computer approach is greatly increased but the methods are suited to large and complex control system applications.

5.1.5 Man-Machine Interface

Man-machine interface plays an important role in process control applications. The interface usually consists of meters, switches, indicators, lights, pushbuttons, and labels which allow a process operator to assess the general situation of the plant at a glance. These are also the means by which he takes manual control of a process load if a situation calls for such action.

Generally, certain man-machine interface functions are provided on the front panel of computer-manual output stations and fallback controllers. Typically, they include the following:

1. Output meter to indicate the current output value.
2. Process meter to indicate the process or measured variable value.
3. Setpoint indicator to show the present setpoint.
4. Mode select switch to select analog/computer or manual operation.

5. Setpoint control to adjust the setpoint manually.

6. Manual output control to adjust the station output when in the manual mode.

7. Alarm light to indicate that an abnormal condition exists.

8. Mode indication light to indicate the present mode of the device.

The process operator console also provides man-machine interface functions for the process operator and process engineer. Those functions normally available are the display and/or alteration of process variable, setpoint, current output to the valve or control element, automatic/ manual mode, and the ability to regulate the output while in manual mode. Digital displays usually are utilized to identify the loop, the process variable, various other functions, and to display entered values. These subjects are considered in detail in Chap. 7.

5.2 SUBSYSTEM CONFIGURATIONS

For a wide variety of applications, the analog-output subsystems can be classified into three different configurations: (1) A single DAC per output channel; (2) a single time-shared DAC with an output memory amplifier for each output channel; and (3) a combination configuration.

5.2.1 DAC per Channel

This is a conventional configuration, as shown in Fig. 5-4. The subsystem contains a number of DACs and a control unit. The control unit provides the subsystem with the means to communicate with the central processing unit (CPU) of the system. It also provides the input data manipulation for each DAC. Under program control, a particular DAC is selected or addressed and the data are loaded into the register of the DAC. The DAC output retains its value until a new digital value is loaded by the control program.

This configuration is usually used for hybrid computation, test automation, and analog display applications which are characterized by the requirement for high speed and high accuracy. If the number of output channels is large, however, the configuration can be quite expensive because of the relatively high cost of the individual DACs.

5.2.2 Time-Shared DAC

Since the cost of the DAC per-channel configuration can be prohibitive, especially if the number of output channels is large, some form of time sharing may be appropriate. A typical time-shared analog-output subsystem is shown in Fig. 5-5. The subsystem contains a control unit, a

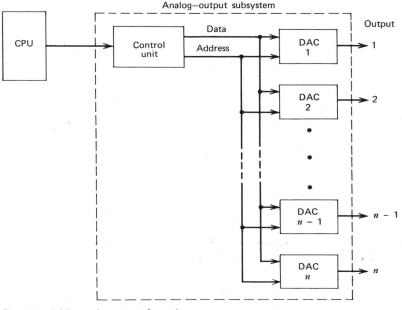

Fig. 5-4 DAC per-channel configuration.

single time-shared DAC, and a number of output-analog memory circuits.

Time sharing involves multiplexing the DAC output voltage into a number of analog memory circuits. As the digital data are received from the CPU through the control unit, the DAC output is switched to the appropriate analog memory circuit. Its output value remains essentially constant until the value is updated by new data. Since some analog memory units exhibit droop with time, the value may require periodic updating even though a new value for the output has not been calculated. The configuration of Fig. 5-5 is commonly used for supervisory control and DDC applications.

5.2.3 Combination Configuration

As dictated by the wide variety of applications, a third type of subsystem configuration can be a combination of a DAC per channel and a time-shared DAC with output memory circuits. The output memory amplifiers may be replaced by computer/manual stations, computer control stations, or back-up analog controllers. In process control applications, the number of per-channel DAC outputs is usually quite limited. On the other hand, the total number of stations serviced by one

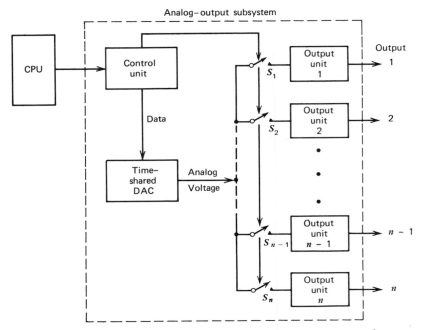

Fig. 5-5   Time-shared DAC configuration. (The output unit can be computer/manual station, computer control station, or backup analog controller.)

or more time-shared DACs may range from 32 to 512 or more, depending on the size of the plant.

The design considerations for the computer/manual station and the back-up analog controller are discussed in Sec. 5.5 and Sec. 5.6, respectively.

5.3   DAC ELEMENTS

Most digital-to-analog converters (DAC) commonly used in process control computer systems consist of the five basic functional elements or building blocks shown in Fig. 5-6: (1) decoding network, (2) analog switch, (3) precision reference supply, (4) summing amplifier, and (5) register and control.

Today, the majority of the DACs used in process control computer applications are built around these five blocks. The major difference between these DACs lies in the manner in which the blocks are implemented and arranged. A thorough understanding of the characteristics

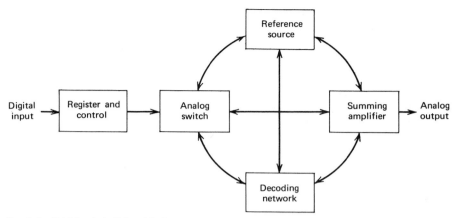

Fig. 5-6   DAC basic building blocks.

and critical properties of each individual block is essential to DAC design.

The design of the register and control is a relatively straightforward logic design task and is not considered here in detail. The primary emphasis of this section concerns the design of the remaining elements, all of which are analog in nature.

Before a design effort is undertaken, the following major performance requirement should be established: (1) accuracy, (2) resolution, (3) conversion time, (4) analog-output voltage range, (5) analog-output current range, (6) output impedance, and (7) temperature coefficient.

Accuracy is the fidelity with which the conversion process is performed, that is, how closely the actual output resembles the intended theoretical value.

Resolution is the minimum change of input data which results in a corresponding output change. It is usually determined by the number of input data bits, excluding the sign bit.

Conversion time is, generally, the interval between the time the data are set into the DAC and the time the output settles to a specified level.

The *analog-output voltage range* is the full-scale output voltage, whether it is a unipolar (single polarity) or bipolar (both polarities) output.

The *analog-output current range* is the full-scale current output capability.

Output impedance is the impedance looking into the output terminals.

Temperature coefficient is a measure of the change in output value as a function of temperature. It is usually expressed as a percentage of the full-scale value per degree Centigrade or degree Fahrenheit over the specified operating temperature range.

Broadly speaking, the accuracy and conversion speed characteristics of various DAC designs are similar and are governed by the precision and response time of each building block used and its behavior with environmental changes. However, some configurations offer certain advantages owing to a unique arrangement of building blocks.

5.3.1   Decoding Networks

The function of a decoding network is to provide an analog voltage or current output which is proportional to the digital input. Generally, the decoding process involves turning on voltage or current sources of predetermined values via the operation of switches and summing the resultant voltages or currents by proper means. Each digit of the number to be decoded is stored in a separate switch and the value of each source is "weighted" to correspond to the value of the digit.

The decoding networks commonly used for DACs can be divided into two main types, weighted resistor networks and ladder resistor networks. These networks can be further classified according to the type of precision source that is used: Either they are voltage controlled or current controlled.

5.3.1.1   Voltage-Controlled Weighted Network.   In the binary system, the most commonly used n-bit weighted network is that shown in Fig. 5-7. When the kth digit is a binary **1**, switch S_k is connected to the reference voltage $+E_r$ which contributes to the output voltage V_o by

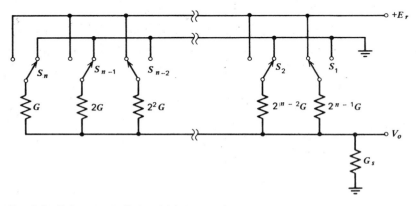

Fig. 5-7   Voltage-controlled weighted network.

a simple voltage divider action. When the kth digit is a binary **0**, switch S_k is connected to ground and E_r does not contribute to the output. In this configuration, $k = 1$ is the most significant bit (MSB) and $k = n$ is the least significant bit (LSB) of the digital input. The values of the conductance in the network are so arranged that the contribution of each succeeding binary digit (bit) to the output voltage is progressively attenuated by a factor of 0.5.

When the kth digit is a binary **1**, the contribution to the output voltage can be calculated as follows (assuming that the reference voltage has zero source resistance and that switch S_k has zero ON resistance).

$$V_{ok} = \frac{G_k E_r}{G_a + G_k + G_s} \tag{5-1}$$

where E_r = reference supply voltage

G_k = the conductance associated with the kth ON switch

G_a = parallel combination of those conductances connected to ground through DAC switches

G_s = summing conductance

The parallel combination of all conductances except the summing conductance is

$$\begin{aligned} G_a + G_k &= G[1 + 2 + 2^2 + \cdots + 2^{n-2} + 2^{n-1}] \\ &= G[2^n - 1] \end{aligned} \tag{5-2}$$

where n is the number of bits. By applying the superposition principle, the general expression of the output voltage for any bit pattern is

$$V_o = \sum_k V_{ok} = \frac{E_r}{2^n - 1 + G_s/G} \sum_{k=1}^{n} A_k 2^{n-k} \tag{5-3}$$

where $A_k = 1$, when the kth digit is a binary **1**

$A_k = 0$, when the kth digit is a binary **0**

The output conductance of the network of Fig. 5-7 is constant and is given by

$$G_o = G_a + G_k + G_s = [2^n - 1]G + G_s \tag{5-4}$$

This network configuration is simple and elegant. It contains the minimum number of precision resistors among the various decoding networks. However, it suffers from the requirement of a wide spread of resistor values which may restrict the resistor network fabrication technique. For an n-bit decoding network, the ratio between the largest and smallest resistor is 2^{n-1}, or 4096 for $n = 13$. Nevertheless, this network is very popular and it is used together with an operational amplifier as a current summing network in the DAC discussed in Sec. 5.3.4.

5.3.1.2 Current-Controlled Weighted Network. The binary weighted network of Fig. 5-8 is used in conjunction with current sources. The same current is applied to every node through its respective switch. When the kth digit is **1**, switch S_k is closed and the reference current I_r is applied to node k. When the kth digit is **0**, switch S_k is open and the reference current is cut off from node k.

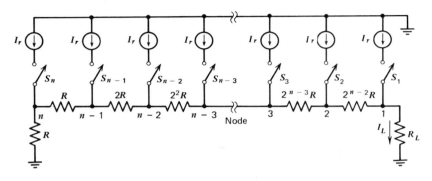

Fig. 5-8   Current-controlled weighted network.

When the kth digit is **1** its contribution to the load current I_L can be calculated as follows, assuming the reference current source has infinite output resistance:

$$I_{Lk} = \frac{R_k I_r}{R_o + R_L} \tag{5-5}$$

where I_r = reference current
R_k = Thevenin resistance to the left of node k
R_o = total series network resistance
R_L = load resistance

The total series network resistance can be expressed as

$$R_o = R[1 + 1 + 2 + 2^2 + \cdots + 2^{n-2}] = 2^{n-1}R \tag{5-6}$$

and the Thevenin resistance to the left of node k is

$$R_k = 2^{n-k}R \tag{5-7}$$

By applying the superposition principle, the load current due to any switch combination is obtained as

$$I_L = \sum_k I_{Lk} = \frac{2^n I_r}{2^{n-1} + R_L/R} \sum_{k=1}^{n} A_k 2^{-k}, \quad A_k = 0, 1 \qquad (5\text{-}8)$$

which clearly shows the binary nature of the output.

5.3.1.3 Voltage-Controlled Ladder Network. The most commonly used voltage controlled binary ladder is shown in Fig. 5-9. The conductance values of the ladder resistors are arranged so that the Thevenin conductance to the left of every node is G. Because of this arrangement, the contribution of each succeeding bit to the output voltage is progressively attenuated by a factor of 0.5.

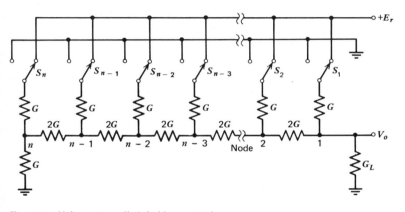

Fig. 5-9   Voltage-controlled ladder network.

Since the Thevenin conductance to the left of node 1 is G, the contribution of the MSB (switch S_1 connected to $+E_r$) to the output voltage is

$$V_{ok} = \frac{E_r 2^{-k}}{1 + G_L/2G}, \quad k = 1 \qquad (5\text{-}9)$$

where G_L is the load conductance. By applying the superposition principle, the general expression for the output voltage is

$$V_o = \frac{E_r}{1 + G_L/2G} \sum_{k=1}^{n} A_k 2^{-k}, \quad A_k = 0, 1 \qquad (5\text{-}10)$$

which again shows the binary nature of the output. The output conductance of the latter is constant and is given by $G_o = 2G$.

The ladder of Fig. 5-9 differs from the weighted network previously discussed in that the ratio of conductance values is only 2, compared with 2^{n-1} for the weighted network. However, the total number of precision resistors used in the ladder is more than that in the weighted network for equal resolution.

A modified ladder network is shown in Fig. 5-10 which performs exactly the same decoding function as the ladder of Fig. 5-9. The conductance values are arranged such that there are four elements $(G, G/2, G/4, G/8)$ in a group, an element of $2G/15$ connected between each succeeding group, and one terminating element of $G/8$. The output conductance of this network is also constant and is given by $G_o = 2G$.

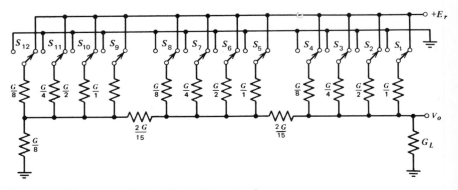

Fig. 5-10 Voltage-controlled modified ladder network.

Table 5-1 shows a comparison between the three different voltage-controlled decoding networks on the basis of the total number of pre-

Table 5-1   Comparison of *n*-Bit Voltage-Controlled Decoder Networks

	Configuration		
Parameters	Weighted (Fig. 5-7)	Ladder (Fig. 5-9)	Modified Ladder (Fig. 5-10)
1. Total number of precision resistors	$n + 1$	$2n$	$n(1 + \frac{1}{4})$
2. Ratio of conductance value spread	$2^{n-1}:1$	$2:1$	$8:1$
3. Output conductance G_o	$(2^n - 1)G + G_s$	$2G$	$2G$

cision resistors, the ratio of conductance values, and the output conductance.

5.3.1.4   Current-Controlled Ladder Network. Another very popular binary ladder network is used in conjunction with current sources as shown in Fig. 5-11. The conductance values are arranged so that the Thevenin conductance of every interior node, node 2 through node $n-1$, is $G/2$ to the left, $G/2$ to the right, and $G/2$ looking down. The resultant Thevenin conductance of every interior node is, therefore, given by $3G/2$. The Thevenin conductance of the two exterior nodes, nodes 1 and n, is $G + G/2 = 3G/2$. All nodes, therefore, have the same Thevenin conductance and every current source sees the same load of $3G/2$, which is also the value of the output conductance.

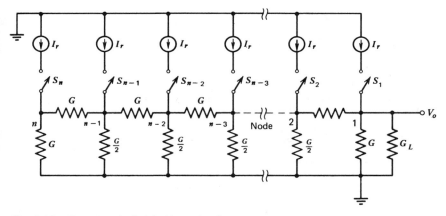

Fig. 5-11   Current-controlled ladder network.

When the kth bit is 1, switch S_k is closed, and a voltage $E = 2I_r/3G$ is produced between node k and ground. Starting with $k = 1$, the contribution of the MSB (switch S_1 closed) is

$$V_{o1} = \frac{2I_r}{3G(1 + 2G_L/3G)} \tag{5-11}$$

where G_L is the load conductance. Because of the ladder configuration, the voltage at node k is attenuated by a factor of 0.5 at node $k-1$, by a factor of 0.25 at $k-2$, and so on. Therefore, the general expression for the output voltage is given by

$$V_o = \sum_k V_{ok} = \frac{2I_r}{3G(1 + 2G_L/3G)} \sum_{k=1}^{n} A_k 2^{k-1}, \quad A_k = 0, 1 \quad (5\text{-}12)$$

Because every current source sees the same load, they can be implemented rather simply, as shown in Fig. 5-12. The desired current is obtained from a voltage reference E_r and is given by

$$I_r = \frac{E_r}{r + 2/3G} \quad (5\text{-}13)$$

where r is the series resistance as shown in the figure. Other types of current source circuits are discussed in Sec. 5.3.3.3.

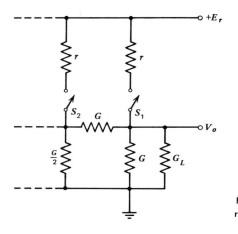

Fig. 5-12 Practical implementation of current source.

5.3.1.5 Bipolar Output.

Bipolar output means that the output voltage or current polarity can be either positive or negative. There are two commonly used techniques to fulfill this requirement: (1) polarity selection and (2) biasing or offsetting. Both schemes are discussed in this section.

The polarity selection technique is illustrated in Fig. 5-13. Reference voltages of $+E_r$ and $-E_r$ are applied to the input of a single-pole double-throw (SPDT) polarity switch. The output of the polarity switch, E_x, is connected to the decoding network and is controlled by the logical control, x.

$$\begin{aligned} E_x &= -E_r \quad \text{when} \quad x = \mathbf{0} \\ E_x &= +E_r \quad \text{when} \quad x = \mathbf{1} \end{aligned} \quad (5\text{-}14)$$

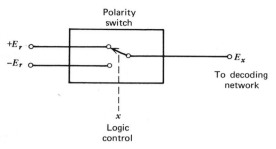

Fig. 5-13   Polarity selection method.

Since the output polarity of the decoding network is determined by the polarity of the reference voltage, this results in a bipolar output capability.

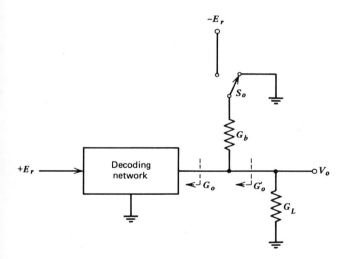

Fig. 5-14   Biasing or offset polarity method.

Biasing or offsetting is another technique for obtaining bipolar output. This method is illustrated in Fig. 5-14. A negative reference voltage $-E_r$ is applied to a bias conductance, G_b, through a sign switch S_o. The conductance G_b can be, and usually is, an integral part of the decoding network. For a positive output voltage, S_o is connected to ground; for a negative output voltage, G_b is switched to $-E_r$. In this mode, the output voltage is biased by a negative potential

$$V_b' = \frac{-E_r G_b}{G_L + G_o + G_b} = -\frac{E_r G_b}{G_L + G_o'} \tag{5-15}$$

where G_o is the output conductance of the decoding network without G_b, and G_o', is the output conductance including G_b. Thus Eq. (5-15) represents the negative full-scale value of the decoding network output. If some of the DAC switches $S_1, \ldots, S_n$ are ON, the equivalent circuit is that shown in Fig. 5-15, where V_o is the open circuit output voltage that would exist if the offset reference $-E_r$ and the conductances G_b and G_L were not present. The output voltage V_o' is given by

$$V_o' = \frac{G_o V_o - G_b E_r}{G_L + G_o'} \tag{5-16}$$

which shows that the output is offset by a fixed potential. The possibility of generating bipolar signals is clearly indicated by the negative offset term. The two's complement number system is used for the digital input code for the negative outputs. This offset technique can be used with any network that has a constant output impedance.

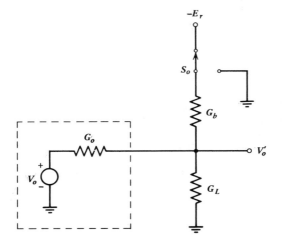

Fig. 5-15 Bipolar DAC equivalent circuit.

5.3.1.6 Precision Resistors. Components suitable for precision resistor decoding network applications can be divided into three major types: wire-wound, film, and monolithic resistors.

For many years, wire-wound resistors dominated precision decoding network applications for accuracies of ±0.1% or better. These resistors can be manufactured to very tight tolerances, have good stability, good temperature coefficients, and high reliability. However, because of their coiled construction, they are limited in speed of response. The best settling time achievable is about 300 nsec for settling to 0.01% of final value.

There are two types of precision film resistors which differ in manufacturing process and composition and which are generally categorized as thin-film and thick-film resistors. Precious metal and Nichrome* films are referred to as thin films, whereas cermet and metal alloys are used as thick films. The substrates on which the film resistors are deposited or fused are ceramic, glass, silicon, or alumina.

Film resistors offer the desirable combination of small size and high-frequency response, together with the accuracy and tight tolerance of good wire-wound types. Film networks are limited, however, in the ratio of resistances that can be realized without complex manufacturing procedures. The settling time of a film resistor network may be better than 100 nsec.

Monolithic resistor networks differ from the wire-wound and film resistor networks in that the network is formed from a common substrate, similar to the manner in which planar transistors are made. The advantages of monolithic networks are space savings, uniform electrical characteristics, and good stability because of identical material and processing for all resistors in the same substrate. This type of resistor network is usually available in both weighted and ladder configurations. Monolithic networks favor the ladder configuration since cost is not influenced directly by the number of elements, and better temperature tracking can be achieved through the repetitive use of two basic resistive values.

Table 5-2 shows the typical characteristics and a performance comparison between ladder networks using these three types of precision resistors.

5.3.1.7   Practical Design Considerations.   A good decoding network design involves many practical considerations, some of which are contradictory. The following are some guidelines for a good and practical design.

STATIC ACCURACY AND RESOLUTION.   In general, the static accuracy of a decoding network should be much better than its resolution. This is necessary to achieve an overall DAC accuracy that is compatible with the resolution. In addition, a certain part of the overall allowable DAC

---

* Registered trademark, Driver-Harris Co.

Table 5-2 Typical Decoding Network Characteristics

Parameters	Wire-wound Discrete	Film Discrete	Monolithic Thick	Monolithic Thin
Number of binary bits (typical)	14	14	12	14
Maximum resistance value	2.5 MΩ	2.5 MΩ	10 MΩ	1.5 MΩ
Maximum input voltage	200 V	200 V	15 V	40 V
Output resistance tolerance	±0.01%	±0.05%	±5%	±5%
Output resistance temperature coefficient (ppm/°C)	±2.5	±5	±100	±5
Output resistance stability	±100 ppm/3 yr	±100 ppm/yr	±1000 ppm/yr	±100 ppm/yr
MSB ratio accuracy*	±0.003%	±0.006%	±0.024%	±0.006%
Cumulative ratio accuracy*	±0.004%	±0.006%	±0.024%	±0.006%
Cumulative temperature coefficient ratio tracking (ppm/°C)	±0.5	±0.5	±50	±0.5
Cumulative ratio stability	±20 ppm/3 yr	±50 ppm/yr	±200 ppm/yr	±50 ppm/yr
Settling time (to $\frac{1}{2}$ LSB)†	300 nsec	< 100 nsec**	< 100 nsec**	< 100 nsec**
Operating temperature range (°C)	0–70	0–70	−20–80	−20–80

* Includes temperature tracking error over limited temperature range (0–60°C).

† Settling time is dependent on network resistance value.

** Limited by measurement instrumentation.

400

errors are accounted for by the precision reference source and analog switches, and so on, so that the network accuracy should be considerably better than the DAC accuracy. Otherwise, the design requirements on the reference source and switches become overly stringent.

In a binary system, the resolution is given by $1/2^n$ where n is the number of bits, or the order of the decoding network. Table 5-3 shows the relationship between the resolution and the number of bits n. As an example, the resolution of a 13-bit system is 0.0122%. The static accuracy of the decoding network, therefore, should be much better than 0.01%.

Table 5-3 Binary Resolution vs. n (resolution $= 2^{-n}$).

n	Resolution
1	0.5
2	0.25
3	0.125
4	0.0625
5	0.03125
6	0.015625
7	0.0078125
8	0.00390625
9	0.001953125
10	0.0009765625
11	0.00048828125
12	0.000244140625
13	0.0001220703125
14	0.00006103515625
15	0.000030517578125
16	0.0000152587890625

NETWORK RESISTANCE INITIAL TOLERANCE. There are two types of initial resistance tolerances which need to be specified. These are the tolerance on each individual resistor, which determines the weight of the respective bit input to the output, and the tolerance on the output resistance of the network.

For a given network configuration and a given initial tolerance on each of the resistors, the worst-case error in the output voltage can be calculated. In this approach the absolute tolerance on each resistor of the network is specified so that the output accuracy is achieved even with a worst-case combination of resistor tolerances. Since the contribution of each input is weighted, the resistor tolerances need not be the same for each resistor. Thus, for example, the resistor for the most significant bit must have a tighter tolerance than the resistor for the least significant bit.

Another specification approach that is preferred over the absolute tolerance method is to specify the ratios or weights of input-bit patterns to the output. It is not pracical, however, to specify the ratio tolerances of all possible input-bit combinations. Based on design experience and analysis of a 13-bit DAC, it was found that a tolerance specification on three key input-bit patterns was sufficient to guarantee successful ladder-type network performance. These three patterns are: (1) the ratio tolerance for each individual bit: (2) the ratio tolerance for the full-scale output, that is, all bits ON; and (3) the ratio tolerance for bits 1–5 turned ON.

For those decoding networks in which the analog switch is connected in series with the network resistor, the nominal switch ON resistance must be taken into consideration in specifying the ratio tolerances. In practice, all ratio tolerances are specified with a dummy resistor for each bit to simulate the switch resistance.

Aside from the ratio tolerance, the cost of a decoding network is quite sensitive to the tolerance on output resistance. Therefore, the tolerance on output resistance should be relaxed as much as the application permits. A DAC configuration in which a tight tolerance on the output resistance is unnecessary should be selected whenever possible.

TEMPERATURE COEFFICIENT, STABILITY, AND TRACKING. In order to maintain overall decoding network accuracy with temperature variations and life, it is mandatory that the resistors track with temperature and time. It is, therefore, not practical to assemble the network using individual resistors purchased at random since they are subject to variations in wire or film properties. It is more practical and economical to order the network resistors in a single package. This allows the manufacturer to use the same spool of wire and/or batch process for all resistors in manufacturing the network, thus assuring good temperature coefficient tracking and stability.

POWER DISSIPATION. Because variation in resistor values due to temperature variations resulting from heat dissipation introduces an error in the output voltage, power derating factors on the resistors in the network should be given thorough consideration. A good practice is to operate each resistor at no more than 25% of its rated power dissipation.

5.3.2   DAC Switches

The DAC switch is the building block connected between the reference supply and the decoding network. The digital-input data are stored by the switch so there is one such switch associated with each bit position. The decoding networks, as discussed in Sec. 5.3.1, can be either current controlled or voltage controlled. Therefore, two types of DAC switches are commonly used: the current switch and the voltage switch.

In the current switching mode, the reference current is applied to the decoding network node when the associated current switch is closed, and the reference current is shut off from the node when the switch is opened. In the voltage switching mode, the voltage switch usually is in single-pole double-throw (SPDT) form. It connects the decoding network to either the reference voltage or the ground potential, depending on the state of the bit in the digital-input data. In some cases, the voltage switch can also be in single-pole single-throw (SPST) form. This is an advantage since it simplifies the switch design.

Semiconductor devices such as diodes, bipolar transistors, and field effect transistors are most suitable for DAC switch applications although electromechanical switches can be used in low-speed applications. Semiconductor devices offer the outstanding advantages of reliability, low power consumption, small size, and high speed.

A perfect DAC switch may be defined as a device that has either zero or unity transmittance upon the application of a proper control signal. It is capable of operating between any two points in a circuit without altering the electrical characteristics of its associated signal and circuitry. Semiconductor switches are not perfect, however, since they contribute some errors to the DAC in both the ON and OFF state. The design problem lies in evaluating and minimizing these errors.

5.3.2.1 Diode Switches. The simplest semiconductor switch is a *pn*-junction diode. It consists of an inhomogeneous semiconductor having a region of *p*-type material and a region of *n*-type material separated by a relatively thin transition region. The static volt-ampere characteristic of an ideal diode is of the form

$$I = I_s[e^{qV/kT} - 1] \qquad (5\text{-}17)$$

where I_s is called the reverse saturation current, q is electronic charge $(1.602 \times 10^{-19}$ C$)$, k is Boltzmann's constant $(1.380 \times 10^{-23}$ J/°K$)$, and T is absolute temperature in °K. The term (kT/q) is approximately 25 mV at room temperature.

The diode characteristic of Eq. (5-17) is shown in Fig. 5-16. For a positive V greater than about 100 mV, Eq. (5-17) becomes

$$I \cong I_s e^{qV/kT} \qquad (5\text{-}18)$$

which is a good approximation for the forward portion of the diode characteristic. The forward dynamic conductance of the diode is obtained by differentiating I in Eq. (5-18) with respect to V.

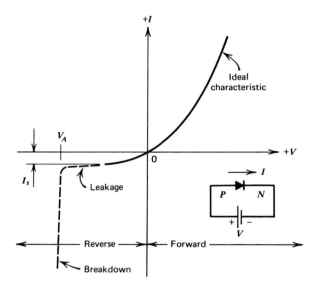

Fig. 5-16   Ideal junction diode characteristics.

$$G_d = \frac{dI}{dV} \cong I_s \frac{q}{kT} e^{qV/kT} = \frac{qI}{kT} \; \mho \qquad (5\text{-}19)$$

The forward dynamic resistance of the diode at room temperature is

$$R_d = \frac{1}{G_d} \cong \frac{kT}{qI} = \frac{25}{I \; (\text{mA})} \; \Omega \qquad (5\text{-}20)$$

For a negative V of approximately -75 mV, reverse saturation is reached and

$$I \cong -I_s \qquad (5\text{-}21)$$

The reverse diode current observed in practice never fully saturates but exhibits an ohmic component in addition to I_s, as shown by the dotted line in Fig. 5-16. This is commonly referred to as the leakage component since it is, in part, the result of surface leakage across the junction. The increase in reverse current due to applied voltage is described by the equation

$$I = \frac{I_s}{1 - (V/V_A)^n} \qquad (5\text{-}22)$$

where I = total reverse current
    I_s = reverse current at low voltage
    V = applied reverse voltage
    V_A = avalanche breakdown voltage
    n = an empirical constant (2 to 4)

It is seen from Eq. (5-22) that as the reverse voltage V increases and approaches the avalanche breakdown voltage V_A, the diode characteristic exhibits a breakdown phenomenon as shown by the dotted line in Fig. 5-16. The breakdown voltage depends on the fabrication and design of the diode but is generally between -10–-30 V in diodes intended for switching use.

The *V-I* characteristic of a diode is highly nonlinear and it is this property that makes the diode useful as a switch. In the forward region, the voltage drop for a silicon diode is 0.6 V. This voltage is temperature sensitive with an average temperature coefficient of -2.0 mV/°C. In the reverse region, the reverse current for a silicon diode is typically 10 nA at room temperature with a reverse voltage bias of 10 V. The reverse current is also temperature dependent. As a rule of thumb, the reverse current *doubles* for every 10°C temperature increase. Both the forward voltage drop and reverse leakage contribute to DAC errors and must be considered in the design.

A basic current switching diode configuration which uses a diode pair for each DAC switch is shown in Fig. 5-17. The precision reference current is either "steered" toward or away from the summing point P, depending on the state of the control signal. A suitable positive going signal representing a binary **1** turns off D_2 and thus steers the reference current into the summing point P through D_1. A negative going signal represents a binary **0** that turns on D_2 and thus diverts the reference current away from P. A similar current mode switch used with a ladder network is shown in Fig. 5-18.

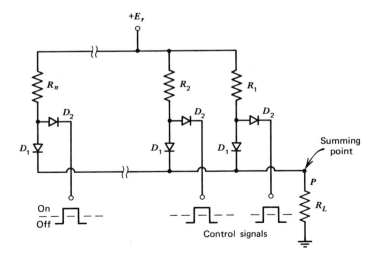

Fig. 5-17 Current-mode diode switch with voltage-controlled weighted network.

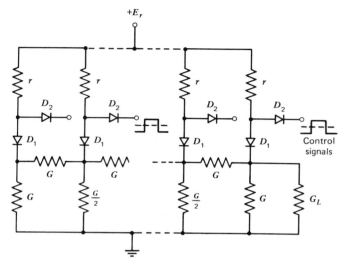

Fig. 5-18   Current-mode diode switch with voltage-controlled ladder network.

With a good diode, the switching speed can be quite high. But this type of switch is not commonly used in high-performance DAC designs because of the following main problems: (1) the forward voltage drop across D_1 and its change with temperature and life; (2) the forward dynamic diode resistance of D_1; and (3) the sum of the reverse currents from normally off diodes and its change with temperature. The first problem means that the reference voltage has to be high enough so that it swamps out the diode voltage drop. For example, an 80-mV change in the forward diode voltage drop can be expected over a 40°C range (-2 mV/ °C). For a required DAC accuracy of 0.1%, with everything else constant, the reference voltage must be at least 80 V. It is noted that this problem exists only with the voltage-controlled decoding networks, however. For current-controlled decoding networks, the diode forward voltage drop and its temperature effect is in series with the reference current source and in no way affects the precision of the DAC. Only the reverse current of the diode switch contributes an error in the output.

As to the diode reverse current problem, the only solution is to select low leakage diodes such that the sum of the reverse currents and its change with temperature has a small influence compared with the least significant bit reference current. If this is not the case, the leakage current causes a change in the DAC output voltage since it flows through the network and its load resistor.

A basic voltage switching diode configuration is shown in Fig. 5-19. It utilizes the clamping properties of the diode. The diode D_1 clamps point P to the reference voltage $+E_r$, and D_2 clamps point P to ground potential. Proper control signals cause the circuit to alternate between these two states in accordance with the state of the data input bits. The forward voltage drop of the diode pair introduces a finite error since the voltage at point P is slightly different from E_r and ground. In addition, this voltage is temperature sensitive. However, reverse current has much less effect in this configuration since it flows through low-impedance circuitry and should be negligible compared with the ON current of the other diode.

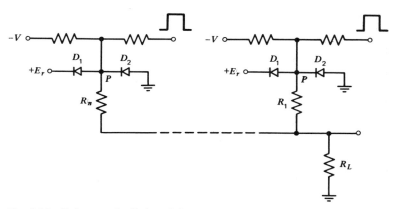

Fig. 5-19   Voltage-mode diode switch.

The forward voltage effect can be minimized by replacing each diode with a pair, as shown in Fig. 5-20. In this manner, the voltage drops across the diode pairs (D_1 and D_1', D_2 and D_2') and voltage changes with temperature ideally cancel each other. Matched diode assemblies are readily available with near ideal clamping properties of forward voltage matching ($V_1 - V_1'$) to within 5 mV and temperature coefficient of diode voltage difference $d(V_1 - V_1')/dT$ to within 10 μV/°C. These matched properties are usually specified at a given diode current.

A disadvantage of this configuration is that there are two diodes between point P and the reference voltage $+E_r$ so that the forward dynamic resistance is doubled. The effect of this extra resistance depends on the size of the network resistors and their location in the network. In bits of lesser significance, the effect is usually negligible.

Table 5-4 Bipolar Transistor Switch Properties

Connecting Method	Normal Connection	Inverted Connection
Circuit		
Equivalent circuit (ON condition)		
Initial offset voltage	$V_{CE} = \pm \dfrac{KT}{q} \ln \alpha_i + I_B r'_E$ $\simeq \pm \dfrac{KT}{q} \ln \alpha_i$ Negative for NPN Positive for PNP	$V_{EC} = \pm \dfrac{KT}{q} \ln \alpha_n + I_B r'_C$ $\simeq \pm \dfrac{KT}{q} \ln \alpha_n$ Negative for NPN Positive for PNP

Saturation ON resistance	$r_{CE} = \dfrac{KT}{qI_B}\dfrac{(1-\alpha_i\alpha_n)}{\alpha_n} + r_C' + r_E'$	$r_{EC} = \dfrac{KT}{qI_B}\dfrac{(1-\alpha_i\alpha_n)}{\alpha_i} + r_C' + r_E'$
	$\simeq \dfrac{KT}{qI_B}\dfrac{(1-\alpha_i\alpha_n)}{\alpha_n}$	$\simeq \dfrac{KT}{qI_B}\dfrac{(1-\alpha_i\alpha_n)}{\alpha_i}$
Equivalent circuit (OFF condition)		

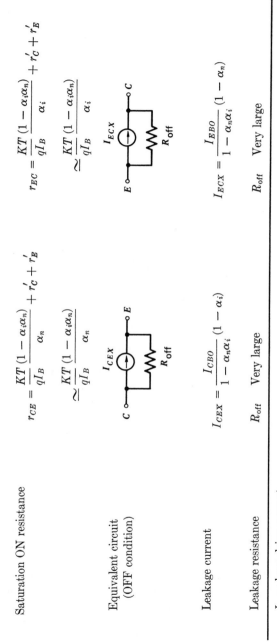

Leakage current	$I_{CEX} = \dfrac{I_{CBO}}{1-\alpha_n\alpha_i}(1-\alpha_i)$	$I_{ECX} = \dfrac{I_{EBO}}{1-\alpha_n\alpha_i}(1-\alpha_n)$
Leakage resistance	R_{off} Very large	R_{off} Very large

I_B = base drive current
α_i = inverted alpha—d.c. common-base current gain
α_n = normal alpha—d.c. common-base current gain
r_C' = collector bulk resistance
r_E' = emitter bulk resistance

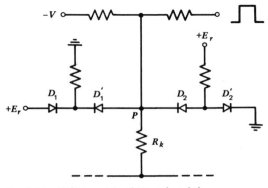

Fig. 5-20 Voltage-mode diode pair switch.

It is interesting to note that the metal silicon hot carrier diodes exhibit a much lower forward voltage temperature coefficient than do the ordinary silicon diodes. Since the temperature coefficient varies with forward bias current, zero temperature coefficient may be obtainable at a specific current for a given type of diode (Hewlett-Packard, 1964). However, the hot carrier diodes are intended primarily for fast switching in high-frequency computers and as a mixer, detector, and rectifier element extending into the microwave region, and they are much more expensive.

5.3.2.2 Bipolar Transistor Switches. The general properties of bipolar transistors used as analog switches have been treated rather thoroughly in Sec. 3.13.2.1 in connection with analog multiplexers. Those properties and parameters pertinent to DAC switch applications are summarized in Table 5-4. In general, these are the same properties important in multiplexer design.

Both the normal and inverted transistor connections are used for DAC switches. The inverted connection is generally preferred because the offset voltage errors are less than those caused by the slight increase in the ON resistance in this connection.

There are two basic types of transistor switches used in DACs: the *current* switch and the *voltage* switch. A common current switching circuit using NPN transistors is shown in Fig. 5-21. When a positive drive voltage $v_i = +\delta$ is applied, the collector current of Q_1 increases and that of Q_2 decreases. The maximum emitter current at Q_1 is limited to the reference source current I_r and the corresponding collector current is $\alpha_n I_r$. In other words, the emitter base junction of Q_2 is reverse biased and all the reference current I_r flows through Q_1. When a negative drive

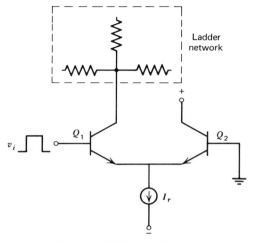

Fig. 5-21   Current switching circuit.

voltage $v_i = -\delta$ is applied, the states of Q_1 and Q_2 are exchanged; that is, the emitter base junction of Q_1 is reverse biased and all the reference current I_r flows through Q_2. Between the drive voltages of $\pm\delta$, where δ equals about 0.1 V, the switch pair has an active region where the emitter current is distributed between the two transistors. If δ equals about 0.6 V or larger, however, the load currents are not affected by the base drive potential. The speed of the circuit of Fig. 5-21 is limited by the amount of charge that must be injected into the emitter depletion layer as the switch is turned on.

The fact that this technique makes use of the diode action of the emitter base junctions in the two transistors suggests the substitution of a diode in place of Q_2 as in Fig. 5-22. This results in a simplification of the circuit.

The error contributions of the switch leakage currents must be considered carefully during the design. When Q_1 is ON, the equivalent current source for the network is the product of α_n of Q_1 and the sum of the current source I_r and the leakage current of Q_2. When Q_1 is OFF, this equivalent current is not completely diverted away from the network since the leakage current of Q_1 is still fed into the network. The cumulative effect of this leakage from all OFF switches in a DAC causes an offset in the output voltage. It is the variations of the current gain α_n of Q_1 and the leakage currents of both transistors, with respect to the environment and life, that are critical to the design since their initial values can be compensated by design and calibration.

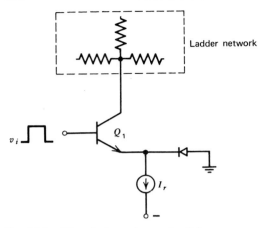

Ladder network

Fig. 5-22 Alternate form of current switch.

A common voltage switch circuit is shown in Fig. 5-23. Transistors Q_1 and Q_2 are a complementary pair and they are connected in the inverted connection. The DAC decoder network resistor R_k is switched either to ground or to the reference voltage E_r, depending on whether the switch is ON or OFF; only one transistor is in an active state at any time. When a positive input voltage of $v_i = E_r + \delta$ is applied, the base drive cur-

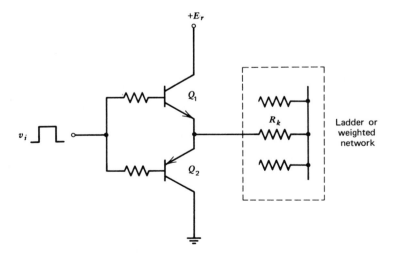

Fig. 5-23 Voltage switching circuit.

rent flows through the Q_1 base collector junction, the transistor is turned ON, and the resistor R_k is clamped to the reference voltage. The voltage δ must exceed the 0.6-V diode potential and, therefore, is typically several volts. It should be noted that the reference voltage source acts as a *current sink* rather than a *current source* since it absorbs the base drive current. This particular requirement of a voltage reference source is treated in a later section on the design of voltage reference sources. When a negative input voltage $v_i = -\delta$ is applied, the base drive current flows through the Q_2 collector base junction, Q_2 is turned ON, and R_k is clamped to ground.

It is the initial offset voltage and ON resistance, as well as their variation with temperature and time, that are of concern in the design of a DAC voltage switch. For transistors connected between the R_k's and $+E_r$, it is the tolerance on the offset voltage and ON resistance that are of importance. The nominal offset voltage value can be compensated by the calibration of the voltage reference, and the nominal ON resistance value can be compensated by making the network resistor equal to the difference between the nominal R_k and the nominal ON resistance of the switch. A typical tolerance for the offset voltage of NPN-type silicon planar transistor is about 200 μV, which results in an error of 0.001% for a reference voltage of 20 V. A typical tolerance for the ON resistance is about 1 Ω, an error of 0.001% for $R_k = 100{,}000$ Ω.

If the decoding network is binary, the cumulative effect of both offset voltage and ON resistance of all ON transistor switches is, at most, *twice* as large as the error caused by only the most significant bit switch. This is because of the successive attenuation in each stage of the network and the fact that $1/2 < \Sigma_{i=2}^{n} 2^{i-2}$ if the offsets and ON resistances of the switches are nominally the same.

The significance of offset voltage error of a transistor switch is proportionally less with higher reference voltages. The greater the reference voltage, however, the larger the requirement for the emitter junction breakdown voltage of the transistor. Since high-speed analog switching transistors can be obtained with an emitter breakdown of about 30 V, a reference voltage of 20–25 V is the usual limit.

For the transistors connected between the R_k's and ground, the cumulative effect of offset voltage of all ON transistors gives rise to a zero offset. Since the voltage reference is isolated from these transistors, some compensation technique must be utilized to minimize the zero offset. A simple compensation circuit is shown in Fig. 5-24. It consists of a compensating resistor R_c which is connected between $+E_r$ and the common emitters of both transistors. The equivalent circuit when Q_2 is ON is also shown. Since the offset voltage of Q_2 is negative, theoretically there exists

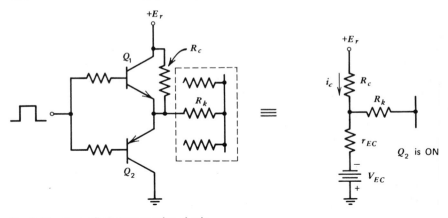

Fig. 5-24   Zero offset compensation circuit.

a compensating resistor value which results in a perfect voltage switch; that is, $V_o = 0$. This value of R_c is

$$R_c = \frac{E_r r_{EC}}{V_{EC}} \qquad (5\text{-}23)$$

where r_{EC} is the ON resistance and V_{EC} is the offset voltage of Q_2. The presence of R_c has no significant effect on the performance of the switching circuit because of its high value.

The leakage current in the OFF transistor in Fig. 5-23 is a less substantial problem. It generates an error proportional to the parallel resistance of R_k and the ON resistance of the conducting transistor. Typically, the ON resistance may be 3 Ω, compared with an R_k of thousands of ohms. Hence as much as 5 μA of leakage introduces an error of less than 15 μV.

The switching speed of the current switch is faster than that of the voltage switch. This is because the input control voltage swing for the current switch is 2δ as compared with $E_r + 2\delta$ for the voltage switch. The switching speed is also dependent on the magnitude of base drive current. In general, the larger the drive current, the faster the speed. The switching speed of bipolar transistors is limited primarily by the minority carrier storage characteristics of the base region. The input or base time constant may be defined as $R_x C_{ib}$ where R_x is the total input resistance in the base circuit and C_{ib} is the input capacitance of the transistor which must be charged to turn the transistor ON. The turn-off time is proportional to $R_L C_o$, where R_L is the load resistance and C_o is the total output capacitance between emitter and ground. Generally, the switch turn-off time is slower than turn-on time.

Another phenomenon related to transistor switching is overshoot transients, commonly called "spikes," which occur during the transition between ON and OFF. The transients are primarily the result of transferring charge from the base drive input to the load through the junction capacitances. They seldom exceed a few hundred millivolts and are of short duration.

5.3.2.3 Field Effect Transistor Switches. The field effect transistor (FET) is another semiconductor device that exhibits characteristics suitable for DAC switch applications. It has the following desirable characteristics: (1) The offset voltage is negligible; (2) the OFF-to-ON channel resistance ratio is high; (3) the capacitance between device leads is moderate; (4) the leakage current is low; (5) the impedance looking into control lead or gate is very high; (6) the driving circuit to the gate may be direct coupled; and (7) the FET switch design, in general, is simpler than bipolar transistor switch design.

The construction, operation, and characteristics of the junction FET (JFET) and the insulated gate (IGFET) or metal oxide semiconductor FET (MOSFET) are discussed in Sec. 3.12.2.2 in connection with their application in analog multiplexers. As with the bipolar transistor, their advantages and disadvantages as multiplexer devices generally apply to their use as DAC switches as well.

The FET switch can be operated either as a current switch or as a voltage switch. A basic current switch consisting of two p-channel MOSFETs is shown in Fig. 5-25. The control voltages V_c and V'_c are derived from a flip-flop circuit. For a p-channel device with typical characteristics, when V_c is zero, V'_c is -10 V; when V_c is -10 V, V'_c is

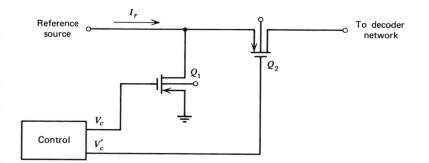

Fig. 5-25 MOSFET current switch.

0. With a V_c of -10 V, Q_1 is turned ON, but Q_2 is OFF, and the reference current I_r is switched to ground. At a V'_c of -10 V and a V_c of 0, Q_2 is turned ON but Q_1 is OFF and the reference current is switched into the summing point or the node of the resistor decoder network.

The current switch has the potential disadvantage that connecting a number of switches to a summing point makes capacitive feedthrough transients significant. On the other hand, the current switch is faster than the voltage switch because the control voltage swing is smaller. The OFF leakage currents in Q_1 and Q_2 and their variation with temperature are also sources of error because they flow into the summing node or decoder network and produce an offset voltage. Therefore, a low leakage device must be used in the most significant bit positions.

A basic FET voltage switch is shown in Fig. 5-26. By applying proper control voltage, only one transistor is turned ON at a given time. When Q_1 is ON but Q_2 is OFF, the decoder network resistor R_k is connected to the reference voltage E_r. When Q_2 is ON but Q_1 is OFF, R_k is connected to ground.

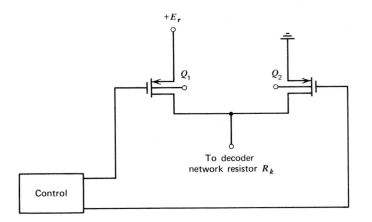

Fig. 5-26   MOSFET voltage switch.

The ON channel resistance and its variations with temperature contribute an error at the output. To minimize the error, the decoder network resistance R_k must be large when compared with the ON resistance R_{on}. The nominal ON resistance of Q_1 and Q_2 may be compensated by considering R_{on} as a portion of R_k. Because of the relative magnitudes of R_{on} and R_k, and the temperature coefficient of R_{on}, it is usually essential to use a device of low R_{on} in the most significant bit positions.

5.3.3 Precision Reference Sources

A precision reference source is one of the basic building blocks for a DAC. It is the source that supplies the reference voltage or current to the decoding network by means of the DAC switches. Because any change in the reference source value results in an equal percentage change in the DAC output, the design of the precision reference source deserves special consideration.

A good precision reference source is characterized by the following properties:

1. Long and short term drift must be low.

2. The temperature coefficient must be small over a wide environmental change.

3. The output noise must be low.

4. The line and load regulation must be very good.

5. The transient response must be very good, and the recovery from step load disturbance or line disturbance must be fast.

6. The range of output adjustment must be adequate and the resolution of the adjustment must be fine.

Some of the desired operating features such as short-circuit protection, reverse voltage protection, and remote error sensing must be incorporated into the circuit design. Generally, the performance and feature requirements of a precision reference source are largely dependent on the DAC with which it is used. The requirements for a reference source to be used with a 13-bit, 0.01%-accuracy DAC are much more stringent than those for one used with a 10-bit, 0.1%-accuracy DAC, for example.

The precision reference source can be either a voltage source or a current source. Fundamentally, both types of reference sources employ the same type of zener reference diode as the reference element and differ only in the output circuitry.

5.3.3.1 Zener Reference Diodes. This section deals with some pertinent characteristics of the reference diodes commonly used with precision reference sources. The discussion is restricted to special types of reference diodes characterized by their ability to maintain a stable output with respect to both temperature and time.

The reference diode is a special version of the silicon *pn*-junction diode and is operated in a well-defined reverse breakdown region. The diode breakdown voltage or zener voltage is usually in the range of 6–15 V. The characteristic of a typical 6-V zener diode is shown in Fig. 5-27.

The forward characteristic of the zener diode is identical to that of

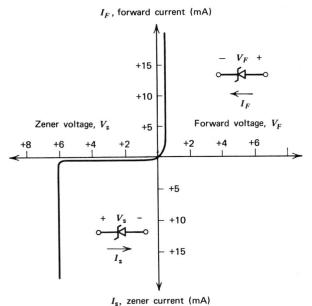

Fig. 5-27   Typical zener diode characteristic.

the ordinary *pn*-junction diode discussed in Sec. 5.3.2.1. When reverse voltage less than breakdown voltage is applied to the zener diode, the behavior of reverse current is again similar to the ordinary *pn*-junction diode. At some definite reverse voltage, depending on the internal structure of the diode, the current begins to avalanche. This is the "breakdown" region where the device is usually biased during normal use as a reference diode.

Temperature coefficient (T.C.), the zener voltage variation due to temperature changes, is the most important parameter of a reference diode. It is defined as the percentage change of zener voltage per unit temperature change, and it is usually expressed in the following form:

$$\text{T.C.} = \frac{M}{V_{z1}} \frac{dV_z}{dT} = M \frac{V_{z2} - V_{z1}}{|T_2 - T_1| V_{z1}} \tag{5-24}$$

where M = a multiplication constant
  V_{z1} = reference zener voltage at T_1
  V_{z2} = zener voltage at T_2
  T_1 = reference temperature (usually 25°C)
  T_2 = any other temperature in °C

Both V_{z1} and V_{z2} must be measured at a specified zener current. Temperature coefficient can be expressed either as percent per degree Centigrade (%/°C) or parts per million per degree Centigrade (ppm/°C). For the former unit, a multiplication constant of 100 is used ($M = 100$); for the latter unit, M is equal to 10^6.

The construction of a reference diode with good temperature coefficient (less than 100 ppm/°C) usually involves temperature compensation. As a result, the temperature coefficient of the temperature compensated reference diode is nonlinear. A typical temperature coefficient distribution for an uncompensated zener diode as a function of zener voltage is shown in Fig. 5-28. The temperature coefficient is positive and increases with increasing zener voltage for zener voltages greater than about 5 V, and the temperature coefficient is negative for zener voltages below 5 V. For example, a diode with a nominal zener voltage of 6 V exhibits a temperature coefficient of approximately 0.03%/°C under normal operating conditions; a 10-V zener has a temperature coefficient of approximately 0.055%/°C, and a 100-V zener, a T.C. of approximately 0.095%/°C. These variations are directly opposite to those found in forward-biased silicon pn-junctions in either conventional silicon diodes or forward-biased zener diodes. Therefore, a good temperature compensation can be attained by utilizing these two opposite effects in diodes connected in series. Temperature-compensated reference diodes are usually available with nominal zener voltages of 6.2, 8.4, 9.0, 9.3, and 11.7 V, depending on the number of forward pn-junction diodes that are used. They are available with temperature coefficients of 100, 50, 20, 10, and 5 ppm/°C over the temperature range of $-55-+150°C$.

There are two commonly used methods of specifying temperature co-

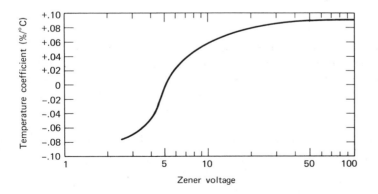

Fig. 5-28   Typical temperature coefficient distribution uncompensated zener diode.

efficient. These are the hourglass method and the box method. The hourglass method, shown in Fig. 5-29, implies that the change in zener voltage always falls within the area bounded by two temperature coefficient lines over the temperature range of interest. The box method is shown in Fig. 5-30. This method involves making zener voltage measurements at various temperatures, such as −55, 0, +25, +75, +100, and +150°C, and then comparing these readings with the maximum allowable voltage change. It is apparent that the hourglass method is more stringent than the box method.

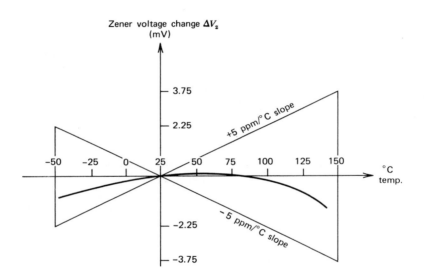

Fig. 5-29   Hourglass method.

The temperature coefficient of commercially available temperature-compensated reference diodes can be improved significantly by selecting an optimum operating current. This must be done for each individual unit. Applying this technique, it is feasible to obtain a unit that has a temperature coefficient of 1 ppm/°C over a restricted temperature range such as 0–50°C.

Another parameter to be considered is the dynamic resistance of the reference diode. This is defined as the slope of the V-I characteristic curve at the operating point, and it determines the amount of zener current regulation that is required in the reference source. For example, the typical dynamic resistance for a temperature-compensated reference diode

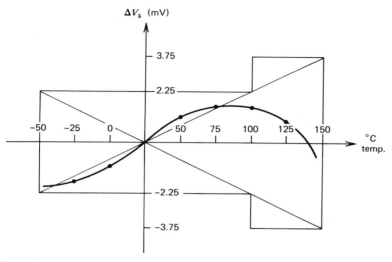

Fig. 5-30   Box method.

is about 15 Ω. Thus a 1-μA change in zener current results in 15×10^{-6} = 15 μV shift in the reference voltage value. The variation of dynamic resistance with temperature is negligible and, therefore, the amount of current regulation needed at 25°C is about the same at other temperatures.

Reference voltage stability, which is the voltage variation with time, is the most critical parameter other than temperature coefficient for reference diodes used in precision reference sources. There are two types of voltage stability to be considered, short term and long term. The time interval used for the short-term stability specification usually varies from 8–40 hr. On the other hand, 1000 hr–6 months is commonly used for the long-term stability specification.

In conjunction with the short-term stability, it is important to know the power-on or warm-up characteristic of the reference diode. How long does it take the diode to reach a stable voltage reading after power is applied? It can be 30 sec, 30 min, or even longer. Immediately after the power is applied and before the voltage has stabilized, the reference voltage value can be either increasing or decreasing with time. This is a prime consideration for the reference diode used in equipment that is turned on and off frequently.

Ordinary temperature-compensated zener diodes typically exhibit a voltage stability of 300–500 ppm/1000 hr. Special preconditioning, such as power aging for an extended period of time and temperature cycling, is necessary to stabilize surface conditions of the diodes. With these techniques, excellent voltage stability, 5 ppm/1000 hr or better, can be achieved. The stability of the device tends to improve with age. For critical applications, reference diodes with certified stability can be obtained from diode manufacturers. The certification often includes the actual stability data for 1000 hr. Knowing this behavior history and the temperature coefficient and dynamic resistance, the user is able to design and calibrate a precision reference source.

It is generally known that uncompensated zener diodes have extremely noisy circuit performance. Since temperature-compensated and long-term stabilized reference diodes usually have much better noise performance, this suggests that good noise performance, good circuit stability, and device reliability are closely related to each other.

A typical noise characteristic curve for a temperature-compensated reference diode is shown in Fig. 5-31. It exhibits a $1/f$ frequency dependence at low frequency and a leveling off at higher frequencies. The high-frequency noise is the result of the basic processes responsible for operation of the device. The low-frequency $(1/f)$ noise results from device "pathologies." The high-frequency noise can be attenuated by connecting a small shunting capacitor across the reference diode.

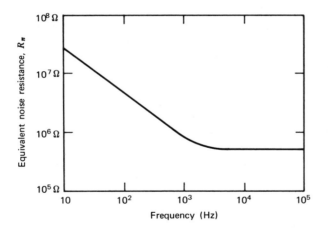

Fig. 5-31   Typical noise characteristics of compensated reference diode.

Generally, noise level decreases as the zener current and junction temperature increase. The change is approximately 20% for a junction temperature change of 25–100°C. Thus the noise rating must indicate bandwidth, zener current level, and temperature.

5.3.3.2   Integrated Reference Amplifiers. The integrated reference amplifier of Fig. 5-32 is a semiconductor device which consists of a zener reference diode and an NPN transistor constructed on a common substrate. It performs the functions of a voltage reference and an error amplifier in a regulated voltage or current supply. The reference voltage of the amplifier is available at the *B* and *Z* terminals.

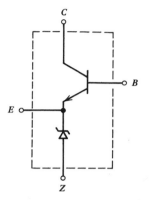

Fig. 5-32   Integrated reference amplifier.

By carefully matching the temperature variation of the zener voltage with the temperature variation of both the transistor base emitter junction voltage and common emitter current gain β, good performance can be achieved. Use of an integrated structure minimizes the error and short-term drift resulting from temperature differentials between the zener diode and the transistor.

As in the case of the zener diode, the most important characteristic of the reference amplifier is the temperature coefficient of the reference voltage. The temperature coefficient must be specified at specific values of zener current, collector current, collector-to-base voltage, and base source resistance. Reference amplifiers of this type are usually available with reference voltages of 6.8, 8.6, 9.5, and 11.0 V. The typical temperature coefficient for a standard unit is about 10 ppm/°C over a temperature range of −55−+150°C. Figure 5-33 shows a typical application of the reference amplifier in a regulated voltage power supply.

5.3.3.3   Basic Reference Source Circuits. The main function of any reference source circuit is to supply a stable output to a given load. The basic performance of the circuit is usually measured by its ability to maintain the output for variations in input voltage, temperature, and

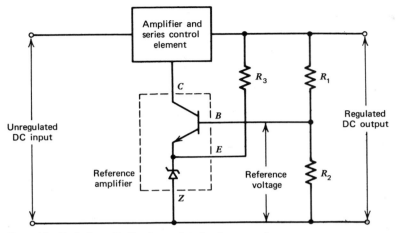

Fig. 5-33 Typical application in regulated voltage supply.

load current. The complexity of the reference source circuit is determined by the types of load it drives and its input and environmental conditions. The circuit can vary from a single regulator to a sophisticated feedback regulator circuit.

A simple regulator circuit is shown in Fig. 5-34. It contains only two components, one resistor and one zener reference diode. The output voltage is

$$V_o = \frac{E_z}{1 + R_z/R_s + R_z/R_L} + \frac{V_{in}}{1 + R_s/R_z + R_s/R_L} \qquad (5\text{-}25)$$

where E_z = equivalent zener source voltage
$\quad R_z$ = zener dynamic resistance
$\quad R_s$ = series resistance
$\quad R_L$ = load resistance

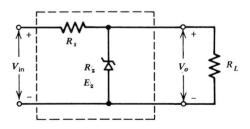

Fig. 5-34 Simple regulator circuit.

For a given zener reference diode and load, the effectiveness of this simple circuit is determined by the ratios R_s/R_z and R_s/R_L. The larger these ratios, the more effective the circuit becomes because the second term is minimized.

For better performance, a cascade arrangement may be used where a first zener diode acts as preregulator for a temperature-compensated reference diode in the final stage, such as shown in Fig. 5-35. The first stage reduces large variations in V_{in} to some relatively low level. The stability and temperature coefficient of the output voltage is then governed primarily by the reference diode in the output stage. Further cascading with a third zener diode stage is not practical because the overall circuit efficiency is rapidly reduced and the consumption of power is greatly increased.

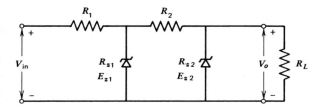

Fig. 5-35   Cascade regulator circuit.

The bridge circuit of Fig. 5-36 is of interest because the output voltage can be made independent of the bridge input voltage. For $R_1 = R_2 = R$ and $R_3 = R_z$ (zener dynamic resistance), the output voltage is

$$V_o = \frac{E_z}{1 + R_z(1/R + 2/R_L)} \tag{5-26}$$

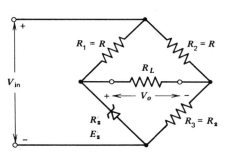

Fig. 5-36   Bridge regulator circuit.

Equation (5-26) does not contain the input voltage V_{in}; the output voltage is therefore independent of the input voltage. In practical design, V_{in} is usually derived from a preregulated circuit. The stabilization factor for input voltage variations can exceed 1000 to 1. However, the stabilization factor for load resistance variations is somewhat degraded owing to the presence of R_1 and R_2. This circuit requires an isolated input voltage supply because the common sides of the input and output cannot be connected together.

It was pointed out in Sec 5.3.3.1 that zener reference diodes require a constant current for temperature stability. Figure 5-37 shows a transistor current source circuit used in a regulator circuit. The current source consists of Q_1, V_{z1}, R_1, and D_1. The emitter current of the transistor Q_1 is approximately equal to V_{z1}/R_1. The collector current is constant and is equal to $\alpha(V_{z1}/R_1)$, where α is the common base current gain of Q_1. Diode D_1 must be the same type of material as Q_1 since it is used to compensate the temperature variation of the emitter base junction of the transistor. The performance of this circuit is about the same as the bridge regulator circuit, but it has the advantages of lower output impedance and a common ground.

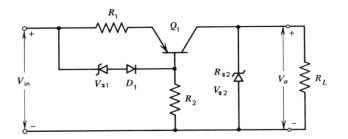

Fig. 5-37   Constant current circuit.

Figure 5-38 shows a complementary transistor current source circuit. It provides two temperature-compensated reference voltages V_{o1} and V_{o2}, one with respect to each input supply line. Each reference diode receives a constant current for which the other reference diode is the reference. This circuit offers an extremely high stability factor for input voltage variations which approaches R_c/R_z, where R_c is the collector resistance of the transistor and R_z is the zener dynamic resistance. Stability factors of 10^5 have been observed (Williams, 1967).

The limit of circuit performance of Fig. 5-38 is imposed by the change

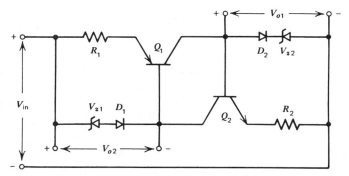

Fig. 5-38   Transistor complementary constant current circuit.

in transistor collector base voltage which produces some slight variation in the reference diode current. This limitation can be eliminated by the introduction of a bipolar transistor FET cascade circuit, as shown in Fig. 5-39. The junction FETs, Q_3 and Q_4, act as self-biased cascade buffers for Q_1 and Q_2, respectively. Neglecting gate leakage current, these buffers have a unity current gain since drain current and source current must be very nearly equal in an FET. This being the case, the transistor collection base maintains an almost constant voltage. The greater part of any input voltage variation is absorbed by the field effect transistors. This results in orders of magnitude improvement in the stability factor. A similar transistor FET cascade arrangement can be used for the same application, where the gate of Q_3 and Q_4 is connected to the base of Q_1 and Q_2, respectively (Williams, 1968). As in the previous circuit, one of the output voltages is floating.

The various reference source circuits, Figs. 5-34–Fig. 5-39, have one serious limitation: They cannot tolerate significant load variations without upsetting the output voltage and the temperature coefficient. Therefore, these reference circuits are most suitable for those applications where the load is constant.

Operational amplifiers can be used together with a zener reference diode to form a series regulator circuit to minimize loading effects. Figure 5-40 shows a typical feedback regulator circuit. A bridge circuit, excited by the output voltage, is formed by R_1, R_2, R_3, and the reference diode. A voltage proportional to the output voltage, $[R_2/(R_1 + R_2)]V_o$, is compared with the reference voltage V_z so that an input voltage change or a load change results in a differential signal at the input of the operational amplifier. The amplifier amplifies this differential signal

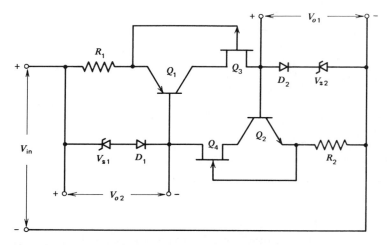

Fig. 5-39   Transistor-FET cascade complementary current source.

and controls the base current of the series transistor Q_1 in such a way that the output voltage across the load remains essentially constant.

The main advantage of this circuit is that it produces an orders-of-magnitude lower output impedance than can be obtained with any zener reference diode, and it improves the load regulation by the same amount. The performance is degraded, however, by the drift and temperature coefficients of R_1 and R_2 and of the differential input stage of the amplifier.

A current reference source can be obtained by removing the zener

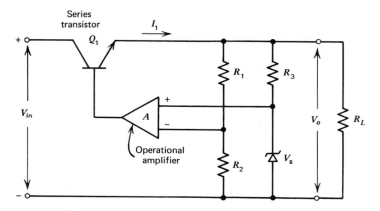

Fig. 5-40   Feedback regulator circuit.

reference diode V_{z2} from the circuit of Fig. 5-37. The reference current (collector current of Q_1) is constant and is equal to $\alpha(V_{z1}/R_1)$ where α is the common base forward current gain of Q_1. The base current of Q_1 affects the accuracy of the reference current, however. One commonly used way to minimize this base current influence is to specify a high α transistor for this circuit. Another technique is shown in Fig. 5-41. A resistor R_2 and a transistor Q_2 are added in such a way that matching of the two base currents, I_{B1} and I_{B2}, is realized. Then the reference current is equal to V_{z1}/R_1. This technique has been used successfully in an integrated 10-bit binary current source DAC (Rudin, et al., 1967).

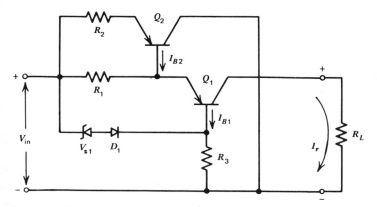

Fig. 5-41   Compensated reference current source.

Another type of current reference source is characterized by using operational amplifiers together with precision resistors (Sheingold, 1963; Newcomb, et al., 1966). The ratio of the resistors is properly selected so that the current source is controlled by a voltage source. Figure 5-42 (a) shows an inverting voltage-controlled current source, (b) a noninverting version, and (c) a modification of (a). The voltage source must furnish adequate driving capability and, therefore, a driving amplifier is usually employed as a buffer. The reference current source can be either positive or negative, depending on the particular configuration and the polarity of the voltage source used.

5.3.3.4 Feedback Regulator Design. This section deals with some specific considerations and practical aspects of the design of a precision voltage reference supply. In DAC applications, most load to voltage references are of a dynamic nature and, therefore, the feedback regulator

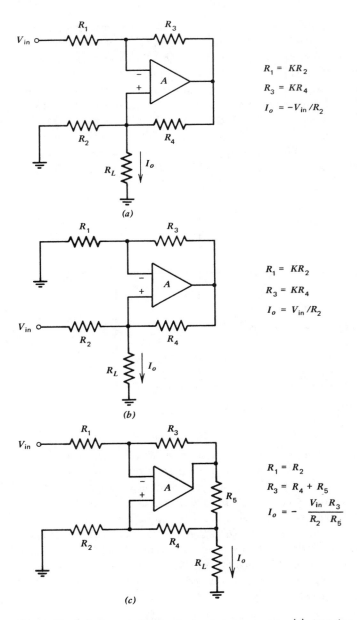

Fig. 5-42 Voltage-controlled current reference sources. (a) Inverting; (b) noninverting; (c) modified inverting.

of Fig. 5-40 is commonly used. It can be shown that the output voltage of this feeback regulator is

$$V_o = \frac{V_z}{\beta} + \frac{h_{22}h_{11}}{A\beta h_{21}} V_{in} - \frac{h_{11}I_1}{A\beta h_{21}} \qquad (5\text{-}27)$$

where V_o = output voltage
V_z = zener reference voltage
β = feedback ratio = $R_2/(R_1 + R_2)$
A = open loop gain of the operational amplifier
h_{11}, h_{21}, h_{22} = common emitter hybrid parameters for Q_1
V_{in} = input voltage
I_1 = emitter current of Q_1

If $A\beta$ is sufficiently large, Eq. (5-27) is reduced to the simple form:

$$V_o \cong \frac{R_1 + R_2}{R_2} V_z \qquad (5\text{-}28)$$

From Eq. (5-27), two stabilization factors can be defined. First, the stabilization factor of the regulator with respect to input voltage variations is defined as

$$S(V_{in}) \equiv \frac{dV_{in}}{dV_o} \frac{V_o}{V_{in}}$$
$$= \frac{A\beta h_{21}}{h_{22}h_{11}} \frac{V_o}{V_{in}} \qquad (5\text{-}29)$$

Second, the stabilization factor with respect to load resistance variations is defined as

$$S(R_L) \equiv \frac{dR_L}{dV_o} \frac{V_o}{R_L}$$
$$= \frac{h_{21}}{h_{11}} A\beta R_L \qquad (5\text{-}30)$$

Another important parameter, the incremental output impedance, is:

$$R_o = \frac{dV_o}{dI_1} = \frac{h_{11}}{h_{21}A\beta} \qquad (5\text{-}31)$$

When the load current is large and the amplifier cannot furnish sufficient base current to the series transistor, a Darlington compound transistor circuit can be used to replace the single-series transistor. Equations (5-29), (5-30), and (5-31) still are valid if the single transistor h-parameters are replaced by those of the compound circuit.

The drift stability of the feedback regulator circuit is determined not only by the drift characteristics of the zener reference diode, V_z, but also by the amplifier A and the resistors R_1 and R_2 in the bridge circuit. If E_a represents the equivalent input drift of the amplifier, the regulator output becomes

$$V_o = \frac{R_1 + R_2}{R_2}(V_z + E_a) \tag{5-32}$$

With some mathematical manipulation, a drift stability equation can be obtained and is as follows:

$$\frac{1}{V_o}\frac{dV_o}{dx} = \frac{1}{V_z + E_a}\left(\frac{dV_z}{dx} + \frac{dE_a}{dx}\right) + \frac{R_1}{R_1 + R_2}\left(\frac{1}{R_1}\frac{dR_1}{dx} - \frac{1}{R_2}\frac{dR_2}{dx}\right) \tag{5-33}$$

where x is a parameter that can be either temperature or time. The first term represents the contribution from the reference diode, the second term represents the amplifier contribution, and the third term represents the resistor divider contribution.

The transient response of the regulator output is predominantly determined by the open-loop frequency response of the operational amplifier and the associated load characteristics. Considering the series transistor as a part of the amplifier, the amplifier with the wider unity-gain-bandwidth product and adequate phase margin ensures a faster settling time.

Remote error sensing is also an important consideration in precision voltage regulator design. Normally, the regulator circuit realizes its optimum performance at the output terminals. When, as often is the case, the load is separated from the output terminals, the regulator performance is degraded at the load terminals by an amount proportional to the impedance of the output leads compared to the regulator output impedance. For example, a 40-mΩ lead (wire and contact) resistance and 100-mA load current creates an error of 4 mV. A remote error sensing scheme is shown in Fig. 5-43. It is a simple modification that brings the inputs of the bridge network out and terminates them separately from the output and ground terminals.

A transistor Q_2 and a resistor R_4 have been added in Fig. 5-43 to provide for output short-circuit protection. When a short circuit occurs at the output, the current in R_4 increases and this is sensed by Q_2. When the voltage drop across R_4 exceeds about 0.6 V, Q_2 turns on and limits the base drive available to Q_1. The output current limit is set by R_4. The transistor Q_1 must have adequate power handling capability under this condition since the total input voltage is across the collector-to-emitter junction of Q_1. When the short circuit is removed, the regulator output returns to normal automatically.

In DAC applications where voltage switching circuits are involved, the precision voltage regulator receives current from the associated voltage

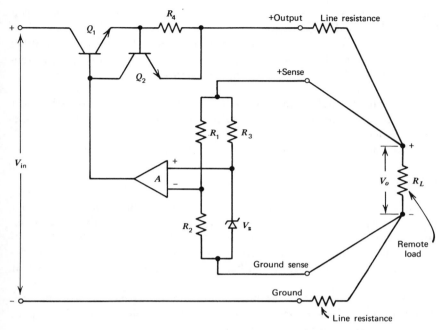

Fig. 5-43   Regulator circuit with remote error sensing and short-circuit protection.

switches. That is, the regulator must act as a current sink and, therefore, a preload resistor is required across the output. When all the associated DAC switches are OFF, the regulator supplies full-scale load current to the preload. When certain switches are turned ON, they supply an amount of current to the preload and, therefore, the regulator output is decreased by the same amount.

5.3.4   Operational Amplifiers

Operational amplifiers are commonly used elements in DAC design. They are principally used as output buffer amplifiers, summing amplifiers, or, as discussed above, in the reference source.

Most DAC decoding networks are relatively high-impedance circuits. As such, they are only useful when driving into much higher impedances such as oscilloscopes and recording instruments. In many applications, therefore, it is necessary to isolate a low-impedance load from the high-impedance decoding network by inserting a buffer amplifier between them.

Operational amplifiers also can be used as a summing amplifier incorporated directly in a DAC design. Various DAC configurations of this type are discussed in Sec. 5.4.

An operational amplifier is a stable, high-gain, DC-coupled amplifier and is usually used with a large amount of negative feedback. It is generally characterized by the following properties:

1. Extremely high open-loop voltage gain, typically exceeding 10^4.
2. High input impedance (typically 400 kΩ) and low output impedance (typically 100 Ω).
3. Very low DC offset and drift with time and temperature.
4. Wide bandwidth, typically the unity gain bandwidth is between 1–10 MHz.

There are three basic operational amplifier configurations: *inverting, noninverting,* and differential. The configurations differ only in the manner in which the input signal is applied and the external feedback elements are arranged. Some of the pertinent closed-loop characteristics for each of these three amplifier configurations are summarized in Table 5-5.

In practice, operational amplifiers are not ideal and require the consideration of various errors introduced by their nonideal behavior. Errors can be classified into four main groups:

1. Static errors due to finite open-loop voltage gain, finite input impedance, output impedance, source impedance, and load impedance.
2. Errors due to initial voltage, current offsets, and drifts caused by temperature, time, and power supply changes.
3. Dynamic errors due to finite amplifier bandwidth.
4. Errors due to noise.

The effect of these errors on associated circuits is not substantially different than those discussed in connection with the amplifiers in the analog-input subsystem (Chap. 3, Sec. 3.14). As in that case, each of these error sources must be considered in the particular context in which the amplifier is being used. The sources of these errors and means of compensating their effects are well documented in the published literature, some of which is included in the bibliography, and will not be discussed in detail.

In general, static errors due to limited open-loop gain can be compensated by initial adjustment of the circuit in which the operational amplifier is imbedded. Changes in the open-loop gain with temperature and life, of course, affect the operation of the circuit but the magnitude of these changes is small if the nominal open-loop gain of the amplifier is sufficiently high.

Similar remarks apply to the errors associated with finite input impedance and source impedance. In most cases these are under the control of the DAC designer and adequate allowance for them can be incorporated in the design. Changes in load impedance and the effects of output impedance are not always under the control of the designer since the amplifier may be used to drive a load external to the process control computer system. However, with proper design, the output impedance can be much less than 1 Ω, and this minimizes most of the errors due to load impedance.

Errors due to initial voltage and current offsets can be compensated by adjustments or compensation circuits. As in the case of the amplifiers discussed in Chap. 3, the variations in these offsets with temperature and life must be carefully considered in the design. The use of matched-pair transistors and balancing circuits (Hoffait and Thornton, 1964) or transistors fabricated on a temperature-controlled substrate (Emmons and Spence, 1966) are effecive in minimizing these drift errors.

Dynamic errors due to finite bandwidth are generally of concern in high-speed DAC design where the objective is to be able to establish a new output level in a few microseconds or less. The settling time calculations in Chap. 3 for first- and second-order systems are useful in determining the required amplifier bandwidth. These calculations, however, assume that the amplifier is linear. In reality, most amplifiers are slew-rate limited owing to current drive limitations in one or more stages of the amplifier. For this reason, the large signal bandwidth, which usually is less than the small signal bandwidth, must be used in obtaining an estimate of the settling time for large output voltage excursions.

One of the most common uses of an operational amplifier in DAC configurations is as an output buffer amplifier. If the output of the DAC is used to control a device in the process or to drive a graphic display device, the buffer amplifier is almost always required. It is, in fact, the exception when the buffer can be omitted.

The primary problems encountered in using a decoder DAC configuration without a buffer amplifier are associated with the output impedance and driving capability of the network. As indicated in the equations for the output of the decoder network, the output value is dependent on the value of the load resistance. Most decoder networks have an output resistance in excess of 1 kΩ and, in some implementations, the output resistance may be as high as several hundred kilohms. Since the load impedance must be considerably greater than the output resistance to avoid an appreciable error due to loading, this requires that the load impedance be very high. For example, in a DAC with an output impedance of 1 kΩ, the load impedance must be greater than 100 kΩ to limit the loading error to less than 1.0%.

Table 5-5 Operational Amplifier Circuit Configurations

	Inverting	Noninverting	Differential

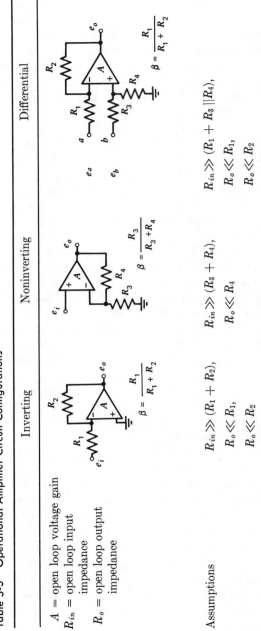

A = open loop voltage gain
R_{in} = open loop input impedance
R_o = open loop output impedance

Inverting:
$$\beta = \frac{R_1}{R_1 + R_2}$$

Noninverting:
$$\beta = \frac{R_3}{R_3 + R_4}$$

Differential:
$$\beta = \frac{R_1}{R_1 + R_2}$$

Assumptions

Inverting:
$R_{in} \gg (R_1 + R_2)$,
$R_o \ll R_1$,
$R_o \ll R_2$

Noninverting:
$R_{in} \gg (R_3 + R_4)$,
$R_o \ll R_4$

Differential:
$R_{in} \gg (R_1 + R_3 \| R_4)$,
$R_o \ll R_1$,
$R_o \ll R_2$

DC gain, (closed loop) $\dfrac{e_o}{e_i}$	$-\dfrac{R_2}{R_1}\dfrac{1}{(1+1/A\beta)}$	$\left(\dfrac{R_3+R_4}{R_3}\right)\dfrac{1}{(1+1/A\beta)}$	$e_o = \left(\dfrac{R_4}{R_3+R_4}\right)\left(\dfrac{R_1+R_2}{R_1}\right)e_b - \dfrac{R_2}{R_1}e_a$ $e_o = \dfrac{R_2}{R_1}(e_b-e_a)$ for $\dfrac{R_1}{R_2}=\dfrac{R_3}{R_4}$
Input impedance (closed loop)	R_1	$R_{in}(1+A\beta)$	At terminal a: R_1 At terminal b: (R_3+R_4)
Output impedance (closed loop)	$\dfrac{R_o}{(1+A\beta)}$	$\dfrac{R_o}{(1+A\beta)}$	$\dfrac{R_o}{(1+A\beta)}$
Characteristics and comments	·Low input impedance ·Single-ended input ·Highest accuracy ·No CM voltage error ·Possible less than unity gain ·Bandwidth is limited if R_2 is large	·High input impedance ·Single-ended input ·Gain ≥ 1 ·Higher bandwidth	·Low input impedance ·Differential input ·Good CMRR ·Drifts same as inverting configuration

Even if the loading error is tolerable, the use of a decoder network to drive an external load without a buffer amplifier complicates the design of the reference source. This is particularly true if the load impedance varies. The reason is that the load current must be supplied from the reference. This increases the required current capacity of the reference source and may require additional reference source circuitry to ensure adequate regulation for load variations.

Another factor in the use of a decoding network DAC without a buffer amplifier in high-speed applications is a possible restriction on the DAC speed of response. Since the time constant of the output voltage is R_oC_L, where R_o is the output resistance of the DAC and C_L is the load capacitance, a high output impedance can restrict the DAC speed when the load capacitance is significant. For an output resistance of 10 kΩ, a load capacitance of only 10 pF results in a time constant of 1000 μsec and a settling time to 0.01% of almost 10 msec.

By using a buffer amplifier, most of these problems can be avoided. The output current is supplied by the amplifier, thus eliminating additional drain on the reference source. Since the buffer amplifier can be designed to have a very low output impedance (typically less than 1 Ω), the DAC risetime and settling time are relatively independent of the load capacitance. The high input impedance of the buffer also eliminates most errors due to decoder network loading.

The buffer can also be a source of errors, however. The most significant of these may be drift and noise. Drift can be the result of drift in the gain of the amplifier and in zero offset drift with temperature and life. Since the gain stability and the gain drift characteristics are dependent on the open-loop gain of the amplifier, these can generally be made insignificant by designing the amplifier with a sufficiently high open-loop gain. Both current and voltage offsets can also affect DAC accuracy. Input current offset results in leakage current through the decoder network, which causes a voltage offset. Similarly, the amplifier offset voltage can cause a change in the amplifier output voltage.

5.3.5 Register and Control

The function and design of the register and control unit for the DAC is relatively straightforward compared with other logic design problems encountered in process control computers. The primary function of the control unit is the digital storage of the value to be converted. This is generally provided by a register of latches or flip flops. The register outputs are used to control the DAC switches, either directly or through appropriate driver circuits.

In addition to the basic storage function, the control unit may provide other functions. Examples include address decoding, error checking, and digital data buffering. Since multiple DACs may be connected to a single digital channel from the CPU, the control portion of the analog-output subsystem may be required to perform address decoding. The address and digital data value are received through the channel from the computer. By decoding the address information, the analog-output value data are routed to the proper DAC register. In essence, the control unit in this application acts as a digital multiplexer.

If a single DAC is used to control a number of analog memory or amplifier circuits associated with current output stations or back-up devices, the control unit may also provide address decoding. In this case, the control unit controls an analog multiplexer which routes the analog output of the DAC to the proper analog storage device.

Another function that may be provided by the control unit is digital data buffering. This is particularly common in specialty applications such as hybrid computers. As an example, it may be required to provide a change in the analog-output voltage of a number of DACs at precisely the same instant. In this case, the control unit includes a digital buffer register for each DAC network in addition to the DAC switch storage register. The buffer registers are loaded with digital data from the computer in a sequential manner; for example, one buffer register might be loaded on each of 10 storage access cycles. When all the registers are loaded, a single signal from the computer transfers the data in these registers into their associated DAC switch storage registers. In this way, the outputs of all DACs change at precisely the same time.

In performing the buffering function, there are two ways in which the DAC switch register can be loaded. Either the data can be transferred on a forced basis in which the storage register latches are changed only if the input-data bit is different, or the storage register can be reset and then loaded with the new data word. This latter method may result in "spikes" in the DAC output since resetting the register causes the DAC output voltage to drop to zero for an instant before the new data are loaded into the register. The severity of these spikes depends on the transient response of the decoder network, the bandwidth of the buffer amplifier, and the time interval between the reset signal and the data transfer. In process control systems, the use of a buffer amplifier is common and the speed of the DAC does not have to be particularly fast. For this reason, the buffer amplifier bandwidth can be limited and the spikes easily controlled. If the DAC is used to drive a fast graphic display device such as an oscilloscope, however, it may not be possible to limit the buffer bandwidth and the spiking problem must be considered. If it is not, the spikes can cause spurious traces and streaks in the display.

The functions of the control unit are the same whether the DAC is of the decoder network type or whether it includes an operational amplifier, as in the case of the DACs discussed in Sec. 5.4.2. The exact requirements are dependent on the particular application for which the DAC and process control computer system are being designed. In most cases, however, the logic requirements are quite minimal and the logic design problem is much easier to solve than the analog design problem.

5.4   DAC DESIGN

The four elements discussed in the previous section are combined in various combinations to form DAC circuits and analog-output subsystems. All DACs contain a decoding network, DAC switches, and a reference source. The operational amplifier is an optional element if its use in the reference supply is disregarded. If it is used, it is either as an integral part of the DAC or as a buffer amplifier to provide an impedance transformation at the DAC output and an increase in the DAC driving capability. In addition to the elements discussed in the previous section, logical control must be provided for the DAC subsystem. This logic provides the circuits which activate the DAC switches and perform other decoding and control functions. In this section, various DAC configurations that use the DAC elements are described. Although not all are commonly used in process control computer systems, there is no fundamental limitation on their use. The selection of a particular configuration depends on the usual analysis of performance versus cost.

5.4.1   Decoder Network DACs

The general configuration of the decoder network DAC is shown in Fig. 5-44. It incorporates the five basic DAC elements introduced in the previous section. Of the five elements, only the buffer amplifier is optional; all others are required to provide the analog-output function needed in process control computer systems. In most practical situations, the buffer amplifier is also included to provide impedance buffering of the decoder network output, however.

The DAC decoder network, DAC switches, and the reference source were discussed in the previous section; this DAC is a straightforward combination of these elements. The manner in which these elements are actually connected is illustrated in Figs. 5-7–5-12 which were used in the discussion of the various types of decoder networks and in subsequent figures used to illustrate the DAC switches. The discussion of the decoder networks, switches, and reference sources also indicated many of the interacting effects that must be considered in DAC design. This includes

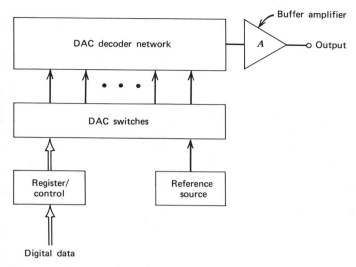

Fig. 5-44 General decoder network DAC.

factors such as reference source transients, switch resistance, and decoder network specifications.

5.4.2 Operational DAC Configurations

This section deals with the characteristics of various DAC configurations using operational amplifiers. The discussions and illustrations are restricted to unipolar output, although bipolar output can be obtained easily by incorporating the schemes described in Sec. 5.3.1.5.

Most operational amplifiers in the various DAC configurations are connected in an inverting configuration and used as summing amplifiers. Because a large amount of negative feedback is utilized, the impedance at the summing point or inverting input terminal is equal to $R_{in}/(1 + A\beta)$, where A is the open-loop gain, β is the feedback factor, and R_{in} is the open-loop amplifier input resistance. This impedance is usually negligibly small compared to the impedances used in the decoding network. It is this property of the amplifier that minimizes the interaction between various inputs and, therefore, makes the use of the principle of superposition particularly attractive for the analysis of the configurations.

Figures 5-45 and 5-46 show two DAC configurations which differ only in the type of decoding network; one uses the weighted type and the other uses the ladder type. The current distribution per bit in the ladder network is more uniform than that in the weighted network and, therefore, the design of the DAC switches is simplified. Because all switches

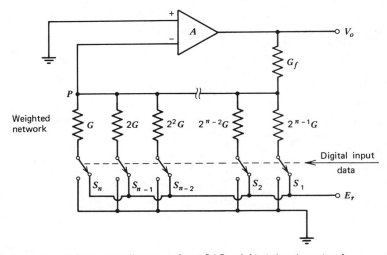

Fig. 5-45   Voltage-controlled operational DAC weighted decoder network.

are connected between the precision voltage reference and decoding network, the voltage reference must be well regulated. This is necessary owing to the significant load current transient that can occur when the switches are actuated. This is especially important when a number of DACs share a common voltage reference.

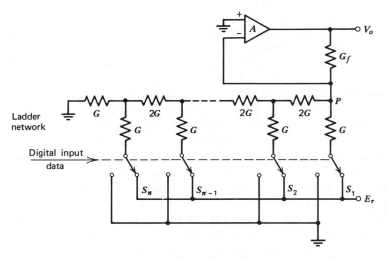

Fig. 5-46   Voltage-controlled operational DAC ladder decoder network.

It is desirable to have the equivalent of a single-pole double-throw (SPDT) bit switch for the weighted network, as shown in Fig. 5-45. This maintains a constant input conductance to the operational amplifier so that the DAC accuracy is independent of the input bit pattern. However, all switches may be replaced by the equivalent of a single-pole single-throw form if the degradation of accuracy is tolerable.

A modified, weighted decoder operational DAC is shown in Fig. 5-47. Note that there are twice the number of precision resistors per bit when compared with the DAC of Fig. 5-45. The switches are of the single-pole single-throw (SPST) form and are connected between the common node of each resistor pair and ground. The main advantages of this configuration are twofold: (1) The switch design is greatly simplified since only an SPST switch is required and (2) the load current transients in the voltage reference are partially minimized.

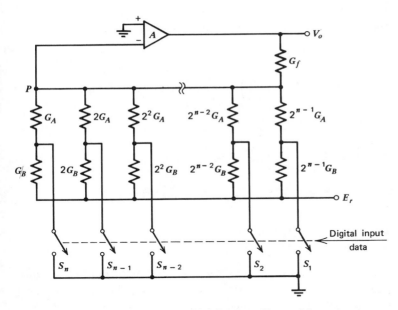

Fig. 5-47 Voltage-controlled operational DAC modified weighted decoder network.

The configuration shown in Fig. 5-48 is of most interest because the current from the voltage reference is constant. This means that the design of the reference can be greatly simplified. Besides, the configuration also employs SPST switches, which are easier to design. It can be shown that the output is

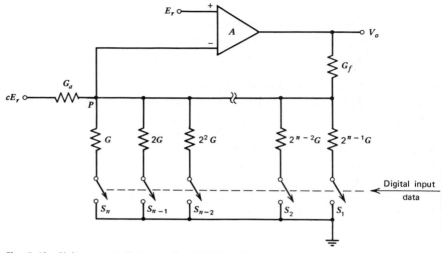

Fig. 5-48   Voltage-controlled operational DAC weighted decoder network.

$$V_o = E_r \left[\left(1 + \frac{G_a}{G_f} - c\frac{G_a}{G_f} \right) + \frac{G}{G_f} \sum_{k=1}^{n} A_k 2^{n-k} \right], \quad A_k = 1, 0 \quad (5\text{-}34)$$

where E_r = reference voltage
G_a = input conductance
G_f = feedback conductance
c = a constant

If $G_a/G_f = 1$ and $c = 2$, then Eq. (5-34) becomes

$$V_o = \frac{E_r G}{G_f} \sum_{k=1}^{n} A_k 2^{n-k}, \quad A_k = 1, 0 \quad (5\text{-}35)$$

which is a binary coded unipolar voltage output. Note that the output voltage V_o and the voltage reference are of the same polarity. This is one of the unique characteristics of this configuration.

If $G_a/G_f = 2$ and $c = 2$, then Eq. (5-34) becomes

$$V_o = E_r \left(-1 + \frac{G}{G_f} \sum_{k=1}^{n} A_k 2^{n-k} \right), \quad A_k = 1, 0 \quad (5\text{-}36)$$

which is an $(n-1)$-bit binary coded, bipolar voltage output. Switch S_1 becomes the sign bit switch and a negative number is coded as a two's complement.

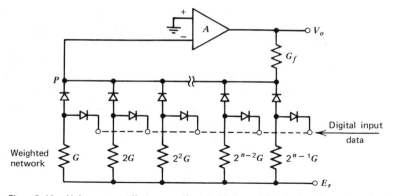

Fig. 5-49 Voltage-controlled operational DAC weighted decoder network with diode switches.

Figures 5-49 and 5-50 show two operational DAC configurations which utilize current mode diode switches. Note that the diode switches are between the summing point P and the decoding network. Figure 5-49 utilizes a weighted decoding network with a voltage reference, whereas Fig. 5-50 uses a ladder decoding network and a current reference. The currents per bit in both configurations are binary weighted. The main advantage of this scheme is the potential of high-speed operation possible with good diodes. The response time of the decoding network is improved because the current in the network is relatively constant regardless of the input bit pattern. The main limitations are the variations of diode forward voltage drop, dynamic resistance, and reverse current with temperature and life.

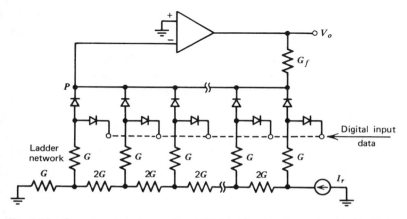

Fig. 5-50 Current-controlled operational DAC ladder decoding network with diode switches.

Figure 5-51 is a variation of the basic configuration shown in Fig. 5-49. All switches are of the SPST form, however, and they are between the summing point and the decoding network. When the switch is closed, the weighted current flows into the summing point P. When the switch is open, the weighted current is diverted from the summing point and flows to ground through the diode. Field effect transistors are well suited for this switch application. Through their use, the problems associated with the diode switch are eliminated. However, the ON resistance characteristic of the FET still deserves careful consideration.

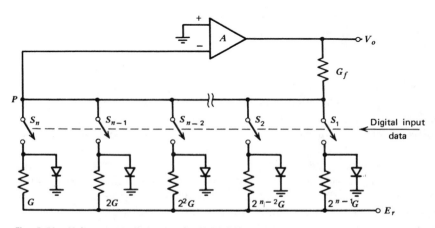

Fig. 5-51   Voltage-controlled operational DAC.

The operational DAC configurations of Figs. 5-45–5-51 have one thing in common: The decoding network and the switches are in the input circuit of the operational amplifier. Operational DACs also can be constructed in which the decoding network and switches are in the feedback path of the operational amplifier, as shown in Fig. 5-52. When all switches are closed, the output is zero. When the kth switch is opened ($A_k = 0$), its contribution to the output is $-E_r 2^{n-k} R / R_1$. The general expression for the output, therefore, is

$$V_o = -\frac{RE_r}{R_1} \sum_{k=1}^{n} (1 - A_k) 2^{n-k}, \quad A_k = 1, 0 \qquad (5\text{-}37)$$

The usefulness of this configuration is limited for the following reasons: (1) All switches are connected in series and, therefore, the resultant errors

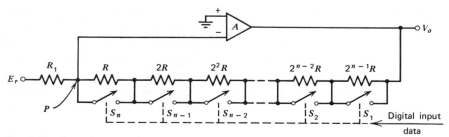

Fig. 5-52   Operational feedback network DAC.

due to each switch are directly additive; (2) the closed-loop band-width of the amplifier is limited by the large resistance in the feedback path; and (3) the accuracy is dependent on the input bit pattern be-cause the offset and drift of the amplifier vary with the feedback resistance.

5.4.3   DAC Error Analysis Methods

Each of the basic building blocks of DACs such as the precision voltage references, decoding networks, switches, and the operational amplifiers contribute to the overall error. The overall error usually consists of the following different components: (1) static error, (2) dynamic error, (3) offset and drift due to temperature, time, and supply voltage changes, and (4) noise. In analyzing each of the error components, it is usually necessary to assume that the remaining ones are negligible because a simultaneous consideration of the effects of all components complicates the problem needlessly. The only result of considering all error sources simultaneously is an estimate of second-order effects. These, however, are usually small compared with the direct errors.

Generally, various error components of the blocks can be calculated by utilizing their respective equivalent circuit models. The calculations are performed by assigning error parameters in the circuit model and then summing each of the contributions at the DAC output. Theoretically, the error calculations should be performed for every possible input bit combination. However, in practice only a number of key input bit pat-terns are usually calculated. For a DAC utilizing the offset method for bipolar operation, these patterns are (1) plus full scale, (2) minus full scale, (3) all bits **1**, corresponding to a slightly negative input, and (4) all bits **0**, corresponding to zero output.

After each of the contributing errors is obtained, the question of how to calculate the overall error still remains to be answered. There are two commonly used methods to obtain the overall error. They are the worst-case (wc) method and the root-mean-square (rms) method. The worst-case method is defined as the resultant overall error, assuming the worst possible combination of each of the contributing errors. Mathematically, it is expressed as $wc = \pm a \pm b \pm c \pm \ldots$, where a, b, c, ... are the contributing errors. The root-mean-square method defines the resultant overall error as the square root of the sum of the squares of each contributing error. Mathematically, it is $rms = (a^2 + b^2 + c^2 + \cdots)^{1/2}$. The worst-case method is rather pessimistic because the probability that all these contributing errors will add in the same direction is very small. It is rarely used in the design of high-accuracy (0.01%) DACs since it imposes an unrealistically stringent constraint on the design.

Another more sophisticated approach to the error analysis problem is a statistical analysis using a digital computer. In the statistical analysis of an electronic circuit, the circuit equations are solved for many different cases. In each case, the circuit component values are selected randomly from within a given range of values according to a defined probability distribution. The statistical results suggest likely behavior limits of a group of manufactured circuits. One such program, ASAP (Automated Statistical Analysis Program), was developed for the IBM 7090/94 computer (Kennard, 1964). It has been used for DAC analysis as well as for the analysis of other circuits such as high-performance amplifiers.

5.5   COMPUTER-MANUAL STATIONS

In direct digital control and noncritical process loop applications, the computer-manual station is perhaps the most widely used analog-output device. The basic circuit elements for the station are (1) an output multiplexing switch, (2) an output memory amplifier, (3) manual controls, and (4) a man-machine interface. A typical station configuration is shown in Fig. 5-53.

During normal operation, the station receives the output command signal from a time-shared DAC or a pulse output and delivers the converted output current to the final control element. The output command signal is the result of a computer calculation that takes into account the status of the process variable and other programmed conditions defined by the control algorithm.

In the event that the computer fails or is taken off-line for some other reason, the station holds the last computer calculated output current value and, therefore, the control element maintains its position. If the computer downtime is prolonged, the station can be switched into a

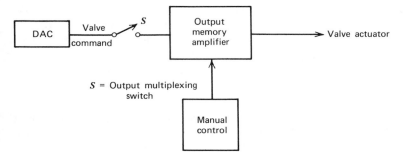

Fig. 5-53 Typical computer-manual station.

manual control mode and the output current can be adjusted manually by the operator. The computer-manual station is sometimes referred to as a current output station because of the nature of the output signal.

5.5.1 Output Multiplexing Switches

The main function of the output multiplexer switch is to switch the output command or setpoint signal from the time-shared DAC to various memory amplifiers; there is one switch for each memory amplifier. The addressing method for the output amplifiers can be either sequential or random and is determined by the programming control.

Generally, the analog-output multiplexing switch is similar to that used for high-level analog-input multiplexing, as discussed in Sec. 3.13. The output multiplexing switch is usually an SPST configuration designed for high-level signals. This is because the DAC output is typically single ended and the output voltage is in the range of 1–5 V or 2–10 V. The switch load may be capacitive because the output memory amplifier is often the capacitor-FET storage type. In this case, the switch OFF leakage resistance is the most critical parameter because it contributes to the resultant output current droop rate. The other important consideration is the operation life of the switch because this type of application generally requires very high reliability.

A reed relay of a special design has been used in a current output station design (Maxfield and Steele, 1967). The OFF leakage resistance of the relay is greater than $10^{12}\ \Omega$ at 25°C and 20–50% relative humidity. The operate time including bounce is less than 1 msec and the release time is less than 0.5 msec. The initial ON resistance is less than 1 Ω. For this low resistance value, a suitable resistor is required between the DAC and the reed relay for current limiting if optimum switch life is to be achieved.

The life specification of the reed relay is 1.26×10^8 operations at its maximum rating of 70 mA. This is equivalent to 4 yr of continuous operation at a station update rate of 1/sec. If the rate is increased to 4/sec, considered to be the upper limit of the rate for the application, the expected life is reduced to 1 yr. However, since the normal operation current flowing through the relay is usually only a few milliamperes or less, the actual life may extend to many more years.

For ultimate reliability and performance, a solid-state switch should be used. MOSFET switches of both p-channel and n-channel type are suitable for this application. Typical ON resistance is about 50–500 Ω and is a function of operating point. The drain-to-source leakage current I_{DSS} is typically 500 pA at $V_{DS} = -20$ V. This current must be greatly reduced in order to achieve a reasonable output current droop rate of 1%/hr. Figure 5-54 shows a low-leakage MOSFET switch circuit in which the leakage is reduced to a satisfactory level (Walton and Leon, 1969). The switch consists of two series FETs. The FET Q_1 is connected to the storage capacitor and must be an enhancement device, but Q_2 can be either a MOSFET or a junction FET. The node between the two FETs is connected through a megohm resistor to a potential equal to that on the storage capacitor so that the potential across Q_1 is very small. The necessary voltage is available at the amplifier output. With the substrate of Q_1 floating, there is a pair of back-to-back diodes between the drain and source. The drain-to-source leakage current obeys the junction diode law and is virtually zero at room temperature because of the low source-to-drain voltage. The leakage is temperature dependent and doubles for every 8–10°C temperature rise. This factor must be taken into account during the design.

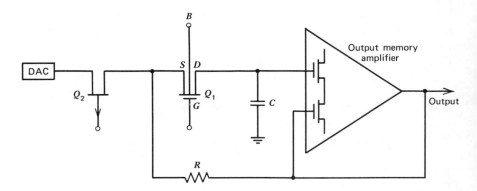

Fig. 5-54 Low-leakage MOSFET switch.

5.5.2 Output Memory Amplifiers

The heart of the computer-manual station is an output memory amplifier which is characterized by nondestructive sensing and analog storage capability. This type of amplifier is sometimes referred to as a track-and-hold or sample-and-hold amplifier, although its characteristics differ from those found in the usual amplifiers of these types. Among the various memory devices discussed in Sec. 5.1.3, the FET and capacitor storage technique is by far the most commonly used for this application. This is mainly because of its simplicity, effectiveness, and high-speed operation capability.

5.5.2.1 Configurations. This section deals with some of the possible configurations for output memory amplifiers of the FET and capacitor storage type. Figure 5-55 shows a simple output memory amplifier without local feedback (Maxfield and Steel, 1967). In track mode, the switch S_1 is closed and the capacitor charges to the command signal from the DAC. The FET, Q_1, forms a high-impedance unity gain amplifier whose output is converted into the proper current level by the voltage-to-current (V/I) amplifier. The arrangement of Q_1, Q_2, and R_s constitutes a source follower so that the output V_o is equal to the input voltage V_i plus the gate-source offset voltage of Q_1.

In hold mode, the switch S_1 is opened and the charge on the storage capacitor C remains essentially constant, as does the output current, due to the very high impedance at the junction between S_1, C, and the gate of Q_1.

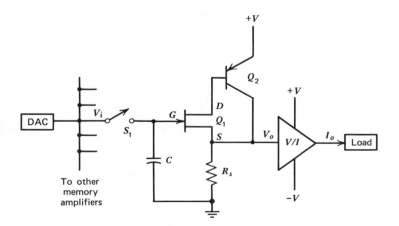

Fig. 5-55   Simple memory amplifier without local feedback.

In this circuit, the temperature coefficient of the gate-source voltage of Q_1 is partially compensated by Q_2 but the overall performance of the amplifier is limited. This is mainly because it is an open-loop design with no overall feedback from output to input. As a result, the output tends to drift with temperature and life. The circuit is also sensitive to power supply variations and, therefore, requires regulated power supplies.

An improved amplifier is shown in Fig. 5-56. The first stage of the amplifier consists of a dual matched pair p-channel MOSFET and a current source I_s. The first stage is followed by additional amplification stages and an output stage, which are all represented by the amplifier A in the diagram. The output current I_o is sensed by the resistor R_F, and the resulting voltage provides feedback to the gate of Q_2. The input voltage is applied to the storage capacitor and the gate of Q_1 through the switch S_1.

This circuit is a conventional operational amplifier configuration with an FET input and current feedback. The configuration offers many advantages, such as high input impedance, better temperature coefficient, superior output current stability, less sensitivity to power supply variations, and single power supply operation.

The load in Fig. 5-56 is connected between the amplifier output and the gate of Q_2 and cannot be ground referenced. This if often acceptable

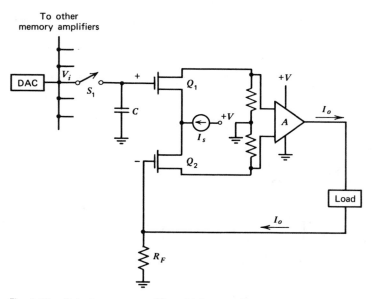

Fig. 5-56   Output memory amplifier with local feedback.

because in many cases the load is a value actuator that is a floating coil. However, sometimes a ground-referenced output is desirable for ground isolation, especially when the subsystem contains a large number of output stations.

Figure 5-57 shows two output memory amplifiers with ground referenced outputs. In Fig. 5-57a, the output stage consists of two PNP transistors (Q_4, Q_5) and one FET, Q_3. The feedback arrangement is the same as that shown in Fig. 5-55. The output current I_o in the load tracks the feedback current I_F. This tracking ability is related directly to the matched properties (V_{BE}, h_{FE}) and operation points of Q_4 and Q_5. The collector voltage of Q_4 follows the collector voltage of Q_5 via the gate-source voltage of Q_3.

In Fig. 5-57b, amplifier A_1 is connected as a voltage follower and the output voltage V_o equals the input voltage V_i. The output stage consists of an operational amplifier A_2, two PNP transistors (Q_3, Q_4), and the resistor network (R_1, R_2, R_F). In this configuration, the output current I_o is

$$
\begin{aligned}
I_o &= \frac{R_2 V_o}{R_F R_1} \\
&= \frac{R_2 V_i}{R_F R_1}
\end{aligned}
\tag{5-38}
$$

The performance of the circuit in Fig. 5-57b is superior to that of Fig. 5-57a because of the feedback incorporated in the output stage design. However, it requires more components and, therefore, is more costly.

5.5.2.2   Practical Design Considerations. This section deals with some practical considerations in the design of the output memory amplifier. The configuration of Fig. 5-56 is used to illustrate some major factors because of its inherent superior performance. The following factors are considered: (1) acquisition time, (2) droop rate, (3) temperature coefficient, (4) static accuracy, and (5) output impedance.

Acquisition time is the time necessary to track the input V_i to the required accuracy. It is often referred to as the minimum sample time or the settling time. There are two time constants in the amplifier circuit. One is the $R_{on}C$ time constant of the input network where R_{on} is the ON resistance of S_1 and C is the capacitance of the storage capacitor. The other time constant is the delay of the amplifier; this is usually negligible. Thus the acquisition time is primarily determined by the input network time constant $R_{on}C$.

The droop rate is the rate of change of the output current when the amplifier is in the hold mode, that is, when S_1 is open. It is usually

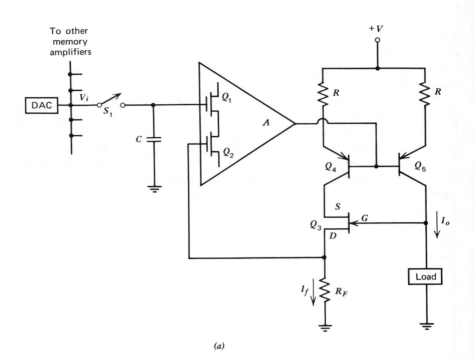

(a)

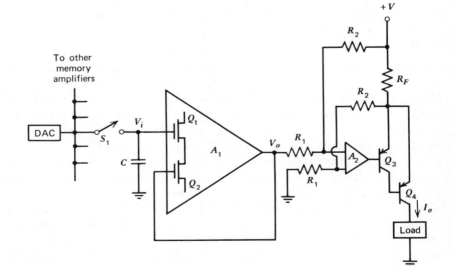

(b)

Fig. 5-57 Output memory amplifier with ground-referenced load.

454

specified as a percentage of full scale per hour. The output current droop is directly related to the rate of voltage change on the storage capacitor C. Leakage resistance at the amplifier input node plays an important role in determining the droop rate. There are four leakage paths at this node: (1) source-to-drain leakage of S_1; (2) input gate leakage of Q_1; (3) storage capacitor leakage; and (4) the leakage through the imperfect insulation material used for mounting or packaging the input node. The droop rate can be either positive or negative, depending on the relative magnitude of the leakage current components.

The voltage change on the storage capacitor is calculated as

$$dV_c = \frac{I_{av}T}{C} \tag{5-39}$$

where dV_c = incremental voltage change
$\quad I_{av}$ = average resultant leakage current
$\quad T$ = hold period
$\quad C$ = storage capacitance

Equation (5-39) can be used to calculate the allowable leakage current for a given storage capacitance and a droop rate specification. For example, to achieve a droop rate of less than 1%/hr for a full-scale input of 5 V and with $C = 1\mu F$, the average leakage current must be less than $(50 \times 10^{-3})(1.0 \times 10^{-6})/3600 = 1.39 \times 10^{-11}$ A = 13.9 pA. This is equivalent to an RC time constant of 100 hr and represents a challenging design problem. Special consideration must be given to the component selection, circuit configuration, and high insulation resistance packaging techniques.

As to the output multiplexing switch S_1, the switch configuration of Fig. 5-54 represents a good solution. The drain-to-source leakage current can be made less than 1 pA at room temperature.

In choosing the storage capacitor, the value of C should be as high as possible, but physical size, price, and quality must be considered. Polycarbonate, polystyrene, and Parylene* capacitors are suitable for this application. The polycarbonate unit has a low temperature coefficient and moderate insulation resistance. The polystyrene unit has high insulation resistance and a predictable negative temperature coefficient of -120 ± 30 ppm/°C. Parylene has exceptionally high insulation resistance and a predictable T.C. (-200 ± 30 ppm/°C). Its dielectric absorption approaches that of polystyrene (0.05% vs. 0.08%). The long-term stability and hysteresis of Parylene capacitors are better than the other film dielectrics. Another advantage is smaller physical size.

* Union Carbide trademark for polyparaxylylene.

The first stage of the amplifier is differential, and either MOSFETs or junction FETs can be used for Q_1 and Q_2. The MOSFET has extremely low gate leakage current, typically less than 0.1 pA at room temperature. However, they are easily damaged by static charge, necessitating special assembly precautions. It should be noted that only Q_1 requires low gate leakage because it is directly connected to the input node. The gate leakage of Q_2 can be tolerated because it does not contribute to the droop rate and, therefore, its gate can be protected with a zener diode. Junction FETs require no special assembly precautions, but their gate leakages are usually orders of magnitude greater and, therefore, device selection for low gate leakage is required.

It is well known that the temperature characteristics of a differential amplifier are predominantly determined by the properties of the first stage. For a good temperature coefficient, it is advantageous to have Q_1 and Q_2 as a matched pair. Typically, a MOSFET matched pair has a 200 $\mu V/^\circ C$ coefficient for the gate-to-source voltage tracking and 50 $\mu V/^\circ C$ for a matched junction FET pair.

The static overall accuracy of the output memory amplifier is determined by the amplifier open-loop gain and the precision of the feedback sensing resistor R_F. An overall accuracy of 0.2% of full scale is easily achieved for an open-loop gain of greater than 1000 and a 0.05% feedback sensing resistor.

Since current feedback is utilized in the amplifier configuration, the output impedance is greatly enhanced. The amplifier approximates a current source and the output current does not depend on the value of the load resistance, provided that the resistance is within the allowable range. The magnitude of the output impedance with feedback is approximately $r_o + (A + 1)R_F$, where r_o is the output resistance of the amplifier without feedback, A is the open-loop gain, and R_F is the feedback sensing resistor.

To maintain the droop rate performance under high humidity conditions, the packaging aspect of the amplifier input node cannot be overlooked. To maintain a time constant of 100 hr with a $1\mu F$ storage capacitance requires that the average leakage resistance at the input node be no less than 0.36×10^{12} Ω. For high insulation resistance, the basic considerations are to provide a high leakage resistance between the input node and the circuit card, to keep water vapor from forming in the input node area, and to protect the area from the accumulation of foreign materials.

It is not advisable to mount the input node directly on the circuit card because the leakage resistance of the card is greatly degraded under high-humidity conditions. A Teflon* standoff may be used for mounting this node. Teflon has a high volume resistivity and a surface on which water does not form readily. It is the most satisfactory and commonly used material for this type of application. The input node area then should be protected by a suitable dust cover to preserve the high leakage resistance.

Potting and encapsulation with high-resistivity resins may be considered as an alternative. In selecting the potting compound, consideration must be given to its volume resistivity, dissipation factor, dielectric constant, and curing temperature and cycle. The curing temperature must be compatible with the temperature rating of the electrical components to be potted.

5.5.3 Manual Controls

Manual control provides the means for the operator or process engineer to manipulate the position of the valve or control element manually. It is commonly used during the start-up of a process or when an emergency condition exists such that the process must be brought to a safe condition without the use of the computer. It is the most direct and simple form of back-up.

Figure 5-58 shows a simple manual control circuit. It consists of a DC voltage source and a single transistor amplifier. This circuit is usually used at the output of the memory amplifier through some switching

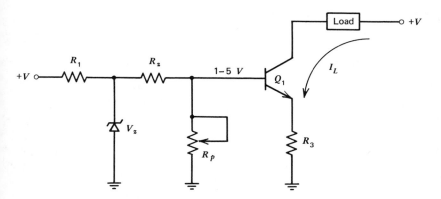

Fig. 5-58 Simple manual control.

* Registered trademark, E. I. Dupont de Nemours.

arrangement. Before the process load is switched to the manual control circuit, a "balance-before-transfer" operation is required for bumpless transfer.

Another manual control can be implemented at the input to the memory amplifier and is shown in Fig. 5-59. It consists of two pushbutton switches and a resistor R. When the "up" switch is pushed, the storage capacitor voltage is slowly increased with a resultant increase in the output current. When the "down" switch is pushed, the storage capacitor voltage is slowly decreased, as is the output current. The rate of charge or discharge is determined by the RC time constant. The transfer to manual control is bumpless since the last value is always stored on the capacitor.

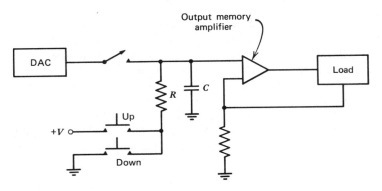

Fig. 5-59 "Up-down" manual control.

5.6 FALLBACK ANALOG CONTROLLER STATIONS

The computer-manual stations discussed in the previous section are satisfactory for most noncritical plant control loop applications. However, they are not adequate for critical back-up applications because they cannot adequately account for dynamic process changes. In the petroleum and chemical industry, for example, 10–30% of the control loops are typically considered critical and, therefore, fallback analog controller stations are required.

Figure 5-60 shows a typical process loop using a fallback controller station. There are three operating modes: normal DDC (direct digital control) mode, fallback-to-analog controller mode, and manual mode.

In the normal DDC mode, the output memory amplifier receives the command signal V via the time-shared DAC. The amplifier produces a

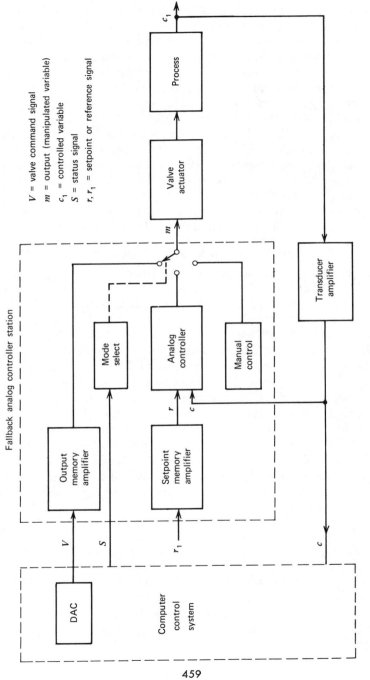

V = valve command signal
m = output (manipulated variable)
c_1 = controlled variable
S = status signal
r, r_1 = setpoint or reference signal

Fig. 5-60 Typical process loop using fallback and controller station.

459

current output or manipulated variable m, which is proportional to the command signal. The output current is sent directly to the valve actuator or final control element and it, in turn, controls the process or controlled variable c_1. This variable is sensed by the transducer amplifier, which produces a proportional process variable or measured variable c. One path of c returns to the computer to complete the DDC process loop. The other path of c goes to the fallback controller.

In the event the computer is not available, a status signal S from the computer initiates the fallback-to-analog controller mode. Automatically, the station output m switches to the analog controller and the controller takes over the process control burden from the computer. The signals r and r_1 are equivalent to a setpoint signal which is the desired setting of the process variable.

The station also can be switched into a manual mode. In this mode, the station output is manipulated manually by the operator.

5.6.1　Analog Electronic Controllers

This section is concerned with the functions and principal types of control provided by analog electronic controllers. The prime function provided by the analog controller is the maintenance of the process variable c at some desired setting, the setpoint r. This task is accomplished continuously by receiving the process variable c and comparing it with the setpoint r, thereby generating an error or difference signal $r - c$ between the setpoint and the process variable. The magnitude of the control action is determined by performing analog calculations on this error signal. This generates a control signal that restores the process variable to the setpoint value.

From this description, it is seen that the controller accepts two input signals, r and c, and produces one output signal, m. There are three basic control actions defined by the relationship between these signals: proportional, integral, and derivative. The proportional control action provides a continuous linear relationship between the output and the input error. Mathematically,

$$m = K_p(r - c) \qquad (5\text{-}40)$$

where K_p is the proportional gain. This control action is seldom used alone since an error must exist if the manipulated variable is to have a finite value. The magnitude of the offset owing to proportional control varies inversely with the proportional gain for a given output.

Another term frequently used to describe the proportional control action is "percent proportional band." This is related to the proportional gain by the expression:

$$PB = \frac{100}{K_p} \tag{5-41}$$

where PB is the percent proportional band.

The second control action is integral control in which the controller output is proportional to the time integral of the error input, namely,

$$m = \frac{1}{T_I} \int (r - c)\, dt \tag{5-42}$$

where T_I is the "integral time" in seconds. Since the output from the integral control action is always changing as long as there is an error, pure integral action is rarely used in practice.

The combination of proportional and integral control action, which is usually called two-mode control, is the most commonly used control scheme. The error or offset is completely eliminated because any disparity between the setpoint and the process variable gives a slow continuous correction in the proper direction to eliminate the offset. In this combined control action, "reset rate" is commonly used to measure the integral response. The reset rate is defined as the number of times per minute that the integral action duplicates the proportional action caused by a given error input signal. In terms of the other constants,

$$\text{reset rate} = \frac{60}{(K_p T_I)} \text{ repeats/min} \tag{5-43}$$

where K_p and T_I are the constants previously defined. The reciprocal of Eq. (5-43) is the time interval in minutes during which the integral action duplicates the proportional action.

The third control action is the derivative action in which the controller output is proportional to the rate of change of the error input signal:

$$m = T_D \frac{d(r - c)}{dt} \tag{5-44}$$

where T_D is the "derivative time" in seconds.

The derivative control action is often called "rate" or "preact" control. This action produces an output signal proportional to the rate of change of the error and in a direction to correct the change. It cannot be used alone because it does not recognize a setpoint. Therefore, it is used in conjunction with proportional control or with proportional and integral control. This latter mode is often called three-mode control.

"Rate time" is commonly used to specify the derivative action. It is the time interval by which the rate action leads the action resulting from proportional control alone. The rate time is measured in minutes and is defined as

$$\text{rate time} = \frac{T_D}{(K_p\,60)} \text{ min} \qquad (5\text{-}45)$$

A general process has inertia, capacity (energy storage), resistance, and transport time lags which determine what the relationship between the process variable and the manipulated variable should be for a certain setpoint. Controllers are equipped with a number of tuning adjustments for each of the control actions to cover a wide range of applications. The typical adjustments available are proportional band—5–300%; reset—0.01–30 min/repeat; and rate time—0.01–10 min.

Virtually all analog electronic controllers are designed with the capability of either "direct" or "reverse" control action. Direct control action is defined as an increase in the manipulated variable when the setpoint is increased. Reverse control action is defined as a decrease in the manipulated variable when the setpoint is increased. These two control actions are necessary to account for the sign of the process gain.

5.6.2   Analog Electronic Controller Configurations

A typical electronic controller consists of three basic circuit elements: the input network, the operational amplifier, and a feedback network, as shown in Fig. 5-61. The input network receives the two input signals, setpoint r and process variable c. The heart of the controller is the low-drift, high input impedance, high open-loop gain operational amplifier. With these properties, the control function is dependent only on the input and feedback network.

Generally, the operational amplifier is one of three distinct types: a magnetic amplifier, a modulator-demodulator AC carrier transistor amplifier, or a DC direct-coupled transistor amplifier. The majority of the present commercially available controllers are of the second type. This amplifier is very complicated and is costly to manufacture. The third type of amplifier offers simplicity and reliability with equal performance

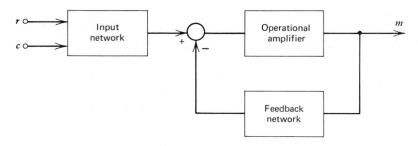

Fig. 5-61   Typical analog electronic controller.

and lower cost. As integrated circuit technology continues to develop, the third type of controller design will result in even more advantages and probably will be used extensively in future designs.

Figure 5-62 shows several controller configurations based on a DC direct-coupled transistor operational amplifier. Figure 5-62a is a conventional design in which two amplifiers are required. The amplifier A_1 is a unity gain differential amplifier, whereas A_2 is connected in an inverting configuration. The output of A_2 has the form:

$$m = -\frac{Z_F}{Z_1}(c - r) \tag{5-46}$$

Different control functions are generated by choosing the impedance functions Z_1 and Z_F. For example, if $Z_1 = R_1$ and $Z_F = R_2 + 1/sC_2$, a two-mode control action is obtained. In Laplace transform notation, Eq. (5-46) becomes

$$
\begin{aligned}
M &= \frac{R_2}{R_1}\left(1 + \frac{1}{sC_2R_2}\right)(R - C) \\
&= K_p\left(1 + \frac{K_I}{s}\right)(R - C)
\end{aligned}
\tag{5-47}
$$

where M = Laplace transform of m, the manipulated variable
R = Laplace transform of r, the setpoint
C = Laplace transform of c, the process variable
$K_p = R_2/R_1$ = proportional gain
$K_I = 1/C_2R_2$ = reset rate (repeats/sec)

Figure 5-62b is a simpler circuit which requires only one operational amplifier. It can be shown by using superposition that the controller output has the form

$$m = \frac{Z_F}{Z_1}(r - c) + r \tag{5-48}$$

Although Eq. (5-48) is slightly different than the conventional control function expressed in Eq. (5-46), it is still applicable for control purposes. If the process variable c is the varying parameter and the setpoint r is constant, then Eq. (5-48) is equivalent to Eq. (5-46) so far as the control action is concerned.

Figure 5-62c shows a controller employing a source feedback technique in which the negative feedback is obtained from the output through the feedback network Z_F to the source of the FET Q_1. In this configuration, the first stage contains two FETs, Q_1 and Q_2, and two separate current sources I_1 and I_2. It can be shown that the controller is also described by Eq. (5-46).

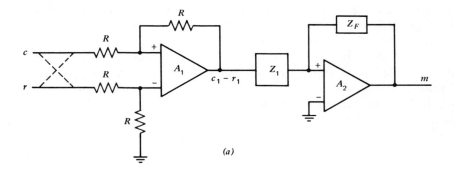

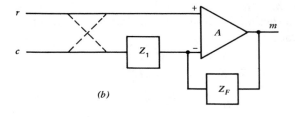

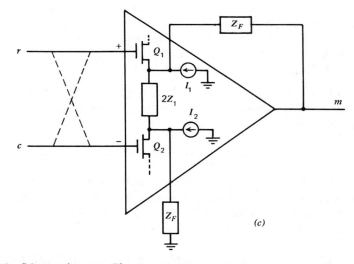

Fig. 5-62 DC operational amplifier controller configurations.

All the configurations in Fig. 5-62 are shown for direct control action. Reverse control action can be obtained by simply interchanging the two input signals as indicated by the dotted lines.

5.6.3   Fallback Analog Controller Station Design

This section deals with some practical considerations involved in the design of fallback analog controller stations. Depending on the feature and performance requirements, the station can take a number of different forms centered around several basic circuit building blocks. Some of these basic circuits such as analog storage, low-leakage multiplexing switches, constant current output, manual control, two-mode analog controller, and man-machine interface are discussed in previous sections. Other requirements such as setpoint memory, setpoint fallback mode, bumpless transfer, and emergency power back-up are discussed in the remainder of this section. Finally, a fallback analog controller station design is given to illustrate the design principles.

5.6.3.1   Setpoint Memory Amplifier.   The setpoint memory amplifier is used to store the setpoint so that it is available to the controller in the fallback mode. The design and storage requirement of this amplifier are similar to those of the output memory amplifier discussed in Sec. 5.5.2 except that unity negative feedback is employed. The amplifier can be the familiar voltage follower configuration shown in Fig. 5-63.

Various setpoint fallback modes are possible: (1) Fallback to a predetermined safe value; (2) fallback to last process variable value; (3) fallback to a computer updated setpoint (setpoint tracking); and

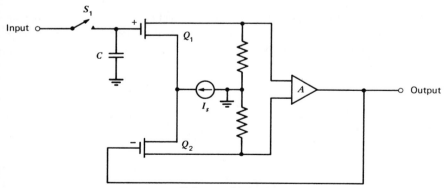

Unity negative feedback

Fig. 5-63   Typical setpoint memory amplifier.

(4) fallback to last process variable initially, and then decay to a predetermined safe value. The choice of the particular fallback mode is dependent on the user's control philosophy, plant operating requirements, process loop dynamics, economic considerations, and other factors.

5.6.3.2 Bumpless Transfer. One of the most important considerations in designing a fallback controller station is how to achieve bumpless transfer between the various operating modes. Bumpless transfer is required in the following mode transitions: DDC to fallback controller; fallback controller to DDC; DDC to manual; and manual to DDC.

A bump is equivalent to a step change or a proportional control action occurring in the controller station output. Such an occurrence is highly undesirable because it can upset the process loop and, in turn, the total process.

In a practical situation resulting from imperfect electronic components and other factors, there is an observable step change in the output at the moment of transfer. It is generally accepted that if the step change is less than 1% of full scale and of short duration, the transition can be considered as bumpless.

5.6.3.3 Emergency Power Back-up. At present, unscheduled power-line failure is a real possibility in most industrial power distribution systems. The consequences of the power failure in the process plant are often serious and mean that back-up must be provided for emergency power.

Generally, the power for process instruments, including fallback analog controller stations, can take either an AC or a DC form. It is desirable to have all circuits operating from a single voltage level, such as 24 V DC, since this results in lower power back-up costs and a higher reliability because fewer electronic components are involved.

Figure 5-64 illustrates a commonly used scheme for providing power under emergency conditions. The battery is maintained fully charged by the trickle charger. Battery voltage regulation is required to account for

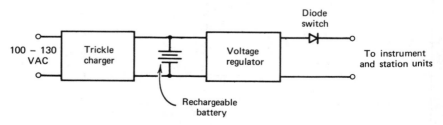

Fig. 5-64 Battery emergency back-up power system.

the load variation. In the event of power failure, the normal supply voltage decreases and the diode switch automatically transfers the load to the back-up battery supply. The battery capacity is typically limited to only a few minutes, but this is sufficient for many cases.

5.6.3.4 Fallback Controller Station Design Example. Figure 5-65 illustrates a fallback controller station design. The circuit design is best understood by describing the main modes of operation.

In the normal DDC mode, the station receives a command signal from the DAC, switch S_1 is closed, and the command signal is stored on capacitor C_1. Diode switches D_1 and D_2 are ON (D_3 and D_4 are OFF) so that the DDC amplifier, Q_1 and Q_2, is connected to the output amplifier A_1. The output current is sensed by the resistor R_F and is fed back to the gate of Q_2. This is the same operation as the output memory amplifier discussed in Sec. 5.5.2.

Under the DDC mode, the setpoint r and the process variable c are continuously applied to the inputs of the differential amplifier A_3. Switch S_2 is in the upper position so both gates of the control amplifier (Q_3, Q_4) see the error signal $r - c$ and the amplifier is prebalanced. Diode switches D_5 and D_6 are OFF (D_7 and D_8 are ON) so that the control amplifier output is disconnected from the output amplifier A_1. The resistor R_1 is the input network, and R_2 and C_2 are the feedback network which generate the two-mode control action. The reset capacitor C_2 is charged on one side to the error signal and on the other side to the command signal v_o. Capacitor C_4 tracks the error signal.

In the event the computer is unavailable, the circuit switches into the fallback controller mode. A status signal comes from the computer and the following events take place simultaneously: D_1 and D_2 turn OFF, D_3 and D_4 turn ON, D_5 and D_6 turn ON, D_7 and D_8 turn OFF, and S_2 is switched to the lower position. As a result, the control amplifier and the output amplifier take over the process control burden and provide two-mode control. Since the capacitors C_2 and C_4 track the respective signals as described, the transfer from DDC to fallback mode is bumpless. The voltage on C_4 discharges slowly through R_4 to ground. The output current I_o holds its last value if the error signal is at null and changes smoothly towards a new value if the error signal is not at null at the moment of transfer. The presence of R_4 and C_4 in no way affects the two-mode control action of the control amplifier.

The introduction of the setpoint amplifier A_2 provides setpoint fallback mode flexibility. It can provide fallback to the computer updated setpoint, to the last process variable, or to a predetermined safe value. Both direct and reverse control action are achieved by interchanging the two

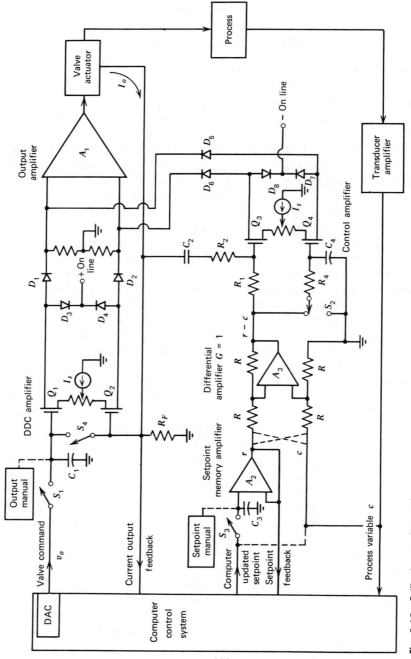

Fig. 5-65 Fallback controller station.

468

input signals to the differential amplifier as indicated by the dotted lines. Current output and setpoint feedback signals are readily available, as shown. Finally, the entire circuit is operated from a single 24-V power supply to ease the problem of emergency power back-up.

REFERENCES

Cundall, C. M., "Backup Methods for DDC," *Control Eng.,* April 1965, pp. 98–103.

Emmons, S. P., and H. W. Spence, "Very Low-Drift Complimentary Semiconductor Network dc Amplifiers," *J. Solid-State Circuits,* September 1966, pp. 13–18.

Hoffait, A. H., and R. D. Thornton, "Limitations of Transistor D-C Amplifiers," *Proc. IEEE,* February 1964, pp. 179–184.

Kennard, A. G., "ASAP—An Automated Statistical Analysis Program," *IBM Tech. Rep.,* TR 00.1142, June 22, 1964, International Business Machines Corp., SDD Laboratory, Poughkeepsie, New York.

Lombardo, J. M., "The Case for Digital Backup in Direct Digital Control Systems," *Chem. Eng.,* July 3, 1967, pp. 79–84.

Maxfield, R. R., and C. J. Steele, "Analog Storage with Field Effect Transistors," *Proc. 1st Ann. IEEE Comput. Conf.,* Chicago, Ill., September 1967.

Morley, R. A., and C. M. Cundall, "The Ferranti System and Experience with Direct Digital Control," *1965 IEEE Int. Conv. Rec.,* Part III, March 22–26, 1965, pp. 90–100.

Nagy, G., "A Survey of Analog Memory Devices," *IEEE Trans. Electron. Comput.,* August 1963, pp. 388–393.

Newcomb, R. W., *et al.,* "Papers on Integrated Circuit Synthesis," *U.S. Gov. Rep. AD653096,* June 1966, pp. 73–85.

Rajchman, J. A., and A. W. Lo, "The Transfluxor," *Proc. IRE,* March 1956, pp. 321–332.

Rudin, M. B., *et al.,* "System/Circuit Device Considerations in the Design and Development of a D/A and A/D Integrated Circuits Family," *1967 Int. Solid-State Circuits Conf., Dig. Tech. Pap.,* 1967.

Sheingold, D. H., "Constant Current Source for Analog Computer Use," *IEEE Trans. Electron. Comput. (Corresp.),* June 1963, p. 324.

Walton, C. A., "The Transfluxor as an Accurate Analog Magnetic Memory," IBM Tech. Rep., TR 02.429, International Business Machines Corp., San Jose, Calif., June 1968.

Walton, C. A., and A. F. Leon, "A Practical Approach to DDC Backup Control," *Control Eng.,* June 1969, pp. 128–131.

Williams, P., "High Stability Direct Voltage Reference," *Electron. Lett.,* February 1967, p. 64.

Williams, P., "Low-Voltage Ring-of-Two Reference," *Electron. Eng.,* November 1967, pp. 676–679.

Williams, P., "A DC Reference Voltage With Very High Rejection of Supply Variations," *Proc. IEEE,* January 1968, pp. 118–119.

"The Hot Carrier Diode Theory, Design and Application," *Hewlett-Packard Appl. Note No. 907,* May 15, 1967.

BIBLIOGRAPHY

Arthur, A. A., "Digital-to-Analog Conversion Ladders," *Electro-Technol.*, November 1964, pp. 60–64.

Baker, R. P., and J. Nagy, Jr., "An Investigation of Long-Term Stability of Zener Voltage References," *IRE Trans. Instrumen.*, September 1960, pp. 226–231.

Banga, J., "Zener Diodes and Their Application in Reference Units," *Br. Comm. and Electron.*, October 1961, pp. 760–764.

Baron, R. C., and T. P. Bothwell, "Performance of a Transistor as High-Accuracy Switch," *IRE Trans. Circuit Theory*, March 1961, pp. 82–84.

Bhola, S. K., and R. H. Murphy, "Low Drift Design of Differential DC Amplifiers," *Electron. Eng.*, October 1966, pp. 659–661.

Blauvelt, D. H., and A. S. Robinson, "Electronic Memory in Automatic Flight Control Systems," *Natl. Autom. Control Conf.*, Dallas, Texas, 1959.

Blecher, F. H., "Transistor Circuits for Analog and Digital Systems," *BSTJ*, March 1956, pp. 295–332.

Bright, R. L., "Junction Transistors Used as Switches," *Trans. AIEE (Comm. and Electron.)*, March 1955, pp. 111 ff.

Colin, D., "High-Speed Techniques in Analog-to-Digital Conversion," *Electron. Instrum. Dig.*, July 1966, pp. 49–54.

Davies, D. L., and S. F. Miles, "The DC Gain of an Operational Amplifier," *Electron. Eng.*, April 1962, pp. 249–251.

Dawson, T. P., and G. C. Taylor, "A Novel Voltage Reference Source," *Electron. Eng.*, February 1967, p. 113.

Demrow, R. I., "Narrowing the Margin of Error," *Electron.*, April 15, 1968, pp. 108–117.

Demrow, R. I., "Protecting Data from the Ground Up," *Electron.*, April 29, 1968, pp. 58–64.

Donaldson, R. W., "Sensitivity Analysis of a Digital-to-Analogue Multiplier and a Weighted-Resistor Digital-to-Analogue Converter," *Electron. Lett.*, October 1967, pp. 447–448.

Eicke, W. G., Jr., "The Operating Characteristics of Zener Reference Diodes and Their Measurements," *ISA Trans.*, April 1964, pp. 93–99.

Eimbinder, J., *Linear Integrated Circuits: Theory and Applications*, Wiley, New York, 1968.

Evans, A. D., "IC's End the 'Driver Gap' in FET Analog Signal Switching," *Electron. Des.*, November 11, 1966, pp. 50–54.

Fattal, L., "Field Effect Transistors as Choppers," *SCP and Solid-State Technol.*, April 1964, pp. 13–18.

Fischer, E. H., "Junction Field-Effect Transistor Characteristics Relevant to Analog Switch Circuits," *U.S. Gov. Rep. AD656889*, April 1967.

Giles, J. N., *Fairchild Semiconductor Linear Integrated Circuits Application Handbook*, Fairchild Semiconductors, Mountain View, Calif., 1967.

Goodstein, L. P., "Two Electronic Analog-to-Digital Converters," *U.S. Gov. Rep. AD446694*, February 1964.

Gosling, W., *Field Effect Transistor Applications*, Wiley, New York, 1965.

Gulbenk, J., and T. F. Prosser, "How Modules Make Complex Design Simple," *Electron.*, December 28, 1964, pp. 50–54.

Hanson, C. C., "Core Memory Stores Analog Signals," *Control Eng.*, September 1967, pp. 87–90.

Harris, J. N., and R. S. Orr, "Input Current Compensation for Transistor Operational Amplifiers," *Simulation*, June 1966, pp. 362–374.

Harris, R. A., "The Computer-Actuator Interface," *Control Eng.*, September 1966, pp. 110–113.

Hendrie, G. C., and R. W. Sonnenfeldt, "Reliability Still Means Backup," *Control Eng.*, September 1966, pp. 131–135.

Ida, E. S., and D. W. St. Clair, "Cutting Controller Costs," *Control Eng.*, October 1964, pp. 86–91.

Hoeschele, D. F., *Analog-to-Digital/Digital-to-Analog Conversion Techniques*, Wiley, 1968.

Jennings, R. R., "Source and Load Impedance: Effects on Closed-Loop Characteristics," *Electro-Technol.*, September 1966, pp. 46–52.

Karplus, W. J., and J. A. Howard, "A Transfluxor Analog Memory Using Frequency Modulation," *Proc. Fall Joint Comput. Conf.*, 1964, pp. 673–683.

Koppel, H. H., "A Solid State Electronic Control System for Industrial Processes," *Electr. Eng.*, September 1961, pp. 682–688.

Korn, G. A., "Progress of Analog/Hybrid Computation," *Proc. IEEE*, December 1966, pp. 1835–1849.

Magigan, J. R., "Understanding Zener Diodes," *Electron. Ind.*, February 1959.

Middlebrook, R. D., "Design of Transistor Regulated Power Supplies," *Proc. IRE*, November 1957, pp. 1502–1509.

Morgenstern, L. I., "Temperature-Compensated Zener Diodes," *Semicond. Prod.*, April 1964, pp. 25–29.

Nagy, J., Jr., "Zener Diode Power Supplies as Fundamental Voltage References," *18th Ann. ISA Conf.*, Prepr. 28.3.63, September 1963.

Pearman, C. R., and A. E. Popodi, "How To Design High-Speed D-A Converters," *Electron.*, February 21, 1964, pp. 28–32.

Richman, P., *Characteristics and Operation of MOS Field-Effect Devices*, McGraw-Hill, New York, 1967.

Rigg, B., "The Characteristics and Applications of Zener Diodes," *Electron. Eng.*, November 1962, pp. 736–743.

Rudin, M. B., "New IC Sparks Low Cost A/D–D/A Conversion," *Electron. Des. News*, May 1968, pp. 50–57.

Schmid, H., "Current Dividers Convert Digital Signals into Analog Voltages," *Electron.*, November 14, 1966, pp. 142–148.

Schmid, H., "Electronic Analog Switches," *Electro-Technol.*, June 1968, pp. 35–50.

Schuck, E., "Nomograph Finds Output Voltage Error," *Electron.*, November 14, 1966, pp. 126–127.

Schulz, D. P., and D. J. Dooley, "Integrating Space Telemetry Systems with Compatible Thin Films on Silicon," *Electron.*, June 26, 1967, pp. 111–122.

Sconzo, S. A., "Reference Diodes—The State-of-the-Art," *Electron. Des. News*, October 1966.

Shipley, M., "FET Low-Level Choppers," *Electron. Equip. Eng.*, February 1964, pp. 62–65.

Shipley, M., "Analog Switching Circuits Use Field-Effect Devices," *Electron.*, December 28, 1964, pp. 45–50.

Soule, L. M., Jr., "Control Instrumentation/Computer Compatibility," *Instrum. and Control Sys.*, May 1968, pp. 111–117.

Stata, R., "Minimizing Integrator Errors," *Electro-Technol.*, October 1968, pp. 46–50.

Stata, R., "Operational Amplifiers, Part I: Principles of Operational Amplifiers and Analysis of Errors," *Electromech. Des.*, September 1965, pp. 48–58.

Stata, R., "Operational Amplifiers, Part II: Inverting, Non-Inverting and Differential Configurations," *Electromech. Des.*, November 1965, pp. 38–43.

Stata, R., "Operational Amplifiers, Part III: Survey of Commercially Available Operational Amplifiers," *Electromech. Des.*, February 1966, pp. 40–51.

Stata, R., "Operational Amplifiers, Part IV: Offset and Drift in Operational Amplifiers," *Analog Devices Appl. Note*, 1966.

Strifle, J., "A Current Mode DA Converter," *U.S. Gov. Rep. AD619778*, August 1968.

Susskind, A. K. (Ed.), *Notes on Analog-Digital Conversion Techniques*, Technology Press, Cambridge, Mass., and Wiley, New York, 1957.

Swartout, C. J., "The Design of a Compatible Electronic Control System for the Process Industries," *17th Ann. Instrum.—Autom. Conf. and Exhibit.*, October 15–18, 1962, New York City, Pap. No. 32.3.62.

Sylvan, T. P., "An Integrated Reference Amplifier for Precision Power Supplies," *General Electric Appl. Note 90.15*, April 1963.

Tobey, G., "Parameter Primer for Op-Amps," *Electron. Des. News*, April 1967.

Todd, C. D., *Junction Field-Effect Transistors*, Wiley, New York, 1968.

Uphill, A. J., "Use of Transistor as a Switch in Precise Analog-Digital-Converter Circuits," *Proc. IEE*, August 1964, pp. 1385–1392.

Wallace, C. L., "Optimizing the Performance of Silicon Reference Elements," *Electron. Ind.*, March 1965, pp. 95–97.

Wenham, D. G., "The Design of Direct Voltage and Current Stabilizers Using Semiconductor Devices," *Proc. IEE*, April 1960, pp. 1384–1393.

Williams, L. L., "An Analysis of a Transfluxor Analog Storage System," *IBM Tech. Rep., TR 02.357*, International Business Machines Corp., San Jose, Calif., August 1965.

Wilken, A. C., "A Precision Analog Memory With Non-Destructive Readout Using Multiaperature Ferrite Cores," *SC-R-65-959*, Sandia Corp., Albuquerque, New Mexico, August 1965.

Young, F. M., "Factors Limiting A-D Conversion—State of the Art," *Data Sys. Eng.*, May 1964, pp. 35–40.

Zalkind, C. S., "An All-Electric Approach—Pulse Duration Control," *Control Eng.*, February 1967, pp. 74–77.

Ziegler, J. G., and N. B. Nichols, "Optimum Settings for Automatic Controllers," *Trans. ASME*, November 1942, pp. 759–768.

"Digital-to-Analog Converter Operates From Low Level Inputs," *Electron. Des.*, July 4, 1968, pp. 108–112.

"Evaluation of MOS Field-Effect Transistors in Analog Gating Applications," *U.S. Gov. Rep. AD474579,* 1966.

"The Memistor," *Electron. Prod. Mag.,* January 1963, pp. 50–51.

"Ultra-Stable Semiconductor Voltage References," *Solid-State Des.,* March 1963, pp. 16–17.

Zener Diode Handbook, Motorola Semiconductor Products, 1967.

Chapter 6

Digital-Input/Output Subsystems

J. PAUL HAMMER

The digital- and analog-input/output subsystems of a process control computer are the primary features that distinguish it from a general-purpose computer. Through these subsystems the process control computer monitors and controls the process to which it is connected. The analog-input subsystem provides the capability of measuring parameters which are represented by analog voltages or currents. The analog-output system provides a means of controlling devices that require a constant or varying electrical signal which is proportional to a numerical quantity stored in the computer.

Both of the analog subsystems deal with quantities that are analog in nature; that is, either the input or output quantity can take on any value within predefined limits. The process, however, is also characterized by parameters that are digital in nature, that is, parameters that take on only a limited number of discrete values. Quite often, these parameters are binary quantities that can be represented by two states such as ON or OFF, IN-LIMITS or OUT-OF-LIMITS, or HIGH or LOW. Non-binary digital quantities such as those represented by a ten-position switch can be coded in a manner appropriate for representation by two-state devices by suitable contact or circuit design. It is convenient to represent digital quantities in the process in binary form as this is compatible with the internal data representation of a digital computer.

To complement the digital parameters that characterize process states, there are output parameters that represent the desired state of the control devices of a process. For example, a motor can be ON or OFF, a damper or valve can be OPENED or CLOSED, and a solenoid can be IN or OUT. Other controls not inherently binary in nature often can be controlled by binary signals. For example, a motor speed controller can be designed such that closing one or more contacts results in one of a set of predetermined speeds. Similarly, binary switches can be used to provide several discrete positions of a damper or valve. As in the case of digital input, the binary representation is convenient because it is compatible with the internal data structure of the digital computer.

In order to control and sense devices that are digital in nature, the control computer must be equipped with special features that interface to the process instrumentation. The digital-input and -output subsystems provide this capability. Digital-input features provide a means of detecting the state or a change in the state of a contact or voltage level as well as a means for counting pulses and determining the duration of pulse signals generated by process equipment or instrumentation. Digital-output features provide electronic or electromechanical switches which may be used to control process actuators and indicators. Pulse output circuits which provide either a fixed number of pulses or a variable pulse duration signal are also provided.

The importance of digital-input and -output features can hardly be overstated when considering a control computer. In typical installations, the number of digital inputs and outputs significantly exceeds the number of analog points. This is because the control of the process and its instrumentation are largely digital, even though the process itself is analog in nature. Digital control is used in the process for the same reason that it is in computers; namely, it greatly simplifies signal handling and provides immunity to noise and to other accuracy problems typically associated with analog control. For example, consider the difference in complexity of a variable speed motor controller compared to a simple ON-OFF contactor.

In this chapter, some of the design considerations associated with the digital-input and -output subsystems are discussed. The circuit design problems associated with sensing circuits, output circuits and logic design problems are considered. System requirements relating to the organization and control of the subsystems are also discussed.

6.1   DIGITAL-INPUT SUBSYSTEM

The basic function of the digital-input (DI) subsystem of a process control computer is to sense events in the process or equipment connected to the computer system. In general, these events are binary in nature or are converted into a binary code by equipment external to the DI subsystem. There are two primary characteristics associated with the digital-input functions. One is the form of the input signal; it may be either a voltage, current, or resistance change. The other is the signal parameter which is of interest in monitoring the process. This parameter may be either the mere existance of a signal, its duration, or the number of events occurring in a given period of time. In providing the basic digital input function, therefore, the subsystem must provide a variety of sensing circuits sensitive to the various signal forms associated with binary parameters in the process.

In this section, the design of the DI subsystem is considered in detail. The general subsystem configuration, the design parameters for signal conditioning and sensing circuits, and the subsystem control requirements are discussed.

6.1.1   Digital-Input Subsystem Configurations

There are a number of subsystem configurations which can satisfy the digital-input functional requirements. The selection of a particular configuration is dependent on the desired performance, the functional characteristics to be achieved, and the design of the computer to which the subsystem is attached. The block diagram of Fig. 6-1, however, represents a common configuration and illustrates the major components of typical subsystems.

The subsystem is connected directly to the CPU or it may be connected via a data control channel. Direct attachment is most common in smaller or special-purpose computer systems. The characteristics of the subsystem control depend on the design of the CPU or channel. In the case of a channel connection, the direct and indirect modes of channel operation discussed in Chap. 2 are often available, with the choice between modes being a user-specified option. The subsystem control performs the necessary functions to maintain communication between the subsystem

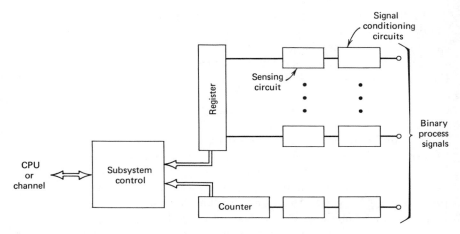

Fig. 6-1   Digital-input subsystem configuration.

and the CPU. It also provides address decoding and other functions associated with the sensing of the digital-input signals. Special functions, such as masking or comparison, may be incorporated in the subsystem control to reduce the computing burden of the CPU.

As noted previously, the two basic functions provided by the DI subsystem are the detection of single events and the counting of multiple events. Both of these functions involve a register: a counting register in the case of counting and a simple holding register for temporary storage of the detection data. As shown in Fig. 6-1, both types of registers may be provided under the control of a single subsystem control unit. In general, the number of bits in the register or counter is equal to the number of bits in the data path of the machine. For example, if the data path is designed for the manipulation of 16-bit words, the DI register or counter is usually 8- or 16-bits in length. Provisions are provided in some DI subsystems to gang two or more counters to provide a greater counting capacity.

The interface between the process signals and the logic signals necessary to control the states in the register or counter involves signal level shifting and/or conversion. The input signals representing the states in the process are typically either voltage levels or switch contact closures. In a few cases, current levels may also be used. The sensing circuit, however, is usually a voltage sensing circuit. Current signals can be converted to voltage signals by using a shunt resistor. Similarly, contact closures can be converted into voltage signals by using a voltage source in series with the contacts and a current limiting resistor.

The sensing circuit is a threshold circuit whose output is a logic level signal representing a binary **1** when the input voltage exceeds a predefined magnitude. Otherwise, the output signal represents a binary **0**. (The relationship between the output and the input can, of course, be reversed with a binary **0** resulting when the threshold is exceeded.) In the simplest circuits, the voltage levels must be unipolar; that is, the input voltages representing the two states must be of a single polarity. Bipolar circuits which respond to signals of either polarity can be provided, however, and they have the advantage of providing greater flexibility in the use of the system.

In the design of the various components of the DI subsystem, the engineer must consider a number of factors relating both to the design of the circuit and to the manner in which it is to be used. Design considerations for the signal conditioning, conversion, sensing circuits, and isolation are discussed in the next section. The subsequent section considers the DI subsystem control.

6.1.2   Signal Conditioning and Sensing Circuits

Since switch contact and current inputs can be converted to voltage inputs through suitable signal conditioning, the voltage sensing circuit is the fundamental circuit design problem associated with the digital-input signals. The sensing circuit is a threshold circuit similar in function to the comparator circuit discussed in Chap. 3. The circuit provides output signals that are compatible with the computer logic signals in response to voltage signals derived from process instrumentation and equipment. The characteristics of the circuit are shown schematically in Fig. 6-2. When the input signal exceeds some threshold voltage V_2, the output corresponds to a binary **1**. If the input is less than a voltage V_1, the output

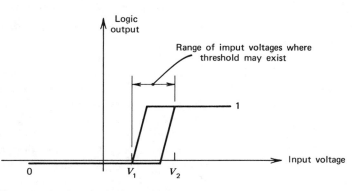

Fig. 6-2   Sensing-circuit characteristics.

is a binary **0**. An input voltage between V_1 and V_2 results in an output signal that is not specified; that is, it may be either at or between the voltages corresponding to the binary **0** and **1** and may not be reliably interpreted by the succeeding logic circuits.

One of the first considerations in the design of the sensing circuit is selecting the design points for the threshold limits V_1 and V_2. The primary consideration is the range of signals encountered in the application. One of the common applications is the sensing of logic level signals in equipment connected to the DI subsystem. For this purpose, it is desirable that the threshold region not include voltages around 0 V. This is because most families of logic circuit utilize a level near 0 V to correspond to one of the binary states **1** or **0**. Thus, V_1 should be selected as a nonzero value, typically about 1 V for common logic families.

Similarly, the voltage V_2 should be selected such that the circuit will detect the other logic state of common logic circuits. Typically, this state is represented by a voltage between 4–12 V. This suggests that V_2 should be about 2 V. With this selection, the threshold region corresponds to an input voltage range of 1–2 V. In contrast to the precise threshold required in the comparator applications of Chap. 3, the existence of this wide threshold region is a result of using less expensive and simpler circuits which nevertheless provide adequate performance.

The above discussion assumes that the logic circuits from which the input signals are derived have nonnegative output voltage levels. Some logic families, however, utilize negative levels so that corresponding negative values of V_1 and V_2 must be accommodated. Thus the designer either must provide two different circuits, one for each polarity, or a differential input circuit where either terminal can be referenced to ground. The latter alternative provides a greater flexibility for the user and minimizes the number of different features that must be provided by the vendor. A differential input circuit is provided automatically when transformer isolation is used. Principles of isolated DI circuits are discussed later in this section.

Logic level signals are only one type of signal likely to be encountered in the application of the DI subsystem. Process equipment often provides output levels in excess of the 12-V maximum signal commonly associated with the logic circuits. Power supplies associated with process instrumentation typically range from 20–45 V so that the maximum signal likely to be encountered is about 45 V. Assuming that the same circuit is to be used for these signals as is used for logic signals, the primary design concern is protecting the sensing circuit from damage owing to these relatively high voltages. A number of design alternatives are possible: Several inputs to the sensing circuit can be provided, each with a different resistive attenuation network to limit the voltage actually applied to the sensing circuit. This approach has the disadvantage of complicating the system installation procedure since either the input signal of each point must be specified prior to installation or a network of resistors selected at installation must be provided for each input point.

A second alternative is to provide voltage clamping which limits the voltages applied to the sensing circuit. In the simplest case, this can consist of diodes connected between the input and voltage sources equal to the desired clamping voltages. Current limiting resistors may be necessary to ensure that the diodes are not damaged by the resultant current flow.

The third and most flexible alternative is to provide a sensing circuit that accepts the necessary range of voltages without damage. The example circuit described later in this section utilizes this approach.

A second consideration in the design of the sensing and signal conditioning circuits for digital input is the control of noise. The sources of process noise and its control are discussed in Chap. 13. The digital-input signals may contain perturbations due to noise sources in the process and this can cause false indications in the DI level-change detecting program or a false count in the case of pulse counting. Low-pass filtering is generally required in the signal conditioning and/or sensing circuit to minimize the noise. Such filtering is also necessary when the circuit is used to sense contact closures since bouncing of the mechanical contacts produces a pulse train input which can result in a false indication or in numerous false counts.

Signal conditioning can be provided by using single-section RC filters at the input to the sensing circuit. The characteristics and design of this type of filter are discussed in Sec. 3.12. The use of a filter, however, can have the disadvantage of providing a variable delay in detection which is a function of the amplitude of the input signal relative to the amplitude of the threshold region. This is illustrated in Fig. 6-3, which shows the output of an RC filter for pulse inputs of different amplitudes. If V_1 and V_2 give the range of threshold levels of the sensing circuit, a large-amplitude pulse is detected sooner than a small-amplitude pulse. Conversely, the fall time of the large-amplitude pulse is such that detection of this change is delayed compared with detection of a small-amplitude pulse.

If the sensing circuit acts simply as an amplifier and signal clipper, filtering the signal after sensing avoids this variable delay. The output pulse from the sensing circuit is a constant amplitude so that the detection delays remain constant for the total input signal range.

The filter time constant depends on the application and the character-

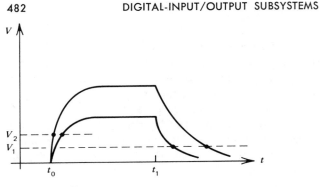

Fig. 6-3   RC filter response.

istics of the signal. For contact closure signals, time constants on the order of 2–5 msec generally provide adequate filtering for contact bounce. In sensing logic level circuits and electronic switches, a much shorter time constant is usually used. Values between 10–100 μsec are typical. If the same circuit is to be used for all types of signals, either a means of changing the filter time constant or different circuits should be provided.

Closely related to the time constant in the case of sensing logic circuits is the selection of the sensing circuit input impedance. To avoid signal delays and transmission line reflections, the input circuit should match the impedance of the input line. For high-speed logic, this usually requires an input impedance between 50–100 Ω. In order to accommodate higher voltage signals, however, a higher input resistance is desirable to limit current flow in the input circuit. An input resistance between 5–10 kΩ limits the current to a reasonable value at 50-V levels. These conflicting requirements can be satisfied either by providing different circuits for the two applications or by giving the customer a means of changing the input resistance at the time of installation.

A sensing circuit can also be designed with positive feedback to provide hysteresis within the threshold region and to guarantee a rapid output transition between logical states. This decreases somewhat the input sensitivity to noise by providing an overlap of the input amplitude ranges corresponding to the output binary states. It does not substitute for input filtering, however, when mechanical contact noise exists. The design example described later includes positive feedback in the sense amplifier to cause it to act as a bistable multivibrator or flip flop.

As in the case of analog-input signals, digital-input signals are also subject to common mode noise. A common mode signal has the effect of shifting the threshold level of the sensing circuit. If the common mode voltage is AC or transient in nature, the shifting threshold value can result in false triggering of the sense circuit. Conceptually, the solution to the common mode problem is the same for the digital-input subsystem as it is in the case of analog-input equipment discussed in Chap. 3 and Chap. 4. A differential sensing circuit corresponding to the differential amplifier of the analog-input subsystem can be used. The common mode potential must be limited to the breakdown potential of the isolating circuit.

Depending on the approach, isolation can be provided at several points in the digital-input subsystem. Referring to Fig. 6-1, the isolation can be provided following the signal conditioning or conversion circuit. This is typified by a transformer-coupled input or a differential sensing circuit. Transformer coupling at this point has the disadvantage of complicating the handling of voltage levels, although contact inputs are easily handled. The approach has the advantage that the input circuits are independent of each other and the polarity of the input signals can be reversed without regard for grounding problems.

Providing transformer isolation following the sensing circuit also provides independent input circuits but may require floating power supplies for the sensing circuit. For independence between channels, a separate power supply must be provided for each channel. Transformer coupling following the register or counter or at the subsystem interface to the CPU also requires floating power supplies. Owing to the sometimes substantial power requirements of the logic circuits, the floating supply may be quite costly. In addition, isolation of these points does not provide channel-to-channel independence unless differential input circuits are utilized. If they are, the channel-to-channel difference in common mode may be limited to a relatively low voltage even though the common mode potential limit between the subsystem ground and the computer ground may be several hundred volts.

Although the design of the voltage sensing circuit is the fundamental problem associated with the digital-input circuits, current and contact inputs also must be considered. Current signals are not commonly encountered in process control applications, but they are easily accommodated by converting the signal to a voltage level by means of a shunt resistor. The value of the resistor must be consistent with the driving capability of the current source and the input voltage range of the sensing circuit. By adjusting the size of the resistor, the effective current detection threshold can be altered.

Contact sense is provided by connecting the contacts between the input circuit and a voltage source, as shown in Fig. 6-4. The voltage applied to the input of the sensing circuit is governed by the attenuator formed by R_1 and the input resistance R_{in}. If the sensing circuit is designed to

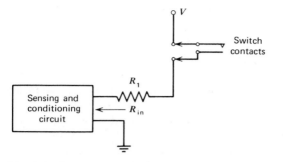

Fig. 6-4   Contact sense circuit.

handle voltages as high as the source voltage V, the external resistor R_1 is not required.

Although the sensing circuit may be designed with a relatively low threshold voltage to accommodate logic level signals, the voltage V is generally quite high to ensure reliable operation of the mechanical contact. The voltage across the contact and the current in the circuit at the time of contact closure must be high enough to break down any surface films or contaminants on the contacts. The voltage level necessary to do this is a function of the contact material and the surface condition. A voltage between 25–50 V at a current level of about 10 mA has been found sufficient for most heavy-duty switches used in process control applications. For light-duty relays, such as dry reed relays, a lower voltage and current level may be satisfactory. Contact protection in the form of an RC network may be required, and the user should be cautioned to consult the relay specifications for information.

The sensing circuit shown in Fig. 6-5 illustrates the design considerations discussed in the previous paragraphs.* A differential input provides moderate common mode isolation and a choice of input polarity and threshold range, without the complexity of carrier-operated transformer isolation. If either input is left open, connecting a DI signal to the other input provides a threshold range centered between $+V_{cc}$ and ground by selecting $R_5 = R_6$. The width of the threshold range, V_1–V_2 in Fig. 6-2, is controlled by the amount of input attenuation $R_3/(R_1 + R_3)$ and the hysteresis generated by positive feedback resistor R_7. The input attenuation is combined with a filter capacitor C to eliminate high-frequency input noise, the value of C to be selected by the user upon installation.

* The author is indebted to Dr. G. A. Hellwarth of IBM Corporation for this circuit example and explanatory information.

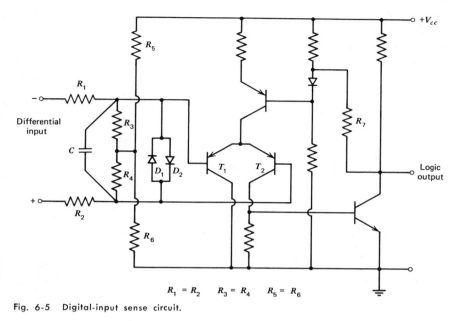

Fig. 6-5 Digital-input sense circuit.

The two shunt clipping diodes D_1 and D_2 aid in providing equal turn-on and turn-off delay when using the filter capacitor C, a problem described in Fig. 6-3, by limiting large-amplitude input signals.

If common mode rejection is required, one of the two inputs should be connected to a reference potential equal to the desired center of the threshold range, this potential being derived at the DI signal source. Thus any common mode signals will be connected on both the signal and the reference line and will be rejected by the differential pair of transistors T_1 and T_2. Either the positive or the negative input line may be used for the input signal, and by choice of reference potential on the other input, either polarity of input DI signal can cause a logical **1** output state. A rapid transition between output logic states is provided by the positive feedback resistor R_7, causing the circuit to act as a bistable multivibrator, and reducing the probability of an indeterminate output state existing at any given time.

6.1.3 Digital-Input Logic Design and Control

The signal conditioning and sensing circuits discussed in the previous section are, essentially, signal converters. They convert the voltage, current, or contact signal derived from the process into a logic signal that is compatible with the logic circuits used in the computer. This logic signal is used to control the state of a single bit position in a register or to control a counter. Under control of the computer program, the data word represented by the states in the register or counter can be transferred to the computer for further processing. In this section, the logical design of the registers, counters, and subsystem control logic is discussed.

The counter used in the digital-input subsystem is designed to count in a manner compatible with the internal data representation of the system; that is, if the computer uses a binary coded decimal (BCD) number representation, the counter is a BCD counter. The design of the counter itself is straightforward, and example designs can be found in most standard references on logic design. The number of bits in the counter is made equal to the number of bits in the computer word or to a fraction of the word length to simplify data manipulation in the subsystem and CPU. Provision for cascading several counters, thereby increasing the maximum count that can be accumulated, is easily provided by using the counter overflow signal from one counter as the input to another counter. The overflow signal can also be routed to the priority interrupt structure of the system. With this option, the occurrence of an overflow is detected automatically by the system and a programmed response can be initiated. For example, the overflow event might result in a constant being added to a location in storage or it might initiate a programmed control action.

One of the logic design considerations is associated with reading the counter. If an attempt is made to read the counter at the same time that a count is received at the counter input, erroneous data may be obtained owing to the transient state of the counter. The error is unpredictable since it depends on the initial and final states of the counter and on the characteristics of the logic circuits. For example, if the initial counter state is **0111**, the resulting data could be anything between **1000** and **0000**. In order to prevent this, logic circuits must be added that inhibit pulses during read operations or that inhibit read operations whenever pulses are received. To minimize interference with the subsystem timing, it is usually easier to inhibit input pulses during the reading operation. It is necessary, however, to store temporarily any counts received during the read operation if an accurate count is to be maintained. Since the read operation generally takes less time than the period corresponding to the highest counting frequency, it is only necessary to provide storage for one pulse. A simple logical delay circuit inserted between the output of the sensing circuit and the input to the counter provides the necessary storage function. The control signal which effects the transfer of data from the counter to the data bus activates the delay function. Following the read operation, the delayed count is added to the counter. If the control signal is not present, input pulses are passed directly to the counter input.

The counter input may be an AC-coupled circuit so that the transitions of the input signal actuate the counter. This type of input can be designed to operate with positive, negative, or positive and negative transitions. The AC coupling also requires a minimum transition time between the two states. The bistable characteristic of a sensing circuit with positive feedback tends to guarantee that the required risetime will be achieved. However, if the input signal passes too slowly through the indeterminant threshold region of a straight amplifier-type of sense circuit, the AC-coupled counter will fail to record the input event.

By using DC coupling, the counter is actuated whenever the input threshold level is exceeded. Depending on the sensing circuit design, positive and/or negative levels can be accommodated. Assuming that the input signals are binary and the sensing circuit is differential, only one polarity option need be provided since the input wiring can always be connected to provide the proper transition or level polarity.

Hardware pulse duration counters are a simple modification of the basic counter. This feature is used to measure the length of an input pulse. By ANDing the output of the sensing circuit and a pulse generator of known frequency, the accumulated count in the counter is a measure of the pulse length. The counter must be read, of course, before the duration of a succeeding pulse can be measured. An indication that the measurement has been completed can be provided by detecting the end of the input pulse and routing this indication to the priority interrupt circuits in the computer.

The register input features of the DI subsystem utilize the same sensing and signal conditioning circuits as those used for the counter input. The difference is that a single sensing circuit is connected to each bit in the register. The data sensed by this arrangement may be bit significant. That is, each bit may represent information independent of every other bit in the register. On the other hand, one or more bits in the register may represent a data entity whose length is equal to that of the basic computer word or a fraction thereof. The data format is generally not considered in the design of the subsystem since the register data are manipulated as an entity. The computer program provides the interpretation of the data according to the format defined by the programmer. Computer instructions and higher level language facilities are usually provided to assist in this interpretation process.

There are several options as to how the register bit is controlled by the output of the sense circuit. The register bit may always reflect that state of the input variable except, perhaps, when the input signal is in the indeterminant region of the threshold characteristic. In this mode, the register bit changes whenever the input state changes. It is possible, therefore, that the input state may change an even number of times between reading operations. In this case, the data transmitted to the CPU show no change even though several may have occurred. It would be possible, for example, that the state of the input might be oscillating and that this condition be undetected because the reading frequency matches the frequency of oscillation. Whether or not these undetected changes are significant depends on the application; the application programmer must be aware of this possibility and select his control algorithms accordingly.

A second alternative is to provide a "latching" register. In this configuration, the state of the register bit changes from, for example, a **0** to a **1** state when the input signal changes state. It remains in the **1** state until the register is reset by the program, even though the input signal previously may have reverted to its original state. Depending on the design of the register, the bit position may latch for an input transition from **1** to **0**, **0** to **1**, or *either* **1** to **0** or **0** to **1**. In any of these cases, the effect of the latching action is to trap the event. When the program reads the DI word, a simple test for zero indicates whether any of the input represented by the bits in the word have changed state since the last reading was taken. If they have, the appropriate programmed action can be initiated and the register reset to its quiescent state. This feature reduces the CPU burden since testing for zero is often provided as a machine-level instruction and is considerably easier than having to compare the data word with a predefined status word to determine whether a change has taken place.

The design of the logic between the register or counter output and the CPU or channel attachment is highly dependent on the functions to be performed, the design of the channel, and the format of the computer instructions. In general, however, this logic performs the interpretation of the instructions, the generation and distribution of timing and control signals within the subsystem, and the synchronization of data transfers between the subsystem and the CPU or main storage.

Primary design factors are the functional modes of operation that are to be implemented. These can range from a single read operation to sequential and nonsequential reading sequences with subsequent storage of data in sequential or nonsequential storage locations. Three modes indicative of the control of a digital-input subsystem are read, read sequential, and read nonsequential. Other modes are, in general, variations or extensions of these types of control.

The read operation is the simplest mode of control. The computer instruction specifies the read operation, the address of the register or counter to be read, and the destination of the data. In some cases, this latter piece of information is not included in the instruction explicitly but is determined by the design of the subsystem. For example, the data may always be returned to an input register or to the accumulator in the CPU.

When issued the Read instruction, the subsystem decodes the address of the register to be read and transfers the register data to an output bus. Once the transfer is accomplished, subsequent instructions either manipulate the data or store them according to data provided by the program. The logical design of the subsystem logic to perform this basic read operation is quite simple. All that is required are the necessary decoders and data paths. If the subsystem is connected to the CPU via a channel, some additional channel control logic may be required. This is generally minimal, however, since indirect channel control (cycle stealing) is rarely used in the execution of a simple Read instruction. This is because the instruction can be executed in one or two machine cycles and the advantages of cycle stealing are minimal.

The Read Sequential instruction allows the user to "scan" several groups of input points with a single command. Two addresses and a count are specified in the instruction and provisions to store these data are included in the subsystem, as shown in Fig. 6-6. The point group address specifies the first group of points to be read. The second address specifies the main storage address into which the first data word is to be transferred. The count specifies the number of groups of points to be read as a result of the instruction execution.

When the DI subsystem is selected to execute the Read Sequential instruction, the addresses and count are transferred into the appropriate registers in the subsystem control logic. This may involve several operational steps if the instruction is a two-word instruction because of the word size in relation to the address size. Once the data are stored, the subsystem decodes the point group address and transfers the DI register data onto a data bus. The data are then transferred to the channel along with the storage address to which they are to be transferred. When the transfer of the data is complete, the storage address and group point address registers in the subsystem are incremented to provide the next addresses and the count is decremented. If the count has not reached zero, the transfer of data from the next input group is initiated.

When the count is decremented to zero, several alternatives are possible, depending on the system design and the program: (1) The DI subsystem can terminate operation and revert to a "wait" state pending further

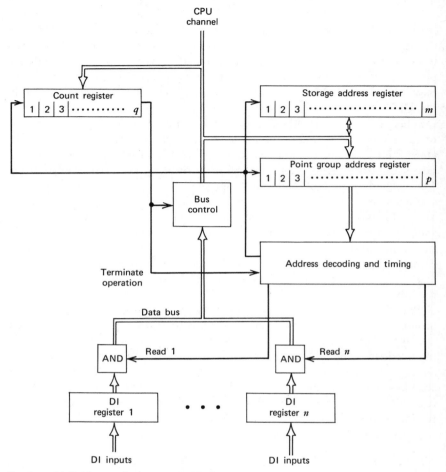

Fig. 6-6   Digital-input subsystem control logic.

instructions: (2) An interrupt signal can be generated to indicate to the programming system that the data transfers have been completed. The subsystem can then go into the "wait" state. (3) The read sequential operation may be automatically reinitiated in what is known as a "chaining" operation. In this operation, two new addresses and a new count are obtained from designated storage locations and the instruction is reinitiated using this new information without program or CPU intervention. (4) The subsystem may generate an interrupt and then chain. In many systems, more than one of these options are implemented. The option is selected by the programmer and is indicated to the subsystem through one or more indicator bits in the instruction word.

The difference between the read sequential and the read nonsequential operation is that in the latter case the point groups may be addressed in a random sequence. It is necessary, therefore, to provide the subsystem with the information needed to control the random addressing. This can be accomplished by using a table of addresses in main storage to specify the point groups to be read. The hardware and the operation are similar to that described above for read sequential. An additional register is required, however, to store the address of the table. Instead of the initial point group address, the initial address in the table is specified in the instruction and is stored in the subsystem. The subsystem then accesses storage and transfers the address of the point group into the point group address register in the subsystem. DI register data are then transferred to storage as described above. After the transfer is completed, the storage address and the table address registers are incremented and the count is decremented. The address stored at the location specified by the new table address is retrieved from storage and the operation is repeated. When the count reaches zero, the execution is terminated with one of the four options discussed previously.

Other modes of control are also possible. For example, data transferred from the subsystem can be stored in random locations specified by addresses in a table instead of in sequential locations as implied in the cases described above. This and most other modes are, however, merely extensions or variations on the basic read, read sequential, and read nonsequential operations.

6.1.4   Digital-Input Error Checking

Since the information associated with the DI function is normally bit significant so that any bit combination is valid, error checking on the basis of code validity is not possible. It is customary, however, to generate checking information in the DI subsystem and to transmit this to the CPU or storage to verify the integrity of the data transfer. This may be accomplished, for example, by generating and transmitting an extra bit for each data word. The additional bit may be used to indicate whether the number of 1's in the data word is even or odd. The extra bit, called a redundant or parity bit, is generated when the data word is read and may be checked either by hardware or by software in the CPU or storage logic. This type of validity checking detects an odd number of bit errors. Other checking methods in which more than one check bit are generated can be used for detecting multiple errors and, in some cases, automatically correcting them.

Each degree of detecting and correcting capability increases the complexity of the system and, therefore, increases its cost. In present generation systems, error correcting codes are not widely used in the digital-input subsystem, although parity checking is often implemented. If an "intelligent subsystem" is considered, the increased cost for checking will probably be evidenced by a larger control or microprogram; that is, by an increase in storage capacity rather than an increase in logic function and circuits. As these types of subsystems become common, increased checking and error correcting capability can be expected.

In addition to methods for checking the validity of data transmissions, hardware can be added to the subsystem to test its basic functions. For example, the various registers in the subsystem can be designed to be loaded with data from the CPU or channel program and these data, in turn, can be read by the system. Thus in the example subsystem of Fig. 6-6, the storage address register can be designed to be read by the system. Data transferred from the CPU or storage into the register and then transferred back can be compared with the original data. This provides a check on both the transmission data path and the register hardware.

If an error is detected, the programming system generally activates a predefined diagnostic routine. The first action usually is to attempt to read the data again. If the error is transient in nature, such as that sometimes caused by a noise pulse, this approach often corrects the problem. However, if several attempts to read the data fail, then the subsystem is assumed to be defective. The program action at this point may be merely to notify the operator of the subsystem status so that a service engineer can be called. In some systems, additional diagnostic programs are executed to isolate the error to a small portion of the subsystem, thereby assisting the service engineer in identifying the defective logic element.

In addition to the error detection provided by the hardware and programming system, the application engineer can include reasonableness checks in his programs. For example, a process may contain over-temperature and under-temperature switches set to close at the limiting conditions. If both switches are found closed simultaneously, an error condition in the system or in the process obviously exists.

6.2   DIGITAL-OUTPUT SUBSYSTEM

The digital-output functions provided in a control computer are complementary to the digital-input functions. The fundamental objective is to provide a means for generating digital control signals and actions for use by process equipment. As in the case of digital input, these digital events are typically binary in nature. The digital-output features associated with the typical control computer differ primarily in the form of the output signal and in the primary characteristics of the output signal. The output may be a current or voltage signal, a semiconductor switch operation, or a contact closure actuated by an electromechanical relay. The output signal or action may be under direct control of the program or, once initiated, it may proceed under control of the subsystem logic. For example, once initiated by the program, the subsystem may hold a contact closed for a predetermined period of time or it may produce a fixed number of contact closures or voltage pulse outputs.

In this section, the functions and design of a digital-output (DO) subsystem are considered. The general subsystem configuration, design criteria for the output circuits, and the subsystem control requirements are discussed.

6.2.1 Digital-Output Functions and Configurations

The basic digital-output function is that of a switch. The switch may control a voltage or current source that is part of the subsystem so that the signal supplied to the connected load is a voltage or current level. Typically, the voltage controlled by this type of digital-output feature is compatible with standard logic circuit signal levels. These output signals are often used to control process or peripheral equipment designed to accept standard logic levels.

The digital-output switch may also control an external source of voltage or current. That is, the subsystem may provide a switch that can be used to control equipment connected to the subsystem. In some cases, several types of switches are provided as features or options in a single process control computer. For example, a semiconductor switch may be provided for high-speed switching of low- to medium-power loads and a relay switch may be provided to handle higher power loads at lower switching speeds.

In addition to the differences in power level, several modes of output control are often provided as hardware or software options. For example, a single computer instruction may result in the output switch's being momentarily closed for a predefined period of time. Alternately, an option may be provided by the program instruction. Since some process instrumentation is controlled by a train of input pulses, a digital-output feature that produces a series of pulses or switch closures can be provided. The pulse width and/or the number of pulses produced as a result of an instruction are specified as parameters in the instruction.

All of the more sophisticated control modes are based on the basic digital-output switching function. Momentary operation, variable duration pulses, and pulse trains can be implemented by suitable programming of the basic switch function, or by hardware included in the subsystem control logic. The choice between the implementations is a hardware/software trade-off that must be made by the system designers on the basis of the performance desired and the relative costs of the alternate implementations.

Figure 6-7 schematically represents the general organization of a digital-output subsystem. The subsystem is connected directly to the CPU or to the channel. In the latter case, the subsystem control logic performs the necessary functions related to the channel operation. The subsystem control logic also provides address and operation code decoding, synchronization and other timing functions, and other control functions such as determining output pulse durations.

The digital-output points themselves are usually controlled on a group basis. To facilitate data manipulation, the number of output points in each group is generally made equal to the number of bits in the computer word or a fraction thereof. Thus, for example, in a computer utilizing a 16-bit word, the number of output points per group is usually either 8 or 16. The number of groups under control of the subsystem

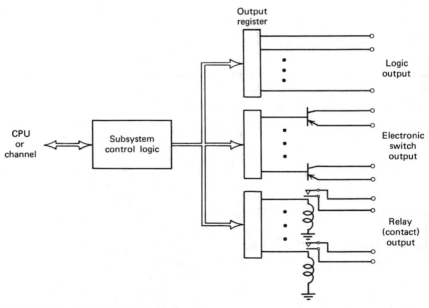

Fig. 6-7   General digital-output configuration.

logic varies depending on the overall design of the system. Eight to sixty-four groups is the common range in current designs, although some systems can accommodate up to 128 groups. Major factors in selecting the number of groups to be controlled by a single subsystem control logic are the addressing structure implied by the instruction format and the physical packaging approach.

In general, the digital-output data are bit significant; that is, each bit in the output word may be used independently to control a process parameter. In some cases, of course, the data represented by a single output group do have word significance. This is the case, for example, when the digital-output group is used to transfer data to a peripheral printer or display device.

Three possible output circuits are represented in Fig. 6-7. The figure shows outputs derived directly from the logic circuits, semiconductor output switches utilizing a transistor, and relay outputs. Although these represent the most common output options, other devices such as silicon-controlled rectifier switches are also possible. A variety of output circuits may or may not be permitted within a single output group or within some other subsystem subdivision. This depends largely on the system architecture and factors relating to the manufacture and sale of the subsystem. The subsystem control functions are largely independent of the type of output circuit so that this is not usually a major consideration. Depending on the output device and the power levels for which it is designed, output termination circuits may or may not be required. When required they generally provide overvoltage protection, arc suppression of relay contacts, or a similar signal control function. These factors are discussed in the next section on output circuit design.

6.2.2   Digital-Output Circuit Design

The output circuits for the digital-output subsystem are typically electromechanical and electronic switches having different power handling capabilities. In general, electromechanical relays provide moderate to high power handling capability at relatively slow speeds. They also have the advantage of total isolation from the subsystem electrical system and can be used to control both AC and DC loads. Semiconductor switches provide higher speed switching. Although most often designed for low to moderate power capability, devices such as silicon-controlled rectifiers and thyristors can be provided for heavy loads. Because of the difficulty of controlling DC loads with these devices, they are most often used for AC loads. Isolation is not an intrinsic characteristic of a semiconductor switch as it is in the case of the electromechanical relay. Isolation can be provided by transformer coupling similar to that mentioned for digital input, however. The degree of isolation is generally dependent on the characteristics of an isolation transformer. In this section, some of the design considerations for these various output circuits are discussed.

The most basic output circuit generally provided is intended for driving low-power logic circuits in equipment external to the subsystem. This feature is sometimes called register output. As an example, the digital-output subsystem can be interfaced to a peripheral printer or other special-purpose display device using the register output feature. In general, the most important design considerations involve output signal levels and switching speed.

The output levels for register output should be compatible with generally available logic circuits. This implies one logic state in the vicinity of 0 V and another in the range of 4–12 V. Output current requirements are generally less than 10 mA in this application. Both positive and negative voltages are used in the various logic families currently available. For this reason, provision must be made for both positive and negative active levels as outputs, and this often requires two different output circuits. If transformer isolation is provided for each output point such that the output terminals can be arbitrarily interchanged, only a single circuit need be provided.

Speed is often an important consideration in this type of output circuit. Logic families currently used in computers are intrinsically fast enough to satisfy most requirements if terminated properly. The termination requirements depend, however, on the load being driven by the circuit. For example, if the load consists of a long coaxial cable, the impedance of the cable should be matched to obtain optimum performance. Since the designer has little control over the user's load, one solution is to provide a few standard loads and a means by which the user can select one of them.

Figure 6-8 illustrates some of the above considerations and the typical application of this type of digital-output feature. The output stage consists of a single transistor driven from the digital-output register. Depending on the logic used to construct the register, the transistor may physically be a part of the register circuit or it may be a separate device driven by the register as shown here. Two optional loads are provided on a termination card and can be selected by using jumpers between the terminals A and B or C. The voltage source V may be provided by the computer system or by the user. The output circuit is unipolar and is designed for positive output levels.

Figure 6-8 also illustrates how the output circuits might be used in typical applications. In the middle circuit, a 90-Ω load is used to match

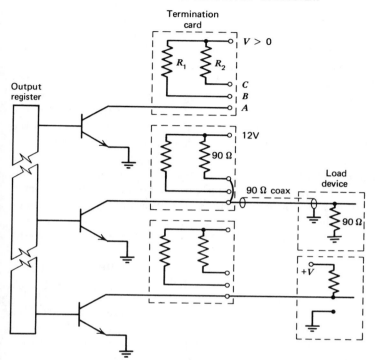

Fig. 6-8   Basic digital-output switch circuit.

a coaxial cable. The lower circuit illustrates an application in which the
voltage source and load resistance are contained in the attached equipment. In both these cases, the loads share the computer grounding system.
As a result, care must be taken at the time of installation to avoid
introducing noise into the computer grounding system since this could
cause malfunction or transient errors.

The basic switching circuit of Fig. 6-8 can also be used to control
medium-power DC loads if an appropriate output transistor is used.
Industrial applications often involve control voltages of up to 48 V and
currents in excess of several hundred milliamperes. Therefore, the output
transistor should be rated for a collector-emitter breakdown in excess
of 50 V and a load current of about 500 mA to satisfy the majority of
the medium-power applications. The design considerations are similar
to those discussed above but, because of the higher voltages and currents
and the nature of typical loads, other factors must also be considered.

Many industrial loads are inductive in nature. Electromechanical relays, actuators, and stepping motor loads are common examples. When interrupting an inductive current, a voltage transient is produced which is proportional to the speed of switching. This transient can easily exceed the load voltage and, if applied to the switching transistor, can cause the transistor to break down. Various forms of voltage clamping can be used to prevent excessive voltage from being applied to the transistor. For example, a back-biased diode connected between the transistor collector and the source (across the inductive load) clamps the collector at the source voltage when the load current is interrupted by the transistor. This approach has the disadvantage of requiring access to the voltage supply. If this supply is provided in the user equipment, it may not be readily accessible. In this case, the diode can be connected to a voltage supply in the subsystem or the user merely may be cautioned that the collector voltage cannot exceed a defined value and that he must provide clamping or other means to prevent damage to the switching transistor.

Because of the higher currents typically employed in medium-power applications and the higher current capability of many industrial sources, consideration of short-circuit protection must also be included in the design. If the user load is shorted, the transistor must carry the short-circuit current of the source. The chance of shorting the load is not insignificant in many applications since wiring changes in the process are often a continuing occurrence throughout the life of the system. Although electronic overcurrent protection can be provided, it is relatively expensive since it must cover each output point. An alternate approach is to provide a fuse in the load circuit. Since metallic fuses are slow compared with the speed at which an overloaded transistor will burn out, some additional protection must be provided during the interval between when the overload occurs and when the fuse melts. This can be done simply by placing a power resistor in series with the collector or the emitter of the output transistor. A third alternative is to allow the output transistor to be destroyed. If overloads are infrequent and the transistor is easily replaced, this is often the simplest solution.

The voltages and currents involved in medium- and high-power digital-output circuits are relatively large compared to those in usual logic circuits, and this can lead to physical packaging problems. The printed circuit techniques common to logic may be unsuitable for the higher voltages and currents. For example, a group of 16 digital-output points each having a rating of up to 0.5 A can result in a ground return current of 8 A. Similarly, the 50-V limit on the voltage in the medium-power output, and even higher voltages in the high-power circuits discussed later, can lead to breakdown problems on the printed circuit boards and connectors. Many of these potential problems are solved merely by changing design parameters such as line widths and spacings for the printed circuits and their board connectors or sockets. Nevertheless, they must be considered by the designer because they can affect package density and other factors such as servicing philosophies. Safety is also a consideration since the usual computer service engineer is not accustomed to high voltages intermixed with logic levels.

Grounding and installation considerations are also important in the design and use of the medium-power output switches. If load current shares conductors that act as logic or analog grounds, the chance of noise being coupled into the computer system circuits is significant. The use of a transformer-isolated digital-output feature that allows totally separate grounding structures is the best solution, although it is more expensive. The added expense to the user, however, is often compensated by the easier installation and the lack of installation delays due to system noise problems. If the digital-output feature is not isolated, the user must be cautioned in his installation of loads and the system designer must carefully consider the grounding structure of the computer to avoid shared ground return paths.

Total isolation of the user and system loads can be accomplished by the use of electromechancial relays. Although slower than semiconductor switches, they have the advantage of being almost perfect switches, as discussed in Chap. 3. They also provide for controlling AC and DC loads, whereas the semiconductor switches discussed above are intrinsically unipolar switching devices. The relay feature is easily added to most process control computer systems. The medium- or low-power switches discussed above can be used as driving circuits with little or no modification. There are, therefore, few unique design considerations in supplying this feature.

Relays that provide a load capability on the order of 100–500 VA satisfy most process control requirements. For heavier loads, the relay output can be used to control an interposing relay or contactor of greater capability. Contact protection is generally required when switching heavy loads in order to minimize pitting of the relay contacts. This protection generally consists of an RC network in parallel with the relay contacts. The values of R and C depend on the characteristics of the load, the voltage, and the current. The user should be provided with design guides for selecting the appropriate value and/or with empirical methods of determining a suitable protection circuit. The design should provide suitable provisions for mounting the contact protection components close to the relay contacts since effective protection is dependent on this factor.

One of the advantages of the relay output is that they can effectively control AC loads. Certain semiconductor devices such as the thyristor are also well suited to this application. The thyristor is a bidirectional silicon-controlled rectifier which passes anode current in either direction following a gate trigger signal. It latches in the conducting state as long as the current through the anode is nonzero. Thus if the gate potential is removed, the device continues to conduct until the anode current passes through zero. In terms of controlling AC loads, this means that there may be a turn-off delay equal to one-half the period of the load current. This is rarely a serious consideration in process control applications.

The thyristor device is available for controlling loads of tens of amperes at peak voltages up to 1500 V or more. A 500-V device with a current rating of 10–20 A satisfies most 115/208-V applications in process control systems. With these types of loads, however, a significant inrush current can occur if the device is turned on when the AC voltage is near its peak value. This current may cause damage to the device and it may also cause noise that can be coupled into other computer circuits. The addition of a zero voltage detector can be used to delay the device turn-on until the load voltage is passing through zero, thereby eliminating the switching overload and the electrical noise. This is indicated schematically in Fig. 6-9. The output of the digital-output register is ANDed with the output of the zero detection circuit. If the register bit changes to a logical 1 while the load voltage is nonzero, the thyristor gate voltage is inhibited by the AND gate, thus preventing the device from turning on until the load voltage is zero.

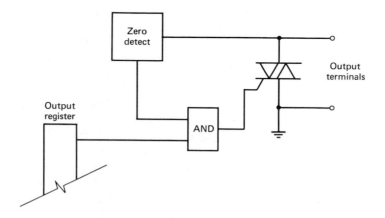

Fig. 6-9   Thyristor digital-output switch.

All the output circuits discussed have assumed a single-ended output configuration; that is, one of the output terminals is common to the computer system grounding structure. This places a restriction on the manner in which the external load can be connected to the system in that the load voltage polarity must be observed and the load grounding must be compatible with the output circuit configuration. A further disadvantage is that the load current flows through a grounding structure that is common to that of the computer system. This can lead to ground loop and other noise problems because of injected noise from the process or the characteristics of the load.

Providing isolated output circuits alleviates the disadvantages associated with single-ended outputs. Output connections can be chosen arbitrarily to account for the polarity and grounding of the load. Since the ground return of the load current is independent of the computer system ground, noise and ground loop problems are minimized.

As in the case of the digital-input subsystem, isolation can be provided on a per-point, per-group, or subsystem basis. Per-point isolation is the most flexible, although generally also the most expensive, since it provides independence between all output points. Isolation on a per-group basis requires that polarity and grounding restrictions be observed within a group of output points. This type of isolation is provided by floating the power source for each output register. By providing a floating power supply for the total subsystem, the subsystem grounding structure can be isolated from that of the computer system. This often requires a substantial amount of floating power and does not remove the polarity and grounding restrictions associated with the single-ended output circuits.

Per-point isolation can be provided by transformer coupling the output of each register bit to the output circuit. Since the amount of power coupled between the circuits is quite small, the components are physically small and relatively inexpensive. An example of such an isolation circuit is shown in Fig. 6-10. The output of the digital-output register is modulated by the carrier signal. The driver provides the power necessary to couple the modulated signal to the full-wave rectifier on the transformer secondary. The output of the rectification circuit, smoothed by the capacitor, drives one of the switch circuits discussed previously.

The degree of isolation provided by this approach depends on the characteristics of the transformer. The breakdown potential of the transformer determines the potential difference that can be sustained between the output circuit and the computer ground system. In general, this should be on the order of 500 V to provide the user with maximum flexibility. The interwinding capacitance of the transformer is a deter-

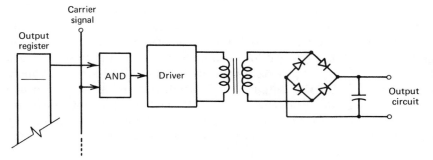

Fig. 6-10 Digital-output isolation circuit.

mining factor in the amount of high-frequency isolation between channels and between the output circuits and the system.

The use of a carrier approach does impose speed limitations on the
output circuits. Since at least one-quarter cycle of the carrier frequency
is required before an appreciable output appears on the transformer
secondary, the action of the output circuit is delayed. For example, if
the output delay is to be less than 1 μsec, a carrier of at least 250 kHz
is required. In practical cases, an even higher frequency is used to allow
for other circuit delays. The use of high carrier frequencies requires that
the designer consider his circuit design and physical layout of the subsystem carefully so as not to introduce excessive noise into sensitive
circuits.

6.2.3   Digital-Output Logic Design and Control

The digital-output subsystem logic provides the control of the individual output points and the functions necessary for communication
between the subsystem and the channel of CPU. The nature of these
control functions depends greatly on the detailed design of the interface
between the subsystem and the channel or CPU and on the control functions implemented in hardware. In this section, the general functions
required and some of the special control modes that can be implemented
are discussed. Although many different functions are provided by currently available subsystems, the discussion centers on the general design
considerations and trade-offs that must be considered by the subsystem
designer.

As in the case of digital input, the digital-output subsystem is generally controlled in terms of groups of points; that is, an address refers
to a group of points rather than to an individual output. For ease of data
manipulation, the number of points in the group is generally equal to

the number of bits in the computer word or to a simple fraction of that length. For example, in a 12-bit computer, the number of outputs associated with a particular group address would be equal to either 12 or 6. The data associated with each address provide the information for the state of the individual outputs in the groups. For example, output data of the form **101111** might indicate that the switches associated with all but the second bit in the output register should be in the closed state.

The basic digital-output control instructions are Write, Write Sequential, and Write Nonsequential. These are analogous to the Read instructions discussed in Sec. 6.1.3, and their hardware implementation is similar. In the Write instruction, one word of data is transmitted from the CPU or a specified storage location to the digital-output subsystem. The associated group address is decoded and the data are transferred to the selected digital-output register. In the Write Sequential and Write Nonsequential instructions, the operation is similar except that subsequent data and, in the case of Write Nonsequential, subsequent addresses are fetched from storage under indirect (cycle steal) channel control. The interrupt on completion and chaining options discussed for the digital-input subsystem are generally available in the digital-output subsystem as well.

The output data transmitted to the subsystem may be interpreted in several different ways. In the simplest case, a **0** would indicate that the associated output switch should be in the open or OFF state whereas a **1** would indicate the closed or ON state. This data format imposes a burden on the system programmer in that every bit of the data must be specified even though only 1 or 2 bits may be altered by the algorithm currently in execution. The program must fetch the storage word representing the present state of the output group, modify only the required bits, and restore the data prior to having them transmitted to the subsystem. This represents a programming burden that can be avoided with other data formats.

One alternative is to associate 2 data bits with each output point. The bits can be coded with the following significance:

00   No change
01   Turn the output ON
10   Turn the output OFF
11   Change the output to its opposite state

A simple decoding circuit at the input to each bit in the output register responds to the 2-bit code to set, reset, or leave the bit state unchanged. With this data format, the programmer need only be concerned with the 2 bits associated with the output currently being manipulated.

Another alternative is to provide a data mask. Two words are associated with each output group, one is a data word indicating which state the output should assume and the other is a mask indicating which bits are to be affected by the data. A mask of **0010** $\cdots$ **0** and a data word of **0000** $\cdots$ **0**, for example, would indicate that the third output in the group should be set to the **0** or OFF state. A modification of this approach, which can be used if the number of bits in the word length is greater than the number of bits in the output group, is to use 1 bit of the data to indicate whether the total word or only a few of the bits are to be implemented by the subsystem logic. For example, if the groups consist of six points each and the data word is at least **7** bits in length, a **1** in the first bit of the data word might indicate that only those outputs corresponding to **1** bits in the word are to be changed. If the data word is **1100100**, for example, the first and fourth outputs are set to the state opposite to their present state. The data word **0100100**, however, indicates that the first and fourth outputs are to be set to their **1** or ON state and that all other outputs are to be reset to the **0** or OFF state.

Similar techniques can be used to simplify address generation in some systems. For example, if a convention is established that a Write Sequential instruction always applies to 16 groups of digital-output points, a mask can be used to sequence only through the desired groups. Using this approach, a Write Sequential (16 groups) instruction might have the following word in the first position of the data table:

$$1000100010000000$$

This is interpreted by the subsystem logic to mean that the second word of the data table is information for group 00 as indicated by the **1** in the left-most bit position. The third word in the table applies to group 04 and the fourth word in the data table applies to group 08, as indicated by the **1** in the ninth bit. This approach conserves data space and compresses the usual address table to a single word at the beginning of the table.

In addition to the interpretation of the data and the address, the subsystem may provide hardware implementation of functions such as momentary or latching output, pulse train output, and pulse duration output. All of these functions can be provided through software in combination with timers provided in the system. Since this can represent a significant computing burden on the system when a large number of outputs are available, however, some or all of these functions may be implemented in hardware. Unique operation codes in the digital-output instructions or tag bits are used to activate these functions in the subsystem control logic.

As with all these features, several alternate implementations are available for momentary operation of digital-output points. For example, when the output point is switched to the 1 or ON state, a monostable multivibrator (single-shot) circuit can be triggered simultaneously. When the output state of the single shot returns to its quiescent state, the transition can be used to reset the digital-output point to the OFF state. Alternately, a counter can be associated with the output point. When the digital output is set to the 1 state, pulses from a constant-frequency oscillator are counted. When the counter overflows or reaches a predetermined state, a signal is given to reset the digital-output point. With some control approaches, such as the use of masking to address individual points, the counter can be shared by several outputs if the restriction is made that subsequent output operations cannot be initiated within the group of points sharing the counter until the momentary action is complete.

The duration of the momentary closure in either of these approaches is a design variable. The type of output and its intended application must be considered. For example, if the output is a relay contact, momentary closure signals of less than 500 μsec are unrealistic since few relays have pick times of less than 1 msec. A common use of momentary outputs is to control stepping motors in analog controllers. Typically, these require a pulse width greater than 10 msec. Semiconductor switch closures used for gating signals in other types of analog controllers, on the other hand, require only 3-msec closures. If single-shot control of the closure duration is used, either a continuous or discrete adjustment of the single-shot time can be provided. If the counter method is used, various counting frequencies and/or counter lengths can provide variations in the momentary closure time.

Variable pulse duration outputs are also used to control analog-output stations that utilize analog memory techniques. In these systems, a constant amplitude, variable duration pulse is used to charge an integrating capacitor. The resultant voltage across the capacitor controls the analog-output voltage or current signal. Durations from a few microseconds to a millisecond are common requirements for this type of equipment. For this feature, the length of the pulse is provided as data to the subsystem logic. A counter associated with the output point is loaded with the data and the count is decremented by pulses from a fixed-frequency oscillator. When the count reaches zero, the digital-output point is reset. Although it is possible to share a counter among several output points, the increased complexity of the logic generally makes this uneconomical. As a result, a counter is usually provided for each output point.

Hardware pulse train outputs also require a counter to monitor the number of pulses produced as the result of a single instruction. Pulses continue to be produced until the count specified in the instruction has been satisfied. Variable pulse width pulse trains of constant frequency can be easily implemented using several registers. The data representing the pulse duration are loaded into a register which is coupled to a counter such that a signal is generated whenever the counter matches the pulse duration data. A signal is also produced when the counter overflows and resets to a zero count. The digital-output point is set to the ON condition when the counter resets to zero and is reset when the counter matches the pulse duration data. The counter overflows are used to decrement another counter which has been preset with the desired number of pulses. This arrangement produces a specified number of variable duration pulses at a constant rate. Typically, digital-output subsystems provide for up to 1000 pulses having a duration of about 1–50 msec.

6.2.4   Digital-Output Error Checking

Error checking techniques suitable for use in the digital-output subsystem are similar to those described in Sec. 6.1.4 for the digital-input subsystem. Thus, for example, the use of parity check bits and the appropriate checking logic in the digital-input subsystem can be used to guard against transmission errors between the CPU or main storage and the subsystem. Detection of a parity error results in an error interrupt and subsequent corrective action as determined by the interrupt service program.

Readback checks can be implemented effectively in the digital-output subsystem. The output register can be designed to be read as well as written. Following the transfer of data from the CPU or main storage into a particular output register, a Read instruction is executed and the resultant data received by the CPU are compared with the transmitted data. If the data are identical, no action is taken; if the data differ, corrective measures are initiated.

Although this technique detects an error within a short period after a digital-output instruction has been executed (one or two instruction cycles, typically), the output register may be in error until the corrective measures are initiated. This situation can be avoided by providing a buffer register for each digital-output register. The buffer register is loaded by the digital-output Write instruction and the transferred data are then read back into the CPU with a subsequent Read instruction. If the data read back agree with the transmitted data, an instruction that transfers the contents of the buffer register into the output register is executed. If an error has occurred in the transmission of the data, the transfer is inhibited and corrective action is taken by the program. Although this technique does not completely eliminate the chance of error since the transfer between the buffer and the output register may be faulty, the transfer operation is relatively simple and the chance of an error is significantly reduced.

An additional error checking procedure that can be used is a "wrap around" check. In this approach, the digital-output signal is connected to a digital-input point as well as to the process load. By sensing the corresponding digital-input point, the programmer can be reasonably certain that the desired output action has been accomplished. If the digital input is not connected directly to the digital output point but, rather, to a transducer connected to the process load, a delay between the output operation and the subsequent input operation used for checking the action may be necessary. For example, if the digital output actuates a motor and the digital input is connected to a centrifugal switch on the motor rotor, a delay equal to the start-up time of the motor must be observed before the digital input is sensed.

BIBLIOGRAPHY

Budzilovich, P. N., "How To Select the Right Relay for Your Application," *Control Eng.,* November 1969, pp. 92–96.

Fredriksen, T. R., "Direct Digital Processor Control of Stepping Motors," *IBM J.,* March 1967, pp. 179–188.

Jacks, E., "The Fundamental Behavior of High-Speed Fuses for Protecting Silicon Diodes and Thyristors," *IEEE Trans. Ind. Electron. and Control Instrum.,* September 1969, pp. 125–133.

Joyce, M., "Understanding Solid-State or Static Relays," *The Electron. Eng.,* August 1969, pp. 43–46.

Kintner, P. M., "Interfacing a Control Computer with Control Devices," *Control Eng.,* November 1969, pp. 97–101.

Talbot, J. E., "Computer Interface Hardware for Process Control Systems," *Instrum. Technol.,* October 1969, pp. 66–74.

"Control Data 1700 Computer System: Analog/Digital Systems General Information Manual," Pub. No. 60187600, Control Data Corp., La Jolla, Calif., n.d.

Control Handbook, Digital Equipment Corp., Maynard, Mass., 1968, 1969.

"IBM 1800 Data Acquisition and Control System: Installation Manual—Physical Planning," Form A26-5922-4, IBM Corp., San Jose, Calif., 1966.

"SEL 80-960A Series Digital Input/Output System," *Tech. Bull.,* System Engineering Laboratories, Ft. Lauderdale, Fla., 1968.

Chapter 7

Man-Machine Interface

DONALD C. UNION

The computer designer sees the computer in terms of logical functions and circuits. He "communicates" with the machine through the media of logic diagrams and the traces on an oscilloscope. His interface to the computer is at the most basic circuit level. The programmer views the machine in terms of logical operations and communicates via the language of the instruction set. His interface to the computer is through the I/O devices such as the card reader and printer.

But what of the process operator and engineer who must use the process control computer for its ultimate purpose? Their interface to the machine cannot be at the circuit level nor, in most cases, at the programming level. These operations personnel require a man-machine interface that provides the ability to guide and monitor the actions of the process being controlled. This need is satisfied through a console or operating panel designed for the particular requirements of operating personnel. The console consists of visual indicators, display devices, and switches or buttons that can be used to enter data, monitor the condition of the process, and perform other operations necessary for the efficient use of the process control computer.

The man-machine interface, however, is more than the physical control panel with its buttons and indicators. The interpretation of operator actions and the responses of the computer and the process must be supported through a suitable set of console control programs. These programs interpret each action of the operator, verify that the sequence of actions is valid, and then initiate the execution of programs to respond to the operator action.

The importance of a well-designed console to provide the man-machine process interface cannot be overemphasized. The console represents the control computer and its programming. The operation of the console must be very comprehensive yet fast, simple, clear, and safeguarded, because through the console the operating people are running the entire plant. Console operations are of prime importance in determining the success of a control computer installation and have sometimes been the deciding issue in the selection of a control computer supplier.

Consoles are for people! People have a right to an opinion on console operations and, consequently, several different opinions are likely to be held, many of them conflicting. Before they are accepted as the basis for a design decision, these opinions must all be weighed in order to determine a solution that will satisfy the typical process operator.

Highly technical people develop control computers and computer programs and highly technical pepole apply them to process control systems. Most of these people are aware of the importance of well-designed man-machine interface consoles and console programs. Their opinions or judgments concerning console operations, however, do not always represent those of typical operating people. The designers of a computer control system must consider the fact that the system will be operated by nontechnical people; this human factor presents the system designers with a very difficult problem that must be dealt with in any effort to design a man-machine interface.

The design of computers and computer programs for process control takes certain identifiable directions if console operations by process operating people are considered. This chapter is concerned with the important and challenging design of these consoles.

The size and complexity of operator consoles spans a broad spectrum ranging from simple consoles that may be rack mounted or placed on a desk to comprehensive control centers involving hundreds of indicators and sophisticated display devices. In some cases the console is integrated into the total design of a control room, whereas in others it is merely considered a supplement to the usual control room equipment. There is, therefore, no standard or typical console that can be described for all the applications in which control computers are used. In this chapter, the discussion emphasizes the design of comprehensive consoles used in centralized control rooms. This approach allows consideration of the complex problems which must be addressed in such a design. The concepts, however, are also applicable to smaller consoles designed for more specialized use. The central display panel discussed in subsequent sections is, in fact, similar to a small console provided for use with the IBM 1800 Data Acquisition and Control System. This console, the IBM 1892 Model 11, is the basis for a number of the figures used in this chapter.

7.1   GENERAL DESIGN CONSIDERATIONS

The process operating people who perform their jobs with the assistance of a control computer have a broad range of skills and responsibilities which are exercised under a range of operating conditions from calm to panic. Their skills and responsibilities are increased by the presence of the process control computer, but the broad spectrum of conditions persists. This makes the design of a suitable man-machine interface very difficult.

The process operating people include process operators, field operators, instrument mechanics, instrument foremen, instrument engineers, operation supervisors, process engineers, and process control specialists. A wide spectrum of information is required by these operating people concerning process and instrument operating conditions and computer system status. Based on this information and their responsibilities, they may have to take several actions. Lists of typical information items and actions are presented in Tables 7-1 and 7-2.

To accommodate the broad range of responsibilities of the operating people, central consoles should be designed for shared operations. By means of a keylock switch or similar device, the central operator's console can be transformed into a console for supervisory and instrument personnel and, by means of another keylock switch, into a process control engineer's console.

There are many ways to effect this transformation. The following example based on the use of keylock switches illustrates the basic concepts. The engineer inserts his key into a keylock labeled ENGR MODE and turns the key switch. The console now has the capability of extended use by an engineer but the engineer must push a button labeled ENGR MODE to utilize this capability. A status light, if available, signals that the console is in the engineer's mode. Pushing the ENGR MODE button at any other time produces a sequence error message SEQ ERROR.

Table 7-1   Console Information Required

Process variable full identification name
Process variable current value
Process variable average value
Process variable last value
Process variable target value
Process variable integrated value
Process variable alarm limits
Process variable control action levels
Process variable states (e.g., In Service, On Manual, etc.)
Associated calculation constants and control constants
Named calculation constants and control constants
Full identification for named constants
Recipe lists
Alarm status logs
Control status logs
History recall logs
Trend logs
Trend plots and displays
Operation logs
Time
Date
Computer system status
DDC loop items

Table 7-2   Console Action Required

Change one of the states of a process variable
Change the target of a process variable
Change a control ratio between process variables
Enter a new laboratory measurement
Change a control constant
Change an alarm limit for a process variable
Switch to a new recipe
Open all supervisory cascades
Open all DDC cascades
Place all DDC loops in MANUAL
Restart the program system
Select a different option in the DDC loop record
Change several values in the DDC loop record

While the console is in this mode, the engineer can address and display the value of parameters and constants that are not normally available to the process operator. In addition, he can alter items such as gain constants, a function not normally included in the process operator's responsibility.

To accommodate operating conditions under shared operations ranging from calm to panic, the process operator must be able to take over the console immediately and to operate it in its normal mode, regardless of from whom he took it over and the conditions of the console at that time. This can be accomplished by pushing a single CLEAR button which clears all the central displays except for the system status indicators and prepares the console for use by the operator. The ENGR MODE key switch may still be turned on but the console reverts to the operator mode immediately because the key switch was necessary, but not sufficient, to put the console into the engineer's mode.

Shared operation of the console among several levels of operating personnel is accomplished by procedures such as these which recognize the importance of quick return to normal operation by a basic operator action. In addition, the higher level operating mode makes use of the normal console function in just a slightly different way so that these extended operations are not evident to the operator. This avoids confusion during normal or panic operations. Other shared or extended functions are discussed in Secs. 7.2.9, 7.2.10, and 7.2.11.

With computer control systems, the central operators have less to do with the control of each process during normal operation, but they may be responsible for the control of more processes because of increased centralization. With the added responsibilities for more processes, operators can become extremely busy during upset conditions. Most of the operator's actions should be made through the console during these times of upsets. The design of console equipment and console programs, therefore, should emphasize near-panic operating conditions. As soon as operators find their console awkward to use during an upset, they stop using it almost entirely and the installation may not be successful, regardless of how much money was spent on the overall computer system, its programming, and the process engineering studies.

An additional consideration in the design of consoles is that once the console equipment and programming are designed, major changes may be extremely expensive. For example, the console program is very dependent on the overall control program design and a major console change could require extensive revision of the whole program package. Because of the magnitude of such a job, this kind of change probably would never be implemented. This emphasizes the necessity of a careful initial design that considers the total information and action needs of the operating personnel.

7.2   OPERATIONAL HUMAN FACTORS

The foundation for console design consists of human factors. The human factors of control room operation in industrial processes have not been well documented, only a few informal studies having been made. The considerations given here are based on the acceptance or rejection of techniques from operating experience with several console designs, discussions with operating people, personal experience, the results of some of the known studies, and studies in related areas.

7.2.1   Naming Processes, Variables, and Constants

Most of the process industries have established a code system for naming process variables according to their service. The Instrument Society of America (ISA) lists the common service code letters for temperature, pressure, flow, and level as T, P, F, and L, respectively. With additional ISA conventions, an indicated temperature variable typically would be named TI-104, a recorded flow FR-120, a controlled pressure PC-48, a recorded and controlled level LRC-26. These codes may be unique only within a single process since they may be duplicated in other processes in the plant. The conventions for naming variables have been altered only slightly with the installation of control computers. Common names for process variables within these computers are T104, F120, P048, and L026. Computer control systems that encompass a number of different processes, however, require a comprehensive naming system for all the directly addressable variables and constants within the plant. A trend for naming them is to use a prefix of two to four letters to designate the specific process, followed by the conventional codes such as T102 for a temperature variable and D028 for a data constant. Thus a code such as ULTF T210 might designate a temperature variable in an ultraformer process.

Names that include several alphabetic letters are much preferred by operating personnel over purely numerical coding because of the association provided by the letters. With such coding conventions, the operating people can memorize an amazing number of code names. As an additional convention, the numeric code that follows the service letters T, P, F, L, and D may be used to designate an area in the process and the number of that type of item in the process unit. When a process variable goes into alarm, an operator knows immediately from its code name the exact location of the alarm condition. If certain areas of a process are more vital, the area code may reflect this fact so that an operator can also assess the importance of the alarm condition.

In general, the use of a prefix process name for all addressed variables and constants helps simplify the entire console operation in a multiple-process control system. In addition to organizing the names of the variables and constants, the prefix provides a simple way to call for displays and typed logs of information for the process and provides a simple way to restrict changes from each console to only certain processes when multiple consoles are used. Thus the process name becomes essentially a separate entity. It is always selected first when any new type of console function is initiated.

Although the code names for variables and constants should be short for quick and easy addressing, they should be expandable for display purposes by an identifying name that is at least 16 characters long. The full name should be as long as is practical within the display capability and computer storage capacity considerations. The identification name provides additional confirmation of the item addressed and need be displayed only upon special request, by pressing a button provided for this purpose, for instance.

Once an addressed process variable is displayed, other associated parameters concerned with its control need to be observed and perhaps changed. These associated parameters should be identified uniquely by their own code names. The same type of parameter, however, is often associated with many controlled variables such as the desired value for the variable, which is often called its target, and the maximum and minimum alarm limits. Consequently, these should be identified by brief alphabetic code names such as TARG for target and MAX for the maximum alarm limit. There are usually so many of these associated parameters that at least four alphabetic characters are required to identify them with reasonable correlations to the full name of the parameter and to ensure differentiation between similar names.

When the process variable is displayed, its code name and its value should be displayed together to minimize identification and recognition problems. All associated parameters displayed concurrently with the process variable should also have their code names and their values displayed together.

A good policy allows operating people to display any associated parameter available to them on their console but allows only higher level operating people to change the value of certain associated parameters. The fact that an operator may not be able to change a parameter need not be indicated in the display until he actually attempts to change the value. Then a simple message such as SORRY tells him that he cannot make the change.

A problem does arise, however, with parameters whose values may be changed under some conditions but not under others. The operator must consider a set of status conditions to determine whether he may change the displayed parameter. If the operator tries to make a change when he should not, a message such as NOTNOW should be displayed to tell him why his actions are not being executed. But he has taken an unsuccessful operating step from which he must recover; such waste of time and annoyance to the operator should be avoided if at all possible.

One solution to this problem is to have alternate names for those associated parameters that cannot be changed under certain conditions. When one of these parameters is displayed, its identifying code name tells the operator whether he can change the value. Examples of this feature are given in Table 7-3. The abbreviations STPT and MAN are

Table 7-3   Console Message Codes

Parameter	Condition	Displayed Code	Displayed Value	Can Be Changed
Setpoint	Cascade closed	CASC	Setpoint value	No
	Cascade open	STPT	Setpoint value	Yes
Output	Output computed to valve	OUTV	Output value	No
	Output computed cascaded	OUTC	Output value	No
	Output computed stored	OUTS	Output value	No
	Output manual to valve	MANV	Output feedback	Yes
	Output manual stored	MANS	Output value	Yes

familiar to the operator as adjustments he has always made when using conventional instrumentation. The other functions such as CASC are not familiar, and this alerts him that he cannot alter their values or states.

7.2.2   Position Orientation of Functions

Operating people are very position oriented. Important console functions should be carefully positioned to distinguish each function or set of related functions. Functions such as addressing a process, variable, constant, or associated parameter, entering or changing values, displaying and changing control states, or displaying and typing alarm conditions should all be distinguished by locating them at different, well-separated, and natural positions on the central panels of the console. A tightly packed grouping of all functions sacrifices clarity and produces a console that takes longer to learn and is very difficult to use under panic conditions. There are times when this packing is justified by the economics of only minor modifications to standard typewriter keyboards to create unique consoles, such as for motel reservation systems. These applications do not have the range of console functions and operating conditions found in process control, however.

A vital concept of console design is to separate functions of different security by position on the console. Operators quickly learn and remember those functions that are safe to perform quickly and those that are not by their position on the console.

The equipment for direct addressing of a process variable by its code name and the resultant display of items concerning that process variable all should be located in the center of the console. The central panel area and particularly the central display of process variable information then attracts the operator's first and primary attention.

The next important functions should be located around this central display and addressing area where they are in the operator's peripheral vision and within easy reach. Functions in this category include computer system status display, status display for the address variable, entry of status changes, selection of associated parameters to be displayed with the process variable and perhaps to be changed in value, entry and changes in displayed value, graphical display of trended variables, alphanumeric display of an information summary, and the selection of these summaries and demand log summaries. The display functions in this category might be grouped around the central display and the selection functions might be grouped around the central addressing function. The data entry function should be aligned as close as possible with the display of the value to be changed.

The most natural position for display functions is somewhat below

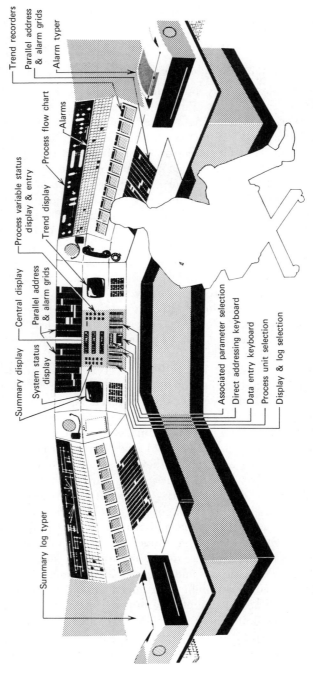

Summary log typer

Summary display
System status display

Central display
Parallel address & alarm grids

Process variable status display & entry

Trend display

Process flow chart

Alarms

Trend recorders
Parallel address & alarm grids
Alarm typer

Associated parameter selection
Direct addressing keyboard
Data entry keyboard
Process unit selection
Display & log selection

Fig. 7-1 Integrated central control console. From IBM Corporation.

518

eye level and facing toward the operator. The most natural position for selection functions is below the display functions on panels that are only slightly inclined from the horizontal.

An artist's sketch of a comprehensive central multiprocess operating console is shown in Fig. 7-1. The central display for the process variable information is in the center of the console and is flanked on the right by process variable status display and entry and on the left by computer system status display and entry. Further to each side of the central display area are two display screens for graphical display of trended variables or alphanumeric information summaries. Immediately below the central display is the data entry function and below that the direct addressing keyboard for addressing process variables and constants within a process. To the left of this addressing keyboard are columns of labeled buttons to select individual processes. The left-most column is reserved for selection of demand displays and logs. To the right of the addressing keyboard are more columns of label tags and buttons that may be dedicated to various parameters associated with the displayed process variable. Above the central area and on the far sides of the console are grids of parallel address and status buttons. These sides of the console are mainly for recorders, special instruments, on-off control buttons for pumps and motors, and process flow charts that may be projected on screens. The two ends of the console are natural positions to mount an alarm typewriter and a logging typewriter.

7.2.3 The Central Display

The primary purpose of the central display should be to display all the information concerning an addressed process variable. Most of this information consists of identification codes and current numerical values. Part of the information, however, consists of status indicators such as those that show whether the process variable is IN SERVICE or OUT OF SERVICE or whether the control loop cascade is CLOSED or OPEN. It takes the operator a certain amount of time to recognize the significance of these status indicators and to evaluate their importance because of their number and the many possible combinations. For quick assessment, the most important status factor should be indicated initially in the central display and the complete status displayed nearby on status display panels. In addition to avoiding confusion in the central display, another reason for separating the complete status is that it should be possible to change each displayed state by the simple and direct act of pushing a button for the desired condition of that state.

Initially, a maximum of information about the addressed process variable should be concisely displayed in the central console display where, ideally, a single glance can absorb all of it without eye movement. The inclusion in this area of the process code name with the variable code is most important since this relieves the operating person of having to look elsewhere for this very important item. The key piece of information to be displayed centrally with the process name and variable code is the current value of that variable in familiar measurement units such as degrees Fahrenheit or in familiar engineering units for calculated values. The measurement or engineering units should be indicated by an abbreviation of perhaps six characters in this same central area of the console. Included initially in the central display for process variables should be the single most important status information which tells the operator the type of variable he has addressed and the degree of control he has over it. It should also prepare him to control the variable in the most direct way if control is permitted. Thus the initially displayed information permits the operator to grasp the significant state of the variable and immediately change the parameter to affect the variable. This is a design for fast emergency operations.

To illustrate these points, an arrangement of information in a central display area is shown in Fig. 7-2. This central display consists of three rows of eleven positions for alphanumeric characters. The left end of the bottom row gives the process name FCC2, meaning fluid catalytic cracker no. 2 in this case. The top row gives the process variable code F017 within that process and its value. The abbreviation for the measurement units is given in the right-hand side of the bottom row. The middle row provides the most important status information which, in this case, is the target value. Complete status information is given by the lighted buttons on each side of the central display.

Table 7-4 lists some of the parameters or messages that might be

Table 7-4   Initial Special Display Information

Condition of Variable	Message	Value Displayed
Out of service	OUT SERVICE	None
Input only	INPT ONLY	None
Time delayed input	INPT DELAYD	None
Primary on manual	OFF CASCADE	None
Output to valve on manual	MAN	Value of output
Cascade open, DDC	STPT	Value of setpoint
Cascade open, supervisory	TARG	Value of target
Cascade closed, DDC	CASC	Value of setpoint
Cascade closed, supervisory	CTRG	Value of target
Miscellaneous data constant	MISC	Value of constant

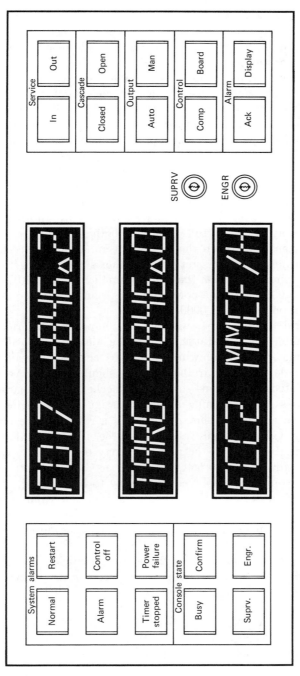

Fig. 7-2 A central display panel.

521

displayed initially in the central display to give the important state information and prepare for immediate operator action. In addition to the choice of items when a process variable is addressed, items applicable to displayed constants are also included in the table.

Numeric values for the process variables and associated parameters displayed in the central display should consist of several significant digits. The number of digits displayed should be kept to a minimum, however, consistent with accuracy and the accustomed ranges of values. This assists the operating people in recognizing the magnitude of the value and in changing it as quickly and as accurately as possible. Experience has shown that four digits with a decimal point and sign is a good practical length. This range from .0001–9999. may be extended to 99999 with the implication that the decimal point is after the right-most digit. Even wider ranges of values are achieved by altering the dimensions from, say, SCF/H to MSCF/H, both of which are commonly accepted dimensions. When the magnitude of a value exceeds 99999, it should be displayed as all asterisks (or other symbols) together with the correct sign. Similarly, when the value is smaller than 0.0001, it should be displayed as .0000 with the correct sign. Absolute zero requires an artificial decimal point such as 0000. .

The central display area, with all the important information for a process variable and its status, is also the ideal place to display associated parameters, some of which might be changed in value. Before a change is made, all the variable's status should be observed and some other associated parameters should be checked. These parameters are usually addressed by pushing a button assigned to each one. The parameter, when addressed, should appear in a row on the central display reserved for value changes and should be identified along with its current value. In Fig. 7-2, the item that may be changed appears in the middle row and is the target, identified as TARG, with a value of +846.0. With an item displayed and ready to be changed, a data entry procedure is followed that produces the desired new value or the desired change in value. While the entry is being made, the current value should continue to be displayed. These entry procedures are discussed in Sec. 7.2.9.

The example console of Fig. 7-2 and the discussion above of its features illustrate an extremely important design concept for the central display: *The central display should be concerned with only one process variable at a time.* The display of all the important information for more than one process variable leads to vast amounts of displayed data. The procedures necessary to indicate clearly which set of information is being changed result in such complex console operations that very few operating personnel can be trained to be competent, and operations can break down completely under panic conditions. In addition, the console program and data areas would consume a large amount of main storage, which is always in limited supply in a control computer.

7.2.4 Console Busy, Clear, and Confirm

As console functions become more comprehensive, some program functions may be queued and occasionally take a noticeable time to be executed. During part of this time, the console where the request was made is busy and is unable to process another request. A CONSOLE BUSY indication near the central display area alerts the operating person that his console is still busy executing his last request.

The provision of a BUSY indicator is an example of the important console design concept of always giving the operating people complete status information; in this case, it is the busy status of the console. The operating people should never have to guess concerning the operating status, nor should they remain uninformed of some special status that may be created by the program design and that may occasionally affect system operation.

If a medium priority request has been made through the console and has placed it in a long busy state caused by a sudden chain of emergency program processing jobs of high priority, the operator should be able to abort his pending request by pushing the CONSOLE BUSY lighted button (or its equivalent). He may then request a high-priority console function, such as displaying a variable, and get an immediate response. There are other situations where an abort function is required, and depressing the lighted CONSOLE BUSY button is a more logical solution than using the CLEAR function.

Low-priority console functions such as demand logs should not tie up the console and therefore should not activate the CONSOLE BUSY indicator. If another low-priority console function is requested while the previous low-priority console function subprogram is still queued, a special message should be displayed to warn the operator of this occurrence. This situation can occur in logging and is discussed in Sec. 7.2.12.

A CLEAR function is another very necessary console function. Depression of a single button (or equivalent) labeled CLEAR should clear the entire central display area of all individual process or variable information. The only centrally displayed information not cleared should be the system status information. Of course, information that is displayed on side panels of the complete console should not be affected by the console CLEAR function.

In order to associate this clear function only with the central display area of the console, it should be located near functions that concern only the central display area. There may be only two such functions: the direct addressing function and the data entry function. Because the clear operation is part of the direct addressing function (Sec. 7.2.5), it should be located in that function area. There are good reasons why the clear operation should not be a part of the data entry function. These are discussed in Sec. 7.2.9.

Another important console function that should be supported by the hardware and software of the console is a CONFIRM action. This feature acknowledges the possibility of a human mistake by all levels of operating people. If an action, such as a change in control status for the entire process, could cause a major change in the plant operation, the program should require a confirmation step before the change is actually implemented.

The CONFIRM function may be implemented by a lighted CONFIRM button (or its equivalent) that is lit when confirmation is required before an action is executed. Depression of the lighted CONFIRM button verifies the preceding step, causes the action to be performed, and extinguishes the CONFIRM light. Actions that sometimes require a confirmation step are discussed in Secs. 7.2.9, 7.2.10, and 7.2.11.

The CONSOLE BUSY and the CONFIRM buttons are often associated with each other and, therefore, are best placed together on the central display panel, as shown in Fig. 7-2.

7.2.5   Direct Addressing

The main objective of process control is to control the process variables. Furthermore, by custom, each control job or loop is identified by the name of the process variable that is being controlled. To permit operating people to communicate with each control job or loop, each process variable measured by any of several means, or calculated, should be directly addressable from the console by a code name such as those described in Sec. 7.2.1. Certain data constants should also be addressable by the same direct procedure and by means of similar code names. These data constants may be obtained from console entries of laboratory measurements which supplement on-stream measurements of process variables. Constants that are in a list to initiate a grade change, for example, should also be directly addressable. In addition, these constants should be addressable sequentially, as described in Sec. 7.2.6.

Direct addressing is accomplished using pushbutton keyboards or their equivalents, as suggested in Fig. 7-3 and Fig. 7-4. The first step in the direct addressing of a variable or constant from the console should be to clear the central display area by pushing a CLEAR button such

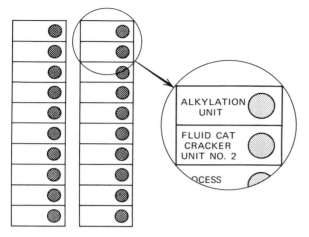

Fig. 7-3   Process select buttons.

as that illustrated in Fig. **7-4**. The second step, in the case of a multi-process console, should be the selection of the process name. The process name should be visible at all times and may be engraved on a legend beside the selection button, as suggested in Fig. **7-3**. The computer pro-

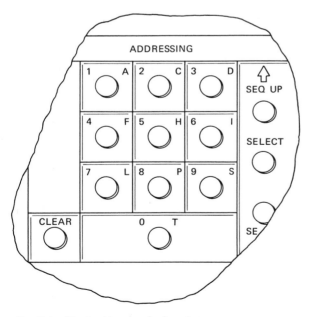

Fig. 7-4   Direct addressing keyboard.

gram should confirm the selection immediately, in milliseconds, by displaying the code name for the process, such as ALKY or FCC2. This process code name should be displayed prominently in the central display of the console since this is where the operator's attention is focused.

The third and final step in direct addressing is to enter the service letter and three-digit code, usually via a ten-key keyboard. Each button or "key" can have a letter and a number, as suggested in Fig. 7-4. The first momentary depression of an address key produces the service letter; the second, third, and fourth depressions produce the code number. The computer program should immediately confirm each character, when keyed, by displaying it in a fixed position in the central display. The display response time should be uniform and on the order of tens of milliseconds so that the operating people may develop a fast and consistent keying action. This part of the computer console program must operate on a high interrupt priority level and must be main storage resident to accomplish this consistently fast response.

As soon as the last digit of the code name is keyed, the computer program proceeds to locate all the information concerning the addressed variable or constant and displays the most important information with a maximum response time on the order of 1 sec. If the variable or constant cannot be located, a simple message such as NOTIN should be displayed to indicate this.

When several variables or constants are to be addressed in the same process, the process name should not have to be re-entered for each direct address. After the first variable or constant is fully addressed, even by another choice of addressing such as sequential or parallel, further use of the direct addressing buttons should be interpreted by the console program as direct addressing of another variable or constant in the same process. As soon as one of the direct addressing buttons is depressed, the central display area is cleared of all information pertaining to the last displayed variable and the service letter on that depressed button is displayed. The process name remains displayed; this differentiates this procedure from the CLEAR operation.

This method of directly addressing several variables and constants in the same process minimizes the number of console operator motions since only the suffix code is entered. As described, this entry is not preceded by an activating step such as the depression of an ENTER ADDRESS button. One suffix code after another can be entered to address directly several variables in the same process. If these addressing buttons were also used for some other function, such as data entry, instead of having a single purpose, it might be necessary to activate their addressing function prior to their use for addressing. This would introduce an extra operation and is thus undesirable.

If there are several similar processes in a plant, such as parallel furnaces in a glass plant, it is often desirable to monitor the same variable in all the processes. In the example console design, this can be done by addressing that variable in one process and then merely pushing the process name button for each other process in turn. As each process button is depressed, the console program combines the new process code name with the displayed variable's suffix code name to form the complete code name for the corresponding variable in each of the processes. The program then displays the most important information concerning the variable, just as if the complete sequence of direct addressing had been followed. These extra features of direct addressing make it as simple to address variables and constants in multiple processes through one central console as it would be to address variables and constants in a single process.

The direct addressing function buttons (or their equivalent), which include the process name buttons and the ten-key keyboard, should all be located in the center of the console just below the central display. The importance of this functional positioning was emphasized in Sec. 7.2.2.

7.2.6   Sequential Addressing

If there are several related variables in a process unit, such as temperature along a reactor, pressure drops up a distillation column, multiple cascaded variables, or grade change lists, sequential addressing is very convenient and quite attractive to operating personnel. The implementation concept here is that a simple console action sequences to the next variable in a prescribed list. The console action can be the momentary depression of one of two buttons for SEQUENCE UP or SEQUENCE DOWN, as shown in Fig. 7-5, so that entry of the code name for the next desired variable is not required.

The operating people may visualize the physical location of the measurements along the process unit and, having directly addressed the first variable, they can sequentially address the others by pushing the SEQUENCE DOWN button, for example, until all related variables along the process unit have been displayed. Similarly, a list of constants, such as those required for introducing a grade change, can be scanned quickly through this simple console procedure. To be general, all directly addressable data should also be sequentially addressable according to sequences established by lists in the console control program.

Experience has shown that the sequential addressing order should be according to the position of the variable in a list and not according to any numerical sequence of code names. This is especially true if the

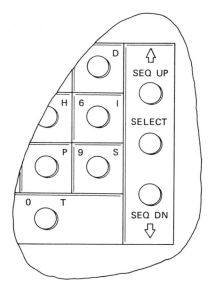

Fig. 7-5   Three sequential addressing keys.

capability is provided for on-line additions and deletions to each list, since this might alter a regular numeric sequence. The position of each name in the list should be for no purpose other than sequential addressing.

Another reason for sequential addressing by position in a list is the sequential addressing of a group of measured variables from which another variable is calculated. These measured variables and the calculated variable rarely have names with the same service letter. In addition, each variable with a different service code letter normally has an unrelated code number. For example, the mass flow F049 of a gas may be calculated from the pressure drop P022 across an orifice plate in the gas pipe, together with the upstream pressure and temperature, P023 and T106. Here are three different service code letters and three different numeric series.

When the names for cascaded variables are placed together in the list for sequential addressing, SEQUENCE UP may be defined as toward the primary variables in a multiple cascade and SEQUENCE DOWN may be toward the secondary variables. This provides an extremely convenient feature for operating people in addressing multiple cascaded loops.

Because sequential addressing is attractive and very useful to operating people, it might be overused and the resultant console activity might become a burden on the computer, especially if full information is displayed for each process variable that is sequentially addressed. This implementation problem may be solved by displaying only the code name for each variable or constant in the list until the operator selects the complete display by momentarily pushing a third button. This button can be included in the set of function buttons for sequential addressing and labeled SELECT, as suggested in Fig. 7-5.

7.2.7 Parallel Addressing

To address a process variable in a single step is an optimum goal to minimize operator motions and to speed up console operations. Direct addressing, as previously defined, took four actions after the process was selected in an initial step or two. Sequential addressing can take one step only if the desired process variable occupies the next position in the list.

A grid of dedicated buttons (or their equivalent), each labeled with the process variable name, permits any of these variables to be addressed in a single step. Such an addressing scheme has been called parallel addressing. The momentary depression of one of these parallel address buttons immediately produces a display of the code name for the process and the variable. It also initiates the selection of data files to complete the initial display of important information concerning the addressed variable. The several steps for direct addressing are simply replaced by the single step of parallel addressing. Physically, the parallel address buttons are often long horizontal rectangles to hold an identification of perhaps 16 characters. The brief code name is included above the identification name, as shown in Fig. 7-6.

In early consoles, the grids of parallel address buttons were only for the quick addressing of the most frequently addressed quantities. Then they became used for an alarm status display for the variable they represented. As this transition took place, the basis for the selection of process variables to be placed in parallel addressing grids expanded to include variables for alarm status display.

With this increased purpose, the size of the grids is bound to increase. As they grow larger, locating a particular button becomes more time consuming. Partitioning the grid areas by process, and perhaps by sections in the process, is helpful to overcome the problem of large grids, as suggested in Fig. 7-6. The partitions should be movable to allow for changes, of course.

The alarm status for each process variable in the parallel address button grid may be indicated in a number of ways. Some solutions have involved two lights behind each button. Since these lights must be con-

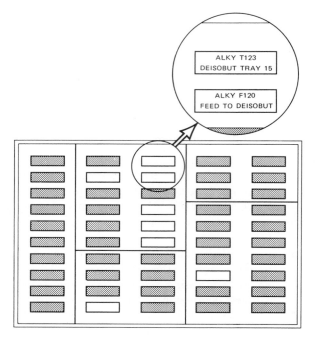

ALKY T123
DEISOBUT TRAY 15

ALKY F120
FEED TO DEISOBUT

Fig. 7-6   Parallel alarm and addressing grid.

trolled by the computer, this is expensive. A less expensive compromise is to use a button that has a single light. When the process variable first goes into an alarm condition, the light is flashed at a constant frequency of about once per second until the operator pushes the parallel address button. Immediately upon depression of the button, the light stops flashing and the central display area shows all the important information concerning the process variable plus its most important alarm condition, as illustrated in Fig. 7-7. Thereafter, the light does not flash but remains lit until the variable passes out of the alarm state.

The parallel alarm display flashes only new alarm conditions and concerns only major or key variables. The operational human factors behind this approach are quite significant. When alarms occur in chain reactions, the parallel display of the major alarms as they occur give the operating people a good feel for what is happening in the process as it happens. All the process variables that remain in alarm may be grouped together for easy analysis in the typed or displayed alarm summaries which may be demanded for each process.

This approach to alarms recognizes the unique value of typed or

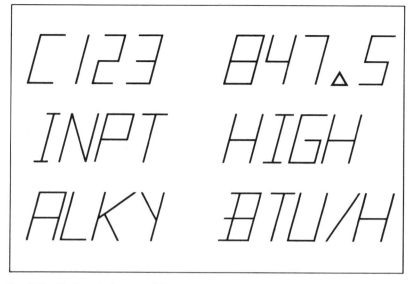

Fig. 7-7   Display of alarm condition.

displayed alarm summaries and the unique features of the parallel alarm status lights. An alarm log has greatest value in summarizing the major and minor alarm conditions in a process for study. The parallel status display of new and old major alarms allows operators to follow what is happening in each process on a real-time basis and to select the important variable for corrective control. They select the most important variable from the display of several new major alarms primarily by its position in the grid. This takes advantage of the position orientation of operating people under panic conditions.

The study of typed alarm summaries to find new alarm conditions is as awkward as the parallel display of all alarm conditions, new and old, major and minor, for a thousand or more variables. The approach just described avoids these pitfalls as well as another pitfall associated with flashing lights. All lights are flashed together at the same frequency. One only has to watch a TV science fiction program with a space ship control room to note the confusion created by console lights that flash at different frequencies. If these frequencies are close together and fairly slow, the lights flash in an infinite variety of changing patterns to the human observer. No more than one flashing pattern conveys any information.

7.2.8   Associative Addressing

Just as associative naming links the process variables and constants with the process, a part of the process, and a service, so does associative addressing reach parameters that are related with process variables and are involved in monitoring or control. In general, the associated parameters are the items most frequently changed to monitor or control a process variable; consequently, these parameters should also be displayed in the central console. When a process variable is displayed, a parameter associated with that variable may be addressed in a single step by pushing a parameter button. If the TARGET button is pushed, only the target for the displayed variable is addressed and displayed concurrently, for example, TARG 846.0 in Fig. 7-2.

Associative addressing may take a single step, with dedicated buttons for each parameter associated with the displayed process variable, or it may involve several steps to enter a general associative code. The single-step approach is best for central operators because of speed, simplicity, and the opportunity to label the dedicated button with a complete description of the associated parameter. Typical parameters for which console buttons may be dedicated are target, average input, minimum and maximum alarm limits, deviation alarm limit, rate of change alarm limit, current deviation, loop gain, nonlinear gain, and the reset and derivative times.

Associated parameters may be thought of as a subset of the sequential addressing list. The use of associative addressing allows the operator to access items easily without searching the total list of sequentially addressed variables. When this kind of addressing involves the entry of a general associative code, the associated item is referred to by position in one or more lists. The number of items, such as density or ratio, in a short list depends on the process variable displayed and might be referred to by multiple depressions of a SPECIAL CONSTANT button. The long lists that are associatively addressed are usually restricted to use by a process or control engineer.

Associative addressing of long lists evolved with DDC. The long list contains every option, address, parameter, and state concerning the DDC loop that controls the displayed process variable. The items in this kind of list make up a data block. To address this information through the console, the engineer must enter the ENGINEER MODE, as described previously.

In the engineer mode after addressing a process variable, further direct addressing is interpreted by the console program as the entry of an associative address code that has a format identical to direct address code. The first letter designates the scaling of the item: decimal, percent, or the engineering units of the process variable. The three numerical digits designate the symbolic number for the item, packed perhaps as a fraction of a machine word in the data block. As this code is entered, it should be displayed similarly to other associated parameters. After the last digit of the code is entered, the value of the item should be displayed to the right of the code.

As long as the console remains in the engineer mode, the address buttons may be used repeatedly for associative addressing of items. This is usually the only special console function that the process control engineer has. His changes to supervisory programs are usually made through another device such as a card reader.

With this approach to switching to the engineer mode, the sequential or the parallel addressing buttons must be used to address another process variable. When another process variable is addressed, the associative code for the item could remain and have its value displayed (if applicable) for the new process variable. In this manner, several loops may be checked for the same item without re-entering the item code number.

If a dedicated button for an associated parameter is pushed while the console is in the engineer mode, the result is the same as for other console modes—except that the name for the parameter will remain displayed as other process variables are addressed in a sequential or parallel manner. The value for this parameter will correspond to each displayed process variable. In this mode, the process engineer can scan the values of an associated parameter for several process variables. A push of the CLEAR button should return the console to the normal operator mode and clear all central displays except the system states.

The associated parameter buttons could be eliminated if an oscilloscope performed the functions of the central display. When the complete information for a variable is displayed, all the associated parameters could be displayed with it. A parameter could be changed by positioning the cursor to it and performing the data entry function. An oscilloscope central display would offer display of all the variables in a process for sequential addressing by moving the cursor to the desired variable and selecting it. Each variable in the list could also have pertinent data for study of them as a group.

7.2.9   Data Entry

To change the value of a parameter is a console function as distinct as any one of the addressing functions just discussed, and it is worthy of even greater design consideration. The data entry function and the status entry function to be discussed next are the only functions that have any direct effect on process monitoring and control. The data entry function, especially, requires safeguards to ensure that changes are made with proper authority, that the correct amount of change is made, and that the new value is actually stored in the active data files. Additional consideration also must be given to the clarity of the operation.

The addressing functions are a means to display all the information that may lead to a data entry. The data entry function is properly prepared for when the displayed information is clearly identified and presented in sufficient detail to give a complete picture of the situation. The necessary detail includes the process name and process variable codes, the current value of that variable, the units used, the complete status of the control loop (if one exists), and the identity and current value of the parameter whose value is to be changed. In some cases, more information may be required before the situation is sufficiently clear. To make certain that the code names are properly interpreted, the expanded name for the process variable should be displayable on request. To supply historical information on how the process variable arrived at its present state and its interaction with other related variables during the past period of time, a historical trend may be required. This can be supplied from permanently assigned trend recorders or from trend displays, which are discussed in Secs. 7.2.11 and 7.2.12.

Previously it was stated that all operating personnel should have the authority to display all the information available through the central console except that information which requires special addressing and interpretation. The authority to change particular values, however, should be assigned to at least three operating levels: process operator, supervisor, and engineer. This authorization may be implemented by means of keylock switches, as discussed in Secs. 7.1 and 7.2.8. Two keylock switches can provide up to four levels of operating authority to change particular parameter values.

The authorization required to change a parameter value may be assigned easily for each process and for each parameter that has a dedicated addressing button. With more programming effort, the authorization can be assigned to each process variable. When authorization is assigned per process, per variable, or per any other set, it applies to all parameter changes within that set. For example, if the second level of authority is assigned for a particular process, only second-level operating people can change any parameter associated with any variable in that process. If the console is in another mode, the first attempt to change a parameter within the process should be answered by a message such as SORRY in the central display.

There are several techniques for entering data. Parameter changes may be made by entering a desired new value. They may also be changed by incrementing or decrementing the present value by a desired amount or to a desired value. Or they may be changed by other techniques that combine the elements of these two fundamental techniques. The first-mentioned data entry method is often called "absolute data entry" because a completely new value is entered. The second technique is called "incremental data entry" because only the desired change in value is entered.

Absolute data entry and absolute addressing are similar in concept. In both operations completely new information is entered. The same keyboard used for direct addressing, which is an absolute addressing technique, can be used for absolute data entry. Several more buttons would have to be added, of course, for a minus sign, a decimal point, and for certain control functions such as starting a data entry and entering the displayed value into working files.

Absolute entry of addresses and data, however, are not as related as they might appear. Their objectives are entirely different. All the addressing techniques previously described, and another to be described in Sec. 7.2.11, use the address keyboard in only slightly different ways. Its use for data entry would constitute an additional use with an entirely different entry procedure. Furthermore, a data format bears no resemblance to address formats, which are very similar for all types of addressing. Finally, the addressing function is not safeguarded; operators should not hesitate to use the address keyboard and to use it quickly! Data entry, however, should be highly safeguarded and it should be approached with caution and care by the operator. An operator should not be expected to approach the same group of console buttons sometimes with rapid motions and sometimes with slow and thoughtful motions. This violates the principle of position orientation discussed in Sec. 7.2.2.

All this discussion leads to the concept that data entry should be done through a distinct set of console buttons. There is no need to share the critical function of data entry with any other function on the console. Its operational category as a potentially unsafe operation should be emphasized by every means available. Positioning the data entry buttons apart from other groups of buttons uses the principle of position orientation to help accomplish the safety objectives for data entry. In addition, the data entry group of buttons should look different from the direct addressing buttons to minimize confusion. Experience has shown that a separate group of unique data entry buttons simplifies console operation with fewer mistakes under panic operating conditions.

Most data entries that process operating people make are changes to improve process operation. They know from experience that a certain amount of change in a particular process variable affects the operation of the process in a certain manner. Thus operators think in terms of a desirable amount of change to make, for example, in the setpoint of the analog controller that controls the variable. Often the amount of change is thought of as so many lines on the recorded chart, not as an absolute value at all.

The most natural method of data entry for process operators, therefore, is incremental data entry, the second fundamental data entry technique mentioned previously. Although the central operators may never make large changes in parameters, other operating people will. Consequently, the incremental data entry function should permit various degrees of change to be made, each with a minimum number of console operations.

A suggested incremental data entry function and button grouping is shown in Fig. 7-8. The buttons are located directly below the displayed

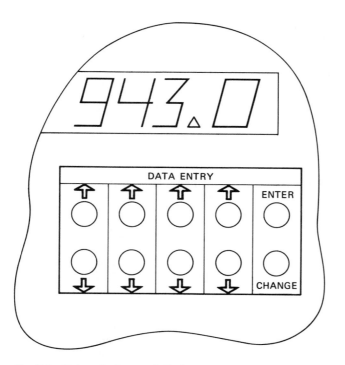

Fig. 7-8 Data entry keys and display.

values in the central display and directly above the direct addressing buttons. The data entry buttons are arranged in four vertical pairs with an up-pointing arrow above the upper button and a down-pointing arrow below the lower button of each pair. Each button pair is associated with one of the four displayed digits that are located temporarily in the bottom row of the central display. The upper button of each pair, when pushed, adds a 1 to its associated display digit. Each time the lower button is pushed, a 1 is subtracted from the associated display digit.

These incremental changes should follow the simple rules of arithmetic. Each upper button should add arithmetically to the displayed value and each lower button should subtract arithmetically from the displayed value. With a positive four-digit value displayed, repeated depressions of the lower button of the left-most pair (the most significant digit) should decrease the displayed value through zero and then make it more and more negative. This should be the quickest way to change the sign of a value, an operation that purposely takes several deliberate steps since operators rarely want to change sign. But when they do, it should be a natural operation that merely takes a few steps.

The simple arithmetic rule of a carry from position to position should also apply. With a displayed value of $+199.9$, one depression of the upper button of the right-most pair (the least significant digit) should produce $+200.0$. Implementing this rule saves many data entry steps.

Finally, the decimal point should move in a natural arithmetic manner as a consequence of incrementing the value. Because only four digits and a decimal point are considered suitable for displayed data, the number $+9.999$ should become $+10.00$ or greater when incremented. As a consequence, all the display positions are used as much as possible and this permits all numbers greater than .01 to be displayed with four digits of resolution. It should also be possible to increment a value from 9000. to 10000 with the decimal point assumed to be to the right of the last displayed digit.

Each depression of the incremental data entry buttons should produce an immediate change in the displayed value. A uniform response time of milliseconds permits the operators to develop a rhythmical and consistent data entry action.

Incremental data entry is the closest digital data entry equivalent for a slight turn of an instrument knob. Only an incremental change is made and the effect on the parameter value is immediately visible before entry to the active files is made.

The data entry operation should have deliberate initial and final steps. In Fig. 7-8, the additional pair of buttons located at the extreme right of the data entry group provides these operations. Depression of the lower button labeled CHANGE initiates the data entry operation. An initial check is made of the authority required to change the displayed parameter. If the operator does not have sufficient authority, the response to his depression of the CHANGE button is a message such as SORRY in the central display. With the proper authority, pressing the CHANGE button activates the data entry buttons, prepares a data manipulation area in the central display and lights the ENTER button. The example shown in Fig. 7-2 displays the parameter to be changed in the middle row. It is the target or setpoint. The data manipulation area can be below the current value where the measurement or engineering units abbreviation is normally displayed. When the data entry CHANGE button is depressed, the value displayed in the middle row is displayed again in this data manipulation area. As the data entry buttons are pushed, the value displayed in the data manipulation area is incremented or decremented.

When the desired change in value has been entered, the final data entry step is taken. In the example of Fig. 7-8, this step is to push the CHANGE button again, which activates and lights the ENTER button, and then to push the lighted ENTER button. If the change is acceptable, the value in the bottom row of the central display replaces the value in the middle row and is placed in the working files of the system. The abbreviation for the units then reappears in the right side of the bottom row and the ENTER button light is turned off. Every successful data entry should be logged by the system on a nearby typewriter to record the date, time, process, item, and new value.

The use of two buttons for the final entry step provides protection against accidental entry of an incomplete data change. Although the ENTER button should be in the data entry group for completeness of the functional group and for ease of operation, its nearness to the data manipulation buttons provides the chance that it will be pushed accidentally. The two-step final entry procedure eliminates that accident without adding to operator inconvenience.

Changes in value greater than some predetermined amount such as 10% may not always be desirable, especially for targets. When a larger change is entered, the CONFIRM button may be turned on to signal the operator that he is entering a large change. If, after checking his entry, the operator believes the entry is safe, he must push the lighted CONFIRM button in order to complete the entry. This confirmation step then becomes the last step of the final entry procedure. When the CONFIRM button is pushed, its light is extinguished and the new value replaces the current one in the display and working files. An entry of a target value that exceeds MAX-MIN limits should be canceled altogether and a message such as TOO HI or TOO LO should be displayed.

There should always be a well-conceived cancel step in a data entry procedure. The main problem to solve is that the final entry step may be skipped by mistake, perhaps because of a momentary diversion, and the next operation begun without realizing that no actual change was made in the working files. The cancel step, as a consequence, must not be merely to skip the final entry steps nor to push a button such as the CLEAR button, since this marks the start of many other console functions that might normally follow a data entry.

The CONSOLE BUSY indicator discussed in Sec. 7.2.4 had the function of indicating a console operation in progress and, as a second function, it provided a means to cancel the current console function. As soon as an authorized data entry is initiated, the CONSOLE BUSY indicator should be lighted and it should remain lit until the final data entry step is taken. The cancel step for data entry, very naturally then, is to push the lighted CONSOLE BUSY status indicator. As an additional precaution, all console buttons except those required for data entry should be deactivated until the entry operation is either completed or aborted.

7.2.10   Status Entry

The status display for process variables was discussed briefly in connection with the central display in Sec. 7.2.3. In addition to this status, there is status associated with each process under computer control that pertains to placing computer control off or on a process. Finally, there is the computer system status which should be displayed in the central display area. The status messages for the computer include NORMAL to indicate the system is operating normally, ALARM to indicate some part of the computer is not operating, TIMER STOPPED to indicate that the computer program is not processing its main control routines, RESTART to restart or reload the program, and CONTROL OFF to indicate that computer control is off all processes. Figure 7-2 suggests a way to group these many status indicators and entry devices around the display on the central console panels.

The indication and entry of status information is best provided by two lighted pushbuttons or their equivalent, either of which may be lit to give the yes/no or on/off status. The alternate unlit button is pushed to change the status. The use of two lighted pushbuttons guards against the burn-out of an indicator light not being detected in time to avoid operating hazards and provides a transitional status between on and off to be displayed by lighting both buttons.

The status entry function requires safeguards similar to the data entry function to ensure that changes are made with the proper authority and that the correct state is changed. In addition, every status entry should be logged by the system on a nearby typewriter to record the date, time, process name, status name, old state, and new state.

The authorization required to make a status entry may be assigned for each process that has a name button and for each status button. Either the SUPRV MODE key, the ENGR MODE key, or no key may be the assigned authority. When a status change requires, for example, the SUPRV MODE, the procedure to place the console into this mode must be followed before the console program acknowledges the depression of the status button and changes the state.

Certain status changes affect the control of an entire process. In these cases, an additional entry step such as CONFIRM should be required to safeguard against inadvertent entry. A lighted CONFIRM button in response to a depression of a status button indicates to the operator that a confirmation step is required. The status change is not initiated until the CONFIRM button is pushed.

7.2.11 Auxiliary Displays and Recorders

Auxiliary displays and recorders, which are included in the example console of Fig. 7-1, add significant flexibility to console operations. For example, recorders are used to show the trend of variables as they are measured or computed. By recording the variable for a long period of time, not only the present trend but also past trends are available for study.

Including several recorders in the central console is a very common practice; some may be permanently wired to important process variables, some may be switched by patchcords to one of several variables, and some may be computer driven and assigned through the central panels of the console. All these recorders may have multiple pens on the same chart.

When recorders are computer driven, a console procedure is needed to select a recorder and pen, to set upper and lower scale bounds, and to choose an item to be recorded. The item to be recorded might be only the current value of a process variable with the upper and lower scale bounds set equal to the MAX-MIN alarm limits. Since this is a particularly common need of the operator, the procedure to initiate this recording should be simple.

As an example of such a procedure, a button on the lower panel is labeled RECORDER. With the process variable displayed, pushing this button clears a portion of the display and causes the letter R to appear to call for the entry of a recorder pen number via the direct addressing keyboard. By assigning two characters for selecting pen numbers 01–99, 33 three-pen recorders can be accommodated. As the pen number is keyed, it appears after the displayed letter R, perhaps in the lower right side of the display replacing the units designation. After the entry is complete, the CONFIRM button is lit and when pushed, the CONFIRM button starts the recording. Until the CONFIRM button or the CONSOLE BUSY (abort) button is pushed, no other button depression on the console should be honored to ensure that the recorder entry function is completed (similar to data entry).

If the RECORDER button is pushed immediately following a CLEAR operation, the letter R appears in the central display area. After a pen number is entered via the direct addressing keyboard, the display fills with the name of the variable being recorded by that pen, the name of the item, and the process name as follows:

T123

PV

ALKY R12

Pressing the SELECT button results in the complete display of all information regarding the recorded variable which may have attracted the operator's attention.

If the selected pen is unused, this is indicated by the display:

UNUSED

R12

If the use of the selected pen is to be deleted, depression of a DELETE button while the recorder pen number is displayed terminates the recording, clears all process variable information from the central display, and replaces it with the UNUSED message shown above.

Operating people with higher authority, such as process engineers, often want to record items other than process variables. By placing the console into the ENGR MODE, the recorder function can be changed to record whatever item is displayed in the middle row of the console display. The parameter to be recorded should be displayed in the normal manner in the central display. A depression of the RECORDER button produces an R in the display, as before, to call for a pen number. After the pen number is entered, an S is displayed after the pen number to call for the scale bounds. A one- or two-digit code is entered to select one of several possible scale bounds. Then the CONFIRM light comes on and must be pushed to start the recording. Later, with a cleared console, the entry of this same recorder pen number produces the display of the variable name, the name or code for the item being recorded, and the process name, as described earlier.

The use of recorders to show on demand historical data stored in the computer produces awkward operations and requires special recorders. These recorders must have high-speed pen movements and high- and low-speed chart movements that are synchronized with the data coming from the computer. Such recorders are available but are higher priced than memory-screen oscilloscopes.

More and more interest is being shown in oscilloscope display equipment for use in central consoles of multiprocess computer control systems. Their capability for displaying many small but legible characters and graphing a group of variables stirs one's imagination. To find a good use for them in central console operations, desirable console capabilities not now generally available and important console functions that can be improved should all be examined.

There remains the genuine need to analyze several control loops at one time to examine interrelated trends and to know precisely how much related variables have been changing. These two important types of information are difficult to obtain from one device. Furthermore, almost none of this information can be obtained from the discrete display of current data for each control loop. This trend information is best given upon demand by a graphical display. The precise changes are best given upon demand by a numeric trend display or typed log.

The pen recorders discussed above have given the needed trend information for many years in the process industry. Their one unique feature is a permanent graphical record, but discrete data records on machine-readable tapes and cards are rapidly replacing the need for permanent graphical records. Graphs and analytical evaluations may be made at any time from the machine-readable records.

Upon demand, a graphical (oscilloscope) display can show interrelated trends among several control loop variables, quickly and clearly, from data in the computer bulk storage. Each variable should be traced versus time because time is the only truly independent variable in a process and it is the one operational people are accustomed to seeing as the horizontal reference or abscissa on recorders. The face of popular compact strip chart recorders shows about 2–4 hr of operation. Consequently, if graphical data traced a 2-hr history of each variable from data stored at perhaps 0.1-hr intervals and continued each trace forward with current data every 0.1 hr, it could be a valuable addition to the central console. In a sense, it would be replacing most of the function of recorders but would be offering capabilities far beyond them as well.

Some recorders for emergency back-up would still be desirable and would be useful on a full-time basis for some applications.

An auxiliary graphical display to show relationships between variables and time need not have much precision. An attempt to provide precision, in fact, leads to one long vertical axis to serve all variables, and this decreases the number of possible variables that can be displayed simultaneously without confusion. Thus as precision is gained, relationships are lost.

Let us assume that the precise changes for each variable may be provided conveniently by another numerical display or typed log. Without precision as an objective then, a graphical display may have a short vertical axis for each variable or group of variables concerned with one controlled variable. Several short vertical axes of equal length can be stacked on top of each other to form several horizontal bands of variable trends with one band for each controlled variable.

The short vertical scale for each band should be linear, of course, but the span can be the difference between the minimum and maximum alarm limits for the process variable and the 0–100 percent range for the control output. A third variable included in each band should be the target or setpoint (when one exists) plotted against the same vertical scale as its process variable.

Each display band should be outlined by low-intensity horizontal lines to represent the assigned limits. A process variable within a band could venture into an adjacent band when it exceeds its limits. The trace of this wayward variable could still follow its own scale, however, just as if the scale extended linearly into the adjoining band.

Each band need be identified only with the code name of the process variable, such as T123, because all recorded items would be from the same process unit. Each of the three traces within the band could be identified uniquely by using dots and dashes for one, dashes for the second, and a continuous line for the third. The noisy process variable would best be shown with the continuous line. These three choices of curve identification are about the limit unless color displays are available.

These display ideas, illustrated in Fig. 7-9, are just an example of the possibilities for computer-driven graphical displays in process control computer central consoles to provide the central operators with an immediate overview of the operation of sections of each process unit. Here is a console function that has been missing and can be provided by current technology.

Another use for auxiliary displays is to provide upon demand numerical trend summaries for each process unit or for sections of each unit more rapidly than is possible with a logging typewriter. The numerical

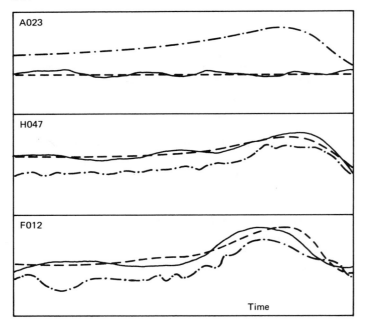

Fig. 7-9   Trend display.

trend summaries would be a valuable complement to the graphical display of process variable relationships. Actually, the numerical trend display may merely list the numerical values used in the graphical trend display. Furthermore, one selection code could call for a group of variables to be displayed graphically on one screen and numerically on another screen, both located in the central panel area of the console.

Control and alarm status summaries around each process convey a great deal of information to the operators. If these summaries could be displayed, operating people might depend more on them and require fewer individual computer-driven indicators for each process variable.

7.2.12   Demand Logs

Periodic hourly, shift, and daily summary logs may be typed nearby central consoles and may provide important operating information, but only the logs typed on the demand of the operator become a functional part of the console. These demand logs should be carefully integrated into the console operational design. Although many demand logs in future consoles may be displayed on auxiliary display screens, as discussed in the previous section, the nature of these logs and the procedure to select them through the console are essentially independent of the device which produces them.

The alarm summary log is a listing of all variables in the selected process that are in an alarm state. A control status summary log lists all variables in the selected process that are in an unusual control state. The exact composition of these two logs is determined by the state of the control system and the particular process.

Demand logs should provide information concerning operation of only one selected process when several processes are under control from a central console. This information is typically the alarm summary, the control summary, one of several operating summaries around parts of the process and occasionally parts of several processes, and trend logs that list values for selected process variables from stored historical data.

For each type of demand log, a console operation sequence is required that should be carefully designed to fit into the general operation of the console. In the previous sections that discussed the addressing of process variables, a basic operational sequence was: (1) Clear the central display; (2) select a process unit; and (3) directly address a variable within that process. A sequence to select demand logs follows this same basic pattern: (1) Clear the central display; (2) select a process unit; and (3) select a demand log. A simple demand log such as an alarm summary would be selected by the depression of a button for that log in step 3 of this sequence. A more complex demand log, such as an operating summary around parts of a selected process, would require a fourth step to select the particular operating summary. The third step, the depression of an operating summary button, would call for a particular operating summary name code to be entered through the direct addressing keyboard as the fourth step. The code to be entered could specify whether the summary is to be typed locally or printed remotely by means of a prefix letter (T = Type, P = Print) and three subsequent digits could specify one of several available summaries. Although a three-digit number provides for more summaries than might be envisioned, the use of a uniform code format for all addressing simplifies the programming and provides uniform operating procedures.

Operating summary demand logs have varied headings and compositions. The user should be able to develop or change these headings and the composition of the log on-line via card entries. This is an on-line RPG (report program generator) capability.

There might be only one trend log on demand per process unit. If this is the case, its composition must be changed easily through the console. It should be possible to delete a variable and add another variable through console entries. When a trend log is first requested by a sequence of clear, select process unit, and select trend logging, only the title and heading might be typed by the computer. At this point either the trend log button is pushed again to receive the log or one of the variables in the heading is directly addressed and a DELETE button is pushed to delete it from the log. Another variable then can be direct addressed and an ADD button pushed to add it to the trend log. This sequence is repeated until the desired combination of variables to be logged is obtained. A second depression of the trend log button would type the new heading and another depression would initiate the printing of the data. This trend log selection procedure could also be used to request the graphical trend display and the numerical summary display discussed in the previous section.

These trend logs by themselves offer another way to show the interrelationships between several variables without the use of trend displays. Columns of numbers may not, at first, convey much trend information to operating people unaccustomed to seeing data logging, but installations where data logging has been operating in the past usually produce operating people who prefer columns of precise numbers to the tangled results of multipoint recorders.

Demand logs have a lower operating priority than other console functions and also may involve typing or printing, which are relatively slow output operations. Demand log programs, consequently, are usually queued to be brought into primary storage for execution from residence in secondary storage. While a demand log program is queued, a request for another demand log should receive a displayed message such as LOGBUSY, but requests for other console functions should proceed as usual. While a demand log is being presented, a request for another demand log should be possible even though it will not be honored until the previous logging operation is complete.

The programs for periodic logs and demand logs have much in common. Either the periodic program request for a log or the console demand request may be loaded into a common log request buffer. A single set of programs services either request.

7.2.13   Alarms

Alarms provided by the computer control system for measured and calculated process variables may be handled sequentially through the central panels of the console and an associated alarm message typer. They can also be handled via parallel alarm and addressing panels. The latter were described in Sec. 7.2.7 in the discussion on parallel addressing. The methods complement each other in that variables of secondary importance may be handled centrally and alarms of primary importance may be handled on the parallel alarm and addressing panels.

All alarm conditions should be typed to provide a permanent hard copy. The typed message should show the date and time of the alarm occurrence, not the time it was typed. The process variable in alarm should be identified by its console name, which indicates the process unit, the service, perhaps a process subdivision, and the item number. The alarm condition, such as ABOVE REF or ABOVE MAX should also be typed.

The acknowledgment of process variable alarms which are not included in the parallel addressing grids may be handled centrally with only one or two buttons. As each alarm occurs, a central panel button labeled ALARM flashes and a horn sounds. Depression of the flashing ALARM button stops the horn from blowing and depression of an adjacent DISPLAY button displays the variable and its alarm condition. If more than one variable has gone into alarm, the ALARM button continues to flash until all new variables in alarm have been displayed and acknowledged by repeated depressions of the DISPLAY button. After all alarmed variables have been acknowledged, the ALARM button remains lit until no variables are in alarm.

When several variables go into alarm rapidly in several process units, the first variable should be remembered and displayed first. Since it may have started a chain reaction of alarms, the next alarm variables to be displayed should be from the same process unit as the first one until all alarms from that unit have been acknowledged. The next unacknowledged alarm variable that a status table search encounters is then displayed, followed by other unacknowledged alarm variables in that process unit. This procedure continues until all new alarm conditions are acknowledged.

The use of parallel alarm and addressing panels for key process variables, as discussed in Sec. 7.2.7, is increasing as larger and multiple process computer control systems are installed. They provide a better feel for what is happening in multiple alarm situations than the sequential alarm procedure described above. They also provide a fast one-step addressing of the process variable.

7.2.14 Console Operation Errors

An operator needs to be told immediately by a message in the display area that he has made a mistake in his use of the console. The immediate error message allows him to recover quickly. The error message should advise the operator clearly what his mistake was. No numerically coded messages that require reference to a book for interpretation should be used! Recovery speed is of utmost importance during panic operations.

The console operation error message given the operator should *never* be printed or stored as a permanent record for management review. Operators would be reluctant to use the console in a self-teaching mode if this were done.

Ideally, operating people should feel a personal bond with their console. The relationship should be warm and friendly, even with occasional humor. A perfect example is a message that appeared on a computer control console at midnight on New Year's Eve: "HAPPY NEW YEAR, GEORGE." The next day, George talked about nothing else!

7.3   COMPUTER AND PROGRAM FACTORS

The design of computers and computer programs for process control takes certain identifiable directions if console operations by process operating people are considered. These directions are discussed in Secs. 7.3.1 and 7.3.2. It is shown that the console program has a larger and more complex interface with the control program than with the computer equipment.

Of all process computer programming, console programming is the most difficult task and the most underestimated. Three inherent console characteristics, discussed in Sec. 7.3.3, are the main cause for console program complexity. All these complex console characteristics have one common origin—man. The console is the interface between the human operators and the entire computer control system. The complexity of the part that the operating people play in the control of the plant is reflected in the design of their consoles. The design must also provide safeguards against console operation mistakes that are sure to happen to even the best operating people. Good console design minimizes the chance of a serious operating mistake, but it can never eliminate all mistakes. The more complex the console design, the more elaborate the safeguards must be.

7.3.1   Computer Interface

The computer interface with the console is mostly a digital signal interface. Digital outputs turn on the lights, command the display tubes with character codes, and command the display scopes if they are digitally driven. To command display tubes and scopes, the digital-output circuits must be fast, on the order of microseconds, and, therefore, solid-state switches rather than relays are used. The display tubes are driven with one common digital word which is routed to the proper display by a digital-output line dedicated to each tube. Oscilloscope displays may be driven by analog-output signals, especially if they have memory screens that utilize long-persistence phosphors. Again, the speed requirement between analog-output values is in the microsecond range.

The more than 100 momentary pushbuttons that may be required per panel are handled very well by digital-input circuits that are latched by the momentary digital signal and that provide a computer interrupt. These pushbuttons may have multiple contacts and their wiring may pass through diode networks in order to code them into one digital-input word with more than 100 bit patterns to identify the button uniquely. For example, a 3-out-of-16 code can serve 128 buttons and provide good error detection. The momentary depression of a single button generates three digital-input signals that most likely will not arrive at the computer at the same instant of time. The issuance of an interrupt signal, therefore, must be delayed to ensure receipt of all three input signals.

With multiple panels, including the parallel alarm and addressing panels and multiple consoles, several latching digital-input words are required, each with its own identifiable internal interrupt signal. Furthermore, because only one program services these panels and consoles, the interrupt signals should go to a single interrupt level. The use of several interrupt levels is an unnecessary programming burden. Not many third-generation control computers have the ability to handle multiple external interrupts per level. Those that do accommodate the multibutton process operator consoles with ease.

An important concept that makes standard central panels possible is that they only interface the computer. When they also interface the process or process instrumentation, their design becomes specialized for each installation.

7.3.2 Program Interface

The total program for process monitoring and control must be designed with a clear understanding of the console needs. Probably the most important design aspect is the grouping of data. The console accesses a large amount of data concerned with each process variable. Consequently, a logical data grouping is by process variable; this is also a natural grouping for processing the variable. In one generalized process control program, for example, 84 data words are associated with each process variable. For computation efficiency, however, these data are split into several data files. The correct group of data in each file is located with a single relative address for the process variable. Furthermore, specific data associated with each process variable may be located by a suffix to its relative address.

For purposes of computation and interprogram communication, relative addresses are completely satisfactory to identify the process variable. But process operating people would be very unhappy with numerical identification. They not only demand alphanumeric codes for the variables, but also they often request additional confirmation upon demand in the form of an expanded identification name, as discussed in Sec. 7.2.1. The data files should be designed to accommodate these descriptive character strings for easy console access and also for minimum computation burden. The lengths of the character strings are partially governed by the economics and capability of the central display that is being used.

Not all process control data are directly associated with process variables. Each data value, however, requires a substantial amount of description for operator understanding and assurance that he is looking at the proper data value. These miscellaneous data must also be organized for easy console access. Again, for computational efficiency, the data and their descriptive character strings would be separated into two files. The complete information for console display of a miscellaneous data value is located in each file with a single relative address.

A logical interface between the two requirements of relative and alphanumeric addressing is a cross-reference table that might be referred to as a name-number table. Each table entry consists of the code name and relative address for a variable or data constant. These entries are ordered according to the sequence desired for sequential addressing and grouped by process unit.

Again, variables and data constants need to be grouped by process unit only for the operating people, certainly not for computation and control needs. In fact, experience has shown that grouping of process variables by process units for computational purposes leads to many problems. Consequently, this console need should not influence the organization of computation data files.

7.3.3 Console Programming

The programming for central consoles is difficult primarily because of the large number of console states, the large number of valid, conditionally valid, and invalid sequence operations from each state, and the maze of sequential links between the console states. A casual observer of a central console in operation would be surprised to learn that a comprehensive central console may have more than a dozen major states from which several unique sequences may be valid.

The concept of console states and sequence operations between them is perhaps best conveyed by a sequence diagram, as given in Fig. 7-10. Most of the console operations discussed in Sec. 7.2 are diagrammed in this figure. Each console state is represented by a darkened circle and the arrows from state to state indicate the valid operating sequences shown in the circles. At any given console state, the initiation of a sequence to another state is made by the console operator as he pushes a button for that operation.

The sequence diagram of Fig. 7-10 has 11 major states at which from 2–8 unique sequences may be valid, 1 may be conditionally valid, and from 12–18 may be invalid. Most of these sequences consist of the depression of one button on the central panels to reach another state. Some, such as direct addressing and data entry, consist of the depression of several buttons with a minor state between each depression until a major state is reached.

A major state is distinguished by noting that the valid sequences after the completion of a particular sequence are different from those that were valid before the sequence was executed. One measure of console program complexity is the product of the number of console states and the maximum number of valid sequences at one state. This measure defines the dimensions of a decision table that might be used to program the console operation logic.

There are three basic states in Fig. 7-10, numbered 1, 2, and 3. The central panel is cleared at state 1; a process unit is addressed at state 2; and a variable or constant in that process is addressed at state 3.

Console programs are simplified when a given sequence leads to only one state regardless of the state at which the sequence is valid, as in the case of the CLEAR operation. When a sequence may lead to a different state as a function of the state at which it is valid, another dimension of complexity is created. Sometimes this is unavoidable; for example, the direct addressing sequence is shown to be valid at three states in Fig. 7-10 and it sequences to a different final state in each instance. In this specific example, the programming complexity is simplified somewhat by having the four-step addressing sequence always identical: a character and three digits.

When a state must be reached by several different sequences, extra care must be taken in console programming to keep all the displayed information including special messages consistent. This problem is hard to show in a diagram. Console messages are used to acknowledge operations and to explain lack of acknowledgement owing to an operating mistake. Care must be exercised to display an error message only when the information replaced by that error message is no longer needed or when it can be redisplayed easily, as, for example, with the push of a single button.

Operating people are not content always to have to start each se-

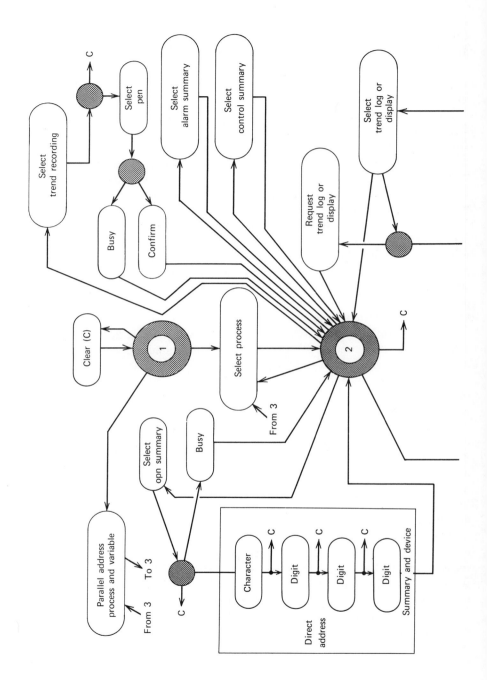

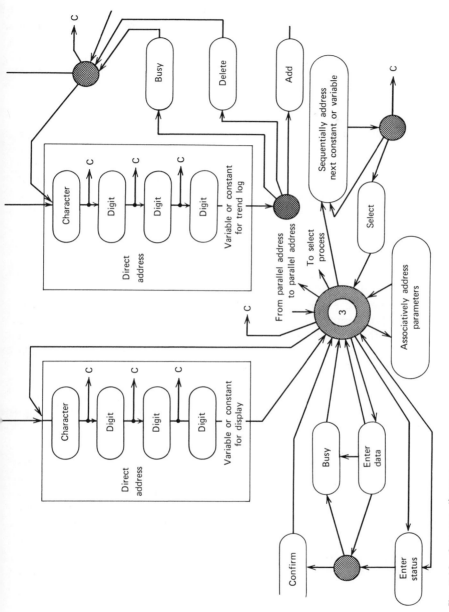

Fig. 7-10 Sequence diagram.

553

quence string from a cleared console. Furthermore, many console operations revolve around states far removed from the original cleared state from which they were initiated. This leads to many sequence loops and loops within loops, as shown in Fig. 7-10. These multiple links between console states along with the many valid sequences from any one state make flow charts for console programs very difficult to follow. A better means for checking the program logic and operating sense of the maze of console sequences is the sequence diagram.

Given sufficient time, operating people make every sequence error possible in the console design. Good design should minimize the frequency of sequence errors and their attendant operator frustration. The number of possible sequence errors, however, approaches the product of the number of unique console sequences and the number of console states. There are 170 invalid sequences in Fig. 7-10, counting only those at the major console states. Before each additional console state is created, alternate console designs should be examined to determine whether a new state can be avoided.

An additional aspect of console programming should be discussed. In Sec. 7.2.5 on direct addressing, the statement was made that a fast and consistent keying action should be made possible by locating a part of the console program on a high-priority interrupt level. The remaining portions of the console program will be at lower levels and some will be near the lowest priority level because of the low priority of some console functions such as demand logs.

The portion of the console program on the high interrupt level responds to all the keying and must display the result of each keying step. Consequently, it must display all the requested information. This portion then becomes the program interface between the console equipment and the main console program, which is at a lower level. This division of tasks on different priority levels causes another degree of console programming complexity.

The further division of the console programs to provide execution of some functions on the lowest priority levels creates a similar programming complexity but not as much as the first division. There are usually fewer branches across this second division and all are usually in just one direction, toward the lower priority level. In both situations, however, a multiword buffer is required to handle the transfer of information between these sections of the console program.

One final aspect of console programming is burden on the system. The console operation must not present more than a small burden because this capacity must always be reserved for it. The console program routines should all execute quickly. Their interference with main storage resident data buffers, however, can result in a significant burden. Since bulk storage data are always accessed in a group, all systems with bulk storage require a main storage resident data buffering scheme to permit changes to adjacent data values in bulk storage by several programs on different interrupt levels. This must be done without excessive use of interrupt masking to prevent the programs from canceling changes made by another program. Care must be exercised to use these data buffers only when data are to be changed by the console since the console program accesses far more data than it changes. If the need for main storage resident data buffers and their complex scheduling schemes is not obvious in programming the control algorithms, the addition of a central console program soon makes the need for these buffers apparent.

If a frequently used portion of the console program is resident in bulk storage, the time to get it into main storage is subject to much more statistical uncertainty than if it is main storage resident with only the data resident in bulk storage. Also, if room must be made for this portion of the console program to be brought into main storage, an exchange burden is created. This is because the displaced program or data must be saved in a bulk storage save area and later returned to main storage.

7.4   CONSOLE EQUIPMENT FACTORS

With the capable, versatile, and convenient console equipment that is available, there are almost no limits to the comprehensive features that can be built into a central console. The main constraints are programming and computer storage. Nevertheless, there has been a strong reluctance to build sufficient capability into the console equipment, and their full development has been delayed.

Reducing the number of alphanumeric display tubes, status lights, and pushbuttons, or replacing them with inferior devices, results in cumbersome and confusing console operation. Buttons become multipurpose and multisequencing, displayed information becomes piecemeal, and important functions are left out altogether. For this reason it is important that the designer not economize by omitting console features. Similarly, the user should not insist on their elimination; rather, users should recognize that the efficiency offered by the comprehensive console may more than offset its increased cost.

The measure of operating convenience in central console design is gauged by operating time and motion studies, by the degree and clarity with which needed information is presented, and by the ratio of operator mistakes to the number of successful operations. The equipment design greatly influences the operating convenience. Of course, the program design may have an even greater affect on operating convenience, but the programming rarely overcomes equipment design weaknesses. The programming can accentuate both the good and bad features of the console equipment. Thus the place to start improving the convenience of central console operations is in the design of the console.

7.4.1 General Design

To cover a broad spectrum of applications, a series of central panel designs, small, medium, and large, may be required. An expansion series with common features and equipment with upward growth is a natural choice. When each member of the series is programmed, however, the programs probably will not be upward compatible. Nevertheless, for manufacturing and maintenance reasons, an expansion series is a wise choice. Perhaps a single central display panel design would serve the needs of the process industries, with other members of the series aimed at the discrete manufacturing industries, laboratory-medical uses, and other applications.

Maximum capability, versatility, and convenience should be the design objectives for the central console equipment, limited, of course, by a given manufacturing cost range. Experience indicates that a general design for the central console for the process industries will be widely accepted if it has enough capability and versatility. Early consoles had limited capability and often were tailored to specific applications. For general acceptance, a central console had to have so much more capability that most control computer manufacturers were reluctant to place such an expensive console on the market. However, the underestimated cost of programming unique consoles for each application has made the more expensive and more capable general console designs widely accepted by experienced suppliers and users. As more general and comprehensive console programs have been developed, the need for even greater capability in the console equipment has been recognized.

As the console equipment is made more capable, care must be taken to keep the consoles as versatile as possible. Grouping of common features achieves a large measure of this desired versatility. The character display in the central panel, when grouped to provide multiple rows of multiple alphanumeric characters with the same number of characters in each row, achieves great versatility. The status displays grouped in one, two, or more groups in a uniform arrangement offer good versatility. The many request buttons may be assigned several types of functions by uniformly arranging them into one or two groups.

Once status and request buttons (or their equivalent) are grouped, their assignments can be made flexible by means of labels and colors. If glued-on plastic labels completely surround the buttons, the user can frame various groups and name each group as well as name each button. The assignment of colors to the status buttons usually denotes degrees of importance or alarm. Color assignments to the opaque request buttons provide division between types of request functions. One color, such as red, may be reserved, however, to mark buttons with critical functions such as ENTER and UNIT OFF CONTROL. These red buttons may also be the only ones that require the confirmation step described in Sec. 7.2.4.

Requests of all kinds are made most conveniently when only one operating motion is required, such as pushing a single button. The present design of consoles uses many momentary buttons to request specific items of information because these buttons are more convenient to operate than other devices such as the rotary switches and dials which were common with computer consoles of the previous generation. These entry devices are discussed in more detail in Sec. 7.4.3. Addressing variables and data entry are also more convenient with keyboards than with rotary switches and dials. This has been confirmed by operating experience in many industries, including the communications industry where the new pushbutton telephones are obviously more convenient to use than the dial type.

Maximum operational safeguards have been stressed in many sections of this chapter and are mostly built into the console programming system. One safeguard, however, should be built into the console hardware. All status lights and displayed information should be controlled only from the computer. No switch should turn on a light directly. This safeguard ensures that when displayed information confirms an entry, the entire flow of information from an entry device to computer and back to the display device was error free.

Other general design factors include colors, contaminants, wear, and panel dimensions. Console background colors are best kept neutral to avoid visual conflicts with the assigned colors for the buttons. The problem of greasy and dirty fingers should also be considered in choosing background colors. Furthermore, the panel finish should have a strong resistance to abrasive wear. All engraved legends should be in black, the color they will eventually become after long use. Dust and dirt are problems that make cleaning difficult when ledges, protruding bezels and rough surfaces are present. Another common contaminant is spilled coffee. The design of panels that are almost horizontal should not leave any space for a cup of coffee! This is an old and well-founded Navy regulation for electronic equipment.

Panel dimensions should be at least 24 in. wide; this is a standard rack size. The narrower 19-in. rack size usually is not adequate. In general, the dimensions for the central panels should be large rather than small. Crowding should be avoided and is not really necessary in most designs.

7.4.2   Display Devices

Individual character tubes and character projection devices have long been used in computer control operator consoles for information display. Many of these devices display only the numerical characters 0–9, and others display only a choice of 12 items which may be short words and may be of different colors. These types are well suited for certain limited consoles tailored to particular applications. However, they lack the information display capability and versatility required for a central console. A device that displays any numeric, alphabetic, or special character is much better suited to central console use. The computer time required to cancel and display each new character on each of these devices must be short, on the order of microseconds, to minimize computer burden. The total computer burden for this function should not add up to more than a smal percentage of the system. The display devices should also maintain the displayed character individually to economize on computer interface equipment and cables. With local character storage a single code bus can be wired to every character display device, and each device can be switched onto this bus to receive a new code, then switched off the bus.

Lighted pushbuttons are popular devices for the display of status information and for the direct entry of status. The relative importance of the status is often emphasized by color, and the status information usually is engraved on each lens. If glued-on plastic labels completely surround the buttons, the user can designate lines to frame the buttons into groups and can label each group. The number of status display and entry devices on the central display panel should be at least 20 for most processes. This provides for an adequate display and entry of system status and a fairly complete display and entry of status for the displayed process variable. More status display and entry devices can be added as auxiliary panels mounted near the central panels. The alarm and parallel address panels are an example of this kind of auxiliary display.

Auxiliary message displays should be capable of showing a comprehensive demand log summary for a process or a portion of a process. Such a summary is read by an operator seated or standing close to the display; consequently, the size of the characters may be as small as 0.25 in. Summaries tend to be rather wide, and this suggests a rectangular display.

At present, two types of CRT displays are available for the central display and the auxiliary message and graphical displays, the short-persistence screen type and the very long-persistence type. With the short-persistence screen, the picture has to be regenerated about 40 times/sec from computer storage or from a special storage associated with the display equipment. This type of display system often incorporates a character generator to form the characters. The characters may be formed by a matrix of dots or with strokes. The stroke or vector method usually provides a clearer display.

With the long-persistence screen, the picture is stored in the phosphors on the face of the tube. The screen is usually driven by analog signals which can form both characters and graphical curves. The analog signals are derived from digital-to-analog converters for each axis; digital signals are used for erasing. The screen may be erased completely or by quarter sections; this is a major difference from the short-persistence screen displays in which a single character can be erased if this feature is supported by the program. The storage screen display equipment is less costly, although at present it has a shorter screen life.

Auxiliary trend displays should be capable of showing messages along with strokes or curves for trending. A set of variable trends displayed without identification in a central console is a severe restriction. However, only an identifying code name is needed for each set of curves and a good character size is 0.5 in.

Strip chart recorders are used in every console, but their numbers have been decreasing as the practice of sharing them between several variables has spread. Each regular size recorder has one or two pens, and the larger size recorders can trend 6–12 variables. The recorders are usually designed to accept the standard live-zero current signals which are converted to a 1–5-V range for 0–100% of scale. Some multipoint recorders are designed for thermocouple inputs. The recorder chart drives are continuous and represent time, usually 0.75 or 1.0 in./hr.

The one- and two-pen regular pen recorders may be computer driven by individual digital-to-analog converters, but the distributed analog-output systems developed for direct digital control (DDC) are more economical. This system, discussed in Chap. 5, uses a single digital-to-analog converter to store analog signals in analog memory circuits associated with each output point. The output of the analog memory circuit then drives the recorder.

In order to send historical trend data conveniently to a recorder, a specially designed high-speed recorder is required and a storage screen CRT is usually a less expensive alternative. The chart drive must be incremental to synchronize chart movement with data retrieval from the computer bulk storage and presentation as an analog signal to the pen. Computer timing between data values varies because this task cannot be given utmost priority in the entire computer control system. To record historical values in a reasonably short time for operational use, the response time of the pen must be better than the common process recorder. Together these factors lead to an expensive recorder.

7.4.3 Entry Devices

The momentary pushbutton has taken over the role of the most common entry device in current console designs. It may be opaque because the effect of the entry is displayed in the central display where the operator's attention should be focused. Its momentary depression should produce a coded interrupt signal to the computer; this is accomplished by multiple contacts on each switch, by a single contact on each switch and a diode code matrix, or by solid state switches and circuitry. Some of these momentary pushbutton switches will be actuated very frequently and a great number of times. As a result, their life expectancy must be long. Their feel to the operator should be a positive snap action with a short movement and the pressure required should be fairly light and uniform from switch to switch. The spacing between the pushbuttons on the console should be a minimum of 0.75 in. to accommodate large fingers and gloved hands. The buttons should be supplied in several colors, and a replaceable label should be provided around every pushbutton to name its function.

The interlocked pushbutton, which remains depressed after being pushed until another in its set is pushed, has many applications in special dedicated consoles. This type of switch, however, has no place on the panel of a central control console. The few instances where such an entry device could be useful are offset by their different computer interface requirements. Because they remain depressed, their contacts remain closed and this eliminates the use of a coded interrupt to sense their position. As a second means to sense them is added to the computer interfacing panels, the cost and complexity of the panels is increased. If this type of pushbutton switch also contains a light or appears to contain a light when depressed, this extra feature should be used only as an aid in identifying which buttons are depressed when their positions are compared with the typed confirming message. The lights or pseudo-lights must always be confirmed if the computer did not turn them on because the computer may not sense the pushbutton switches accurately.

Rotary switches are used on many consoles, including computer maintenance and programmer consoles. On central control consoles they largely have been replaced by the popular momentary pushbuttons. Rotary switches are slow to use. They often are hard to turn and hurt the operator's fingers because long-life design requires that they be stiff. The names for the switch positions either must be crammed onto the moving dial and mostly read sideways or must be engraved on a label which surrounds the dial. Occasionally, a numbered dial is placed behind the panel such that only the number for the current position is visible through an opening. This design is unsatisfactory for dials with letters or names since all but one letter or name are hidden. When several rotary switches are required to select one function from a great many functions, the desired rotary switch position must be set and the other rotary switches set to a "home" position before a console read button is pushed. This is a clumsy operation. The computer interface equipment for these switches is similar to that required for the interlocked pushbuttons. This is why one or two of them are not added to a central console with many momentary pushbuttons. Nevertheless, when only a small number of console functions are to be provided and no display is needed, a rotary switch to select the function, one to four more to enter data, and a pushbutton to request a read action constitute a very practical console for limited applications. A small printer or a typewriter without a keyboard can provide confirmation of data entries and answers to requests for information.

Thumbwheel switches are similar to rotary switches with their dials behind the panel, and they have the same limitations. They do have the advantage of saving space if this is an important consideration.

Toggle switches are an excellent means for entering binary or on-off information and requests. They are commonly used in large quantities on computer maintenance and programming consoles. Their use on operators' consoles would probably be limited to use on auxiliary side panels to turn on equipment such as pumps and motors, very likely without going through the computer at all. Toggle switches are subject to inadvertent flipping by sleeves and glove cuffs and are not suitable for important console functions. In addition, binary information is rarely entered through the operator console so that decimal entry devices are required.

A dial switch, such as that used on a telephone, can be an extremely inexpensive and reliable console in itself for some very limited applications. A small printer or a typewriter without a keyboard can provide confirmation of dialed data entries and answers to information requests. The computer interface wiring is very minimal for one dial switch, and the computer program burden from repeated interrupts from the dial-generated pulses is negligible. Of course, a dial switch has no place on the central operator console except on the operator's telephone, and even here a pushbutton phone with a keyboard similar to the direct addressing keyboard on the central panels is more appropriate. The dial switch is slow, limited to code and data entry, and it hurts the operator's fingers if used for long periods of time. These problems are not faced in very simple and limited applications where the dial switch is an appropriate device.

A light pen is often available as an entry device with advanced CRT displays. The computer receives a coded signal from the console that gives the coordinates of the point on the CRT screen where the light pen is pointing. A momentary foot-switch or a switch in the light pen activates the pen position sensing system. Acceptance of this type of entry device by process people for central operator consoles has been limited so far. The pen is usually fragile, has a signal wire attached to it, and, in general, is foreign to plant operators. The concepts represented by the light pen, however, have many possibilities and future console equipment developments probably will be based on its use. Basically, a large array of momentary pushbuttons represents a grid to which the operator points with his finger. By displaying the grid on the display screen, the entry procedure may be tailored to each type of console function by changing the displayed format of the grid. The operator merely points the light pen at the desired function to initiate the console action. The possibilities are indeed exciting. Central operator console equipment and programming definitely comprise a technology whose growth is just beginning.

BIBLIOGRAPHY

Bailey, S. J., "Pushbutton Programming for On-Line Control," *Control Eng.*, August 1968, pp. 76–79.

Bernard, J. W., and J. W. Wujkowski, "Direct Digital Control Experiences in a Chemical Process," *ISA J.*, December 1965, pp. 43–47.

Brittan, J. N. G., "Nature and Structure of Man-Computer Communication in Team Control Systems," *Comput. J.*, November 1967, pp. 219–226.

Callisen, F. I., "Control Rooms of the Future," *Chem. Eng.*, June 2, 1969, pp. 107–111.

Gandsey, L. J., "Two Way Communication Between Operator and Computers," *Chem. Eng. Progr.*, 61(10), 93(1965).

Grimsdale, R. L., and M. Barraclough, "Cathode Ray Output for Digital Computer," *J. Br. IRE*, June 1961, pp. 497–501.

Hubbarth, W. F., "Shrinking the Man-Computer Interface," *Control Eng.*, August 1965, pp. 63 ff.

Kern, J. L., "The Computer-Operator Interface," *Control Eng.*, September 1966, pp. 114–118.

Laird, R. F., Jr., "The DDC Control Center Buffer Zone Between Man and Machine," *ISA Prepr.* 30.2-3-65, 1965.

Larsen, M. J., and G. L. Glahn, "A User Designed Operator Console," *ISA J.,* April 1968, pp. 45–51.

Machover, C., "Information Display and Data Acquisition Systems," *Pap. 68-704, Proc. 23rd ISA Conf.,* 1968.

Mahan, R. E., "Data Display: Part I: System Characteristics," "Part II: Systems Hardware," *Instruments and Control Systems,* March 1970, pp. 81–84; April 1970, pp. 131–135.

McCormick, E. J., *Human Factors Engineering,* 2nd ed., McGraw-Hill, New York, 1964.

Robertson, D., "Graphical Display for Process Communications," *Case Western Reserve Univ. Rep. SRC-68-4,* June 1968.

Rose, G. A., " 'Light Pen' Facilities for Direct View Storage Tubes—An Economical Solution for Multiple Man-Machine Communications," *IEEE Trans. Electron. Comput.,* August 1965, pp. 637 ff.

Roth, J. F., "Output Devices for Industrial Computer Systems," *Ind. Electron.,* July 1963, pp. 511–514.

Smith, C. F., "Graphic Systems for Computers," *Comput. and Autom.,* November 1965, pp. 14 ff.

Stark, D. E., and R. A. Sylvester, "A Multi-Station Remote Control and Data Handling System with Flexibility," *Proc. ISA,* (18)2, 1963, pp. 6 ff.

Thode, E. F., and W. J. Sewalk, "Talking to Computers," *Chem. Eng.,* November 22, 1965, pp. 147 ff.

Tolcott, M. A., A. N. Chambers, P. Preusser, and staff, "Human-Factors Engineering Design Standards," *Electro-Technol.,* August 1961, pp. 103–109.

Union, D. C., "Addressing Methods for a Process Control Computer," *Instrum. Technol.,* May 1969, pp. 48–50.

Weisberg, D. E., "Man-Machine Communication and Process Control," *Data Process. Mag.,* September 1967, pp. 18–20 ff.

Weisberg, D. E., "Graphic Displays; Matching Man to Machine for On-Line Control," *Control Eng.,* November 1968, pp. 79–82.

Welbourne, D., "Alarm Analysis and Display at Wylfa Nuclear Power Station," *Proc. IEE,* November 1968, pp. 1726–1732.

Williams, T. J., "Direct Digital Control Computers—A Coming Revolution in Process Control," *Proc. Texas A&M 19th Ann. Symp. Instrum. for the Process Ind.,* January 1964, pp. 70–81.

Woodson, W. E., and D. W. Conover, *Human Engineering Guide for Equipment Designers,* 2nd ed., Univ. Calif. Press, Berkeley, Calif., 1964.

Chapter 8

System Design Factors

RICHARD E. MACH

Compatibility between the component units that make up a computer process control system is fundamental to good system design. The component units described in previous chapters must be carefully coordinated if overall system performance objectives are to be achieved. These units, categorized as the processor, the data-processing input/output (I/O) equipment, the process I/O hardware, and communication-based terminal units, are not inherently compatible. The processor, including the central processing unit, main storage, and data flow control channels, is composed of digital logic and shares properties common to high-speed digital circuits. The data-processing I/O equipment, however, is primarily electromechanical in nature and is relatively slow compared with electronic logic. The process I/O and communication units consist of both digital and analog circuits. The analog circuits are not as fast as the logic but they are considerably faster than the electromechanical devices. This incompatibility in speed, as well as in other characteristics, is not unique to digital process control computer design but is part of any computer design problem. Compatibility is achieved in the system structure by use of techniques such as buffering, interrupt, overlap, and cycle stealing.

Compatibility between component units is not sufficient however. The component units also must be compatible with the environment, with the process inputs and outputs, and with the needs of the human interface.

Although the environmental and human interfaces suggested in Fig. 8-1 are required in any digital computer system, the process control application imposes unique requirements. For example, the environmental operating limits often must be extended beyond those of the usual business or scientific computer to cope with the industrial environment. The human interface must also be more extensive to satisfy needs ranging from those of the semiskilled operator to the sophisticated system programmer. In contrast to these interfaces, the process I/O interface is a totally unique feature of the process control computer. Its existence is a result of the necessity for a direct interface between the computer and the real world of the process installation. The process I/O require-

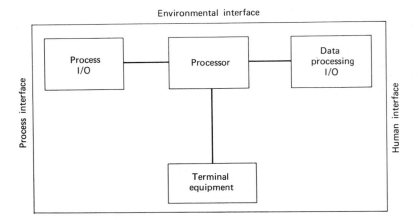

Fig. 8-1   System component interfaces.

ments are many and varied and place unique interfacing requirements on the system designer.

The compatibility of the computer system and the process often depends on the particular installation as well. For example, if the designer is utilizing existing process hardware, adaptation of old and/or new hardware may be necessary to meet overall system needs. Even if the design is totally new, the need for future expansion or alteration must be given thorough consideration in developing a satisfactory total solution.

Since many of the component units have unique attributes which require special consideration when applied to process control, every aspect of design and application must be evaluated. For example, one of the most restrictive limits on data-processing I/O relates to environmental restrictions on the I/O media itself. This has special significance in process control system applications because office-type environments cannot be tacitly assumed. Punched cards and printed paper forms are sensitive to variations in relative humidity. If the air is too humid, the paper becomes limp; if it is too dry, static electricity may be a problem. This is only one example, but it illustrates the need for careful examination of compatibility in a digital process control computer.

Many of the interface compatibility requirements are discussed in the previous chapters dealing with specific subsystems. In this chapter, attention is focused on design factors that are common to most of these subsystems or that are related to the compatibility between subsystems or between the process and the total system. The objective is not to provide an exhaustive design guide, but rather to make the reader cognizant of many of the important areas that are sometimes overlooked in the early stages of system specification. Consideration of these areas is essential to good system design.

8.1   SYSTEM DESIGN REQUIREMENTS

Compatibility between the component units which constitute a digital process control computer is assured by careful consideration of overall design factors. The total system is, in a sense, the sum of its subsystems plus the effects of a thorough overall system design. In addition to the coordination of the components units of the system, the overall design must also include considerations common to all subsystems such as packaging and power. These factors, considered mundane by some engineers, are extremely important in the effective implementation of a process control computer project.

8.1.1   System Architecture

Architecture, according to the dictionary, has several meanings: The first is, ". . . the art or practice of designing and building structures . . . ," and the second is, "formation or construction as . . . the result of a conscious act." Both definitions are appropriate when applied to the design of a digital computer system. The digital computer control system is a structure that must be designed and built as the result of a conscious act. It is the responsibility of the system architect to define the structure of the total system and to ensure the compatibility of the subsystems that make up the computer system. His concern is not with the detailed design of the subsystems but rather with the interrelationship between the several subsystems. This includes the interaction of the hardware and software as well as interactions between the hardware component units. Therefore, the system architect must have a thorough understanding of all aspects of process control computer design, including the specialized knowledge involved in unique subsystems such as those for process I/O functions.

Much of the system architecture of the process control computer involves considerations common to any digital computer. For example, the basic speed incompatibility of electronic and electromechanical devices exists in both types of system. Similarly, the trade-off between hardware and software is a design factor in the general-purpose computer as well as in the process control computer.

There are, however, unique architectural aspects in the design of the process control computer. One of these is the requirement for customizing the system to satisfy the needs of each particular application in the broad spectrum of the control system market. The design of a completely customized system for each particular application is economically impractical since the benefits of mass production are negated. Thus one of the first questions that must be addressed in the design of the system is which capabilities are to be provided as standard features, which are to be available on a special-order basis, and which are not to be offered by the vendor. The answer to this question involves both business and engineering considerations and the system architect must be a key participant in the decision.

Independent of which features are offered as standard and which as specials, the application always arises in which an unavailable capability is required. The architecture of the system must be such that additional features can be incorporated in the system without major redesign. For example, a common requirement is the attachment of equipment which is not manufactured by the system vendor. The architectural requirements for such an attachment must be anticipated early in the design if the ultimate marketability of the system is not to be degraded. Since the total range of these requirements cannot be anticipated, the system architecture should stress interface structures that allow the attachment of asynchronous devices. Related requirements include provision for extra power, space for special circuits, and flexibility of the programming system.

Even with standard features, the application requirements pose unique problems. In the business and scientific computer the number and types of standard features are limited because the computer system is relatively self-contained. In the process control computer, the range of features is typically much greater and the number of possible system configurations may be quite large. For example, in one analog-input subsystem, a limited number of features resulted in over 75 unique configurations. This complexity is increased even more when all the subsystems are considered.

The architectural significance of this range of features is a complex subject and only a few aspects of the problem are mentioned here. Many of the considerations are economic and competitive. For example, the system must be competitive in both large and small configurations. Packaging features with small unit modularity generally results in an economical system for small configurations but an expensive system when a large configuration is required. This is because the cost of the package does not increase at the same rate as the physical space provided by the package. Similarly, it is sometimes less expensive to provide logic controls for 1000 analog-input channels as a standard feature, for example, than it is to provide less control capability plus the ability to expand the function. The trade-off is between extra logic and increased complexity in the specification and construction of the machine.

Another architectural consideration is the result of the characteristics of the process plant being controlled. In many industries a process installation is expected to last for 20 years or more because the process technology evolves very slowly. Computer technology, however, is changing rapidly and every few years sees a marked decrease in the price/performance characteristics of the central processing unit. The peripheral subsystems, such as process I/O, do not undergo as rapid a change and, therefore, may be suitable and economical for many years. It may be desirable, therefore, to have the capability of installing a new central processor without altering the process I/O subsystem and the process. The ability to do this requires careful architectural design to ensure that the subsystems are compatible with future generation processors. Although it is unlikely that the subsystems will last for the entire life of the plant, it is quite possible that they can be used with several generations of processors.

A closely related consideration is expansion of the system after initial installation. Most processes change as a result of improvements and the computer must often expand to provide increased capability. In addition, new control philosophies may require modification to the initial capability of the system. Thus the design of the computer control system must be such that features can be added or deleted with minimum disruption of the hardware and software. This requires careful planning with particular emphasis on interfaces, modularity, power, and programming system design.

These are only a few of the unique architectural considerations in the design of the process control computer system. They illustrate that the overall system design factors must be specifically addressed if compatibility between subsystems and between the system and the process is to be achieved.

8.1.2 Packaging

The physical packaging of the electronic and electromechanical components in a process control computer is an important aspect of the overall design. The packaging problem is more difficult than in the case of the usual computer because of the unique requirements of modularity, environment, customer access, and special terminations. One of the main impacts of modularity is the cost of the package. The size of the physical package can increase many times with only a nominal increase in cost. For example, a small cabinet enclosing about 30 in.^{3} of volume costs about \$1 as an off-the-shelf item. A cabinet of 950 in.^{3}, more than 30 times larger, costs only four times as much. Thus a small packaging modularity can be very detrimental to the cost of large systems. On the other hand, the competitive nature of the market may preclude the use of a larger package for a small system. A compromise based on the anticipated distribution of large and small machines in the market must be negotiated.

The packaging generally used for computer logic consists of printed circuit cards which are plugged into backpanels or boards with printed circuit or hardwire connections between card sockets. The boards are often mounted in hinged or sliding gates. This approach results in a economical package that is compatible with the requirements of high-density logic circuits. Since the processor and some of the subsystems in the process control computer system are predominantly logic circuits, this type of package is suitable and is often used. The decision to use it, however, must be reviewed in the light of the unique requirements imposed by the industrial application.

One of the unique characteristics of the application is that the usual computer packaging concept is not familiar to the process control user. Most industrial instrumentation is available in small modular packages suitable for rack and panel installation. Typically, these units are designed to be compatible with the standard 19 or 24 in. rack and panel assemblies. They provide for easy expansion of the system as the needs of the application change. The higher cost of the small package is partially offset by the availability of many off-the-shelf unit packages in a wide variety of shapes and sizes.

The environmental requirements for the package are also an important consideration. The usual computer package is designed primarily for office or laboratory environments. The unpredictable industrial environment is often more easily controlled in a small modular package that can be sealed and/or purged with instrument air. The environmental requirements of packaging are discussed in more detail in Sec. 8.2.

The peripheral subsystems of the process control computer often require special packaging. The analog I/O subsystem, for example, has unique requirements for high insulation resistance, the ability to handle high common mode voltages, isothermal areas, and shielding. The requirement for terminating many input lines and providing easy customer access to the terminations is common to both the digital and analog I/O subsystems. A package designed for high-density logic is often not compatible with these requirements. These considerations are also discussed in Chap. 3 and in Sec. 8.3.

In addition to these functional criteria, the package must satisfy the aesthetic standards of the modern marketplace. Function and aesthetics are not incompatible and the proper objective of industrial design is a product that is both functionally correct and aesthetically pleasing. Examples of areas in which the engineer and the industrial designer must cooperate include panel design, selection of finishes, and overall package shape and appearance. Panel design must include the human factor considerations of Chap. 7 which tailor the function to the needs of a wide variety of skilled and unskilled operators. The selection of surface finishes and package shapes is often affected by the industrial environment. Rough, pebble-like surfaces suitable for the clean environment of an office may not be acceptable in the industrial control room where equipment must be easily cleaned because of the dust-filled atmosphere of the cement plant or blast furnace. Similarly, packages that include crevices and dirt-catching protrusions may be aesthetically pleasing but functionally deficient.

8.1.3 Power

As in the previous sections, many of the power supply design factors in the process control computer system are the same as in the business and scientific computer. For example, power supply capacity, sequencing, and overvoltage protection must be considered. However, the unique requirements of the process control computer must also be evaluated. The modularity and expansion of the system requires a decision as to whether total power supply capability is included in the basic system configuration or whether it is provided with each unit of modularity. The former is most economical in large systems because of the base cost of a power supply. That is, the capacity of the supply can be increased considerably for only a nominal increase in cost. In addition, the need for power supply sequencing and control circuits often adds additional cost when modular supplies are used.

The unique power requirements of the peripheral subsystems must also be considered. The digital I/O subsystem requires instantaneous power surge capacity to supply the needs of relays and solenoids adequately. This is similar to the requirements for electromechanical devices such as printers and punches.

The analog I/O subsystems require carefully filtered supplies to avoid crosstalk, feedback, and noise interference. In critical cases, the analog supplies cannot be shared with electromechanical equipment because of transient noise considerations.

The need for careful planning of the analog subsystem grounding structure is discussed in Chap. 4. This same consideration must be included in the overall architecture of the system. Unless the grounding structure is consciously designed, the performance of the analog subsystem can be severely degraded owing to ground loops and other coupled noise sources.

The reliability of many industrial primary power sources is of concern in systems used for real-time control. The economic trade-offs between occasional loss of control resulting from a power failure and the cost of back-up motor-generator sets or batteries depends on the particular application. The system design must be such that emergency power sources can be incorporated easily and effectively. One example of the situations that must be anticipated is a dip or momentary loss of power. Such a transient must not result in control devices being reset or altered since this could cause a hazardous condition in the process. Thus the power supply stability and the method of switching over to emergency power is an important design consideration.

8.2 ENVIRONMENTAL INTERFACE

The physical environment is very important in most control system applications. Environmental parameters of temperature, humidity, contamination, shock, and vibration are usually much more severe and more variable than in the business office environment. A discussion of these environmental factors and necessary measures to ensure satisfactory system performance is presented in the following sections.

8.2.1 Temperature

This parameter varies over significantly wider ranges in industrial process installations than it does in office situations. In some cases, equipment is installed in sheds with little or no temperature control, or it is installed in close proximity to furnaces or other sources of heat. The need for equipment operation over the temperature range in which a man can be expected to work presents itself quite often. This may mean a temperature spread of 40–120°F without even taking outdoor installations into consideration. Wide temperature variations affect both mechanical and electronic components of a process control computer system.

The impact on mechanical devices, such as the data-processing I/O units, can be especially severe. These devices usually are designed primarily for office environments. Wide temperature variations result in performance degradation as a result of such factors as changes in precision fits and lubrication problems. Although the temperature capability of this type of equipment has gradually improved, it often requires special consideration. In cases of severe environment, devices such as data-processing I/O units are installed in special rooms where the environment can be controlled. Another possible approach is to enclose the unit in a protective cover that serves to maintain a micro-environment. This approach is most applicable for smaller units such as typewriters and small printers.

The impact of temperature on electrical/electronic devices is more subtle. Operation and performance are often degraded rather than completely inhibited. Temperature variations alter component values and this causes changes in circuit parameters. Both low and high extremes induce changes. Furthermore, operation at high temperature for extended periods of time shortens component life.

The effect of temperature variations on digital circuits is manifested in power supply variations and component value shifts. This results in reduced fan-out (ability of one circuit to drive many load circuits) and reduced ability to differentiate between "up" and "down" signal levels. For analog circuits, the supply and component value variations affect total accuracy of the subsystem. Specifically, the sensitivity, zero offset, and linearity can be affected by these variations.

The change in electronic performance resulting from temperature variations is minimized in one of two ways. One approach is to guarantee parameter temperature stability through the use of components which exhibit only a minimal change in properties with temperature and to use circuit designs which are inherently less temperature sensitive. Components with known temperature coefficients are sometimes used in this approach to balance out temperature-induced changes in other components. For example, resistors having a positive temperature coefficient may be used to compensate the effect of the negative temperature coefficient of copper conductors.

A second approach is to provide temperature-controlled enclosures for sensitive components. In some cases, it may be necessary to enclose an entire subassembly in an oven. Often, however, the temperature characteristics of a circuit are determined by a few components and only these need to be maintained in the oven.

The concept of thermal resistance is useful in considering cooling approaches. Components dissipate heat. This heat is dissipated to some cooler region, a heat "sink." The thermal resistance of the path between the heat source and the heat sink determines how high the temperature of the power-dissipating components rises above the sink temperature. Generally, the ambient air is the ultimate heat sink for a component. Drawing an analogy between thermal and electrical phenomena, as shown in Fig. 8-2 where temperature corresponds to voltage and heat

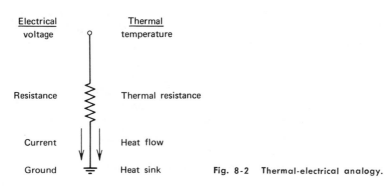

Fig. 8-2   Thermal-electrical analogy.

flow to electric current, as the thermal resistance increases, the temperature of the heat-dissipating component increases above the ambient. If the ambient is already high, the temperature of the component becomes excessive, and component life may be seriously reduced. The designer must keep the thermal resistance between each component and the ambient air as low as possible.

In most cases, the primary mechanism for removal of component heat involves convection, with air as the heat transfer medium. Thermal resistance is minimized by providing a free flow of air between the component surface and the ambient air heat sink, as shown in Fig. 8-3a. The industrial process environment complicates this situation. Contaminated atmosphere and/or outdoor installations often require sealed enclosures, which precludes the possibility of providing a direct free-flow air path between the component and the enclosure ambient air. In this case, the heat content of the air inside the enclosure must be transferred to the walls of the enclosure via convection and from the walls to the outside ambient air via convection, as in Fig. 8-3b. The additional convection transfer paths greatly increase the thermal resistance between the component and the ambient. Cooling fins and the use of fans to circulate air inside the enclosure are beneficial but add to the cost. Also, the reliability of fans operating at the high temperatures that may be experienced inside a sealed enclosure under high ambient temperature conditions is generally poor.

Alternatives to air convection cooling exist. Unfortunately, most of these introduce significant technical and/or economic problems. Examples of such alternatives include refrigeration, liquid convection, and conduction cooling.

Refrigeration may be applied at the unit enclosure level or it may be incorporated via cooling of a special equipment room. In the latter case, atmospheric contamination often remains a problem, since the contaminants are gaseous in nature and are not readily removed from the room air by filtration. A source of clean air must be ducted to the room air inlet. Refrigeration can also cause servicing problems since condensation may occur if the cooled enclosure is opened in a hot, humid, ambient environment. Applied at the unit level, refrigeration costs are not insignificant, running about $400 and up for a typical unit enclosure, depending on the heat load.

Liquid convection cooling techniques substitute a liquid for air as the internal heat transfer medium. The thermal resistance associated with liquid transfer is considerably less than that of air. This approach has been common in the design of power transformers for many years. When the concept is extended beyond the basic transformer materials and

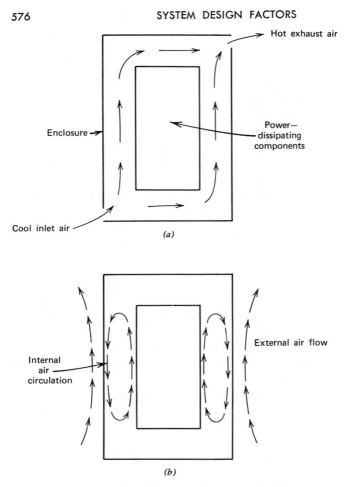

Fig. 8-3 Enclosure cooling. (*a*) Free-air flow enclosure; (*b*) sealed enclosure.

service requirements, however, the selection of suitable coolants at reasonable costs becomes quite difficult. A wide variety of materials are used in present-day electronic fabrication. The coolant selected must be inert with respect to all these materials. Furthermore, if the unit package is other than a throw-away module, the liquid coolant residue remaining on circuit boards and wires after removal from the enclosure must either be readily removable with some inert solvent or it must not interfere with unsoldering and replacement of faulty components. Specially formulated liquids are available which meet or come very close to meeting these requirements. However, they are costly, typically $20–$50/gal.

An additional problem with liquid cooling is the maintenance of adequate cooling during troubleshooting when the unit is removed from the liquid for access.

Conduction cooling is a third alternative. In this approach no attempt is made to facilitate convection cooling as in the previous cases. Rather, a direct path of solid materials is provided between component and a heat sink such as the enclosure frame. If the solids are good heat conductors, as are most metals, low thermal resistances can be achieved. The practical limitations of conduction cooling include cost and serviceability, as in the case of liquid cooling, but for different reasons. Conduction cooling of solid state components, such as transistors and integrated circuit modules, depends on the formation of a low thermal resistance boundary between the component and the thermal path member, as shown in Fig. 8-4. This means a solid mechanical contact must exist between the thermal path member, usually a metal, and the component case and/or leads. This is not readily achieved in a manner that is serviceable. Pressure plates to press against the component case are limited by varying component sizes and shapes. Potting or encapsulation in plastic does not result in a serviceable unit below the potted assembly level. However, it is used in some cases with thermally conductive fillers to promote conductivity. One approach that has been applied in some military applications is to improve the thermal conductivity of the printed circuit board by cementing the back of the card to an insulated metal heat sink. This sink forms the main path for heat removal via the component leads between the component and the card. Again, the limitation is the level of repairability.

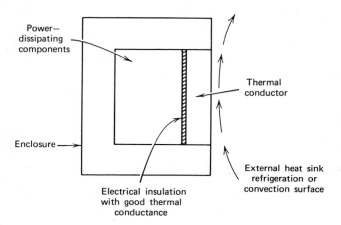

Fig. 8-4   Conduction cooling.

An additional problem in conduction cooling is the requirement for electrical isolation but thermal transmission. Since most good thermal conductors are also good electrical conductors, the choice of a suitable insulating barrier material is limited. A thin layer of a material with high dielectric strength is a common solution. Anodized aluminum is also suitable, although care must be exercised in assembly to ensure that the anodized surface is not scratched. In addition, the anodized barrier is hygroscopic and electrical leakage under high humidity conditions may be a problem.

Independent of the cooling approach, certain aspects of the thermal problem remain unchanged. When assembling packaged components in an enclosure or rack, the cumulative heating effect of the units mounted below a given unit must be considered when determining the ambient temperature. Even in a well-ventilated enclosure, the temperature of the air passing the top units may be 20–30°F above the air temperature at the bottom of the enclosure. Furthermore, the heat dissipated by fan motors is an additional heat burden on the cooling air.

Radiant energy is another heat load to be considered in the thermal design of enclosures. Outdoor installations are always faced with this problem. The temperature rise due to solar heating of sealed enclosures can easily be 40–50°F in the summer time, depending on the shape of the enclosure, its color, and its geographic location relative to the horizon. Although light colors, such as white, minimize radiant energy absorption, enclosure color should not be depended upon for radiant energy absorption control since the enclosure may be repainted a darker color at some future time. Radiant energy from nearby sources, such as furnaces or rolling mill stations, can also be present, even in indoor locations. The radiant load can be even more severe than that produced by solar radiation.

One method of controlling radiant energy input is by shielding the enclosure with a protective panel which blocks the radiant energy but allows free-air circulation between the panel and the enclosure. For outdoor installations, a simple panel over the top of the enclosure is often sufficient. However, field installations of sun covers or sheds to protect the enclosure adds to system cost.

The effects of high temperature on a working system are twofold, degraded system performance and shorter component life. In any outdoor installation, maximum temperatures do not exist continuously. Relatively high air temperatures are only present for a portion of the day, and then usually only during the summer. Expressed as a percentage of the whole year, the maximum temperature situation may constitute about 5% of the total time, depending on the location. If degraded performance can be tolerated for these relatively few periods, a worst-case design is not necessary. In addition, component life, although shortened by exposure to high temperature, is not degraded to the extent that it is if exposure is continuous.

8.2.2 Atmospheric Conditions

In the last section, the effect of sealed enclosures used for protecting electronic and electromechanical hardware from weather and contaminated atmosphere was discussed relative to its impact on thermal design. This section considers protection from these elements and a third agent, humidity. The atmospheric conditions present a common problem to the designer in that extended exposure results in degraded electrical performance and mechanical/electrical failure due to corrosion and wear.

Humidity and gaseous atmospheric contaminants result in oxidation and corrosion which can degrade or inhibit the proper operation of electromechanical devices. Airborne dust causes wear due to abrasion of moving parts. In addition, the life of bearings in high-speed equipment can be shortened appreciably.

Although inert dust does not affect electrical performance *per se*, dust in combination with humidity can result in low-resistance leakage paths in electrical circuits. If the dust is conductive, short circuits are possible. Similar situations occur when condensed humidity causes the ionization of salt residues between adjacent conductors. Salts and other corrosive agents also attack unprotected contacts and conductors. Corroded contacts can cause intermittent system operation which is difficult to diagnose and repair. In severe cases, the oxide products of the corrosion can bridge adjacent conductors with resultant leakage paths or short circuits.

Humidity effects can be minimized through the use of nonhygroscopic encapsulants and coatings which protect the components. This technique is not useful in protecting moving contacts, of course. Hermetically sealed switches, such as mercury-wetted or dry contact reed relays, and gas-tight connectors are sometimes used. However, hermetic sealing on a large scale, such as an entire computer enclosure, is very difficult to achieve in practice. All openings to the enclosure must have pressure-tight seals. The reason is that temperature and barometric pressure variations result in considerable pressure differences between the interior of the "sealed" enclosure and its environment. Even if the enclosure is initially sealed with dry air inside, the breathing action due to pressure/temperature variations causes the enclosure to gradually infuse moist air from its surroundings. The result is further aggravated if the air temperature inside the enclosure later drops since the internal relative humidity rapidly reaches saturation because of the restrictive sealing and condensation occurs. In effect, a cloud is formed and it may even rain inside the enclosure.

Because most enclosure designs can be expected to breathe to some extent, a package of desiccant is often placed in the enclosure to absorb moisture from the air. The use of a desiccant also provides a solution to the problem of guaranteeing low internal humidity when the enclosure is resealed after being opened for servicing. Of course, the desiccant eventually becomes water-logged and must be replaced or rejuvenated by heating. The time schedule for this depends on the amount of desiccant, the unfilled volume inside the enclosure, the hygroscopic characteristics of materials in the enclosure, the quality of the seals on all openings, and the external atmospheric conditions.

A second approach, in lieu of a desiccant, is to maintain the temperature inside the enclosure at a high level. Air entering the enclosure is then reduced in relative humidity even when the outside air is saturated. The key to this approach is to keep the interior warm enough that even when the outside air temperature is falling rapidly, the inside air humidity is maintained well below its saturation point. If ambient temperature variations are not too great, it is sometimes possible to depend upon equipment power dissipation to achieve an adequate interior temperature. However, most enclosures installed outdoors require a thermostatically controlled heater to maintain a suitable interior temperature. The reliability of the heater and the effect of opening the enclosure for servicing must be included as design factors.

Humidity is just one aspect of weather that must be considered in outdoor installations. Weather manifests itself more directly in the form of rain, hail, snow, and perhaps even salt spray. These agents must not be allowed to act upon electromechanical components and electronic circuits if continued operation and performance within specification is expected. The National Electrical Manufacturers Association (NEMA) has published a standard, I.S.-1.1, which describes the requirements for enclosures that can withstand this exposure (NEMA, 1966). Several manufacturers produce off-the-shelf enclosures designed to meet the requirements of this specification. Their use is recommended for process control computer system components when outdoor installation cannot be avoided.

Some form of contamination is usually present in an industrial process plant environment. Splashed liquids such as lubrication oils, machining coolants, and plating baths are potential contaminating agents. Protection against liquid contaminants is fairly straightforward. A closely fitting cover is generally adequate to prevent splashed liquids from reaching the protected equipment. Airborne gases present a much more difficult problem and little information is available regarding what concentrations to anticipate for design purposes. Their presence is not always obvious and their effect can be severe. Airborne contaminants have access to any area that is not air-tight, so the preceding discussion on designing with sealed enclosures is applicable. Even if the enclosure is truly sealed, opening it for servicing in the presence of airborne contamination leads to some subsequent deleterious action by the contaminating agent after the enclosure is resealed.

Typical airborne contaminants are gaseous derivatives of sulfur and chlorine, such as hydrogen sulfide, sulfur dioxide, chlorine gas, and the mercaptans. In combination with moisture, these contaminants form acids which readily corrode electronic components and mechanical hardware. In addition, they may cause deterioration of electrical insulation material.

Because so little information is available regarding typical or maximum concentrations of these agents in various industrial process environments, the design of equipment for operation where the presence of these agents reasonably can be expected is best guided by two rules:

1. Use materials that are as resistant as possible to attack by these gases.

2. Seal the equipment in an enclosure that minimizes exposure to gases.

In view of the paucity of information regarding the concentration which these gases reach in practice, a pragmatic approach is to use the human toxicity limits as a design guide (Sax, 1963). Certainly, one would not expect most equipment to be installed in areas where service personnel were required to work while wearing gas masks.

8.2.3 Explosive Atmospheres

In addition to corrosive contaminating agents, potentially explosive atmospheres are not uncommon in petrochemical and other installations. Gases such as methane, hydrogen, and vapors from various petroleum distillate products can be present in sufficient concentrations to produce an explosion if ignited. Although the subject is quite complex and beyond the scope of this book, the reader should be aware of some of the basic concepts and design problems associated with the operation of electronic equipment under these conditions. Sources of more detailed information on this subject are noted in the references at the end of this chapter.

The National Electrical Code, in considering explosive or potentially explosive atmospheres, establishes a classification system to define different types and degrees of explosion hazards. Three classes define the nature of the hazard: gases and vapors, class I; dusts, class II; and ignitable fibers, class III. These classes are further subdivided into groups A–G. The groups define the type of gas or dust more specifically.

Groups A–D are concerned with gases and vapors, whereas E–G cover metal and organic dusts. Finally, two division designations are used to describe the probability than an explosive concentration actually will be present in a given location. A location where hazardous mixtures of gases or vapors are normally present is a division I location. If the hazardous mixtures are not normally present, but could be present owing to some fault or accident, it is a division II location.

Typically, electrically operated process control equipment cannot be used in division I or II locations without special design precautions. Three basic approaches are used to qualify electrical equipment for use in these hazardous environments. One approach is to use an explosion-proof enclosure. The other approaches are to purge the enclosure with fresh air or to employ the concept of intrinsically safe electrical design.

The purpose of an explosion-proof housing is to prevent the ignition of the hazardous environment surrounding the housing even if the electrical equipment ignites the gases that have seeped into the enclosure. The intent is not to protect the electrical equipment inside the enclosure; on the contrary, the equipment is sacrificed in order to protect the surrounding environment from destruction. Explosion-proof design must consider the stresses caused by the burning gases and the sealing of all entries to the enclosure. In some cases, complete sealing is not possible. For example, the armature of an explosion-proof motor must be free to turn and, therefore, there must be clearance around the shaft. Cable entries must be sealed but drains must be provided to release condensation from humid air. In general, clearances and other openings are provided as long narrow passages. The design is such that the flame front from an ignited mixture inside the enclosure is extinguished before it can propagate through the passage to ignite the outside environment. Relatively long, narrow passages accomplish this extinction by making the flame front pass through an area of large surface-to-volume ratio. The heat content of the burning mixture is rapidly dissipated to the relatively cool walls and the temperature of the mixture drops below the combustion point. The use of explosion-proof housings is typically limited to relatively small enclosures since the material thickness necessary to withstand the explosive force is the limiting factor.

If the physical size of the equipment is too large to employ explosion-proof housings, or if protection of the equipment as well as its surroundings is desired, purging may be employed. A purged equipment housing is supplied with a source of clean, fresh air. A positive pressure differential is maintained between the enclosure and the outside environment. Air leakage is outward from the enclosure and explosive vapors or dusts cannot enter the housing; thus the hazardous situation created by sparks within the enclosure possibly igniting contaminated air is alleviated. The system design must include protective devices to deactivate equipment if the source of purging air is interrupted. Clean air usually must be piped from a remote location outside the division I or II area. The cost of piping the air must not be overlooked.

In some cases, the purging air can perform double duty by cooling the equipment within the enclosure. However, if the equipment produces a high heat load, the volume of air necessary to provide adequate cooling may complicate the purging action. To maintain positive pressure differential with a large volume of air may require unacceptable large, and noisy, air leakage velocities.

The newest concept pertaining to the safe operation of electrical equipment in hazardous locations is that of intrinsic safety. Intrinsic safety is concerned with how much energy is required to ignite a combustible mixture and with electrical circuit design which precludes the release of that energy. Approved intrinsically safe equipment can be used in division I locations. In essence, an intrinsically safe circuit must be designed so that it is incapable of releasing sufficient electrical energy to ignite a combustible mixture under normal and abnormal circuit conditions. A related approach, the use of barrier circuits, stresses design in which hazardous conditions in connected equipment such as transducers and actuators are minimized even if a system fault occurs. This is done by limiting the energy that can be delivered through the input and output terminals of the subsystem. The system itself is not located in the hazarous areas, but this approach allows it to service devices located in division I and II locations (Redding and Towle, 1967). Safe operation in these approaches is achieved though the inherent design of the equipment rather than through adding protective devices after the equipment is electrically designed.

A less stringent concept, that of nonincendive design, is applicable to division II locations. In this case, the equipment must not ignite an explosive mixture under normal equipment operating conditions only. Consideration of what constitutes intrinsically safe and nonincendive design is quite complex. The types of explosive mixtures, ignition energy levels, definition of normal and abnormal conditions, and testing techniques are among the important design factors. This is not a design problem for the neophyte. Design reviews by professional experts and testing by experienced laboratories is strongly recommended.

Although design considerations for division I and II locations are a valid concern for the process control computer designer, these systems are rarely installed in such locations. They are, however, installed in purged control rooms in hazardous environments. This is the most common practical approach since no system vendor at present offers an approved division I or II computer system as a standard product.

8.2.4   Shock and Vibration

The ability of control system equipment to withstand vibration is very important for industrial installations. Vibration is transmitted to the equipment enclosures through the floor and other structural members from nearby sources, such as heavy rotating machinery. Because 50- and 60-Hz electrical equipment is commonly used, vibration at these frequencies and their harmonics sometimes dominate in certain applications. Other frequencies exist, however, depending on the particular environment in which the equipment is installed, the characteristics of the vibration source, and other factors such as building structural resonances. The enclosure and the equipment within the enclosure should not have any appreciable resonant responses at 50 and 60 Hz and their harmonics. However, good design tends to minimize all resonances. Cable clamps and cable lacing tie points should be provided to minimize wire movement in the presence of vibration. Excessive movement can result in broken or loose connections. For the same reason, connectors often require hold-down clamps. Hinged gates, sliding racks, and other movable assemblies within the enclosure should be provided with secure latches to prevent unintentional motion.

Generally speaking, resistance to mechanical shock is not as important as resistance to vibration for industrial environments. One area where shock resistance becomes significant, however, is in the transportation of equipment to the installation site. A primary example is that of oil-field data-gathering terminals. In this case, the equipment may have to be transported over rough, unimproved roads to a remote site. The designer must decide either to incorporate the necessary structural strength into the basic design or to provide a special shipping container.

8.3   THE PROCESS I/O INTERFACE

The interface between the control computer and the various process sensors and transducers is an area where compatibility is very important and often inadequate. Compatibility in this instance refers to the mechanical aspects of the interface, not to electrical factors such as signal level and impedance matching. An adequate design of the computer/ process interface must consider the basic questions of isolation and cabling.

8.3.1   Cabling

Wires coming from sensors and running to transducers are connected to the control computer at some type of termination facility. This usually consists of a series of terminal strips that provide screw-down termination for the wires. The terminals may be mounted in the cabinet housing the central processing unit or analog subsystem equipment or in a separate enclosure reserved for termination. The use of a separate enclosure for termination has advantages and disadvantages which must be carefully considered.

For the vendor, a separate enclosure is usually more expensive, and this means an increase in total system price to the user. However, the use of a separate enclosure provides a definite interface between the process and the computer. Process maintenance and installation personnel do not require access to the main computer cabinets when changes in instrumentation are required. This may be an important factor if the computer system is leased rather than purchased. It may also be important in establishing the responsibilities of trade union personnel who are involved in the installation and maintenance of the equipment.

A separate enclosure allows it to be shipped to the installation site prior to the shipment of the computer system. With the termination facility on hand and installed, instrumentation personnel can complete the installation of sensors and transducers and proceed with the checkout of the process wiring independent of the computer installation. This results in a smoother computer installation because it minimizes many of the wiring errors and instrumentation problems that might otherwise occur that complicate the installation and debugging of the computer system.

Whether or not the termination facility is a separate enclosure, attention must be given to physical access to the terminal screws. Most processes and experimental instrumentation are subject to periodic changes because of the addition of new equipment and changes in the process or experiment. With poor access to the terminals, error and accidents are more likely to occur. Observance of the following guidelines promotes good termination area design.

1. It should be possible to see the terminal screw while attaching a signal wire to it.
2. Sufficient clearance should be provided so that the use of normal electrical tools and test equipment is not subject to interference due to other wiring, doors, panels, and so on.
3. Termination panels should be stiff enough that pressure applied in

securing screws does not cause appreciable flexure. Failure to observe this rule can result in broken connections at other terminals.

4. Cable or wire ports should be designed so that removal of existing lines is not required when rewiring or adding new wires.

Depending on the type of instrumentation and process wiring used, there may be one, two, or three wires per input signal. Thermocouple wire is almost aways solid; wire for other lines may be solid or stranded. Stranded is preferable for ease of handling. Wire gage generally ranges from AWG 20–AWG 14, although AWG 12 thermocouple wire is occassionally used. This necessitates a terminal or screw large enough to handle a wire of up to 64-mil diameter or a corresponding size terminal lug. A minimum screw size of 8-32 is recommended.

Wire access ports to handle the number and type of input/output lines must be provided in the enclosure. The wires may be cabled but, since uncabled wire requires more space, conservative design practice dictates that this be assumed.

Within the process, signal lines are usually installed in cable trays or in conduit. Thus conduit fittings should be available for access ports. Insulated fittings may be necessary to prevent grounding the system through the conduit. Grounding, other than at specific points in the system, can lead to ground loops that result in analog subsystem errors. System grounding and its importance are discussed in Chap. 4 and Chap. 13.

An AWG 14 solid wire is relatively stiff and not easily formed. Therefore, sufficient room must be allowed inside the enclosure to route these wires to their appropriate terminal in a nonoptimum fashion. This should be accomplished in such a way that interference with other wires and with the use of tools and test equipment is minimized. It is customary to label wires with adhesive labels or with hang tags. Adhesive labels are usually wrapped tightly around the wire and do not require additional space other than that required to read the label; additional space must, however, be provided for hang tags.

Because alteration to the wiring is sometimes required while the system is in operation, the mechanical design must guard against accidental grounding or shorting of input/output lines. When common mode potentials are present, shorting may damage both the transducer and the system. System design should be such that system damage is restricted to a single point or at most, a few points. The possibility of damage to time-shared equipment should be minimized since this results in losing information from many input points. It is important that the system user recognize the fact that, even though the equipment vendor attempts to anticipate every reasonable fault that can occur, he cannot economically protect against every conceivable failure. Users should therefore take all necessary measures in the design of their process instrumentation to minimize the possibility of a hazardous or serious condition.

8.3.2 Isolation

At some time in the life of the computer process control system, it will be necessary to separate the control processor from the process inputs and outputs. This is very likely necessary during the initial debugging steps and probably later in tracking down and isolating system faults. In either case, it is very desirable to be able to present known test level inputs to the process controller and have the controller exercise output lines without disturbing the process. It should not be necessary to unscrew process I/O lines to provide breaks between system components for fault isolation. Besides being time consuming, new problems can be introduced due to wiring errors made during subsequent reconnecting operations. Provision for plug-in elements in the electrical path between the controller and the process can be used to facilitate isolation. However, the addition of connectors and contacts may reduce overall system reliability compared with screw-down or soldered connections. The trade-offs between reliability, serviceability, and hardware costs must be considered in determining the best approach for each case.

8.4 HUMAN INTERFACE

The interaction between working personnel and the computer control system occurs on two basic levels. First, there are those who program and operate the system. Second, there are the personnel who must install and service the equipment. The interface requirements for programmers, operators, and, to a certain degree, service personnel are design considerations for the process operator console (POC). The POC requirements and design considerations constitute a subject in themselves and are covered separately in Chap. 7.

The interface requirements for service personnel are basically that the component system units be accessible, serviceable, and safe to work on. This section is concerned with these three aspects of the system hardware.

8.4.1 Accessibility

The subject of accessibility is closely related to packaging. In general, system units may be packaged in wall-mount enclosures, rack-mountable enclosures, or in free-standing, walk-around configurations. The latter are common to most computer I/O units such as tape drives, printers, and files. These units usually require three- or four-sided access for servicing.

The need for sealed enclosures in some process control applications typically means that single-door enclosures are desirable. Wall-mountable and free-standing units are both common in single-door access designs. Single-door construction minimizes the sealing problem but, at the same time, greatly restricts access. Equipment installed in single-sided access enclosures is made accessible for service in one of two ways. Either the component package is made serviceable from the front face, or techniques such as chassis slides, hinged subassemblies, or plug-in subassemblies with extender frames are used. In the latter situation, the subassembly is unplugged from a connector fixed at the back of the enclosure and the extender frame is inserted into the connector. The subassembly is then plugged into mating connectors at the front of the extender, as shown in Fig. 8-5. The function of the extender is to act as a jumper and to relocate the subassembly in front of the enclosure to provide five-sided access. Unfortunately, the very act of unplugging the defective assembly to install the extender sometimes clears the fault. In addition, the need for an extender complicates some aspects of on-line servicing. Both this approach and the use of chassis slides have a cost disadvantage when compared with a hinged-gate assembly. In one case it is the cost of the extender hardware and, in the other, the cost of the slides.

Particular attention should be paid to the accessibility of items commonly involved in servicing operations. These include termination areas, test points, fuses, and connectors. Because these are areas that are checked first in almost any service procedure, they should be readily accessible. Easy access to clearly identified customer replaceable items such as lamps and fuses can significantly reduce service calls since many service calls are completely eliminated.

8.4.2 Serviceability

An important aspect of serviceability is the definition of what is expected of the service personnel. In some cases, such as those involving the computer processing unit, service personnel are highly skilled and equipped with the necessary test equipment such as oscilloscopes and voltmeters. The seviceman is expected to determine the exact cause of the malfunction and repair the defective unit. At the other end of the spectrum, servicing may consist of simply replacing a faulty plug-in module with a new one. The fault in this case is either predetermined by using diagnostic procedures at the control panel, or localized to a general

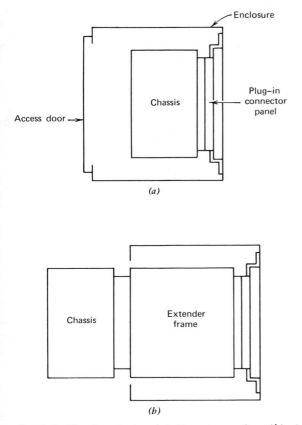

Fig. 8-5 Chassis extender. (a) Normal operation; (b) chassis extended for service.

unit or area, and the exact fault is found by cut-and-try methods. Whether the service approach is either of these two extremes, or somewhere in between, the equipment designer must know what level of serviceability is required and design accordingly.

The question of what level of hardware assembly should be repaired, and where, must receive thorough consideration. As complexity increases, so does cost. But hardware cost must be traded against repair labor and time delay. Also, the cost of stocking replacement assemblies becomes considerable if they are very complex.

Some degree of field servicing is required in any event, and several points should be kept in mind. Servicing can be facilitated by providing the serviceman with adequate test points. What is adequate can, of course, be debated. The design trade-off is the cost of test points against the savings in service time. This is a difficult trade-off to evaluate in quantitative terms. The designer must consider the significance of each test point in terms of what is provides the seviceman, how it relates to other test points, and how it may affect customer use. If possible, these and other service aids should be evaluated by service personnel during the system design and testing phase.

Another aspect of servicing process control computer systems concerns adverse environments. The performance of the serviceman in these environments, although for only short time periods, can be severely degraded. This can be especially significant if dealing with outdoor terminal hardware. The question of servicing in adverse weather and the provision for protective sheds or umbrellas to shield personnel from the elements should be considered.

8.4.3 Safety

Personnel safety is a design factor in any computer system. In the case of process control computer systems, however, the safety of the controlled process is also a consideration. Generally, safety hazards can be identified and restricted to confined areas of the machine. Access to these areas should be restricted to trained servicemen by locked doors, protective shields, and warning labels and signs. These techniques should be employed judiciously, however, if they are to be effective. For example, safety shields should not be added as an afterthought but should be carefully integrated into the overall design. Safety shields that are difficult to remove and awkward to store after removal can be self-defeating. They may be either reinstalled inproperly and inadequately or left off altogether.

Voltage levels in the process control computer system normally do not exceed 50 V, with the exception of the primary power and utility outlets which are usually 115 or 220 V AC. The location of these voltages is well known to the designer and adequate safety measures generally can be provided. In the process I/O termination facility, however, these conditions may not be satisfied. Dangerous potentials in excess of 50 V can exist at the terminals owing to the nature of the transducer employed or owing to a fault in the process instrumentation. The location of these voltages cannot be predicted by the vendor because they vary with each installation. Furthermore, personnel provided by the vendor may not be aware of dangerous areas in the termination facility because of unfamiliarity with the details of the signal wiring. An additional safety hazard is that many times the transducer power cannot be disconnected during servicing because the process, and perhaps the computer, are still operating.

The user should tag or otherwise identify dangerous potentials in the termination facility. Training of personnel should emphasize the potential danger of high voltages and should provide basic training in first aid for shock victims. Access to the termination facility should be restricted to trained personnel. In the final analysis, however, there is no substitute for adequate training and safety awareness on the part of all personnel.

REFERENCES

Abbot, A. L., and F. Stetka, *National Electric Code Handbook,* McGraw-Hill, New York, 1963.

Carroll, J. M., *Mechanical Design for Electronics Production,* McGraw-Hill, New York, 1956.

Magison, E. C., *Electrical Instruments in Hazardous Locations,* Plenum Press, New York, 1966.

Redding, R. J., and L. C. Towle, "Barrier Method of Ensuring Safety of Electrical Circuits in Explosive Atmospheres," *Control and Sci. Rec.* (IEE), September 1967.

Sax, N. I., *Dangerous Properties of Industrial Materials,* Reinhold, New York, 1963.

"Electrical Safety Practices," *ISA Monogr. 110,* Instrument Society of America, Pittsburgh, Pa., 1965.

"Electrical Instruments in Hazardous Atmospheres," *ISA Recomm. Pract. RP 12.1,* Instrument Society of America, Pittsburgh, Pa.

"Enclosures for Industrial Control," *Publ. IS-1.1-1966,* National Electrical Manufacturer's Association, New York, 1966.

"Instrument Purging for Reduction of Hazardous Area Classification," *ISA Recomm. Pract. RP 12.4,* Instrument Society of America, Pittsburgh, Pa.

"Intrinsically Safe and Non-Incendive Electrical Instruments," *ISA Recomm. Pract. RP 12.2,* Instrument Society of America, Pittsburgh, Pa.

Chapter 9

Programming Systems

ALEX J. ARTHUR

A programming system or operating system* is a suite of generalized programs for a particular computer which includes, and operates under the surveillance and control of, a master program. This master program generally is given a name such as executive, supervisor, or monitor. The application programs are not generally thought of as part of the programming system. They are, of course, essential to the total application system and are run under the control of the programming system, as discussed in Chap. 14.

The individual programs that operate under the control of the executive perform functions likely to be required by a large percentage of the users of the particular computer family. Typical of these functions are mathematical routines such as those for the calculation of the sine, cosine, tangent, and square-root functions; resource management programs for functions such as assigning input/output (I/O) units to application programs; and language processors for converting source code into executable object code.

The need for an operating system in the application of a control computer stems from two major considerations: (1) efficient use of computer resources such as central processing unit (CPU) time and storage and (2) time responsiveness to process events.

The first of these, efficient use of resources, is also the major consideration in the use of programming systems in general-purpose computing systems. The computing speed of modern computers is such that having the central processing unit remain idle for any reason during the execution of a program can reduce the throughput of the system significantly. For example, during the few seconds that it might take an operator to ready data cards in a card reader, the computer could have executed several hundred thousand program steps. It is desirable, therefore, to make the execution of any program as automatic as possible,

---

* A distinction is sometimes made between an operating system and a programming system. The usual distinction is that the operating system is an integrated set of programs whereas the programs in a programming system may not be interrelated. For the purpose of this chapter, this distinction is not important.

allowing operator intervention only for those actions and decisions that cannot be made on the basis of a predetermined algorithm. Since it is almost impossible for the user programmer to know the state of the machine at every instant and, therefore, to optimize the efficiency of his program, the programming system must automatically provide optimizing functions such as scheduling and assignment of I/O devices. With the majority of the decisions made at electronic speeds by the executive, the production of the machine is enhanced.

Time responsiveness is related to efficiency in that minimizing time delays increases efficiency. In the control computer, however, time responsiveness is a separate consideration because the computer must respond to events occurring in the process being controlled. An example demonstrates the role of the programming system in providing timely response. Suppose that the computer is executing an optimization routine and an event requiring special corrective action suddenly occurs in the process. This might be, for example, a signal indicating that a vat of liquid is about to overflow. Without a suitable programming system or its equivalent to support external interrupts, the optimization program would continue execution until completed or until the program itself released the CPU. The CPU then would poll all process inputs, discover the overflow signal, and notify the operator via a message on a typewriter. The operator would survey the state of the plant, enter the appropriate commands on a console, and the computer would load the program that responds to the overflow condition. This sequence of events would take considerable time and the vat might have overflowed before corrective action was taken. This is obviously an untenable situation; the control computer must be able to respond to events in the process quickly if the proper corrective action is to be effective.

The programming system in conjunction with special hardware features supplies the means to provide time responsiveness to process events. It can terminate the program currently in execution, monitor a number of inputs to determine the state of the process, load the correction program, and effect the necessary control. All of this takes only a fraction of a second and is accomplished without any operator intervention. It is reasonable to state that computer process control could not be used effectively if it were not for the functions provided by the programming system and the special hardware which supports it.

In addition to providing the control and other necessary functions for efficiency and time responsiveness, the programming system performs many other useful functions for the application programmer in an efficient and convenient manner. Examples of such functions include error handling, program loading, diagnostic aids, and input/output device control. These functions are provided as part of the programming system because they are needed in most applications. The net effect is that the application programmer can concentrate on the application problems rather than on computing problems. In a sense, he programs a more elaborate system which is formed by a real machine and a programming system combined. By using the programming system he benefits from the expertise of the programming systems vendor in these specialized areas and also reduces his programming expenditures.

The use of the programming system, of course, has a price: the storage and execution overhead imposed by the programming system. That is, the programming system itself occupies storage space and requires computing time to perform its functions. In addition, all or portions of the system may be subject to an initial and/or monthly fee charged by the computer or software vendor. Most users, however, find that the value of the programming system is well worth the cost. Although a user might consider the development of a specialized programming system for his patricular application as a means of reducing the overhead, the design and implementation of such a system requires a significant effort and specialized knowledge which can rarely be justified for a small number of projects.

A typical control computer programming system used at present in a variety of data acquisition and control applications is the Multiprogramming Executive (MPX) system for the IBM 1800 Data Acquisition and Control System (IBM, 1969). This programming system allows for time-shared execution of batch programs when the processor is not executing the real-time control programs. It also allows for several programs or portions of programs to be resident in main storage and to be executed concurrently, hence the name multiprogramming. The major components of MPX are illustrated in Fig. 9-1.

The Executive Director, a component of the System Executive, coordinates and controls the multiprogramming services provided by the system. Control is passed to the Director as a result of either an interrupt or a call from program. It provides services such as loading routines into main storage, coordinating the use of main storage and the execution of programs, identifying interrupts and routing program control to the proper interrupt service routine, maintaining timers and the real-time clock, and determining when the processor facilities are available for use by the Batch Processing Supervisor.

Executive I/O, another component of the System Executive, includes input/output (I/O) routines for the control of typewriters, reading and writing the disk storage, error detection and recovery programs, frequently used interrupt routines, and certain utilities such as dump and

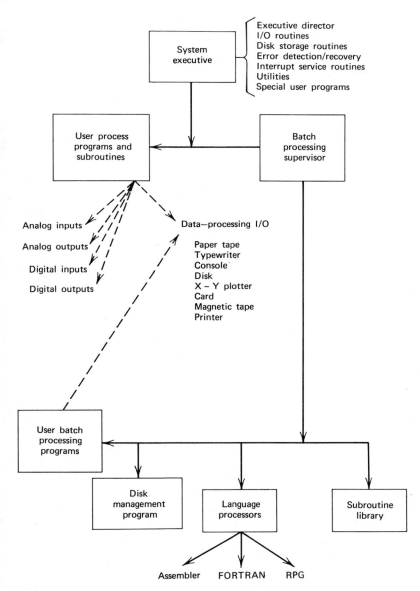

Fig. 9-1   IBM 1800 multiprogramming executive programming system.

597

trace programs. Further routines, such as the printer routine, may be optionally included in the Executive I/O. User-written programs may also be incorporated into the user area of the System Executive at the time the system is generated. These might include special arithmetic subroutines, interrupt servicing programs, and frequently used application programs.

The Batch Processing Supervisor is another executive program in MPX which controls the nonreal-time execution of programs run in the "background." That is, these programs are run on a time-shared basis whenever the System Executive determines that processing time is available. The Batch Processing Supervisor provides the necessary job-to-job control of the execution of these "batch processing"* programs. It also provides control for the other components, shown in Fig. 9-1, that may be used for preparing or executing programs.

The Disk Management Program consists of utility routines that assist in the efficient use of disk storage. Some of the functions included are repacking data on the disk to eliminate unused areas, adding and deleting programs from disk storage, copying one disk onto another, and providing for other file maintenance functions.

The language processors provide the necessary translation from source code into executable object code. In the case of MPX, language processors are provided for an assembly language, FORTRAN, and RPG (Report Program Generator).

The subroutine library provides a variety of routines that are available for use in application programs. Typical of these are programs for evaluating arithmetic and logical functions, conversion routines for converting between various I/O and CPU data codes, and input/output routines for both process and data-processing I/O devices.

The user batch processing programs are the user-written programs run on a time-shared basis under control of the Batch Processing Supervisor. These might include new process control programs that are being debugged and tested or routines to perform nonprocess-oriented tasks.

In addition to the components of MPX shown in Fig. 9-1, the programming system includes a number of other routines. These are primarily associated with the generation of the system. That is, they assist the user in tailoring the basic programming system to his particular applications. This process, called system generation, includes defining

---

* In programming terminology, "batch processing" refers to the processing of groups or "batches" of programs. It should not be confused with the process control concept of processing batches of materials rather than a continuous stream of material.

the size of main storage partitions, determining which subroutines are included in the System Executive, and defining the interrupt structure for the particular system.

The next section discusses the environment, which dictates many of the design requirements for a control computer programming system. Subsequent sections consider design problems of particular importance in the design of such a system. Finally, a brief section speculates on future directions and developments in programming systems.

9.1 PROGRAMMING SYSTEMS ENVIRONMENT

Process control computers present the programming system designer with a major challenge to his ingenuity. Many applications require capabilities in relatively small computers that are generally available only in large-scale systems with very high-performance CPUs. These unusual and sometimes onerous demands arise from the requirement to support unique I/O devices, the need for real-time response, the significant fluctuations in processing load, the wide range of possible system configurations that must be supported by one basic programming system, and the unique reliability requirements imposed by control applications.

Many of these requirements are no longer unique to control computer programming systems, however. With the advent of intercomputer communications using telephone lines and terminals allowing direct data capture and man-machine interaction, many of the requirements for data processing and scientific applications are beginning to seem remarkably similar to those for control. Thus many of the features of real-time programming systems for data acquisition and control applications are being incorporated in more general-purpose programming systems. Nevertheless, these requirements normally must be more carefully considered by the control computer programming system designer.

9.1.1 Unique I/O Devices

The first obvious special features of a control and data acquisition computer are input/output (I/O) subsystems with characteristics very different from the I/O devices normally found on a data-processing computer. These consist of the analog and digital I/O subsystems, the hardware details of which are discussed in previous chapters. To the programming system designer, this equipment presents some unusual problems. First, they are essentially random address devices; that is, a user program typically reads a number of process inputs which, because of the characteristics of the instruments and their connections to the system, are often physically separated. The addresses are, therefore, often nonsequential, but they must be read in such a manner that the user program organization is not affected by this separation. This differs substantially from conventional data-processing input devices in which data typically are acquired and stored in a sequential order.

Also unusual is the large number of I/O addresses. Each of the analog and digital I/O subsystems may provide for thousands of I/O connections to the process with each connection having a unique address. Except for direct access secondary storage, no conventional data-processing I/O device has anywhere near this many addresses associated with it.

Another unusual aspect of the process I/O subsystems is that their performance may be affected by the manner in which they are controlled. For example, there is often a limitation on the rate of reading a single analog-input point to achieve the stated measurement accuracy. Normally, no rate interlocks are provided in the hardware, as some users may want to operate at the maximum rate to achieve a wide input bandwidth, even though measurement accuracy is degraded. These users, however, are usually the exception, and the programming system normally provides an optional interlock to assist the average user.

A unique problem is that digital I/O devices are frequently used on a single-contact basis in application programs. That is, each individual bit in the digital I/O word may have significance independent of every other bit. As a result, the programmer frequently must manipulate single-bit data in a word. The programming system must make this easy and safe and yet cannot interpret the digital I/O bit patterns for significance (a common data-processing (DP) I/O error-checking procedure).

Another unique data manipulation characteristic is the data transfer to and from the process I/O devices. These often consist of single-register transfers such as reading the data from one analog-input point. If a register contains 16 bits, this is equivalent to reading an input record consisting of only two characters, a very rare procedure in either data-processing or engineering calculations. I/O operations for these applications typically involve much longer sequences of words or characters and, indeed, on many DP I/O devices such short records are treated as noise records.

Lastly, the state of these input devices is dynamic and normally independent of the computer system. This precludes the use of such standard DP I/O error recovery techniques as automatic reread.

9.1.2   Real-Time Response

A major consideration of the system design for a control computer is its need to react to events in the external world within a time interval dictated by the application. This "real-time responsiveness" has implications going deep into the basic design of both the hardware and software systems. The fastest response time required, for a given set of hardware, affects the algorithms selected by the system designer for most components of the programming system.

In general, control computers need response times that are some rather ill-defined function of the highest control frequency required by the process. The response times are normally related to the type of device generating the external event signal; for example, high responsiveness is required for events generated by electronic equipment, lower responsiveness for relay events, lower yet for mechanical, and lowest of all for human-initiated events. The reason for this is that the events are commonly assigned to the signaling device on the basis of the required service rate or responsiveness.

There is a limit to the responsiveness obtainable from a given computer and programming system. This limit is approached when the overhead time of the computer and programming system dominates the total service time. Since the hardware overhead is fixed early in the design, the major area of flexibility is in the programming system. By careful consideration of the requirements for responsiveness in the detailed design of the programming system and all its components, maximum responsiveness can be achieved by the programming system designer at very little increase in implementation cost.

9.1.3   Processing Load

The processing load on a control computer has some unique characteristics when compared with that of a typical data-processing or scientific computer. The average control computing load often is not high, and, indeed, is frequently very low by the standards of data-processing and scientific applications. Yet there may well be a few control functions that require the total available computing power. The decision as to how powerful a machine is needed is based on the performance required to do these worst-case functions, *not* that required to do average functions. It is under worst-case conditions that a control computer may pay for itself.

Even on a scheduled basis, there may be major peaks in the activity that are time dependent. For instance, the duration of a major load or production rate change is a critical period in many continuous process plants. If the process is not tightly controlled, the plant may produce useless material during this period. To provide adequate control, the control inputs are read much more frequently, special control algorithms are used, and the computer is operating near its peak capability. Once at a steady load, however, the plant may be quite stable and may need only light control. Since load changes may occur only every 6 hr and last only 10 min, the average computer load is light, no matter how heavy it is during these 10 min.

The design of the programming system must recognize these unique load characteristics of the control environment. In particular, the efficiency of the system under peak load must be emphasized since it is often this peak activity which justifies the use of the control computer.

9.1.4   Multifunction Processing

The processing load characteristics discussed in the previous section result in a large peak-to-trough utilization ratio. A technique to minimize this ratio and thus to lower the cost of implementing computer control of any one control function is to use the computer in a multifunction manner—that is, to use the computer for several different functions, some of which may not be directly related to the control task. Using what is basically a serial device, the CPU, to do multiple functions causes some performance degradation on any single function compared with the performance when the computer is totally dedicated to just that one function. Assuming that this slightly degraded performance still satisfies the application requirement, however, using the computer for other tasks when its control load is light can considerably decrease the total cost per function.

A common function added to the processing load in a multifunction system is the maintenance of the control programs themselves. This maintenance aid may be only a special loader that allows the computer temporarily to disassociate itself from the process and load a new set of control programs. In larger programming systems, however, the added functions are typically much more sophisticated. Most systems provide various resource management functions such as the disk management program discussed in connection with Fig. 9-1. Language processors are another common feature provided in such multifunction systems. These allow the development and debugging of programs concurrently with the control function. Some systems provide for the execution of programs not even related to the control problem. For example, some control computers are used for conventional data-processing applications, such as computing payrolls, concurrently with the control application.

Since the control application is of primary importance in a multi-function environment, these additional functions must not be allowed to interfere significantly with the control task. This is guaranteed by a combination of hardware and software which allows for the interruption of a program whenever a program of higher priority must be executed. Most control computers today are provided with hardware to provide preemptive priority interrupts. Many current programming systems for control computers utilize this hardware to provide some form of "multiprogramming."

Multiprogramming is a generic term covering a number of techniques that allow the execution of two or more programs, effectively simultaneously. As an example, suppose that two programs (or tasks) are to be executed, one concerned with the control of the process and the other a low-priority noncontrol job. The control task is executed until it has been completed or until there is a delay in its execution resulting from, for example, an I/O request. The CPU resource is then free and is reassigned to the noncontrol task. This task continues in execution until the control task is again ready for execution (e.g., its I/O operation is complete) or another high-priority control task is initiated as a result of an interrupt. In either case, the programming system preempts the CPU resource from the noncontrol task and assigns it to the control task.

There are a number of different multiprogramming control algorithms that can be used to control program execution. All these algorithms, however, depend on some sort of prioritizing which may or may not be under user control. Even in the simplest computers and programming systems, there are priority hierarchies implicit in the hardware *and* the software. In the simplest case, error recovery procedures have priority over any other function. This, in turn, means that the I/O interrupt servicing routines actually have priority over the user program; they interrupt it, check for errors, clear a busy indicator, and return to the point where the user program was interrupted.

Control computers must, however, give the user considerable control over the assignment of relative priorities between the various programs in the computer, including the I/O interrupt service routines. In the more complex systems, priority queuing is possible on virtually every system resource with only a few predetermined priorities, these being essential to keep the machine running.

In summary, many complex techniques are available to allow use of "spare" computing capacity. These must be implemented so as to allow maximum utilization with minimum impact on the control responsiveness of the computer. Some of the problems associated with this requirement are discussed in more detail in later sections.

9.1.5 System Generation

The programming system is typically provided by a vendor, either the computer vendor or a software vendor. Because of the spectrum of applications to which the computer system can be applied and for reasons of economy, the programming system designer attempts to write a programming system that can be used in many different applications. This requires that the programming system design be flexible in terms of handling different hardware configurations, different assignment of priority functions, and other special user requirements. The required adaptability generally is provided by allowing the user to "build" a customized programming system for his application from components supplied by the programming system vendor. The procedure involved in adapting the general programming system for a particular application and hardware configuration is referred to as *system generation.*

System generation involves describing to the programming system the characteristics of the hardware on which it is to run, the software options that are to be incorporated into the programming system, and the modification of particular program modules to correspond to the selected hardware and software options. In the control computer programming system, a very important system generation requirement is the assignment of relative priorities between program routines and the allocation of programs to specific storage devices.

The basic objective of system generation is to lower the demands of the programming system on the system resources and thereby increase its effectiveness in the particular application. Effectively tailoring the programming system to the application requires considerable application information, and, as the application changes, so must the system. The programming system designer cannot, therefore, do the actual tailoring of the system to the application. He must, however, provide the user with the facilities to accomplish the task himself.

The approach of tailoring what is basically a general-purpose system probably will not result in the most efficient system possible. The effort required to write a custom programming system that provides maximum efficiency is considerable, however, and rarely can be justified by a user. By careful design, the programming system designer can provide a flexible system with high efficiency suitable for many applications.

System generation has often been a source of user dissatisfaction despite the equally common user demand for customized programming systems. Some of the criticism is the result of insufficient or inefficient system generation procedures and is well founded. Many of the complaints, however, are unjustified. System generation for a sophisticated and complex computer control system is itself a complex and often lengthy procedure, involving 100 or more crucial decisions. In some cases, a totally satisfactory system is not achieved on the first try and additional iterations are required. In addition to its inherent complexity, system generation is only performed a relatively few times in the life of the system and is, therefore, an unfamiliar procedure. This increases the chance of error and the need for subsequent corrective procedures. It is, however, the only effective means yet devised for vendors to provide efficient, tailored systems on a basis consistent with the mass production of hardware and software. System generation is an important consideration in control applications and is discussed in more detail in Sec. **9.4.**

9.2 PROGRAMMING SYSTEMS COMPONENTS

A programming system breaks down naturally into functionally distinct groups of programs and subroutines. The elements of these groups are normally referred to as components. In a control computer programming system, four categories of components may be readily defined: (1) standard subroutines and utilities; (2) man-machine communication facilities; (3) diagnostic routines; and (4) resource management routines.

The functions included in these categories are relatively self-explanatory and their interrelationships are illustrated in Fig. 9-2. Many of them are similar to their counterparts in data-processing programming systems. In this section, the first three of these categories are discussed briefly with emphasis on requirements unique to the process control application. Because of its importance in this application, resource management is discussed separately in Sec. **9.3.**

9.2.1 Standard Subroutines and Utilities

A subroutine library containing a reasonable selection of scientific, engineering, and data-processing function routines is a part of every programming system. Although process control applications have many unique facets, many of them require standard engineering calculations or data manipulation functions. Routines for these functions must, therefore, be available just as in noncontrol programming systems.

Most current control programming systems have been developed from a background of scientific and engineering support systems and so have relatively powerful subroutines in these areas. For example, subroutines are available to perform most standard algebraic and trigonometric functions, curve fitting, solution of differential equations, and matrix arithmetic.

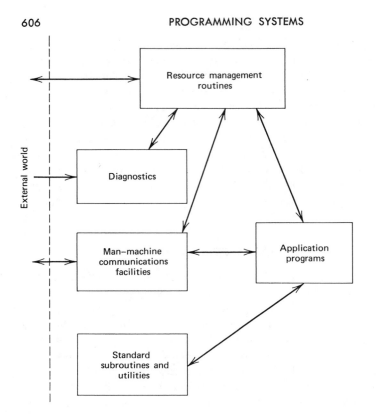

Fig. 9-2   Major programming system relationships.

The support of data-processing functions is typically much weaker in current subroutine libraries for dedicated control computers. This can lead to problems in the use of control computers for data acquisition either for experimental purposes or as a by-product of the control functions. A very large volume of data is potentially capturable in either of these application modes. For example, if 400 analog inputs are read once per minute, 576,000 analog-input records are accumulated in one day. This is a very large data file and requires processing functions similar to those used, for example, in processing an inventory movement file for a large plant. Current subroutine libraries for process control computers, however, do not generally make available to the user more than a few of the functions available in regular data-processing libraries. To compound this problem, most control programmers are engineers by training and are unaware of established data-processing concepts for the building and processing of large data files. Recent indications, however, are that the need for more extensive data-processing support in control applications is being recognized (Minutes: Workshop, 1969). Programming systems for future control computers probably will provide increased subroutine and language support for the necessary functions.

With the ready availability of higher level languages and the regular publication of algorithms (in, for example, the Communications of the Association for Computing Machinery), the need for the more abstract functions in the basic subroutine libraries is lessening and special-purpose libraries are emerging. For instance, the subroutine library may support only basic matrix arithmetic functions; it is assumed that those users who require the more complex matrix functions have access to descriptions of the necessary algorithms. Using a higher level language, they may readily program the precise function they require and thus create a special subroutine library.

Subroutine libraries are normally expandable by the user and are often limited only by the number of unique subroutine names possible and by available storage. The subroutines are stored in relocatable format so that they may be readily appended to any program. In addition to the routines themselves, a directory containing the subroutine name and its location on the storage device must be maintained.

There are relatively few unique considerations in the design of subroutines for process control computer programming systems. One particularly important consideration, however, is the multiple use of subroutines. In all real-time applications that are event driven, there is high probability that the execution of a subroutine will be interrupted by a higher priority task. As in the case of any interrupted program, intermediate results and working areas must be preserved to allow resumption of the program following the interrupt service routine execution. However, since the interrupting program may require the interrupted subroutine, special precautions must be taken to ensure that subroutine data integrity is preserved—that is, that intermediate results and work areas are not destroyed.

Conceptually, the simplest way of guaranteeing data integrity of a subroutine is to ensure that the subroutine is never interrupted during execution. This can be done by masking all interrupts during execution, thereby preventing the recognition of an interrupt. The first action upon entering the noninterruptable subroutine is to set the interrupt mask and the last action is to reset the mask. Thus even though a higher priority interrupt may occur, the subroutine is allowed to complete execution before the interrupt is recognized. This approach is acceptable when the delay due to the subroutine execution time will not prevent the necessary system time responsiveness to the highest priority interrupt. It may, indeed, be the preferred technique for short subroutines where the bookkeeping associated with other methods consumes more time than the execution of the subroutine.

Another technique to provide for the integrity of a subroutine is to append a copy of the subroutine to each of the programs that invoke it. By assigning unique parameter storage and working areas along with the copy of the code, the possibility of destroying data in the interrupted subroutine is avoided. Providing multiple copies has the disadvantage of increasing storage requirements since many copies of a subroutine may be resident simultaneously in storage. It also increases the time required to load a program from secondary storage since the subroutine must be loaded even though it already may be resident in main storage. A third disadvantage stems from the maintenance requirements: If the subroutine is ever altered, all programs in the system must be searched to ensure that all copies of the subroutine have been updated.

Another common technique minimizes these disadvantages but increases the complexity of the subroutine. This technique involves a type of subroutine, known as a *reentrant* subroutine, which is written in such a way that all data and intermediate results are separate from the actual code of the subroutine; that is, the code itself is essentially read-only code. The data and intermediate results are addressed via a base address. If the routine is interrupted and the interrupting program requires use of the subroutine, the base address is altered to provide unique data and working area addresses for execution. This is illustrated in Fig. 9-3. To establish addressability for level 1 on interruption from level 3 to level 1, the programming system saves the address currently in BASADR and puts LV1WA into BASADR. On restarting level 3, the interrupted value of the base address is restored to reestablish addressability for level 3. Some systems use a general working area for all the reentrant subroutines and by convention will load LV3WA into the base address, thus eliminating the requirement to save the contents of the base address on interrupt. Others have a dynamically allocated area associated with each routine when needed.

Although this approach temporarily creates multiple data and working areas, only one copy of the routine is resident at any time. This results generally in less total storage even though the subroutine itself may be larger. It may also reduce the loading time of the interrupt service routine if the required subroutine is already in main storage. Maintenance problems are eased since only a single copy of the subroutine need be updated. Balancing these advantages is the increased complexity of the subroutine and, because of the continual need to

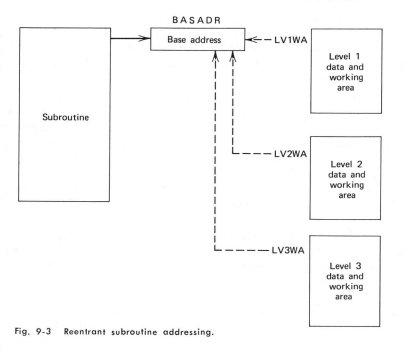

Fig. 9-3 Reentrant subroutine addressing.

reference a base address, the potentially increased execution time.

Still another approach is to use a "gated" or "locked" subroutine. Here, an indicator is set when the subroutine is first entered. If the subroutine is interrupted and is required by the interrupting program, use of the routine is denied to the interrupting program. This is called a *gate conflict* or *lock*. A number of techniques are used for resolution of such locks. They all imply a suspension or delay of the interrupting program or "requestor" and in all the interrupted program (or "owner") is allowed to execute; that is, the subroutine completes the request it is currently executing. The most common technique used in control computer programming systems is to look then at a queue of requestors and essentially interrupt out of the current owning level to the level of the highest priority requestor.

Another, more elaborate, technique sometimes used is that of "priority push through." In this scheme, the owner is temporarily elevated to the level of the highest requestor until his use of the subroutine is complete. The owner is then demoted back to his own level and the subroutine given to the highest level requestor.

If the subroutine is lengthy or if it contains extensive looping procedures, the execution of the interrupting program may be significantly delayed pending completion of the subroutine execution. For this reason, this technique is limited to use in those cases where the probability of a lock conflict is quite low and the resultant occasional delay is not detrimental to the application.

Each of these four approaches to the serial reuse of a subroutine has its advantages and disadvantages, as summarized in Table 9-1. For this reason, a programming system generally contains subroutines utilizing all four approaches. In some cases, subroutines providing a given function are available in more than one form. The user selects the particular technique that best satisfies his needs at system generation time. The decision as to which is to be used is based on storage availability, execution time requirements, and the anticipated degree of use of the subroutine.

A second unique consideration in process control programming systems is the manner in which the subroutines are handled during assembly/compilation and execution. In some data-processing systems,

Table 9-1 Comparison of Techniques for Subroutine Integrity

Technique	Advantages	Disadvantages
Noninterruptable routine	1. Simple implementation 2. Low overhead in speed 3. Easy space maintenance	May seriously affect responsiveness of system everytime subroutine is used
Multiple copies of subroutine	1. Simple implementation 2. Fast execution 3. No impact on response	1. High space requirement 2. Slower program loading (increased size) 3. Difficult maintenance (multiple updates)
Reentrant routines	1. Good responsiveness 2. Reasonable space requirements 3. Easy storage protection 4. Easy maintenance	1. Code enlarged somewhat 2. Normally slower execution (dependent on hardware architecture)
Locked routines	1. Little impact on responsiveness and space when no conflict	1. Lock resolution routine normally elaborate 2. Serious impact on requestor when lock occurs

subroutines are made available at execution time by a loader program that scans the program being loaded, identifies the required subroutines, extracts them from the subroutine library, and appends them to the program. It is usually important in control computer systems that this resolution not be left until execution time since time responsiveness may be seriously degraded by the searching procedures. Instead, the output of the compiler or assembler is passed through a preloader component (sometimes implicit in the compiler or assembler, otherwise variously called storage load builder or linkage editor). This component typically does much of the same work otherwise done by the loader. In particular, it normally does all the searches and builds a special copy of the program that will require little processing by the loader itself at load time. These special copies are typically stored in a special library or secondary storage and create some quite severe update and maintenance problems.

In addition to the subroutines, programming systems have utility programs that are available to maintain the subroutine and other libraries, that is, to provide for the addition, deletion, and reordering of programs in the library. Many other utilities are possible, such as storage dump to printer, data file dump to printer, data device initialization, and data file copy. The system designer must restrain himself in the provision of utilities, however; it is sometimes easier for the user to write his own specialized utility program than to use the utility provided by the system.

The design of utilities for the process control programming system involves no significant considerations different from those found in normal data-processing system design. They are generally used less frequently in the process control situation since, once the system has been debugged and is on-line, the number of changes and the use of debugging aids decrease. Nevertheless, it is preferable that the utilities operate in a multiprogramming mode if this capability is provided by the programming system. This eliminates the need to take the system off-line to perform necessary updating and housekeeping chores, even though these may be infrequent requirements in normal operation.

9.2.2 Man-Machine Communications

Although the direct communication between the computer and the process is a unique characteristic of the control computer, extensive interaction between the computer and the human may also exist. In many respects, this interaction is similiar to that encountered in the application of scientific and data-processing computers. There are, however, some unique man-machine considerations in process control applications.

One important consideration is the diversity of users encountered in the typical process control application. The basic man-machine communications are between the computer and one of four people: (1) the programmer, (2) the process engineer, (3) the computer operator, and (4) the process operator. These may be four different individuals, the same individual at different stages of the job, or any combination of these. The facilities required by the user operating in each of these capacities is distinct, whether the user is a single individual or four separate people.

The programmer is a highly trained individual whose job is to prepare and maintain the programs required in the control of the system and the process. Although he may be very knowledgable about the process, his primary concern is directed toward the computer. The man-machine facilities available to him must allow him to instruct the machine to do every function it is capable of performing. Since the programmer is a specialist, the communcations also may be very specialized. For example, languages consisting of abstract symbols and complex rules are acceptable. Similarly, machine responses such as error messages may also be coded in an unnatural language. Computer languages and their associated language processors are the primary communications media between the programmer and the machine. The I/O devices used for this communication are typically standard computer devices such as typewriters, card reader/punches, video displays, and printers.

The process engineer is concerned primarily with the process, not with the computer, and his background and education often are not related to computer science. The highly abstract languages and complex rules frequently used in programmer-machine communications generally are not acceptable for effective use by the process engineer. Higher level languages that are closely related to either the control problem or to ordinary mathematics are acceptable. In addition, special job control languages or limited command sets for controlling the execution of programs, making minor modifications to the programming system on a component basis, and similar functions are useful communication facilities.

The computer operator does not generally require the same degree of control over the computer functions as does the programmer and process engineer. The computer operator must be able to initiate and control the execution of programs and to maintain the operation of the programming system. In general, he deals with programs as entities and is not required to understand their internal functioning. For example, he may be expected to respond to an error condition by reinitializing a program or by aborting its execution; generally he is not expected to isolate the source of an error below the program component level.

In many control applications, a specially trained computer operator is not available. Instead, his responsibilities are fulfilled by the process operator. The process operator is concerned primarily with the process. Frequently he has a limited mathematical background and is apprentice trained in the operation of a particular process. Although trained in the general concepts of computer control, he is only expected to interact with the computer in a limited manner. Typically, this interaction involves requesting a limited amount of process-related data, such as setpoints and process parameters, or requesting that the computer perform a specific function, such as producing a log. Abstract languages are usually impractical owing to the amount of training required to use them. Rather, a limited set of commands consisting of ordinary English words entered through a typewriter or the activation of switches assigned to specific functions are the usual communication media between the operator and the machine.

The diversity of characteristics between the professional programmer and the process operator is a significant consideration in the design of process control programming systems. This diversity is greater than that found in many computer applications, although a similar spectrum of characteristics is becoming more common in user-oriented terminal systems. As indicated above, the primary communication media for satisfying these diverse needs are a variety of programming and job control languages and specialized I/O devices.

9.2.2.1 Programmer Communications. A variety of languages may be used by a programmer in controlling the operation of a computer. These normally are categorized as either assembler or compiler ("higher level") languages.

Assembler language is normally the lowest level of language used by the professional programmer. It allows him to specify exactly which instructions the computer is to execute but relieves him of some of the tedious bookkeeping tasks by allowing instructions and data to be referenced by mnemonics or assigned names. The assembler scans the input from the programmer and builds various tables from which it can resolve the names into addresses relative to an origin specified by a special control card. It translates mnemonic instruction symbols into their corresponding machine language instructions or into short predefined sequences of instructions.

Effective utilization of an assembler language requires detailed knowledge of the internal operation of the computer and usually involves extensive training. It has the advantage, however, of giving the programmer complete control over the detailed machine operation. Although assembler languages are used extensively by programming systems programmers and, to a lesser extent, by experienced application programmers, current emphasis is toward minimizing the use of assembler language in favor of higher level languages.

The higher level languages are today the dominant program preparation facilities used by application programmers, both professional and casual. They have the advantages over the assembler languages of being relatively machine independent and largely self-documenting. The languages are generally procedural oriented, using symbols and phraseology already familiar to the casual user. The compiler must thus go through a fairly complex analysis of the programmer's input to convert it to actual machine instructions (Lee, 1967).

To do this analysis, two different approaches are possible in compiler design. The first is to do the analysis completely and to produce directly executable machine language code. Code is only produced for those functions required by the particular program, and the compiler is not involved at application execution time. The storage space and execution speed of the compiled code, normally called the object code, are thus dependent only on the user's input and the efficiency of the compiler. The second approach is to build the input code into a table. At execution time this table is scanned by an interpreter program, which uses the table entries only to decide which portions of itself to execute. An interpreter thus must be able to execute any possible function allowed in the language and at execution time must decode which function is required. The storage space and execution speed of the object code are mainly dependent on the size and speed of the interpreter.

Both compilation schemes are used extensively, the second being very attractive for programs that are compiled only once and for which the execution speed and storage requirements are not critical. The gain in compiler speed may well outweigh the loss of execution speed, and the interpreter can easily be written to execute multiple programs simultaneously. The direct compilation scheme is often the more attractive in control computers where the execution speed and storage size of the user's program may be more important than the compilation time required. Most compilers are, of course, a mixture of the two types, the mix being somewhat dependent on the language definition.

Probably the best known higher level languages are ALGOL, COBOL, FORTRAN, and PL/I. Of these, FORTRAN is the only one that has found extensive acceptance in control computer applications. It is scientifically oriented, relatively easy to learn, and widely taught. The COBOL language is rarely used for scientific applications, being more oriented toward business data-processing applications. PL/I provides facilities oriented toward both scientific and data-processing applications but is a relatively new language. It is probably the best balanced of the three languages and contains powerful functions for both applications. In addition to these languages, specialized languages for process control applications have been proposed and some have been implemented. Among these are AUTRAN, PROSPRO, and RTL (Gaspar *et al.*, 1968; Markam, 1968; Schoeffler and Temple, 1970). These are relatively new languages and have not been widely accepted at this time.

FORTRAN has been implemented by most control computer vendors. It does not, however, support all the required process control functions in its standardized versions (American National Standards Institute, 1966). As a result, vendors have extended FORTRAN to support the specialized I/O equipment of a control computer and to provide other functions, such as bit manipulation, which are not a part of the standard language. These functions are primarily implemented through the subroutine CALL procedure in the FORTRAN language, although some vendors have provided syntactical extensions. Although functionally similar, the implementations provided by the various vendors vary in detail.

At the time of this writing (mid-1970), programming languages for process control computers are a subject of considerable discussion (Weiss, 1968). Several efforts toward the standardization of process control language have been suggested, and one activity has been relatively active (Weiss, 1969; Minutes: Workshop, 1969). The emphasis of this group has been toward machine independence, ease of learning to reduce the required level of skill, and inclusion of data-processing functions to facilitate such procedures as creating and maintaining data files. Although it is too early to predict the outcome, it is unlikely that a standard will be developed quickly and an implementation before 1973 or 1974 seems unlikely.

In addition to demands for a clearly defined higher level language for control computers, some current users accustomed to assembler language are requesting a degree of efficiency which is difficult to attain in languages that deliberately exclude machine dependency. It is very doubtful, however, if much of the attainable efficiency of assembler coding is, in fact, realized in practice without extensive testing. This is, therefore, a questionable goal. Techniques are being devised for optimization of the compiler output code, and these may well satisfy all but those users who have particularly stringent efficiency requirements.

The design of a compiler for control computer applications differs little from that for other data-processing applications. As discussed in Sec. 9.2.1, means for the multiple use of subroutines must be provided. Reentrancy, masking, gating, and multiple copies of subroutines are all employed to provide this capability. The compiler also must support extensions to the standard languages. This includes support of the specialized I/O subsystems, timers, other real-time functions, and bit manipulation to facilitate handling bit significant data. Some specialized control languages have introduced key words and descriptors that closely relate to process terminology—for example, a few support statements such as CLOSE(VALVE 5) AT (1) MIN.

In addition to these languages which are relatively general in nature, more specific application-oriented languages designed for the solution of specific problems have recently been introduced. These are oriented toward use by process engineers and are exemplified by PROSPRO and BICEPS, which utilize a "fill-in-the-blanks" programming technique to reduce programmer training requirements (Markham, 1968; McCurdy, 1969). These application-oriented languages, although often applicable to a wide range of problems, generally are oriented toward a particular process or mode of control, such as direct digital control (DDC) or supervisory control. As the field of computer control matures, other specialized problem-oriented languages undoubtedly will be developed in those areas common to a sufficient number of applications to justify the development of the language processor.

9.2.2.2 Computer Operator Communications. A computer operator controls the flow of jobs (a job is a group of related programs) and data through a computer. He regards the program, or even complete jobs, as an entity. Similarly, he controls data flow by manipulating physical data storage entities (e.g., disk packs, magnetic tape reels, spools of paper tape, decks of punched cards). A special language, frequently called the *job control language,* is used by the computer operator to control the job and data flow by communicating information and instructions to the programming system. The programming system responds with special messages requesting operator action or indicating program or machine errors. The job control language is also used by the programmer to indicate to the computer detailed information that he does not wish to build into the program and/or that may be overridden by the computer operator, if necessary, to get the job run. The job control language facilities are vital to the operation of data-processing installation with a full-time operator. The flexibility offered by the job control facilities is also very valuable in a control computer installation. However, to the casual computer operator frequently encountered in control computer installations (e.g., an engineer adding another control program to the system), the facilities provided by the job control language are sometimes a source of confusion. However, since it is often desirable to have these facilities available for installations running both control and regular data-processing programs, the system designer must seek ways of reducing their impact on the casual computer operator.

One method of simplifying the appearance of the job control language to the casual operator is by the use of "default" options. For certain parameters, unless otherwise specified, the programming system uses a predefined, or default, value of the parameter. The default value may have been built into the programming system, specified at system generation, or specified by a special job control language facility. The more sophisticated operator can override the default option by appropriate parameter specification through the job control language. For example, the programming system may assume that a program being compiled is to be run on the machine on which it is being compiled and that any secondary storage allocated need not be preserved after this compilation job is complete. Through the job control language, these default assumptions could be overriden. For instance, a job might well be performed to compile a program for another computer with a main storage of different size. In this case, the parameter specifying main storage size would need to be specified, yet normally it would not need to be. The default options, therefore, allow the casual operator to use the system effectively with only a minimum knowledge of the system. Through the job control language, however, the more sophisticated operator can invoke the full capabilities of the system.

9.2.2.3 Process Operator Facilities. Process operator communications with the computer range from using a typewriter to using elaborate custom-built consoles with multiple typewriters, many pushbuttons, thumb switches, display lights, video displays, and trend recorders. There is at present little agreement in the industry on what functions should be performed by a process operator rather than by a computer operator or process engineer. In some installations the process operator is restricted to performing only the simplest functions through a specialy designed console. In other installations, process operators are trained to perform almost all the duties expected of a full-time computer operator. Until some clearer picture emerges from the industry, it is difficult to predict what specific process operator provisions a control computer programming system should contain except support for the required I/O devices.

The considerations relating to process operator communication are not, however, soley concerned with software. As discussed in Chap. 7, the use of custom consoles has been quite common in control applications. Often these are integrated with control panel functions unique to the particular plant being instrumented. The uniqueness of the console dictates unique software support. As a result, this portion of the programming system generally is not provided by the vendor. Rather, it is considered as an application program and is designed by the user or software vendor to operate under the programming system.

Since most custom consoles are interfaced to the process control computer through the digital-input and -output subsystems, providing adequate support of this equipment through the language processors and other programming system functions is about all that can be done by the programming system designer. If and when vendors and users agree on standard console hardware, vendor-written software to support console functions undoubtedly will be developed.

Some vendors are providing a minimal process operator console and associated software support. In general, however, these are inadequate for a broad range of applications and must be supplemented by custom hardware and software. It is important, therefore, that adequate software support of the digital I/O subsystems be provided in the programming system.

Although the software support of process operator communications is an important consideration, this is only one facet of the design task. Many human factor considerations must be included in both the hardware and the software designs. A detailed discussion of the man-machine interface is the subject of Chap. 7.

9.2.2.4 Summary. Although Sec. 9.2.1–Sec. 9.2.3 discuss separately the functions of man-machine facilities for the programmer, process engineer, computer operator, and process operator, many users do not require separate individuals to perform these functions, except at initial installation and plant start-up time. Thereafter, typically, there is no full-time computer operator, and these duties are shared by the process engineers and operators. Again, some of the process engineers typically will become quite good programmers. In a well-tested, smoothly running computer control system it is easy to lose sight of the value of programmer and computer operator man-machine facilities. The first emergency requiring good control over the computer, however, reminds one forcibly of their importance.

9.2.3   Diagnostics

The diagnostic components in a control computer programming system are significantly different from those in data-processing or scientific programming systems because of a difference in emphasis. The primary objective in the diagnosis of a data-processing or scientific computer is to enhance serviceability—that is, to isolate and verify the fault as quickly as possible, thus minimizing machine downtime and/or the service engineer's time. On the other hand, the primary goal in the diagnosis of a control computer is to maintain system availability—that is, to continue to offer the user some minimum degree of surveillance and control of his plant even though service time may be increased. For example, in a data-processing application it may be preferable to abort all currently executing jobs so that the total system resources can be utilized in the diagnostic procedure. In the on-line, real-time process control application, this approach is often unacceptable. If at all possible, the system must maintain control of important plant parameters. Ideally, this "limp-along" operation should continue while the service engineer diagnoses and rectifies the system fault.

There are three different typse of diagnostic components involved in most programming systems: error detection, fault location, and calibration components. The error detection or error recovery programs are invoked or initiated by other components of the programming system on the occurrence of an *apparent* error. The error recovery programs first validate the error; that is, they establish whether the error is transient or solid. If transient, the error recovery program has achieved just what its name suggests: error recovery. If the error is solid, then it must be brought to the attention of appropriate personnel and information about the error must be saved to help the service engineer in selecting which fault location component to use. The error recovery program typically then transfers control to the resource management components for disposal of the involved resources. The fault location components are used by the service engineer to isolate the fault to a replaceable or repairable module. To assist in the repair of modules, the service engineer may well be able to initiate calibration or exerciser programs, which can repeat required operations many times and give him an insight into the precise adjustments required.

The total diagnostic picture is that of a tree with each level being used to select the component to use in the next level, as illustrated in Fig. 9-4. The boundaries between the levels are not rigidly defined across all programming systems but are, instead, a function of the diagnostic philosophy of the particular system.

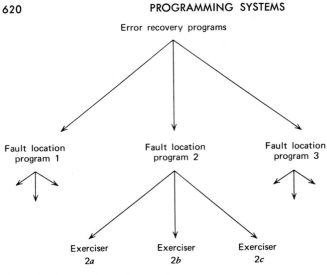

Fig. 9-4   Overall diagnostic procedure.

A source of some confusion in the discussion of diagnostics is that the basic concepts, and particularly the error recovery procedures, are commonly identical for hardware and software errors. The reasons for this are many. The majority of errors on a modern computer are programming errors and are corrected by changing the programs. Only a small number of the possible hardware errors can be detected unequivocally on first occurrence since many hardware errors manifest themselves as an uncorrectable program error. Also, the immediate consequences of many errors are essentially independent of their cause. For example, an occasional double-bit failure in a storage word (which, of course, will not be detected by the hardware parity check feature) is just as serious as if a program were setting these two bits to incorrect values.

In a control computer programming system, the requirements placed on the resource management procedures by the diagnostics are to isolate the effects of the error or fault to the smallest possible set of resources and, also, to provide adequate facilities for the product user to protect his plant and process against possible malfunctions. This is done in a number of ways. First, the error recovery procedures normally are quite conservative. That is, they are biased towards declaring errors to be solid rather than transient. Second, they are normally quite specific in naming an apparently failing resource. This enables the resource management routines to disconnect a small portion of the overall system logically. Third, a special user program is frequently involved to allow for possible plant shutdown, process operator alarms, and other limp-along functions. Fourth, the logically disconnected resources typically must be logically reconnected to the system by some special program invoked by the service engineer.

The phrase "logically disconnected resource" is used to emphasize that normally the fault location and calibration programs can still use the resource but no other program can. Any attempt to do so by other programs may be treated in itself as an error. However, many systems allow back-up facilities for many resources. That is, the user may desig-nate one resource as the back-up or fallback resource for another. This is most commonly allowed only for similar resources; for example, the output for one printer may be redirected automatically to another. If this is done, then obviously when the service engineer logically recon-nects the resource, the back-up resource is no longer required as such.

In control computer programming systems there is a growing tend-ency to associate much of the fault location program logic with the error recovery procedure. This is done for two basic reasons. First, it enables the resource management functions not only to control single resources, but it may also allow control over which functions can be performed on that resource. For example, if the resource is an analog-to-digital converter with five program-selectable ranges and it fails in such a way that only one range is faulty, then it is valuable to continue to use the other four ranges.

Second, the service engineer typically is not on-site for even a quite large control computer system and must be called in by the operations staff. It can be very helpful to him to know considerable detail as to what the error is. He may well be able to bring special tools and parts with him or even to justify bringing a specialist service engineer with him, as he typically has much more confidence in his standard fault location program than in any other programs.

In summary, diagnostics are a critical function for real-time on-line computers. It is vital that they be designed and implemented with the possible consequences of this behavior clearly in mind. Their differences from the diagnostics on nonreal-time computers are normally individu-ally quite small, but the sum of all these differences produces a different effect that leads ultimately toward the "fault-tolerant" computer system —that is, a complete, applied computer system that can and will con-tinue to do some useful work so long as it has some reliable resources and functions available and will allow repair or replacement of faulty resources (hardware and software, systems programs, and applications programs) with little disturbance to the useful work it is doing. Only some military and space systems currently approach even remotely being truly "fault tolerant," and these normally do so at prohibitively high cost and only under specialized conditions. True fault tolerance is, however, an obvious long-term goal of all control systems designers.

9.3   RESOURCE MANAGEMENT

Resource management is a very important function in control computers and involves many design complexities. Although the majority of the resource management functions are found in the executive or monitor portion of the programming system, resource management is a consideration in the design of almost all system components.

The four resources available to a computer are CPU time, storage, I/O devices, and software. Resource management is concerned with the assignment of these resources to programs that are contending for their use. It involves the scheduling of tasks, the allocation of storage and I/O devices, and the control of the use of software resources.

The CPU time is apportioned among the various contending programs by either scheduled or unscheduled task switching. In this section we use the word "task" to mean a single contender for CPU resources, rather than continuing to use the rather vague term "program." A scheduled task switch is one resulting from an action either of the programming system or of a task presently being executed. For instance, when an I/O operation is completed, control is transferred to the programming system. It then decides to which task control is to be returned after doing any necessary processing of the I/O information (e.g., checking for errors). An unscheduled task switch is the result of an event that is not under the control of the programming system and is not, in general, synchronized with the actions of the currently executing tasks or the programming system. Such an event might be, for example, an interrupt generated in the process being controlled. In response to the interrupt, the programming system must stop the task currently in execution, normally in such a manner that it may be resumed later. The programming system must then accept the interrupt and start the execution of the task associated with the interrupt. This normally must be done within a defined response time and thus has important effects on all components of the system.

The efficient apportionment of primary and secondary storage is a complicated and difficult problem that can have a significant affect on the viability of an application. From a technical viewpoint, storage management is an important factor in determining the time responsiveness of the system. Economically, inefficient use of storage increases the cost of computer use if additional storage facilities must be purchased or rented to satisfy the application requirements.

Input/output management is required in control computers or, indeed, in any multifunction programming system. Without techniques such as overlapping I/O operations and program execution, the disparate speeds of the CPU and the I/O devices result in the CPU's remaining idle for significant periods of time awaiting completion of I/O operations. The problem of CPU idleness resulting from the relatively slow speed of I/O devices is common to all computers. In the control computer, the characteristics of the special I/O devices and subsystems lead to additional I/O management considerations. The large number of I/O addresses, the randomness of their use, and the amount of data involved require sophisticated management routines. In addition, the unique availability requirements of many control applications make necessary special management procedures for error recovery, such as the automatic reassignment of output messages if an output printer malfunctions.

Software management is an often neglected function in programming system design and yet provides major benefits for the user if it is done well. To the software management routines every program in the computer system is a resource, just as to other management routines each piece of hardware is a resource. The programming system maintains status information for each program relative to its availability, location, and usage within the system. A suite of management utilities are generally available to the user for manipulating programs and at the same time maintaining the accuracy of this status information. Examples of such information include lists of subroutines required by a mainline program for execution, lists of sharable program modules, and directories of program modules on secondary storage.

Many of the management problems are little different from those involved in the management of hardware resources; for example, finding a subroutine and appending it to a program is analogous to opening a data file for a program. However, rather different solutions may arise for some cases. Consider, for example, a library subroutine and the problems arising from its use by multiple tasks. This was discussed in Sec. 9.2.1, and the alternatives of masking, gating, reentrancy, and the use of multiple copies were introduced. These solutions and the decisions relating to their use often have no direct analogies in hardware management where architectural features of the machine have predefined the available alternatives.

Software management methods are dependent on the precise market sought by the system designer and the detailed design of the CPU. The ease of implementation and the performance differences of the various methods are largely dependent on this last item. For instance, if the CPU has base addressing registers accessible by every instruction at little storage or performance penalty, reentrant routines are readily achievable and are probably little different in performance from standard routines.

The necessity for sophisticated resource management in a control computer arises from both economic and technical realities. As discussed in Sec. 9.1.4, the multifunction use of the computer can significantly reduce the effective cost of computer control through the use of otherwise wasted computing capability. Effectively utilizing this excess capability, however, requires the scheduling and other control services provided by the resource management components of the programming system. From a technical viewpoint, the ability of the programming system to handle scheduled and unscheduled functions with a minimum of operator intervention provides the necessary time responsiveness for real-time control applications.

9.3.1 Basic Management Considerations

In a control computer, the basic managerial problem is created by the contention among many possible functions for the available resources. Programs are contending for CPU time, for the use of particular I/O devices and software, and for primary and secondary storage. Despite fundamental differences in the characteristics of these resources, the problem of contention is common to all of them.

The criteria against which the solutions to the contention problem must be judged include the efficient use of the available resources, the priority of the contending functions and the closely related requirement for time responsiveness, and the logical consistency of the solution. If economy and efficiency were not a consideration, contention could be eliminated by providing separate resources dedicated to each computer function. Thus, for example, multiple processors and I/O devices could be provided so that no resource need be shared by multiple functions. This is obviously not an acceptable solution for the general problem of contention because of the cost of the hardware and the resulting inefficient utilization. It may, however, be acceptable, and even preferable, for particular contention problems. For example, providing separate typewriter output devices for routine logging and for alarm messages may represent a logical and economical solution.

The priority of contending functions must be considered carefully. Assigning resources on a "first-come, first-served" basis can severely degrade the time responsiveness of the system to a high-priority function if a low-priority job is initiated first. On the other hand, this approach may be acceptable in scheduling two functions having such a low probability of contention that the resulting degradation of the higher priority function can be tolerated.

The criterion of logical consistency has several facets. One is the external appearance of the system to the user. It would seem inconsistent and be confusing to the user, for example, if the printing of an operating log were interspersed with routine messages concerning relatively unimportant events in the process. Such a procedure would be justified only if the messages concerned an alarm condition in the process and alternate output devices were not available for alerting the operator to a potentially hazardous condition.

The second aspect of logical consistency concerns the relationship between two contending functions. If the functions are mutually exclusive and of equal priority, either might be logically allocated a given resource. If, however, one program depended on the output of the other, logical consistency would normally dictate that the resource be allocated to the independent function first.

The simplest management algorithm for the solution of contention problems is one that ensures that contending functions will be run in a mutually exclusive manner; that is, through the design of the program itself or the executive routine, a program requiring a particular resource will never be initiated unless the resource is exclusively available to the program. In many executives this approach is implemented by requiring that a program executing on a particular priority level be completed before another program may be initiated on the same level. In this case, assigning all requests for a resource to a single level will avoid contention problems since the programs will be run to completion before a contending function is initiated. For example, if all printing routines are assigned to the same priority level, the designer has ensured that any given message will be completely printed before the printing of a new message is initiated. In this particular case, the algorithm also guarantees logical consistency for the user. It may or may not satisfy the additional criteria relating to the effective use of the resource and the application priority. In general, the use of this management algorithm is acceptable if the probability of contention between two functions is low, that is, if the two contending functions are rarely executed simultaneously. (It should be noted that few real-time systems support absolute priorities, the common misconception to the contrary. The priorities generally used are all relative and are affected by many other factors outside the control of the programming system.)

Another possible algorithm is to reject any attempt to use the resource if it is already in use. The immediate question that arises is, of course, What can the application programmer do then? The system must have some means for him to recover. One technique is to allow him to set up a time delay on his function and, when the function restarts after the delay, it can again try to use the resource. In other cases, however, the programmer need not take a specific action to recover. If, for example, the function is run on a cyclic basis, a new request will be initiated on the next cycle. This approach can be troublesome if the two functions contending for the resource are in some way synchronized. In this case the request may never be honored.

A third alternative is to establish a waiting line or queue for the use of the resource. Each time the resource becomes free, it is assigned to the next item in the queue. Queuing is used extensively in process control programming systems and in other applications involving multifunction processing. The nature of the resource being controlled and the requirements of the application are the primary factors in determining the algorithms that control the processing of the queues.

There are four basic queuing disciplines in common use. Other approaches are usually a combination or a variation on these four types of queues. The queues and their basic characteristics are as follows:

- *Push down stack* or *last-in-first-out* (*LIFO*) *queue*. In this queue, items to be processed are added and later removed from the head of the list. It is analogous to the processing of railroad cars on a railroad siding or spur. The last car entering the siding must be the first to be removed.
- *Sequential* or *first-in-first-out* (*FIFO*) *queue*. This queue is similar to the waiting line at a theater box office. Entries are processed from the head of the queue and new items are added at the end of the queue.
- *Prioritized queue*. In this queue, each item is assigned a priority and items are processed in the order of this priority independent of the order in which they entered the queue. A simple example of this type of queue occurs at the intersection of a highway and a railroad track. The intersection is the resource being controlled and, primarily for historical reasons, a train has priority over the automobiles passing through the intersection.
- *Any-sequence queue*. Items are always added at the end of this type of queue but may be processed from any place in the list. The order of processing is generally determined by factors other than the nature of the items in the queue. Many quality control procedures involve this type of queue. Products are placed on a con-

veyor belt (the queue) at a single loading point and defective items can be removed at any inspection point along the length of the belt.

In a programming system, the items in the queue consist of blocks of data, programs, or control information relating to the pending request. Physically, these items are stored in the computer main or secondary storage. The arrangement of the information in storage is not necessarily contiguous. That is, the queue items may be stored in widely disparate sections of the storage device depending on the availability of space. The items in a particular queue are linked together either by means of control words added to the storage block containing the item or by means of an index or directory maintained by the system executive program. When items are added to the queue or removed for processing, the executive updates the necessary control words or directory entries

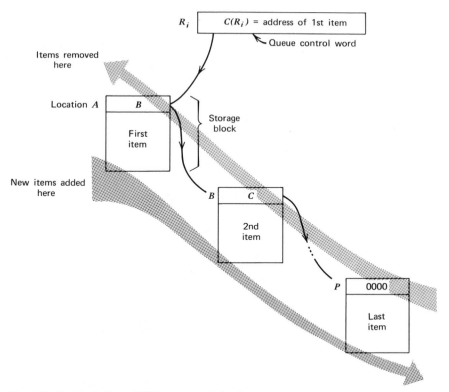

Fig. 9-5 Last-in-first-out (LIFO) queue—chained.

to maintain control of the queue. Examples of the algorithms involved in this procedure are discussed in subsequent paragraphs.

Although the priority and FIFO queue structures are the most commonly used in control computer programming systems, each has its particular advantages and all should be considered by the system designer. In the following discussion, each of the queue types and examples of their associated queue management algorithms are examined. In addition, their characteristics and examples of their use in a control computer programming system are considered.

The push down stack or last-in-first-out (LIFO) queue is a simple structure both from a conceptual and from an implementation standpoint. Both the control word and the indexing method of linking items in the queue can be utilized. As illustrated in Fig. 9-5, the queue consists of a series of items located in blocks of storage that may be scattered throughout main or secondary storage. In this figure, the use of control words for linking the items is illustrated. A control word that

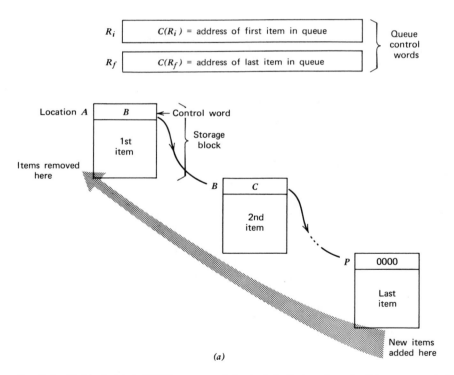

Fig. 9-6   First-in-first-out (FIFO) queue—chained. (a) Storage layout; (b) queue control algorithms.

contains the address of the first location of the following item in the queue is affixed to the beginning of each item. Thus, for example, the first item in the queue is stored starting at location A. Location A is a control word that contains the address B of the word that contains the first word of the second item in the queue. A queue control word containing the address A is maintained in storage to provide the programming system with the initial link into the queue.

The basic queue management algorithms for an LIFO queue that is chained through control words are conceptually simple. Assume, for example, that the first item in the queue is stored at location A so that the queue control word R_i contains the address A. To add a new item that is stored at location $Z + 1$, the address A is transferred from R_i into

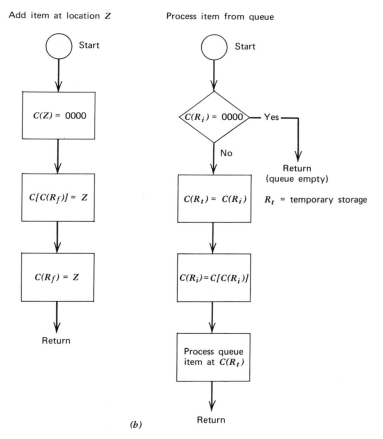

Add item at location Z

Process item from queue

(b)

Fig. 9-6 (b)

location Z. The queue control word R_i is then loaded with Z to provide the initial link into the queue. To remove the first item from the queue, the address of the second item is loaded into the queue control word R_i from the pointer word in the first item. That item is then removed and processed as determined by the controlling program. When the queue is empty, the location R_i is loaded with 0000 or some other distinctive data.

It is also possible to control the LIFO queue with an index or directory. In this case, control words are not associated with the blocks of storage representing the items in the queue. Instead, the address of the start of each item is maintained in a table and a control word containing the number of items in the queue is maintained. Adding an item to the queue is accomplished by inserting its beginning address in the next available location in the table and updating the control word. To process an item, the address of the first item in the queue is obtained from the table using the queue control word which, in turn, is then decremented by one. If the table is of finite length, the queue may become full, and

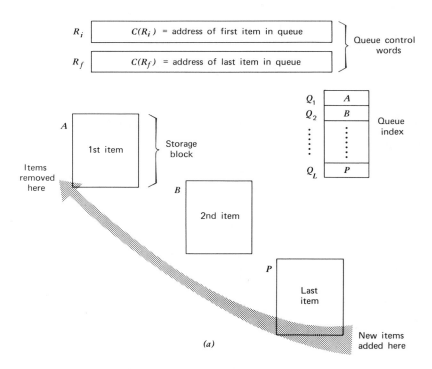

Fig. 9-7 First-in-first-out (FIFO) queue—indexed. (a) Storage layout; (b) queue control algorithms.

Add item at location Z to queue

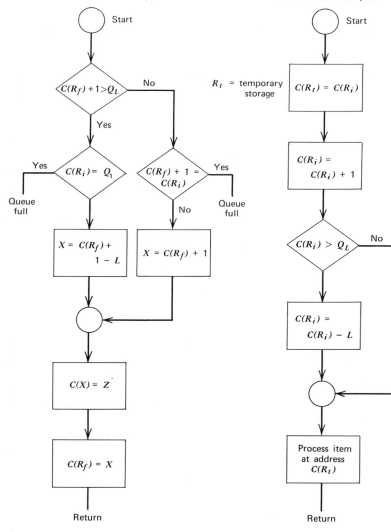

Process item from queue

R_t = temporary storage

(b)

Fig. 9-7 (b)

this eventuality must be considered in designing the queue control algorithm.

The LIFO queue can be used in cases where the order of processing items is of no importance from either a logical or a priority standpoint. For example, it can be used to control unused storage blocks. In this case, all that is needed by the programming system is a record of where the unused storage is located and how many storage blocks are unused. The order in which the blocks are selected for use is not important, so logical consistency and priority are not controlling factors. Neither is the fact that in a LIFO queue there is a finite probability that some items in the queue may never be processed.

The sequential or first-in-first-out (FIFO) queue is commonly used in control and other multifunction programming systems. In this approach, the items in the queue are processed in the order of their entry into the queue. The arrangement of an FIFO queue is illustrated in Fig. 9-6a. As in the case of the LIFO queue, the queue can be chained together by means of control words affixed to the queue items, as indicated in the figure. Two queue control words are used in the management of the queue. One contains the address of the first item in the queue, and the other contains the address of the last item. The queue control algorithms, illustrated by the flowcharts in Fig. 9-6b are examples of how items are added or removed from the queue. Adding an item to the queue is relatively simple. If the initial location of the item to be added is designated as Z and its contents by $C(Z)$, the algorithms consists of setting $C(Z)$ to 0000 to indicate that it is now the last item in the queue and updating the next-to-the-last queue item to provide chaining into location Z. The queue control word indicating the last item in the queue must also be updated so that it contains the address Z. Except for a test to determine whether the queue is empty, the algorithm for removing an item from the queue is very similar, except that the queue control word containing the address of the first item in the queue is used.

Indexing can also be used to manage the FIFO queue, as illustrated in Fig. 9-7a. As in the previous case of indexing, the control words are not affixed to the storage blocks containing the queue items but instead are concentrated in a separate index list. The algorithms for adding and removing items from the queue are somewhat more complicated since the index list is usually fixed in length. Two alternatives are possible in controlling the index list. Every time that an item is removed from the queue, the control words associated with the other queue items can be shifted up so that the first word in the index list always contains the address of the item at the head of the queue. Another alternative is to allow the index to rotate or wrap around; that is, when the location in the index list is filled, the next entry is made in the first location of the index, assuming that the item represented by the first entry has already been removed from the queue. The example algorithms for controlling an indexed FIFO queue that are illustrated by the flow diagrams of Fig. 9-7b assume this latter mode of operation.

The FIFO queue is used extensively in computer control programming systems for items that are not dependent on priority. For example, it can be used to queue noncontrol jobs run on a multiprogrammed system or for the control of routine maintenance and diagnostic programs run at periodic intervals. This type of queue can also be used for many of the data-processing I/O equipment programs such as printing routine logs or punching cards.

Because of the extensive use of prioritized functions in the computer control programming system, the prioritized queue is probably the most important queuing technique in these systems. In the prioritized queue, items are processed, not in the order of their arrival as in an FIFO queue, but rather in the order of a priority assigned to the item. If several items of equal priority are in the queue, they are processed on an FIFO basis. The prioritized queue is, in essence, a series of FIFO sub-queues chained together in which processing of a lower priority subqueue is not initiated until the higher priority subqueues are empty. Figure 9-8 illustrates the logical arrangement of a prioritized queue. This example assumes that the items in the queue are chained by means of control words, although the indexing method is also possible. In the prioritized queue, the management algorithms are more complex since each new item must be added at the end of one of the subqueues, depending upon its assigned priority. In order to facilitate this, a second control word or other indicator may be added to each storage block, indicating the priority of the item stored in the block. Alternately, the number of queue control words may be increased so that there are two words associated with each priority level. The first word in each pair contains the address of the first item in the subqueue corresponding to a particular priority level, and the second word contains the address of the last item.

The management algorithms for adding or removing items from the queue are similar to those described and illustrated for the FIFO queue, except that additional steps must be added to ensure that the item is added or removed from the proper priority subqueue. If separate queue control words are maintained for each priority level, this is not par-ticularly difficult and the FIFO algorithms described earlier may be used with the additional selection of the proper control word correspond-ing to the item being added or removed from the queue. If the control word chaining management method is used, however, the algorithm

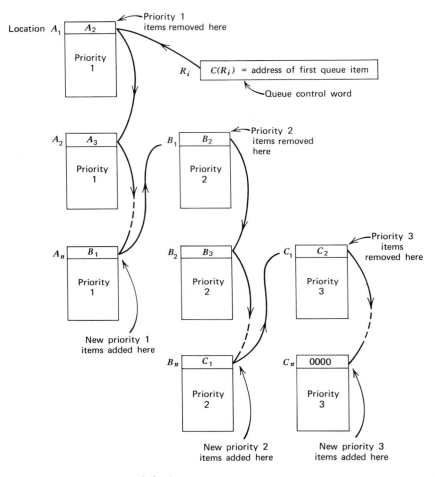

Fig. 9-8   Priority queue—chained.

must contain a search technique which traces through the queue until the appropriate boundary between items of different priorities is found. Once it is found, the queue control algorithms again closely resemble those used for the FIFO queue.

In the any-sequence queue, new items are always added at the end of the queue, but items may be removed from any location. The concept of the any-sequence queue is illustrated in Fig. 9-9 for the case of a queue linked via control words affixed to each queue item. In this case, however, two control words are utilized, one indicating the address of the preceding item in the queue and the second containing the address

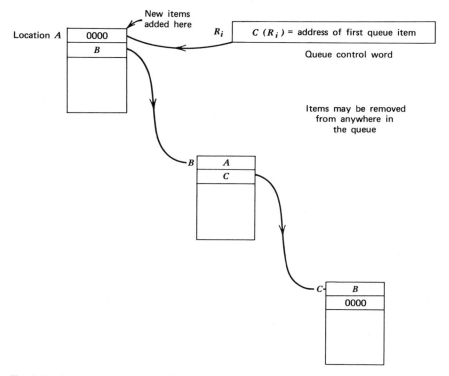

Fig. 9-9 Any-sequence queue—chained.

of the next item in the queue. Only one queue control word is required to indicate the address of the first queue item to the programming system. Adding items to this type of queue can use an algorithm essentially identical to that used for the LIFO queue when the items are linked by control words. The removal of an item from the queue is somewhat more complicated, however. The control word in the preceding queue item must be modified to contain the address of the item following the one that is to be removed. Similarly, the control word in the item following the item being removed must be modified to link it to the item preceding the removed queue item. A queue index or directory can also be used to manage the any-sequence queue.

The any-sequence queue is not used extensively in control computer programming systems, but there are cases where it is advantageous. Many of these cases involve optimizing the use of computer resources. For example, the any-sequence queue of pending jobs may be searched for a job that requires a program already resident in main storage. Similarly, an any-sequence queue for the analog-input subsystem may be searched for requests requiring the same gain setting of the amplifier or a disk queue may be searched to determine whether requests are pending that can utilize the present position of the disk access arm. In these examples, the purpose is to minimize time delays in processing of tasks by eliminating the need to reload a program into main storage, to change the gain of an amplifier, or to reposition the disk access arm, respectively. Such techniques are useful in enhancing the time responsiveness of the system if the execution of the search algorithm is short compared with the time delay involved in servicing the queue requests.

Although the queuing techniques described in the previous paragraphs can be considered basic, variations or combinations can be used. For example, in a prioritized queue, the subqueue for each level may be structured as an any-sequence queue. This might be done to optimize the utilization of hardware and software resources without destroying the requirement of processing items on a priority basis.

As is evident from the examples given in the descriptions of the basic queuing techniques, the selection of a queuing method is dependent on the purpose of the queue as well as on the characteristics of the hardware and programming systems. There are additional considerations involved in the design of the queuing structures, however. These include such things as the lengths of the queues and whether the queues and/or queue indexes should be maintained in main or secondary storage. It is difficult to predict the length of the queues unless considerable detail is known concerning the characteristics of the particular application. Queuing theory can provide a first-order approximation of the probable length of a given queue, but care must be exercised in the use of these results. One common problem is that the theoretical queuing calculation must assume a rate at which items are added and deleted from the various queues. This is often assumed to be a distribution function such as a Poisson distribution, and this may not accurately represent the actual situation in a given application.

Simulation of the application and the programming system can also be used to predict the probable length of the queues. As before, however, the simulation results must be used only as guidelines unless an exceptionally accurate simulation model is available. The best estimates of queue lengths are obtained by actually observing the operation of the programming system as it is controlling the process. This is, of course, not possible in the initial design of the system, so it is important that the programming system be designed such that queue lengths can be easily changed as operating experience is accumulated. For the purpose of storage allocation design, it may be necessary to limit the space available for the sum of all the queues in the system. However, if the lengths of the individual queues within this total space allocation are easily altered without extensive programming changes or a new system generation, the installation of the system can be greatly simplified.

The fundamental limitation on the length of the sum of all queues is the availability of storage. If control words are used to chain the items in the queue, this is the only limiting factor. If indexing is used, the finite length of the index table represents an additional limitation on the number of items. In either case, however, there is always the possibility that one or more queues may be filled. This must be recognized in the queue management algorithms, and appropriate recovery procedures must be defined. An example would be to transfer some lower priority queue items from main storage to secondary storage.

In the management of any given resource, any one or any combination of the above techniques may be used. In the following subsections, their application to specific groups of resources is discussed. The concepts are also expanded to include multifunction contention for multiple resources.

9.3.2   Function Switching

One consequence of contention in the multifunction real-time environment of the control computer programming system is function switching. This is the means by which the CPU and other computer resources are shared among the various programs requesting their use. In gross terms, function switching can be categorized as either scheduled or unscheduled. A scheduled function switch occurs, for example, when the program currently in execution requests that a function be run or when a task execution has been completed and the programming system examines a queue to determine which function is to be executed next. On the borderline between scheduled and unscheduled function switches are those initiated as a result of an interrupt from a timer or the real-time clock. These are scheduled in the sense that they are preplanned; however, they are unscheduled in the sense that the timer or clock interrupt is asynchronous with respect to the task currently in execution. Other interrupts generated within the computer itself and in the process being controlled result in unscheduled function switches. For example, an internal interrupt indicating that a requested process I/O function has been completed may result in the temporary suspension of the currently executing function while the I/O device is provided with data for the next task. Similarly, an interrupt from the process may require the loading of a routine to deal with an alarm condition.

In general, a function switch requires that the programming system perform some analysis as to the need for a function switch, some book-keeping necessary to monitor and control the overall operation of the computer, and the loading or branching to the new function to be executed. Two primary considerations in the design of the function switching procedures are the time response requirements imposed by the application and the relationships that may exist between the various programs in the system.

It is possible to divide the response requirements into the following general classes, which are illustrated in Fig. 9-10.

CLASS I. Simple functions requiring the fastest possible response. Response requirements are typically of the same order of magnitude as the CPU and programming system overhead time so that only simple servicing routines stored in main storage are acceptable. Function switching for this class must be accomplished in less than 200–500 μsec.

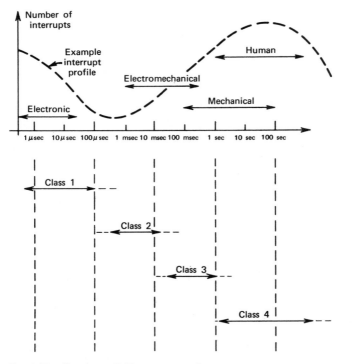

Fig. 9-10   Function switching response classes.

An example of a class I requirement is counting nuclear events from an array of detectors and building a histogram of the events.

CLASS II. Complex functions with required response times considerably greater than the CPU and programming system overhead time. The required response, however, precludes accessing secondary storage for servicing routines. The required response time is in the range of 0.5–20 msec. Sequence-of-event recording during a power plant emergency is an example. On the occurrence of any one of a related group of events, a log of the subsequent occurrence of other events must be initiated with the time of the occurrences recorded to within a precision of approximately 1 msec.

CLASS III. Complex functions but with lower response requirements which allow secondary storage access for servicing routines. In general, the response times for this class are on the order of 20 msec to about 1 sec. A power failure alert in the computer may initiate a function switch in this class. On a computer powered by a motor-generator set, if the main power is down for more than 0.5 sec, either an automatic switchover to back-up power or a shutdown of all the application programs in an orderly manner must be initiated. The motor-generator set might be capable of sustaining computer operation for 1.5 sec after the loss of main power, and the switchover or shutdown must be accomplished in this interval.

CLASS IV. Very complex functions, but only low responsiveness is required. This class is essentially satisfied by the automatic initiation of a normal job. The servicing routine, therefore, may generally be inserted in the regular job stream for execution since the fastest response time required is on the order of 1 sec or more. For example, after midnight a file of cumulative operating data must be processed to prepare a management report for use in the morning.

There is, of course, considerable overlap between these classes. In addition, the classes and their boundaries are technology dependent. The boundaries stated above were selected on the basis of current hardware and software capabilities. For example, the class III function generally requires access to secondary storage, and a practical response time for secondary storage devices normally used with control computers may be on the order of 100 msec. As hardware speeds increase and as new hardware and software architectures are introduced, the problem of satisfying the requirements of a given class are eased. Simultaneously, however, the greater capability of the computer makes its use in new and more demanding applications possible. This, in turn, creates new classes with more demanding requirements.

In the design of a multiprogramming system, the relationship between the various programs also must be considered in function switching. In any large group of programs, there are many relationships that can exist between the programs over a given time span. However, between any two the relationship is essentially simple and reduces to whether or not they are mutually exclusive. If they are, they must run in the sequence determined by their interdependence. If the programs are mutually exclusive but otherwise independent, then their execution sequence is insignificant. For instance, two programs, both of which use most of the primary and secondary storage, are obviously mutually exclusive but yet may be run in any order. When two mutually exclusive programs are not independent, this is normally because one of them must complete a function (e.g., build a data file) before the other can be executed (e.g., prepare a statistical report on the data in the file).

Those programs that are not mutually exclusive can be run "simultaneously" but, since the CPU is a serial device, a decision algorithm is needed to determine on which program the computer should concentrate. The simplest approach is a first-in-first-out (FIFO) algorithm in which the programs are run in whatever order they were requested. As discussed in the previous section, this is an elementary algorithm to implement and is often used in simple data-processing computers. It can also be used frequently and quite satisfactorily in control applications to handle functions with class IV response requirements.

The next simplest algorithm for simultaneous execution is dependent on the interprogram dependence being essentially nil and is called a time-slicing algorithm. After a fixed time interval, the program currently running is interrupted, its status saved, and another program is started or restarted. After each program has had a time slice, the cycle is repeated, starting with the first program. This basic algorithm is well suited to the support of human users working through terminals. In this application, the user requires only that the computer maintain a reasonable conversation rate at each terminal. In fact, too fast a response may well be intimidating for the inexperienced user.

9.3.2.1 Interrupt Handling. As indicated in the previous section, most unscheduled and some scheduled function switches are initiated by an interrupt. In a typical control application, the programming system must respond to tens or hundreds of thousands of interrupts every hour. As a result, the time responsiveness of the computer and its ability to satisfy the four classes of function switching requirements are highly dependent on the manner in which the hardware and the programming system respond to interrupts.

In most modern control computers special hardware—the priority interrupt feature—has been incorporated. The design of this feature is discussed in Chap. 2. The feature generally operates on the basis of preemptive elevation of the CPU from level to level within the interrupt structure. That is, the CPU resource can be preempted by a program having any level of priority higher than that associated with the function using the resource. The fine detail of the programming system supporting this feature is dependent on the exact nature of this interrupt structure, but the basic concepts can be discussed without reference to a particular hardware design. Many data-processing, scientific, and engineering programming systems have implemented a very similar function in software so that they may readily restart functions that have been suspended while awaiting I/O transfers.

On the occurrence of an interrupt at a level higher than the one currently in execution in the CPU, the hardware normally transfers control to a unique address for the interrupting level. To effect a complete transfer of control from the function currently in execution to the new function, however, the system status at the point of interruption must be preserved. Following the execution of the new function, control is returned to the interrupted program.

Before examining the servicing of interrupts in detail, it is instructive to consider some general aspects relating to the utilization of the system when a multilevel priority interrupt structure is employed. The first important point is that the term "priority" is often misleading or, at least, misunderstood by many designers. It is often assumed that devices or functions assigned to higher priority levels are more important than those assigned to lower levels. That is, the term "priority" is associated with the importance of a function rather than with the concept of precedence. As demonstrated by the example on page 903, the assignment of seemingly unimportant functions, such as typewriter control, to the higher levels of priority sometimes can improve the overall efficiency of the system without significantly degrading the performance of functions that are of greater importance to the application or the programming system. Thus it is important to realize that priority interrupt systems establish a precedence for the execution of functions and that this may be independent of the importance of the function to the system.

In a system having more than two levels of preemptive priority, functions executing on the lowest level may be interrupted for the execution of a program on the next higher level. The execution of this function, however, may also be preempted by a function operating at an even higher level. Thus the programming system must maintain the necessary data to allow these several levels of preemption and a return to each of the interrupter programs in their defined sequence of decreasing priority.

The data in Fig. 9-11 illustrate the level of interrupt activity encountered in a particular application and the degree of preemption that occurred at the various levels. These data were collected from a system being used for supervisory and DDC control in the petrochemical industry. In this particular application, nonreal-time batch processing is also performed on level 5, the lowest priority level. During the approximately 18 min when these data were collected, the processor load was approximately 50% and in excess of 58,000 interrupts were processed. Thus the average rate was on the order of 50 interrupts/sec. During the same period of time, approximately 10%, or 5800, preemptions occurred. As would normally be expected, the number of preemptions increased with decreasing priority level. The exception to this on level 5

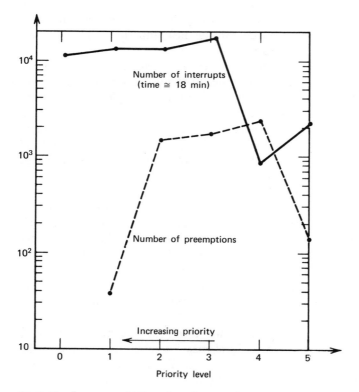

Fig. 9-11 Frequency of interrupts and preemptions.

results from the fact that the nonreal-time processing on this level is performed when real-time process activity is low. In a system more heavily loaded with process-related jobs, one might expect the number of preemptions on the lowest level to be the greatest.

Closely coupled with the number of preemptions in a multilevel priority interrupt system is the length of time required to service interrupts on the various priority levels. Figure 9-12 shows data derived from the same system which provided the data for the previous figure. The interrupt service time is the length of time between the initial occurrence of the interrupt and the completion of the task associated with the particular interrupt. Because of periodic preemptions and, perhaps, a delay in recognizing the interrupt due to a function of higher priority, the total service time includes wait time during which the task is inactive. As Fig. 9-12 indicates, the wait time generally decreases with increasing priority. This, in conjunction with more complex or lengthy tasks associated with the lower priority levels, results in a general increase in the total service time with decreasing priority. As in the previous case, the data in the figure show some exceptions to this general trend owing to the particular application and assignment of devices to interrupt levels.

In the particular system from which these data were derived, the

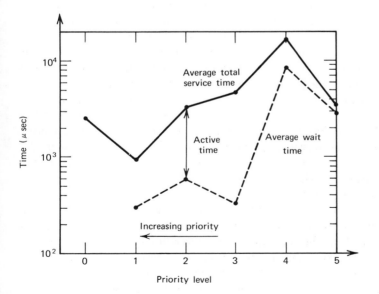

Fig. 9-12 Interrupt service time.

activity on the various priority levels is relatively constant on the four highest levels. This, as well as the number of interrupts, the number of preemptions, and the service time, is the result of the particular devices assigned to the various levels and the characteristics of the application. Thus the data should not be considered typical nor should they be used as a design guide. They are presented here merely to illustrate the interrupt characteristics of an actual process control application.

In examining a general multilevel priority interrupt structure and its associated servicing, the following nine distinct steps can be identified:

1. Complete the execution of the current instruction.
2. Inhibit or "mask" all interrupt levels.
3. Store the address at which the executing program is being interrupted.
4. Record the status of the CPU by storing the contents of some or all active registers.
5. Reset the interrupt indicator and unmask higher interrupt levels.
6. Determine the exact source of the interrupt.
7. Determine the action to be taken and transfer control to the appropriate service routine.
8. Execute the service routine.
9. Initiate the return to the interrupted program.

Of these actions, 1, 2, and 3 are typically performed automatically by the hardware in a modern control computer. The execution of the service routine in step 8 is a programmed function. Depending on the design of the hardware and programming system, the other steps may be either hardware, software, or a combination of hardware and software functions. In addition, the hardware and software design may cause steps 3–6 to be performed in a sequence other than that indicated in this list.

The interaction between the hardware and the programming system is apparent by examining this typical sequence of steps in the interrupt processing routine. Interrupts generally cannot be recognized during the execution of an instruction since this would severely complicate the problem of preserving the status of the CPU for the subsequent return to the interrupted program. Since individual instructions are typically noninterruptable, the execution time of the individual instructions is one factor contributing to the time responsiveness of the system to interrupts.

The second action, the masking of all interrupt levels, is necessary to prevent an interruption during the subsequent steps in which the status of the system is being stored for later use. If the interrupts were not masked, a higher level interrupt could occur during the status storage action and the necessary status information would not be available for the return to the interrupted program. In most machines, the masking of all interrupts for at least one instruction cycle following the recognition of an interrupt is provided as a hardware feature. If only a single instruction is automatically masked, the instruction executed during this masked time is usually a masking instruction to inhibit interrupts during subsequent steps.

One critical piece of status information to be saved is the address at which the interruption occurred. As a result, the hardware generally provides for the automatic storage of this address. This is done in conjunction with the interrupt masking of step 2 and the branching action to the interrupt service routine. In a system having a hierarchy of priority levels, the branching action results in a transfer of control to an address associated with the particular interrupt level that is active. In many cases, the interruption address is stored in the location to which control is transferred. In effect, the interrupt causes a Branch and Store Instruction Counter instruction to be executed.

The content of the instruction counter is generally insufficient status information to provide for resumption of the interrupted program. As a result, the next action is to store the status of other CPU registers for later use. At the time of interruption, these registers contain intermediate results of computations or data manipulation and other data necessary to reinitiate execution of the interrupted program. Typically, the contents of the accumulator, other arithmetic registers, the index registers, and the interrupt status register must be preserved. If the storage of these data is a programmed function, the programmer may be able to increase the time responsiveness of the system by saving only a partial status of the CPU. This is acceptable either if the subsequent routines are known not to disturb the contents of a register or storage location or if the status can be reconstructed in some other way upon return to the interrupted program. If the saving of status information is provided as an automatic hardware function, the status of all operating registers generally will be preserved, since the subsequent action of the program cannot be predicted easily by the hardware function.

Once the status of the CPU has been stored, the occurrence of a higher level interrupt will not prevent the successful return to the previously interrupted program. Therefore, the programming system or, in the case of a hardware status save, the hardware can unmask interrupts on the levels higher than the level presently in execution. Lower and equal level interrupts remain inhibited to preserve the preemption property implied by the interrupt hierarchy.

Depending on the hardware and software design and the assignment of devices and subsystems to the interrupt structure, the automatic branch to the location assigned to the level may or may not be sufficient to determine the exact cause of the interrupt. For example, if multiple devices are connected to a single interrupt level, it is necessary to determine which device initiated the interrupt before the proper service routine can be selected. As another example, in some hardware designs, one word of data is associated with each interrupt level. The individual bits in the word can then be used to represent sublevels to which individual devices can be assigned. In this latter case, the particular bit causing the interrupt may be identified through hardware so that a subsequent branch to a unique storage location can be initiated automatically. Or, the identification of the active bit may require the execution of a software routine. In either case, step 6 in the interrupt processing procedure must be accomplished.

Once the exact cause of the interrupt is determined, control is transferred to the interrupt service routine associated with the exact function requiring execution. It should be noted that during this and the preceding step, the higher level interrupts are not inhibited and the program execution may be preempted. This was illustrated with the data of Fig. 9-12.

At the completion of the interrupt service routine, program control is transferred back to the executive to initiate the return to the originally interrupted program. As in the case of responding to the interrupt, this return may involve a combination of programmed or automatic hardware actions. Several differences should be noted, however. Masking of all interrupt levels is, in general, not required in restoring the CPU status because, if the restoration routine is itself interrupted, the original status information remains stored and is available for later use. In addition, rather than exiting directly to the interrupted lower level program, it is often worthwhile to check for pending interrupts of equal priority before existing from the level. This may prevent the immediate interruption of the program to return to the level just serviced. This latter point, however, is often heavily influenced by the particular hardware design, and other return algorithms may be preferred.

Interrupt processing is illustrated in Fig. 9-13 along with terms that are useful in discussing the procedure. The lockout, resolution, interrupt program, and response time correspond to one or more of the nine steps discussed previously. The lockout time is equal to the execution time of the instruction immediately preceding the occurrence of the interrupt. If the level associated with the interrupt had been masked previously, the masking time becomes part of the lockout time. The lockout time, therefore, is the time interval during which hardware or program constraints inhibit the recognition of an interrupt.

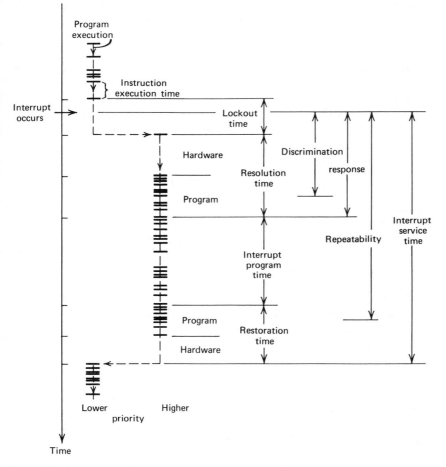

Fig. 9-13 Interrupt servicing.

In terms of the previous discussion, the resolution time is the interval between the completion of the last instruction of the interrupted program and the initiation of the first instruction in the interrupt service program. It is during this interval that the system performs the actual function switch and determines the exact cause of the interrupt. The response time is the time interval between the occurrence of the interrupt and the start of the interrupt service program. Thus, at most, the resolution time and the response time differ by the lockout time.

A closely related concept is that of discrimination. The interrupt discrimination is a measure of the ability of the system to recognize two successive occurrences of the same interrupt. It is, therefore, defined by the time interval between the occurrence of the interrupt and the resetting of the interrupt indicator. The discrimination time is also the maximum lockout time for a higher level interrupt that occurs after a lower priority interrupt has initiated the hardware function switch. The discrimination and response time differ by the execution time of the program that determines the exact source of the interrupt. If this determination is made by the hardware automatically, the discrimination and response times are essentially equal.

The interrupt program time is the time required to execute the routine that services the device that initiated the interrupt. The restoration time is the sum of the program and hardware time required to transfer control from the interrupt program to the interrupted program. The total interrupt service time is the interval between the occurrence of the interrupt and the completion of the restoration process.

The repeatability is most closely related to the discrimination. Whereas the discrimination is a measure of the time interval between two interrupts that will be recognized as separate events by the system, the repeatability is a measure of the rate at which the same interrupt can be serviced. This rate is the reciprocal of the time interval between the occurrence of the interrupt and the time that the interrupt level is reexamined for pending interrupts. This reexamination generally occurs during the program portion of the restoration procedure. If the interrupt program time is significant, it is obvious that the discrimination might be considerably better than the repeatability.

All these time intervals and interrupt processing characteristics are determined by both hardware and software factors. Except for his participation in the original definition of the hardware, the system programmer has little control over the hardware factors contributing to the interrupt servicing performance of the programming system. By knowing the characteristics of the hardware, however, he may be able to devise algorithms to minimize hardware limitations that might otherwise degrade the time responsiveness of the system in handling interrupts.

The only portions of the total interrupt service time that are under the direct control of the system programmer are the program portions of the resolution and restoration time. He may also have control over the interrupt program time if the routine is provided by the programming system vendor. If it is user written, of course, he has no control over its effect on the time responsiveness of the system. As a result, careful consideration should be given to those portions of the interrupt servicing procedure that are under control of the designer.

The portion of the response time contributed by the programming system is normally associated with saving the environment at the time of the interrupt and determining which service routine is to be used. To decrease the response time, it may be possible to save only part of the environment and still establish the entry interface to the interrupt service program. For example, if it is known that the decoding algorithm used in selecting the service routine utilizes only two of seven available index registers, then only the contents of these two registers need be saved. If the service routine requires the other registers, however, it must save and later restore their contents. Since the programming system has no general way of returning to the interrupted program if the environment is not saved, the programming systems designer should leave to the application specialist the responsibility of by-passing the save in those few applications in which it can be tolerated.

The use of interrupts in a control computer system significantly enhances the efficiency and time responsiveness of the system. On the other hand, certain functions must be designed as noninterruptable to ensure the integrity of the system. For example, the saving of the CPU status during the processing of an interrupt must not be interrupted if the ability to return to the program is to be guaranteed. In addition, masking or inhibiting interrupts is a resource management technique available to the programmer for the control of particular functions that are not serially reusable. However, the excessive use of masking, either hardware or software, can largely negate the advantages offered by the priority interrupt concept. For this reason, masking must be considered carefully by both the programming systems designer and the application programmer in the design of programs.

As a specific example of masking in one current computer, the shortest instruction takes 1 storage cycle, the average instruction 4 storage cycles, the mask and unmask instructions 3–5 storage cycles, and the longest noninterruptable instruction 22 storage cycles. In this machine, the program can mask, perform 16 instructions, unmask, and still have been noninterruptable for no longer than the time that could arise from the execution of one instruction. It seems reasonable to state as a general aim that the programming system should not mask all levels for longer than the execution time of the longest noninterruptable instruction. Only a few systems currently approach this criterion, and these are machines with extremely wide variations between their average and longest instructions.

Excluding error procedures for fundamental system resources, masking basically should be used only to process simple resources or to restructure the interrupt handling. By careful design of the resource management algorithms and other programs, this basic approach usually can be implemented. For example, consider an FIFO queue with the queue elements linked by control words and a queue control block pointing to the element currently being serviced. If an element is to be added to the queue, the queue is searched without masking to find the last element in the queue. Interrupts are then masked and a check is made to see if it is still the last element. If it is not, interrupts are unmasked, and the search is started again. If it is still the last element, the queue has not been altered and the new element is added by setting the queue control word for the new entry and performing the necessary linkage. The interrupts are then unmasked. Although this algorithm is more complex than one in which the interrupts are simply masked during the total procedure, the approach results in minimum masking time and in minimum degradation to the time responsiveness of the system.

9.3.2.2 Time Scheduling. Most of the load on a control computer consists of scheduled functions; for example, every 5 sec the system may scan and alarm 200 inputs, and every 15 min it may print their average value over the previous time interval. The overall system schedule is determined by four basic sources: timers or clocks, queues, interrupts, and the computer and process operators. Of these, the timer- and operator-initiated scheduling are the only functions concerned exclusively with real or relative time. The timers provide the basic load, as most functions are executed on a cyclic basis and much of the process operator interface is concerned with initiating and stopping the cyclic execution of functions. Even though concerned with real time, the operator actions are unscheduled in that they occur asynchronously with respect to the internal system operation. As a result, they are handled by means of the interrupt structure and servicing procedure.

The programming system must support the hardware timers in such a fashion that reasonably accurate time periods of a wide range may be used. For example, some programs run once per second, some once per week. Care must be taken to prevent "creep" in the time periods owing to the delay that may occur between the expiration of the appropriate timer and the execution of the associated function. As a result, high-frequency timers normally appear as one of the highest priority devices on the control computer. In addition, the programming system timer support algorithms must include procedures to compensate for instruction execution time during the update and servicing procedures.

An important use of timers in sequential control programs is for introducing time delays between two portions of a user's task. For example, the sequence "fill fermentation tank, allow to ferment for seven days, and then test if alcohol content is greater than 10%" might well be a part of a sequential control program for a distillery or brewery. As another example, most mechanical devices have start-up and settling delays for any movements, and their position cannot be read accurately during these times. Thus the control sequence for them is "set movement relay on, wait N time units, then read position." It may be important for the computer and programming system to allow N to be specified accurately since the user may wish to compute the movement rate of the mechanism for troubleshooting purposes.

A number of different types of timer algorithms are possible, although for easy user comprehension it is sensible to use a method similar to that used by the hardware timers. The hardware timer can be of a "count to zero" or a "clock comparison" type, both of which are designed to give a CPU interrupt at the expiration of the required time interval. A third type, rarely used, is where the software must make all tests, no specific timer interrupt being given.

In the "count to zero" timer, a positive (or negative) count is loaded into the timer when it is started, and this count is decremented (incremented) at a fixed rate. A unique interrupt is given when the count reaches zero.

The "clock comparison" timer works by comparing the current clock value with a clock setting recorded for comparison, usually in an associated clock register. This comparison is done every time the clock is updated and, when it compares equal, a unique interrupt is given.

The degree of timer support and its exact nature are highly dependent on the hardware timer and clock. Important considerations in determining the programming support are the number of hardware timers, their resolution and capacity, and the resolution of the real-time clock. In general, most control computers provide more than one interval timer. The programming system must allocate and control these timers for the various programs requesting their use. If the number of timers is insufficient, software timers derived from the hardware timers must be provided.

If the capacity of the interval timers is insufficient to provide the measurement of long timer intervals directly, the programming system must provide extensions to the hardware timers. This is easily accomplished by a software counter that is incremented (decremented) each time the hardware timer reaches zero and causes an interrupt. The software extension timers can be chained so as to produce longer and longer time intervals, although the resolution of each subsequent timer in the chain becomes less. This usually is not a problem since in most applications the required resolution becomes less as the time interval increases. If this is not the case, the requesting program can utilize more than one of the chained counters to obtain the required resolution.

If insufficient hardware interval timers are available to the application programmer, the programming system can provide additional timers. These are implemented in a manner similar to the chained counters. The difference is that all the timers are updated at each expiration of the hardware counter.

The software extension of the hardware timers can be serviced in two ways. The programming system can either test all software timers on each expiration of the hardware timer or use each one as a base for the next. That is, it tests successive timers only if the preceding timer has expired. The first method provides a larger number of independent timers and is a simple and small algorithm. It can, however, be very time consuming if a large number of timers are simultaneously in use. The second method requires more code but normally executes much less code than the first and allows much longer time intervals on a fixed word length machine for the same table space. Combinations of these methods, of course, can be used.

The problem of creep exists in servicing both the hardware timers and their extensions. Even if the timer routines are assigned to the highest priority level, some delay occurs between the expiration of the timer and the initiation of the servicing routine because of the response time of the interrupt handling procedure. Unless the response time is small compared with the resolution of the timer, a specific procedure is required to compensate for creep. If the timers are assigned to a lower priority level, such a procedure is almost always required since the response time is unpredictable. A simple method of avoiding creep is to allow the timer to continue to operate after the expiration interrupt occurs. When the interrupt service program actually begins execution, it can read the current value in the timer and provide the appropriate compensation.

9.3.3   Storage Management

Two resources are basically involved here, main and secondary storage. Some computers offer a deeper hierarchy of storage levels, but the basic management principles are similar. The most common type of main storage is magnetic core storage, although indications are that future systems will make extensive use of monolithic semiconductor storage. Secondary storage is most often provided by direct access storage devices (DASDs) such as drums and disks. Sequential access devices such as magnetic tape are rarely used in control applications, although they may be used in data acquisition applications for data storage.

The primary reason for having two or more levels of storage is economic. Typical cost/bit ratios between magnetic core and disk storage are 100-1 or more. On the other hand, the access time for a DASD secondary storage device is typically more than 10,000 times greater than that for main storage devices. This relatively slow access time can degrade the time responsiveness of the system unless carefully considered in the programming system design. The programming system must, therefore, be designed to economize on main storage as far as can reasonably be done within the system performance constraints. If necessary, secondary storage can be handled relatively inefficiently to achieve this goal.

9.3.3.1 Main Storage Management. The organization of main storage is dependent on the details of the programming system and the hardware it supports. In general, however, it is partitioned into a number of sections, one of which is usually assigned premanently to the system executive. Other sections may be assigned permanently to other routines, or they may be available for transient use by routines currently in execution. Figure 9-14 illustrates a main storage map for the IBM 1800 multiprogramming executive operating system (MPX) and serves as an example of main storage utilization. The executive I/O partition contains the routines and tables necessary for the control of the data-processing I/O devices in the system. It includes INSKEL COMMON, which is used for communication between the various types of storage loads used in the system. The executive director is the heart of the executive system and provides the services and control necessary in a multiprogramming system. The USER ROUTINES area of the system executive contains user-written routines which, for reasons of frequent usage or rapid time response requirement, are permanently resident in main storage. These four components constitute the system executive and are permanently resident in main storage.

The MPX system also provides special storage area configurations called SPAR. These are, in essence, extensions of the system executive and may contain frequently used servicing routines that may be subject to periodic modification. By building these into a SPAR, rather than building them into the system executive, modifications can be made without the necessity for a complete system regeneration.

In the MPX there are 23 fixed-length partitions provided for the multiprogramming execution of programs. The length of these partitions is defined by the user at system generation time. Storage loads, stored on secondary storage devices in storage load image form, are loaded into

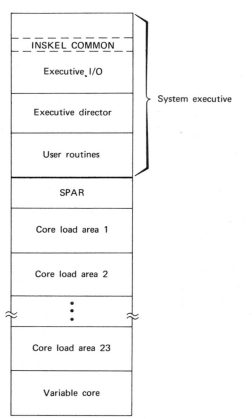

Fig. 9-14   Example main storage utilization. From IBM, 1969.

these areas for execution. The VARIABLE CORE provides working storage for both the system executive and the programs loaded into the various partitions.

The efficient use of main storage is dependent on two factors. The first is the efficiency of the code itself. That is, by careful coding and the use of efficient language processors, the storage requirements of the program components can be minimized. Second, the efficient use of main storage is dependent on the management algorithms used to control the program load areas and the variable storage used by the programs.

The basic concept employed in main storage management is to treat it as a very scarce resource. Among the techniques useful to implement this concept is to try to allocate storage space at use time and only for the time it is in use. The limitation on this is that the allocation algorithm itself has to be main storage resident so that some compromises have to be made in a practical case.

As an example of storage allocation, consider the storage required to read analog values from an analog-input subsystem. Ideally, the programming system should allocate this space for the duration of the read operation and subsequent processing of these data only. Since the programming system cannot readily detect when the total processing of the data is complete, this might seem impossible. However, analysis shows that the problem consists of three parts:

1. Read the digitized millivolt values as an input image into main storage.

2. Move the values from the input image into the variables specified by the user program.

3. Process the variables.

There are two allocations required in this procedure: The first is the allocation of storage space for the input image and the second is the allocation for the individual variables.

The first allocation is required before step 1 is performed, and the space may be freed after step 2 is complete. The second allocation is needed before step 2 is initiated, and the space may not be freed until after step 3 is performed. Both these allocations can be done although the second allocation is normally done for steps 1–3 in many current compilers. Routines for the first type of allocation have been written for a FORTRAN compiler. They allow dynamic allocation of a card image, for example. Similarly, the second type of allocation also has been done automatically, again for a FORTRAN compiler (Lowry and Medlock, 1969). Some languages contain powerful statements to assist in the implementation of such functions (IBM, PL/I, 1969). There are indications that emphasis will be placed on similar facilities in new languages designed for control computer use (Minutes: Workshop, 1969).

System routines normally should be designed to be independent of their data areas so that both the routines and the data areas can be allocated independently. Some system routines should also be designed so that they can be permanently resident in main storage and the resident copies used by all programs requiring that routine. Thus only their data areas need be dynamically allocated. This can be very economical for multiprogramming systems with a large number of possibly concurrent routines.

To assist in main storage management, it is advisable that programming system components be written with the main storage management algorithms and their consequences in mind. For instance, it may be possible to capitalize on algorithms that are particularly efficient in loading large blocks of code from a DASD or on the known characteristics of groups of subroutines. Specifically, by classifying the subroutine library into mutually exclusive groups, it may be possible to implement relatively efficient transient procedures whereby the use of one member of the group makes all other members available in main storage. However, until one member is used, none are main storage resident. The savings in storage space is made by overlaying a group with a new group; the mutually exclusive groups should, therefore, be of approximately the same size.

Another storage management consideration arises from the nature of many application programs in a control computer. Specifically, on most runs of the program only a small percentage of the code is actually executed. It is a tautology in programming that although the basic algorithm takes little space or effort, the exceptions and possible errors require a vast amount of code even though these events rarely occur. Some large data processing systems (Meyer and Searight, 1970) address this problem by the dynamic allocation of storage to user programs on a page (i.e., block) basis. The page size is chosen carefully to be some reasonable compromise between excessive "page turning" and excessive unused storage per page.

The programming system preferably should allow the user to override the system selections on storage allocations (e.g., force main storage residence of certain routines), but it should also allow the storage allocation to be transparent to the user. That is, if the allocation is not specifically stated by the user, it should appear completely automatic. This allows the user to be concerned with the allocation problem only when he wants to be. An example provides insight into the reason for this approach: Contrast the user's trying to control one portion of a pilot plant using a process not yet reasonably understood. Computer efficiency does not matter particularly to him and he would rather not deal with the details of efficient allocation. He is concerned probably with only one loop at a time, is changing the control algorithm frequently, and he specified the biggest and most powerful computer he could get. He was not particularly concerned with storage costs. Now consider the same user two or three years later when he is controlling the same portion of a production version of this plant. He now has a good understanding of the process but is concerned with minimizing the load on the computer since he has specified the computer size on an economic basis—that is, the smallest size he thought could simultaneously handle all the loops. In the first case, it is not worth his while to consider possible inefficiencies in the use of the computer. In the second case, it is very important to get maximum efficiency from the computer.

9.3.3.2   Secondary Storage Management.   This management problem is concerned primarily with allocation of secondary storage to programs and with the maintenance of data or programs resident on secondary storage. The thrust of conventional data-processing programming systems with regard to secondary storage management has been toward device-independent storage management. That is, the trend has been to allow programs to be written without reference to, or concern for, the particular secondary storage device that might be required by the program. Such programs have been faced with trying to delay the decision of where to read or write data until just before execution time so that regularly run jobs would not be delayed due to the failure of, for example, a particular tape drive.

On the other hand, scientific and engineering programming systems have not, in general, used large data files nor have they been required to run programs on a regular fixed schedule. As a result, this type of feature has not been pursued very strongly. As computer control programming systems have generally been developed from such systems, they too have not emphasized this type of capability. This is despite the fact that they actually may need more powerful abilities than data-processing systems because of the real-time nature of the application. In addition, since a computer operator may not be present, the programming system must be able to recognize the unavailability of a secondary storage unit and to compensate for it without operator intervention.

A control computer will always have programs resident on secondary storage devices if such devices are available on the system. The first important technique is to design the programming system in such a way that it not only allows the user programs to be independent of any one secondary storage device but also permits the programming system itself to be independent of any particular device.

To allow device independence for user programs, a simple technique can be used. If the primary task is resident on one device, the user is allowed to name a back-up task which is resident on another device or even permanently resident in main storage. The system automatically transfers control to this back-up task if it cannot access the primary task. A reasonable degree of secondary storage device independence can be attained quite readily using this technique.

Having allowed for the loss of a task residency device, it is advisable to provide for the automatic reintegration of the device into the system when it again becomes available. For example, a power surge occurs that is sufficient to trip out one of the secondary storage devices but nothing else; after 10 sec recovery time, it becomes available again. All references to it should be treated with care until some checking has been done to make sure that the storage medium (e.g., disk pack) was not replaced or altered. A simple check is to compare the media labels. If different, the new medium is not considered usable unless this decision is overriden by either the user program (e.g., it calls for the use of a pack with the new label) or the computer operator (e.g., if he wishes to replace what was fundamentally a scratch pack). In essence, techniques similar to those used in data-processing systems for the manipulation of data media should be used. That is, if a device goes down, all data files or program files associated with it should be "closed" and the user program should have to "open" them again when the device again becomes available.

File organization is another consideration in the management of secondary storage. The organization of files can affect the effective utilization of the storage media and the ease of accessing data. Effective utilization is a factor in system cost, and ease of access can affect the time responsiveness of the system. Both of these are important considerations in control computer applications. The efficiency of various file organizations and access algorithms have been studied in detail (Lefkovitz, 1969).

The two major types of data file arrangement are sequential and random. Sequential is the simplest type of arrangement and is characterized by the order of the records in the file being in the same sequence as some characteristic of each record. For example, if analog values are read from a chromatograph and written to secondary storage as they are read, the file of values is sequential by time.

A random file has the data arranged in a not necessarily sequential manner and the data may not need to be processed sequentially to find a particular record. The word random does not mean necessarily that the data are completely unordered, as then the only way to find a particular record would be to search the whole file. It may simply mean that there is an algorithm available to calculate the position of a particular record. This allows direct access to the data more economically than sequentially processing the whole file until the record is encountered.

In a random file there may well be vacant records to allow for a simple algorithm. For example, consider a file associated with each analog-input point. The file normally would be arranged by multiplexer address and a record assigned for each possible address. If the record size and the position of the first record are known, the position of any other record can be calculated from its multiplexer address. Using current data-processing programming systems, the processing of such a file proceeds as follows: The file is "opened" and all the file characteristics are provided by the using program (e.g., file length, the record size, and position of key fields, such as the multiplexer address in this case). These data are compared with those specified in the file label which was created at the time the file was originated. If they do not match, it is not the right file and an error recovery routine is initiated. If they do match, thereafter the program need only reference the records by the multiplexer address in order for the programming system to fetch the correct record.

A variant on this technique is "indexing" the file. That is, instead of using an algorithm to find the records, another file is created that contains nothing but the key field values and the corresponding record number or location. This approach is useful for small files, for variable-length records, or for files in which the key field values are very irregular. If a file is to be processed sequentially on a certain field, rather than create a separate index field, it may be economical to incorporate the index into each record. Thus each record contains an index pointing to the next sequential record. Indexing is the basic method used by programming systems of all types to maintain their program libraries. The index file, typically called a directory in this case, contains the name, location, and size of each program module.

Indexed files create some interesting problems for the programming system designer. To update such a file, both the index file and the data file may have to be updated. During this process certainly the record and its index, and possibly the whole file, become serially reusable resources. On early versions of this type of support, a power failure at a critical moment was a disaster. Today, however, indexed file systems are effectively used.

9.3.4   Input/Output Management

The input/output devices in a control computer system can be grouped into two classes: process I/O and data-processing I/O (DP I/O). These are characterized as follows: The information available to the system from a DP I/O device, or vice versa, is stable with time. It is under the control of the computer system and the complete job environment can be duplicated readily by reinitializing all media. The information available from process I/O devices is dynamic with time and is not necessarily related to the state of the computer system. It may be impossible to duplicate the job environment except approximately.

These differences between the two classes strongly influence the basic handling of the I/O devices by the programming system. Most current control programming systems handle DP I/O in a manner similar to that used in current engineering and scientific programming systems. They process requests for execution of a function on a DP I/O device and execute that function with little or no provision for multiple buffering, I/O overlap, or request look-ahead. They leave to the user most problems of device recognition and file verification.

Some of the latest systems do support more sophisticated features such as (1) automatic retry on a device error, (2) automatic use of alternate devices, (3) queuing of multiple requests for use of a device, (4) error indicator passback to the user program calling for the device use, (5) media verification, and (6) some limited file protection and device recognition. These are fairly easily implemented. Features 1, 2, and 4 are driven from an interface between the device interrupt indicators and the user program. This interface analyzes the device status and, if an error has occurred, a retry operation is attempted if this is possible on the particular device. As an alternate, the complete operation can be tried again on a different device, or the error indicator can be converted to some meaningful form and passed back to the user program. If the alternate device is of the same type, it can be very simple to switch the complete operation to it. Few systems, however, have done much to address the case of different types of devices.

The hardware error indicators may be widely different for different device types, so it is useful to standardize them for the user programmer and also to integrate them with the software error indicators. For example, an improper calling sequence to an I/O routine is just as serious an error as a parity check on the device. In each case the user program may have to take some recovery action over and above the system action.

Queuing the requests for I/O is, of course, virtually essential in a multiprogramming system. Ideally, queuing should be transparent to the user program and should be prioritized. A useful method of queuing user I/O requests is to link them in a chain and put the task making the request into suspension until the requests have been completely serviced. This and other queuing techniques are discussed in Sec. 9.3.1.

Media verification and control are also important aspects of I/O management. Despite all the precautions exercised by the manufacturers, the magnetic coatings used on secondary storage devices such as tapes, disks, and drums are not always perfect. Most of the imperfections are detected at the factory before shipment, but others show up only with use and, indeed, may develop owing to the conditions of use. The latest programming systems, therefore, provide techniques for finding the imperfections and avoiding use of that portion of the medium. One way of doing this is for the system or maintenance engineer to use the statistics gathered from the automatic device retry functions.

On writing magnetic tape, the device retry function typically assumes that, if it cannot write, a portion of the tape is bad and it moves the tape and tries to write again. Only after a fairly large number of such trials does it assume that the tape unit is faulty. On disks and drums a special program can be used to write and read so-called "worst-case patterns" on the storage medium. If faults arise on these operations, the maintenance engineer may adjust the mechanism slightly, or he may allow the programming system to flag the record(s) as unusable. In this latter case, the system ignores this particular record in all subsequent operations, perhaps automatically substituting a physically different record when the user program tries to access the faulty record. Some of the latest data-processing programming systems invoke this media analysis program as part of their automatic device retry, but no current process control programming systems do so.

Process I/O devices are, as noted before, unique in that the data they present to the computer, or that they expect from it, are time dependent. This has two rather interesting facets. First, the data may be critically time dependent; that is, it may be vital to read or write the data within a very short time span. For example, a plant emergency may require reading the status of the five digital inputs, two temperatures and a pressure, and deciding which valves to close and which motors to stop. These actions may have to be initiated in, say, 5 msec.

The second interesting facet is that the data may be input to a digital filtering algorithm. Depending on the characteristics of the filter, it may not really matter if an occasional reading is missed or is in error.

As a result of these factors, the process I/O management routines are very different in many details from DP I/O management routines. It is rather difficult to discuss specific algorithms independent of the actual application as many of the possible management algorithms strongly influence the accuracy and quality of the whole control system. Some general approaches, however, can be suggested.

The appearance of the process I/O to the user program can be made similar to that of the DP I/O if queuing of requests, prioritizing of requests, and error passback are supported. In addition, some way normally is required to provide for the immediate execution of a process I/O function regardless of the operations currently in execution.

For the usual process I/O options used on control computers, three management techniques are commonly used: The first is to do the I/O on the basis of a specific request from the user program, just as for DP I/O. The second approach is to do all process I/O with a master I/O routine which executes on a fixed time interval basis. The status of each I/O point is maintained in a table that is updated once during each interval. The user requests for I/O then become transformed into requests for status data stored in primary storage. The third technique is similar to the second, except that the master routine runs on a continuous basis.

These last two techniques have some interesting pros and cons when compared with the first technique. They may result in data that are less timely for any particular point. However, they do make checkpoint/restart operations simpler since a complete set of status data is available. In addition, they normally have much less execution overhead than the first technique when a large number of I/O points is involved, but they do require a large quantity of storage space for the current value of the points. The third scheme ensures maximum use of the I/O hardware but may actually overuse it. Some types of relays, for example, may well last only a few months at their highest rated operating speed.

9.4   SYSTEM GENERATION

System generation is the final stage in the production of a programming system. Just as an automobile designer has to allow for a vast number of different options on what is considered by most users to be one type of car (e.g., color, transmissions, engines, radios, power accessories, body styles, differences in state regulations), so a programming system must be able to support a wide range of environments and user requirements. The major difference, of course, is that in a programming system this final stage can be carried out by the user and may be readily repeated if necessary. During this stage, the system generation problem, the user "builds" a programming system tailored to his particular application by selecting components and architectural alternatives which are to be included in his programming system. In this section, the environment (the machine) and the user requirements (the components and their tailoring) are discussed, and finally some comments on the uses and abuses of system generation are made. In applying a control computer, very careful consideration must be given to system generation by someone with a good knowledge of the application. Indeed, it should be considered just as carefully as the machine selection was considered.

9.4.1   Machine Description

A programming system must be adapted to its operational environment. Most manufacturers offer an array of different combinations of boxes under the designation of one system name. These combinations can be used to create many distinct system configurations. Most users expect, not unreasonably, that the complete range of configurations can be supported by one programming system. If the range is too large, however, this may not be desirable as excessive distortions may then be required in many of the algorithms in the programming system. For example, the programming system used during the system generation process is normally designed to operate on a minimum hardware configuration and is designed to be relatively insensitive to the environment. To do this, it normally has to be very inefficient in its use of the machine by, for example, not allowing any overlap of I/O operations. During system generation for a more than minimum configuration, these limitations are removed.

The first thing normally specified during system generation is the type and main storage size of the object CPU. These specifications are, of course, checked for validity of support under the programming system. The storage size information is used to indicate to the various components of the programming system how much space is available for table building and other storage allocations. The information is also used by the programming system in its control of the system. Subsidiary uses may include use by address verification routines included in user-written programs to prevent errors caused by storage overflow.

The next information provided to the programming system may concern some of the optional features on the CPU such as timers, clocks, and floating point hardware. This information is preserved by the system to assist in management of these features.

The rest of the environment description consists of a description of the peripherals, or I/O units, and their interface to the CPU. The interface normally requires fairly detailed description, but the I/O units themselves usually can be classified by type. It is fairly rare for the system generation procedure to require submodel identification. For example, the vendor may well offer a low-speed analog-input subsystem in a variety of somewhat different submodels. Most of the differences, however, will not affect the programming system support.

The differences that do require unique support normally need only a small number of different instructions. Although these extra instructions could be deleted, it is not common to do so. The reasons are to minimize the cost and complexity of the sytem generation procedure, to keep the number of unique programming systems reasonably within the grasp of the service engineer and, also, to allow some reasonable degree of portability of the programming system between slightly different hardware configurations.

Also required for system generation is the number of channels that must be supported and an identification of which features are connected to each channel. The system must also know which interrupts represent each device and, lastly, the type and location of its own residency device currently and on the object system.

In addition to a description of its environment and as part of the total system generation procedure, the system may require the initialization of various media such as disk surfaces. The programming system may need parameters for some devices, characterizing their application or performance. This is particularly true for teleprocessing and process control I/O devices. The parameters required might be, for example, the number of telephone lines and their data rates, or the size of the analog and digital I/O multiplexers. This information is used subsequently for device management and error detection.

9.4.2 Component Selection

In producing the programming system, the vendor prepares program components for the support of all equipment and features that can be included in the computer system. In addition, he may provide alternate routines which, although they perform similar or identical functions, differ from one another in performance or other details. For example, both standard and reentrant subroutines may be provided for inclusion in the subroutine library. Few users require all the components provided by the vendor and, as part of the system generation process, the user must specify which of the available components are to be included in his particular programming system.

The normal user of a programming system needs at least one language processor, a link editor (or builder), and a library of macro routines or subroutines. The user system designer, normally at system generation, indicates which of a number of possible languages (and even a number of possible processors for one language) he wishes to retain in his system. He also indicates which libraries and which macro routines and subroutines in the library he wishes to retain. The system generation routines then simply copy these processors, libraries, and routines from the master system DASD or tape to the new system residence media.

The reasons the user may wish to delete parts of the system are threefold:

1. He foresees no probable use in his application for them.
2. To increase system efficiency since the unnecessary components require space on the system residence media. In addition, their presence may affect the system performance if, for example, a disk seek has to be performed regularly across the component.

3. He may wish to restrict the facilities available to his application programmers, to enforce a one-language standard, for example, or to simplify the appearance of the system for casual programmers.

In small systems component selection is usually very easy. If the system components are provided as card decks, the user selects the appropriate decks and loads them using a system builder program. On systems generated from tape, the components are preceded by a header record containing the name of the component. On DASD systems, these header records are grouped together into a directory; that is, a single file, the records of which contain the name and location of all data sets on the DASD media. The user specifies the program components he wishes to build into the system by providing program names to the system builder through the card reader or typewriter. For both tape and DASD, the search routine for the individual components is quite simple and the components are simply copied onto the system residency media. For individual routines within a library, the library itself typically contains a directory (or index) of its contents that can readily be searched for routines specified by the user.

In addition to searching for and copying system components specifically requested by the user, the programming system builder may automatically delete unnecessary programs. This is done on the basis of the machine description provided to the system. Thus, for example, if the machine description does not include a high-speed printer, the control routines for such a printer are automatically deleted.

9.4.3 Tailoring

Component selection provides for the adaptation of a very general programming system into a more specialized system for a particular application. The adaptation is relatively gross, however, since the components provided by the vendor must still be designed for a wide range of features and other system environmental factors. Tailoring or "fine tuning" of the selected components is a procedure that further adapts the system components to the particular application in which the programming system is to be used. Tailoring involves the selection of alternatives which have been included in the design of the system components and the specification of parameters required by the various components.

The need for tailoring arises simply from economic considerations. It may be nearly impossible for the programming system market research team to identify the possible breakdown of a system beyond some fairly gross component level. Beyond this level, it is valuable if the development team provides options or techniques to allow the user to fine tune the programming system to the application. Although tailoring takes time and effort by the user, it is an aid in achieving higher performance or lower storage cost.

One obvious fine-tuning function concerns the specification of variable length tables that are used dynamically by the system and that are normally resident in main storage. It is important that such tables be controlled by the user. The application may well be such that only certain tables ever really get large. By allowing the user to define the length of such tables, the reservation of a large quantity of main storage which rarely gets used can be avoided. However, this requires insight into the particular application and the habits and experience of the application programmers. This is why these decisions cannot really be made by the programming systems designer. An example of such a table is any kind of queue table that can contain a maximum of N entries. For most users, a small value of N may well be adequate (Martin, 1965, p. 47). For some users, however, a small value of N may destroy much of the value of queuing for that resource. By making N user selectable, the programming system can be used by both users with, perhaps, no other real differences.

Selection between architectural alternatives is the most important tailoring function. This arises when there are different algorithms available for doing a system function, all of which are presented to the user for decision. An example of this might be whether typewriter messages from application programs and background programming systems programs should be buffered in secondary storage or in main storage. There is a trade-off between these two algorithms in the total main storage used, but their effective performance is virtually the same. The buffer-to-secondary storage scheme can handle much larger numbers of larger messages, but its own code is, of course, much more extensive. Standardizing on one or other by the vendor would either penalize the user with a small number of short messages or would preclude the use of the programming system by a user with a large number of lengthy messages.

However, apparently more important than this type of decision is the one concerning the overall appearance and layout of the system on both main and secondary storage. Analysis of this decision shows that it is really a series of decisions. That is, the user must select between the various allocation, residence, and management algorithms provided for each portion of the application support. These choices are particularly important for those algorithms supported by the permanently main storage resident portions of the programming system, since storage space and efficiency can be significantly affected.

Another very useful tailoring function common in data-processing programming systems, but not often used currently in control computer programming systems, allows the user to modify the most commonly used system routines for his particular application. It may be that because of the ground rules established for the application programmers, the use of certain functions in the system are precluded. For example, if floating format is not allowed on reports or input cards in a system using FORTRAN application programs, the "E format" need not be supported by the FORTRAN I/O routines. By removing the code relating to this format from the FORTRAN I/O routines, they may be reduced substantially in size. This can be easily accomplished if the programming system programmer included conditional assembly statements through his code which can be activated or deactivated by changing a single parameter and then reassembling. The practical limit of this technique is when the elimination of such functions affects the size of the routine by only a small percentage. If conditional assembly statements or a similar technique are not provided, the elimination of functions such as this may require the skills of a relatively sophisticated programmer.

Although the information for tailoring the system must be available for system generation, it is frequently preserved by the programming system in such a way as to be alterable later without severe consequences. This is generally easy to do for tailoring techniques involving the specification of parameters such as table or storage partition lengths. The architectural decisions, however, are usually embedded in the system and cannot be changed without redoing much of the system generation procedure.

A fascinating function that can be performed readily through system generation is to alter the apparent configuration of the complete hardware system so that the application programmer may ignore differences between machines. This is valuable in process control as it may well allow a machine in a laboratory or pilot plant environment to run application programs the same as those run by a machine having a different configuration and located in the actual plant. Both the CPU and the I/O devices and subsystems can be included when using this technique. The programming system simply maintains equivalence tables to relate the real and simulated configurations. The machine or unit simulated may even be a simpler, or at least more readily comprehensible, machine for the application programmer to use. For example, it makes sense to simulate magnetic tape on a DASD for a wide range of programs. In this case, the magnetic tape support routine has to have some indication as to which are the real and which are the simulated tapes.

9.4.4 Implementation

The implementation of many current system generation procedures lacks the sophistication inherent in the programming system itself. This is particularly true in the control computer area. Three reasons for this may be discerned:

1. The fact that the current control computer programming system often has been derived from the much simpler scientific and engineering programming systems. These systems, simple and often thought to be computer bound, traditionally have had a minimal system generation procedure. It merely assumes a certain set of I/O and associated assignments and allows little specific tailoring of system routines.

2. The common limitations of limited programming system development time and money. The available time and funding is usually directed toward more basic problems.

3. The assumption that the user doing the system generation is a sophisticated programmer. The user is assumed to be quite capable of making whatever adaptations of the system may be required to tailor it efficiently to his application. Although many user programmers are indeed capable of modifying the programming system, many more are not, and their needs must be recognized by the system generation designer.

The system generation procedure is thus frequently a conglomeration of different bits and pieces, sometimes even requiring immediate operator input of system parameters. Only a few control computer programming systems have a carefully designed single job stream for system generation that can be run readily by an operator with the results returned to the system generation analyst/programmer for verification of its accuracy before release as a usable system.

Even when a well-designed generation procedure is used, however, some of the problems encountered when installing the computer system will be traced to problems the solution of which will require repeating some or all of the system generation procedure. Since system generation is typically a complex procedure, it may take considerable time to rebuild all the application programs. As a result, techniques are needed to minimize this impact. There are at least five such techniques:

1. During initial system generation, three things are needed: (a) The master system generation media must be preserved. For example, if the total programming system is delivered on magnetic tape, the first thing to do is to make a working copy that can be used as the control system generation input medium. (b) A copy of the final object system must be

preserved. (c) The records of the actual system generation should be saved for later analysis to see if problems can be explained and thus avoided in subsequent system generations.

2. The system generation procedure can be repeated on-line when multiprogramming is supported. Very few current control computer programming systems allow complete on-line regeneration, but large portions of the system regeneration normally can be accomplished in this manner.

3. The programming system may be patched on-line to solve, or at least to ease, certain types of problems. If the wrong algorithm was selected and the desired algorithms can be inserted easily, for example, this technique can be quite useful. Care must be taken to disable all references to the affected function until the switchover has been successfully accomplished. The most modern control computer programming systems have powerful facilities to do this type of on-line modification. This capability is required, at least to a limited degree, for installing temporary fixes for bugs found in the programming system.

4. As previously discussed, it is very useful if the system can allow dynamic on-line alteration of many parameters specifying items such as table lengths, buffer sizes, and the number of buffers. The programming system designer must, of course, balance the use of this technique with the difficulties implicit in using such dynamic parameters. Dynamic alteration of such parameters is more easily implemented if the affected storage areas in the executive are grouped together. Hence, as long as the total required space remains the same, the distribution between the various tables and buffers may be altered readily under the control of the user programmer or, in some cases, the operator.

5. It may also be possible to change the distribution of I/O units without system regeneration. For instance, if the programming system supports two types of devices on a single channel, it could be easily designed such that, as long as the total number of devices remains the same, the distribution between the two types can be adjusted dynamically. This can be achieved by the use of two tables for each device. The first is associated with the device type rather than with a specific device. It is used to check whether there is a device of the requested type at a given position on the channel. It might also contain a pointer to the device table for the particular device at that position. The device table contains the device characteristics, including a specification of the device type. If the device tables are a standard size, it is easy to manipulate the information in them and the pointers in the device type tables. The channel control table contains a reference to each possible position which will point to one or other device type tables. For example, Fig.

9-15 shows the tables for a system that may have 12 channel positions and a maximum of 8 DASD type A and eight DASD type B. It is easy for the programming system to manipulate the pointers dynamically, as shown by the change from the solid line to the dotted line in the figure. The only real overhead in this approach is some extra space in the device type tables. The total number of device tables remains the same.

The decision as to which of these techniques to implement is, of course, dependent on the details of the particular programming system. It also requires intelligent consideration by the user system designer. He should try to take advantage of these facilities but, particularly in the matter of on-line (or on-the-fly) patching, great care should be exercised. Failure to observe proper care may make matters very much worse and, conceivably, could destroy the programming system with resultant loss of control over the process.

In conclusion, process control programming systems are introducing many people previously accustomed to scientific and engineering programming systems to many problems such as system and file integrity that are common to data-processing systems. Frequently, errors have an even more radical effect on the company's economics. Thus it is important that process control programming systems be carefully tailored to their application through the system generation procedure. System

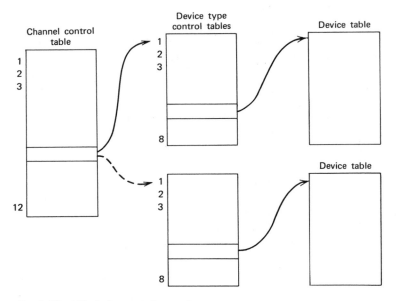

Fig. 9-15   I/O device control example.

generation is an important and powerful tool but, like all powerful tools, it must be controlled carefully and used with intelligence and foresight.

9.5   GENERAL CONSIDERATIONS

Previous sections have covered major design considerations involved in the development of a control computer programming system. Emphasis has been placed on the design aspects of particular importance or uniqueness in such systems. In addition, however, topics common to the development of any programming system must be considered since these often can have a significant effect on the design or use of the system. This section briefly describes some of these more general topics. Specifically, program testing, program performance, documentation, system updating, and conversion problems are discussed.

9.5.1   Program Testing

Programs must be thoroughly tested before they are released for use by their ultimate user. This maxim is all too easy for the control computer application programmer to ignore. His previous programming experience has often been in an engineering or scientific "job shop" environment. Programs are developed in close association with the engineers requiring the program and, in many cases, the programmer and the engineer are the same person. The programs often have served their purpose when they first run completely successfully. The control computer environment swings to another extreme, frequently even much further than the regular data-processing environment. The programs are used repeatedly and it is possible that an error can cause irreparable damage to the process being controlled. Table 9-2 illustrates the operator, I/O, and failure environments of the three types of programming systems.

The control computer programming system can help the application programmer test his programs in a number of ways: (1) by providing support for tracing the execution of his program; (2) by simulating the process to his program; (3) by being well debugged and tested itself; and (4) by being well maintained.

Program tracing is a well-known program debugging aid. Fundamentally, the ability to trace the program execution requires that certain occurrences cause the display of the machine status. Following the display, the program execution is resumed. This is almost an exact description of interrupt processing, and many machines now have a trace interrupt with the trace support portions of the programming system executing on the interrupt level. The hardware is responsible for recog-

Table 9-2 Programming Systems Environment Comparison

Programming System Environment— Function	Scientific and Engineering	Data Processing	Control Computer
Repetitive use of any one program	Rarely	Daily, weekly, monthly	Seconds, minutes, hours, days
Operator	Programmer/ Engineer	Trained computer operator	No computer knowledge
Cost of program failure	Low, much knowledge still gained	Computer rerun	Process failure; cost of ruined materials/equipment can be very high
Input/Output			
Quantity	Minimal-medium	Medium-high	Medium-very high
State	Static	Static	Dynamic
Type	Simple	Complex	Complex

nizing the events on which to interrupt the program execution. Typically, these hardware trace features are rather restrictive in the events they recognize, so they must be supplemented by additional software in the trace package. Various levels of tracing can be provided:

INDIVIDUAL INSTRUCTION TRACE. Each instruction of the program under test is displayed along with the contents of any registers whose contents have been altered.

BRANCH INSTRUCTION TRACE. When executed, the branch instructions in the program under test are displayed along with the condition states they are testing. The programmer can thus examine the logical flow of his program.

HIGHER LEVEL LANGUAGE TRACE. This is readily implemented by the user by inserting source statements for the display of intermediate results during program execution. Once the program is thoroughly tested, these statements are removed and the source program is recompiled.

Trace packages, although useful for the verification of the program logic and very safe if used properly, have a major drawback for debugging real-time programs: Because of their very nature, they interfere with the timing performance of the application program and so may disguise potential bugs caused by timing interactions.

Simulation, the second testing aid, is very powerful in discovering timing interactions as well as other problems. There are two common types of simulation: (1) simulation of the computer and programming system, and (2) simulation of the process. Both are very useful, although they frequently are not used to maximum advantage by application programmers.

The first type of simulation can be used in conjunction with a trace package to overcome the timing interaction problem. Better yet, however, the simulation of the programming system and the process can be combined to provide a powerful debugging aid. This is utilized by running the application program in a simulation mode during which all interfaces between it and the programming system are made to the programming system/process simulator. For example, an application program calls for an analog-input function using the conventions defined by the programming system analog-input routine. However, as it is in the simulation mode, the programming system analog-input routine has been replaced by a routine which generates a value according to the specifications for that analog input. The alternate routine generates an analog-input reading according to a predefined algorithm, performs the same checks as the programming system routine, and provides the proper delay before returning to the application program. At the expense of some storage then, the application program can be made to react just as it would have in real life. As an added benefit, all of its inputs and outputs may be preserved for later analysis to verify its performance.

As the value of process equipment and material flows continue to increase, it would seem that such simulators would be very popular, but this is not yet true. Although conceptually simple to implement, the simulation programs are also very large and quite expensive. They tend to be complex, difficult to operate, and demand a degree of knowledge of the process rarely possessed by the application programmer. They have been used with great success in the space program and some military systems (Gibson, 1969; Day, 1969; Witzel and Hughes, 1969). Their use in control applications probably will increase as modeling techniques and the design of simulation programs are refined.

The third aid to the application programmer, a well-tested programming system, is very important. The system should also be forgiving. That is, it should recognize a wide variety of possible errors, identify them as precisely as possible, and prevent their propagation through the system. The techniques for testing the programming system are no more than variants of those available for application programs, and so tight management control is required to ensure adequate testing. A common technique is to have an independent test audit group whose functions are the following:

1. To examine the specifications of all tests run by the implementation team and express an opinion to higher management concerning their adequacy and efficiency.

2. To select a subset of the tests for rerun against the system to verify that the system does pass them.

It should be remembered, however, that it is virtually impossible for the vendor to test the whole system exhaustively. The system generation procedure that allows many options to be specified can result in a huge number of unique programming systems, all of which would require extensive testing. An application programmer's job, in addition, is to come up with essentially unique combinations of functions. These factors, combined with the number of possible states of a typical control computer, if faulty hardware is allowed for, result in a virtually inexhaustible set of conditions which would all have to be tested to guarantee an error-free system.

Closely coupled with a well-tested system as a testing aid is a well-maintained system. This consideration is discussed in Sec. 9.5.4.

9.5.2   Program Performance

The question of programming systems performance is a controversial subject in the industry at the present time. The controversy stems from two major questions: (1) What is a meaningful measure of the performance of a programming system? And (2), assuming that performance criteria can be defined, what are meaningful techniques for determining the performance of the system? The first question is similar to the problem of defining criteria for the comparison of different hardware systems. Performance can be measured in terms of execution time, time responsiveness to particular interrupts, storage requirements, and a variety of other factors. The importance of each factor depends greatly on the particular application for which the system is being used and, in addition, complex interdependencies exist between many of these factors. Suppose, for example, that two programming systems can be used on the same CPU. One provides faster interrupt response than the other, but the second is more tolerant of system failures such as main storage parity errors. Which one can be said to have the "best" performance? Thus it is very difficult, at best, to define performance criteria that are meaningful for the comparison of programming systems. For a series of articles concerning performance measurement, see the *IBM Systems Journal*, Vol. 8, No. 4, 1969.

There are three different approaches currently being used to assess programming system performance, none of which is restricted to control computer programming systems. The first is the purely theoretical approach in which a number of simplifying assumptions are made. An attempt is then made to predict the performance of different algorithms for different functions of the programming system. This approach so far has failed to yield more than minor benefits in terms of design criteria for higher performance systems but has confirmed a few of the empirically available results.

A more promising current technique is simply to simulate the actual algorithms under controlled situations, vary the parameters, and compare the performance given by the simulations. This is quite a powerful technique and has invalidated many common assumptions made by systems programmers. The limitations of this approach are, of course, the degree to which the environment can be controlled and the degree to which the performance under controlled conditions can be related to the performance in an actual application.

The third technique is to try to measure the performance of existing programming systems under known conditions. This is sometimes implemented through the use of a "benchmark" application program which is run using various sets of test data. The design of a suitable test program and data sets is very difficult, and the resultant figures are open to a wide variety of interpretations. On some large data-processing systems, it is not unknown for the timing on a user "benchmark" problem to vary over a 30% range when executing under apparently identical conditions. The reason for this variation is the statistical nature of every parameter in the overall computing system.

As a general rule, therefore, the programming system's overall performance is normally characterized only under very artificial conditions. For example, interrupt response times are quoted that assume the system is completely quiescent awaiting the interrupt, that no cycle steal is taking place, and that all other interrupts are masked. These statements are not very harmful as long as they are recognized for what they are; they are only useful for relative ranking of programming systems, not for absolute evaluation.

These three techniques for assessing the performance and the general conclusion discussed in the preceding paragraphs are primarily related to the *a priori* prediction of performance. Once a particular system has been installed on a specific application, however, there are useful techniques to measure the performance of the overall hardware and software system during specific time intervals and under the actual operating conditions existing during this interval. Although these measurements are *ex post facto* in terms of selecting the system, they can be very useful in identifying modifications which should be investigated to improve system performance.

Both hardware and software techniques can be used to gather the data for analysis of system performance. In the case of software, a program is provided which records the number of times specific events occur or the time interval between two events. For example, it may record the number of interrupts occurring on a given level or the access time for data stored on a DASD. The measurement program runs under control of the programming system executive and is contending for the use of system resources. Therefore, it can affect performance and this factor must be considered in its use and in the interpretation of the data.

In the hardware approach, an independent monitor is attached to the computer system. The monitor typically consists of counters and timers actuated by signals in the computer being monitored. The computer signals are monitored using sensing circuits that do not disturb the normal operation of the computer circuits or the programming system. For example, by attaching a sensing probe to a particular interrupt signal, the number of times that it is actuated can be counted. Similarly, the access time of the DASD or the activity in particular partitions of main storage can be measured. The raw data represented by the values in the counters and timers then can be analyzed to provide statistical measures of particular aspects of the system performance. The data presented in Fig. 9-11 and Fig. 9-12 were collected using this type of monitor and are indicative of the data that can be obtained.

As in the case of any measurement results, some care must be exercised in deriving conclusions. If the measurement interval did not represent a period of "typical" operation for the system, for example, or if a dependency between two measurements is not recognized, the conclusions derived from the data may be erroneous. Properly used, however, the data may suggest modifications to the programming system, application programs, or hardware that can significantly increase the efficiency and/or performance of the overall system in the particular application.

9.5.3 Documentation

This is a requirement crucial to the use of a programming system. The system must have its features thoroughly documented. The amount and nature of the documentation may vary with the particular vendor. In general, however, a well-documented system is accompanied by a variety of manuals and other supporting publications.

There is normally a pamphlet or small manual giving some explanation of what the programming system does for its user and what kind of algorithms it uses for various functions. There is also a manual, or perhaps manuals, specifying how to use the system, how to generate it, how to write the interfaces to its components, how to operate it, and what its messages mean and what to do about them. In addition, another series of manuals typically is prepared but often is restricted in circulation, owing mainly to the cost of keeping them up-to-date. These are the manuals specifying how the components actually work. These manuals are of most interest to those either debugging or modifying the programming system. They usually contain a detailed description of how the modules operate, their storage layout, the layout of their tables and control information, and a flow chart.

The most basic documentation remains, however, the source program listing. Without this, the foregoing manuals are of no great assistance to anyone trying to explore the internals of the system. But even the source listings may be of limited value unless certain precautions are taken. A primary responsibility of the programming system implementation management must be to enforce reasonable coding standards or the source listings may be virtually useless to anyone except the original programmer. The standards should include such things as meaningful labels, no absolute numbers except if used in parameter definition cards, no "trick" codings such as the use of coincidental no-op instructions instead of the standard system no-op, and above all, plentiful and meaningful comments and remarks. Every piece of data and every table should be carefully and clearly explained. So should all branches, subroutines, and unusual pieces of code. The module listing should be prefaced by comment statements giving a good description of its purpose and its interfaces to other modules. The language processor used by the development team should automatically apend cross-reference lists to the module source listing. With these standards and supplementary information, the source listing becomes a valuable and useful document for the experienced programmer.

Another useful document is a performance specification manual listing the speed and size (where possible) for each component under a reasonable selection of realistic operational conditions. Few current system descriptions provide more than the size of each component and some sketchy timing figures. If they are provided, it is important for the user to recognize that the data are valuable only as guidelines. As discussed previously, performance is difficult to define and is highly dependent on the use of the component.

Another useful document is a technique manual wherein a number of hints and tips for the use of the system are given. The best of these manuals include a sample application system showing a range of inputs and outputs and how the individual programs were written. The manual describes how the programs relate to one another, the programming system, and the machine. Thus the application programmer may be able to use already tried and tested techniques that need no more than minor adaptation to his own application.

Finally, each manual has to be oriented carefully to its readers. For example, two versions of some manuals may be required: one for the experienced programmer and one for the casual programmer.

9.5.4   System Update

A programming system requires maintenance. Software does not wear out nor does it age with use, but no programming system is perfect on original release to the application user. As discussed in Sec. 9.5.1, a programming system cannot be tested exhaustively and so users find combinations of system functions that do not operate as they apparently should. In this event, the programming system vendor normally operates in the following manner. His field operative validates the problem with the user and classifies it according to its impact on the application. For example, if it prevents the use of a programming system component essential to the user operation, it is classified as severe; if it concerns a component of little use or importance to the user, it is noted for information purposes and a description is forwarded to the programming system maintenance group for consideration in future versions of the component.

For severe problems, the vendor's field engineer can normally get immediate access to a maintenance coordination group to determine whether the problem has been reported previously and whether a solution or "fix" has been developed. If not, two alternatives can be considered: (1) Develop a by-pass of the problem for the user by, for example, disabling some portions of a specific component, or (2) develop a temporary fix. The distinction between these alternatives is that the by-pass merely avoids the problem (typically with some resulting degradation in performance and/or function), whereas the temporary fix actually solves the problem. The fix is considered temporary since it has not been thoroughly imbedded in the system. A permanent fix is one that, with the rest of the system, is submitted to a regression test against all the tests the system previously passed plus a new test for the specific problem area.

In either case, the programming systems maintenance group now must decide whether to provide a permanent fix, put a restrictive statement into the documentation for the system, or simply distribute the temporary fix to those users requesting it. This decision is made as a result of estimating the value of the fix to all users and the cost of the various alternatives.

As a result of this corrective process, however, the programming system maintenance group accumulates a number of permanent fixes that must be developed for the programming system. To get these developed, proved, and to the user, a common vehicle is the programming system "release." This is simply a package of all the latest fixes, document changes, and programming system improvements. The fixes may consist merely of patches to the previous version of the component or they may be complete replacements. It is usual to make releases on some announced schedule so as to facilitate user planning.

When a new release is issued, the user or field agent must have tools available to apply this release to the programming system; that is, to do the system update. In a control computer programming system, it is desirable to do this system update on-line so as to minimize the impact of the update on the application. Before considering on-line update, however, we discuss the less complicated case of an off-line update to illustrate the functions that must be performed.

In the case of an off-line update, no user program is running concurrently with the system update. The update is normally done by a special program which scans the update input provided in the new release. For each component to be updated, there is a record in the input stream indicating the component name and the type of modification (add, delete, or replace) either for parts of the component or for the whole component. The system residence device is then searched for the component, and when found the designated update is performed. If the named component is not found, the update program assumes that the component has been deleted from the system previously. After logging this information to the update operator, the program proceeds to the next component name record in the update input stream.

After the update of the system has been completed, the next step is the update of the user programs. The procedure for doing this depends on the manner in which the user programs are coded. If components of the programming system such as subroutines are imbedded in the user programs, the update procedure may be quite lengthy. Fundamentally, a search of all user programs for references to the affected programming system components must be made. To reduce the time taken to do this, many different techniques are used. Most of these depend on keeping available some of the information which was used to create the program but which, however, is not needed for its execution.

Alternatively, the user programs may be programmed such that the necessary programming system components are selected every time the user program is executed. To ease the update procedure in this case, a list of the names of programming system components used can be attached to the executable version of the user's program during assembly or compilation of the source program. As system update proceeds, a list of all components updated can be generated and this list compared with the lists appended to the user's programs. All programs that use an updated programming system component then can be identified. Their executable versions can be flagged as no longer valid and the operator informed of this action. After verifying that the updated component is suitable for use in the user program or after making the required modifications, the user program can be reinstated in the program library.

In a real-time programming system, the same updating functions are required. When the system update is done on-line, however, the procedure is made more complex by the random nature of the demands for the application programs. It is important that a version of the user's program be available at all times. The replacement version must, therefore, be completely prepared and ready for use before the old version is flagged as invalid. As a result, the user's program library fills up with invalid copies of his programs. As part of the on-line system update, therefore, the user program library must be purged, as must the system library. A convenient and commonly used scheme for doing this is to use the space for a just-deleted component program to hold the replacement copy of the next program to be updated. A number of variants of this algorithm are used. Each offers some reduction in the amount of space wasted by components or programs that are smaller than the space available, but they retain the major virtue of not always requiring that the other programs or components in the library be moved physically.

A major problem may arise in attempting to update executive program components: Two copies of the executive can become simultaneously resident in the system and, at some point in time, the new copy has to be used rather than the old one. The safest technique currently available is to demand that the actual switchover be completely under operator control. The reason for manual control is that, since this is a very drastic change to the system, it seems unreasonable to switchover automatically at some arbitrary time unrelated to the state of the process.

9.5.5   Conversion Problems

Computer technology is changing rapidly and, in particular, cost for a given performance is dropping very rapidly. Many users thus find that, when the total workload on a system has begun to exceed its capacity or major alterations of the plant are anticipated, it may well be cheaper to install a completely new computer with much more capability than to expand the old system. One of the biggest problems, and frequently the highest cost exposure, is in the conversion of programs and data files from the old system to the new one. Very few process control users have yet had to face this problem since the present generation of control computers is the first to be installed in fairly large numbers. The problem has been addressed in more conventional data-processing applications, however, and information applicable to control installations can be derived from this experience.

A major task is the conversion of programs to the new machine, new I/O devices, and new performance constraints. Ideally, they should be completely rewritten to maximize efficiency, but this is not normally practical for some substantial transition period. If programs are written in a higher level language that is supported on the new system, they can be recompiled and, normally can be used quite successfully with few changes. They may not make efficient use of the new system, but this is generally acceptable for a reasonable transition period.

The very critically time-dependent programs are usually the first to be rewritten for the new machine and, therefore, do not present any particular conversion problem. Those programs written in machine language but whose timing is not critical present a difficult conversion problem as they may not justify a rewrite early in the conversion period. A number of different techniques can be used. If the old and new machines are similar in machine language, it may be possible to run a copy of the program through a simple language conversion program (LCP) which detects those places in the program where the old and new languages differ. If possible, the LCP does the conversion itself and, if not possible, it flags portions of code for rewriting by a programmer.

If the machine languages are not very similar, then either emulation or simulation may be used. For simulation, a program is provided that runs on the new machine and simulates the old machine. The old programs are run under control of this program. As simulation is relatively slow, a technique used on data-processing machines is emulation. For this, special hardware is added to the new machine to perform some of the more commonly used functions of the old machine. A simulator program can then be written that is smaller and runs faster than the equivalent simulator without these special features. This combination of hardware and software is called an emulator. Very little use has yet been made of this technique in computer control systems.

The second problem, data conversion, can be very costly and time consuming for a user if a large volume of data is involved. Many programming systems provide facilities for accessing data from a wide variety of media so that the programming system may not need to make specific conversion facilities available. It is also common to use a third machine as an intermediary if it has two different kinds of transportable media attached to it, one of which is acceptable by the old machine and the other by the new machine. Data can then be transferred to the new machine media via processing on this third machine.

Sometimes the data storage media may be acceptable between the two machines, but the new programming system will not accept the data formats of the old programming system. It is common to provide special conversion routines to handle such a case.

9.6 FUTURE DIRECTIONS

There are a large number of major trends discernible in present-day computer control systems which will probably dominate the next generation of machines and programming systems. The experimental period is over. Computer control is a technical success and, increasingly, an economic success. The process engineer increasingly will regard the computer as a very powerful and extremely flexible controller. The market for large special-purpose hardwired controllers may very well disappear, as may the relay room except for power-handling purposes. The plant engineer and his management will start to explore the economic advantages of truly centralized control and easy data collection. Many large-scale loops involving multiple individual processes will become automatic for the first time; for example, on-line recalculation of all plant controller parameters based on product mix required and current production yields with the new parameters then applied directly to all control loops. By coordinating information over a whole plant, many pieces of process equipment will be reduced in relative size. For example, large mix tanks are not required if the component variations in the process flow can be compensated automatically downstream.

These developments will lead to complex control computer systems and will place a strong emphasis on good programming systems. The programming systems will exploit the computer efficiently, yet clearly show the application systems designer the results of his decisions.

The process engineer will turn increasingly to the mini-computer for noncentralized control. This will include experimental work, back-up control, and remote control and data collection. To help him exploit the mini effectively, program preparation, debug, and simulation facilities for the mini will be available on larger computers. Software vendors may well offer special application program packages that can be readily generated on a large computer and then loaded into a mini to do a reasonable variety of standard functions. These methods will allow a simple programming system to suffice for the mini yet take advantage of the efficient I/O devices and large storage of a larger computer.

Simple as it is, the programming system for the mini is likely to be some reasonable subset of the system available for a larger machine so as to ease programmer and operator training requirements.

For some plants, mini-computers and other components will be used to create distributed control systems. Such systems will be popular in plants where the product flow goes through a number of discrete steps which, for process, safety, or even historical reasons, are rather poorly integrated.

Other plants will rely on a centralized control system focused on a large central computer. Such a system will be attractive to many continuous flow plants with well-integrated, physical compact, 24-hr processes. Many such plants will be run from a centralized control room by a very small operations staff. In an emergency, manual control from this room is essential so that little remote automatic control is possible.

The difference between these two approaches is that in the first, the complexity tends to be external to the individual computer, whereas in the second it is internal to the computer. Thus the second type of system requires a more complex programming system that can, however, balance the computational load over the whole plant much better. The economics of each type of system, and the many in-between compromises possible, are constantly fluctuating. Computer costs are dropping rapidly, as are the costs of remote digitizers. New plant wiring schemes are becoming attractive. Large computers, however, are still offering enormous gains in computational power each year.

The centralized system normally offers the lowest computing costs for a given control system so that other factors will influence whether to centralize the control system. These include wiring costs, operational convenience, and rate of changeover to computer control of the various sections of the plant.

The systems engineering to design control systems for distributed plants is going to be complex. The programming systems designer will be under pressure to support a vast array of possible computer networks as well as means of attaching instruments. The programming system will need to be adaptable and very flexible since the relative advantage of any one configuration will change quite rapidly. The alert systems engineer will constantly be factoring new capabilities and new cost/performance ratios into his design.

The total computational capability of the control computer(s) in a plant will, of course, be governed by the peak load conditions, and the normal huge amounts of space capability will be utilized for many noncontrol functions. Thus programming systems will be required that are much more general purpose so that they can efficiently handle a mix of process control, teleprocessing, scientific and engineering, and other types of jobs.

There will be an increasing degree of sophistication in the use of computer languages and in the languages themselves. Knowledge of certain languages will become imperative for all control engineers, regardless of their basic area of interest. In addition, users will become more willing to trade off computer efficiency for language capability and ease of use.

Extensions will probably be made to some of the current programming languages to satisfy the special requirements of control computers. It does not seem reasonable to create a special language only for control computers, as a typical control program has only a few functions in it that are not present in other types of programs. The development of relatively general application programs having many of the characteristics of specialized languages seems likely. Some presently available programs utilizing "fill-in-the-blanks" techniques for data and configuration specification are indicative of this approach.

REFERENCES

Day, E. C., "IBM System/4 Pi—EP/MP Computer System Simulation," *Rep. No. 69-825-2337*, IBM Corp., Owego, New York, June 1969. (Presented at IEEE Comput. Group Conf., Minneapolis, Minn., June 17–19, 1969.)

Gaspar, T., V. V. Dobrohotoff, and D. R. Burgess, "New Process Language Uses English Terms," *Control Eng.*, October 1968, pp. 118–121.

Gibson, A. L., "Functional Simulation of System/4 Pi," *Rep. No. 68-825-2345*, IBM Corp., Owego, New York, 1968. (Presented at IEEE Comput. Group Conf., Minneapolis, Minn., June 17–19, 1969.)

Lee, J. A. N., *The Anatomy of a Compiler*, Reinhold, New York, 1967.

Lefkovitz, D., *File Structures for On-Line Systems*, Spartan Books, New York, 1969.

Lowry, E. S., and C. W. Medlock, "Object Code Optimization," *Comm. ACM*, January 1969, pp. 13–22.

McCurdy, P., "BICEPS in Computer Control," *Pap. 69-511, 24th ISA Conf.*, 1969.

Markham, G. W., "Fill-in-the-Form Programming," *Control Eng.*, May 1968, pp. 87–91.

Martin, J., *Programming Real-Time Computer Systems*, Prentice-Hall, Englewood Cliffs, N.J., 1965.

Meyer, R. A., and L. H. Searight, "A Virtual Machine Time-Sharing System," *IBM Syst. J.*, Vol. 9, No. 3, 1970, pp. 199–218.

Rosen, S., *Programming Systems and Languages,* McGraw-Hill, New York, 1967.

Schoeffler, J. D., and R. H. Temple, "A Real-Time Language for Industrial Process Control," *Proc. IEEE,* January 1970, pp. 98–111.

Weiss, E. A., "Needed: A Process Control Language Standard," *Control Eng.,* December 1968, pp. 84–87.

Weiss, E. A., "Conflict Rages over Process Language Standards," *Control Eng.,* July 1969, pp. 77–81.

Witzel, T. H., and J. S. Hughes, "Flight Software Laboratory," presented at IEEE Comput. Group Conf., Minneapolis, Minn., June 17–19, 1969.

"American National FORTRAN," X3.9-1966, American National Standards Institute, New York, 1966.

"American Standard Basic FORTRAN," X3.10-1966, American National Standards Institute, New York, 1966.

"IBM System/360; Disk and Tape Operating Systems, PL/I Subset Reference Manual," Form C28-8202, IBM Corp., White Plains, N.Y., January 1969.

"IBM 1800 Multiprogramming Executive Operating System Introduction," Form C26-3718, IBM Corp., San Jose, Calif., 1969.

"Minutes: Workshop on Standardization of Industrial Control Languages," Purdue University, Purdue Laboratory for Applied Industrial Control, February 17–21, 1969.

BIBLIOGRAPHY

Abegglen, P. C., W. R. Faris, and W. J. Hankley, "Design of a Real-Time Central Data Acquisition and Analysis System," *Proc. IEEE,* January 1970, pp. 38–48.

Aiken, W. S., "Programming Digital Computers for On-Line Process Control," *ISA J.,* November 1959, pp. 62–65.

Anderson, J. P., *et al.,* "D825—A Multiple-Computer System for Command Control," *Proc. FJCC,* 1962, pp. 86–96.

Aron, J. D., "Real-Time Systems in Perspective," *Proc. Real-Time Syst. Semin.,* IBM Corp., Houston, Tex., November 1966, 26pp. (Also: *IBM Syst. J.,* Vol. 6, No. 1, 1967, pp. 49–67.)

Bennett, W. F., "An Approach to the Design of Command and Control Systems," *Anthology of Technical Papers,* IBM Federal Systems Center, Gaithersburg, Md., 1965, pp. 7–16.

Biles, W. R., and H. L. Sellars, "A Computer Control System," *Chem. Eng. Prog.,* August 1969, pp. 33–35.

Bouman, G. H., and J. J. Eachus, "A Real-Time Digital Computer Control System for Industrial Applications," *ISA Proc. Summer Inst.—Autom. Conf. Exhib.,* May 1962, 10pp.

Brereton, T. B., "Real-Time Computer Systems: Bibliography 24," *Comput. Rev.,* August 1970, pp. 483–489.

Brereton, T. B., and W. J. Lukowski, "An Automatic Test System for Software Checkout of Naval Tactical Data Systems," *Proc. Joint Conf. Autom. Test Syst.,* Univ. of Birmingham, England, April 1970, pp. 375–385.

Brown, K. R., and G. W. Brown, "Real-Time Optimal Guidance," *Proc. Real-Time Syst. Semin.*, IBM Corp., Houston, Tex., November 1966, 11pp.

Buchman, A. S., "The Digital Computer in Real-Time Control Systems, Part 5: Redundancy Techniques," *Comput. Des.*, November 1964, pp. 12–15.

Burrows, J. H., "Program Structure for Military Real-Time Systems," The Mitre Corp., for ESD, AFSC, USAF Rep. No. ESD-TDR-64-161, January 1965, 24pp.

Canning, R. G., "Progress in Fast-Response Systems," *Syst. & Proced. J.*, July–August 1967, pp. 20–24.

Casciato, L., "The Control of Traffic Signals with an Electronic Computer—A New Application of Real-Time Data Processing," *Proc. IFIP Cong.*, 1962, pp. 231–235.

Cassimus, P., "Design Considerations for Real-Time Systems," *J. Syst. Manage.*, December 1969, pp. 23–28.

Chang, W., "Single-Server Queuing Processes in Computing Systems," *IBM Syst. J.*, Vol. 9, No. 1, 1970, pp. 36–71.

Chang, W., and D. J. Wong, "Analysis of Real Time Multiprogramming," *J. ACM*, October 1965, pp. 581–588.

Chapman, J. W., and L. Danner, "A Real-Time Telemetry Data Collection and Processing System," *Proc. Real-Time Syst. Semin.*, IBM Corp., Houston, Tex., November 1966, 12pp.

Chorafas, D. N., *Control Systems Functions and Programming Approaches*, Academic Press, N.Y., 1965.

Clough, J. E., and A. W. Westerberg, "FORTRAN for On-Line Control," *Control Eng.*, March 1968, pp. 77–81.

Connelly, M. E., "Real-Time Analog-Digital Computation," *IRE Trans. Electron. Comput.*, February 1962, pp. 31–41.

Coppey, P., "Real-Time Operating Systems," *Proc. 4th AFIRO Cong. Comput. and Inf. Process.* (in French), April 1964, pp. 229–242.

Cormack, A. S., "Interaction Between User's Needs and Language-Compiler-Computer Systems," *Comput. J.*, April 1965, pp. 8 ff.

Couleur, G. F., R. R. Smith, and D. L. Bahrs, "Techniques for Multi-Level Programming with Real Time Contraints," *Proc. ACM Natl. Conf.*, September 1962, 14pp.

Coyle, R. J., and J. K. Stewart, "Design of a Real-Time Programming System," *Comput. and Autom.*, September 1963, pp. 23–34.

Critchlow, A. J., "Generalized Multiprocessing and Multiprogramming Systems," *Proc. FJCC*, 1963, pp. 107–126.

Cunningham, R. J., et al., "A Language for Real-Time Systems," *Comput. Bull.*, June 1966, pp. 202–212.

Darga, K., "On-Line Inquiry under a Small-System Operating System," *IBM Syst. J.*, Vol. 9, No. 1, pp. 2–11.

Davidson, L., "Operating Systems for OLRT," *Datamat.*, October 1967, pp. 27–29.

Davidson, W. G., "Real-Time Programming and Athena Support at White Sands Missile Range," *Proc. FJCC*, Part 1, 1965, pp. 871–877.

Day, R. F., and C. A. Hobbs, "A Real-Time Digital Computer for Radar Control and Data Processing," *Proc. 6th Natl. Conv. for Mil. Electron.,* June 1962, pp. 82–85.

Doede, J. H., and L. Blanc, "Dynamic Memory Protection in Interactive Environment Programming," *ISA Trans.,* Vol. 7, No. 1, 1968, pp. 6–10.

Donegan, A. J., "Some Aspects of Programming Real-Time Problems," *ISA Proc. Summer Inst.—Autom. Conf.,* May 1960, 8pp.

Donegan, J. J., "Programming Real-Time Process Control Computers," *ISA J.,* November 1961, pp. 46–49.

Donegan, J. J., C. Packard, and P. Pashby, "Experiences with Goddard Computing Systems During Manned Spaceflight Missions," *Proc. ACM 19th Natl. Conf.,* August 1964, pp. A2.1-1–A2.1-8.

Donegan, J. J., C. Packard, and P. Pashby, "A Real Time Computing System for Supporting Manned Space Flights," *Comput. and Autom.,* October 1964, pp. 22–25.

Douglas, R. M., "Digital Computer Achieves Real-Time Flight Simulation," *Data Process.,* April 1965.

Eckert, G. M., "Control and Programming Problems," *Proc. 1968 Int. Data Process. Conf. and Bus. Expos.,* Vol. XIII, *DPMA,* June 1968, pp. 309–325.

Elmore, W. B., and G. J. Evans, Jr., "Dynamic Control of Core Memory in a Real-Time System," *Proc. IFIP Cong.,* Vol. I, May 1965, pp. 261–266.

Everett, R. R., C. A. Zraket, and H. D. Benington, "SAGE—A Data-Processing System for Air Defense," *Proc. EJCC,* December 1957, pp. 148–163.

Feldman, J., and D. Gries, "Translator Writing Systems," *Comm. ACM,* February 1968, pp. 77–113.

Flores, I., *Computer Software: Programming Systems for Digital Computers,* Prentice-Hall, Englewood Cliffs, N.J., 1965.

Forbes, J. M., "An Introduction to Compiler Writing," *Comput. J.,* February 1965, pp. 98 ff.

Frank, W. L., W. H. Gardner, and G. I. Stock, "Programming On-Line Systems," (2 parts), *Datamat.,* May 1963, pp. 29–34; June 1963, pp. 28–32.

Frost, D. R., "FORTRAN in Process Control," *Pap. 68-708, Proc. 23rd ISA Conf.,* 1968.

Gass, S. I., "Project Mercury Real-Time Computational and Data-Flow System," *Proc. EJCC,* December 1961, pp. 33–46.

Gaylord, C. V., "Multiprogramming and the Design of On-Line Control Systems," *Data Process.,* May 1968, pp. 26 ff.

Gertler, J., "High-level Programming for Process Control," *Comput. J.,* February 1970, pp. 70–05.

Ginsberg, M. G., "Notes on Testing Real-Time System Programs," *IBM Syst. J.,* Vol. 4, No. 1, 1965, pp. 58–72.

Grebert, A., "A Working Real-Time System in the Textile Industry," *Comput. Bull.,* June 1966, pp. 32–33.

Greco, R. J., "Evaluation Techniques on Real-Time Computer Systems: A Survey of Practices," *Rep. GITIS-69-11,* School of Information Science, Georgia Institute of Technology, 1969, 28pp.

Green, W. K., and A. Peckar, "A Real-Time Simulation in Project Mercury," *Proc. EJCC,* December 1961, pp. 54–65.

Gross, J. L., "Real-Time Hardware-in-the-Loop Simulation Verification of the Gemini Guidance Computer and Its Operational Program," *Proc. Real-Time Syst. Semin.,* IBM Corp., Houston, Tex., November 1966, 28pp. (Also: *Simulation,* September 1967.)

Habermann, A. N., "Prevention of System Deadlocks," *Comm. ACM,* July 1969, pp. 373–377.

Hansen, M. H., B. Y. Manghan, and J. D. Cornell, "Scanning and Processing Plant Variables," *Chem. Eng. Progr.,* August 1969, pp. 36–41.

Head, R. V., "Testing Real-Time Systems—Part 1: Development and Management," *Datamat.,* July 1964, pp. 42–48; "—Part 2: Levels of Testing," August 1964, pp. 54–57.

Head, R. V., "Real-Time System Configurations," *IBM Syst. Res. Inst. Pap.,* April 1963.

Head, R. V., "Programming the Real-Time System," *J. Data Manage.,* February 1964, pp. 22–31.

Head, R. V., "Seven Configurations for Real-Time Computer Systems," *Control Eng.,* June 1964, pp. 104–108.

Head, R. V., "Real-Time Programming Specifications," *Comm. ACM,* July 1963, pp. 376–384.

Hess, H., and C. Martin, "TACPOL—A Tactical C&C Subset of PL/I," *Datamat.,* April 1970, pp. 151–157.

Hoover, E. S., and B. J. Eckhart, "Performance of a Monitor for a Real-Time Control System," *Proc. FJCC,* 1966, pp. 23–35.

Hoover, W. R., A. Arcand, and T. B. Miller, "A Real-Time Multicomputer System for Lunar and Planetary Space Flight Data Processing," *Proc. SJCC,* 1963, pp. 127–140.

Hosier, W. A., "Pitfalls and Safeguards in Real-Time Digital Systems with Emphasis on Programming" (2 parts), *Datamat.,* April 1962, pp. 43–47; May 1962, 68–72.

Hovorka, R. B., *et al.,* "The Control System in Action," *Chem. Eng. Progr.,* August 1969, pp. 46–52.

Jacobs, J. H., and T. J. Dillon, "Interactive Saturn Flight Program Simulator," *IBM Syst. J.,* Vol. 9, No. 2, 1970, pp. 145–158.

Jensen, H. F., and J. Muller, "Understanding the Problems in Real-Time Systems," *Proc. 1969 Int. Data Process. Conf.,* Vol. XIV, DPMA, 1969, pp. 129 ff.

Johnstone, J. L., "RTOS-extending OS/360 for Real-Time Spaceflight Control," *Proc. SJCC,* 1969, pp. 15–27.

Johnstone, J. L., "A Real Time Executive System for Manned Spaceflight," *Proc. FJCC,* 1967, pp. 215–230.

Joslin, P. H., "The AN/FPS-85 Real-Time Multiprogram Monitor for System 360," *Proc. Real-Time Syst. Semin.,* IBM Corp., Houston, Tex., November 1966, 30pp.

Kahn, P. G., and M. E. Fuller, "Program Conversion: A Discussion of Techniques," *Data Process.,* November 1969, pp. 28–31.

Kelly, G. N., "Project Gemini Mission Support Programming System," *Anthol. Tech. Pap.,* IBM Federal Systems Center, Gaithersburg, Md., 1965, pp. 25–31.

Kipiniak, W., and P. Quint, "Assembly vs. Compiler Languages," *Control Eng.*, February 1968, pp. 93–98.

Kornbluh, M., "Design Techniques—With Emphasis on Real-Time," *Data Process.*, Vol. XIII, DPMA, 1968, pp. 103–116.

Lafferty, E. L., "The Role of Simulation and Data Reduction Programs in the Development of Real-Time Systems," *U.S. Gov. Rep. AD609500*, December 1964.

Landeck, B. W., "MPX—An Operating System for Process Control," *Pap. 68-709, Proc. 23rd ISA Conf.*, 1968.

Lasser, D. J., "Productivity of Multiprogrammed Computers—Progress in Developing an Analytic Prediction Method," *Comm. ACM*, December 1969, pp. 678–684.

Lechner, R. J., "MILDATA—A Study of Post-1975 Command Control Information Systems," *IEEE Trans. Aerosp. and Electron. Syst.*, July 1966, pp. 434–445.

Lewis, R. J., "Real-Time Computerized Expressway Traffic Control in the Chicago Area," *Proc. ACM 20th Natl. Conf.*, August 1965, pp. 325–335.

Liang, L., "A Real-Time Programming Language and Its Processor for Digital Control of Industrial Processes," *Proc. SJCC*, 1969, pp. 843–848.

MacGowan, B., "1800 Data Acquisition System," *Comput. Bull.*, June 1966, pp. 87–89.

Mack, E. B., "Programming Real-Time Computer Systems," *Data Process.*, May–June 1965, pp. 143–145.

Martin, T. J., *Design of Real-Time Computer Systems*, Prentice-Hall, Englewood Cliffs, N.J., 1967.

Martin, T. J., "Program Testing in Real-Time Systems," *Comput. and Data Process. Manage.*, November 1964, pp. 28–31.

Maxwell, M. S., "An Automatic Digital Data Assembly System for Space Surveillance," *Proc. EJCC*, December 1961, pp. 257–263.

Matherne, R. J., "The Central Computer Approach," *Datamat.*, April 1970, pp. 140–142, 146.

McCreight, W. R., "The Real-Time Revolution in Medicine," *Comput. Bull.*, May 1968, pp. 8–10.

McIrvine, E. C., "Planning Software for a Manufacturing Line," *Control Eng.*, April 1968, pp. 100–103.

Mensh, M., "Organizing Process Computer Software," *Autom.*, February 1968, pp. 57–60.

Mueller, J. H., "Aspects of the Gemini Real-Time Operating System," *IBM Syst. J.*, Vol. 6, No. 3, 1967, pp. 150–162.

Mueller, J. H., "The Philosophy of the RTCC Control Programs," *Proc. Real-Time Syst. Semin.*, IBM Corp., Houston, Tex., November 1966, pp. 32–40.

Muntz, R. R., and E. G. Coffman, Jr., "Preemptive Scheduling of Real-Time Tasks on Multiprocessor Systems," *J. ACM*, April 1970, pp. 324–338.

O'Conner, J. T., "Eastern Air Lines' Real-Time Computer System," Vol. IX, *Proc. 1965 Fall Conf. and Bus. Exp.*, DPMA, 1965, pp. 296–309.

Palm, F. R., "A Real-Time Operating System for the Saturn V Launch Computer Complex," *Proc. Real-Time Syst. Semin.*, IBM Corp., Houston, Tex., 1966, 8pp.

Parker, R. W., "SABRE System," Vol. IX, *Proc. 1965 Fall Conf. and Bus. Exp.*, DPMA, 1965, pp. 310–317.

Parker, R. W., "The SABRE System," *Datamat.*, September 1965, pp. 49–52.

Peavey, R. D., and J. E. Hamlin, "Project Mercury Launch Monitor Subsystem (LMSS)," *Proc. EJCC*, December 1961, pp. 68–78.

Pfaff, A. M., and C. C. Rawlins, "NASA-MSC's Real-Time Computer Complex," *Proc. Real-Time Syst. Semin.*, IBM Corp., Houston, Tex., November 1966, 22pp.

Pike, H. E., Jr., "Process Control Software," *Proc. IEEE*, January 1970, pp. 87–97.

Pinder, H. G., "A Set of Programs for Control of Sequencing Systems," *Pap. 68-706, Proc. 23rd ISA Conf.*, 1968.

Pistole, D. G., and J. R. Erickson, "Acceptable Standards for Process Control Software," *Pap. 68-731, Proc. 23rd ISA Conf.*, 1968.

Puser, W. F. C., "System Timing for On-Line Computer Control" (2 parts), *Instrum. Tech.*, December 1968, pp. 41–46; January 1969, pp. 51–56.

Reardon, R. S., "IBM System/360 as a Real-Time System," *Comput. Bull.*, June 1966, pp. 90–92.

Reardon, R. S., "The Problem of the Peak in Real-Time Systems," *Proc. 1966 Br. Joint Comput. Conf.*, Eastbourne, England, 1966, pp. 188–195.

Rosin, R. F., "Supervisory and Monitor Systems," *Comput. Surv.*, March 1969, pp. 37–54.

Rothstein, M. F., *Design of Real-Time Systems*, Wiley, N.Y., 1970.

Sargent, W. H., "Real-Time Wagon Progress Control," *Comput. Bull.*, June 1966, pp. 27–31.

Schroeder, R., "Input Data Source Limitations for Real-Time Operation of Digital Computers," *J. ACM*, April 1964, pp. 152–158.

Schoeffler, J. D., "Process Control Software," *Datamat.*, February 1966, pp. 33, 39.

Schrage, L., "A Mixed-Priority Queue with Applications to the Analysis of Real-Time Systems," *Oper. Res.*, July–August 1963, pp. 728–742.

Schrage, L., "Analysis and Optimization of a Queueing Model of a Real-Time Computer Control System," *IEEE Comput. Group Conf. Dig.*, June 1969, pp. 34–36.

Scott, M. B., and R. L. Hoffman, "The Mercury Programming System," *Proc. EJCC*, December 1961, pp. 47–53.

Sellars, H. L., *et al.*, "Program Scheduling and Data Handling," *Chem. Eng. Progr.*, August 1969, pp. 42–45.

Shafritz, A. B., A. E. Miller, and K. Rose, "Multi-Level Programming for a Real-Time System," *Proc. EJCC*, December 1961, pp. 1–16.

Shaw, C. J., "JOVIAL—A Programming Language for Real-Time Command Systems," *Ann. Rev. Autom. Program.*, Vol. 3, Pergamon Press, N.Y., 1963, pp. 53–120.

Shertell, A. V., "On-Line Programming," *Datamat.*, January 1965, pp. 29–30.

Spitler, R. H., and B. K. Kersey, "A Research Laboratory for Processing and Displaying Satellite Data in Real-Time," *Data Process.*, January 1965.

Stanley, W. I., "Methods of Dynamic Analysis of System Performance Developed for the Gemini/Apollo RTCC Project on the IBM 7094," *Anthol. Tech. Pap.*, IBM Federal Systems Center, Gaithersburg, Md., 1965, pp. 33–40.

Stanley, W. I., and H. F. Hertel, "Statistics Gathering and Simulation for the Apollo Real-Time Operating System," *IBM Syst. J.*, Vol. 7, No. 2, 1968, pp. 85–102.

Stevens, R. T., "Testing the NORAD Command and Control System," *IEEE Trans. Syst. Sci. and Cybern.*, March 1968, pp. 47–51.

Stewart, D. H., D. H. Erbech, and H. Shubin, "A Computer System for Real-Time Monitoring and Management of the Critically Ill," *Proc. FJCC*, 1968, pp. 797–807.

Stimler, S., and K. A. Brons, "A Methodology for Calculating and Optimizing Real-Time System Performance," *Comm. ACM*, July 1968, pp. 509–516.

Teager, H. M., "System Considerations in Real-Time Computer Usage," in J. E. Coulson, (Ed.), *Programmed Learning and Computer-Based Instruction*, Wiley, N.Y., 1962, pp. 273–280.

Temple, R. H., and R. E. Daniel, "Status of Minicomputer Software," *Control Eng.*, July 1970, pp. 61–64.

Theis, D. J., and L. C. Hobbs, "Mini-Computers for Real-Time Applications," *Datamat.*, March 1969, pp. 39–61.

Tiffany, P. C., "Software Design for Real-Time Systems," *U.S. Gov. Rep. AD803389*, January 1970, 28pp.

Trapnell, F. M., "A Systematic Approach to the Development of System Programs," *Proc. SJCC*, 1969, pp. 411–418.

Turner, G., "On-Line Programs from Control Logic Diagrams," *Control Eng.*, September 1968, pp. 102–104.

Villiers, J., "Concept of a Real-Time System, Automating Air-Traffic Control," *Proc. IFIP*, May 1965, pp. 297–304.

Weaver, S., "Importance of Manufacturer Software," *Control Eng.*, January 1968, pp. 60–64.

Webb, K. W., "Real-Time Management of Range Operations," *Proc. Real-Time Syst. Semin.*, IBM Corp., Houston, Tex., November 1966, 21pp.

Weiler, P. W., R. S. Kopp, and R. G. Dorman, "A Real-Time Operating System for Manned Spaceflight," *IEEE Comput. Group Conf. Dig.*, June 1969, pp. 137–142.

Weingarten, A., "Analyzing a Real-Time System," *Proc. ACM 21st Natl. Conf.*, 1966, pp. 179–182.

Weingarten, A., "The Analytical Design of Real-Time Disk Systems," *Proc. IFIP Cong. 1968, Booklet D*, pp. 860–866.

Wilkes, M. V., *et al.*, "Panel on Priority Problems in Computer Systems," *Proc. IFIP Cong.*, 1962, pp. 711–715.

Williams, J. M., "Design of the Real-Time Executive for the UNIVAC 418 System," *Proc. ACM 21st Natl. Conf.*, 1966, pp. 401–413.

Willmot, T. L., "A Survey of Software for Direct Digital Control," *Pap. 68-834, Proc. 23rd ISA Conf.*, 1968.

Witt, B. I., "Dynamic Storage Allocation for a Real-Time System," *IBM Syst. J.*, Vol. 2, No. 4, pp. 230–239.

Yourdon, E., *Real-Time Systems Design,* Information and Systems Institute, Inc., 1967, 205pp.

"Advances in Fast Response Systems," *EDP Anal.,* February 1967.

"A Language for Real-Time Systems," *Comput. Bull.,* December 1967, pp. 206–212.

"Analysis of Some Queuing Models in Real-Time Systems," Form F20-0007, IBM Corp.

"A Survey of Airline Reservation Systems," *Datamat.,* June 1962, pp. 53–55.

Computer Software for Process Control, BISRA, 24 Buckingham Gate, London, 1969.

"Real-Time Control at BEA," *Data Process.,* January–February 1968, pp. 14–21.

"The IBM Approach to Real-Time Systems," *Comput. Bull.,* June 1966, pp. 86–87.

Chapter 10

Project Organization and Management

H. WAYNE VAN NESS

As new technologies are introduced into the business world, management personnel are faced with the problem of making decisions on projects in areas that have developed since their formal training was completed. They must resist the inertia that would allow the status quo to prevail and cost their company a chance to stay abreast competitively. Successful management can lead to significant strides forward in society's efforts to be more productive, allowing more time and effort to be devoted to creativity.

This chapter addresses the problems that confront the management of a plant when it is proposed that a computer control system be installed on one of the plant units. Two major areas are considered: the control problems of management and the personnel problems. Basic techniques are discussed for project control, and the types of disciplines that are required for a successful installation are defined.

10.1   PROJECT PHASES

The process control computer project is not appreciably different from any other project that involves the design, installation, and operation of process equipment. The spectrum of talents required is broader because the successful implementation necessitates knowledge of computer technology and, in most cases, advanced control concepts, in addition to the usual process design knowledge. The project may be more complex simply because the control scheme itself is more detailed and complex. Nevertheless, the usual process control computer project consists of the same basic phases involved in conventional control projects.

Figure 10-1 illustrates these program phases and indicates the relationship between the hardware and software considerations. The hardware consists of the physical process plant to be controlled, the computer, and the control equipment. The software consists of the computer programs and, implicitly, the control strategies used in the control of the hardware. The diagram is, of course, a greatly simplified representation of the control computer project. In an actual project the phases overlap and the differentiation between the hardware and the software is often difficult to define. Although these complexities are discussed in more detail later in this and subsequent chapters, the simplified diagram of

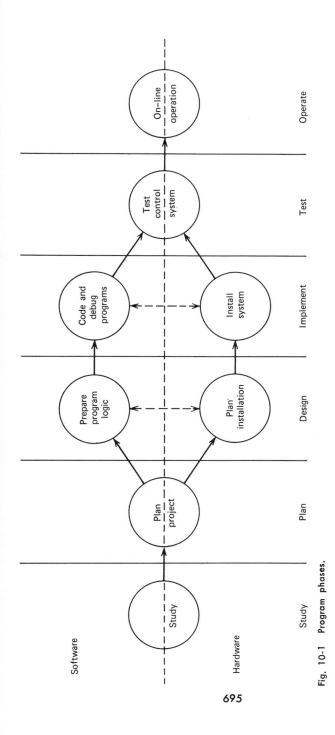

Software

Hardware

Study   Plan   Design   Implement   Test   Operate

Study

Plan project

Prepare program logic

Code and debug programs

Test control system

On-line operation

Plan installation

Install system

Fig. 10-1   Program phases.

Fig. 10-1 provides the basis for an overall examination of the process control computer project and its implications to management.

10.1.1 Study Phase

Most process control computer projects begin with a study to determine which of several processes should be considered for computer control, the feasibility of controlling a given process with a computer, the economic return on the investment, or some combination of these objectives. Merely proving feasibility of control is a less common study objective than it was a decade ago; control computers have been applied to such a wide variety of applications that their capability for controlling many industrial processes is well documented in the literature. This does not mean, however, that the feasibility of control should not be examined in the study. Clearly, this fact must be established or the other objectives are academic. Feasibility is often easily proved on the basis of prior experience, however, and the emphasis of the study is usually directed at one of the other objectives.

The study team which performs the analysis of the plant and its control takes different forms depending on the magnitude of the proposed project, the experience of the company in computer process control, the resources available, and many other factors. In a large experienced company, the nucleus of the team may be a permanent group assigned to the corporation engineering department. As the result of experience and study, this team probably has most of the knowledge required for the analysis. They know what questions to ask, who to ask, and how to evaluate the significance of the answers. For the smaller, inexperienced company, the team is often formed as a task force drawn from affected departments in the company. There may be only two or three permanent members of the team, with part-time assistance obtained from the other departments only when necessary. If at all possible, the permanent members of the team should have some experience with computer control, and they should have a thorough knowledge of the overall plant operation and the particular process being considered. These are necessary elements of their knowledge since a valid analysis of a computer control project involves factors that affect the entire plant, as well as the specific process. A detailed consideration of the disciplines required in the total project is found in Sec. 10.3. Most of these same disciplines are required to some extent during the study phase as well.

The most common objective of the study phase is an estimate of the economic return on the computer investment, since increased profit is almost always the ultimate reason for the installation of a control computer. This increased return may be the result of greater throughput, less waste, the use of new process techniques, or any one of many other factors. Using their knowledge of the process characteristics and the control concept assumptions as a basis, the members of the study team must examine each major source of increased profit and estimate its magnitude. In addition, they must consider the initial cost of the installation, its cost of upkeep, and the relationship between the improved process performance and overall business factors. Through a comparison of the costs and the returns, the wisdom of the control computer investment can be evaluated. The many facets of economic justification and some of the applicable study techniques are considered in detail in Chap. 11. If the objective of the study is to select which of several processes should be considered for computer control, the decision is usually still economic in nature, and most of the same facets must be examined in the study. The several candidate processes must each be studied and the estimated return on investment compared to determine the best investment.

Management's role in the feasibility study is to guide the effort toward selection of a process with obtainable goals that have a realistic economic yardstick applied to them to measure the anticipated payout. All too frequently, these studies are brief, with the process selected on something other than an engineering basis and with goals poorly defined. In such instances, there is no means to evaluate the project at its conclusion to determine whether the feasibility assumptions were correct. As in process control, this feedback is necessary; without it, one cannot reasonably expect improvement in future feasibility studies.

The first major management decisions concerning the study are those involving the selection of the study team itself. As noted above, the nucleus of the study team should consist of individuals with a broad knowledge of the plant and its operation. If part-time assistance is required from other departments in the plant, management should ensure that these people are available at the right time and that they are free to devote the required time to the study project.

Whatever the objective, the study is equally concerned with software and hardware since both contribute to the cost of the installation and to its success. A study biased toward either consideration may overlook an important factor and negate the results of the study. For this reason, management must insure that the proper disciplines are represented on the study team and that the investigation is thorough and unbiased.

A factor related to the selection of the team is how the results of the study are to be passed on to the installing team so that the installation objectives are consistent with the selection criteria. One solution is to have the same people responsible for both the installation and the study. This is sometimes impractical, for the installing team is often drawn from those with knowledge of a specific process, and involving them in the selection could bias the objectivity of the study. If the study is a justification, rather than the selection of a specific process, it is very practical to have the same people participating in the installation and justification. This not only gives the desired continuity but also assists in ensuring that realism is applied to the justification.

The first action of the manager and his study team should be to define the objectives of the study and to establish the ground rules under which the study is to be performed. The objectives must be realistic and attainable in the available period of time. They should call for conclusions that are consistent with the accuracy and precision of the information available. They should also include an assessment of the importance of the assumptions that are required to reach the conclusion. Another factor, often forgotten, is who is to make the final decision, since this may affect the emphasis of the study.

Once the objectives have been defined and a course of action is implemented, it is the study team manager's responsibility to guide the efforts of the group toward the objectives in a timely manner. The lack of firm objectives or management control can result in a long, drawn-out study which either fails to obtain conclusive results or produces answers more precise than required. The manager must continuously monitor the progress of his team and carefully justify extensions to the schedule or content of the study. Every additional day of study means a delay of the installation, and this is translated directly into dollars and cents through the loss of the increased profits attributable to the computer.

On the other hand, a hasty study can result in an incorrect decision and an unnecessary capital investment. The manager must analyze each conclusion to ensure that all important factors have been considered and that the data are accurate. Often, periodic review sessions with other departments in the plant are a valuable means of critique and guidance.

Once the study is complete, its results must be presented to the management groups or individuals responsible for the final decision. This is a very important step since even an excellent study can be negated by a poor presentation. The assumptions, objectives, conclusions, and estimates of risk should be clearly and succinctly stated. Unnecessary detail should be avoided. For example, technical details should not be belabored if the management is business, rather than technically, oriented.

10.1.2 Planning Phase

If the decision is made to install a computer control system, the next major phase of the program is planning the installation. Planning is the most important contribution made to a project by those responsible for its management. It must be recognized that planning is not a one-time task but continues throughout the life of the project. A plan, at any given time, represents the project management's estimate of the job ahead—how and when the objectives will be attained. As the time to perform a particular task approaches, a planner sees this task more clearly than he sees a task that is later in the project. Consequently, close-in tasks are the most accurately portrayed. As the project progresses, understanding of future tasks improves and the plans should be modified to reflect this additional knowledge. These plans can serve as a means of measuring progress on the project and a vehicle for reporting progress to others involved with the project.

Three fundamental areas must be addressed during the planning phase of a project. These are personnel, fundamental systems design, and physical hardware requirements. Many of the decisions made at this initial stage of planning have dramatic effects on the future progress of the project. Details relating to these areas are discussed in later sections of this chapter and in subsequent chapters.

With proper planning, a project manager can avoid a situation where he is responding only to crises rather than preparing operations to avoid problems. The amount of detail that is incorporated in a plan must be sufficient to track the progress so as to allow detection of trouble soon enough for correction. Excessive detail, however, is expensive and can cloud events to the point where significance is lost.

An important management function is pulling together those people required to design and implement the objectives established by the study group or project team. The timing for these people to become involved with the project must be integrated with the overall plan for the project and their other assignments. One of the largest expenses incurred on a project is labor, so this must be controlled from a cost, as well as from a performance, standpoint. Also, the types of talent required are in high demand and, consequently, should not be improperly utilized because of poor planning.

Another topic to be settled early in the planning phase is the question of which philosophy of control is to be employed for the installation. By "philosophy of control" is meant the conceptual way in which the theories of control are implemented for a particular process. The concepts range from merely gathering data for off-line computer analysis to total control of the process as an integral part of a computerized company information system. Between these limits, there are a variety of control techniques that can be implemented. Each technique has its advantages and disadvantages which must be matched to the characteristics of the process.

The concepts and implementation of these control philosophies are discussed in other chapters, but their importance in planning must be emphasized here. Different approaches to the control problem can yield significantly different economic returns for a given process while consuming varying amounts of resources such as time and equipment. Thus an early decision must be the selection of the control concepts to be employed. This does not require that all the details of the control scheme be determined at this early time, but the broad outline of the approach must be established. Sometimes the philosophies are not mutually exclusive but may be implemented incrementally during the life of the project. For example, data logging and monitoring are natural predecessors of closed-loop control.

Concurrent with the development of the philosophy of control, there must be preliminary planning on the equipment to be employed. A given control philosophy may be dependent on the availability of instruments that can provide a necessary signal to the computer. If a particular instrument does not exist, then some other control philosophy or measurement technique must be sought that is not dependent on that device. Technology in instrumentation is advancing very rapidly, so the status of new developments should be checked to determine what has been tested and proved operational in an industrial environment. A project should not be placed in a position where it is dependent on an instrument that is in the prototype stage, however; if it does not prove out, the entire project may become unfeasible. Also, all too often projects have been delayed while an instrument is corrected or redesigned to meet its performance requirements.

As illustrated in Fig. 10-1, the hardware and software aspects of planning are virtually inseparable. The programmers require the hardware to develop and debug the control programs. But hardware without programs cannot control the process. Close coordination of programming and engineering is essential if delays in the project are to be avoided. Although not shown explicitly in the figure, similar coordination with the computer vendor and contractors responsible for the construction of physical facilities is also important.

10.1.3 Design Phase

Following the development of the basic plans and schedules for the installation comes the actual work of designing the overall control system. One of the first activities during this phase is the selection of the particular computer to be used in the installation. With this question answered, the hardware and software designers can proceed with the specific hardware and software design activity necessary to prepare for the installation and use of the system.

As indicated in Fig. 10-1, the hardware and software tasks can be defined as relatively separate activities during this phase. Although this has the advantage that the activities are easier to control, there are some inherent dangers. The primary one is that the hardware and software groups might proceed independently of each other and not be aware of decisions that affect the other group. During this phase, it is very important for management to ensure that communications between the groups are maintained at adequate levels. Failure to do so can result in excessive delays during the implementation and test phases while inconsistencies between the hardware and software systems are resolved.

The consideration that must be given to the selection of the computer is discussed in Chap. 12. In this section, some of the management questions that must overlay this decision are discussed. Aside from the hardware, software, equipment maintenance, and special equipment that are supplied, there are the support services described later in this chapter to consider.

It is difficult to come up with an accurate, impartial scale that adequately provides a means of comparing competitive offerings. Unfortunately, the system selection sometimes degenerates to allowing emotions to override sound engineering judgment. Statistics representing the basic characteristics of various computers are readily available, but they do not necessarily tell the user what the relative performance of a system will be on his particular job. For example, one of the first specifications analyzed for a system is the storage speed, the time required to execute 1 cycle of access to main storage. This does give a basic indication of the speed of the system to perform computations, but it must be evaluated in terms of how many cycles are necessary to fetch the necessary data and instructions and to perform the desired computation. In addition, it does little good to have a nanosecond memory on a system that spends a disproportionately large amount of time waiting on an input/output device such as analog input. With considerations such as this, the analysis becomes more involved, for one must look at the performance of the entire system and the interactions between the various components.

One solution to this comparison problem that is evolving is the use of simulators on existing computers that give an estimate of system throughput for a given job mix. As with any simulation, this can be only as valid as the accuracy of the job stream and system specifications submitted to it. If properly implemented, however, simulation yields a quantified relative performance index that is useful for comparisons.

Another advantage of simulators for analyzing job streams is that computer systems are very modular in construction and, consequently, with a given system there are alternatives to perform a given function. Very common alternatives, for example, are the ways in which floating point computations can be performed. They can be done by subroutines that take many instruction cycles to execute, or hardware can be added that performs the same computation in one instruction cycle. The latter alternative gives faster results but costs more. The question is: Is the added speed necessary for this application, considering the additional costs? There are many such alternatives that must be evaluated between specific hardware additions or programming. A properly constructed simulator can evaluate various configurations and show which one yields the best throughput for the cost.

To expand the above example one more step, consider one dimension of the subroutine versus the hardware approach. The demand for space in main storage for programs tends to grow beyond original estimates as programs reach the completion stage. Although a subroutine may perform a computation at less cost, it takes more space in main storage, and thus may be less economical in the long run. Allocation of storage requires close attention throughout the life of the project for it dramatically affects the performance of the system.

The selection of the computer system may include submission of specifications to several computer vendors. Two fundamental approaches are available in preparing the bid request. One is to prepare an explicit set of hardware specifications that precisely define what equipment is to be bid. This has inherent disadvantages since, in the early stages of a project, the system design is in an embryonic stage and the requirements are not fully defined. In addition, there are many good systems on the market today, each with its own characteristic capabilities. How a desired end-performance is achieved may vary widely from system to system, yet each does the job. To define equipment precisely in a bid request may rule out vendors equally qualified to be responsive, only because they have a different approach. On the other hand, with a fixed specification, the selection judgment can be based on dollars alone, with the assurance that comparable systems are being evaluated.

The alternate approach is to prepare specifications that state performance requirements and allow the vendors to define a system configuration that meets those requirements in the most competitive manner. This approach requires that the engineers making the choice must be prepared to apply technical judgments as to what is being offered and its adequacy in meeting the requirements. Unfortunately, performance specifications are often ambiguous in their statement of requirements, and this places a large burden of judgmental evaluation on the individual or group making the selection. Nevertheless, this approach can lead to a more economical system.

One point that must not be forgotten when specifications are being developed is that, as the system design progresses, change is inevitable. Allowances for change must be made in preparing the specifications, contract, and budget. For example, this allowance may take the form of uncommitted process I/O for the computer or other system features. Similarly, if the system is to be leased, the cancellation clause should be examined to see whether what may prove to be superfluous features (such as extra analog-input points, digital inputs, and interrupts) can be returned to the vendor for credit.

Once the vendor has been selected and the working relationship defined, it is helpful if both parties, supplier and purchaser, view the installation as a team effort in which both have a vested interest in the system's success. The vendor is anxious for the project to gain success with the accompanying brighter outlook for future sales; the purchaser, of course, is interested in the return on his investment.

Once the computer system is selected and the purchase order submitted, the hardware and software designers can begin the detailed design and implementation of the hardware and software systems. Although they are familiar with the general capabilities of the system as a result of the study and specification activity, they must now learn the system in detail if they are to produce an efficient and profitable installation.

On the software side of the operation, the programmers must completely understand the programming system. This includes learning which features and subprograms are a part of the programming system, what additional application programs are available from the vendor, and how new programs can be incorporated into the programming system. Often the computer vendor can provide formal instruction on the programming system and its use which accelerates this learning process.

The next step is the design of the application programs required for the control of the process. In cooperation with the system engineer, the overall control strategy must be decided upon and the general program logic must be designed. During this phase it is important to recognize unique operating situations that must be identified and handled by the program. Interrelated alarm conditions and secondary dependencies in the process are examples of such conditions. Often seemingly unimportant aspects of the system operation can prove to be extremely important in the overall acceptance of the system. For example, the operating log has little to contribute to the actual gathering of data and control of the process, but it must be given a great deal of thought at design time. To many on the operating floor, the typewriter on which the log is published *is* the computer. Operators form their opinions on the basis of how the logging device functions and the usefulness of what it produces. If printing is assigned a low priority on the interrupt servicing scheme, it may be slow in completing a message. The operator who asked for the message can only assume that the system is correspondingly slow. On the other hand, the quick response of a system in which keyboard functions are assigned a high priority often provides the operator with the feeling that the computer is a valuable control tool under his command.

Following the design of the general program logic, most programmers utilize flow charts to refine further the structure of the required programs. These flow charts detail the program logic for each of the major subdivisions of the overall logic. Although some programmers prepare flow charts only after the program is developed, use of such charts is to be encouraged in the early program design phase since often they can help identify logical problems or program redundancy. In addition, they form the basis for the final documentation of the program and can significantly reduce the documentation effort. The resulting program logic flow charts should be reviewed with the hardware designers and other personnel affected before any significant coding is done. These reviews act as an effective coordination technique to ensure that both the hardware and software designers are operating under the same assumptions. For example, the software designer might assume the existence of a particular signal, whereas the hardware engineer had not planned the installation of an instrument to measure the parameter. Similarly, the instrumentation engineer might find that the conversion routine being used by the programmer for the linearization of a variable is incorrect or is not accurate enough for the particular calculation being attempted.

Paralleling the development of the programs, the design of the instrument system, computer installation, computer room, communications, and other physical facilities is taking place. This involves a complete understanding of the computer-process interface. It must be verified that the instrumentation signals are compatible with the characteristics of the computer system. Accuracy and precision requirements must be defined. The detailed design of the physical installation also must be done. This involves the design of the computer room (if necessary), the design of the cable routing for signals lines, the assignment of terminal addresses, and many other details. Here again, the vendor often can provide assistance in the form of assignment forms, check lists, and installation advice.

For management, the design phase is one of review and coordination. The decisions being made by the hardware and software designers must be reviewed by other groups involved in the project. The manager must continually review his plans to make sure that no aspect of the project has been neglected or is falling behind schedule. In this and other matters, he must coordinate the activities of his group with the vendor to ensure that the physical installation is complete and satisfactory when the equipment is delivered.

10.1.4   Implementation Phase

After the design has been prepared for a project, it is time to take on the task of translating the design into being. The programmers must develop the actual control programs that ultimately analyze data from the process and provide control signals to the process. The program logic has been defined by laying out flowcharts that are a graphic representation of the decisions which must be made by the computer in controlling the process. After these flowcharts have been reviewed and approved, coding, the generation of the actual instructions that are executed by the computer, must be done. Several programming languages are available that can be used by the programmers to translate their logic into the language that is understood by the computer. One such language that is receiving wide usage today and is understood by many is FORTRAN. FORTRAN, however, is a relatively old language that has some disadvantages in real-time applications. Some of the newer languages, including those specifically designed for industrial control, should be considered as possible alternates. After the initial coding is completed, the programs must be tested or "debugged" to make sure that the program performs the desired logic and to solve problems that develop as various sections of code are brought together. This testing can take place on computers that are provided in test centers to simulate the process, if these are available. Although they may not represent exactly the situation encountered when the final system is installed, they are a good aid that allows the program preparation to progress efficiently.

While the coding and debugging is taking place, work is proceeding to prepare the hardware side of the installation for the computer. This involves the building of the computer room and air treatment equipment (if this is necessary), installation of instrumentation and signal wires between the process and the computer room, and installation of control devices. In an existing plant, the devices that directly interface to the process may have to be installed without disturbing production, and this requires close scheduling with purchasing and those who operate the unit. If at all possible, instruments and their signal leads should be installed and monitored while the process is in operation, well in advance of the arrival of the computer. Although not a guarantee, this procedure usually expedites the computer system installation because attention can be focused on the computer with minimum concern for instrumentation problems.

10.1.5   Test Phase

At some time, the day of truth must come when the control system takes on the job of running the process. There are many techniques for testing and simulating programs prior to the installation of the equipment, but no simulation can truly represent the actual process. The principal shortcoming of simulation is that most methods of simulation are designed by the programmer to have the simulated process respond to the control system as he expects it to. These are the same criteria on which he developed his programs. Unfortunately, it is rare for these assumptions to be absolutely accurate. Thus thought must be given to the errors that could be latent in the programs and how they might be countered if they arise.

Prior to going on stream, a check should be made of the input and output signal connections to ensure that all wires are connected as planned. The best way to perform this check is by program, with someone observing the process-end to ensure that the complete data path is correct through the entire system. This path involves instrument terminals, leads, transducers or transductors, computer terminals, programming addresses, and a similar route to the controller or control element. The programming could be perfect but lead to chaos if a pair of leads is crossed.

When first starting to try the system as a whole on live data from the process, it is best to do it gradually and systematically. This can be done by adding one control loop or function at a time and not advancing to the next until assured that the previous one is operating as desired. Sometimes this cannot be done because of the interaction of the various loops, with the requirement that they be handled together. Another approach is to plan to be on-line for short periods of time and then to back off to review the results. This latter approach may run into difficulty if the control scheme is based on an inferential algorithm which requires a significant amount of history on which to operate. In this case, it is sometimes possible to go on stream with the data collection but have the control functions go to a typewriter, or some other means of indication, so that they can be monitored before they are enacted.

Again, it should be emphasized that the cutting-in of the control system must be closely coordinated with the operating people. An ideal situation is to have operating people on the process, feeding information back to the control people at the computer. Since the computer and process usually are geographically remote from each other, communication aids such as walkie-talkies or sound-powered phones are often helpful.

It is very normal, in the preliminary test stages, for people to ask for every variable to be logged every time it is observed or computed. This is a predictable overestimate of the needs. Many of these variables actually have little or no meaning to the operating people. However, by logging them, two things happen: (1) Operations gets a feel as to what is taken into consideration when making control computations, and (2) they select those they want maintained in the log as the ones that have meaning to them.

If the log is printed too frequently, the reader cannot readily extract the information that is needed to run the process. One acceptable way is to log on an exception basis. When an upset (i.e., a disturbance beyond stated bounds) occurs, the offending variable is logged. By the same token, the shift foreman can "demand" a log when he wants it so that only those variables that are out of a normal range log out.

There is legitimate reason for programmers to ask for data to be logged while the system is still being checked out since they need to know what conditions were when the control theory went awry. One characteristic of these data is that if all goes well, one doesn't need them. They are necessary only if something goes wrong. A technique that has been successful in some instances is to use auxiliary mass storage on the system, such as disks, to record records containing the pertinent variables. These can be recorded every time the cycle time calls for a control computation. When the end of the available space has been reached, the new record can then be written over the oldest record in the file. In this way, the file cycles and always contains the latest data available. The data, however, have been retained long enough to be recalled if an upset occurs in the control. These data are also very useful when reviewing performance and when trying to determine how to improve the system.

Following a session on stream, it is well to solicit comments from all involved as to what they observed, the reactions to the approach, and suggested improvements. There are many changes at this stage of the project, and some of them cause what can appear to be a complete relapse. In some cases, the programmers must go back, make changes to the code, recompile, and then go through the debugging procedure again, or the hardware designers may have to install a new instrument. On some changes, it may take weeks to get back to the position held prior to the initiation of the change. Sometimes several modifications are pending to be implemented at one time. If it is desired to evaluate the effect of a change, only one change should be made at a time and then it should be allowed to run long enough to evaluate its merits. Otherwise, whether the net result is good or bad, it is difficult to attribute the result to the correct change.

10.1.6   Operation Phase

After the system has been tested to the extent that it is considered ready to be used as a regular operational tool, it should be brought on stream and turned over to the operating personnel. They should have been involved in the design and testing so that, by this time, they are familiar with the system and are comfortable when using it. When this transfer takes place, it would be incorrect to assume that there are no more errors in the programs. At the same time, there will be errors reported that are not errors but are merely misunderstandings of how the system operates. One good technique at this stage is for the operation foreman to keep a diary of what he observes on the use of the system. This can then be reviewed regularly by the project team who can correct the error or explain how the system works in that given situation.

Once the system is operating, one of the remaining duties of the project manager is the assessment of the success of the project. This should be done at a time when the system is ready to be judged. There will be pressure for the system to be operational from the day that it is installed, but it is a disservice to the project if evaluation is allowed to take place prematurely.

The process of evaluation is an obligation of the project team and falls under the jurisdiction of the project manager. Evaluation involves the personnel from operations that have been associated with the project as well as someone from accounting who has the cost information, not only of the project, but also of the changes that have occurred in production expenses or product qualities since installation. The results of the evaluation should be submitted to the group that conducted the feasibility/selection study for comparison with their anticipated performance. Differences should be explained whenever possible. An objective viewpoint is most important; the evaluation is intended as a learning tool, not an indictment for errors.

Generally, the improvement or incremental change is the figure that management is interested in seeing and, consequently, it is necessary to have established a well-defined base from which the improvement is calculated. The base should be a statistically sound representation of what is typical for the process unit before the control system is added. The improvement can take the form of higher quality product, less quality give-away as the result of production that is closer to the limit on bounded specifications, more productivity per unit of resource or raw material, added safety, or many other areas.

The installation of the computer is frequently accompanied by the installation of additional instrumentation and control equipment. The point is sometimes made that some of the improvement could have been achieved with the new instrumentation without the computer. When the equipment is installed as a system under one project, the results should be evaluated as a function of the total system. To try to allocate part of the benefits of the installation to some fraction of the total installation is beyond the accuracy of the evaluation since it usually requires estimates that are highly subject to bias.

The test period over which the evaluation is performed should be comparable to the base period so that it provides a fair comparison. The duration of the evaluation period should also be consistent with the properties of the process. For example, if production is to be compared before and after the installation, the test period must be long enough to reflect and correct any changes in quantity produced by the amount of in-process inventory fluctuations, since these inventories can greatly offset production quantities in some processes.

It is highly unlikely that some parameters have not changed since the feasibility study was made, and account must be taken of these changes. These may be independent of the new control system. In a paper mill, for example, dramatic changes in production quantity and quality can take place when the pulp comes from a different source. If the base for the evaluation was established while the mill was running on a given pulp and, at the time for evaluation, a new pulp or bleaching unit has been installed, the project team must make judgmental allowances for such changes.

Rarely does a project come to a definite conclusion where a well-defined end is reached; there is usually more that can be done. The scope of future applications for expansion of the system will have been learned throughout the development of the project. In preparing the evaluation, future potential should be pointed out so that judgments can be made as to whether this project should be enhanced or the resources diverted to new projects.

10.2 PROJECT ORGANIZATION

Process plants generally are organized to operate the process on a continuing basis. When a project is started that has a definite duration with the accomplishment of explicit objectives concluding the project, it sometimes calls for a different type of management structure. The installation of a process control computer is often such a case. As can be seen by the shading in Fig. 10-2, practically every department in a plant is involved in the project before it is completed. The project team is often composed of members of these various departments, yet each remains responsible to his own department. A project of this type is usually on a tight time schedule and cannot afford the luxury of waiting for all managers to concur on all points to gain the necessary project direction. One person must have the responsibility for the installation and be the focal point for the activities of the total project. As a project manager, he may not have direct supervision responsibility over the installation

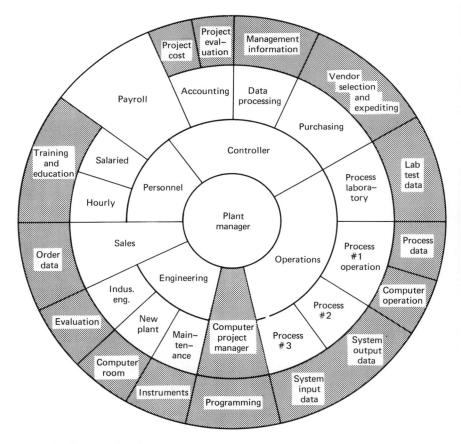

Fig. 10-2 Organization chart.

team, but incentives should be placed at his disposal to ensure that people respect his authority, indirect as it may be.

A problem that arises is how a staff of this type should be worked into the plant hierarchy. Consideration should be given to what the planned life is for the group. If they are to defy Parkinson's Law and disband at the conclusion of the installation, then it is unnecessary for them to be integrated into an existing operating department. It has been found successful in some installations to have the project manager responsible directly to the plant manager, eliminating jealousies between departments that are affected by the installation. The hypothetical organizational chart shown in Fig. 10-2 illustrates this type of structure.

Another approach that has been used successfully in this type of project is shown in Fig. 10-3 (Lecture, 1968). The project team shown here is the study team that studied the feasibility of the project, selected

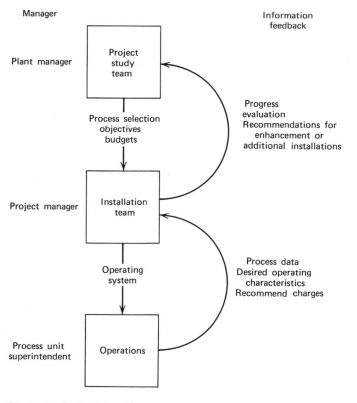

Fig. 10-3   Project interaction.

the process, and established the objectives for the project, including the anticipated returns on the estimated cost of the project. This group is probably made up largely of those department managers directly under the plant manager. The involvement of top level management has been found to be a necessary ingredient for successful computer installations. It has been stated that the chief executive plays the key role in systems management (McKinsey, n.d.). There are five things he should do to ensure the success of the project:

- Approve corporate systems objectives and project priorities.
- Decide on the most effective organizational arrangements of systems and user personnel.
- Assign responsibilities to the using organizations.
- Insist that the detailed systems plans be included in the operating plans and budgets.
- Follow up to see that the achievements match the plans.

The project study team is chaired by the plant manager and produces the project objectives and budgets for the project. A member of this group should be the project manager who has the responsibility of the installation. Once the project is under way, this group should periodically review the progress of the project through the reports they receive from the installation team.

The installation team is composed of those people who perform the various tasks necessary to bring the system to an operating state. As discussed in later sections of this chapter, each brings to the project his individual talents as well as the viewpoint of his particular department. Their objective is to bring the system to an operating condition within the time and cost budgets allotted to them. After the system has completed whatever shake-down testing is deemed necessary, it is turned over to the department responsible for the operation of the particular process unit. Throughout the testing phase, those who will ultimately be operating this system should be involved, giving their recommendations to the installing team as to what changes should be made to improve the installation. Some member of the operating group should be on the installing team with the understanding that he will have the responsibility for maintaining the programs and making whatever additions or corrections are desired after the system goes on stream.

10.3   PROJECT PERSONNEL

Today, the performance of hardware systems is dramatically improving while the price for a given function is going down. The investment made in manpower for an installation of a control computer has become one of the more significant items of expense. With the improved performance comes the capability to do more sophisticated control functions with the system. It logically follows that to take advantage of this capability requires the efficient use of technical know-how that may or may not be present at a given site. The project must draw on all the disciplines that exist in the plant design, operation, maintenance, and accounting departments. The size of the group that is required to perform the installation task is a function of the established objectives and the talents possessed by the various members of the team. From a control and efficiency standpoint, the team should have no more members than are necessary to accomplish the objectives. Ideally, the team size would be only one person with all the necessary know-how and capabilities. Practically, such an individual does not exist, and the magnitude of the job calls for more people. Every additional member requires that much more management and, even more critically, that much more coordination, however.

The pressures on the project team build to a crescendo that far surpasses the run-of-the-mill type of work that is experienced from day to day. When a person is approached for a job on the installation team, career opportunities beyond the successful installation of this particular system should be explained to him. These opportunities must be adequate to see him through the long hours that are required to get the system on stream. In other words, do not allow an individual to feel "locked-in" on a project.

On the other side of the coin, however, there is a danger in building up an individual's expectations to the point where he insists on moving on to a new area without taking advantage of the knowledge gained from this project. The success of this installation means that top management is going to be interested in making more installations on other processes at other plants, and they should be able to expect to benefit from the investment that has been made in training, both formal and in the school of hard knocks, on the first installation. In short, beware of guaranteeing that the individual will not be involved in future installations. This warning may prove unnecessary. The thrill of being involved in development of a system that is on the threshold of a new industrial era (i.e., computer control) gets in the blood of many, and they would not want to be in any other section of the business.

For many on the installation team, the assignment to the project need not be a full-time task since their contributions are needed only at various stages during the life of the project. It is very probable, for example, that the instrumentation selection and procurement activity is very similar to the work normally done by the plant instrument engineer. The instruments should be installed and operating before the computer arrives so the demands on his time diminish in the latter part of the project. Conversely, the programmers are involved from the start of the project until the system is deemed operative. Consequently, planning and scheduling is critical if the right talents are to be available at the proper time to complete the project on schedule without wasting manpower.

10.3.1 Project Manager

The most important selection is that of the project manager. The project manager should realize that in putting together such an installation, he is given an opportunity to view intimately the workings of the company while producing significant benefits to his company through the proper use of new technology. The price for this opportunity is the assumption of a formidable task that requires creativity and unique management methods. It is necessary for him to be successful in soliciting cooperation from practically all branches of the plant, for each has its contribution to make to the project. Personnel relations consume a significant percentage of his time, and it is inevitable that the project manager must recognize plant politics and play them to his advantage. As each department makes its contribution, whether large or small, it has the potential of making or breaking the project. The technical aspects are equally challenging, for in a control system some of the latest theory is employed or derived and then given the final test in the cold, realistic world of actually running the process against the economic stop-watch of dollars returned on the investment.

The project manager obviously must possess all the characteristics of an accomplished manager. He must be forceful, decisive, and have the respect of all with whom he is working. Throughout the life of the project there are repeated occasions where alternative ways of accomplishing a task are presented, each with its corresponding attributes. The manager must be prepared to evaluate these alternatives and render a decision that is explicit and consistent with the objectives. A computer does not tolerate "maybe" for an answer, and the conclusion must be implemented with all ambiguity removed.

In addition to universal management characteristics, the project manager also must be technically conversant with a good percentage of the project and, on the balance, reinforce his shortages with a staff that can assist in the decision-making process.

The responsibility for the success or failure of the project is directly assigned to the project manager. Accompanying this responsibility must be the necessary authority to accomplish the goals of the project. When this authority cannot be given, for any of several reasons, he must be able to enlist top management support in enacting some of the activities that are required. For example, when the system is ready to interface first to the process, some planned process excursions (experiments) may be the most expeditious means to test out the model and control theory that have been implemented. Operations may be reticent to perform these since they do not want existing operations disturbed by intentional upsets. These planned experiments may disrupt the productivity of the process but can save money in the long run. In addition, the operating personnel gain confidence in the system when they see it handle upsets successfully. Obviously, such an experiment requires careful, thorough planning. To achieve the necessary cooperation in this type of venture, however, frequently involves plant management.

John S. Baumgartner, in his book *Project Management,* offers the following summary: "Project management consists of the actions involved in producing project deliverable items on time, within contemplated costs, with required reliability and performance at a profit to the contractor. The purpose of project management is to insure achievement of these objectives through functional organizations and over their specialized interests. The project manager's role basically is one of planning, controlling, and motivating the project team."*

10.3.2 Operations

Operations personnel, from the superintendent down to the hourly operator on the production floor, must participate in the installation for practical and psychological reasons. From the practical side, they have been running the process and will continue to run the process; consequently, they have a wealth of information on their current method of operation. On a new plant, of course, this is not true, but their experience on related process units is often very valuable. Sometimes some very fundamental control concepts are revealed through the homely rules of thumb that are the operating criteria of the operator. All through the life of the installation, a close relationship must be maintained with operations, for they must ultimately accept the system and make it a tool working for them. Without this acceptance, the system will never reach its ultimate success. Close coordination with operations is particularly required at any time the installation of the control computer would interfere with the process or hinder production. This can occur

---

* Baumgartner, 1963, by permission of the publisher.

when new instruments are added, planned experiments are run, or the initial tests of the control system are planned.

If it is ultimately planned that the system be turned over to operations, someone from that department should be associated with the project from the beginning. In that way, they have an understanding of how the control logic evolved. This proves very valuable as new approaches to the control theory or plant modifications are proposed.

10.3.3 Instrumentation

The interface between the computer and the process comprises the instruments that provide signals for the computer and the equipment that receives control commands from the system. For an existing process, the individual selected to design, procure, and install the necessary instruments must be aware of the instrumentation already on the process, its capabilities and shortcomings, and how—or whether—it can be brought up to a condition of being of service to the system. For a new system, he must understand the instrumentation requirements and the availability of equipment.

The most likely candidate for this work is the engineer responsible for instrumentation in the plant. His involvement must come early in the project since he must brief the project team on what is installed and what new instrumentation is available that could be added. Instruments are frequently a long delivery item that can become the controlling activity when scheduling the installation, so the instruments that are going to be added must be selected and ordered early to avoid future delays. The instrument engineer must become familiar with the computer sampling system so that he can assure that the components are compatible. This includes voltage levels, signal ranges, filter designs, and noise characteristics of the connecting wires. By the time the computer is installed, the work of the instrument engineer is essentially completed and his time demands by the project diminish rapidly.

10.3.4 Process Engineering

In addition to the practical aspects of operations, it is necessary to have process engineering represented on the project to define the function of the process in engineering terms and relationships. It is this basic process knowledge that eventually evolves into the process model, regardless of the complexity of that model. It is the process engineer who transforms control theory into practice through his practical knowledge of the physical or chemical phenomena that are taking place in the process.

The project manager looks to the process engineer to assist in defining the critical variables that must be sampled, the frequency at which they must be sampled, and their relationship to the process objectives. The process engineer works closely with the instrument engineer on the design and selection of the proper instruments and controllers. Between them, they must procure instruments having the necessary accuracy and precision for the control that is planned. At the same time, there is no need to pay excessively for accuracy that is not required. This balance must be maintained through the entire loop of the system, for no system can be expected to control better than the data that it receives or the capabilities of the controllers under its command.

The process engineer must learn the functional capabilities of the computer so that he is able to take maximum advantage of these capabilities during the control logic design. His ingenuity can greatly enhance the performance of the system.

The demands for process engineering capabilities are concentrated at the beginning and end of the project. In the early stages, the process engineer is intimately involved with the control logic. During the implementing stages, he has far less to do, unless he is doing some of the coding. At the end, the process engineer is needed to assist in the test runs that are made to prove out the control theory. With his knowledge of the process he is able to interpret results of the tests and suggest improvements or solutions to problems.

10.3.5   Plant Laboratory

Within most plants there is a laboratory that has the responsibility for quality control analyses and grade determination. Laboratory personnel are involved at two stages of the installation. While the relationships between variables are being established in the design phase, observations can be made by the lab staff with temporary instruments or special lab tests. Often these are not planned to be carried on a containing basis but are used only for logic development. The involvement of the lab personnel at this stage is dependent on how well established the knowledge of the process variables is.

There are certain variables necessary in a control scheme that cannot be directly instrumented and fed into the computer system. For example, a quality test that requires the physical destruction of a sample may be one of the design objectives toward which the system is to control. Since this can only be determined in a lab, the computer system can be designed such that this parameter is tested and entered manually into the system. To coordinate this type of normal operational procedure properly, someone from the lab should be involved with the installing team to integrate the lab operation into the system.

The demands on the time of this group are generally small except for the coordination aspects and those special observations that they ask to have run. However, the system can have significant effect on their day-to-day operations after the installation is completed.

10.3.6   Statistics

In essence, a digital control computer takes many observations of process variables, analyzes them, and, from the analyses, determines what control action is necessary to keep the process on target. Inherent in this analysis are two basic problems:

1. Differentiating between variances in the observations that should trigger control action and noise that may exist on the signal.

2. Determining the interaction and relationship between variables to determine how much control action is required and to which variables it should be applied.

A statistician is often helpful in both instances. The latter problem requires the analysis of data that have been gathered during the design phase of the installation. Techniques such as regression are readily available to determine the correlation between observations and results but, without the proper understanding of the underlying statistical principles, false interpretations can be developed.

How sensitive the system should be to variances in observations is one of the most critical areas in the design of the control logic. Almost every signal has noise stemming from the minor process fluctuations, noise introduced by the instrumentation, and noise pick-up in the signal leads back to the computer. If the system is too sensitive to variances, it may take control action on meaningless noise. On the other hand, a system can be developed that damps out the signal to such an extent that it does not recognize a valid change in the process and fails to take the desired control action. The degree of filtering applied to a signal can be provided by the instrument engineer through use of electronic filters and by the statistician through a mathematical filter in the program.

The installing team should have the consultation, if not the direct involvement, of a statistician from the design phase through the implementation phase.

10.3.7   Mathematics-Operations Research

On many digital computer process control installations, the final objective is to have the system optimize the performance of the process with respect to certain variables. The process engineer can determine the relationship of those variables and state them in the form of equations. Given these relationships, an algorithm can be developed that selects the optimum from the alternatives that are available during a given time interval. There are several such techniques available that can be applied, and it is the task of the mathematician to select and assist in the implementation of the best one. In some situations that prove to be unique, it is necessary to develop a special optimizing algorithm for a process.

The services of the mathematician are required during the design phase while the programming flow is being laid out. He should also be available at the time the system is tested to assist in the final tuning of the system. Most likely, after the design phase, this individual need not be a full-time member of the team but merely be available for consultation.

10.3.8 Programming

Whenever a digital computer is involved, there are programmers who translate the logic of the control scheme into the specific instructions that are executed by the computer. Programming is a relatively new profession, having come into being with the development of computers and their broad usage. Programming requires a thorough knowledge of the computer, its components, and how to take advantage of the specific configuration that is being installed.

Programming represents the principal expenditure of manpower in a control system installation. The manpower is a direct function of the degree of sophistication that is being attained and the tools that are available to do the job, such as the hardware and programming systems. The programmers are involved in the project from its planning phase until the system is completely operative. They must be familiar with all of the other activities being carried on since most decisions directly affect their coding in one way or another.

10.3.9 Physical Planning

The installing and housing of the equipment that is included in the control system requires that engineering consideration be given to how the system adapts to its environment. Where it is inadequate, corrections must be made. For example, certain contaminants in the air can be highly damaging to the internal components of a computer. Contaminants often can be removed by filtering the air supply to the computer room and maintaining a positive pressure within the room. The engineering required for this type of adaptation is referred to as physical planning.

This work must be done early in the design phase and should be completed by the time the computer arrives at the start of the installation phase. Most of the physical planning activity falls within the domain of plant engineering and, if outside contractors are required, this department is familiar with factors relating to dealing with contractors.

10.3.10   Accounting

Frequently, the role of the plant accounting department is overlooked in setting up an installation team. While the feasibility study is in process, accounting is involved in the financial analyses used in deciding among the various candidate processes. Following the selection, accounting personnel help establish the budgets and objectives. Their involvement is vital from several standpoints. First, a process control computer system is a significant investment and its benefits are subject to an economic evaluation at some point in time. Accounting can better participate in this evaluation if it has been acquainted with the project during its entire life.

Second, production data are available within the system that accounting is probably currently developing from some other source. These financial data are available on a real-time basis, not after the fact, and can be used for assistance in making management decisions. Consequently, accounting may want to feed data into a plant information system that it either has developed or is planning to develop. This can be one of the fringe benefits received from the system that should be considered in its evaluation.

Third, when a control decision is made, whether part of an optimization scheme or not, the economical consequences of the alternatives must be considered. Cost accounting information is necessary in establishing this logic.

As can be seen, accounting's participation is concentrated at the start and end of the project, with only part-time participation required during the other phases.

10.4   SOURCES OF SKILLS

Few plants, except the very largest, have manpower available that offers all the talents required for a control computer installation. Either the particular skills needed do not exist, or, more likely, they are already committed to programs that take advantage of them. Three alternatives are available for staffing a project.

10.4.1   In-house

The most desirable situation exists when the team can be drawn from plant personnel where the system is going to be installed. A major advantage of this approach is that the team has previously established knowledge of the process and the political atmosphere surrounding it.

Another item to be remembered is that a project of this nature does not draw to a clear, discernible conclusion. The term "maintenance" is used in reference to that work that takes place on the system after it has been deemed operative. This takes the form of program modifications, improvements in the model that is being used to represent the process, and the addition of further control functions. Once the time of acceptance is past, those who are not resident locally are assigned new duties and are no longer available for improvements and maintenance. Consequently, a local cadre of knowledgable people is necessary to ensure the continued growth and successful operation of the system and process that it controls.

Another advantage of developing and using in-house skills is the potential for installing future systems on other processes throughout the company. The competitive world in which most organizations operate today demands that companies take advantage of any technological edge they may have. If they have in-house capability for developing and installing control systems, this advantage should be exploited to its fullest.

Obtaining or developing programming talent is particularly important. Not all people have the aptitude for programming, so tests have been developed that give an indication of an individual's abilities along this line. The tests are designed to indicate whether an individual possesses the logical characteristics that are required in programming. To avoid frustrations and disappointments, these tests should be given prior to assigning anyone to programming responsibility. By getting the right person assigned, the work progresses in a much smoother and more efficient manner. It should be emphasized that a lack of programming aptitude in no way reflects on the abilities of an individual to do other technical work.

Another source of talent is the corporate research or engineering center. These people have the mission of being abreast of new developments and spearheading their application to areas that can be profitable for their company. Frequently they already have done work on the concepts that must be considered during the system design stage of the installation. At the same time, they can enhance their backgrounds by having an opportunity to put some of their theory to test. Their influence on what data are to be gathered from the process often allows those data, in turn, to be helpful in research projects as new techniques and processes are developed.

10.4.2   Computer Supplier

Most of the suppliers of digital computer systems offer the services of their technical people. The extent to which an organization wishes to use this help can vary over a wide range from job to job. Obviously, the vendor is not always going to have first-hand knowledge of the specific unit that is to be put under control. On the other hand, he may have been involved in work on a similar process for another site. Areas in which suppliers usually have talents that may not be available from other sources are statistics, mathematics, and optimization. System engineers working for computer manufacturers are involved in application development where they are applying mathematics to the control of industrial processes through the use of computers, specifically the product of their company. Their contribution to the project can take the form of an algorithm or technique that has already been developed and may even be programmed.

The computer supplier is expected to bring to the project thorough knowledge and understanding of programming and real-time executive programming systems in general and specific knowledge, in depth, of the system being installed. In addition, the supplier can be expected to assist in the layout of the equipment, specifications on power and air requirements, and other problems related to physical planning needs.

The policy for the use of these services varies from vendor to vendor with respect to what is included in the cost of the equipment and what is available on a fee basis. To avoid misunderstanding during the installation, the extent of the participation on the part of the vendor should be clearly defined and an explicit commitment made for performance. If this is not done, severe misunderstandings can develop and the installation may be delayed. It must be remembered that these people are highly skilled and must be used effectively by their employer. Their time should be scheduled as accurately as if they were employees of the installing company.

Another point to determine at an early stage of the project is what degree of information security is desired concerning the installation. This point of clarification grows in importance when a great deal of research goes into the system design. It may be desirable to keep developmental information from competitors. On the other hand, it is only natural for the manufacturer of the computer to be interested in proving applications that he can market to other companies with similar applications. If any process information is proprietary and the vendor is going to be involved actively in the development, a clear understanding must be reached as to how the security of this information is to be maintained.

In some instances, a cost savings can be realized in a joint venture where the burden and benefits of some new research can be shared mutually by the outside organization and the installing company. For example, a computer manufacturer may have developed a control theory for a specific process that he wishes to market for use on his system and he would like to prove the theory by employing it on a process. By joining forces, the user gets assistance in the installation and the computer manufacturer gains a proven application for his marketing efforts.

10.4.3 Consultants

In the past few years, consulting organizations have been formed that, for a fee, make their services available to clients to assist in the installation of control systems. The offerings of these organizations vary greatly across the spectrum of capability. It cannot be overemphasized that a clear definition of their role is a must for smooth installation. It is possible that a "turn-key" contract can be negotiated where a consulting firm takes total responsibility from design, through installation, to acceptance. The major disadvantage here is that when the job is completed, much of the information necessary for maintenance of the system leaves the project with the contractor.

It is quite natural for an installation to involve talents under varying contractual arrangements. For example, it is not unusual for a computer company system engineer and a consultant (in some instances called "software house") system engineer to be working in concert on a project with the installing team's programmers. The project manager has the responsibility of ensuring that all contacts, both supply and service, do not overlap while yielding coverage for all aspects of the installation.

10.5 TRAINING

If a computer control system project is one of the first installations to go into a plant or company, the training requirements are extensive. This education is not limited to the installing team. Everyone that is involved with the project, directly or indirectly, should have some degree of knowledge as to what is going to take place and what to expect in the way of results.

10.5.1 Plant Management

Plant management should be given training on what to expect from a control computer, how it interfaces to the total job of running the plant, the potentials offered by the system, and what returns can be expected in a given time period. Without some background and understanding in new technologies, it is difficult to judge what to expect in the way of returns on investments. This is not to imply that top plant management should get involved in the details of the installation, but they are better prepared to monitor the progress of the project if they thoroughly understand the basic concepts and considerations that are involved.

This type of training can be provided in several sections by different groups with different intents. Computer vendors often offer executive classes that teach the concepts of computer control in a concentrated course directed to management-level people. In some instances, these classes are directed toward a specific industry or application. Generally, the objective is to provide a base for the students on which to build their understanding. When a specific installation is undertaken, occasional seminars should be held for plant management to talk about the specific application and generally inform them of how the system functions.

10.5.2   Programmers

Programming is the generation of the instructions of the control logic in a language that can be executed ultimately by the computer. There are those who are professional programmers, making a career of programming. For the most part, these people have devoted most of their time to commercial systems that are more numerous. Today, with higher level languages available (e.g., FORTRAN), it is far easier to teach someone who has process knowledge how to program than it is to teach an experienced programmer how to control a process. However, in each shop, there must be at least one person who learns the computer programming system in reasonable depth. This is necessary to have the ability to do system generation, maintenance, and, most importantly, to integrate programs into the total process control system. This type of in-depth training is most efficiently taught by the system supplier who has the responsibility of authoring and maintaining the programming system.

Two levels of programming language are available to those doing the coding. The higher level languages are directed at making coding easier for the programmer at the sacrifice of a small increment of computer efficiency. These languages are taught by the computer supplier and general courses are available in books, technical schools, and universities. If a program is going to be executed frequently in the system so that efficiency is important, or, if the programmer wants to deal with machine instructions directly, coding can be done in assembler language that is very close to actual machine language. This type of programming is generally taught by the computer supplier since it is unique to each computer.

10.5.3 Engineers

The digital control computer has functional hardware characteristics that must be taught to the engineering members of the installation team. These include, for example, the hardware features and their physical specifications which are important in interfacing to the process. By necessity, this training must be given by the supplier of the computer. The information is required early in the project, immediately after the selection of the vendor, so that matching equipment can be ordered that has compatible physical features. It may even be required before the vendor is selected if the overall installation costs are affected. This area covers such items as characteristics of analog-input filters, digital signal forms, and interrupt signal durations and recognition. They must be fully understood by the engineer responding for physical planning and by the instrument engineer.

10.5.4 Operations

Although mentioned last in this section, the training of operating personnel is one of the most important requirements for a successful installation. The operating personnel often view the installation with a great deal of reserve and even hostility. Until they are assured that the magical "black box" is not going to put them out of a job, they cannot be expected to be very enthusiastic. As mentioned earlier, the cooperation of the operating personnel is essential, and they can be very helpful in the system design and start-up.

The training of operating personnel should begin early and include seminars that acquaint them with what to expect and what the objectives are for the installation. This can best be taught by the installing team, and it yields the side effect of developing rapport with these personnel. Any aura of mystery should be dispelled by giving introductory courses on the fundamentals of computer control systems and relating them to the specific process. This early training should dwell more on concept than on details. Unfortunately, computers have been given a leading role in science fiction and have been raised to a position of awe in the eyes of laymen. This sometimes makes it difficult to bring computers back to where they can be thought of as applicable to the industry with which they are associated.

As the system design progresses, the seminars should go into how the process operation will change and what the operator's new role will be. Before the system first goes on stream for its shakedown, the operating people involved should be briefed thoroughly on the start-up procedures. It should be pointed out to them at this time that the system is new, that it does involve instructions executed in a predictable order, but that it may have errors. Their help is to be solicited in detecting and correcting errors as they are found. Operators often tend to be quite critical of the manner in which the system runs "their" process. Getting them involved also has the psychological effect of making them feel more a part of the installation.

It is human nature for people to remember the errors and mistakes

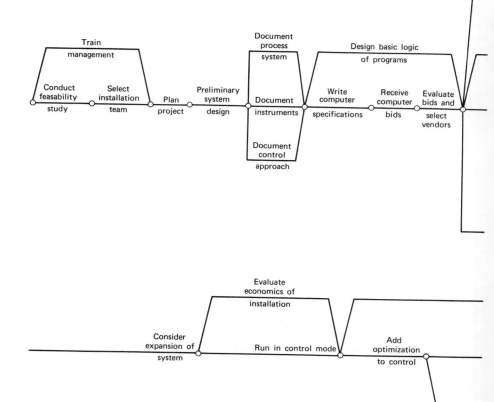

Fig. 10-4　Project control chart.

that are made but to overlook the success. The case of computers is no exception, so the way the operating people are prepared for an installation is often critical to its success.

10.6 PROJECT CONTROL AND SCHEDULING

The job of a manager is to get a system operative on schedule while meeting the performance objectives set forth at the start of the project.

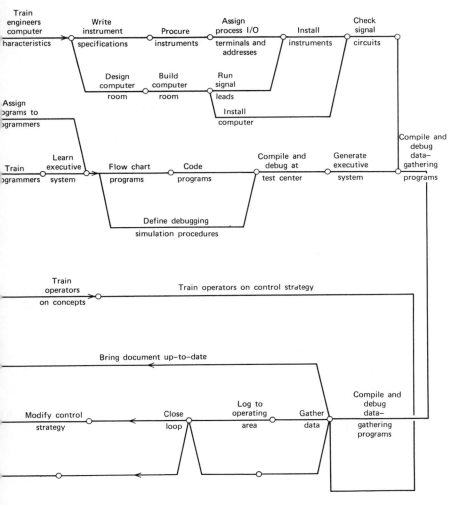

Fig. 10-4 (Continued)

This must be accomplished within a budget of time, manpower, and money. Consequently, planning for the project must include developing estimates of available resource requirements and then checking these estimates throughout the life of the project.

Two commodities that warrant particularly close tracking are time and money. The demands for each of these will be heavy. At the outset, as budgets are prepared, remember that change is a necessity to allow the systems engineers to come up with a good design. Allow for it in the budget.

From the time standpoint, even experienced individuals tend to underestimate the time required to complete a task. Experience shows that this is particularly true in estimating the time to take a program that is coded and initially debugged to the ultimate point where it is operative and ready to take over control. This must be kept in mind as schedules are defined and also when programming changes are requested after the detailed work is under way.

One technique for planning and controlling a project of this nature is through the use of critical path scheduling (often referred to as PERT). This technique can utilize the computer to assist in evaluating various options that present themselves. A representative diagram for a computer installation is shown in Fig. 10-4. When time is assigned to the various tasks during planning, it can be determined whether the overall time objective can be met and, if not, where emphasis needs to be placed. Throughout the life of the project, information developed during planning can be used to analyze progress by substituting actual times for the estimated times. This gives an analysis of how the project is progressing relative to the plan.

It has been estimated that PERT adds from 1.5–5.0% to the cost of a project, depending on the amount of detail that is carried in the analysis (Baumgartner, 1963). There are other methods that cost less and can offer adequate control for a project. One such method is discussed in *Project Management* by Baumgartner in the chapter "Project Control: The Status Index." Basically, a status index on tasks within a project at any given time is computed by the formula

$$\frac{\text{progress}}{\text{scheduled progress}} \times \frac{\text{budget}}{\text{actual expenditures}} = \text{status index number}$$

The closer the index is to 1.0, the closer the task is to performing according to plan.

Time and budget considerations are given equal weight in the status index. Thus, for example, if the progress is better than expected but the expenditures are greater than expected, the index still may be close to

1.0. If time and money are not equally important, different weighting factors could be applied to the two factors in the index. In any case, an index greater than 1.0 indicates performance is exceeding that expected. Conversely an index of less than 1.0 indicates that the progress is less than was expected for the money being spent.

By plotting the index with respect to time, various conclusions can be drawn concerning the project progress. Figure 10-5 illustrates some of the possible results of such an analysis. In general, this method of

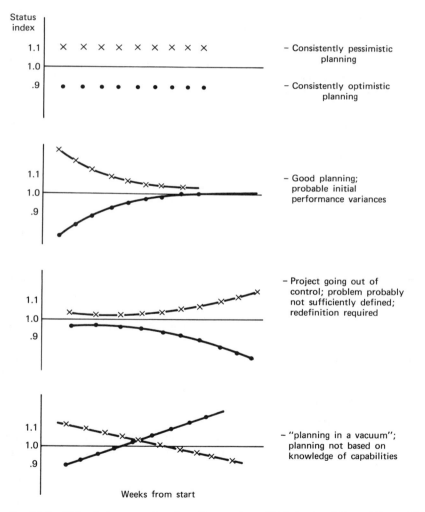

Fig. 10-5 Status index examples. From Baumgartner, 1963, by permission of the publisher.

project control provides an indication of the time/cost performance of the project in relation to the planned progress. It also provides the basis for projections of progress. By using the technique for subprojects within the overall activity, the method can provide a means of identifying and ranking actual or potential problem areas. As such, it also provides a means of reallocation of resources from areas that are progressing very well to those that are lagging.

It should be carefully noted that the status index does not provide a plan since the planning schedule is a factor in calculating the index. Nor does it require detailed planning, as other methods such as PERT do. It is, however, an effective method of tracking project progress toward a defined objective.

Another interesting method of portraying project slippage described by Baumgartner (1963) was developed by J. D. Burke of the Jet Propulsion Laboratory, California Institute of Technology. It consists of plotting the actual date versus the scheduled date for various activities during the project. If the project reaches its objective on schedule, the plot terminates at the 45° diagonal.

A horizontal line on the graph indicates that the project is proceeding without slippages, whereas a vertical line indicates a revision to the schedule. The technique and an illustration of three project histories are shown in Fig. 10-6. In the project represented by curve D, problems are recognized early in the project and a number of short revisions are initiated. The project finally maintains its schedule and meets its objective, as indicated by the horizontal line intersecting the diagonal. In project E, however, revisions are not made until barely a month before the schedule date. Following a number of short revisions, the project terminates without reaching its objective. Curve F illustrates a project whose objective is scheduled for a date which has already passed. It has, therefore, passed the diagonal and the objective cannot be achieved on schedule.

One of the oldest and most familiar means of representing plans and progress graphically is the Gantt or bar chart. If properly prepared, it can be understood quickly by everyone who has occasion to study it, and yet it requires a minimum amount of time to prepare.

All methods of measuring progress are dependent for accuracy on the integrity of the reporting individual. Project reporting procedures are not the place to hide problems that an area manager may feel are denigrating to his section's capabilities. Discretely used, progress reporting can be used not only as a guide for control but also as a lever for the project manager to use in inducing better performance from the sections on the project. For one thing, pride can be fostered in the staff that is

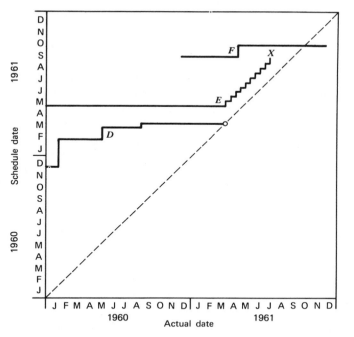

Fig. 10-6   Project tracking graph. From Baumgartner, 1963, by permission of the publisher.

on target, and extra effort is often put forth to keep on target. Conversely, no group wants to be shown as the task holding the bottom position when the various groups are ranked.

A word of caution might be entered here. The project should not be allowed to become so engrossed with keeping track of progress that progress itself is deterred. A balance should be struck between what information is needed to control a project and the effort required to gather those data.

10.7   PROGRESS REPORTING

The installation of a control computer is a major project and employs some of the best technical talent in a company. Top management expects and deserves to be kept abreast of the progress of the project. This reporting should not only discuss the progress to date but prepare management for the events to come in the near future. One principal objective of project reporting is the elimination of surprises for plant management and the project team.

The time between when the system is installed and when it is truly in control is a period during which it is hard for those not closely involved with the project to understand what is taking place. This is when the operating system is being generated, programs are being recompiled as the debugging process takes place, and the final integration of the individual programs into the total system is taking place. In reporting progress during this phase, it is better to report a conservative estimate of completion since this activity is very easy to underestimate. The story is told of a major programming undertaking that was reported 99% complete for two years with a sizable programming staff working on it all the while. Remember that any reporting scheme has the job of selling the project by factually presenting the status of unanticipated problems, how they have been surmounted, and the project position relative to the forecasted schedule and budget. If management has had the necessary training mentioned earlier, reporting is much simpler since management has a basic understanding of the system and appreciates the importance of significant problems and accomplishments.

In reporting progress, do not allow the project to carry the R&D connotation too long or it can lose drive and favor with those responsible for directing resources. During the course of the installation, it is well to note savings that are being realized by the partial implementation in the reports to management. Sometimes some of the least sophisticated savings can be the most significant. If there is one relatively easy to obtain objective that could be accomplished soon after installation and that would pay the cost of the system, reaching this objective greatly relieves the pressure for results and allows the necessary time to develop the more difficult applications.

10.8 DOCUMENTATION

There is no excuse for not having the development work on the project properly documented. However, this is the hardest item to administer since most engineers and programmers are "too busy chasing the pigs to build a fence." Documentation should not be left until the conclusion of the project since, all too often, it never gets done. Even if it is completed, it has lost part of its usefulness since documentation is necessary for the effective communication between the various sections of the project. Wiring diagrams, instrument lists, flowcharts, sample logs, terminal assignments, and the like should be distributed to those responsible for the installation for their approval and information. This helps to ensure that everyone works from a common base. The use of engineering drawings on vellum, with blue-line copies made for distribution, is an efficient vehicle for disseminating current information while maintaining the master copy intact. As logic or physical details change, the drawings should be revised, dated, and reissued so all can be current. Dating and numbering revisions to drawings are a necessity. Without this, effort unwittingly may be put into work based on information that is outdated.

Program coding is developed from flowcharts that portray the logic of the control scheme. When a program is compiled, a listing should be made. Part of the program is header information that is used only in identifying the listings. These headers should include information such as the name of the programmer, date and number of the revision, and which flowchart and revision the program was taken from. The small amount of time invested to keep listings current pays off in greater efficiency throughout the life of the project.

At the close of the project, all drawings, listings, and flow charts should be brought up-to-date in an "as-built" set of documents. These should reflect the total system as it currently stands. Without these, any future maintenance or extension of the system requires a great deal of unnecessary effort to determine what has already been done. Each individual should be responsible for completing the documentation of his area before he is allowed to disassociate himself from the project.

REFERENCES

Baumgartner, J. S., *Project Management,* Irwin, Homewood, Ill., 1963.

Lecture Notes from Arnold E. Amstutz, Assoc. Professor of Management, Sloan School of Management, Massachusetts Institute of Technology, 1968.

"Unlocking the Computer's Profit Potential," McKinsey, New York, N.Y., n.d.

BIBLIOGRAPHY

Ast, P. A., J. C. Cugini, and R. S. Davis, "Trends in Management of Industrial Control Projects," *Pap. 518 1969 ISA Ann. Conf.,* October 1969.

Bemer, R. W., "Economics of Programming Production," *Datamat.,* September 1966, pp. 32–34, 39.

Bohus, C. P., "Computer Systems Planning," *Mech. Eng.,* August 1968, pp. 17–21.

Bond, W. F., "Project Management for Computer Process Control," *Pap. 519, 1969 ISA Ann. Conf.,* October 1969.

Brooks, G. M., "Interaction of Process Control Programming and Instrumentation," *Proc. Texas A&M 21st Symp. Instrum. for the Process Ind.,* January 1966, pp. 15–18.

Chestnut, H., *Systems Engineering Methods,* Wiley, New York, 1967.

Donnelly, J. F., "Hybrid Mock-up Prepares Instruments and Controls for Refinery Start-up," *Instrum. Technol.,* August 1967, pp. 56–59.

Drewry, H. S., "Project Planning Pays Off," *Control Eng.*, September 1966, pp. 91–94.

Drewry, H. S., and J. R. Howard, "For Successful Computing Control . . . Start Simple," *Control Eng.*, August 1965, pp. 76 ff.

Jackson, R. R., "Men, Machines, and Management Controls," *Proc. Texas A&M 18th Ann. Symp. Instrum. for the Process Ind.*, January 1963, pp. 1–11.

Kay, R. H., "The Management and Organization of Large Scale Software Development Projects," *Proc SJCC*, pp. 425–433, 1969.

Lane, J. W., "How On-Line Computer Projects Go Wrong," *Control Eng.*, June 1969, pp. 111–114.

Lumb, A. C., "Computer Control of an Alkyl Amine Plant—A Project History," *Pap. 68-716, Proc. 23rd ISA Conf.*, 1968.

Moore, J. F., "Guidelines for Computer Control in New, Existing, and Multiple-Unit Plants," *21st Ann. ISA Conf.*, New York, October 24–27 1966.

Phister, M. Jr., "Checklist for Large-Scale On-Line System Planning," *Control Eng.*, November 1967, pp. 87–89.

Plumer, F. J., "Total Construction Project Management," *Iron & Steel Eng.*, March 1970, p. 107.

Stout, T. M., "Instrumentation Project Management—Its Effect on Costs and Payout," *ISA J.*, May 1966, pp. 46–49.

Weiss, E. A., "Working with the System Supplier," *Control Eng.*, September 1966, pp. 122–125.

Weiss, M. D., "Organizing for Systems Engineering," *Control Eng.*, May 1963, pp. 89–94.

Weiss, M. D., "Getting the Control Systems Engineering Done," *Proc. ISA*, 1962, pp. 6 ff.

Chapter 11

Economic Justification

LYNN S. HOLMES, JR.

The term *control* is not easily defined and the definition often depends on who provides it. For the control theorist, control is seen as a mathematical relationship between process parameters; for the process operator, control is viewed in terms of its operational aspects. Management ultimately defines control in terms of process economics. To them, control is certainly not manipulating the variables to draw nice smooth lines on a control chart. It may not even be causing the process to produce the maximum amount of saleable product. Make no mistake about it, the most common object of control is to produce for the company the maximum amount of net long-term profit consistent with the safety of the operating personnel and plant. Regardless of how much money is available for new projects, there is never enough to go around. A control system, like any other investment, must stand on its own feet and meet the profit criterion. The days when a control computer could be installed primarily for prestige are past, if they ever existed at all.

This chapter is devoted to a discussion of the many practical aspects of identifying the need for, and justifying the installation of, a process control computer system. The material is based primarily on data and experience from the process industries. However, because of the general nature of the subject, most of the information applies equally well across industry lines. Improvement in the control of product quality, for example, is an important consideration in steel manufacturing, petroleum refining, chemical manufacturing, and most other industrial endeavors. One of the purposes of this chapter is to point out some of the places where justifications can exist irrespective of a process industry orientation.

The discussion considers the general characteristics of processes for which computer systems can usually be justified. There are exceptions, of course, but these general guidelines are useful in identifying potential applications. Specific areas of justification common to many different processes are then considered. The subsequent section considers how these and other sources of justification may be related to the control philosophy planned for use in the process.

The areas of potential justification result, hopefully, in economic credits. The debit side of the ledger is equally important in establishing the need for a computer control system, however. These debits consist of implementation and operating costs directly attributable to the computer control system. Having identified both credits and debits, the discussion considers the tasks that must be accomplished by the justification team and the factors that must be considered in calculating the potential payout of the system. The last topic, postimplementation studies, stresses the importance and value of evaluating the accuracy of the justification after the system has been installed and is operating.

The topics discussed have been carefully chosen to provide a realistic guide to proving or disproving the economic justification for the installation of a computer control system. Logical and well-directed economic justification studies are very important because, if the need for a control computer cannot be recognized and justified, all else is academic.

11.1 IDENTIFYING POTENTIAL APPLICATIONS

Not all processes are candidates for computer control systems. In fact, even identical processes can exhibit widely different benefits from the use of a control computer. For example, distillation tower B in a refinery can justify a control computer whereas an identical tower A in the same plant cannot. The difference is that tower A runs the same material all the time and its optimum operating conditions can be very nearly fixed and maintained manually. Tower B, however, is used to process many different charge stocks which require significant and frequent adjustments in operating conditions. In this case the computer is paid for by reduction in losses experienced in changing the charge stocks.

This example illustrates one process characteristic common to many processes that can justify a control computer: the frequency and economic effects of process disturbances. Other general process characteristics that identify an application as a candidate for computer control are the size of the process and the value of the products, and the complexity of the required control. As with any generalization, however, there are exceptions, but the guidelines discussed in this section are useful in the early identification of potential applications.

11.1.1 Process Size

A control computer involves a significant investment and cannot be paid for by a process that does not produce sufficient gross profit. Size, in terms of dollars of plant investment, is one criterion that is closely correlated to the gross profit. It is not, however, the only criterion; operating cost, product value, and value added must be considered in assessing whether or not a process is of sufficient size to justify a control computer. Viewed in a broad sense, however, process size is one process characteristic that can be used to identify a process as a candidate for computer control.

Size is of importance because of the relatively large investment in the control computer and the relatively small percentage process improvement that is attributable to the computer control system. Although the improvement depends on the type of process and many other factors, a guideline of 0.5–6.0% improvement, which was originally proposed for the chemical industry, has been widely accepted (Williams, 1961; 1965). The minimum return of 0.5% is largely based on the intangible benefits provided by a computer control system, such as better documentation and increased process knowledge. If the improvement exceeds the 6.0% level, Williams suggests that the process was not optimally designed and that significant return can be realized by modifying the process without a control computer.

These guidelines, although useful for rough approximations, should not be applied blindly to any process in any industry. Judging from reports in the literature, the improvement achieved in many cases has exceeded these guidelines. This is an indication that either the criteria are conservative or that many processes are less than optimally designed. In addition, significant improvements make good news in trade publications whereas less spectacular results, or even losses, may not be announced publicly. It must, however, be concluded from published reports that computer control has resulted in substantial improvement in many industries. This has been achieved through the effective implementation of computer control projects which, in general, were preceded by soundly based economic justification studies.

If two years is accepted as the payout period, the improvement attributable to the computer control system must increase the profit by an amount equal to half the system cost per year after taxes, or, roughly, an annual amount before taxes which is equal to the cost of the system. Although it is difficult to state a "typical" installed system cost because the definition of a "typical" system is equally difficult, a $300,000 system provides for the control of a relatively complex process. Based on the 0.5–6.0% guideline, the plant investment should be from $5,000,000–$60,000,000 for a 2-yr payout.

In addition to these guidelines, evaluations based on economic factors and on gains realized in actual installations have provided justification guides based on the product value in various industries. One such compilation, shown in Fig. 11-1, covers a number of different processes in several of the industries in which computer control has been extensively employed. The data also illustrate that the market conditions can affect the basis of justification drastically. For example, in a production-

Industry Process	Values and Costs per Product Value	Unit of Product Raw Material	Minimum Size for Computer System
Chemical			
Ethylene	4–5 cents/lb	1–2 cents/lb	600,000 lb/day
Ammonia	$70–$90/ton	$10–$20/ton	300 tons/day
Styrene-butadine rubber	15–18 cents/lb	8–10 cents/lb	200,000 lb/day
Ethylene oxide	12–15 cents/lb	5–6 cents/lb	200,000 lb/day
Vinyl chloride	8 cents/lb	4–5 cents/lb	400,000 lb/day
Petroleum refining			
Crude distillation		$2–$3/bbl	60,000 bbl/day
Catalytic cracking	$3–$4/bbl		40,000 bbl/day
Hydrocracking	$4/bbl	$3/bbl	20,000 bbl/day
Iron and steel			
Blast furnace	$60–$65/ton	$35–$40/ton	1,000 ton/day
Oxygen steelmaking	$75–$80/ton	$50–$70/ton	3,000 tons/day
Hot strip mill	$105–$160/ton	$80–$90/ton	2,000 ton/day
Cement			
Rotary kiln(s)	$3–$4/bbl	$1–$2/bbl	6,000 bbl/day
Pulp and paper			
Continuous	$115–$130/ton	$30–$40/ton	200 ton/day
Bleaching	$140–$160/ton	$50–$60/ton	150 ton/day
Paper machine	$400/ton	$200/ton	75 ton/day
	$200/ton	$125/ton	200 ton/day
Electric power			
Steam plant(s)	1 cent/kWh	0.4 cent/kWh	200 MW

Fig. 11-1 Justification based on product value. From Stout, 1966, by permission of the publisher.

limited plant in the pulp industry, a 10% increase in production results in a gain of about $2000/day. On the other hand, if the plant is market limited, a 10% decrease in raw material cost realized through reduced losses results in a savings of only $600/day. This, and similar examples, illustrate that a production-limited plant is almost always a better candidate for computer control than is a market-limited plant.

11.1.2 Process Disturbances

Another characteristic common to many processes for which computer control has been justified is that the process is subject to frequent or significant disturbances which result in economic penalties. All processes are subject to a wide variety of disturbances which occur as the result of such things as changes in ambient conditions, raw material properties, catalyst characteristics, load conditions as in the economic dispatch of electric power, equipment performance, and many others. A disturbance may not even be a physical phenomenon; it can be economic or seasonal in nature.

Some disturbances are more significant than others in both their effect on the process and their frequency of occurrence. In general, if a process is subject to a number of moderately high-frequency disturbances, it is a candidate for further investigation. Even if the disturbances are somewhat infrequent, on the order of once each day or two, but the consequences are severe enough, the process may be able to justify a system. However, if the rate of disturbance is higher than the ability to compensate for the effect of these disturbances, computer control may be ineffective. This situation is unusual and normally indicates that the process should be modified whether or not a computer control system is installed.

The type and frequency of the disturbances determine the relationship between the frequency and value of control action. Figure 11-2 illustrates this point by comparing two processes, A and B. Process A is characterized by a very rapid rise in profitability when the time for adjusting the unit setpoints to compensate for disturbances is increased from once a

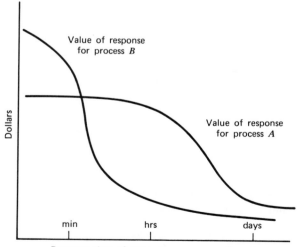

Fig. 11-2 Process disturbance characteristics (after Savas, 1965; by permission of the publisher).

day to once every few hours. This process can be optimized off-line by a conventional computer. Hence most of the profit for better control can be realized without an on-line system. Credit should not be taken for increased profit of this type to justify computer control. Process B, on the other hand, is insensitive to infrequent optimization but shows a rapid rise in profitability when adjustments are made every few minutes. This type of process is a very likely candidate for computer control.

Although disturbances are usually considered to be uncontrollable factors, some industries are characterized by what can be considered man-made disturbances. For example, some paper mills concentrate on the production of small lots of specialty material such as fine papers or coated board. In this situation, grade changes may occur as frequently as once or twice a day. Using computer control to sequence start-up operations and to bring the product within specification in a shorter period of time can result in substantial benefits by increasing the throughput for a fixed capital plant investment. It has been reported that computer control of a paper machine has reduced the time for changing paper grades by 20% (*Chem. & Eng. News*, 1966).

The benefits that can be obtained through elimination of disturbances may be either easy or difficult to identify, depending on the nature of the process. As an example, batch processes involve charging a vessel with a material or recipe of materials and controlling heat, pressure, agitation, recipe additions, or combinations of these four along a time-level profile to a specification end-point. If there is time lost in charging and draining the vessel because the operators cannot monitor everything rapidly enough or be in several places at the same time, throughput gains can be easily identified. However, if improved turn-around time requires recognition of the end-point according to a mathematical model or more efficient control of the time-level profile, the expected gains can be more difficult to prove.

11.1.3   Process Complexity

Processes that are technologically or operationally complex are candidates for computer control. Technological complexity is usually the result of many process variables, multiple interactions between the variables which require control action that is not obvious, a large number of product specifications, or a combination of these factors. Operational complexity is usually a result of technological complexity and is often characterized by the requirement that the operator perform a complicated sequence of operations according to a strict timetable and within a defined set of constraints. Both types of complexity can be further increased by the need to remember process history for relatively long periods of time. This is typically the case in processes with large residence times, such as glass furnaces, blast furnaces, and cement kilns.

Manual control of a complex process often requires the simultaneous adjustment of interacting control variables with a good chance that the manipulations will cause a period of upset. In processes of moderate complexity it is seldom possible manually to manipulate more than two variables at a time. To paraphrase a quotation attributed to Napoleon, "Most process operators can change one variable successfully, an experienced supervisor can probably simultaneously manipulate two variables adequately, but only a genius can correctly adjust three variables at the same time."

The alternative to manipulating a number of individual variables simultaneously is to move them one at a time. This can cause long periods of off-specification operation, which is probably as undesirable as a mild upset.

Computers operating with even a simple mathematical model can simultaneously move multiple variables successfully. Hence a source of justification in complex processes is reduction of the time required to go from one operating level to another and also reduction in process upsets caused by adjustments that compensate for process disturbances. As a specific example, a cement plant experienced a 75% elimination or reduction in process upsets with a control computer (Pit and Quarry, 1966).

Another important source of justification in complex processes involves the ability of the computer system to accomplish control actions easily that are not usually considered for conventional instrumentation. Some complex loops, for example, can be better stabilized through the use of feed-forward algorithms, multiple cascade control, or other advanced control approaches. In addition, computer control strategies can sometimes be based on variables that are difficult, or even impossible, to measure using conventional instrumentation. These derived variables are calculated by the computer using a mathematical representation of the phenomenon and data from conventional instruments. For example, an index of catalyst activity or data from analytical instruments can be incorporated easily into the control algorithm. This is usually possible with little, if any, increase in the cost of the computer system. It is generally true that the cost of computer control increases much less than the cost of conventional control for a given increase in control complexity. This is the result of the stored program and the time-shared input/output equipment utilized in the control computer.

11.1.4   Other Process Characteristics

There are a number of other characteristics important in identifying candidate processes for computer control. One of these is that the process should have a long life expectancy. This assures that the company not only recovers its investment but makes an attractive long-term profit. In other words, it is not only important to present the payout economics, but also to consider the profit position of the investment over a period of years. In determining this position, it is imperative that anticipated changes in product price, demand, and competitive situations be included in the computation. For example, the development of a new process and its implementation by several manufacturers can result in a price change which makes an older plant uneconomical, even with computer control.

The degree of instrumentation in the existing plant is also an important consideration in identifying potential applications. The computer must have adequate data from the process instrumentation if computer control is to be effective and profitable. Poorly instrumented plants in which the operator controls the process on the basis of intuition may require substantial capital for new instrumentation. This affects the return on investment provided by the control computer.

Adequate knowledge of the process technology is also a requirement if a rapid payout is required to justify the computer system. Effective control requires that the process be adequately described so that appropriate control algorithms can be incorporated in the control computer program. As an example, processes based on chemical reactions usually are good candidates for computer control because the chemical reactions are generally well understood and can be described easily in algorithmic or mathematical terms. Similar concepts apply to processes in which reactions are based on thermodynamic principles that are adequately understood. Processes that involve mechanical operations such as grinding or pulverizing are often more difficult to describe because the relationship between these operations and the amount or quality of production is not always well understood. This does not mean, however, that computer control should not be considered for such a process. The data acquisition and computational capability of the control computer coupled with mathematical modeling techniques provides the means for developing an empirical model which can lead to substantial improvement in process operations. A lack of process knowledge merely means that the payout may be delayed owing to the expense of developing the model empirically. These factors must be considered in the justification computations.

Whether or not adequate process knowledge is readily available, the availability of technical people familiar with a process can make that process a good candidate for computer control. The planning, installation, and operation of a computer control system comprise a technically complex task and require a well trained staff. If such a staff is available, a process that otherwise might be a poor candidate may provide the basis for justification. For this reason, large companies with extensive engineering resources may consider computer control for a process that would not be considered by a smaller company.

The lack of such a staff, however, does not preclude the possibility of computer control. The services of consultants and vendor technical services can often provide initial technical assistance. The payout period is often extended, however, since the effective installation and use of the computer require that the company develop a technical capability, and this takes time and money.

11.2   GENERAL JUSTIFICATION SOURCES

The process characteristics discussed in the previous section provide guidelines for identifying potential control computer applications. Merely identifying the process, however, is rarely enough to convince management to make the substantial investment required for the implementation of computer control. An adequate justification for the investment requires that specific sources of savings and payout be estimated. In this section some of the sources of justification are examined and the bases for estimating their value are discussed. Not all of these sources are found in every process, but experience has shown that these are often lucrative and should be considered in any justification survey. The sources discussed here are relatively independent of the control strategy planned for the plant. The effect of strategy on sources of justification is discussed in Sec. 11.3.

11.2.1   Better Control

If control is defined in terms of the profit criterion discussed earlier, all sources of justification result in better control. In this section, however, control is viewed in terms of the operational definition that control is the maintaining of the process status as closely as possible to predefined operating conditions. These conditions are determined by the process engineer on the basis of his knowledge of the process and its dynamics. He may utilize process set-up tables, manual calculations, or a computer program for optimizing the process status. The process operators implement his instructions and then modify the process adjustments to maintain operation at the desired operating points in spite of process disturbances.

The operator's efforts usually are not sufficient to maintain the process at its best operating point. This may be a result of his lack of knowledge of the process dynamics, or it may be that important process conditions, such as feed stock composition, have changed since the process engineer determined the desired operating points. In addition, the speed or complexity of the process may be such that manual control of set-points cannot be performed fast enough to maintain the process at the desired operating points. Even the most skilled operator has difficulty maintaining process operation at optimum levels for sustained periods of time.

Less than optimum performance in the process is usually manifested in process operation which occasionally reaches the optimum operating condition but then falls back to a lower point. In essence, the process oscillates about an average value that is usually less than the optimum. Whatever the control criteria are, consistent operation of the process at, or nearer to, its optimum operating point results in an important source of justification. If the control criterion is throughput, the higher average throughtput realized through smoother control of the process results in a net return to the company. Similarly, the savings in additives, the recovery of valuable by-products, and the reduction of losses are all important benefits of better control action.

The control computer provides better control through its ability to respond to disturbances more quickly and to provide control which anticipates the effects of interacting adjustments. Although these two control computer capabilities are often sufficient to justify the system, further benefits may be realized by utilizing advanced forms of control not easily implemented with conventional control equipment. This includes techniques such as on-line optimization, adaptive adjustments, feed-forward, and control based on calculated variables that cannot be directly measured. These techniques can be applied to reduce process upsets and loss of production owing to upsets caused by adjustments made to compensate for disturbances.

Specific estimates of the value of smoother control can be obtained through a variety of techniques. Research papers relating the effect of the independent variables on performance are a good source of information. Supplementary to this, the engineer can gather data from historical logs or from tightly controlled test runs to establish performance variations actually experienced on the process. In the latter case, the effort must be organized to provide uniform conditions during the test and to apply manual control that approaches computer control as closely as possible.With data collected from this type of test, statistical comparisons of the performance fluctuation of the test run results and normally observed variations can be made and appropriate conclusions drawn.

It is not always feasible to conduct test runs on operating units. When this is the case, historical log data must be examined for periods of lined-out operation. The worst periods of performance fluctuation then can be compared with the best periods. Using the hypothesis that computer control can at least equal the best periods of manual control, credits can be established.

In some processes this theoretical knowledge can often be used to establish a maximum possible return from the process. For example, a chemical process involving the combination of two elements has an upper bound determined by the laws of chemical equilibrium. If this theoretical maximum return is not sufficient to justify the installation of a control computer, the project must be abandoned or other sources of justification must be found. Theoretical upper bounds on process performance are most useful in disproving a source of justification since performance rarely approaches the theoretical limit. If the process control computer is not justified under theoretical conditions, it can never be justified in a practical installation.

Justification in new installations is particularly difficult because of a lack of historical data and experience. If the new process, or portions of it, is similar to other processes in the plant, then experience on these related units sometimes can be used to arrive at estimated performance. These data can often be supplemented by controlled tests on the similar unit.

Mathematical simulation also provides a powerful tool for studying possible improvements in an old process or performance in a new process. Off-line optimization studies using mathematical models in conjunction with simulated process dynamics can provide insight into control problems and the feasibility of their solution with a control computer. With an adequate model, actual process operators can be used to control the simulated process. The efficiency of their control can be compared to computer control of the same model responding to identical process disturbances. Estimated returns of computer control can be determined from this comparison.

The techniques available for investigating sources of justification, therefore, include the use of historical data, controlled tests on existing processes, and simulation techniques. In addition, the literature can provide information concerning the success of other companies in the same industry (Washimi and Asakura, 1966).

11.2.2 Consistent Quality

Most processes have the single objective of producing a given product of a specified quality at a profit. Quality can be measured by off-line tests or by on-line instruments. The quality of the product varies with process disturbances, and chemical kinetics or other factors usually determine the amplitude of these variations. Variations in the quality of the product may or may not be economically significant. If they are, better quality control provides an important source of justification.

Quality variations which result in products that are out of specification are clearly an economic liability. They result in product being discarded or reprocessed, both of which cost money. If off-specification material is shipped to the customer and rejected, the additional loss in transportation costs may be significant.

A less obvious fact, however, is that producing product which is considerably better than the specification is also expensive. The superior material may cost more to produce and yet it returns revenue that is less than should be expected for a product of this quality. Thus the justification of a control computer may be based on improving the consistency of the quality, not on improving the quality.

Control computers can provide more consistent quality as a result of smoother control. The speed of the computer and its ability to apply advanced control techniques are usual bases for the improvement. The possibility of improving quality and the consistency of quality has been recognized by industries and their customers. In one case, a steel supplier reported that his customers were specifically requesting sheet steel that had been rolled on a computer-controlled mill. In another rolling mill, a company reported a significant decrease in the standard deviation of sheet gage (Brower, 1966).

One word of caution should be noted in working with product quality as the source of justification. All laboratory procedures are subject to testing deviation. It is necessary to establish the standard deviation of the test because it is not possible to reduce quality variations below the deviation of the test unless blind replication of routine analysis on samples is standard procedure. This is not the usual practice in a production laboratory.

11.2.3 Manpower Efficiencies

The displacement of manpower is in many cases a poor source of justification for a control computer. There are several reasons that process manpower cannot always be replaced. First, there are normally just enough process operators assigned to a unit to handle emergencies that cannot be controlled by the computer. These emergencies include such things as pipe rupture, stuck valves, power failure, steam failure, and instrument air failure. Second, if process manpower is replaced, this is usually offset by the accompanying increase in the technical manpower needed to program and maintain the computer. Third, there may be union contractual agreements that restrict job elimination.

There are, however, situations where manpower reductions can be used for computer justifications. These instances normally occur when a computer is a necessary requirement for control room consolidation. There is an accelerating trend in the modern processing plant either to combine a number of control rooms during modernization or to provide consolidated control rooms for new units that are being added in a plant expansion.

Another instance where manpower reduction is a primary source of control computer justification is in the area of automatic control and monitoring of laboratory instruments. A laboratory technician normally handles several instruments for both control and computation. If the computer eliminates the manual calculations and assists in control, one operator can service more instruments and manpower can be reduced. Likewise, laboratory clerical efficiencies can be effected because the computer can be used to collate test results and to prepare the printed report. Laboratory chromatograph monitoring and mass spectrometer monitoring are excellent examples of this type of justification (Briggs, 1967).

As a specific justification example, a laboratory technician can service from 8–10 chromatographs without a computer. With a computer a technician can control 20–30 instruments. Therefore, in a 30-chromatograph laboratory, two men can be eliminated per shift. If the laboratory is operated three shifts per day, seven days per week, this means a savings of $8\frac{2}{3}$ men.

Computer-based laboratory information systems offer opportunities for other clerical savings. This subject is discussed in Sec. 11.3.4.

11.2.4 Maintenance

Improved process control can have significant effect on high-cost maintenance items, and this source of payout should be investigated in every justification survey. This source of justification is most evident if the need for maintenance can be associated directly with the effectiveness of control. That is, if the interval between maintenance action or the cost of the maintenance is affected by how well the process is controlled, savings can be credited to the control computer. In addition to the direct maintenance savings resulting from decreased maintenance requirements, process throughput is increased since the process is available for production. This is particularly important in a production-limited plant since it may eliminate the need for capital expansion of the facility.

As specific examples, heat exchanger fouling can often be decelerated through tighter control, and corrosion can be reduced by better removal of contaminants. In processes involving rolling, cutting, or grinding operations, wear sometimes can be reduced by better control of the speed and cutting rate. In addition, computer control in rolling mills has minimized "wrecks" in which the slab damages a roll because of the improper setting of a stand. Because the roll is so costly, avoiding only a few wrecks can provide substantial justification through reduced maintenance costs.

Better control has led to noteworthy reductions in shutdown frequency for kiln refractory repair for one cement manufacturer (*Pit and Quarry*, 1966). Most processes which involve control of burning zones in kilns or regenerators can reap benefits if it can be shown that improved control reduces hot spots or lowers average burning temperatures. However, refractory replacement savings may not be a short-term benefit because the life of the material may be a number of years. Thus savings may be of the deferred type.

11.2.5 New Installation Savings

New units can provide several sources of justification not found in updating existing process plants. By programming the computer to simulate the process, operators can refine their techniques even before the process is operational. Process disturbances can be simulated and emergency procedures can be tested and learned. This type of training for the operator often contributes to a faster start-up and more efficient control in the early months of operation (*Baytown Briefs*, 1967).

Having the computer on-line in the early phases of a new unit start-up also provides other savings. The tighter control afforded by computer control allows the unit to be brought up to full production in a shorter period of time. This provides additional months of productivity for the process and the increased revenue is a credit for computer justification. Engineering cost is also reduced since the unit may be transferred to the operators for control at an earlier date.

By getting the process lined-out earlier, the characteristics of the unit are learned sooner. Control and equipment deficiencies are identified and can be corrected. This effectively increases the throughput of the plant and may even result in reduced maintenance since damage to the unit caused by a deficiency is avoided. The increased knowledge of the plant at an earlier date also allows corrections to the control strategy to be implemented sooner. The measurement and computational capability of the control computer are particularly important in capitalizing on these benefits since they provide better data and the means for performing the time-consuming data reduction required in analyzing the operation of the new plant.

Although these are largely one-time savings, they should not be neglected in a justification survey. The increased revenue of a process that attains full production several months earlier can provide a substantial justification credit.

11.2.6 Process Knowledge

One common result of computer control is increased process knowledge. This is an intangible benefit and it is difficult to assign a monetary credit for the purposes of justification. It should, however, be included as an intangible factor in a justification study. As the result of better process knowledge, improvements in the control strategy can often be effected. Examples of better knowledge include an understanding of interactions between process variables, discovery of process parameters dependent on variables which cannot be measured but can be calculated, and determining the effect of more timely control.

Increased process knowledge generally stems from two sources. The first is the justification survey itself. An effective survey requires a good understanding of the process if justification credits are to be estimated accurately. The study of the process during the justification survey is often the most comprehensive study ever made of unit operation. The results often provide the basis for improvements that can be realized partially without computer control. If the improvements can be implemented with the existing control hardware, these benefits cannot be included as credits for computer justification. Usually, however, control can be further improved by the computer, and these increased gains are valid justification credits.

The second source of increased knowledge is directly creditable to the installation of the computer. Through techniques such as data smoothing, correlation, and other statistical methods, dependencies between process variables can often be identified and utilized for implementation of more effective control strategies. The computer, properly programmed, is ideal for these types of studies since it can process large quantities of data in a relatively short period of time. From these data, effects can be uncovered that are not obvious in a mere scanning of historical records. For example, subtle differences in operating conditions between day and night can be detected, and these might lead to more effective control of diurnal variations. In addition, the tight control afforded by the computer during test runs often results in more accurate data for further analyses. Historical trends based on these more accurate data provide additional information on process dynamics that is useful in refining control techniques.

Another advantage of computer monitoring of processes is the ability to continue the data collection under extreme upset conditions. It is the job of the process engineer to analyze events just before, during, and after major upsets either to prevent or to minimize similar upsets in the future. It is not unusual for the engineer to try to troubleshoot an upset condition by first examining the log sheet for the period in question. He normally finds that the log is blank for the upset period. The reason is simply that the operators are too busy trying to regain control of the unit to take the time to enter values on the log when the unit is out of control. The computer, however, continues to collect the data. After analyzing the data, the process engineer can implement changes to the control program which minimize the chance of a similar upset. This results in smoother control, one of the major sources of justification for computer control.

11.2.7 Malfunction Detection and Safety

Another intangible source of justification which is one of the major potential benefits of computer control is malfunction detection and safety monitoring. Trouble anticipation through the computer reduces the number of unscheduled down periods with resultant increases in throughput. In addition, it is not out of the realm of possibility that the computer can take action to prevent major damage to the unit during an emergency. Many process units have complex systems of mechanical and electrical interlocks that prevent such things as the mixing of hot oil and air, reverse flow of fluids in differential pressure-sealed systems, and other conditions that could have disastrous results if they were to occur. Most of the interlocks are relatively simple cause-and-effect relationships, and it is usually not economically feasible to allow for all circumstances. Many of these, however, are easily implemented through the computer program. Often this provides more reliable detection since computers, unlike humans, have no fear and continue to operate in the face of imminent danger.

A computer can be programmed to scan process inputs and to examine possible trouble conditions every few seconds. This type of constant surveillance is utterly impossible for the process operator to accomplish, and yet it is a simple matter for the control computer. The early detection of malfunctions often results in significant economic benefit.

A less spectacular but more routine safety application is the use of the computer for monitoring interrelated alarm conditions. For example, a gradual rise in the liquid level in the bottom of distillation towers is not nearly as significant by itself as it is if it is accompanied by a gradual decrease in the suction pressure of the bottoms pump. A control computer can be programmed to check such interrelated conditions routinely and to issue routine messages which warn the operator of impending trouble. The computer, of course, only recognizes those alarm situations that have been included in the program by the process engineer. Thus the computer can prevent the recurrence of a catastrophe; it cannot, however, prevent a catastrophe that has not been anticipated by the process engineer.

11.2.8   Multiprogrammed and Time-Shared Computing

Another justification area is the use of the control computer in a time-shared or multiprogrammed mode. Time sharing involves the use of the computer to perform noncontrol jobs during the periods when it is not required for control actions. This mode assumes that the computer works on the noncontrol job whenever it is free, even if only for a fraction of a second. Multiprogramming, on the other hand, involves sharing the computer resources among several jobs that are simultaneously resident in the main storage of the computer. These modes of operation are discussed in more detail in Appendix B, but their relationship to justification is considered in this section.

A process control computer is characterized by the fact that under normal conditions an appreciable percentage of the computing capacity of the machine is not used. The only time that the entire capacity of the machine may be utilized for the control program is during periods of upset or under emergency conditions. Because of this, the computer can be used to do off-line engineering calculations in its spare time. The reason engineering calculations are mentioned specifically, and not data processing in general, is that much engineering work is written in a high-level language such as FORTRAN. A high-level language provides greater system protection to prevent an off-line program from destroying an on-line program as a result of programmer error.

This latter point is an important consideration and deserves detailed discussion. Conventional data-processing programs are written either in an assembly language or in a high-level business language such as COBOL. Control computers, however, are not usually supplied with compilers for the high-level business languages. Therefore, by the process of elimination, business applications for the control computer usually must be written in a low-level language. This introduces the distinct possibility for programming system destruction by human error during program debugging. Although some control computers have the ability to hardware-protect their storage from destruction, it is seldom possible to protect everything from inadvertent overlay. Because of this, there is an element of risk involved in using a control computer for off-line work on programs that are written in low-level languages. One can argue that an assembly language program can be debugged when the computer is off-line or on another machine, and that this would elimi-nate the risk. Experience has shown, however, that computer programs may operate flawlessly for years but can "blow up" as the result of some unanticipated set of circumstances. An additional hazard is that the input data of assembly language programs may not be read under con-trol of a supervisory monitor program that checks for keypunch errors. This can cause inadvertent modification of the executive programming system. For these reasons, a careful study should precede any decision to implement conventional data-processing applications in an on-line control computer.

There is one engineering application where the off-line use of a control computer is highly profitable, and this is in model development, up-dating, and tuning. The control computer provides volumes of data that can be analyzed to produce more accurate and more detailed models. If only a sketchy model exists, the free time of the computer can be used to regress these data to provide for expansion of the model.

11.2.9   Other Justification Areas

Many other sources, some tangible and some intangible, can provide justification for a control computer. For example, reduction of utility costs can provide many opportunities as a source of control computer payout. Reduced process variations often lead to more efficient heater operations, or, in the case of electric furnaces used in steel manufactur-ing, proper control can reduce fluctuation in electric power utilization and thus reduce surges that cause demand penalty payments. This illustrates another important item. Contractual commitments can often influence payout. These agreements are usually called "take or pay" or demand contracts and are characterized by clauses which specify that a company is obligated to pay for a minimum amount of material whether or not it is processed. Some additionally require that the customer pay a premium for material usage in excess of a specified amount. The agreement can be further complicated by limiting the maximum delivery rates over fixed time periods without penalty. As can be seen, take-or-pay contracts can be a two edged sword. On one hand, they can degrade incentives if they are long-term agreements and the payout is based on reducing raw materials below the minimum specified in the contract. On the other hand, the computer can be used for scheduling delivery and for predicting future use. The increased accuracy of the prediction can often eliminate penalty payments.

More efficient use of chemical additives can also be an important by-product of improved process control. Since the additives can be sig-nificantly affected by the efficiency of the process, the more efficient operation under computer control can result in justification credits. Lubricating oil manufacture and motor gasoline production are examples of such processes. For instance, there is a rather delicate economic balance between the optimum severity levels of catalytic crackers and reformers and the tetraethyl lead level required to produce specification gasoline. Optimum control of catalytic crackers and reformers is successfully accomplished though control computers and the savings in tetraethyl lead provides part of the justification.

Operating efficiency improvements of mechanical equipment offer opportunity for justification credits in computer control. Compressor efficiencies in recycle or refrigeration subsystems can be improved through proper control (Koch and Schildwachter, 1962; Farrar, 1965). Even simple compensation of ambient temperature effects from daylight to dark for refrigeration systems should be included in payout investigations.

Another place that can provide returns is control of steam throttling in process unit subsystems that include single- and multistage turbine drives. In these systems steam at one pressure level is distributed to turbines which discharge it at a lower pressure into a gathering system that supplies steam to other low-pressure turbines. Here control reduces unnecessary throttling that is caused by changes in steam demand at varying system loads. This is accomplished by a horsepower/steam balance to determine which turbine should be on each system. The payout is enhanced if an admission-extraction turbine that permits finer interstage adjustments is in the system.

11.2.10 False Justification

Control computers can be solidly justified on the basis of legitimate economics. When this is done and the control strategy is well designed, both technically and operationally, the installation is a success. If, however, the justification is built on false premises, no matter how well the system is designed, it will be a failure and the entire status of computer control within a company can be jeopardized.

False premises can take several forms. Extrapolation of statistically based data is extremely difficult and can lead to erroneous conclusions. Justifications that assume the ability to operate the process in untried areas are questionable without extended confirmation testing.

An example of how a false premise might be used to justify a control computer is useful in illustrating this point. Suppose that the tighter control possible with the control computer allows the recovery of more of a valuable product from a less valuable product. This can be a lucrative source of payout and is one of the best and most easily proved ways of justifying a computer. This justification can all but disappear if the removal of the valuable product causes the viscosity of the less valuable product to increase to the point where it cannot be pumped. The more valuable product must be blended back into the process to reduce the viscosity and the gain is lost!

Another error that can be made is to base justification of control computers on product price structures that contain artificial constraints which are violated by the computer. For example, payouts have evaporated in cases where reduction in off-grade material was the major source of gain from the computer control. The reason is simply that in many markets there exists a demand for two distinct grades of the same product. One is called by its trade name; the other is simply referred to as an "off-spec" product and is sold at a reduced price. When improved process efficiencies reduce or eliminate the source of this off-spec material, the sales department may protest strongly because they have customers who do not require the tight specifications of the actual product. A manufacturer can find himself in the position of running his process intentionally to produce out-of-specification material. Thus it is important to examine the structure of the marketplace both from a product distribution and from a price structure point of view before using justifications of this type.

A similar error can be made if incorrect assumptions are made concerning the character of the market. There are usually two distinct economic cases that provide the basis under which process control is evaluated. These are the market-limited case and the production-limited case. If justification is evaluated under the production-limited assumption, when in fact it is a market-limited case, the entire payout calculation can be wrong. The production-limited case assumes profit on incremental production, whereas the market-limited case derives its payout primarily through reduction in raw material and utility costs. In addition, postponed capital investment for equipment replacement can sometimes be discounted in time and used for justification in the market-limited situation. As noted earlier, the production-limited case usually provides greater possibilities for control computer justification.

Another serious error is to justify computer control out of the context of the entire plant. For example, if throughput is to be increased by better control, it is obvious that the unit feeding the process must have the capacity to supply the increased input or, if raw material is the feed, that the supply is available. A not so obvious error can be made for a process that is producing an intermediate product. Here the entire processing flow must be examined to determine the effect of improving the process in question. The justification is false if the only effect of computer control is to modify operations on a preceding or succeeding unit with no material effect on the end result of the processing.

Although there are other examples of false justification that can be discussed, these examples illustrate typical errors. The important point is that justification claims must be scrutinized in the cold light of reality from every point of view. Supply and marketing, as well as technical considerations, must not be overlooked.

11.3 JUSTIFICATION FROM CONTROL STRATEGIES

In the previous section general sources of justification were discussed. These potential sources were considered without specific reference to particular applications or control strategies. Although many of these justifications can be found in any process, some sources are more closely associated with particular processes or control strategies. Because of the broad spectrum of control computer application, it is not practical to discuss the most probable sources of justification for each industry or application. This information is available in the literature, mostly in the form of specific examples. The number of control philosophies or strategies, however is much smaller, and justifications can be discussed in terms of the broad categories of control philosophies. In this section, several of the major philosophies are described and probable sources of justification are discussed.

11.3.1 Monitoring, Alarming, and Logging

Strictly speaking, monitoring, alarming, and logging are not examples of true computer control since the system takes no control action. However, the same basic control computer is used, and in many cases monitoring, alarming, and logging comprise a preliminary step to closed-loop computer control.

The names "monitoring," "alarming," and "logging" are relatively self-explanatory. The control computer is used to collect data from the process and to monitor process status on a real-time basis. These data are used to provide alarm messages to the operators if isolated or inter-related conditions indicate the necessity for an alarm. The data are also used to provide information for off-line studies such as model building. Information is also stored, often after some degree of data reduction, to provide historical operating records for the process. The logging function consists of periodically reducing pertinent data and providing a printed log to the operator, process engineer, or management. Logs may be produced on a predefined time schedule or on demand. They provide a means of evaluating the total operation of the process and, in addition, may be used for accounting or billing purposes.

There are three sources of payout in installations which use the control computer for monitoring, alarming, and logging. First, there may be manpower savings in two areas. It is the practice in many process plants to use yield clerks to tabulate and calculate many items which measure the operating efficiency of the process. These include such things as the quantity of product per pound of steam consumed, quantity of product per kilowatt hour, and similar indices of efficiency. With its data collection and computational capability, a computer can eliminate this manpower need for a particular unit.

The second source of manpower savings is the result of the trend toward control room consolidation. Here the instrumentation and controls for several process units are brought together in one control house. Although it is usually impractical for one operator to monitor all units with conventional instrumentation alone, the computer can provide the necessary exception reporting which makes single-operator control possible. Thus manpower savings are realized through this relatively simple computer application.

The second major source of justification is the result of the monitoring function. The function of the yield clerk discussed above is to provide management with a means of measuring the performance of the operating crews, to provide insight into the mechanical state of the process, and to provide status information. When a computer is added to the process, these calculations can be done as often as required, not just once per shift. Therefore, the computer detects inefficient operation much sooner than is possible through manual calculations. In addition, efficiency indices that are too complex to consider for manual calculation are easily implemented with the control computer. The indices are also more accurate and timely because of the data acquisition characteristics of the control computer. All of this leads to higher overall operating efficiency.

Finally, the data collected by the computer can be the source of information to be used in the development of mathematical process models. As discussed in Sec. 11.2.6, significant process improvements can result from a better understanding of the process even if the model is never used in direct control of the unit.

There are, of course, a number of intangible benefits that accrue through the use of the control computer for monitoring the process. For example, the major jobs of a MARC (monitor alarm and result calculations) system in a power-generating station is sequence-of-events recording. Here the operators are interested in determining the sequence in which relays opened when an outage or upset occurred, usually with millisecond resolution. The sensing and time interval measurement capabilities of the control computer easily provide this function. In continuous processes, event recording can answer questions such as which parameter indicated the upset first and how the upset propagated through the process. Thus the monitoring function leads to the development of better control strategy and provides information to reduce losses caused by upsets.

The logging function is also an intangible benefit because it provides more timely, accurate, and consistent information than is available by manual methods. How often have you heard the comment from the process engineer that log sheet information is all but useless for analytical purposes? If this is the case, and my personal experience indicates it is, the mere act of producing an accurate log of operating information has value in reducing test runs and process engineering effort.

11.3.2 Supervisory Control

Supervisory control produces all of the justification that is attributable to monitoring, alarming, and logging, and, in addition, provides the benefit of producing higher profit to the company through more efficient operation as the result of frequent optimization and/or control calculations.

Supervisory control implies that the computer manipulates setpoints of the independent variables. This can take two forms, either closed-loop control or open-loop control, the latter often being called *operator guide* control. In the former case, the computer actually manipulates the process instruments, whereas in the latter case, the computer instructs the operator to make the indicated move.

In both cases, the computer calculates the optimum set-points through the use of mathematical models and optimization techniques which utilize methods such as linear programming and gradient projection (Wrathall, 1966; Sauer *et al.*, 1964). Through the computational power of the computer using optimization, the process can be held against more of its limits simultaneously than is possible with conventional control. An additional technique which often improves process performance under supervisory control is the use of the computer to calculate derived parameters that cannot be measured directly. Often these derived parameters are more indicative of process status, and their availability provides the operator with information for better control decisions. Supervisory control usually provides a lucrative source of computer control justification which can be as much as an order of magnitude greater than any of the other sources.

11.3.3 Direct Digital Control

Direct digital control (DDC) is a recent and exciting application of electronic computers to the regulation of process control loops. DDC is the substitution of time-shared computer I/O capabilities and logic for the analog functions normally supplied by conventional instrumentation. Originally, there was hope that DDC could be justified on the basis of direct instrumentation replacement through the use of the computer. This has not proved to be the case, and the reasons are quite simple. Although computer reliability has been proved very high, there still exists the possibility of failures, and process operators are reluctant to operate without conventional instrumentation back-up since some computer malfunctions could cause the simultaneous loss of all control (Griem, 1965).

Another reason that direct instrument replacement does not provide complete justification for DDC lies in the actual hardware costs. Interfacing a control computer with a control element involves converting the digital value produced by the computer to an analog voltage or current which is used to actuate the valve or control element. These digital-to-analog converters (DAC) are still rather expensive, costing as much as $1000 each. To reduce this cost, a DAC can be multiplexed to several sample-and-hold amplifiers or analog memory circuits, as discussed in Chap. 5. Even with this approach, however, each analog output may cost $500 or $600 without back-up functions. This, coupled with the $100,000–$200,000 base cost of the medium-sized control computer, precludes direct instrument replacement as the sole means for justifying DDC.

The foregoing discussion should by no means be taken to indicate that there is no economic justification for DDC. On the contrary, there are many justification sources but, like many new concepts, DDC is not a cure-all for every situation. The primary criterion for choosing DDC over conventional instrumentation is the regulation problem which is not easily handled by the conventional approach.

In general, DDC can be applied in a control computer dedicated to the implementation of DDC, or it can be combined with supervisory control. In both cases, the justification results from the wide flexibility afforded by the control computer. For example, in an application of DDC to the manufacture of fiberglass, there was very little difference between conventional and computer control for individual loops. The improvement came from the ability to use control techniques not possible with standard analog instruments (Griem, 1965).

DDC offers the control engineer a wide choice of techniques that either were not possible before or were possible only in specialized situations. For example, cascading of loops can be altered automatically without hardware modification, tuning constants can be adapted to loop disturbance characteristics (Bakke, 1964), control can be based on calculated variables such as BTUs and residence time, feed-forward control can be easily implemented, automatic ramping is possible, and dead time effects can be compensated.

In recent installations there has been a growing trend to combine supervisory control and DDC in the same system. This has many advantages and generally allows the justification for both types of systems to be added. As a matter of fact, it often provides even more sources of justification. The reason for this is that the dedicated application of DDC and supervisory control with separate control computers results in a duplication of hardware functions and adds control complexities. A single-control computer can implement both control strategies with little additional hardware and software.

11.4 PROCESS INFORMATION SYSTEMS

The decrease in the cost/performance ratio of new control computers has presented industry with the opportunity to widen the horizons of computer applications to include systems which economically provide all levels of management with timely information on which to base day-to-day decisions. Although these systems, often called information systems, are not control systems in the sense usually associated with process control, they often include a control computer as a basic element. In addition, the theory on which information system application is based is often similar to that used in process control.

In this section, two of these information systems are briefly considered. The first, the laboratory information system, is very similar to a process control system in that the computer controls the operation of various types of instruments, collects data, and provides information to the operator. The second, the management information system, consists of an interconnected system of computers. The control computer serves as one of the information input devices for this computer system complex.

11.4.1 Laboratory Information Systems

The application of control computers for monitoring laboratory instruments is a rapidly growing field. Control computers are used to monitor a wide variety of these instruments such as mass spectrometers, chromatographs, infrared analyzers, and distillation apparatus. In fact, any instrument that outputs a continuous signal that is recorded or graphed for interpretation is a candidate for computer monitoring.

Generally, a laboratory instrument detects composition or characteristic changes in the sample being analyzed by using some type of detector whose output is a continuous analog signal that varies as the composition or characteristic varies. In conventional use, the output signal is either amplified or attenuated to serve as an input to the recording device which produces a graphical output trace. Digital data are obtained from the trace by manually selecting data points or by using mechanical devices such as planimeters. The analog signal conditioning and recording equipment is subject to noise and mechanical limitations that can add damping or lag.

The computer, however, senses the output signal directly, applies digital filtering, and often detects information features that would be lost as a result of limitations of the conventional equipment. In addition, the direct conversion of the output signal into digital data eliminates the errors associated with the manual measurement and recording of information from the output trace.

In addition to providing more accurate data, the computer provides a more accurate analysis. The computational techniques that can be implemented in the computer are more precise, and usually more accurate, than the techniques employed in manual calculations. Techniques that are too time consuming to be considered for manual analysis are easily provided with the use of a computer. The consistency of the analysis is also better since the programmed decision rules in the computer always provide the same decision for the same set of data. This is often not the case when human interpretation of data is involved.

Finally, the computer calculates results much more rapidly than is possible with manual analysis. The printed reports produced by the computer are more legible than the typical laboratory log sheet, and they are easier to use. The timeliness of the reports often contributes to better process control since product variations can be detected and corrected sooner.

The payout for the application lies in two areas. As discussed in Sec. 11.2.3, manpower reductions can be significant in a large installation because the analysts are relieved of the burden of determining cut points and attenuation factors as well as the calculations involved in the analysis.

A second and even more significant payout results from improvement in precision analysis. For example, one large chemical company reports that it was not out of the ordinary for computer-run chromatograph analyses to have a standard deviation of 0.02% compared with 0.2% in conventional analyses (Briggs, 1967). In many plants the ability to produce a product that is 0.1% closer to specifications is worth thousands of dollars.

In some cases, hardware savings also can be credited to the use of the computer. In some analytical instruments, electromechanical equipment is used to sequence phases in the analysis procedure. For example, the equipment may control the admission of the sample, the injection of a carrier material or gas, and the cleaning of the instrument after completion of the analysis. The computer can provide this sequence control through the process I/O subsystems. In addition, amplifiers, attenuators, and recording equipment often can be eliminated.

11.4.2 Management Information Systems

Management information systems consist of an interconnected complex of several computers arranged as a hierarchy beginning at the lowest level and continuing upward through all facets of a business. In general, control computers form the first, and sometimes the second, echelon of the information system. The process control computers perform their conventional functions and, in addition, pass information to the second-level computers. This echelon uses the data for computing optimizations, product blending, and unit scheduling, as well as for preparing summary reports for mangement. The second-level computer also provides reduced data for use in higher levels of the information system hierarchy (Hodge, 1965).

Although the detailed justification of management information systems is beyond the scope of this book, the basis for the justification is the timely analysis of the operating data for the entire plant. The analysis provides management with the information required to make operating decisions and to formulate policy. In sophisticated management information sytems, control theory principles are applied to business decisions to optimize the total profit of the company within its policy, operation, and market constraints. The contribution of the control computer to the management information system is an intangible benefit that can be a justification credit in the decision to install a control computer.

11.5 SYSTEM IMPLEMENTATION COSTS

The justification of a control computer involves examination of both credits and debits. The credits are the sources of payout discussed in the previous sections and the debits are the costs of implementing the computer control system. There are many items that influence the system cost. Some of these costs are easily identified, whereas others are not so apparent and may remain hidden for several years. This section discusses the more obvious items and attempts to highlight some of the important, but obscure, factors that may be significant in particular applications.

11.5.1   Hardware Cost

It would seem that the most easily identifiable item of cost is that of the computer system itself. Vendors can supply detailed itemized tallies from the outset. However, through no fault of the vendor or the buyer, the bid price of a system is usually less than the installed price. The reason is that bid specifications are seldom explicit enough to define an application in sufficient detail for an exact estimate. This is particularly true for the first application. If the same application is installed a number of times, later installations can be estimated very accurately.

There are a number of common errors which lead to inaccurate bid prices. First, main storage estimates tend to be low because additional programs are often added to cover applications or situations not anticipated in original estimates. Allowing a generous contingency is ultimately more economical because main storage is one of the least expensive items of a computer. Although the immediate reaction to this statement is often one of disagreement since it can account for 25–30% of the cost of a computer system, the real cost of inadequate main storage must be considered. When programmers are faced with a very tight storage limitation they spend considerable time coding and recoding programs to make them as small as possible. Furthermore, compromises are usually made in program flexibility. This leads to the expenditure of more manpower during system design and programming and the expenditure of more manpower than necessary when the system requires modification after installation. If limited main storage causes only one man year of extra programming effort, the effort can be more than paid for by the installation of 16,000 words of incremental main storage which costs about $25,000. It is quite true that computer programs, like gases, expand to fill the available space. But another analogy is also true, the larger the space, the lower the pressure.

The second item that is subject to inaccurate estimation is the analog and digital "front end" of the computer. It is rarely possible to specify exactly the number of inputs that computer system will be required to sense or the number of switches and outputs that the system requires for control. These items can be estimated, but they are dependent on noncomputer items such as operator consoles, analog instrumentation, and other equipment which has been specified but whose interface requirements vary from one manufacturer to the other. These devices are usually out for bid at the same time the computer is being bid and, therefore, the exact requirements are not known.

Another significant factor that can influence the cost estimate is the modularity characteristic of the control computer front end. For example, digital-input points are most often available as a group of points. Each group may contain 10, 16, or 24 points, depending on the computer manufacturer. In addition to the groups of points, a control unit that can handle only a specified number of groups is required. If the system expands during the implementation phase to the point where an added group of inputs also requires another control unit, the effective per-point cost of these additional points may be very high. Bid specifications try to anticipate these costs by requiring the vendor to list the incremental cost of adding inputs. However, the vendor usually lists these costs with the cost of the control unit evenly distributed over the maximum number of groups that the unit can control. Individual prices on the control unit and groups of points, coupled with information concerning the maximum number of groups that can be attached to the control unit, provide a better picture of the cost of additional groups of points.

A similar situation exists if the addition of a group of points requires more mounting cabinets or racks. Familiarity with the rules for configuring the control computer and individual prices are the best guarantee for elimination of errors due to this and similar restrictions.

Another hardware cost that is often underestimated is the cost of signal conditioning. Sometimes instruments and control elements have electrical or noise characteristics that require elaborate interfaces for successful use with a control computer. This can be very expensive, particularly if these interfaces are not off-the-shelf items and must be developed as custom features.

11.5.2   Software Cost

Of all the items that must be evaluated in computer control applications, the most difficult to estimate is the cost of programming the computer. Programming is a difficult and time-consuming task, and estimating a programming effort remains more an art than a science. If a user were to install a computer with absolutely no programming system and do all the work necessary to provide a flexible, sound programming system that supplied all the needs of a closed-loop computer control application, the programming cost would vary from the cost of the hardware to several times this amount. Although users rarely do this amount of programming, it is very important that vendor-supplied programming systems be carefully evaluated during the justification study since additional programming can represent a significant implementation cost.

There are essentially three categories of programs necessary in a control computer application: programs to control the computer hardware, programs to control other programs, and programs to control the process. The first two categories are generally supplied, at least in part, by the vendor as part of the system. These are often called the *programming system* for the computer and are discussed in Chap. 9. The third category, programs to control the process, are called application programs and may be provided by the vendor. Uusually, however, the user must write some application programs for his particular process. The considerations involved in writing an application program are discussed in Chap. 14. In this section, these categories are discussed only briefly and their effect on implementation cost is emphasized.

Computer control requires a programming system that controls the hardware and software capabilities of a computer functioning in the real-time environment. These programming systems are called by various names, such as real-time monitors, time-shared executives, free-time systems, and multiprogramming systems. Regardless of the name, they must provide the essentials necessary for the effective use of the computer hardware in the control application. Some of the functions that must be provided by the executive programming system are the following:

1. Program and event sequencing.
2. Job to job control.
3. Control of the analog and digital I/O subsystems and their interactions.
4. Servicing of interrupts on a priority basis.
5. Diagnosis of hardware and software errors and initiation of corrective action whenever possible.

Besides programs to provide these functions, the executive usually includes programs that allow the computer to be programmed in assembler language and one of the higher level compiler languages, such as FORTRAN. Both of these are important to the user since they provide an efficient means of writing the specialized application programs that are required.

Another group of closely related programs is sometimes called the "utility program library." These programs perform simple jobs such as listing or reproducing a deck of punched cards, copying one disk onto another, or more complex tasks such as modifying programs stored on disk, relocating and loading programs, and other functions that are required in the everyday use of the computer system. Often, the first-time user of control computers overlooks the value of such programs and does not thoroughly investigate their availability with the ordered system. If these programs must be developed by the user, time is lost in programming effort in addition to thousands of dollars.

The executive and related programs are absolutely essential for a computer control system and any company that is contemplating such a system must insist on this type of program from the computer vendor. Regardless of how inexpensive a computer may be, if it does not have a programming system of this type, it is probably too expensive to be considered for a complex real-time, closed-loop control application.

To illustrate the costs involved in producing this type of programming system, assume that there are 50,000 instructions in the executive, exclusive of constants, comments, buffers, and so on. A common rule of thumb for estimating the manpower required to write, debug, and document a program is five completed instructions per man day. If it is assumed that there are 22 working days per month, the total effort required to produce 50,000 instructions is almost 38 man years. At a conservative $20,000/man year, the total cost is over three-quarters of a million dollars and a highly sophisticated programming system costs several times this amount!

The application programs may take many forms from simple scan programs to highly sophisticated "fill in the blanks" supervisory programs (Ewing et al., 1967). Some of these programs are available from the vendor and can be included in the programming package supplied with the computer. In addition, the vendor often can supply the optimization program or the subprograms required to develop an optimization program for a particular application.

Generally, however, the user must plan on supplying some of the application programs for his particular process. The user may supply the manpower to develop these programs or he may contract for their development by a consultant, a firm specializing in software development, or the computer vendor.

The type and sophistication of the programming systems available from the various vendors cover a broad range. These programming systems must be carefully evaluated during the justification study to determine their applicability to the particular process being considered. The effort required to modify vendor programs and to develop new programs should be estimated for each vendor system and included in the implementation cost. In addition, the cost of developing unique application programs must be considered in the justification survey.

11.5.3 Interface Equipment

The equipment required to interface the computer to the process must be carefully evaluated to avoid unexpected costs. The important points to consider are the electronic compatibility, the programming compatibility, and the mechanical service factor.

The design engineer must be thoroughly familiar with the electronic and electrical characteristics of the computer, its I/O interface, the sensors, and the actuators that are used to pass information to and from the process. Overload characteristics of the computer or the instrumentation can require elaborate interface circuits if the consequences of certain types of transients cause a failure in the computer. Similarly, the stability of instrument calibration is quite important because the quality of control must not be a function of instrument stability. This can lead to elaborate and expensive calibration procedures that have not been anticipated.

Programming compatibility is a factor sometimes overlooked, particularly on the first system installed by a company. For example, if each setpoint station does not move a consistent amount in response to each pulse received from the computer, the program may have to use different coefficients for each station. If, in addition, there is a hysteresis effect, the number of coefficients can be increased further. This increases the requirement for data storage in the computer. The signal-to-noise ratio of interface equipment is also important because noisy signals can require electronic filters that limit scan frequencies or elaborate digital filters that impose an unnecessary computing burden on the system.

The mechanical service factor of interface equipment must not be ignored. In many cases equipment that is satisfactory for normal manual operation is unsatisfactory when interfaced to a computer. In general, this is because a man normally makes adjustments only a few times per shift, and this imposes no mechanical strain on the equipment. The computer, on the other hand, generally adjusts the equipment every few minutes or so, and in this type of service relay points may fail, rotor inertia can cause errors in positional information, and mechanical components can fracture owing to the number of operations.

From the above discussion it is quite obvious that a computer vendor should be selected prior to the selection of interface equipment; to do otherwise can lead to costly errors that can quickly eliminate the apparent price advantage of one vendor over the other.

11.5.4 Vendor Services

The considerations in selecting a control computer are discussed in Chap. 12. Since selecting a system is equivalent to selecting a vendor, vendor selection is also discussed in that chapter. The selection of the vendor is also a consideration in establishing the implementation costs of a computer control system, however, and this facet of the problem is discussed here.

One of the major cost implications of vendor services has been discussed in Sec. 11.5.2. This is the evaluation of the adequacy of the vendor-supplied programming system. There are, however, other vendor services that can have significant effect on the implementation cost of the computer control system.

Technical assistance from the vendor is one of these factors. Most vendors provide some degree of technical assistance to the user during the design and implementation phases of a computer control project. These services may be included as a part of the system price or rental, or they may be purchased under a separate contractual arrangement. Even after implementation, the proximity and availability of technical service engineers is important. Throughout the life of the system there will be new programming systems, hardware improvements, or modifications of hardware and software that are requested by the user. In all of these cases the availability of vendor technical engineers is very desirable. As before, the costs of these services may or may not be included in the basic purchase or rental cost. If they are not included, they should be considered as a separate cost item in the justification survey.

The competence of the vendor in computer control applications is a factor that can affect the cost of designing and installing the system. This is a difficult factor to identify quantitatively in the implementation cost of the system. It should, however, be included as an intangible item in the justification study.

One other factor to consider in evaluating vendor services is the maintenance organization that is provided and the availability and proximity of spare parts. Control computers must operate almost continuously, and when a failure occurs, it is very important that service facilities and spare parts be close at hand. If, for example, a large stock of spare parts must be purchased by the user, then this consideration must be included in the cost of the system. The loss of production or the loss of production gains during the periods when the computer is inoperable can also be considered a cost if it has not been accounted for in establishing the justification credits. The availability of trained maintenance personnel is a major factor in determining the magnitude of this expense.

Some users service their own systems, which yields some advantages. However, the cost of training the service employees must be included in the payout calculation. There is one disadvantage to the use of in-house maintenance that should not be overlooked: Control computers have fairly long runs between failures and it may be months between times when the serviceman sees any kind of machine trouble. As a result, he gets little practice and repair times may be significantly increased with corresponding increase in production loss because of the unavailability of the computer. Since the vendor service personnel usually maintain several control computers and have considerably more experience, they often can repair the system in a more timely manner.

11.5.5   Other Costs

Other costs that must be examined in a thorough justification study are the system operating and installation costs. This includes items such as additional training programs for personnel, maintenance and spare parts costs as discussed above, modification or construction of control rooms, supplies such as typewriter ribbons and blank forms, and the cost of maintaining the necessary technical staff. In all cases, these costs are not unique to the computer control system. For example, paper forms are required in any system, as are the control room, staff, and spare parts. There may be, however, differences in these costs when the control computer is installed and these must be considered. For example, the paper forms for printers and typewriters are generally more expensive than the forms used for manual logging operations. Similarly, some of the control room modifications, such as raised flooring, are directly attributable to the installation of the computer.

The increase in these costs must be added to the cost of the system. Similarly, savings in these items which are the result of the computer installation must be taken as credits. The main factors in this evaluation are quite obvious and a comparison between operations before and after the computer is installed is usually sufficient to identify the major items. A brief study by the justification team of computerized and conventional installations is a useful approach in initiating this phase of the study.

11.6   THE JUSTIFICATION SURVEY

The identification of potential applications of control computers and the detailed examination of justification credits and costs should be implemented through an organized justification study or survey. Control computer justification surveys take many forms, ranging from a brief meeting of a management committee to long, expensive, mathematical model developments that consume many man months of effort and cost more than the computer being justified. Neither of these two extremes is desirable. Ideally, the survey should not require more than six weeks to two months, and it should be conducted by a well-managed team that includes representatives from the management, operations, technical services, financial, and marketing departments.

The length and detail of the survey depends on many factors. If this is to be the first installation of a control computer, the survey may be lengthened by the inexperience of the team in identifying the important facets of computer control justification. In later justification studies, they can more quickly focus on the major payouts that can be achieved. Similarly, if the installation is in an industry or is for a process that has not been previously computerized, the justification cannot benefit from the experience of other companies that have reported their results.

Another factor that affects the length and detail of the justification survey is the attitude of the person who requests the survey and the person who makes the final decision. An executive who is positive toward a computer control or who, through independent study, is convinced of its value requires much less information than the one who is skeptical of the total idea. In the case of the positive executive, the justification team may have to expend considerable energy convincing him that a control computer for a particular process is not a profitable investment!

The first step to be taken is the organization of the team and the establishment of the ground rules under which the survey is to be conducted. Some of the important factors that should be defined include the following:

1. The basis of the justification premise; is the process production-limited, market-limited, or a percentage of each?

2. The definition of manpower credits; for example, is the design engineering a direct cost or is it considered as an overhead expense for the total company operation?

3. The definition of intangible credits; for example, is the value of increased process knowledge to be included as a credit and, if so, how is its value determined?

4. The values of the product, raw material, utilities, and other constants that are required in payout computations.

After deciding on these basic assumptions, the team should establish a study guide that systematically outlines the parameters that are to be evaluated. This ensures the completeness of the study and provides a means of measuring the progress of the team. The guide can take many forms, depending on the type of process being studied and the detail required in the justification. Table 11-1 is an example of such a guide and illustrates some of the considerations to be included in the survey.

The first section of the table outlines the analysis of process disturbances. It lists the sources of various disturbances, their magnitude and frequency, and the compensation that is required to eliminate their effect. Finally, the cost of the disturbance is estimated to provide a basis for the justification credit. The second section of the guide is concerned with the control of the process, independent of the disturbances. In this particular case, the effect of measurement accuracy is considered. The economic value of increased accuracy of computer control is estimated as a credit for justification. The third section considers economic factors such as product value and production cost. Both incremental and fixed costs are listed. The fixed costs are important when plant expansion is anticipated since computer control can effect some fixed-cost savings. The incremental costs are useful in the production-limited plant when the basis for the justification is increased throughput.

This example is, of course, limited, and an actual study guide outlines many of the other sources of justification and cost discussed in previous sections. In any case, however, the data gathered in completing the guide provide the basis for determining a profit function for the process. Detailed study of this function and its sensitivity to process and control parameters leads to a decision concerning the profitability of a control computer investment.

At this point, a number of actions are possible if the potential payout is attractive. Either a preliminary mathematical model can be developed for a more detailed study of the process, specifications can be developed and issued to vendors with requests for price quotations, or a computer can be ordered and an implementation team established.

The alternative chosen depends on the objectives and experience of the company. If wide application is expected and the company has little experience with computer control systems, it is probably best to begin by acquiring experience through the development of a mathematical model of the process. As a word of caution, however, detailed and technologically precise mathematical models are sometimes difficult to develop and may require considerable manpower and time. Experience has shown that less precise models are often adequate and produce almost the same results as more detailed models. As a matter of fact, several companies have found that a model will control a process even if the values of the model coefficients are badly in error but their signs are correct.

Smaller companies that cannot justify a staff dedicated to computer process control may wish to issue specifications to hardware and software vendors or consultants for implementation of the control system. This is quite practical. However, every company that is considering the installation of a computer control system must plan to have at least one engineer at the end of the project who is experienced in the application and installation of computer control systems. There are several reasons for this, but the primary one is to maintain continuity following the actual installation. When the changes and modifications which are sure

Table 11-1   Process Control Study Guide Outline

I. Disturbance Analysis	Disturb-ance	Fre-quency	Magni-tude	Compen-sation	Cost
A. Inputs					
1. Raw materials					
a. Feed stock					
.					
.					
2. Additives					
.					
.					
3. Catalyst					
.					
.					
4. Utilities					
a. Fuel gas					
b. Fuel oil					
c. Steam					
d. Electricity					
B. General Process					
1. Ambient changes					
2. Equipment fouling					
3. Grade changes					
C. Others					
1. Analytical accuracy					
2. Safety factors					

to come are required, the user must be in a position to carry out the tasks. Otherwise, long expensive delays may result.

Finally, companies that are staffed to implement control systems may order the computer hardware and set about developing the models and control strategy. The organization of implementation teams is varied but should include at least one trained control engineer who is responsible for the instrumentation and process/computer interface. An experienced programmer who is responsible for the implementation of the programming tasks is also a requirement.

11.7   PAYOUT COMPUTATIONS

One of the recommended members of the justification team is a representative of the finance department. This recommendation is based on the fact that the computation of investment payout is not only complex,

Table 11-1 (Continued)

II. Variables Analysis	Measured Accuracy	$ Variation within Accuracy	Total Effect
A. Independent controlled			
1. Feed rate			
2. Reaction temperature			
B. Independent uncontrolled			
1. Raw material quality			
2. Ambient conditions			
a. Temperature			
b. Humidity			
C. Dependent variables			
1. Yield			
2. Quality			
3. Quality distribution			

III. Economic Factors	Fixed Cost	Incremental Cost
A. Production costs		
1. Raw material		
2. Utilities		
3. Labor		
4. Maintenance		
5. Investment		
B. Product values		
1. Production		
2. Off-grade		
3. By-products		
4. Quality give-away		

it is also dynamic and can change from time to time based on a number of financial considerations. The engineer is sometimes content to use the time-honored formula

$$\text{profit} = \Sigma \text{ products} \times \text{value} - \Sigma \text{ raw materials} \times \text{cost} - \Sigma \text{ additives} \times \text{cost} - \Sigma \text{ utilities cost} - \text{fixed costs}$$

The result is corrected for federal income tax, and the number of years necessary to recover the investment is calculated. There is considerably more to it than this, however, in a thorough evaluation, and the advice of the financial expert is required.

The first decision that should be made is whether to purchase or lease the computer equipment. This decision is influenced by several factors. The current cash position of the company is important. If cash is short and interest rates are high, it may be advantageous to lease the equipment so that it can be considered an expense item instead of a capital expenditure. Another alternative is to lease the equipment for a period of time with the option to purchase and apply part of the lease price to the purchase price.

There is another choice available, and this is the execution of what is sometimes referred to as a "third-party lease." The user purchases the computer from the computer company, resells it to the leasing company, and leases it back for a guaranteed number of years. The advantage of this arrangement lies in a lower lease cost since leasing companies are usually willing to take a lower return on their money in exchange for a fixed-term lease.

After the decision is made relative to equipment purchase, the user must consider how to handle the programming costs. Some companies take the position that programming is similar to research and development and the cost is expensed. Others consider programming as a capital item and depreciate its cost. Finally, some companies ignore the in-house programming costs and consider them as an overhead burden since programming is usually accomplished under a technical services department.

The next step is determining the depreciation period of the equipment and related capital expenses if it is decided that purchase is attractive. This decision then leads to the problem of which type of depreciation to take—straight line, sum of digits, double declining balance, or a combination of methods. This decision is again strictly financial and must be based on corporate policy.

Another decision to be made is the type of dollars to be used in computing the payout. The decision can be to ignore the variation of the dollar with time or, as so many companies are doing, to use a "time value of money" approach, sometimes called "discounted cash flow."

With these basic financial decisions made, computation of payout can proceed. There are many accepted methods of computing payout. One of the popular methods is through what is called an equation form of payout (Eliot and Longmere, 1962). This method takes the formula shown earlier for dollars profit and formalizes the various components of each of the summation terms into what are called economic factors. The payout time is calculated as the ratio of the fixed investment to the sum of the net profit corrected for federal income taxes and depreciation allowance. The influence of the time value of money can be incorporated by a stepwise calculation over given periods with summing of the cash accumulations on a discounted basis. This technique is an excellent method of computing payout because it pinpoints the various factors that exercise major influence on the payout period for given types of processes.

11.8 POSTIMPLEMENTATION STUDIES

Postinstallation activity for conventional data-processing computers is a continuing requirement. Volumes change, government reports change, payroll procedures change, and so forth. Although changes occur in processes, the control computer is not subject to as severe fluctuations in applications and requirements as is the commercial or scientific computer installation. The reasons for this are twofold. First, the control computer's primary job is to control the process, and after the first year most of the operating problems have been solved. If the system is properly designed and if it is accomplishing its mission, there is very little incentive for doing the job faster if nothing utilizes the excess computing capacity. Second, the application is on-line and real time. Operators hesitate to permit unnecessary disruption of control to allow engineers and programmers to debug or modify control schemes. In other words, once a control computer is operating properly, there is great resistance to anything that "rocks the boat."

There should, however, be a postimplementation study of the control computer installation. This study is often omitted because of lack of time or interest, but management should demand a complete evaluation of every application. Such a study verifies the basis of justification and the accuracy of the assumptions and payout calculations. This information is valuable in guiding future justification studies for other processes. Often as not in initial installations, the real payout of a control computer is found to be other than that predicted by the justification study.

The postimplementation study can also lead to further improvement in control. With the smoother control that is possible with the control computer, process interactions not detected before the installation may be discovered. This can lead to changes in the control strategy which can be implemented at little cost and which increase the payout of the application.

REFERENCES

Bakke, R. M., "Adaptive Gain Tuning Applied to Process Control," *Prepr. Instrum. Soc. Am., 19th Ann. ISA Conf. and Exhib.*, October 1964.

Briggs, P. B., "Computer-Controlled Chromatographs," *Control Eng.*, September 1967, pp. 75–80.

Brower, A. S., "The Spread of Computer Control in Steel Mills," *IEEE Trans. Ind. Electron. and Control Instrum.*, April 1966, pp. 16–23.

Eliot, T. Q., and D. R. Longmere, "Dollar Incentives for Computer Control," *Chem. Eng.*, January 1962, pp. 99–104.

Ewing, R. W., G. L. Glahn, R. P. Larkins, and W. N. Zartman, "Generalized Process Control Programming System," *Chem. Eng. Progr.*, January 1967, pp. 104–110.

Farrar, G. L., "Computer Control Registers Big Gains," *Oil and Gas J.*, October 25, 1965.

Griem, P. D., Jr., "Direct Digital Control of a Glass Furnace," *20th Ann. ISA Conf. and Exhib.*, October 1965.

Hodge, B., "Company Control via Computer," *Chem. Eng.*, June 7, 1965, pp. 177–180.

Koch, D. A., and J. C. Schildwachter, "How To Predict Compressor Performance," *Hydrocarbon Process. Petrol. Refiner*, June 1962, p. 151.

Sauer, R. N., A. R. Colville, Jr., and C. W. Burwick, "Computer Points the Way to More Profit," *Hydrocarbon Process. Petrol. Refiner*, February 1964, pp. 84–92.

Savas, E. S., *Computer Control of Industrial Processes*, McGraw-Hill, New York, 1965.

Stout, T. M., "Computer Control Economics," *Control Eng.*, September 1966, pp. 87–90.

Washimi, K., and M. Asakura, "Computer Control and Optimization," *Chem. Eng.*, November 21, 1966, pp. 131–136.

Williams, T. J., "Studying the Economics of Process Computer Control," *ISA J.*, January 1961, pp. 50–59.

Williams, T. J., "Process Control Today," *ISA J.*, September 1965, pp. 76–81.

Wrathall, E., "Linear Programming for Optimum Blast Furnace Burdening," *Blast Furnace and Steel Plant*, May 1966, pp. 399–404.

"Closing the Loop at Northwestern States," *Pit & Quarry*, May 1966.

"Computer Teams Up To Help Operators at Fuels Control Center," *Baytown Briefs* (company newspaper of Humble Oil and Refining Company), January 13, 1967.

"More Paper by Computer Control," *Chem. and Eng. News*, June 6, 1966.

BIBLIOGRAPHY

Andrews, A. J., G. D. Stacy, and P. U. Webb, "A Control Engineer Looks at the Impact of Plant Computers," *Proc. Texas A&M 20th Ann. Symp. Instrum. for the Process Ind.*, January 1965, pp. 15–22.

Bean, K., "The Forgotten Costs of Installation," *Autom. Data Process.*, August 1962, pp. 20–23.

Cline, R. P., "Keep a Proper Perspective When Considering Use of Computer Control," *Oil and Gas J.*, January 17, 1966, pp. 72–74.

Curry, R. B., "The Significance of Computer Investment Decisions," *Comput. and Autom.*, September 1962, pp. 8–10.

Fox, E. C., "Nature and Purpose of Computer Process Control," *Autom.*, October 1963, pp. 48–54.

Fraade, D. J., "A Review of Closed Loop Control of an 85,000 Barrels Per Day Crude Unit," *Proc. IFAC-IFIP Conf.*, Stockholm, September 1964.

Fried, L., "The Post-Implementation Feasibility Study," *Datamat.*, January 1966, pp. 47–50.

Gillings, D. W., "Process Control Economics," *Instrum. Pract.* (a series of six articles published from July–December 1964).

Jakubik, R. F., D. Kader, and L. B. Perillo, "Consider All Potential Savings When Justifying Process Control Computers," *Autom.*, March 1964, pp. 81–84.

Janes, J. D. W., "Measuring the Profitability of a Computer System," *Comput. J.*, January 1963, pp. 284–293.

Lasher, R. J., "Computer Controls Blending of Motor Gasoline," *Oil and Gas J.*, December 4, 1967, pp. 78–82.

MacMullan, E. C., and F. G. Shinskey, "Feedforward Analog Computer Control of a Superfractionator," *Control Eng.*, March 1964, pp. 69–74.

Mears, F. C., "Success is Word for Computer-Controlled Chromatograph Unit," *Oil and Gas J.*, December 18, 1967, pp. 90–93.

Moore, J. F., "Guidelines for Computer Control in New, Existing, and Multiple-Unit Plants," *21st Ann. ISA Conf. and Exhib.*, October 21, 1966.

Shirley, D., and B. J. Webber, "Choosing a Computer—Rent or Buy?" *Data and Control*, January 1966, pp. 24–27.

Skalnik, C. R., "Centralization + Modernization + Computers = Profits," *Proc. Texas A&M 21st Ann. Symp. Instrum. for the Process Ind.*, January 1966, pp. 9–14.

Stout, T. M., "Management Aspects of Process Control Computer Installation," presented at Ann. Conf. TAPPI, February 24, 1966.

Stout, T. M., "Economics of Computers in Process Control; Part I: System Justification and Evaluation," *Autom.*, October 1966, pp. 82–90.

Tivy, V. V., "Direct Digital Control in the Process Industries," *Proc. Texas A&M 21st Ann. Symp. Instrum. for the Process Ind.*, January 1966, pp. 5–8.

Webb, M. S., "Justification for Control Computers," *Chem. Eng. Progr.*, October 1965, pp. 83–86.

Williams, T. J., "What to Expect from Direct-Digital Control," *Chem. Eng.*, March 2, 1964, pp. 97–104.

Willis, V., "Computers in Process Control," *The Eng.*, November 6, 1964, pp. 759–762.

"Computer Gives Quick Payback in Bleach Plant Control," *Pulp & Pap.*, October 23, 1967.

"A Computer to Cope With Catastrophe," *Eng.*, October 9, 1964, pp. 450–451.

"Hot Strip Mills—Computer Control Begins To Pay Out," *Control Eng.*, December 1963, pp. 24–25.

Chapter 12

System Selection

I. C. (JIM) TRAUTH, JR.

The selection of the control computer is a difficult and crucial task in any computer control project. The characteristics of competitive offerings must be compared and evaluated in terms of their ability to satisfy the requirements of a particular application. The problem is compounded by the many features that are available, the typical requirement for special features, and the lack of common specifications and terminology.

In general, the selection process consists of three basic steps: First, the system must be carefully defined to ensure that both the prospective vendors and the person(s) responsible for the selection have established a common base of understanding. In many cases, this system definition will, in fact, take the form of a definition of the process and the control problem to be solved. The second step in the selection process is the collection of data concerning the available systems. The estimated price

779

of the system is but one of the necessary pieces of data. Information concerning the projected performance of the system, functional descriptions, specifications, and other information such as the results of software comparison tests must be compiled.

Once the system information is available, the basis and the method of comparison must be established. The basis for comparison is a matter of defining and weighting the importance of various machine capabilities and features. In addition to the information concerning the system itself, other factors must be considered. These include such things as the financial situation in the company, its long-range plans, and intangible factors relating to the system and the vendor.

In this chapter the selection process is discussed in terms of these three basic steps. Since there is no single "right" selection method, the chapter considers important factors that are sometimes overlooked without attempting to define a single selection procedure.

12.1   SYSTEM DEFINITION

To select a system, it is first necessary to prepare a reasonably detailed description of the job and computer requirements. The preceding chapter discussed the justification for computer control. While preparing the justification, the investigating team would of necessity have considered the control task and the computer requirements. As a result, their study and system description can be used as the basis for the specification of the computer system.

The manner in which the initial control and system descriptions are transformed into selection criteria depends, in part, on the experience of the user. For the experienced company that has developed a cadre of computer control personnel, the control task description can be used as the basis for writing a specification for the control system. The performance requirements that form a portion of this specification are, in fact, the selection criteria.

For the inexperienced company, the task is somewhat more difficult. In general, the best approach is to concentrate on describing the control task to be accomplished. Then, with the aid of technical personnel from prospective vendors or a consulting firm, this information can be transformed into a specification on equipment requirements. If the control job or the process is proprietary, special precautions are necessary to ensure adequate information security.

An important aspect of the system specification is the definition of inputs and outputs, both analog and digital, which form the interface between the computer system and the process. The price of the computer features necessary to monitor the inputs and control the outputs often represents a substantial portion of the total computer price. In addition to the system price, there are also other costs associated with these inputs and outputs. These costs may vary from those for a single resistor that converts a current signal into a voltage, or the cost of wiring a signal to the computer, to the case where an instrument must be installed to transduce a pneumatic signal into an electrical signal, or to where a special instrument must be designed and built for the particular application.

A list of the probable inputs and outputs should be prepared. For analog inputs, the first entries on the list should probably be those inputs to be used for control. In addition to the input from the primary process variable, an input should be included for setpoint feedback. Since process dynamics enter the picture and the results of control action are usually delayed before they can be sensed from the process variable, setpoint tracking can be done using the setpoint feedback. Those analog instruments required for material balance should also be added. After the control variables, the next most obvious variables to include are those which indicate alarm conditions, that is, those inputs required to indicate that a variable has exceeded its normal or safe operating limits. It may be necessary to correct some flow measurements for temperature and/or pressure, and an appropriate indication of these variables should be available to the computer. If the temperature and pressure are subject to frequent variation, they should be on the analog-input list. If, however, the variation is slight or infrequent, the operator can enter the information into the computer only when significant changes have occurred. Finally, those variables to be used for logging, special reports, and operator information should be added to the list.

A form such as that shown in Fig. 12-1 may be useful in defining the analog inputs to be read by the computer. This serves the function not only of giving an accurate count of the inputs required but can also be used by the personnel defining the subsequent models and conversion routines for the analog-input subsystems. In addition to merely identifying the various analog inputs, information concerning their characteristics should be noted in the table. As a minimum, this should include the required scanning rate, the signal level, and special conditions such as the presence of a common mode voltage. This information is useful when the vendor offers several analog-input subsystems having different characteristics. By grouping the inputs according to scanning rate, it may be found that it is possible to use a small high-speed analog-input sub-

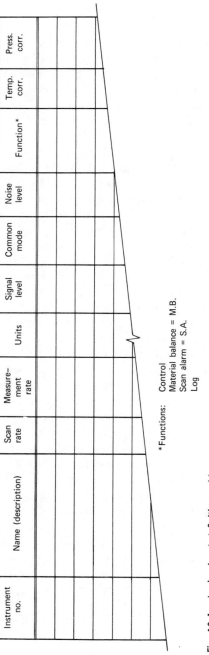

*Functions: Control
Material balance = M.B.
Scan alarm = S.A.
Log

Fig. 12-1 Analog-input definition guide.

system for those inputs requiring a high scanning rate, whereas the remainder of the inputs can be read using a slower, but less expensive, subsystem.

The second job is to define the digital inputs to the computer. These can probably be separated into three categories: signals that are continuous and are to be read by the computer at regular intervals; signals that require immediate response upon closure, such as alarm contacts; and event items that are to be counted by the computer. Machinery rotation, turbine flowmeters, unit counters and so on, fit into this category. In the case of these pulse-type instruments, the maximum pulse rates should also be noted since the vendor may offer features with varying speed capabilities.

The digital inputs from the process that require monitoring at regular intervals include process valve limit switches, setpoint "on-" and "off-computer" control switches, and contacts to indicate the position of multiposition valves such as those associated with multistream analyzers. In some case, it may be more economical to indicate a variable as multiple discrete digital inputs using devices such as level and pressure switches rather than as an analog input to be interpreted by the program.

The second category of digital inputs, those requiring immediate response, usually are the process interrupt signals. These include level, temperature, and pressure alarms. As another example, a signal from a chromatograph equipped with a peak-picker can signal the computer when the peak is ready to be read. In the case of events that happen less frequently than every second, such as traffic counting, it may be more economical to use interrupts than pulse counters.

In addition to these inputs derived from the process, the digital-input requirements for man-machine communications must be enumerated. As discussed in Chap. 7, many of the operator actions are conveyed to the computer as digital-input signals. For example, status requests and operator response to an alarm condition are often indicated by a switch position. In order to minimize operator annoyance caused by a delay in response, many of these digital inputs should be handled as high-priority interrupt signals. Thus the needs of the man-machine interface should be defined in order to estimate the digital-input signals that must be handled by the computer. A form similar to that shown in Fig. 12-2 is helpful in identifying and listing digital-input signals.

The next area to consider is analog output. The first task in this area is to list the process variables to be controlled by the computer which require output signals. If the computer was justified on the basis of better or more optimum control, those control variables used in the justification will probably be the first on the list. Following this, the

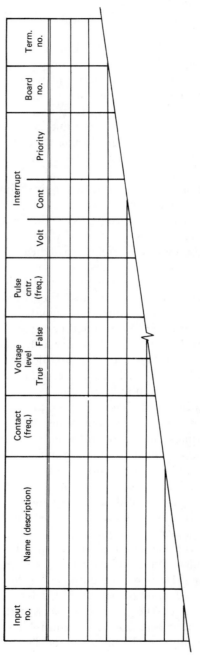

Fig. 12-2 Digital-input definition guide.

784

overall plant should be studied again to determine the effects of changing these optimizing control variables. If these changes necessitate
changing the control point of other control variables upstream or downstream, these other control variables must be added to the output list.
Although many variables do not affect the economics of the plant directly, they must be adjusted regularly to maintain smooth operation of
the plant (for example, flow out of intermediate storage vessels) and
these should be considered for computer control.

While making the output list, additional information should be collected. Does the existing controller accept a remote setpoint? If so, what
level and type of signal does it require: pneumatic, voltage, or current?
Many controllers manufactured today include stepping motors or the
provisions to add them. Others include, or have provisions for, memory
amplifiers which may be set remotely. A stepping motor is a small pulse
motor that is attached to the setpoint shaft of the controller and allows
the setpoint to be changed with a pulse train. As discussed in Chap. 5,
a memory amplifier accepts a short duration analog-output signal from
the computer and holds the signal, thus allowing many controllers to be
set from a single analog-output circuit in the computer. The digital
output that may be required to select the appropriate memory amplifier
should not be overlooked.

In addition to these analog outputs primarily concerned with the
direct control of the process, other outputs are frequently required for
man-machine interface and communication. These signals are used as
inputs to graphic display devices which assist in describing the status
of the process to the operator. Typically, the graphic devices used for
this function are strip chart recorders and oscilloscopes. Because of the
difference in the speed of response of these devices, the analog output
requirements can differ appreciably. For this reason, the characteristics
of the output signal, as well as the number of signals, should be
determined.

The other form of computer output used to control the process and to
provide man-machine communication is digital output. As discussed in
Chap. 6, digital output provides the capability to close relays in the
process or, if the relays are included in the system, to turn process loads
on and off. These outputs are frequently required in process control to
actuate devices such as motors, tap selectors, and indicator lights. In
addition to the number of required outputs and their frequency of use,
the voltage, current, and transient characteristics of the load should be
noted. These factors can be significant data since, for example, heavy
motor loads may require interposing relays or contact protection circuits,
and these can contribute to the system cost.

Other forms of digital output that may be required in the process include register output and pulse duration or pulse train signals. Register outputs are used to transfer multibit words of information between the system and another system or between the system and peripheral equipment. For example, it might be used to transfer data from the computer to a special-purpose strip printer or to communication equipment which links the process control computer to a centralized computing center. The signal levels, speed, terminations, and matching requirements for these outputs should be defined since they may require special matching circuits or other modifications to the standard computer system.

Pulse duration or pulse train signals are required by certain types of control elements. For example, motor driven setpoints frequently require pulse duration signals for actuation. In some computers, pulse duration and pulse train signals are provided by programming the standard contact operate feature in conjunction with a real-time clock. If a hardware pulse duration and pulse train feature is available, then the trade-off between the hardware and software implementation of this function must be considered. In any case, parameters such as speed, level, and termination characteristics are important factors.

As seen from the above discussion, the definition of the process input/output equipment is primarily a matter of identifying process and man-machine interface signals and describing their characteristics. This must be done carefully, however, since the cost of process input/output functions often represents a substantial portion of the computer system cost. The prudent planner usually includes a contingency factor in his estimates. For a fairly well-defined installation which, perhaps, is modeled after an existing unit, the contingency may be less than 10%. In a new or poorly defined process, the contingency may be as high as 30%. As the control problem is refined in subsequent preinstallation negotiations and investigations, this level of contingency can be reduced. It is a rare installation, however, that does not require additional process inputs and outputs up to, and through, the installation and checkout phases of the project.

In addition to the process input/output requirements, factors relating to the computer performance must be defined. These include computer speed, storage capabilities, peripheral device requirements, and programming system capabilities. In some ways, these are even more difficult to define than the process input/output features because they involve more than merely counting the required inputs and outputs. As pointed out in the various chapters on programming, estimating programming effort is very difficult. As a corollary, estimating the size of a program is equally difficult and many of the required characteristics of the computer depend on the size and frequency of execution of the control programs.

For the experienced user, the definition of the control problem may take the form of a detailed description of the computer control system itself. This might include, for example, definitions of the storage size, access times, instruction set size and content, and the existance of special features such as hardware storage protection and high-efficiency compilers. This type of definition is difficult to prepare because it requires detailed knowledge and interpretation of computer design and use. In addition, such a description may be unduly restrictive in that it often presupposes a solution to the control problem and prevents the vendor from suggesting a solution that might be more economical.

An alternative to this type of machine description is a statement of the control problem to be solved and the functional requirements and constraints that must be satisfied. In this approach, the required calculations can be grouped by function and identified with respect to required performance. For example, control requirements might be described in terms of the number of loops that are to be controlled using a two-mode control algorithm and the frequency of the control calculation. Similarly, the data to be stored can be described along with their retention time, access frequency, and other pertinent factors. As another example, the relative importance of alarm conditions can be defined along with an indication of the corrective action required for each type of alarm.

A functional description of this type has several advantages. First, it allows a description of the control problem in the language that the user probably knows best, the control language of the process. It leaves the description of the machine details to the computer vendor, an expert in describing computer systems. Second, and equally important, it allows the computer vendor some flexibility in satisfying the control requirements. As a result, he may be able to take advantage of unique characteristics of his system and provide a more economical solution to the control problem.

As a specific example, a specification may be written to provide a multiplexer for 500 analog inputs and a scan rate of 600 points/sec. This would require a relatively expensive high-speed multiplexer for 500 points. The control technique, however, may actually require limit scanning of 50 critical variables at 10 times/sec and all 500 variables every 5 sec. Thus the high-speed multiplexer need only have 50 points, and the other 450 points could be connected to a less expensive low-speed multiplexer. Similarly, a digital-input scan rate of 1000 words/sec may be specified to determine when a switch closed within 1 msec. A more satisfactory solution might be to connect the switch to a process interrupt.

This description of the system definition is, of necessity, brief and general. In a situation where hundreds of characteristics and features of a control system must be considered and matched to the requirements of a particular control problem, it is not possible to provide a cookbook answer to system definition. The key points are that the user must understand his process problem and identify crucial parameters that must be observed and controlled. In addition, he must remain flexible in his concept of the computer system to allow the vendor to assist him in defining the best solution to his problem. Defining a computer control system is an iterative process in which, ideally, the general requirements of the control problem are refined in a series of discussions between the user and one or more prospective vendors. The effective interplay of these parties can result in a computer control system that provides an optimum cost/performance trade-off for the particular process to be controlled.

12.2   SELECTION METHODS

The criteria for selecting a control computer may be simply stated as the requirements for the best system to do the required job. There are many excellent companies that offer sound hardware and software. The real problem is to select the vendor who offers the equipment and services that best fulfill the requirements of the particular control problem. Emotions have no place in selecting the vendor, nor does the microsecond speed of the central processing unit alone. One computer may compute twice as fast as another, yet the advantage is lost if the speed cannot be utilized because of an I/O limitation. Similarly, a system can have overall hardware advantages that are all but worthless if programming systems are unavailable to utilize these features.

Several different methods may be used in selecting a vendor and equipment. One method is to invite several vendors to describe their equipment and to discuss the specific control requirements. If convenient, these discussions might be held at the vendor's plant site since this might allow inspection of his manufacturing and test procedures. During the discussions, or shortly after, the vendor should be asked to supply an approximate equipment and software list and prices. Applicable manuals, brochures, and seminar training courses should be requested. The vendor should also be asked to supply the names of several installations in the area that might be visited. The list may not be very long because many users do not care to discuss their computer installation for proprietary reasons; others fear becoming tour agencies for computer vendors. Often these visits can be more easily arranged between the companies involved, especially if some kind of trade agreement exists. Some few companies gladly show their computer installation for the advertising value they get from such demonstrations. Another source of information about a vendor's equipment, software, and services are user and professional society meetings.

After evaluating the data presented, the user should be able to list the vendors in order of preference. The selected vendor can then be contacted to work out details of system configuration, final price, implementation plan, and contractual arrangements.

Another approach is to prepare a general specification and send it to vendors for preliminary bids. On the basis of these bids, a single vendor can be selected for further discussions on a detailed system specification. Alternatively, several vendors can be requested to present their system details before a decision on the specific vendor is made.

A method often used in selecting a vendor is to solicit competitive bids in response to a detailed specification. Stout (1966) includes a discussion on specification writing. A necessary prerequisite to writing a specification is the complete description of the system discussed in the previous section. The main purpose of the specification should be to convey to the vendors the job to be performed by the system, the ground rules for performance, and the interface for information transfer. With this information, the vendors may then bid their equipment which best meets these requirements. If hardware details are to be specified, there should be a statement in the specification preamble to the effect that, if the hardware and functional specifications conflict, the functional specification should govern the response. The specification should be worded in such a way that narrative rather than strictly numerical responses are welcomed. This effectively gives the vendor the opportunity to state his case as to why he proposes a particular solution.

Still another method that a company can employ in selecting a vendor is to hire the services of a consultant to study the job and recommend the equipment. There are both highly qualified consultants and uninformed consultants; there are both objective consultants and biased consultants. The selection of a consultant is almost as difficult as the selection of a computer system. However, if a company has confidence in a particular consulting firm and the projected payout can stand the overhead, their services can be invaluable.

In all the selection processes described above, the user should be trying to identify many factors. Is the hardware being considered capable of doing his job? What are the characteristics of the central processing unit and its auxiliary storage, process I/O, and peripheral devices? Is the system flexible enough to be changed should the job definition change? What are the environmental and space requirements? What software is available for the system? What is standard and what is provided for an additional fee? Is the proposed system in operation? If not, when will it be available? What training is available? Where and at what cost? What vendor services such as system and installation planning, programming, analysis, and maintenance are available and at what cost? Does the vendor have a history of installation problems? If so, an acceptance test may be required. Does the vendor consistently meet his delivery commitments and availability quotes? If not, penalty terms may be required in the contract.

In the final analysis, it is the purchaser who must make the decision. Any of the methods mentioned above can be used. However, there is no substitute for first-hand knowledge of the vendor, his equipment, his programming systems, and his services because each of these four items is an ingredient of a successful installation. Even with the mechanical selection techniques such as decision tables and benchmark problems, no vendor will be unquestionably superior to another on every point, and the decision will require intuitive judgment. Thus one of the most useful aids to equipment selection is an informed, objective evaluation. The best way to get the background necessary for this type of evaluation is through education and surveys of working installations. The employees doing this must remain aloof from sales pressure, ask penetrating questions, and be able to recognize meaningful demonstrations.

12.3   COMPARISON TECHNIQUES

The previous section discusses the general procedures that can be used in selecting a vendor and, as a result, a system. It does not, however, provide guidance as to how to reach the decision; that is, it does not supply a technique for comparing competitive solutions to the control problem. The problem solution does not consist solely of hardware, or software, or vendor capability. It is, rather, a complex mixture of these and other factors. As a result, there is no simple method for comparing the proposed solutions. Following a brief discussion of specsmanship, this section discusses two comparison techniques. One, although quite common, is not highly recommended because it oversimplifies the procedure. The second is a modification of the first and attempts to weight the factors involved in the comparison in proportion to their importance in the total solution of the control problem. Neither is presented as being the only method, but the discussion should assist the user in developing a comparison method and avoiding the pitfalls inherent in some approaches.

In Chap. 4 there is a section entitled "Specsmanship." Specsmanship is the art of writing and interpreting specifications. The use of such a term is not meant to suggest subversion or deception, as might be inferred from its similarity to the often maligned term "gamesmanship." Although vendor and user alike are sometimes accused of specsmanship, a detailed investigation almost always reveals that the source of the problem is a difference in the definition being used by the opposing parties, rather than any attempt to deceive. The source of the misunderstanding is primarily the result of a lack of standard definitions and specification format in the computer and data acquistion field. Although some minimal standards have been accepted by various professional and standards organizations, their general acceptance is quite limited and can be expected to remain so for some years.

Because of the lack of standardization, anyone who attempts to compare competitive offerings is cautioned that the comparison must be made using the same definitions for equivalent specifications. As a specific example, the accuracy specification for an analog-input subsystem may be quoted as 0.03% of full scale on the 50 mV range; a competitive system may be specified at an accuracy of 0.05% of full scale. Whether one system is superior to the other requires further investigation and definition of terms. The 0.03% specification, for example, may refer to a mean accuracy, whereas the specification for the other system may be a total accuracy figure which includes the effects of noise and limited temperature excursions. Sometimes these differences can be determined only after detailed discussions between the vendor and user.

Software is not immune to specsmanship. The terms "monitor," "executive," and "scheduler," for example, can refer to programs that perform the same basic function, or a scheduler and monitor may be only portions of an executive. Similarly, a compiler that requires an additional 2000 words of storage or an extra 2 sec of compilation time is not, *per se*, inferior, since it may produce more efficient object code which results in significantly faster object program execution.

Although difficult, it is of paramount importance that differences such as these be identified if an accurate comparison of competitive solutions to the control problem is to be made. Experience is extremely valuable in identifying these factors, but a thorough study of all aspects of computer control systems provides the base for the inexperienced user and supplements the experience of the sophisticated user.

One of the most widely used methods for evaluating hardware is the use of a large table which lists characteristics as row entries and in which a column is provided for each computer type (Pardo and Vance, 1966). This technique is a good first step but it cannot possibly provide an adequate evaluation because it is impossible for a single entry, such as "yes" or "no," to adequately compare computer characteristics.

Consider, for example, one entry which appears in every table: "Number of index registers?" One might think that if one computer has three index registers and another has only one, that the former computer has a definite advantage over the other. This may or may not be the case. If the index registers are main storage resident in the system with three registers and are implemented in hardware in the one-register system, there may be little difference since the indexing with main storage resident registers takes appreciable time whereas the hardware register indexing is essentially time-free. Another consideration is how the index registers are used. If the executive program of one system requires more dedicated index registers than the other, a numerical advantage can be lost. On the other hand, if the dedicated use of more index registers reduces the system overhead, the advantage may be regained.

Another question that often results in misleading answers is "Number of wired instructions?" Basically, the maximum number of wired instructions in a machine is determined by the number of bits in the operation code portion of the instruction (4 bits = 16, 5 bits = 32 codes, etc.). The other bits in the instruction word typically produce modifications to the instruction (indirect addressing, indexing), operand address, and execution information (number of shifts, jump or skip conditions, etc.).

Sometimes a special form of an instruction may be as important as the standard form. For example, a special form of the Modify Index instruction may be used to modify a storage location, thus allowing direct modifications of an indirect address or allowing any storage location to be used as a counter. A special form of the Branch or Skip instruction may be used to reset conditions or interrupt masks as the test, or after the program, is completed. Similarly, some computers have an instruction for each input/output device action, whereas others use a single instruction to call input/output control words.

Depending on the interpretation the proposal writer applies to the question, a machine with a 5-bit operation code could have from 20–150 instructions when special instruction forms and the effect of modification bits are considered. The only effective way to compare the machine instruction sets is not by number but rather by evaluating the functions performed and determining their usefulness in a particular application.

Another area where a table answer provides minimum information is the number of registers: arithmetic, control, buffer, general purpose, and so on. If the evaluator hopes to gain any useful information from this type of data, he must study the computer data flow.

The questions "Priority Interrupt: yes/no; Number of Levels = _____?" provide minimum information concerning the interrupt structure of a machine. The term "priority interrupt level" probably has more definitions than any other three-word phrase in use today, not only between vendors but even within a single vendor's organization. To determine the effectiveness of the interrupt structure of a machine, the evaluator must study the descriptive material concerning this feature and relate it to the particular control problem being considered.

Another "yes/no" entry that does not tell the whole story is "cycle stealing." This entry is meant to compare different systems as to their ability to process I/O information through stealing cycles from the normal computational operation in progress. The real question is how many cycles are required to pass a piece of information from the input to the computer memory or from the computer memory to an output device. These important aspects of channel operation are lost in simple yes/no questions.

The table normally includes some entries related to programming systems but the number of questions is typically very small. In general, these questions are broad in scope and not definitive. For example, the yes/no reply to "Real-time monitor?" cannot possibly answer such questions as the amount of main and bulk storage used by the monitor, the philosophy of its design, the flexibility of time-sharing or multiprogramming features, or the many other attributes that monitors may or may not have. The important question of software comparison is considered in more detail in a subsequent section.

Table entries can provide quick answers to some quantitative questions such as storage capacity and speeds available and/or bid, I/O device speeds, maximum number of process I/O points available, the speed of the multiplexer, and the instruction execution time. In most cases, however, the table entries must be supplemented with operational data if they are to be used for comparisons.

An alternate comparison technique which relates the system parameters to the specific control problem is a refinement of the table discussed in the previous paragraphs. The primary considerations in the comparison are divided into categories such as vendor, hardware, software, services, and miscellaneous. Each category is assigned a weighting value which represents its estimated effect on the successful completion of the project.

Once these weighting values are assigned, the categories are subdivided as much as possible. For example, the vendor category includes reputation, stability, experience, and location. Hardware includes such factors as main storage size and speed, bulk storage characteristics, instruction set size and speed, I/O capabilities, and so on. The software category includes executive characteristics, I/O subroutines, time sharing, multiprogramming, application programs, and so on. Services is subdivided into training, installation, programming assistance, maintenance, and so on. The miscellaneous category accounts for those items that do not fit into one of the other major categories. The weighting assigned to each major category is then prorated among the subdivisions according to its effect on the control problem solution.

Each vendor and his proposed equipment are rated in each subdivision using an arbitrary scale. For example, if application programs available from the vendor satisfy 80% of the anticipated requirement, a value of 80 is assigned to this subdivision of the software category. Similarly, the complete lack of a feature results in a 0 entry. The sum of the products of the ratings times the subdivision weighting values provides a single weighted average index which is a fair indication of the proposal effectiveness. If this effectiveness index is divided by the cost of the system, the evaluator has an effectiveness per dollar figure to aid in his system selection.

Although the use of weighting values and assigned ratings appears somewhat arbitrary, this technique has been used effectively by users. The seemingly arbitrary nature of the procedure is modified by the averaging effect of the many factors that can be considered and the weighting factors selected. As previously indicated, no single system is obviously superior in every respect, and this technique reduces the competitive evaluation to a single index which is related to the application through the weighting factors.

The method has another inherent advantage which, in part, accounts for its effectiveness as a comparison technique. The assignment of meaningful weighting factors requires careful consideration of both the application and the system characteristics. This naturally tends to focus attention on heavily weighted factors, and these are generally assigned relatively accurate weights as the result of the thorough study. Thus the weighted average index accurately reflects the factors most important to the success of the project. Unimportant factors, even though they may be assigned grossly incorrect ratings, tend to lose significance in the overall index because of their low weighting factors.

The basic selection techniques discussed in this section are not the only ones that are employed. More sophisticated techniques such as decision tables and trees are also effective. The discussion emphasizes two important factors, however, that must be considered by the user in any method of comparing systems. First, the simple yes/no comparison technique is rarely adequate because of the many factors that must be considered and the fact that no single system is superior in all aspects of performance. Second, it is important that the comparison procedure reflect the relationship between the particular control problem and the proposed system solution.

The discussion of this and previous sections provides information concerning the three basic phases involved in the selection of a vendor and system for the solution of a particular control problem. In the remaining sections of the chapter, specific considerations and factors of concern in the comparison of proposed system solutions are discussed. These include hardware-software trade-offs, the place of software tests in the selection procedure, the ability to expand the system, the question of buy versus lease, and the intangible selection factors.

12.4 HARDWARE-SOFTWARE TRADE-OFFS

The design of a digital computer involves many decisions concerning whether a function should be implemented in hardware or whether it should be performed by means of a series of program instructions. Every function provided by the software can be implemented directly in hardware; similarly, almost every hardware function can be provided through software written in terms of a few basic instructions. The wired program machine represents one end of a spectrum at which the hardware is used for the implementation of the total system function. An analog controller is an example of this type of computing device. The two- or three-mode control algorithm calculation is determined solely by the interconnection of electronic components and is not alterable except by rewiring or adjustment of components. At the other end of the spectrum, a computer can be built in which the only executable instructions are the basic logical functions such as AND and OR. In this computer, all arithmetic instructions, for example, must be reduced to a program written in terms of these few basic instructions. A more realistic example of a machine near this end of the spectrum is a digital computer in which there is no hardware adder. All addition, subtraction, and multiplication operations are performed by a table look-up or a subroutine.

Between the end points of this hardware-software spectrum, there is a continuum of trade-offs which must be considered by the computer designer. Each of the trade-offs is associated with a performance factor and a cost factor. For example, the use of program subroutines to implement floating point arithmetic operations results in a performance penalty in that the execution time is generally greater than if the operations were implemented in hardware. On the other hand, the hardware implementation is relatively expensive. The benefit of increased speed performance and the elimination of the need for floating point subroutines must be compared with the hardware cost.

Included in the evaluation of these various trade-offs are other factors such as design philosophy and architecture of the computer, compatibility between the computer and other computers, its intended application, and the distribution of manufacturing and development costs within the vendor's organization. Thus the answer to each trade-off question is rarely obvious and different vendors, or different groups within a single vendor organization, are likely to make different choices. For this reason, the comparison of competitive solutions to a given control problem often requires consideration of hardware-software trade-offs. Using the floating point example, the user must decide whether a machine offering floating point hardware is justified on the basis of the required performance for his application. In this section, some of the more common hardware features of current generation computers having equivalent software implementations are briefly considered.

Since a process control computer spends a good percentage of its time communicating with the outside world via analog and digital inputs and outputs, one of the first areas to be considered is how this information is brought into, and transmitted from, the computer. Three methods are generally used today. The first, which requires the maximum amount of hardware, is a storage channel where many input or output values are transferred directly between main storage and peripheral devices as a result of only one or two instructions. This is generally known as a cycle stealing or direct storage access channel. A second alternative, requiring slightly less hardware and more programming, is a method to read each point or group of points directly into storage. The third alternative is for all access to and from storage to be through the accumulator.

The channel is basically a path to and/or from the main storage of the computer and its associated control circuits. In some cases an individual channel is subdivided into subchannels. The relative speed of the channel and the interference it creates on the normal processing is dependent primarily on the channel control circuit configuration. Specifically, part of the channel controls must include a storage address register whose contents indicate where the information being transferred is to be stored. In a high-speed channel, these storage address circuits are implemented using high-speed circuits. Thus the address is available when the data are ready for transfer. In the case of lower speed channels, the storage address register may itself be stored in the main storage. In this case, additional storage access cycles are required to access this address register, increment it to prepare it for the next word, and re-store it. Although the programming aspects of both types of implementations remain roughly the same, it is important to determine the degree of interference created with the program running in the processor. Many channels also have word count registers associated with them, and the same consideration must be given to these registers as is given to the address register.

The use of the direct storage access channel allows many inputs to be read with one program instruction. In addition, this type of channel may allow the individual readings to be brought into storage as a result of external stimuli or sync signals. When the channel has completed its operation, a program can then be run to perform whatever processing is desired. Thus this type of channel can relieve the programmer of having to issue commands for each input and output operation. Direct storage access channels are particularly useful when many input and/or output values are to be transferred to and from storage or where the information transfer is to be done at high speed. This channel may also be useful, however, on slow-speed devices where many readings are to be collected before a program is run to process the data.

The second method of getting information from a peripheral device is by means of a channel which provides a direct Read or Write from storage. In this type of channel, an instruction is required each time storage is accessed, but many pieces of information can be read into storage before stopping to process the data. While this method of data transfer requires programming steps for each data transfer, it may allow better organization of the program. To be fully effective, the reading of data directly into memory should not affect the accumulator or arithmetic registers.

The third method of bringing information into the computer is through the accumulator. This requires additional programming steps to load or store each piece of information into storage. This method does offer the advantage, however, of having the data in the accumulator ready for processing.

The next area of consideration is the method by which inputs are read; that is, are they always, or can they be, read randomly and/or sequentially? Random addressing allows the reading of inputs in any order; thus inputs may be read in the order that is most convenient for processing and, in addition, the sampling rates can be adjusted for each particular point. This method does require that the address of the point be set for each reading, which calls for additional programming steps and storage space. The ability to read inputs in random order is probably most important in the area of analog input since the frequency of processing generally varies with time and from input to input.

An alternate method is to read the inputs in sequential order. When all inputs have been scanned, the accumulated table of values must then be scanned to select the values to be processed. Modified sequential scanning arrangements, in which a scan of a predetermined number of points is the result of a single program instruction, are also offered. The best solution is probably to have a system in which both options are available under control of the program.

Another area of concern in most process control systems is data comparison. The data from analog-input points are generally compared against a high and low limit value. One method of accomplishing this is to have hardware threshold circuits on each input point. These circuits generate an interrupt every time the process variable crosses the limit. This method has the disadvantage that hardware is required for each input point and, in addition, it may burden the interrupt system of the machine. Another method is to accomplish the comparison by means of Compare instructions, but this involves transferring each piece of data into the accumulator.

An alternate method is to have a hardware digital comparator incorporated in the analog-input subsystem control unit in which the comparison limits are changeable by the program. This allows the comparator to be shared by all analog inputs under program control and also allows the limits to be manipulated according to current process conditions. Since the accumulator need not be involved in the comparison operation, this method reduces the computation and data-handling burden of the central processing unit.

In addition to limit scanning, the digital comparator offers other programming economies. Since a process variable must be converted into engineering units before it can be utilized by most application programs, the analog-input values are usually converted each time they are read. However, if the analog input is staying close to its previous value, the raw data can merely be compared with the previous reading using the comparator and the conversion program run only when the data have changed by a predetermined amount. The comparator can also be used to indicate when certain control programs are to be run since control action is only necessary in response to process changes.

A software method of analog-input comparison is through the use of a Compare instruction. This instruction causes the program to take one branch if the value in the accumulator is greater than the value in storage with which it is compared, another branch if it is less, and a third branch if it is equal. If the Compare instruction is not included in the computer instruction set, limit testing is done by subtracting the analog value from its upper limit and then branching if the result in the accumulator is negative. If the result is positive, the lower limit is subtracted from the analog reading, followed by a branch if this result is negative. This requires more programming steps than the use of a Compare instruction. Both methods, however, require some programming steps and access to the accumulator for each comparison. The advantage of the method is that special analog-input hardware is not required as it is in the previous cases.

In the case of digital inputs, a different type of comparison is generally required. It is usually desired to know whether any digital inputs have changed state—that is, whether any inputs that were previously "on" have switched to "off," or vice versa. Detecting the change of state in a group of digital inputs can be done with a logic function called "exclusive or" (EOR), which can be implemented in either hardware or software. When an EOR function is performed between a previous input pattern and a new pattern, those inputs that have changed state are uniquely identified. When using software, the program must determine whether the change resulted from a switch closing or opening. This is accomplished by an additional logical function called an AND or EXTRACT. When the AND is performed with the bit pattern that resulted from the EOR operation and the new input pattern, those inputs that have changed from "off" to "on" will have their bits "on" since the AND instruction results in "on" bits if the corresponding bits are "on" in both operands. To determine which inputs have switched to the "off" state, the AND instruction is performed with the bit pattern that resulted from the EOR operation and the old input pattern.

When the inputs that have changed have been isolated and the type of change has been determined, the program to be run in response to the change must be identified. Several methods are available to perform this identification. One method is to have a table in storage containing a series of words with a single bit in each available location. A series of instructions ANDs the input pattern with each bit position and continues on zero or branches to the appropriate service routine when the bit is located. An alternate method is to shift the bit to an identifiable position such as the lowest order bit (odd) or the highest order bit (negative sign). To implement this, the program would consist of a series of shifts and branches. The use of an index register reduces both methods to a loop operation. If the instruction set includes a Shift and Count instruction, the bit that has changed can be identified quickly through the use of this instruction. This instruction generally operates in conjunction with a shift count register or index register, so the handling of these registers must be considered in program estimates.

The above discussion involves handling the digital inputs in groups. Some computers offer instructions which allow the addressing of individual digital inputs and operations on individual bits in the accumulator. Where these instructions are not available, the individual bit can be obtained from the group by the AND instruction and a series of mask words.

An appropriate question at this point is, "Why not provide hardware to compare for change of status similar to the analog comparator discussed earlier?" Such hardware, in essence, provides the same function as that provided by the interrupt feature on most computers. The interrupt generally provides this function without the interference of cycle stealing. In addition, the interrupt circuit latches (or "remembers") the input state so that it can be serviced even though it may change state again before the servicing program can be executed. This approach requires additional hardware, whereas the software method of identifying digital-input status requires additional programming. Since both methods are generally available in a control computer at different prices, the user must make the hardware-software trade-off decision when he assigns inputs to the system. If immediate response to a status change is required, then the higher cost of the interrupt point probably can be justified. On the other hand, the standard digital-input feature in conjunction with software generally provides adequate performance for most status indicators.

The use of external (i.e., not a part of the computer system) interrupts is unique to process control computers and offers additional opportunity for hardware-hardware, as well as hardware-software, trade-offs. Basically, the interrupt hardware suspends operation of the normal sequence of instructions to allow special instructions to be executed. The instructions executed by the interrupt hardware must, by necessity, be those that do not disturb the registers of the central processing unit or those that cause the contents of these registers to be saved. For example, some computers having the interrupt hardware execute an Add One to Storage instruction. Another interrupt response instruction can be a transfer between storage and an input or output device implemented such that the accumulator or other data registers are not involved.

Most frequently, however, the interrupt hardware forces a Branch instruction and saves the contents of the instruction register. The interrupt hardware may also save the contents of some of the arithmetic and control registers, but this operation is generally left to the interrupt response program. Similarly, the restoration of the register content after completion of the interrupt program is included in the program. The branch address in the Interrupt instruction establishes the level of interrupt and, if the levels have an assigned priority, lower priority levels must be masked so that they cannot cause interrupts until the higher priority program is completed. The masking of lower priority interrupts may be included in the interrupt hardware and/or it may be a requirement in the interrupt response software. If the hardware does not mask the lower priority levels until the interrupt service program is complete, the first action of this program is to save the old mask and set the new mask.

Each input that can cause an interrupt may be connected to an individual level or the level may be shared by many inputs. If the level is shared, the interrupt response program must have a word or register available so that it can identify the particular source that caused the interrupt. Since interrupts are generally more expensive than standard digital inputs, the user may wish to connect the individual inputs to digital inputs if the latching function is not required and then connect the group of digital inputs to a single interrupt input through a diode OR circuit. This system requires programming to determine which specific input caused the interrupt, but it may provide economy in detecting infrequent changes which require immediate attention by reducing the number of interrupt circuits in favor of the less expensive digital inputs.

In addition to the hardware-software trade-offs in the input/output portions of the process control computer, there are the more often discussed trade-offs in the central processing unit. For example, the implementation of arithmetic operations and floating point arithmetic can be done in either software or hardware. These particular cases were discussed briefly as examples in the introductory paragraphs of this section. The central processing unit trade-offs in the process control computer are not substantially different than those that must be considered in selecting a computer for other applications. For this reason, detailed discussions of these trade-off examples are not presented here. It is noted, however, that they must be considered in the comparison of competitive solutions to a proposed control problem.

For the designer, the hardware-software trade-offs represent decisions which must be made to optimize the performance of the computer for a spectrum of process control computer applications. For the user, the trade-offs selected by the designer must be evaluated in terms of the particular application being considered. In the comparison of competitive system solutions, therefore, the absence of a hardware feature in one machine should not result automatically in a low rating. If the missing feature is, or can be, provided by the programming system and this does not result in significant burden, the machines may, in fact, be comparable. The necessity for considering the importance of hardware-software trade-offs in this manner further emphasizes the inadequacy of the simple yes-no comparison table discussed in the previous section.

12.5 SOFTWARE TESTS (BENCHMARK PROGRAMS)

A thorough evaluation of the software provided with a control system is too often neglected when developing ratings for the comparison of competitive systems. Hardware specifications are more easily quantified, and this probably accounts for the emphasis on hardware in comparative studies. For example, almost all specifications require a quote on the computer storage cycle time or require that the cycle time be less than some specified number of microseconds. All too often, this specification is weighted more heavily in the comparison process than is justified by its effect on overall system performance. Storage cycle time does provide an indication of the state of the art of the computer hardware and it is the basis for most of the other operations in the computer system. However, its effect on the problem to be solved is dependent on the number and types of instructions available, the I/O design, and the programming systems design.

The overall performance of the computer system is a function of all these factors. The importance of the software in this performance should not be underestimated. It has been stated that, ". . . but process control applications, in our experience, do not stand or fall on the vendor's hardware. . . . More important in my mind, in retrospect, is to assess the vendor's software with respect to significant differences in programming costs it will produce, . . ." (Johnson, 1968).

The software and its interaction with other factors in the computer system can be evaluated with the use of software test programs, sometimes called benchmark programs. This is a program that provides a solution to a defined sample problem. The program is run using defined input data and the execution time is compared with the time required to solve the same problem on another computer. Due to the noncompatibility of programming languages for various computers, the benchmark problem must generally be reprogrammed for each computer involved in the comparison. If the problem is a test of a higher level language, the reprogramming effort may be minimal since a certain degree of standardization exists between some high-level lanaguages. In assembly language, however, the significant noncompatibility of the languages usually means that the reprogramming represents a significant effort for all except the most trivial problems.

The definition of the benchmark problem can have a considerable effect on the results obtained. An evaluation conducted for the Air Force (Joslin and Aiken, 1966) compared four computers with four different problems and produced the results shown in Table 12-1. Several general

Table 12-1 Benchmark Program Comparison

	Computer			
	A	B	C	D
Compilation times:				
Problem				
W	2.53	2.47	1.27	1.00
X	3.19	2.67	1.00	1.30
Y	2.57	2.29	1.11	1.00
Z	3.19	2.79	1.00	1.03
Execution times:				
Problem				
W	1.00	1.10	2.10	2.57
X	1.36	1.00	2.09	2.00
Y	1.00	1.32	2.76	1.35
Z	1.12	1.00	1.12	4.05

observations can be made from this table. The selection of any particular problem would rearrange the performance rating of the four computers. Similarly, basing the selection on compilation time in some cases produces results different from those obtained when it is based on program execution time. In general, it can be concluded that small differences in benchmark timings have little or no significance.

To provide an accurate evaluation criterion for a computer control system, the benchmark problem should contain an appropriate amount of process input and output operation. In addition, it should result in an instruction mix and programming system burden similar to that which will be encountered under typical operating conditions. Definition of a benchmark problem that represents the actual job to be performed is a real challenge to the specification writer. The user best prepared to do this is probably ordering a second system similar to one already installed or a replacement system.

When such an effective benchmark problem is prepared, the question is how many vendors have the equipment available and can afford to expend the programming effort in a bid situation. The problems must, of necessity, be rather trivial or the vendor cannot afford to program them. Similarly, if the user prepares the programs, he must be able to expend the programming effort for each system to be compared. Furthermore, he must ensure that each system is programmed with the same degree of expertise if the results of the benchmark tests are to be meaningful. Probably the main thing that can be said for benchmark programs is that they can prove the existence of hardware and software and may point out obvious inadequacies in either.

Benchmark problems may be run as simulations on equipment other than the computer being bid because the vendor does not have a prototype of the computer or a working programming system. If this is not acceptable to the user, however, the vendor can be required to bid only equipment that is proven. This, however, imposes a limitation that a purchaser may not find desirable: He may not be able to consider the most up-to-date system available because the vendor cannot bid it. This is not to say that requiring the existence of proven hardware is necessarily bad because some companies have bought "paper tigers" that never quite lived up to their billing. When dealing with a vendor of proven performance and reputation, however, the paper tiger machine is not usually a problem and the hardware existence requirement limits the user's choice of systems.

12.6 EXPANSION

When selecting the system, the user must also consider whether the initial system can be expanded in the future and what is involved in such an expansion. When preparing to install a new system, the user may feel that the equipment specified will be fixed for the life of the system. The purpose of this section is to discuss some of the reasons it may be desirable, or necessary, to expand an installed system and how this expansion might affect the selection of a system.

As pointed out in the previous chapter, control computers are generally cost justified; that is, it is shown that it is profitable to install one. The same is true for other types of process equipment. It is also true that most companies have a limited amount of money to spend on new process equipment, expansion, or improvements. Although results of a computer feasibility study may indicate the profitability of a digital process control computer system, an analysis of the financial and/or personnel position may indicate that it is not possible to undertake a full-scale project immediately. Since the profitability picture indicates that the project should be started, two paths are open to the project manager: (1) Buy a small system which provides only part of the functions of the required job; or (2) buy a small amount of the required equipment on a system that is expandable at a later date to the full system requirements. The user can get more initial equipment for his investment in a small system but, when expansion is required, he is faced with installing a second system and programming the systems to work together or with exchanging the small system for one that meets the total project requirements. A reduced version of a larger system costs more money because a portion of the cost of the total system capabilities is included in the basic price of the system. This extra investment can be realized at the time of system expansion, however, since it will not be necessary to pay for a second basic system, its programming system, or the programs to coordinate the operation of two systems. The equipment already installed and functioning can probably continue to operate during the expansion except for short interruptions to install the additional equipment and programs.

Expansion still plays a part in system selection when the user is not faced with these financial problems and can install the full system from the beginning. The justification was based on operating data which, in the case of existing plants, are generally obtained from operating log sheets or special data collection runs. These data are compared with more optimum operating conditions as obtained from textbook unit operation, pilot plant studies, special test runs, or best operating records. The installation of the computer provides the user with the best tool available for process definition. The computer can collect data at regular intervals with an accuracy not previously available. Basically, it provides a new dimension in operating process definition. The natural result of this new definition is generally a changed and expanded role for the process control system. This often leads to expansion of the system originally installed.

An operating process is not static; it is a dynamic phenomenon. Equipment wears out and must be replaced, operating parameters change, bottlenecks develop, and certain portions of the process must be expanded. These changes require corresponding changes in the control system, and the installed system must have the flexibility and expandability in hardware and software to meet the changing process requirements.

Often within an operating plant several units may be close together or even may share a common control room. The process control system may be justified and installed on one of these units but, if the control system is expandable, it may be relatively inexpensive to extend computer control to these nearby units. Although not included in the original system, this planned expansion to other units may even be used as a part of the system justification.

When evaluating system expansion, several areas must be considered. Probably the first area of consideration is the central processing unit storage size. The user should determine the main storage expansion increments, the cost of installing each increment, and the maximum expansion available. If bulk storage devices such as drum, disk, and tape are available, the cost of adding or expanding these should be evaluated since the project often can be expanded through the use of bulk storage, rather than main storage. This decision will be based on the resident main storage requirements, the frequency of storage exchanges, and the time required to make the exchange, including the access time of the bulk storage. In addition to size, speed should also be considered. If several main and bulk storage speeds are available, some savings may be realized by starting with the slower speeds as long as these speeds can be increased if needed. In addition to storage speed, other optional hardware features may affect the time required to complete a certain job. Among these features are hardware multiply and divide, floating point hardware, input/output channels, and faster input/output devices. The availability of these features should be considered during the initial system selection procedure.

Since most expansions require additional process inputs and outputs, the availability and prices of these features should be established along with system maximums. The ability to add a line printer and additional typewriters should be established as well.

A final area to consider under expandability is the ability to interconnect the computer being considered with other computers. The connection may be to a similar computer on an adjacent process or in the laboratory, or to another computer in a larger management information system. The types of connections available (channel-to-channel, through digital input/output features, or over a telephone line) should be determined and evaluated.

12.7 SPECIAL FEATURES

Process control computers are designed with a relatively large number of standard input and output options. The manufacture of these features on a production line basis is one of the reasons that control system prices are considerably lower than they were several years ago. Since it is impossible to include all application areas in the system design, however, most control computer installations include a certain number of special features designed for the user's particular system.

Special features generally cost more than their equivalent standard feature because of the special engineering and production required and the relatively low market quantity. These special features should be identified in the vendor's proposal along with their cost. While the proposed special features will be required in some cases, it may be possible to alter the process or its equipment in such a way that standard features can be used. On the other hand, the use of special features may reduce the total system cost by providing a reduced capability where the full standard system capability is not required. These decisions represent a standard versus custom hardware trade-off.

Special features can fall into many categories, including modification of standard features to adjust their timing or voltage levels, the inclusion of special devices not normally offered on the proposed system, or even special responsibility to connect several pieces of hardware together. In many of these cases the vendor may be uniquely capable of delivering the complete system. In others, he may simply act as the user's purchasing agent by purchasing the equipment from another vendor and shipping it to the user with an appropriate price markup for handling. In this case the user may investigate purchasing the equipment directly from the supplier, saving the markup, and having a direct service relationship should problems develop.

Along with standard hardware, most vendors provide standard software to support the standard features. Special programs may be required to support special hardware and may or may not be supplied by the vendor. Bid specifications often contain the requirement for special programs not normally supplied with the system. The costs for these programs should be listed as separate items in the proposal, and the user should obtain some independent bids from companies which specialize in programming.

12.8   BUY VERSUS LEASE

Computer control systems are installed to increase the monthly profit and/or decrease the monthly expenses by more than their monthly cost. In the case of a purchased system, the computer system is expected to increase the operating profits sufficiently to recover the system purchase cost in a specified number of months or payout period. The purpose of this section is to discuss some of the aspects of leasing versus purchasing the system. These are important factors which must be considered in the comparison of competitive solutions to the control problem.

One of the first areas of consideration is the company budget. Most budgets have separate categories for capital improvements and operating expenses. If funds are available for capital improvements or if the cost of a control computer has been included in the cost of a new process unit, this might allow purchase of the system. The lack of these capital funds may dictate leasing the system. As an alternative when the user wishes to purchase the system but does not have capital funds available, he may lease the system initially so as not to delay the project while waiting for the funds to be included in the capital budget. Other areas discussed later may also indicate the desirability of an initial lease period before purchase. If this alternative is considered, the user should establish what amount of his lease payments, if any, may be applied toward the subsequent purchase of the equipment.

As a third alternative if the user wishes to purchase the equipment but does not have the total capital available, a time payment purchase can usually be arranged through a commercial lending institution or through the vendor.

In evaluating purchase or lease options, the user should be aware of what is included, the available options, and his commitments under each option. Lease contracts vary between vendors. The main areas to compare are the lease period and whether maintenance is included in the lease price. Lease periods generally vary from three months to five years, with reductions for the longer term contracts. Long-term lease contracts lose some of the advantage of leasing the system, generally cost more than a time purchase contract for the same period, and do not result in the user's owning the system at the conclusion of the lease period. If the lease or time payment contracts do not include maintenance, the maintenance costs should be added to obtain the actual monthly operating expense. In addition, some lease contracts do not cover full-time use of the system or all of its components. The use of those parts subject to overtime charges should be estimated, and the charges should be added to the monthly operating cost.

At the current rate of development of computer technology, the user must consider that improved systems will be available in the future. The desire to have the most modern system is often balanced, however, by the fact that a large part of the cost of a control system is in the installation and programming effort which would have to be expended again if the system were exchanged. Many of the original process control systems are still installed and performing quite satisfactorily after nearly 13 years. Probably the main reason for changing the system is that the process requirements have outgrown the system expansion capabilities and/or maintenance, and costs of spare parts have increased or these parts have become difficult to obtain.

The expansion capabilities of the system discussed in Sec. 12.6 may also affect the decision to buy or lease. There is often a difference in the price of expanding a system that is leased and a system that is purchased. This difference is usually small, however, unless it entails the replacement of a major component of the system.

Another related experience is to determine that a device purchased on the original system is no longer needed. These possibilities may require consideration of installing the system under a lease contract and purchasing the system after the requirements have been confirmed. An alternative is to purchase part of the system and lease those items which are required only during the system implementation phase. In all these cases, an area of consideration in the selection procedure is the difference that may exist in the vendor commitment to leased and purchased systems.

One other factor is taxes. In the case of purchase, the user may depreciate the equipment at an appropriate rate and may also be eligible for investment incentive tax credit. On a time purchase plan the interest is also deductible. As the owner he is liable for the real property taxes on the system. In the case of a lease, the payments are operating expenses and some vendors may pass on to the user any allowable investment incentive tax credits. Since the vendor retains ownership of the equipment, he is subject to the real property taxes assessed against the equipment.

12.9 INTANGIBLES

When the system requirements have been defined and the vendors have responded to these requirements, there are generally some items that cannot be assigned a direct cost or value. These should be qualitatively evaluated in the selection process as intangibles, however, and given adequate consideration.

The area of vendor services was discussed in the previous chapter relative to the specific requirements of the project. The location or proximity of vendor technical personnel, for example, has a bearing on their availability to the project team. Often days can be lost in an implementation effort that could be saved by a one or two hour visit with a vendor technical representative. Similarly, all of the people associated with the project require some training on the system to be installed, varying from a one- or two-day summary overview of the system hardware and software, to a knowledge of the system FORTRAN and/or assembler languages, to a complete knowledge of the monitor or executive system. Thus the availability and location of several levels of system training has an effect on the project.

The merits of vendor maintenance services were discussed in the previous chapter, but the user should also consider the effect of the proximity of similar systems. Although training and diagnostic aids are important factors in maintenance proficiency, the relative high mean time between failures of current systems results in minimum use of troubleshooting techniques. The maintenance engineer who has responsibility for several systems and/or experienced assistance in his immediate locality may maintain a higher level of proficiency which results in more efficient servicing techniques. This is ultimately translated into less downtime for the user and, therefore, increased system availability.

Another area to which it is difficult to assign a value is the availability of user organizations which include the system being considered. These organizations provide sounding boards for problems and may change the complexion of a problem from a unique case to a general hardware or software deficiency. Similarly, they may result in solutions to common problems. These organizations and their meetings should be controlled by the users to be effective. The vendors should encourage the use of the organization and attend the meetings to provide answers to questions, where possible, and to record complaints. In addition to discussing problems, the users generally discuss some of their applications that they are willing to make available to other users.

The use of digital computers for process control is relatively new and, even after revision of some of the early projected growth figures, is still one of the fastest growing markets for computers. The natural result of such a market is for many companies to offer systems and/or services to fill the market needs and also for some companies to be formed specifically to address this market. It is equally natural for some companies to fail or withdraw from the market as a result of the competitive pressures or underestimation of the requirements for success. The user should attempt to assess the stability of the vendors he is considering in his system selection. This is not to say that he should select a vendor merely because he has been in the market the longest. Some of the early vendors withdrew after a considerable tenure. More important are the vendor's current marketing practices. Some of the most common problems result from underpricing and overselling. Beware of too low a price. At the present time, there are no nonprofit organizations marketing computer control systems. Some vendors bid their early systems low to establish themselves in the field. If they have the financial stability to do this, the user may get a bargain, but the effects of the learning process on the implementation schedule must be anticipated. In the area of overselling, the user must satisfy himself that the vendor has bid sufficient equipment to do the job and that he has the staff to provide the proposed services. One might assume that this problem can be solved by getting the vendor to sign a performance contract and letting him worry about these problems. Such problems generally return to the user, however, in the form of reduced quantity or quality of hardware or services. They may even result in having to start negotiation with a new vendor late in the project cycle or even after the system has been installed.

In evaluating a proposed system, a value generally can be assigned to the required programming systems and their features, but programs that provide capabilities not required for the specified job should also be considered. Some users require the computer system to have the ability to do off-line work between process jobs in order to justify the system installation. Other users say that the computer is going to be programmed for its process job and left alone. In most installations, programming the computer for its process job is a continuing activity. First of all, the installation of the computer provides a large step in knowledge of the process and therefore the best method of control will probably change. Second, the process requirements and equipment generally change and require program changes. Thus the ability to debug and change programs on-line has a value even in the dedicated applications.

The executive system is generally delivered with the system and does not have to be changed except for revisions and corrections by the vendor, in which case the vendor would supply a new executive program. The user should inquire about how he can make changes to this executive should they be required. Although such changes are not anticipated at the time of system selection, the ability to make them on the proposed system has a value. If this cannot be done easily, the method and cost of getting these changes made should be established and included in the system evaluation.

REFERENCES

Johnson, A. E., Jr., "Significant Aspects of Computer Project Management," *Proc. 3rd Ann. Workshop Use of Digital Comput. in Process Control,* Louisiana State Univ., February 21–23, 1968, Instruments and Control Systems, Philadelphia, Pa., 1968, pp. 50–53.

Joslin, E. O., and J. J. Aiken, "The Validity of Basing Computer Selection on Benchmark Results," *Comput. and Autom.,* January 1966.

Pardo, V. A., and J. M. Vance, "Guide to Computer Specifications," *Control Eng.,* September 1966, pp. 99–104.

Stout, T. M., "Selecting a Process Computer Vendor," *Control Eng.,* August 1966, pp. 49–54.

BIBLIOGRAPHY

Anderson, A. H., and G. E. Kaufer, "How to Select the Best System," *Control Eng.,* June 1967, pp. 93–98.

Boutwell, E. O., Jr., "Comparing the Compacts," *Datamat.*, December 1965, pp. 61–69.

Brandon, D. H., "Standardizing Computer Selection," *Comput. and Autom.*, October 1965, pp. 32 ff.

Bromley, A. C., "Choosing a Set of Computers," *Datamat.*, August 1965, pp. 37 ff.

Feldman, R., "Selecting Input Hardware for ddc," *Control Eng.*, August 1967, pp. 75–79.

Florkowski, J. H., A. A. Kusnik, and O. R. LaMaire, "How To Select Random Access Storage Devices," *Control Eng.*, January 1966, pp. 91 ff.

Hendrie, G. C., and R. W. Sonnenfeldt, "Evaluating Control Computers," *ISA J.*, August 1963.

Hillegass, J. R., "Standardized Benchmark Problems Measure Computer Performance," *Comput. and Autom.*, January 1966, pp. 16 ff.

Huesmann, L. R., and R. P. Goldberg, "Evaluating Computer Systems Through Simulation," *Comput. J.*, August 1967, pp. 150–156.

Parker, N. H., "Selecting the Best Vendor," *Chem. Eng.*, August 19, 1963.

Chapter 13

Installation and Physical Planning

RICHARD N. POND*†
SHANE L. SHANE*

*S. L. Shane authored Sec. 13.1–Sec. 13.5 and R. N. Pond contributed the remaining sections in the chapter.
† Deceased.

A computer control system is often installed in an industrial environment. This environment is usually more severe than that found in the typical office location. Ambient air temperature and humidity vary over a wide range and frequently reach levels appreciably greater than the outdoor conditions at the plant locality. Various manufacturing processes increase the solid, liquid, and gaseous contaminant level of the local air supply. Large quantities of electrical power are consumed and can induce appreciable electrostatic and electromagnetic fields at the computer site.

All these factors have direct and varying effects on system performance, design, and installation requirements. Consideration of these effects by the system design engineer throughout the design cycle helps him choose the best trade-offs in system hardware, system programming, and system installation costs while still preserving the required overall performance. Similarly, it is essential that the user's facilities design engineer be familiar with the environmental limitations of the computer system so that he may provide an adequate installation at an economical cost.

Factors included in the installation and physical planning for a process control computer are the following: Selection of the installation site; design and erection of necessary building structural modifications; design and installation of the power supply system, including the building wiring; design and installation of heating, ventilation, air-conditioning, and filtration systems; design and installation of the signal and control interface between the computer and the instrumentation systems, including the interconnecting cabling; and provisions for the control of damage caused by fire, flood, earthquake, and other forms of catastrophic damage to the process control system. This chapter discusses these factors and their importance in terms of the operation of the process control computer system.

13.1 COMPUTER SITE PREPARATION

The preparation of the site for a new process plant is probably a familiar task for the facilities engineering department of most companies. It starts with the selection of a site on the basis of factors such as transportation facilities, utilities, and the source of raw materials for the process. These considerations are not altered appreciably when a process control computer is to be installed to control the process. There are, however, additional considerations included in the site selection which are attributable to the characteristics of the computer. Similarly, the computer may be a factor in the preparation of the site since recognizing the unique requirements of the computer and incorporating them into the design and layout of the process can often enhance the performance of the system.

The modern process control computer system is specially designed for use in industrial environments. As such, it can operate under conditions which are intolerable for the commercial or scientific computer. Nevertheless, the control computer is a complex electronic system and its performance is affected by the conditions under which it must operate. Careful attention to the preparation of the computer site ensures more reliable day-to-day operation and increases the useful installed life of the system. For these reasons, careful site preparation is well worth the required effort and expense.

In this section, the particular site requirements that are unique to the installation of the process control computer are considered. These factors must be merged with those that have traditionally been considered in preparing for the installation of new process equipment.

13.1.1 Site Selection

Several locations often offer significant advantages as an installation site for the computer system. The best site is seldom obvious and, therefore, evaluation of the effects of each possible location on the performance and the installed cost of the system must be made. The more common factors influencing the selection of the installation site are the following:

1. The amount of floor space required for the process control computer system and its associated supporting equipment. Storage of necessary supplies, office space for personnel directly associated with the system, and the actual machine units all contribute to the total amount of space needed.

2. Structural requirements of the building. This includes allowable floor, beam, and column loading and the amount of modification needed to isolate the installation from the surrounding environment.

3. The interrelationship of various personnel groups. For example, it is often desirable to locate the process control system in, or adjacent to, the major process control room as a convenience for both the process operators and the computer system personnel. Similarly, if engineering or business computations are a significant portion of the computer work load, these affected personnel should also have ready access to the system.

4. Availability of required utilities such as power, heating, ventilating and air-conditioning equipment, water, fuel when required, and lavatory facilities.

5. Technical and cost factors associated with providing the signal interface to the process variables for control monitoring purposes. Once again, site selection in or adjacent to the main control room often offers significant advantages since many of the signal and control functions are readily available and can be connected to the system easily and economically.

6. The relative availability of capital and operating monies. Even though a near ideal site is available, it is sometimes necessary to select a different location simply because sufficient capital equipment funds are not available, even though an increase in future operational cost of the system is incurred.

The first step usually taken by the facilities engineering group is to review the general plant site and its existing structures. From this general review, a number of potentially acceptable sites can be identified. Some of these locations will have significant disadvantages when reviewed by other affected personnel, and they can be disgarded after only this preliminary analysis. The remaining sites must be examined in terms of system layouts and economics before the final decision is made. This normally requires a preliminary layout of the computer system and its supporting equipment to determine whether sufficient space is available and to determine the extent of necessary building structural modifications. With these data, the analysis can be completed and a site selected.

13.1.2 Layout

Figure 13-1 is an example of a process control computer room layout. The best layout is that orientation of machine units which provides maximum operation efficiency and system serviceability. Both system performance and reliability are affected to some degree if sufficient space is not provided. The number and types of units making up the process control system varies between user installations and within vendor product lines. In addition, the physical aspects of the building structure, such as the length-to-width ratio of the room and the allowable floor and beam loading, place varying restrictions on any layout. For these reasons, it is not possible to provide a specific set of criteria which can be followed in making the layout. However, the following rules-of-thumb and consultation with the appropriate vendor representatives can be used to prepare an acceptable layout. For convenience purposes, these general rules are separated into two broad categories: those primarily affecting system operational efficiency and those which have their major affect on system serviceability.

13.1.2.1 System Operational Efficiency. The typical process control system is made up of several different types of units. Some require direct operator access on a continuous basis, some only infrequently, and others only on rare occasions and then principally to ascertain whether the unit is functionally ready for operation. Table 13-1 is a listing of the more common units found in a process control system, the amount of attention required by the operator when he is in attendance, and the general type of attention required.

From the amount and type of attention that the operator needs to give to various system units, it is apparent that some units should be installed adjacent to the main processor (or the operator console if it is a separate unit) and that other units can be effectively placed out of his main traffic pattern. This is illustrated in the example layout of Fig. 13-1. The computer operator's principal work area is bordered by the central processor with its console, the printer, the card reader/punch, and the console typewriter unit. All these units require frequent attention from the operator. The magnetic tape units are near the work area but are slightly removed from the more frequently used equipment. The disk drive and the process I/O units require little attention and are furthest from the operator's work area.

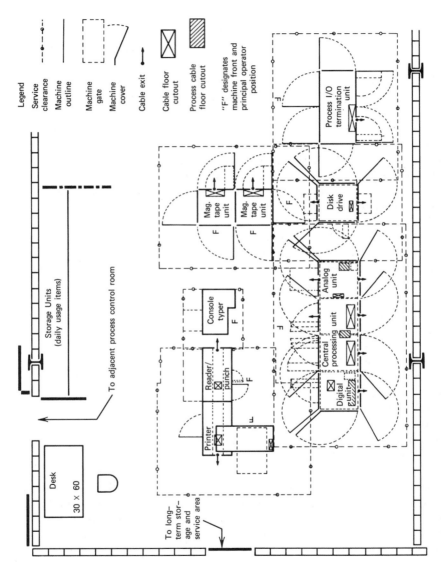

Fig. 13-1 Example computer room layout.

Table 13-1   Operator Requirements

Unit	Amount of Operator Attention Required	Type of Operator Attention Required
Main processor and system console*	Frequent or continuous	Monitor displays. Set control functions.
Process operator console	Frequent or continuous	Monitor displays. Set functions. Enter data or program changes.
Card read/punch	Frequent	Load & remove cards. Operate feed control.
System output printer	Frequent	Load & unload paper.
Remote printer(s) (often equipped with keyboard for direct entry of data)	Infrequent	Load & unload paper. Read instructions. Normally provided for operator guidance at remote locations.
Magnetic tape drive	Infrequent	Load & unload tape reels.
Paper tape read/punch	Infrequent	Load & unload tape.
Disk storage drive	Infrequent	Load & unload disk packs if removable.
Auxiliary main or other storage	Seldom	Check function switches and/or lights.
Analog/digital input/ output	Seldom	Check function switches and/or lights.
Other control	Seldom	Check function switches and/or lights.
Communication adapters	Seldom	Check function switches and/or lights. Make communication system connection when provided.
Power and power distribution	Seldom	Check function switches and/or lights.
Signal interface	Seldom	Check function switches and/or lights and wiring.
Miscellaneous input/output switches and displays	Frequent	Data entry and manual logging.

* Some vendors provide a separate system console. Others build the console into the main processor. The system console should be considered as the principal operator work station unless a separate supplemental process operator console is installed.

In addition to the actual machine units, significant quantities of card and paper stock, preprinted forms, programs and subroutines on cards, magnetic tape, paper tape, and magnetic coated disks must be stored in the machine room and convenient nearby areas. Enclosed fire resistant cabinets are usually provided to protect the vital records and to reduce the potential fire hazard of the unused stock. Space for the items required in normal daily usage should be provided at a location near the machine.

Operating and programming personnel often spend a major portion of their working day in, or at least near, the system machine room. The layout must provide space in the office area or machine room for necessary furniture such as desks, chairs, work tables, filing cabinets, office supply cabinets, coat racks, and other office equipment.

13.1.2.2 System Servicing Efficiency.   Those layout factors that affect system servicing are more a function of machine design than are those that influence system operation. The following basic considerations should be followed unless conflicting specific recommendations are furnished by the computer manufacturer.

1. At least one input and output unit is usually required to load a system program diagnostic routine. These I/O units are often specified as part of the minimum system configuration offered by the system vendor. They should always be ordered and installed unless special arrangements have been made with the vendor. These I/O units usually require frequent operator attention during normal operations and should be located near the operator console.

2. Many units in the system are electronically controlled by some other unit in the system. System malfunction can be more quickly diagnosed if the display lights and function switches are visible from the primary servicing position of each unit. In addition, electromechanical devices such as card and paper readers and punches, printers, disk drives, and plotters are located such that the motion of the mechanical mechanisms can be viewed by servicing personnel from the primary servicing position at the controlling unit. The portions of the mechanical mechanisms that need to be observed are usually those normally visible to the operator when the unit is in normal use.

3. Space for servicing each unit must be provided. Internal electronic and mechanical subassemblies are typically mounted on movable structural members within the machines. These structural members, commonly referred to as "gates," are attached to the basic machine frame with hinges or slides. During servicing operations, access to both sides of the gate is often necessary. In the absence of specific vendor recom-

mendations, a minimum clearance of 12 in. beyond the open gate position and 6 in. beyond an open cover position should be provided.

4. Servicing operations on the system frequently require access to two or more units at the same time. Usually the units that require this kind of action are the control unit and connected input/output unit(s), and the processor and control units. Conflicts of servicing operations can be minimized if gate and cover swings for related units are not overlapped.

5. Service clearances for adjacent units can usually be overlapped without seriously affecting serviceability. The greater service clearance requirement of either unit should be used as the minimum spacing for the two units.

6. Even though the spacing of units is sufficient for normal operator actions, it may not be adequate for servicing personnel and test equipment. The layout should provide sufficient space for oscilloscopes and other test equipment adjacent to each unit which may require this type of equipment for servicing operations. In addition, space for the use of logic books and other documentation required during servicing should be provided.

7. Some of the units installed in the system interface directly to instrumentation and control devices and it is occasionally necessary to change or add wiring in this area. Access to these areas is normally infrequent, however, and it is seldom practical to relocate the unit when a change is necessary. Thus service clearance areas similar to those provided for other units in the system should be provided.

8. Electrical equipment such as power service equipment is often present in the computer room. System units should be located so that normal service clearance for this type of equipment is maintained.

9. A complement of spare parts, servicing tools, test equipment, servicing diagrams, and instructional material is usually provided with the process control system. Space for the storage of these materials should be provided in or near the machine room. Providing an enclosed area enhances the appearance of the installation. Space required for this storage function and to provide an off-line servicing area usually varies between 70–150 ft², depending on the size of the computer system.

Most preliminary layouts are made by placing templates on a scale drawing of the area. Templates are sometimes provided by the system vendor but, if they are not, they can be easily made from light-sensitive plastic sheets. Each machine unit template should include gate and cover swings, caster and leveler locations, interconnecting cable entry locations, and minimum service clearance in addition to the basic machine outline. This is illustrated in Fig. 13-1, which was prepared using this type of template. Layouts prepared by this method are easily reproduced on several types of duplicating equipment or by normal drafting methods. The dimensional accuracy of the templates is sufficient for most planning functions and for use as the basis of final working drawings.

13.1.3   Structural Requirements

The process control system is made up of a number of separate units interconnected by signal, power, and control cabling. Floor space requirements vary from that required for a single relay rack to several hundred square feet, depending on the size and complexity of the process control computer. The more common medium-sized systems usually require 300–500 ft² of floor space including the necessary service and storage areas.

Most vendors recommend that the computer be installed in an environmentally controlled area. This may mean that the user must construct a room in which to house the computer equipment. There are two primary reasons for this recommendation: First, even though the computer is designed for use in severe environments, the input/output equipment, such as disk drives and card punches, is often comprised of the same units supplied in the vendor's standard computer product line, and these cannot withstand long exposure to severe environments. Second, a controlled environment generally results in significantly lower maintenance and longer machine life. Although the construction of a computer room seems to be a disadvantage for the user, studies have shown that in most cases this is more economical than the design of specialized input/output equipment for industrial environments.

Design and fabrication of remote terminals for use by process operator personnel does not follow this general rule since the cost of controlling the environment at the process operation station can easily exceed the cost of the terminal. Since each terminal location is a special case, it must be considered separately and conditions acceptable to both the user and the vendor must be established well in advance of actual unit installation. The design of the main computer room, however, is a more general case and basic construction parameters can be established.

Selection of construction materials for a computer room depends primarily on local government codes and the user's common practice. Masonry, wood, metal, and other materials can be used as desired. However, because of the relative importance of the system to the user's business, fire resistant materials should be used as much as possible.

Probably the single most important difference in room construction is the provision for an effective barrier from the surrounding environment. Many industrial processes use corrosive and abrasive materials in manufacturing operations. In others, explosive and combustible mixtures are often present in the atmosphere. The building structure must be designed so that the computer system is effectively isolated from these conditions. This can be readily accomplished by using vapor barriers in exposed structural elements such as walls, floors, and ceilings. In some severe environments, it may also be necessary to provide air locks and special floor mats or grates at main entry doors, especially if personnel traffic into the area is high. Extensive use of weather stripping at all doors, windows, and other openings and the sealing of pipe, duct, and conduit entrances is a necessary measure in these cases. Often a positive air pressure of 0.1–0.2 in. of water is maintained in the room as an aid in controlling infiltration of dirt and potentially corrosive or explosive materials. This slight pressure can usually be carried by normal building construction materials unless unusually long unsupported spans are used.

The weight/size ratio of typical process control units is such that the floor loading rating normally found at the installation site is rarely exceeded if the recommended service clearance around each unit is provided. Although most systems can be installed on a floor structure rated at 50 lb/ft^2, a detailed structural analysis of the floor should be made whenever (1) the floor rating is less than 100 lb/ft^2, (2) the analysis is required for compliance with local codes, (3) the minimum specified clearance between units cannot be met, or (4) the existing floor structure has undergone significant modification since originally constructed.

In the absence of specific data, the following factors can be used in making this analysis:

1. A partition allowance of 20–25 lb/ft^2 is customarily added to the floor load rating unless objectionable to the local building department.

2. The individual machine unit weight can be assumed to be uniformly spread over the total floor area occupied by the machine unit and the service clearance. The maximum amount of service clearance used in this calculation is customarily limited to 30 in. in any one direction from the unit.

3. Raised or false floors, as illustrated in Fig. 13-2, are often used for protection of interconnecting cables, for supplying conditioned air to the machines or area, and for personnel safety. An allowance of approximately 10 lb/ft^2 must be applied for the weight of the false floor.

4. A 20-lb/ft^2 allowance for each square foot of service area used in the computation should be assumed as live load.

5. If multiple machine units are installed adjacent to each other, no allowance should be taken for end clearance provided in the layout.

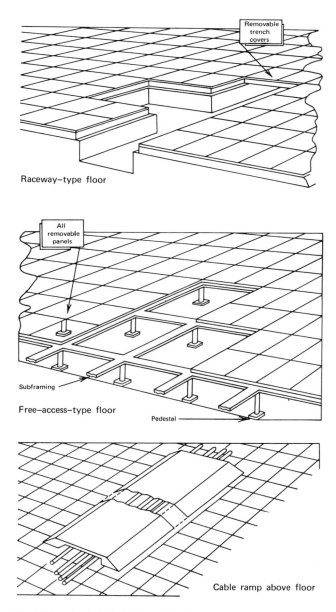

Raceway–type floor

Free–access–type floor

Cable ramp above floor

Fig. 13-2 Cable installation methods.

13.1.4 Cabling

System cables are often separated into several general categories based on their unique physical characteristics, the method of procurement, and the responsibility for the design, installation, and maintenance. The most common categories are (1) main computer system interunit cables, (2) cables between the computer and remote terminal units, and (3) cables between the computer system and the process instrumentation and control devices.

13.1.4.1 Procurement and Functions. Cabling between units of the computer system is normally provided, installed, and maintained as part of the computer system. It is used to provide both AC and DC power to the system units and for the transmission of data and control signals within the system. Electrical parameters of the circuits connected to these cables are usually such that the total cable length must be restricted. The maximum length limitation is specified by the vendor for each unit connected to the system. Typical values seldom exceed 100 ft and lengths of less than 25 ft are not uncommon. This restriction seldom causes a severe problem in layout since most system units in the main computer room are installed close to each other for operational efficiency.

Cabling between the computer and remote terminal units is normally procured, installed, and maintained by the user since it is often considered as building wiring by local code. Cable construction is similar to that used for computer unit interconnection since the remote terminal is similar or identical in design to a terminal installed in the computer room. Suitable cable for terminal connection can usually be purchased from the computer vendor. Cable specifications concerning the electrical and mechanical characteristics of the cable which affect the computer or terminal operation are provided by the system vendor. The vendor may also provide general recommendations concerning the cabling specifications affected by the installation environment. However, specific criteria are usually defined by the user.

Cabling between the computer and process instrumentation and control elements is normally designed, procured, installed, and maintained by the user. Specifications for this wiring depend primarily on the function of the instrument or control device and the electrical characteristics of the computer interface. Basic recommendations concerning wiring practices are available from the computer, instrument, and wiring vendors. Most of the voltage and current levels used with this type of wiring are low level. Thus most wiring can be classified as low-voltage secondary signal and control wiring, and open wiring methods are usually acceptable to local code authorities unless hazardous locations are encountered.

In general, however, open wiring for signal and control cables is not recommended by the computer vendor. Recommendations on wiring practices, their affect on system performance, and the effect of environment on this wiring are discussed in Sec. **13.6.**

13.1.4.2 Cable Construction. System interunit cables are generally fabricated from multiconductor, jacketed cable designed for hard service. Each cable assembly is usually terminated on one or both ends with one or more multicontact connectors. Commonly used insulations for individual conductors and for the outer jacket are polyvinyl chloride and polyethylene. The completed cable assembly often includes one or more braid or foil shields. Coaxial, twisted pair, planetary lay, and bunched cable constructions are used individually and in combination. Dimensions of these cables vary with the application; cable diameters of 0.25–2 in. are common, whereas the majority average 0.75–1.5 in. Bending radii of the finished cable also vary with a typical maximum radius of 4–6 in. for the larger cables. Maximum length limitations for each cable are dependent on circuit parameters of the interconnected units. Since many computer unit time constants are extremely small (nanosecond, microsecond, and millisecond time constants are not uncommon), cable length can also be restricted as a result of signal propagation time. Table **13-2** lists typical length limitations for current equipment designs. The

Table 13-2   Typical Cable Lengths

Unit Type	Typical Maximum Cable Length
Communications adapter (long line)	Virtually unlimited
Communications adapter (limited distance)	3–10 mi
Remote terminal (multiwire)	1000–2000 ft
Magnetic tape unit	100–200 ft
Buffered channel units	200–400 ft
Low-speed input/output units	50–100 ft
High-speed input/output units	25–50 ft
Very high-speed units	Less than 25 ft

classifications shown in the table are arbitrary, at best, and are only intended to indicate the time dependency of cable length. Other variables which could increase or decrease these cable lengths must be considered and the vendor should be consulted to determine specific restrictions.

13.1.4.3 Cable Installation. Cables between units in the main computer room are normally laid on the building floor. Most multiwire cable assemblies (AC and DC power, signal, and control) are terminated with connectors. Removal of the connectors, suitable protection of each wire termination, reassembly of the connector, and repair of possible damage usually precludes pulling of cable assemblies through conduit or floor duct. The use of cutouts in the building floor and stringing cables in wireways or racks below the floor is also impractical because of the limited access. Cables can be laid beneath a raised false floor effectively or, if traffic is light, simple cable covers placed in traffic aisles provide sufficient protection. Figure 13-2 illustrates these possibilities.

Cabling to remote terminal locations is normally installed as open wiring or in conduit, depending on local practice and code requirements. Similarly, process instrumentation and control wiring is usually installed in conduit, busduct, wireways, cable trays, or as open wiring. The choice again depends on local practice, code requirements, and the environment. Generally, signal wiring enclosed in conduit or busduct is less susceptible to noise pick-up, and this method is usually recommended by the vendor.

13.1.5 Vibration

The internal structure of the modern computer is a complex mechanical combination of independent masses interconnected via a series of beams and columns. Almost all parts of the structure and even the component masses have some elasticity. This structure is subjected to vibration caused by process equipment. In addition, the electromechanical devices such as card readers, magnetic tape units, printers, drum and disk storage units all convert electrical energy into mechanical motion, and this provides sources of vibration in the unit itself. The result of this complex structure is a system that may be sensitive to externally induced vibration. This is particularly true if the external vibration contains a frequency component which matches a resonant frequency of a substructure in the machine unit.

Failure in the unit caused by vibration can be a permanent structural failure or a temporary change in operational performance of the machine. For example, the electrical characteristic of an electronic component may be temporarily or permanently altered. However, the failure of principal concern is a permanent structural failure where the machine is seriously damaged or at least temporarily removed from service while repairs are made.

Typical resonant frequencies of concern are in the range of 5–100 Hz. Most systems are designed to withstand relatively severe vibrations in the basic axes at these frequencies. They also have excellent shock resistant capabilities. Specifications furnished by the vendor must be consulted, however, since significant differences exist between different units.

The three terms commonly used to specify allowable vibration intensity are frequency, amplitude, and acceleration. The conventional units used are Hertz and inches. Thus, for example, a mathematical expression used to specify the allowable vibration might be $G = 0.103\ AF^2$, where A is the amplitude of the vibration in inches, F is the frequency in Hertz, and G is the acceleration in in./sec². Typical vendor specifications are based on either peak-to-peak or rms values.

Vibration specifications vary widely between systems. Typical values found in the current literature range from 0.1 G sustained and 0.25 G intermittent to 5 G sustained. Shock limits typically vary between 1–50 G.

Vibration in environments where personnel work continuous 8-hr shifts is rarely a problem since even peak accelerations in excess of 0.1 G sustained are rarely encountered. Some unusual conditions, such as mobile van mounted systems or systems installed immediately adjacent to undamped forging, rolling, or other heavy equipment, may require the installation of isolation equipment. Isolation mounts can be specially tuned semi-elastic pads, pneumatic dampened reaction masses, or simple spring mounts. Selection of the most suitable method of isolation depends on the relative economics and the magnitude and characteristics of the vibration.

13.1.6 Lighting

General office lighting practices should be followed in the process control computer system installation area. Illumination levels in the range of 60–100 ft-c measured at desk height level generally prove most satisfactory since personnel must read considerable quantities of printed material.

Some units in the system may contain cathode ray tubes or low-intensity back-lighted displays. In addition, maintenance personnel utilize oscilloscopes extensively during servicing operations. Lighting levels in the range of 30–40 ft-c are most acceptable in the immediate vicinity of display units and where oscilloscopes are in use. On-off control of alternate rows or alternate lighting elements by providing separate power circuits is a method of achieving good control of light intensity levels.

13.1.7 Safety and Fire Precautions

Safety is a vital factor in planning for a process control system installation. This consideration affects the choice of the installation site, the building materials, fire prevention equipment, air-conditioning, and the electrical supporting equipment design.

13.1.7.1 Computer Room. The process control computer system should be housed in a noncombustible or fire-resistant building or room. If it must be located near areas where flammable or explosive materials are stored or processed, precautions must be taken to minimize the hazard of fire and explosion. The most commonly used method is to seal, pressurize, and air purge the computer room since explosion-proof packaging of the computer units is usually economically less practical than declassification of the installation site. Selection of an installation site outside the hazardous area also may be a practical alternative. However, operating procedures, instrumentation and wiring costs, and other considerations often dictate that installation at the process site is mandatory.

Walls enclosing the computer area should extend from floor to ceiling and should be of noncombustible materials. Where false or hung ceilings are used, they should also be made of noncombustible or fire retardant materials. Similarly, raised false flooring, cabinet work, and storage units in the area should be constructed of similar materials. When other materials are used, they should be protected in accordance with local safety and building code regulations.

13.1.7.2 Fire Prevention and Control Equipment. Both fires in ordinary combustible materials such as card and paper stock and fires in electrical equipment can occur in the computer area. Fires from either source have been rare to date, especially those originating in the computer equipment itself. However, some risk does exist and design of the computer facility should minimize this risk. The following considerations should be observed:

1. Portable fire extinguishers of suitable size and number should be installed in the computer room. Carbon dioxide is the nonwetting agent most generally recommended for electrical equipment (class C hazard). The most practical size for these portable extinguishers is 15 lb. Local codes and plant safety regulations should be used in the establishment of required inspection and maintenance procedures.

2. Standpipe or hose units should be installed within effective range of the computer facility for use as a secondary extinguishing agent for a class A hazard.

3. In some installations, local building codes or ordinances, local plant safety practices, or insurance conditions require the installation of automatic water sprinkler systems. The following recommendations should be considered if they conform to these requirements:

(a) *Pre-action sprinkler system.* This system is of the "dry pipe" type. Water is introduced into the piping only by valving controlled by high-temperature sensing devices located within the room. The use of such a system can preclude damage to equipment and supplies in the computer room should mechanical damage to the sprinkler heads or piping occur.

(b) *Sprinkler heads.* Heads rated for an intermediate temperature range (175°F) should be used. The use of these higher temperature heads provides some additional protection from premature operation of the sprinkler head.

4. A fire detection system should be installed to monitor continuously both the computer room and the storage area containing critical master data and program files. The detection system should be designed such that both the input power and air-conditioning systems can be shut down. An audible and/or visual alarm system is normally provided in conjunction with the fire detection system. The alarm system should be designed to alert key personnel immediately at the plant site.

13.1.7.3 Air-Conditioning Systems. In most installations, heating, ventilating, and air-conditioning equipment is installed to support the process control computer area (and the process control room, if a combination installation is provided). If the computer system operates unattended for extended periods of time, it is essential that air conditions be monitored on a continuous basis. Temperature, humidity, and room pressure (if air purging declassification methods or control of corrosive substances requires) should be monitored. Sensing apparatus should include connections to remote alarm stations to alert personnel to unusual conditions in the computer room.

Air-conditioning switches should be located at both the air-conditioning equipment and in the computer room. At least one set of controls should be accessible to authorized personnel outside the room for use in an emergency. Air ducts run through the computer room should be equipped with fire-stop fusable link dampers at each wall and at such other locations as prescribed by local code and/or insurance regulations.

13.1.7.4 Electrical Systems. Safety and fire protection considerations in the design of the electrical power distribution system should include the following:

1. The main line circuit breakers for the computer system should be of the remote operable type. Emergency disconnect stations should be installed near the main computer operator primary work station and, as a minimum, near one main exit door from the computer area. Also included near the exit should be a disconnect station for the air-conditioning unit.

2. When emergency standby power is provided for use in case of main power system failure, switchover from normal to standby power should

be accomplished by automatic means. An indication of which power system is in use should be provided.

3. Special battery-operated lighting systems should be installed even if they are not required by local code.

4. Lightning arresters should be installed on the secondary power distribution system.

5. Waterproof plugs and receptacles should be installed under raised floors, in trenches, or at any location that might be subject to submersion in liquids.

13.2 PRIMARY POWER

The modern digital process control computer system is a highly sophisticated combination of electronic and electromechanical devices which requires reasonably stable and well-regulated electrical power. In the industrial plant, however, large amounts of electrical power are switched off and on intermittently, arc-producing electrical apparatus is used, and repair, revision, and damage to the plant power systems are common. All of these produce electrical disturbances which can affect the operation of the computer. Although the digital computer power system has excellent stability and line isolation characteristics which minimize the effects of plant electrical disturbances, severe disturbances can result in computer malfunction. For this reason, the installation planning for the computer site must consider the provision for a suitable source of primary power for the computer system.

The industrial power systems engineer must consider the computer system to be a critical electrical load which may require isolation from the remainder of the plant electrical system. This isolation can be accomplished using power-line filters, isolation transformers, motor- or engine-driven alternator sets, or the newer solid-state inverter/converter power systems. The decision on the degree and type of isolation depends on how much process operating improvement can be attained by the increased capital investment for electrical isolation. Although this is a difficult trade-off to evaluate, the starting point is easier to define. The design engineer must match the electrical specifications of the computer with those of the plant electrical power system. At the time of this writing (early 1971), two groups are working on standards related to primary power sources for computer systems. These are the committees SP-54 in the Instrument Society of America (ISA) and Subcommittee 3 (SC3) of the Business Equipment Manufacturers Association (BEMA). The recommendations of these activities will probably be available sometime after 1972.

13.2.1   Electrical Specifications

Electrical specifications provided by the system vendor define the nominal voltage, nominal frequency, and the number of required phases. Specifications also include voltage and frequency tolerances and the amount of power required, expressed as kilovoltamperes, kilowatts, or amperes. Unfortunately, power distribution frequency and utilization voltages typically vary both by country and by local power company practice. In addition, the amount and type of load that may be connected to the power net at the utilization voltage is sometimes regulated by local law or practice. It is generally impractical from an economic viewpoint to provide units or systems that can be connected directly to any common utilization voltage or frequency. Thus each system is usually designed for connection to a specific source of power. If connection to some other source is required, the equipment modification may be available from the vendor as an optional feature or on an extra cost basis.

Small computing systems requiring a power source of less than approximately 2 kVA are usually designed for connection to the lowest readily available voltage normally found in the typical plant. These computer systems are commonly single-phase units. Typical nominal voltage levels are 115 or 120 VAC on 60-Hz power systems and 195, 220, or 235 VAC on 50-Hz systems.

Intermediate and large-scale systems require a much larger amount of power and vary in load from 3–5 kVA to more than 30 kVA; the more typical systems average 10–15 kVA. These systems are normally designed for operation from three-phase power sources, although individual units in the computer system are often single phase. Nominal voltage levels are either 208 or 230 VAC for 60-Hz systems and 195, 220, or 235 VAC for 50-Hz systems. Fifty-Hertz systems are sometimes provided with internal autotransformers or internally connected phase to neutral so that the common 380 and 408(415) VAC utilization voltages found in some countries can be used.

A voltage tolerance specification of ±10% is typical for most equipment. The specification is based on steady-state conditions and must be assumed to include transient effects, although most vendors differentiate between transient energy and electrical noise. The ±10% voltage tolerance can usually be assumed to apply to rms voltage measurements including subcycle variations but not necessarily electrical noise.

Frequency tolerance specifications vary from ±0.5–±3 Hz from the nominal 50- and/or 60-Hz value. Poor frequency control affects internal computer power supply regulation and motor speed. If the plant power source is commercial, power frequency rarely presents a problem. However, an industrial engine-driven generator set designed to supply heating, lighting, and motor loads is seldom equipped with better than $\pm 1-\pm 3$-Hz control. Although within the specified limits for some systems, electrical specifications are only one set of variables affecting computer system operation and tighter control provides a safety factor for proper system operation. Good frequency and voltage control pays dividends in lower downtime and better computer system utilization.

13.2.2 Power Sources

Both commercial utility company power and user-supplied, prime-mover-powered generation systems are used as primary power sources for computer control systems. Public utility sources have two major advantages. As stated above, they inherently have excellent frequency control characteristics. In addition, voltage stability at transmission and subtransmission levels is generally quite good because the amount of generation capacity is large compared with connected load changes.

The main disadvantages of utility line power are the following:

1. Generation capacity is seldom dedicated to one customer. As a result, voltage regulation at one customer location is affected by the type and amount of loads switched at other customer locations. The effect of these load-oriented voltage disturbances can sometimes be mitigated by utilization of preferred feeder lines where a dedicated line is provided to a specific customer and other customer loads are limited to continuous duty service.

2. Some portion of the distribution system generally is run on poles above grade. This method of distribution is particularly sensitive to lightning activity in the area. Even though the lightning strike is on some other portion of the distribution system, operation of line protection devices can cause large voltage disturbances on the entire transmission system. Thus power interruptions of several cycles to several minutes duration can occur at any time.

3. Distribution systems are subject to physical damage. Lines, insulators, and poles are broken and cause power interruptions and voltage disturbances on the system. Above-ground distribution systems usually have a higher incidence of failure than do buried systems since they have greater exposure to weather and man.

In some installations a prime mover independent of the commercial utility system is used to power local generation equipment. Steam, diesel, gasoline, natural gas, and other forms of engine drive are used to run these plant generation systems. These systems are not subject to the weather exposure and the mechanical power distribution system failures found in the public utility system. However, they have disadvantages of their own. Engine and generator mechanical failures occur and, generally, maintenance costs are higher than those for power purchased from the public utility. In addition, frequency control is seldom as good as the public utility source. This is simply because the normal utilization equipment for this type of power seldom requires close control of frequency.

Both public-utility- and local-plant-generated power can be used successfully as the power source for the computer system. The decision on which should be used involves the cost of power, its availability, and, in some cases, the question of emergency or back-up power sources.

There are basically three situations in which emergency power sources may be justified. If the plant normally continues operation despite interruptions in the primary power source and if the computer is necessary for continued operation, an emergency power source must be provided. This may already be available if the plant requires some electrical power for sustained operation. In this case, the computer may be operated from the emergency source provided for plant operation, assuming, of course, that the source has sufficient capacity to power the computer system. If the plant does not require electrical power for continued operation, a back-up source of electrical power for the computer system must be provided.

The second situation in which emergency power sources are required is the case in which a power interruption forces the plant operation to be suspended and the computer is required to effect the shutdown procedures. Although the electrical requirements of the back-up source are the same as in the previous case, the emergency power source need not have the capacity for sustained operation. In some cases, only certain subsystems, such as analog output, need be powered from the back-up source.

The third situation in which electrical back-up power sources may be advisable is when the economic value of the process dictates that it be operated continuously even though it could be shut down during a power interruption. If this is the case, it is usually required that the computer also continue to operate in order to maintain plant operation at optimum levels. As before, emergency power may be supplied exclusively for the computer system or it may be derived from existing sources used for plant operation.

The possible sources of emergency power and their cost are generally known to the plant power systems engineer. He must, however, compare the specifications of the emergency power source with the electrical specifications of the computer to determine the adequacy of the back-up source. As noted previously, computer systems may require better voltage and frequency control than is needed for other plant equipment. In addition, provisions for switching from the primary to the back-up source without causing a computer malfunction must be provided. Some vendors offer hardware to effect this transfer as an optional feature. In any case, the vendor can provide the information required to design a suitable switching circuit.

13.2.3 Power Distribution

Plant or building wiring for the computer system follows normal construction practices for low-voltage (less than 600 V) electrical wiring. Wiring can be run in metallic or nonmetallic conduit, busduct, wireways, or as surface cable. Metallic conduit or busduct is preferred because of their better electromagnetic isolation properties.

The typical computer system consists of several units which connect to the building power system. All machines in the main computer system room or area should be connected to a single power distribution panel with each connected unit powered by a separate branch circuit. The riser or feeder servicing the computer room panel should not serve other loads. Both single and three-phase units can be powered from the same building source. Remote units generally can be connected to any convenient power source near their installation site and, since most remote units require small power quantities, multiple units can be connected to one branch circuit.

The three-phase computer system ordinarily can be connected to either a wye or delta connected power system. Since most systems utilize both 120- and 208/240-VAC power in the same unit, some vendors use a step-down transformer in the computer to provide the 120-VAC power. Other vendors rely on the plant distribution system for both the 208/240-VAC and the 120-VAC power. In either case, the plant distribution wiring is identical except for the grounded neutral conductor in the wye connection.

Two standards are commonly used by the computer system design engineer as a source of design data for system AC powering. These are the "National Electrical Code," NFPA #70, published by the National Fire Protection Association in the United States, and the "Canadian Electrical Code," published by the Canadian Standards Association. These standards are used since most state and local code-making groups use these documents as a basis for their laws governing electrical construction. Some local codes do differ from the national codes; however, the difference is usually small and does not appreciably affect facilities design.

A second standard affecting both computer and installation facilities design has been published in the United States. This is "The Standard for Protection of Electronic Computer Systems," NFPA #75, prepared by the National Fire Protection Association. This standard provides general information and classification of computing equipment with respect to design considerations as a function of fire hazard. It also provides information concerning interconnecting cable design and protection and facilities construction practices necessary for the protection of personnel, records, and the computing equipment. While not yet generally incorporated in local construction codes, it is nevertheless a useful document in the design of a process control computer installation.

13.2.4 Computer System Grounding

Three basic grounds are required for the computer system. These are the DC, AC, and equipment safety ground. Both the DC and AC grounds are used to provide a voltage reference point for the computer system circuits. As discussed in Chap. 4, the DC ground may be further subdivided into separate digital and analog grounds. Both AC and DC grounds are used to provide a voltage reference point for the computer circuits. The safety ground is used to limit frame potential to a safe level with respect to building and other conductive material to ensure personnel safety.

Many computer systems are designed for connection to the building power system through an attachment cord and plug assembly. Common design practice in these machines is to provide a separate insulated grounding conductor as part of this assembly. This grounding conductor is used to provide the physical connection between the internal computer system grounds and the building grounding station. When this grounding system is used, it is essential that the building branch circuit and power feeder wiring include a separate ground conductor which provides a permanent low impedance at joints and fittings, since conduit is not recognized as a suitable grounding conductor by the computer vendor.

In addition, the modern computer system often includes some type of filter in the AC input circuit to control conducted line noise in the radio frequency range. These filter circuits consist of both inductance and capacitance, with the inductance connected in series with the circuit conductors and the capacitance connected between the circuit and grounding conductors. If the building wiring grounding conductor is used as the shunt path to earth for the power-line filter current, current flow must be restricted to 1 A or less to meet most testing laboratory listing requirements and code restrictions. Because of more restrictive limitations in some countries, current in the grounding conductor is often restricted to less than 15 mA. These restrictions impose a definite size limitation on the amount of capacitance used and, therefore, on the effectiveness of the filter circuit.

Although it is not possible to provide a grounding system which guarantees that a zero potential ground reference is always available to the computer and instrumentation systems, practical grounding can be accomplished by adhering to the following rules:

1. The connection to ground (earth) should be as short as possible to minimize the AC and DC impedance to ground.

2. The ground resistance measured between earth and the grounding electrode must be as small as possible, preferably less than 1 Ω. Many local codes consider resistance between the grounding electrode and earth to be satisfactory if it is less than 25 Ω. Values of up to 250 Ω are not uncommon in actual practice and, when encountered at an installation site, require that additional grounding be provided.

3. The difference in ground potential between the instrument and computer system grounds should be less than the voltage level of the data signals.

4. The resistance between the power system ground and the computer system ground should be low enough to allow sufficient current flow for operation of fault protection devices such as fuses, circuit breakers, and relays in the event of a system fault.

5. Computer system ground should be isolated from grounding systems used for electrical noise generating equipment. If both grounding systems must be interconnected to meet local code or safety requirements, the connection to the earthed conductor should be located between the computer system and the noise generating equipment.

6. The grounding connection between the computer system and earth should be limited to a single path to minimize the chance of a ground loop.

Normally, the computer system grounding connection proves satisfactory if the following conditions are met:

1. All grounding conductors from individual computer units are carried back to a common grounding bus in the computer room power distribution panel. Units located in remote locations outside the computer room can be grounded at their individual locations.

2. The grounding bus in the computer room power panel is electrically insulated from the power panel neutral (if present) and from the power panel itself.

3. A single grounding conductor, at least equal in size to the feeder circuit conductors, but not less than AWG 6, is connected between the power distribution panel grounding bus and the nearest building grounding station.

4. The isolation transformer case and neutral grounding conductor are connected to the same building grounding station as the computer room power distribution grounding conductor.

5. All grounding conductors are electrically insulated to prevent inadvertent grounding at other than desired locations.

6. All mechanical joints, grounding electrodes, and grounding conductors are periodically inspected both visually and electrically to ensure that proper circuit continuity exists.

In some cases, it may be necessary to provide an additional high-frequency grounding system. This requirement often exists when the grounding method results in a lead length in excess of 50 ft and high-frequency electrical noise, such as that produced by arcing contacts, is present on the power-line conductors. The high-frequency ground can consist of a conductive material of approximately 15 ft² or larger installed in direct contact with building steel or masonry. This provides a capacitively coupled ground for high-frequency noise. A low-impedance connection of 5 ft or less in length is installed between the power distribution panel grounding bus and the conductive material. An alternate solution is to eliminate the effect of the noise by additional filtering and/or shielding at the noise source.

13.2.5 Plant Power System Grounding

In recent years, the trend in low-voltage power distribution system grounding has been toward the solidly grounded method. That is, the power system is directly connected to earth with a grounding conductor. Usually this grounding conductor is carried to the building service entrance grounding station and it is mechanically attached to a cold-water piping system, grounded building steel, or a man-made electrode consisting of driven rods, buried plates, or other conducting material in direct contact with permanently moist earth. Other practices still followed occasionally are ungrounded (except through unintentional distributed capacitance between conductors and the raceway or conduit system), resistance grounded, and reactance grounded. These methods are not recommended for use with process control computer systems.

13.2.6   Electrical Environment

Computer system operation can be affected both by changes in the industrial power system and by radiated electrical energy present in the ambient environment. The total electrical environment can be separated into three basic categories when one considers its effect on computer system operation. These are power-line steady-state voltage or frequency disturbances with a time duration in excess of 1 or 2 cycles, power-line conducted electrical noise or transient effects with a frequency range from the power-line fundamental to megahertz, and electrostatic or electromagnetic disturbances transferred directly to machine frames, covers, and circuits from sources other than the power-line connection.

All computer systems are designed to allow some fluctuation of line voltage and frequency from the rated nominal values without causing a system malfunction. The more common tolerance values are $\pm 10\%$ for voltage and ± 0.5 or ± 1 Hz for frequency variations. If the line voltage varies in excess of the specified value, protective circuitry is energized and the system disconnects itself from the power source. It is then necessary for the operating personnel to restart the computer system manually.

If the voltage change does not exceed the setpoint of the protective circuitry, a different type of failure can occur. The normal use of AC power is for motor loads such as cooling fans and drive motors for electromechanical devices. Although AC voltage fluctuations have little effect on the operation of forced-air convective cooling systems, speed control of the electromechanical drive motors is often critical. Excessive variation in AC line voltage can cause these motors to run fast or slow, and this may cause data recording or reading errors. In addition, since DC power supplies depend on line peak voltage levels for output regulation, DC regulation cannot be maintained if the input line AC voltage is outside of specification. This can also cause a system malfunction. Line frequency change effects are similar to those caused by voltage variations. Poor motor speed control and/or DC power supply output regulation problems result, and system shutdown or an excessive error rate can be expected.

Poor voltage or frequency control often can be traced to an intermittent overload on some portion of the power generation and distribution system. The overload condition is usually a direct result of switching a relatively large load on the industrial power system. For example, an overload can result from the inrush current required to start a large electrical motor or from the effects created by steady-state loading of the building load center transformers above their rated capacity. If the power capacity problem exists in the plant power system, the corrective measures are probably obvious and easily accomplished. If the capacity problem is caused by overloads on the serving utility company distribution system or generation equipment, the solution becomes relatively more difficult. Fortunately, the problem usually exists in the industrial plant power system and is easily solved by providing some electrical isolation for the computer system.

The most common and practical isolation is provided by a separate step-down transformer, since the usual line disturbance is line voltage fluctuations. This parameter is most affected by the *relative* change in the connected load. Thus the effect is reduced at the higher distribution voltage levels simply because the ratio of available capacity to the switched load requirement is higher. For this reason, the ratio of primary to secondary voltage of the isolation transformer should be as high as economically practical.

The second basic broad electrical environment category is conducted power-line noise and transient effects. Most computer systems contain electrical filters to decrease their sensitivity to spurious voltage signals transmitted through the input power lines. The most common approach is to use capacitance to ground as an AC short-circuit for transient energy. In addition, inductance is sometimes added in series with the AC input lines to increase the filtering effectiveness. It is not economically practical, however, to supply the degree of filtration which eliminates all sensitivity to this type of electrical noise. This is especially true when a standard system is provided for use in many different process industries because each installation environment differs in significant respects from every other installation. For this reason, the user must consider noise control measures when installing the computer system.

The more usual sources of electrical noise are interruptions of inductive loads, arc-producing equipment such as brush-type motors and welders, and similar equipment which produces electrical transients. The typical disturbance is an exponential voltage pulse appearing between phase conductors or between the power system grounding conductor and the building structural elements, especially the computer room floor. Voltage levels for pulse trains induced by interruptions of a simple relay coil inductive load can be as high as 15,000 V, and 500–1000-V levels are common. Frequency or pulse repetition rates for these disturbances vary from very low to very high, with the most common being in the high- and ultra-high-frequency range. The propagation rate for the voltage disturbance depends primarily on the electrical characteristics of the transmission line. In the typical computer system industrial power network, the propagation time is on the order of 1.5 nsec/ft.

Since this propagation time is comparable with logic circuit delays, reflections of transient pulses can increase the effects of these pulses on the system. The industrial plant secondary power distribution system contains many branch circuits of different lead lengths, and each branch circuit differs in characteristic impedance. The electrical load connected to each set of branch circuit conductors also varies from circuit to circuit. The amplitude and shape of high-frequency pulses are altered by these different electrical parameters. Since the loads and the transmission lines are not matched, reflections of the original pulse are produced in the distribution system. The voltage division caused by the multiple paths present in the secondary distribution system, the presence of high-frequency short-circuits (capacitance), and resistive loads provide significant attenuation of a noise pulse. However, the parent pulse remains on the transmission line (phase to phase, phase to ground, or phase to neutral) until the negative reflection appears on the same transmission line. Thus even if a perfect capacitor of zero impedance were connected between an infinitely large ground and the noise generator, the voltage between the conductor and ground would be at the noise signal level until the negative reflection returned to the voltage measuring point. For example, a noise pulse remains on a 40-ft length of grounding conductor for approximately (1.5 nsec/ft) $\times$ (2 $\times$ 40 ft) = 120 nsec. This is an appreciable period of time compared with the speed of computer circuits and could cause a malfunction.

Most computer systems are designed to reject high-frequency noise pulses. Typical design conditions for noise rejection are 600–1000 V in the megahertz frequency range. These design limits are satisfactory for most industrial environments. If additional noise immunity is required, the cost is usually lower if the user installs the necessary isolation apparatus in the plant power system rather than purchasing additional filters from the system manufacturer.

The principal methods used for control of conducted power-line noise are source elimination, attenuation of the signal on the transmission path, and desensitization of the computer system.

Source elimination is the only complete solution since this removes the entire problem. This is usually impractical, however, and some electrical noise is always present. A practical solution usually requires a combination of control measures. The following procedures generally provide adequate control:

1. Remove or suppress arc-producing equipment.

2. Use inductive or mechanical coupling when power systems common to both noise generation and computing systems must be used.

3. Provide short low-impedance grounding paths to earth ground for both the noise generating apparatus and the computer system. If these grounds must be common, make the electrical connection to earth between the computer system and the noise source.

4. Make the computer system grounding conductor as short as possible. Use building steel, cold-water piping systems, or other grounded structural building components as a supplemental grounding method.

5. Establish maintenance and control procedures that ensure continued integrity of *both* the power and grounding systems.

The third factor in the electrical environment is electrostatic and electromagnetic radiation. The effect of this radiation is a direct coupling of noise energy into internal computer circuits. Some of the more common radiation sources are radar, radio, television, other communication media, and large magnetic fields caused by large currents.

Typical field strength measurements of the ambient electrostatic field range from a few millivolts per meter in residential and light manufacturing areas to several volts per meter in isolated locations, such as the near field of a "pencil" radar beam. Attenuation of the electrostatic field can be accomplished by a simple grounded reflective screening. Effective screening can be in the form of a thin solid reflector such as a conductive foil, or a reflector consisting of wire screen with a conductor spacing of less than $\frac{1}{8}$ wavelength of the radiant energy source. Attenuation to less than 1 V/m is usually satisfactory when the computer system is in its normal operating condition. During periods of computer system maintenance when unit enclosure covers are open and internal circuit components are directly exposed to the environment, additional attenuation may be required. Thus screening of the computer room should limit the actual radiant energy level to a field strength of 300–500 mV/m.

Allowable electromagnetic field strength limits are more difficult to define. The computer system uses some circuit components that are magnetic in nature. Although magnetic fields may affect these components directly, they can also have a secondary effect, such as influencing circuits which, in turn, influence basic computer functions. An example of this phenomenon is coupling and resultant current flow in signal lines between the computer and process instrumentation.

A more direct effect of electromagnetic fields is the alteration of the characteristics of magnetic devices such as relays, coils, core arrays, and magnetostrictive delay lines. Of less importance, considering the typical magnetic field strength encountered in the industrial environment, is the effect of stray fields on magnetic recording media such as magnetic tapes, disks, and drums. Typically, the limiting field strength for DC fields is in excess of 50 G if only the effects on recording material needs to be considered. However, test equipment such as an oscilloscope is necessary for servicing the computer. Scope trace distortion with non-shielded equipment can occur at intensities of less than 25 G.

Control of electromagnetic fields can be simple or quite difficult. As with conducted electrical noise, control can be accomplished by elimination of the source, attenuation in the transmission path, and desensitization of the receiver. The most effective method is the elimination of the source. This is often feasible simply by relocation of either the source or the computer room. The second effective method is the use of a magnetic material as a shield. Fortunately, normal industrial environments have very limited areas in which intense magnetic fields exist. Thus the solution in most cases is simply to select a computer site where little or no exposure exists.

13.3   AIR-CONDITIONING

The typical control computer uses ambient air as the primary coolant for internal circuit components and electromechanical devices. Air from the ambient environment is introduced into the unit enclosure, routed to the internal component to be cooled by natural or forced convection, and is then exhausted to the room or computer area. Both the ambient air and any airborne contaminants are thus brought into intimate contact with the internal functioning components of the computer system. Control of the basic air properties within the computer system tolerance levels is, therefore, essential if satisfactory system performance and life are to be attained.

For the purpose of this discussion, air-conditioning is defined as the control of specific air quality including its temperature, water content, and contaminant level. Usually, however, vendor specifications include criteria for only the first two air properties. In addition, these data are presented on the basis of maximum tolerance limits with little or no data indicating possible increased performance attainable if closer control is provided by the user. These limited data are generally sufficient for the user's purposes unless optimum operation of the computer on a real-time basis must be provided. In this case, additional data are needed. Typical specifications presented in vendor literature are summarized in Table 13-3.

Table 13-3   Typical System Environmental Specifications

System Unit Type	Condition of Operation	Temperature Range °F	Relative Hum. Range %	Max Wet Bulb Temp °F	Filtration %
Pure electronic	Power on	40–120	8–95	85	20
		32–130	5–95	85	20
Pure electronic	Shipping or long-term storage	−40–140	Maximum 95	—	—
Card, paper, and magnetic tape input/output units	Power on	50–110	10–80	78	20
		65–80	20–80	78	20
		60–90	20–80	78	20

13.3.1   Effects of Air Properties

Each of the basic properties of the air found in the industrial environment can affect the performance of the process control computer. Some of the effects are temporary and reversible, whereas others can cause permanent damage that must be repaired by the replacement of components or subassemblies. In this section, the possible effects of temperature, humidity, and airborne contaminants are discussed. The next section considers air-conditioning systems that can be used to maintain air properties within acceptable limits.

13.3.1.1   Temperature Effects. Both electrical circuit parameters and dimensional stability of mechanical components are affected by changes in ambient temperature. These effects are reversible when specified temperature limits are not exceeded; that is, each component functions in a specific manner when exposed to a given temperature. If the limits are exceeded, however, permanent alteration of operating characteristics can result, and in some cases structural damage to the component occurs.

Electromechanical devices are made up of combinations of materials selected as the best economic compromise to obtain a desired operating characteristic. Since many materials are used, different coefficients of expansion exist within the same mechanical device. For example, the read/write heads of a magnetic recording device could be mounted on a stainless steel supporting bracket attached through a fixed or movable mounting unit which, in turn, is bolted to a basic frame casting. The recording media could be nickel-cobalt plating on a steel drum rotating on bearings mounted in retainers attached to the same base casting. Or, the recording media could be a magnetic material in a ceramic or plastic base mastic coated on an aluminum disk substrate clamped to a spindle which is supported by bearings and bearing retainers clamped to the unit basic casting.

With recent advances in recording technology, the recording density on the magnetic material results in a requirement for tight dimensional tolerance between data tracks. Thus allowances for thermal expansion differences in the mechanical mechanisms are small. Differences in temperature of the mechanism between recording and retrieval of the stored data can cause a change in the alignment of the mechanism and affect data recovery.

Thermal stress also has a characteristic effect on electronic circuits. The most prevalent is a change in circuit impedance which in turn causes alteration of signal characteristics and the voltage or current output from the circuit. In analog circuits, temperature changes can result in parameter variations that affect system accuracy. As in the case of mechanical effects, the design of the computer system is such that these variations do not cause a malfunction within the specified temperature limits. Since the process control system has both analog and digital capabilities, compensation for temperature variation of critical input/output circuits can be accomplished mathematically by the computer program if actual circuit temperatures can be measured or computed. This is, however, an additional burden on the programming system and this approach is rarely used.

Perhaps the most significant variable of concern to the normal process system supplier and user is the inverse relationship between component life and temperature. As with vacuum tubes, operation of solid-state circuit elements well below their critical temperature limits results in a significant increase in the useful life of the equipment. In any computer application, this fact is important to the user since operation at lower temperature levels decreases the amount of time the system is out of service for maintenance.

13.3.1.2 Relative Humidity Effects. Effects of relative humidity of the ambient air are usually subjunctive; that is, the cause and effect relationship is usually indirect. Electrically, changes in relative humidity alter the absolute value of circuit components. This change is usually the result of a change in the capacitance or resistance of circuit conductor insulating materials. This characteristic has little practical effect unless the change is exceptionally large, on the order of 50–75% or more.

Of more direct importance are sustained humidity levels in the range of 0–20% and/or 60% and higher. In the 0–20% range, dissimilar materials build up static electrical charges whenever contact is broken or friction exists between the two surfaces. For example, the operator's shoe soles rubbing on the tile floor can cause the generation of a static charge on the operator's body and clothing. The same condition can exist when chairs and other furniture are made from insulating materials. The amount of charge on the operator or his chair can easily result in potential differences as large as several thousand volts. When contact is made with the grounded machine frame, this stored energy is dissipated by a current flow through the machine covers or frame to system ground. The net result can be system malfunction because of internal coupling of the spark energy into an electronic circuit.

A second manifestation of static charge is improper feeding of cards, paper, and, on occasion, magnetic tape. Jamming of feed mechanisms is caused by the electrostatic attraction of adjacent material surfaces. In addition, static electrical discharges as the material passes through the machine can result in system errors.

Relative humidity levels in the 60–95% range have a more permanent effect on system operation. Many airborne contaminants are either soluble in water or act as desiccant materials. If the contaminant absorbs water and forms a salt solution, it can cause a chemical reaction between certain system components; usually this reaction is the corrosion of some metal part of the system. The presence of high relative humidity does not in itself cause the corrosion; its presence is necessary for the reaction to occur, however, and the rate of corrosion is much more rapid at higher humidity levels.

A second failure mechanism which is an indirect result of high relative humidity is actually caused by condensation. If system components are at a temperature lower than the dew point of the air parcel, condensation forms on exposed parts. Entrainment of water in the convective cooling air stream can expose internal circuits to physical damage. Damage from this source usually occurs on inactive components such as conductors, connectors, and component housings where circuit heat dissipation is minimal.

High relative humidity also tends to increase the amount of moisture associated with a dust or smoke particle present in the atmosphere. An increase in the moisture content of the particle often increases its adhesive characteristics so that small particles tend to stick together. This increases the mass of the particle and, thus, its inertia. As a result of the increased inertia and the inherent increase in adhesive characteristics, more dirt is collected by impingement on moving parts than would be the case if the particles were dry and small.

Probably the most irritating effect of relative humidity for operating personnel is the change in the characteristics of card and paper documents used with the input/output system units. Card stock exposed to variations in relative humidity in excess of 20% require a conditioning period prior to use on the machine to allow all portions of the card to reach the same moisture content. This minimizes distortion which otherwise can cause card jams or other equipment malfunction. A different condition exists with paper stock. This material simply loses tensile strength when exposed to excessive moisture, and destruction of paper forms can occur during printer operation.

A direct result of excessive humidity is degraded common mode rejection performance. As shown in Chap. 3, the common mode performance of the system is dependent on certain leakage impedances in the analog-input subsystem. Excessive humidity can reduce these leakage impedances to the point where the common mode rejection ratio is appreciably lowered.

13.3.1.3   Airborne Contaminants. Two basic categories of airborne contaminants are of primary interest to the computer user. These are particulates and corrosive gases. Particulates can be abrasive in nature and they can become a problem because they block air passages and filter systems, or in combination with water they can cause corrosion of system components.

Magnetic recording devices such as drum, disk, or tape storage units typically use an air-bearing principle to maintain separation between read/write heads and the recording media. The spacing between the head and the recording surface may be less than 100 μin. Although relatively high-efficiency air filtration is used to remove potentially damaging particles for the air stream, small particles can and do tend to come in contact with moving parts. Materials such as smoke, tar, oil, and other heavy liquids tend to collect these small particles. If the build-up of these materials becomes excessive, read/write heads can come in contact with the recording media and cause complete deterioration of both the head assembly and the recording media.

A second mechanism of failure is that induced by relatively hard particles being imbeded in the recording media by impact. During successive passes of the read/write head assembly over the imbeded particle, abrasion of the metal from the assembly can cause changes in air bearing clearances with resultant "crashing" between the assembly slider and the recording media. Both types of failure result in data loss and excessive maintenance.

Corrosive gases are also of concern since many industrial plants use corrosive materials in their manufacturing process. Acids are used in the cleaning and plating of metals; sulfur and sulfur compounds either are used as reducing agents or are released as a product of combustion; chlorine, fluorine, bromine, and other members of the halogen family are used in bleaching and etching processes. Each of these materials can combine chemically with several of the metals used in computer construction. The resultant compound can be a nonconductive film which insulates relay or other contact points; the reaction can also cause a conductive salt or base compound to bridge between conductors, creating short-circuit conditions. If sufficient quantities of the corrosive agent are present, complete mechanical failure of system components or conductors can occur. Any of these failures increases maintenance requirements and, therefore, reduces the availability of the system.

All computer vendors are vitally interested in the corrosion problem and system component design includes provision for reducing system susceptibility. For example, noble metals such as gold, silver, and platinum are used extensively as contact plating materials; mechanical bonds between circuit wiring and circuit elements are formed as gastight joints; zinc, nickel, chromium, and other similar materials are used extensively as protective coatings for metallic parts. Plastic or natural rubber base materials are often used in relatively thin films as a protective seal for exposed parts. In addition, electrical insulation materials are selected both for their electrical characteristics and their resistance to moisture and corrosive agents.

Nevertheless, the typical industrial process environment must be considered hostile to the computer and instrumentation systems. Control of the environment within acceptable operating limits must be provided by the user. Only in this way can the user realize the full potential of computer control.

13.3.2   Air-Conditioning Systems

Air-conditioning was defined as the control of temperature, moisture content, and contaminants in the ambient air parcel. The preceding section discusses the reasons for control. This section reviews the more common methods of providing the required degree of control. Many of the usual control methods used for one parameter also have application to the other two. Conversely, however, some of the control methods for one characteristic can increase the difficulty of controlling the other two.

13.3.2.1   Refrigeration Equipment.   Cooling of the ambient air parcel is usually accomplished through the use of standard comfort cooling refrigeration systems. Engine or electric motor-driven compression systems of either the piston or centrifugal type are used. Gas absorption or steam ejector units also have application in the field. In some unusual processes, an abundant supply of cold water, brine, or other liquid may be available in adequate quantities to provide the necessary heat sink.

The actual cooling coil used can be either the direct expansion type or the chilled water or brine type. The direct expansion system has the advantage of lower initial cost but this type of system is somewhat more difficult to control. The use of chilled water or brine also offers the advantage of simpler modification when changes in design are required.

Compressor, condenser, and fan coil unit capacities are computed in the same manner as for a comfort cooling system for personnel. Sensible and latent heat computations should be based on the accepted comfort air optimum design condition of 70°F at 50% relative humidity. A lower relative humidity value of 30–40% should be used if corrosive materials are present in the environment. There should be adequate capacity to provide control to within ±3°F and ±5% relative humidity with the ability to adjust the basic setpoints within the 20–50% humidity and the 60–85°F temperature range.

Sizing of the condensing unit is usually predicated on computed sensible and latent heat loads. Either air- or liquid-cooled condensing units can be used equally well. The choice is usually made as the result of the normal computations of capital and operating costs for the two different methods.

Sizing of the cooling and reheat coils can differ from that usually used in the normal comfort cooling system. The computer system units usually use ambient air as the basic cooling medium. The air is introduced to the computer unit cabinet near the floor line, routed through the unit by natural or forced convection, and discharged to the surrounding area near the top of the unit. In order to minimize operator discomfort and to ensure effective cooling of the individual components, the temperature rise within the computer cabinet is quite low. Values for this temperature rise are typically between 5–10°F and seldom exceed 15–20°F. Design of most commercially available fan coil units considers a temperature differential across the coil of from 20–30°F, although a value of 40°F is sometimes used. This apparent incongruity need not be of concern to the refrigeration design engineer since the problem can easily be resolved in several ways. The most common is to change the room air flow patterns, as described in a later paragraph. A second method that can offer significant advantages in control of relative humidity consists of a slight modification of the air handling unit and cooling coil. This relatively minor change consists of either increasing refrigerant temperature in the coil or increasing air flow through the coil. Either change reduces the temperature differential of the air stream across the coil and requires an increase in the coil surface area to provide the same cooling capacity.

Standard comfort cooling air-conditioning units can be used as the basic cooling unit in the air-conditioning system. These units have the distinct advantage of lower initial cost and can be easily replaced when damaged. Single units of 2–15-ton capacity are readily available with either air- or liquid-cooled condensing units. Typically, they are self-contained and only require connection to primary AC power sources and a plumbing attachment to drains and cooling water tower (if used). Their main disadvantages are that most have no provision for introduction of water into the air stream for humidification, and particulate filtration efficiency is minimal. Thus modification of the unit is desirable to improve dirt extraction efficiency and humidity control.

Small window units are also available commercially. These units can be used effectively in some small applications. Their most common use is to cool areas remote from the main computer room where small remote terminals are installed for operator guidance and only supplemental cooling is required.

13.3.2.2 Air Distribution. Probably the most difficult problem facing the air-conditioning engineer is air distribution within the computer room or area. Typically, the size of the area is small and the total heat load is quite high. Operator and maintenance personnel are normally in attendance on a full-shift basis, so air noise should be minimized. Air flow patterns are subject to change simply because machine and storage unit layout configurations can change radically several times during a year. In addition, air pressure differentials in the room are not only affected by the air-conditioning system but also by the forced convective patterns caused by the individual computer unit fans.

A process control computer system may convert as much as 10–15 kVA of electrical energy into heat. The total floor space required for such a system is approximately 400–600 ft². If a 10-ft ceiling height and an allowable temperature differential between supply and return air of 15°F are assumed, the air change rate in the room must be between 30–45 complete air changes/hr for just the sensible heat load of the computer system. Even assuming a much higher temperature differential between supply and return air, it is obvious that an unusual air change rate exists in the machine room. Experience has shown that the actual air change rate in a digital computer facility varies from a high of about 90 to a low of approximately 20 changes/hr. The typical rate is probably nearest the low end of the range, between 20–30 changes/hr. The high air change rate need not cause any unusual design or operational problems. It usually can be taken care of by design of the air distribution system using static pressure differentials rather than the inertia of a high-velocity air stream. Good air diffusion is the key to both the potential problems of noise and the short-circuiting of the cold wet supply air either directly into a computer unit or back into the return air system.

Good air diffusion or mixing can be accomplished in several ways. Perhaps the most effective is to do the actual air mixing in the air-handling unit or the duct work connecting the unit to the machine room. In this way, part of the return air can be routed through the cooling coil, cooled to or below the dew point, and mixed with by-pass return air at the air-handling unit. The use of face and by-pass damper systems controlled as a function of relative humidity in the machine room is common. When better temperature and humidity control is desirable, reheat coils can be used.

Air mixing can also take place in the actual machine room. This system is the most economical, but control of both temperature and humidity is more difficult. Its main disadvantage is that individual supply air diffusers must be readjusted when heat and air flow patterns change within the machine room.

Supply and return air can be routed into and out of the computer room or area in many ways. The most common are the following:

1. Installation of the air-handling unit directly in the machine room. No duct work of any kind is used except for cooling air to and from the condensing unit if it is air cooled.

2. Overhead ducted supply and return.

3. Overhead ducted supply with ceiling plenum return.

4. Floor-level ducted supply with a ceiling-ducted or plenum return.

A large computer system is often installed on a raised floor, as illustrated in Fig. 13-2. The space between the raised floor and the plant floor makes an excellent supply air plenum in addition to its primary function as a protective device for computer system interconnecting cable assemblies. When the underfloor space is used to distribute supply air, conditions of the supply air should be changed from those used for overhead supply. Temperatures less than 65–70°F can cause operator discomfort from cold flooring. In addition, both the temperature and relative humidity of the supply air should be near computer system specifications since the supply air can be introduced directly into computer cabinets with little or no mixing with room air. In every case, the temperature of the underfloor air must be above the room dew point to eliminate the possibility of condensation occurring on computer component parts. The optimum settings for this type of air supply system vary between 60–70°F at 50–70% relative humidity. Higher temperatures and lower relative humidity values can be used provided that mixing with room air takes place prior to introduction of the supply air into the computer system cabinets. Heat transfer characteristics of the floor structure and the resultant effects on operator comfort must be taken into consideration if the supply air is appreciably below 65°F. Return air for this type of air distribution system can be ducted, carried back to the air-handling unit by ceiling plenum, or carried directly to the air-handling unit. With a room size in excess of 25–30 lin ft in one direction, some form of ducted or plenum return is desirable to maintain a more even temperature distribution throughout the machine room and to minimize short-circuiting of supply and return air.

Independent of the type of air distribution system used, some make-up air is required. The usual functions of make-up air are odor and oxygen control for personnel. A second function also important to the computer user is computer room pressurization. Pressurization assists in the control of both corrosive gases and particulate air contaminants. The pressure differential between the room and surrounding environment need not be excessive. Typical values used for this type of dust and gas control vary from 0.1–0.2 in. of water. Higher values can be used if significant air velocity is required. However, in the general case it is only necessary to maintain air flow out of the machine room.

The actual quantity of make-up air required for personnel comfort and room pressurization varies with each specific site. For oxygen and odor control, most users establish fresh make-up air quantity on the basis of 15–20 ft³/min for each person normally in the machine room. The quantity needed to overcome losses due to exfiltration through doors, windows, and other openings varies considerably with construction of the room. Usually the practice of using 90% recirculated air and 10% fresh air make-up proves to be an effective compromise.

Relative air pressures in the environment and the air distribution system arranged in descending absolute pressure levels then should be: supply air, room air, outside air (fresh air make-up), and return air. Alterations of this sequence to exchange the relative positions of fresh air and return air can be accomplished by using a fresh air make-up fan system. The use of a positive pressure in the return air system is desirable so that all portions of the room and air-handling systems can be kept at a positive pressure with respect to the surrounding environment.

13.3.2.3 Humidity Control. Most mechanical refrigeration cooling coils are operated at surface temperatures of 45–55°F. This temperature is normally below room dew point and, as a result, some water is condensed from the ambient air stream. If load calculations include sufficient coil capacity for a high latent heat load, all the air passing through the cooling coil can be cooled below dew point and the air moisture is in a saturated state. Reheat of this air moisture is required to provide adequate humidity control. Reheat can be accomplished by the direct addition of heat from an external source or by mixing the cold saturated air mass with higher temperature air. Either method provides excellent control of the dehumidification process since the relative air quantities or the amount of heat added can be controlled by a humidistat monitoring the supply air condition.

Addition of water to the air parcel presents a slightly different problem. In this case a source of moisture must be provided. This can be a liquid or steam source and can be introduced in the air stream at any point in the air cycle. The usual practice is to add moisture in the form of water on the upstream side of the cooling coil if dew point control or water wash design concepts are being used. Spray and centrifugal water atomizers are commercially available and can be installed in supply air ducts. The pan-type humidifiers can be installed in either the return or supply ducts, but usually they are used to add moisture directly to the computer room ambient air parcel.

13.3.2.4   Air Filtration. Most processes used in the manufacture of a product also tend to generate waste materials. These wastes can be particulate, gaseous, or liquid in form. Although modern industry must reduce the amount of waste product dumped into the surrounding atmosphere for both legal and economic reasons, it is seldom possible to remove all the waste products. Therefore, any ambient air parcel contains some undesirable products of both industry and nature. Many of these products cause wear, corrosion, or malfunction of computer system components. The concentration of these materials present in the air environment must, therefore, be reduced to levels consistent with effective computer system operation.

The simplest control measure is filtration of solids from the make-up and recirculated machine room air supply. Two basic filtration methods are effective for this purpose. These are mechanical filters, usually of the impingement or adhesive type, and electrostatic filter systems. The mechanical filter has a lower installation cost than the electrostatic but a higher replacement cost. The electrostatic system can also be more susceptible to corrosion and is limited to use with particles that can be electrically charged.

The dirt loading of any air mass is made up of particles of many different materials, sizes, and shapes. Eighty-five–ninety percent of the total weight of a typical airborne particulate sample is made up of particles that are $5\,\mu$ or larger in size. Conversely, however, 90% or more of all particles in the air sample are less than $5\,\mu$ in size. Filter efficiency ratings are a measure of the filter's ability to remove dirt from the air sample. Filter efficiency ratings can be based on weight of particulate removed, number of particles removed, or a combination of the two. A further refinement of efficiency rating is possible if a "standard" dust sample graded according to particle size distribution is used. The most effective filter efficiency rating method from the computer manufacturer's viewpoint is the one based on both weight and particulate count methods. The United States National Bureau of Standards has defined such a test. Several computer system suppliers base their recommendation for filter efficiency on this NBS test method. A typical specification listed in vendor literature is: "Mechanical filter system minimum efficiency rating of 20% when clean and when tested by the National Bureau of Standards Discoloration Test method using atmospheric dust."

The equivalent filter efficiency rating for an electrostatic precipitator filter system using the NBS test method would be 80% efficiency since the electrostatic filter system operates on a different filtration principle.

Special ultra-high-efficiency filter systems are available for special-purpose applications. Mechanical filter media which remove all solids above $0.3\,\mu$ in size are commercially available. The major disadvantage of these filter media is a relatively high pressure loss across the filter. The pressure drop can exceed 1 in. of water.

Whenever a high-efficiency filter medium is used in the air distribution system, a prefilter of lower efficiency should also be used. Most prefilters are low enough in cost that they can be disgarded when dirty. Even a 1-in.-thick glass fiber filter is effective for large particles. Collection of the large particles by an inexpensive prefilter allows the more efficient main filter to funtion for a much longer period of time between filter changes and thus reduces total filtration costs.

Most electrostatic precipitators should also be used in conjunction with a mechanical prefilter since this method of operation increases filter effectiveness. To control highly corrosive solids, and to minimize dirt carryover into the supply air stream, a mechanical after-filter can also be installed downstream from the electrostatic filter.

13.3.2.5 Corrosive Gas. Control of corrosive gases is much more difficult than control of particulate matter. Some gases are soluble in water and can be controlled by installation of a water wash system. Control of soluble salts and gases is also a by-product of a water wash or spray coil refrigeration system.

Purging for gas control is a common method of dilution of gas concentration levels. This concept presupposes that a clean air source is economically available and can be used for computer room fresh air make-up. Where the quality of the fresh air source is not adequate, special filtration can be used. Both activated charcoal and other materials with specific chemical properties can be used for this purpose.

Probably the most effective method for corrosion control is a combination of several control measures. These include the following:

1. Pressurize the computer room to minimize infiltration.
2. Maintain low ambient relative humidity levels at the machine unit. Values between 20–30% usually prove to be the most effective compromise with other considerations.
3. Reduce particulate contaminant levels by the use of high-efficiency filters.
4. Reduce gas concentration levels by the use of activated charcoal or chemical filters in combination with water wash if the gases are soluble.

13.4 HAZARDOUS ENVIRONMENTS

The "National Electrical Code," NFPA #70, Article 500, defines the various types and degrees of hazard present when the industrial plant or process includes combustible liquids, gases, or solids. Additional data on the definition and control of this facet of the industrial environment can be found in most technical society publications. Of particular interest to the facilities design engineer are those areas where the primary product is the combustible.

Although solid-state switching techniques are used extensively in process control computer signal circuits, some exposed switching contacts are used. So are elements whose surface temperature can exceed 80% of the ignition temperature of several of the common explosive mixtures now in common use. To encapsulate the modern computer system totally or to make it explosion proof is not yet economically practical. Thus it is necessary for the user to provide the control of hazardous environments.

As with particulate control, the most advantageous method of control is purging. Purging does not eliminate the combustible material, it simply reduces the fuel-air (oxygen) mixture below the explosive limit. As with corrosive gas control, the source of fresh air for purging should be as free from the combustible material as practical, and a positive air pressure must be maintained in the computer room at all times when power is on.

Since most control rooms in the process are normally classified as nonhazardous areas and the computer system also needs a similar environment, consolidation of the control and computer rooms can result in significant construction and maintenance cost savings.

13.5 FACILITIES COSTS

The actual costs for facilities construction vary with each selected site and the amount and quality of the facilities already present at the site. These variations in cost can be quite significant. However, it is possible to use some basic cost data to arrive at an approximate budget estimate before proceeding with a complete preliminary design and specification generation. The following data can be used as average values and have sufficient accuracy for use in preliminary construction estimates for evaluating alternate site proposals.

Primary Power

Transformer supply	$100/kVA + $100/circuit
Motor alternator set	$150/kVA + $100/circuit
Engine alternator set	$175/kVA + $100/circuit
Solid-state inverter/converter	$1000/kVA + $100/circuit

Air-Conditioning

Window air-conditioning unit	$200/ton
Self-contained package unit	$300–$50C/ton
Built-up package unit	$500–$800/ton
Complete package system	$500–$800/ton
With good temperature, humidity, and particulate control	$1500–$2000/ton

Space

Raised flooring (installed)	$3–$5/ft²
Partitioning—studwall	$5–$8/lin ft
Partitioning—movable	$10–$15/lin ft
Partitioning—movable steel panel	$25–$35/lin ft
Ceiling—dropped acoustical	$3–$5/ft²

13.6 INSTRUMENTATION AND WIRING PRACTICES

Instrumentation and wiring requirements for process control computer systems are similar in many respects to those required in a well-engineered analog process control installation. However, the increased accuracy, resolution, and speed of response available with a digital process control computer generally require that process, transducer, and signal transmission noise be at a much lower level than is necessary in lower bandwidth analog control systems. As a result, some of the economical wiring installation short cuts that prove satisfactory for analog systems only invite trouble when used on digital systems. Good instrumentation and wiring practices contribute to an economical system by providing prompt start-up through the reduction of installation troubleshooting and rework.

Physical planning of the instrumentation subsystem is the preliminary planning required to ensure good installation practices. This phase of physical planning should include the following:

1. Careful consideration of the application of each instrument, including an understanding of the type of instrument required and the location of the instrument in the process.

2. Detailed specifications for mounting sensors to minimize measurement errors that can be introduced by process noise, structural vibration, ambient temperature changes, sampling lags, and other factors.

3. Planning of signal wiring to minimize potential errors introduced in signal transmission circuits. This includes consideration of wire routing, use and placement of junction boxes, and the types of supports for wire and cable.

4. A thorough understanding of the computer control system input and output interfaces.

Items 1 and 2 are not unique to digital process control computers, and they have been thoroughly discussed in the literature (e.g., Considine and Ross, 1964). Since these factors are common practice in all process control systems, they are not discussed in this section. However, items 3 and 4 have specific considerations peculiar to digital process control computer systems, and these are discussed in this and subsequent sections.

The major concern in the installation of instrumentation for use with process control computer systems is the maintenance of signal fidelity. Signal fidelity requirements are much more stringent than those for the normal analog control system; transmitted signals that are quite acceptable for an analog installation may be totally unsatisfactory for a digital system. The characteristic differences between analog and digital control systems which contribute to the high signal fidelity requirements for a digital system are summarized in Table 13-4. The most important differences affecting installation and physical planning are the increased accuracy and resolution available with the computer control system. In order to utilize this capability, however, noise sources in the process must be carefully controlled. The increased bandwidth of the process con-

Table 13-4   Control System Characteristics

Analog	Digital
Continuous data which permits averaging of ripple or noise	Sampled data system which may read peak value of ripple or noise voltages
0.5–0.25% accuracy available	0.05% or better accuracy available
0.1% resolution available	0.001% or better resolution available
Frequency response flat to 5 Hz or less provides attenuation of ripple and noise	Frequency response may be flat to several thousand Hertz and this provides only minimal noise and ripple attenuation

trol computer analog-input subsystem makes the system more susceptible to random noise errors than is true in the lower bandwidth analog control systems.

It is essential that the physical planning for the wiring between the sensors in the process and the computer system be such that the actual installation does not result in deterioration of the sensor signal. This section discusses potential error sources and suggests installation techniques which minimize their effect on process control computer systems.

13.6.1   Loop Resistance and Source Impedance

When current output sensors (transmitters) are used in process control computer systems, the overall loop resistance for each sensor must be considered to ensure that the sensor is capable of working into the planned load. Computer-controlled systems have the following three characteristics not normally found in analog systems:

1. Longer transmission lines due to centralization of control equipment.
2. The possibility of an additional receiving element since one sensor can be feeding both the computer and an analog controller.
3. High speed of response so that transmission line time constants may become critical in high-speed data acquisition applications.

The desire to be economical on long transmission runs by utilizing small-diameter wire can introduce excessive resistance in the loop and limit the useful range of the sensor. Practice has also shown that the use of small-diameter wire is not economical since many start-up delays are the result of broken wires and connections. This experience dictates that transmission wire size should be selected primarily for the strength required to withstand the rigors of installation, modification, and maintenance. The minimum suggested transmission wire suggested by some industry users is AWG 16, 7-strand copper wire for single-pair cable and AWG 20 stranded copper for multiconductor cables of 10 or more pairs.

The use of series receivers may also result in excessive loop resistance since some transducers are designed to supply only one analog control loop. This is especially true in the case of transducers designed for use in intrinsically safe systems. The choice is to waive the intrinsically safe system requirement or to utilize two parallel transmitters. Because present-day process control computers are not designed to be part of an intrinsically safe system, the first choice is often the more practical.

Since high transmission loop resistance can increase the transmission line time response to an unsatisfactory level, it is necessary to consider this effect if measurement lag might degrade system performance. Although this is uncommon in most process control applications because of the relatively long process time constants, it must be considered in high-speed data acquisition applications.

Loop resistance can also be a limitation owing to the common mode rejection characteristics of the analog-input subsystem. In order to obtain the specified performance, the total loop resistance is usually restricted to some maximum value. In addition, the unbalance between the two sides of the transmission line may be limited to a specified value, typically 300 Ω–1 kΩ. Exceeding either specification can result in degraded common mode rejection performance.

The series combination of the sensor source impedance and the transmission line impedance, in conjunction with the input impedance of the measuring system, forms a voltage divider which attenuates a voltage sensor signal. This attenuation is the same phenomenon discussed in Sec. 3.12.3 in connection with signal conditioning attenuators. In most cases this attenuation introduces only a small error since most analog-input subsystems are designed to have an input impedance of at least 100 kΩ, and most have an input impedance of 10 MΩ or more. The source and transmission line impedance of a sensor would have to approach 50 kΩ to cause a 0.5% maximum error in a system with a 10-MΩ input impedance. This percentage error is approximated from the formula

$$\% \text{ error} \cong \frac{100R_s}{R_{in}}$$

where R_s = sensor source resistance plus transmission loop resistance
R_{in} = computer subsystem input resistance

In high-performance systems or in systems with low input impedance, the error may be appreciable, however, and should be considered in planning an installation. If it is a problem, the error can be compensated

by recalibrating the sensor after installation in the system or by calibrating the sensor with a load which simulates the system input
impedance.

13.6.2 Electrical Noise Pick-up

Since industrial plants contain power lines and equipment such as
electric motors, generators, and relays which can introduce electrostatic
and electromagnetic noise into signal transmission wiring, it is important to reduce the effect of such error voltages. In process control computer installations, the following general rules should be observed:

1. Design or modify the noise generating equipment so that it produces minimum noise.

2. Route instrument signal wiring so that it avoids electrical noise
generating areas.

3. Select instrumentation signal wiring designed to protect against
pick-up of electrical noise.

These general rules are similar to those discussed in Chap. 4 for the
control of noise in analog-input systems. The basic concepts of noise
control are the same whether applied to a subsystem of the control
computer or to the total plant installation.

13.6.2.1 Electrostatic Noise. Many electric fields such as those
radiated by arc-producing equipment are present in an industrial plant.
These fields can be the source of noise in the signal transmitted to the
process control computer if they are allowed to be coupled capacitively
to the signal wires, as shown in Fig. 13-3a. The alternating noise signal,
being superimposed on the transmitted signal by the capacitive coupling,
is known as electrostatic noise.

Since the value of the capacitance is a function of the distance between the noise source and the signal lines, spatial separation is a partial
solution to electrostatic noise problems. Signal wires should not be
routed in the vicinity of noise producing equipment, even though this
may result in longer wiring runs. A minimum separation of five feet is
recommended between the noise source and a parallel run of cable trays
carrying signal wiring. If the signal wire is in solid conduit, the clearance can be reduced to 2.5 ft.

In addition to spatial separation, an effective method of breaking the
capacitive coupling is the shielding of the individual signal pairs and
grounding the shield. The capacitive coupling of the external noise source
is then to the shield and not to the signal wires, as illustrated in Fig.
13-3b. In order to be effective, however, the shield must be grounded at

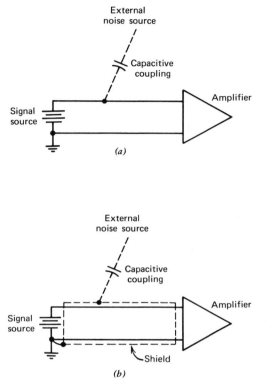

Fig. 13-3   Electrostatic coupling. (a) Without shielding; (b) with shielding.

only one point, and this ground point should be at the zero or reference potential of the signal system. If the shield is not grounded, it is much less effective and it may even accentuate noise coupling. If the shield is grounded at more than one point, creating a ground loop, the resultant current flow causes a voltage gradient along the shield; this is undesirable. The shield should be covered with a thick outer insulating jacket to prevent accidental grounding of the shield at an unwanted point.

Figure 13-4 illustrates two grounding schemes for a grounded thermocouple. The approach shown in Fig. 13-4a is *incorrect* because shield currents may flow along the thermocouple lead wire to ground, introducing an error signal into the signal conductor. Figure 13-4b is *correct* because the shield is grounded at the point considered to have zero potential in the signal environment and the unwanted currents do not flow in the signal conductor.

A recommended electrostatic shielding for signal wires is a 2-mil

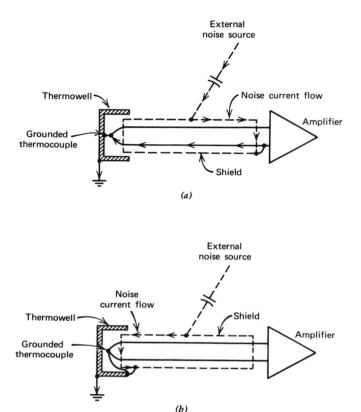

Fig. 13-4 Grounded thermocouple shield installation. (a) Incorrect grounding of shield; (b) correct grounding of shield.

Mylar* tape with a 1-mil aluminum foil helically applied with a 25% overlap, plus an 18-gage standard drain wire in continuous contact with the aluminum foil. The foil is an excellent shielding material and the drain wire reduces the shield resistance to a low value. The combination is an ideal construction for reducing electrostatic noise. In addition, it also reduces installation costs since the design permits faster termination than is achievable with braided shields.

Shields should be continuous and no junction boxes should be used. For this reason, in addition to others, individual twisted and shielded pairs are recommended in preference to the use of multiconductor cable.

* Trademark, E. I. DuPont de Nemours & Co.

Field experience has shown that the overall cost of installing individual pairs is not higher than multiconductor cable. The pulling costs are offset by the extra connections and breaking out of the pairs. Also, fewer initial mistakes are made.

If junction boxes must be used, carry the shield as close as possible to the signal terminals, as shown in Fig. 13-5a. No more than 1 in. of signal wire should be exposed and the shield should be connected to adjacent screw terminals. If connectors are used, the shield should be connected to the pins adjacent to the signal wires to provide maximum possible shield continuity and effectiveness. This is illustrated in Fig. 13-5b.

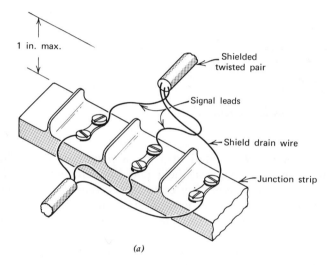

(a)

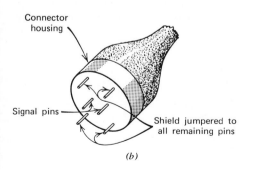

(b)

Fig. 13-5 Recommended shielding practices. (a) Recommended junction box wiring; (b) recommended connector wiring.

13.6.2.2 Electromagnetic Noise. Processing plants not only have many sources that radiate electric fields and cause electrostatic noise, they have many sources that generate magnetic fields of erratic and varying strengths. These magnetic fields can cause what is known as electromagnetic noise in transmission signal conductors.

The AC current flowing in plant power lines and equipment produces a magnetic field in the vicinity of the conductors. When transmission signal conductors are subjected to this magnetic field, current is generated in the transmission loop to oppose the magnetic field. Figure 13-6

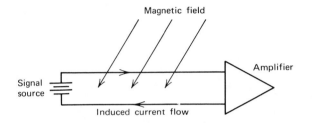

Fig. 13-6   Generation of electromagnetic noise.

illustrates this effect on parallel signal leads in a magnetic field. The diagonal arrows represent the magnetic field and the solid arrowheads on the transmission signal conductors illustrate the current flow induced by the magnetic field. This spurious current flow in the signal conductors creates transmission signal errors identified as electromagnetic noise.

This electromagnetic noise can be minimized by twisting the transmission signal conductors. Figure 13-7 illustrates the canceling effect of current flow in twisted conductors. The diagonal arrows represent the magnetic field and the solid arrowheads on the twisted signal conductors

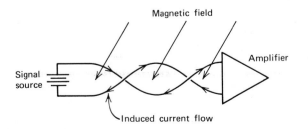

Fig. 13-7   Effect of twisting conductors.

illustrate the direction of current flow induced by the magnetic field, as in the previous figure.

Twisting the signal leads forms a series of loops instead of just one large loop. Hence, canceling, rather than additive, currents flow in each wire. The shorter the lay (i.e., the more loops/ft), the better the magnetic noise rejection. A 2-in. lay is about the optimum from both a cost and noise rejection standpoint. Many refineries, however, specify 8 crossovers/ft, or approximately a 1.5-in. lay, to obtain somewhat better rejection.

Additional protection from electromagnetic noise can be gained by running the twisted and shielded signal leads in conduit or in properly designed cable trays. Care must be taken in installing conduit or cable trays to ensure that their shielding qualities are fully realized and that other noise problems are not introduced. It is very important that conduit or cable trays themselves do not carry a potential. Ideally, both conduit and cable trays should be grounded at *one and only one point*. In addition, good electrical continuity must be maintained for the entire length of the run. In many cases it is extremely difficult to achieve the ideal of single-point grounding. An alternate approach is to install a ground grid throughout the entire plant site. This generally consists of an 00 AWG or larger conductor bonded to the cable trays, conduit, and other signal ground conductors at frequent intervals. Because of the low resistance of this grounding grid, extraneous ground currents flow through it rather than through the trays, conduit, and other signal ground conductors.

Cable trays should have essentially solid bottoms, sides, and a tight cover. Ventilation and drainage slots are required, of course, but the shield coverage should be no less than 85%. To take full advantage of the shielding, the covers should make good electrical contact with the sides. In extreme cases, additional shielding can be achieved by wrapping the insulated shielded pairs with several layers of a high-permeability magnetic tape. The high cost of this tape, however, precludes its general use in field installations of long transmission runs.

In addition to these practices that apply to the signal wiring, similar procedures can be applied to the noise source. Magnetic shielding can reduce external fields from power lines and equipment. Installation of power lines in steel conduit provides such magnetic shielding. In addition, twisting the power conductors reduces the resultant magnetic field.

Spatial separation of magnetic field producing equipment and signal wiring is also an effective practice. Thus the same precautions applicable to wiring practices for the avoidance of electrostatic noise also apply here. An often suggested separation between signal wires in conduit or

Table 13-5 Recommended Power and Signal Line Separation

Power Wiring Capacity	Minimum Separation
125 V 10 A	12 in.
250 V 50 A	18 in.
400 V 200 A	24 in.
5 kV 800 A	48 in.

in grounded cable trays having 85% coverage is given in Table 13-5. In addition to maintaining a suitable separation, parallel runs of signal and power wiring should be avoided. When crossovers are necessary, they should be made at right angles with a separation of no less than 1 ft.

13.6.3 Common Mode Noise

Common mode voltage has been defined and its effect explained in describing the analog subsystem designs in Chap. 3. There are many sources of common mode potentials with which one must deal in applying a digital process control computer. These sources generally can be divided into two categories. The first category covers common mode potentials caused by the nature of the signal source. The second encompasses sources due to unintentional phenomena, such as noise or ground loops.

Some common examples of the first category are grounded thermocouples, bridge circuits, and current output transducers. Common mode potentials of any of these can range up to 50 V or more. These potentials are much higher than those of the unintentional sources, but their effect on the system performance is often much less. This is because they are known and they are relatively well-behaved sources. Thus their effect can be anticipated and, if necessary, suitable steps can be taken to compensate for their effect through hardware or software techniques.

As a specific example, bridge circuits are common in many different measurement systems. They are used in transducers for the measurement of strain, pressure, temperature, and resistance. Figure 13-8a shows an installation where a common mode potential exists. Assuming that the measurement system has a high input impedance so that the current flow between A and B is negligible with respect to the current flow through the bridge resistors, the common mode potential V_{cm} is

$$V_{cm} = \frac{V_B R_2}{R_1 + R_2}$$

where V_B is the bridge excitation voltage. The common mode voltage,

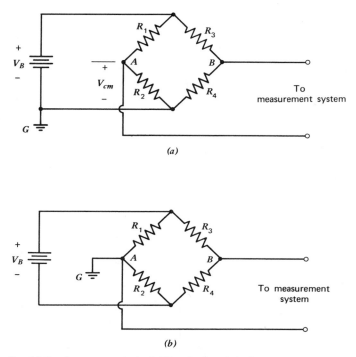

Fig. 13-8 Common mode in bridge circuits. (a) Common mode potential present; (b) no common mode potential present.

therefore, is between $0-V_B$ V. Typically, the bridge excitation is in the range of 0–20 V. If the normal mode signal between A and B is 10 mV, a 10-V common mode voltage results in an error of 0.1% for a system with a common mode rejection ratio of 120 dB. Depending on the application requirements, this error may or may not be tolerable.

Figure 13-8b illustrates how this common mode potential can be eliminated. The ground G in Fig. 13-8a is removed and transferred to point A in Fig. 13-8b. Thus Fig. 13-8b shows a ground-referenced input signal to the measurement system. Figure 13-9b shows that this solution is not possible if the power supply provides excitation for more than one bridge. The bridges are connected to ground at A and A' to provide ground-referenced output signals V_o and V_o', and both bridges are driven by an ungrounded power source V_B. In this configuration, however, the bridge resistors R_1 and R_1' are in parallel, as are R_2 and R_2'. As a result the characteristics of each bridge are altered. In addition, if either R_1

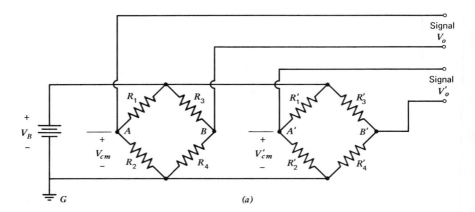

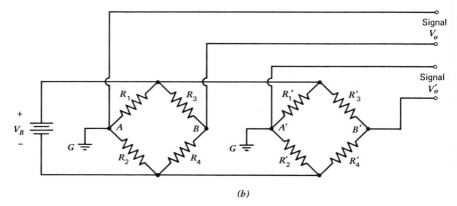

Fig. 13-9 Multiple bridge circuits. (a) Common mode potential present; (b) no common mode but bridge inoperable.

or R_2 is a variable element in the bridge, changes in the bridge cause a change in both output voltages. This is unacceptable. Thus there are two alternatives:

1. Rely on the common mode performance of the system and install the bridges, as shown in Fig. 13-9a, without a ground-referenced output signal.

2. Supply each bridge with individual excitation, as shown in Fig. 13-8b.

Another common source of common mode voltages is associated with current output devices such as transmitters and controllers. These voltages are normally of much greater magnitude than those encountered in the bridge circuits previously discussed. In some controllers, the common mode voltage exceeds 50 V. The general case is illustrated in Fig. 13-10. This represents the process control computer monitoring the

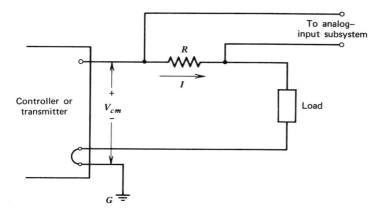

Fig. 13-10   General current output device.

current output I of an analog controller or transmitter. The current I flowing into the load is determined by measuring the voltage across R. The common mode potential is essentially equal to the source voltage provided by the controller. If the current from the controller returns to a ground point, the common mode can be eliminated by placing the resistor in the grounded leg, as shown in Fig. 13-11. However, in prac-

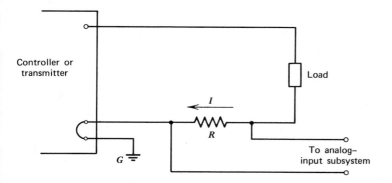

Fig. 13-11   Elimination of common mode.

tice this is not always possible because of the particular design of the current source or because of noise considerations.

As an illustration, a typical current output circuit is shown in Fig. 13-12. The output current is maintained constant by the feedback action

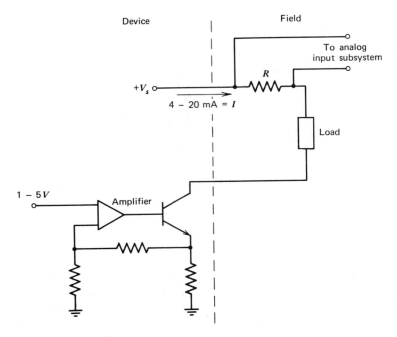

Fig. 13-12   Typical current output source.

from the emitter resistor to the amplifier input. As the amplifier input is varied from 1–5 V, the collector current and, therefore, the load current vary from 4–20 mA. The common mode potential at the resistance R is equal to the voltage V_s. This voltage varies with the system design and may be as high as 65 V; the lowest value presently available is 24 V. The common mode potential is reduced slightly by placing the measurement resistor on the other side of the load in series with the collector of the output transistor. The common mode cannot be eliminated, however, unless the output stage is modified or unless an output tap is provided on the emitter resistor for monitoring the output current.

If the common mode potentials due to the nature of the signal exceed the common mode limitations of the analog subsystem or are considered to be hazardous to personnel, steps must be taken to isolate them from the process control computer. Isolation can be achieved by modifying the transducer or measurement technique or by using an isolating device that can withstand the common mode voltage.

The modification of the transducer might consist, for example, of re-wiring the power supply to provide a ground-referenced measurement signal, as was suggested for the bridge circuits of Fig. 13-8. In the case of common mode due to the measurement technique, perhaps another measurement can be substituted and the desired parameter calculated from it.

The common mode potential may also be caused by the process characteristics. This might be the case where a bonded thermocouple is used to measure the temperature of an object which is at some potential with respect to ground. For example, measuring the temperature of an electrolytic refining process often results in a common mode potential. In this case, the measurement technique itself might be changed; for example, a bonded thermocouple might be replaced by one that is insulated from the thermowell.

If these techniques cannot be employed, an isolator can be used. These devices are per-point amplifiers or voltage-to-current transducers designed to operate with high common mode voltages. Some isolators can be used with common mode potentials up to 700 or 1000 V. Their disadvantages are that they are used on a per-point basis, and this may represent a significant capital expenditure if a large number are required. In addition, they contribute to system inaccuracy, which may be significant with some devices, particularly those based on the use of magnetic amplifiers.

It is extremely important to identify high common mode voltage sources prior to the installation of the computer control system. Each transducer and measurement should be reviewed during the system planning phase to determine whether common mode voltages are present. This is particularly important if the analog-input system has a 15 or 20 V common mode limitation, a common situation with some types of solid-state multiplexers and amplifiers. Failure to identify these sources can result in system damage and/or system installation delays.

Care should be taken in the installation of the instrumentation so that common to normal mode conversion is minimized. As explained in Chap. 3, a primary cause of this conversion is imbalance in the instrumentation lines and in the line-to-ground leakage impedances. Although lines and leakages can theoretically be balanced, this is difficult to achieve in practice because of the distributed nature of the line and leakage impedances. For this reason, the installation should avoid the introduction of unnecessary imbalance in the lines.

Twisting the signal conductors tends to maintain equal capacitances to ground. Similarly, selecting wiring with good insulation minimizes imbalance resulting from leakage between the signal conductors and ground. Junction boxes are a source of leakage and imbalance and should be avoided if at all possible; if necessary, however, the terminal boards used in the boxes should be selected for high leakage resistance properties. Terminal board materials with this property, plus fair mechanical strength and excellent fire resistance, are mineral diallyl phthalate and polycarbonate.

The second category of common mode sources, those due to unintentional phenomena, is often the most troublesome even though the common mode potentials rarely exceed a few volts. The problem stems from the unpredictable and often sporadic nature of these sources. Unintentional sources can be characterized by the means through which they produce the common mode potential. Specifically, the common mode voltage may be attributable to electrostatically coupled currents, electromagnetically induced potentials, or conducted currents caused by unintentional grounds.

Electrostatic and electromagnetic noise have been discussed in the preceding sections of this chapter and in Chap. 4. These discussions have emphasized normal mode noise, but common mode noise is generated by essentially the same process. The only difference is that both conductors are treated as a single conductor. For example, flux passing through the loop formed by two signal conductors results in normal mode noise. Common mode noise, however, is generated by the flux passing through the loop formed by the transmission pair and the ground return circuit. Similarly, capacitive coupling of noise into both conductors of a transmission pair can result in common mode noise.

The elimination of common mode voltages from such unintentional sources follows the concepts presented before. The most effective is the elimination of the noise source either by modification of the noise producing equipment or by appropriate shielding techniques. Spatial isolation of transmission lines and the noise source is also effective. Thus the precautions of routing discussed earlier are applicable in the case of common mode as well. The other technique, electrostatic and/or electromagnetic shielding, also helps to eliminate the effects of common mode potentials. By the use of properly connected low-impedance shields, the common mode potentials are shorted to ground and do not interfere with the measurement signal.

13.6.4 Crosstalk

Crosstalk is the coupling of one signal into another transmission line pair. AC transmission signals from such devices as AC strain gages, magnetic flowmeters, conductivity meters, and linear variable differential transformers (LVDTs) are one source of crosstalk. Another source is pulsating DC signals, as found in telemetry systems, turbine flowmeters, and low-level solid-state alarm systems. These on-off or oscillation signals can be superimposed on adjacent low-level transmission signal lines, creating noise errors in the transmitted signals. Crosstalk is primarily electrostatic, rather than electromagnetic, in nature.

Routing is one approach in reducing crosstalk effects. Segregate the various types of signals and run them in separate conduit or trays. This means the AC signals should be separated from other types of signals and run together. For example, all turbine meter transmission lines run together and all LVDT conductors run together. This reduces the coupling into DC transmission signal lines, but crosstalk can still exist between the turbine meter lines and the LVDTs.

The AC lines can be "balanced" to reduce the effect of crosstalk. "Balanced" lines are individual pairs of AC signal transmission lines isolated by a pair of transformers, one at each end of the transmission line. Both these transformers have their center tap connected to the same common ground. This is quite easy to accomplish for LVDT installations and for some magnetic flowmeters, but it is hardly practical or economical for most other AC applications.

Actually, shielding and grounding the shield are very effective. Both the noise generating pairs and the receiving pairs should be shielded and grounded. In addition, maximum spatial separation between different types of signals should be maintained.

13.6.5 Ground Loops

A ground loop is a conductive path between two grounded points in a system. Ground loops cannot be tolerated in signal transmission conductors or shields in a digital system, any more than they can be tolerated in analog systems (see Sec. 4.3.3.3.). Current flowing in the ground loop can introduce normal or common mode noise into signal circuits, depending on the particular current path. Careful initial planning of the system grounding structure is required to avoid ground loops in the signal conductors, their shield ground, and the conduit or cable tray grounds. Some of the more common installations and suggestions for avoiding ground loops are discussed in this section.

The key point in the avoidance of ground loops is that a ground loop is a conductive path between two or more grounded points in a system. If the ground system contains only one grounded point or, if there is more than one ground point, the ground points are at precisely the same potential, then no current can flow in the ground path. Actually guaranteeing that this is the case, however, can be difficult in a system that contains hundreds of signal transmission lines and transducers. Only careful planning and installation can provide assurance that all ground loops have been eliminated.

Experience has shown that shields are often involved in ground loop problems. For example, a shield is inadvertently grounded at both the transducer and the process control computer system. As a result, a ground current flows between the two ground points through a signal conductor and thereby introduces a noise signal into the measurement path. As in most ground loop problems, this interference often shows up as a lack of repeatability in the system readings. To check for such an unwanted ground, remove the shield from the point at which it should be grounded and make a continuity check to system ground. If a short or low resistance is present, this indicates that another point in the system is grounded. Carefully tracing the routing of the shield and the condition of its insulation and termination usually reveals the unwanted ground. This same procedure can be used for detecting suspected ground loops caused by other paths as well.

In addition to inadvertent connections of the shield to ground, there are a number of other common sources of multiple grounds. If the output of a transducer is single ended and referenced to the transducer case, a ground loop may be the result of an inadvertent circuit through the signal conduit which is connected between the transducer case and the computer system grounded frame. In this case, the transducer cases may remain grounded but a short section of insulating conduit should be used to isolate the transducer from the computer system. Similar problems exist when signal conduit or busducts are connected to the computer frame and to the building steel. Isolating the conduit from the system or building steel is one solution to this problem. Another is to provide a low-impedance connection between the building steel and the computer system ground to ensure that they remain at precisely the same potential. Some users of this approach specify that the resistance between any two points in the grounding system must be less than 3Ω.

In planning the system installation, the grounding system should be carefully considered. Each transducer should be examined to determine its internal grounding arrangement. With this knowledge, a circuit diagram of the grounding structure can be prepared and possible ground loops can be identified. During installation, continuity checks should be made to ensure that unwanted grounds have not been added.

Even after these precautions, however, it is sometimes found that ground noise interferes with system operation. This is largely due to the distributed nature of the wiring system which precludes a complete understanding of the grounding system on a discrete basis. Thus a certain amount of empirical engineering may be required. It may be found, for example, that a particular transducer works best when the shield is grounded at the computer system rather than at the transducer, the preferred grounding point. These cases could be explained if a sufficiently detailed analysis were performed. Usually, however, the empirical solution is accepted and the unexplained phenomenon is attributed to the "art" of grounding. Except in extreme cases, this is probably a reasonable approach.

Sometimes, after months of trouble-free operation, the computer system operation may suddenly become erratic owing to noise problems. The source of this noise may be any one of several things. A ground loop may have been created as a result of modification of the instrumentation or the installation of a new instrument. Another common cause is deterioration of the insulation, which results in an unwanted ground in the system. Or new equipment that is a source of noise may have been installed in the process.

Finding and eliminating the case of such noise can be a long and painstaking task. Any recent modifications to the process should be examined to see if they might account for the interference. Similarly, changes in the operating characteristics of process equipment should be noted. This might reveal, for example, a motor in which the brushes are arcing from lack of maintenance. If possible causes such as these are not found, the analog-input system and the instrumentation wiring should be checked for unintentional grounds resulting from damage or deterioration of the wiring. Often a cut-and-try method of disconnecting certain inputs is a useful technique for isolating the noise to a particular area of the process or computer system. Once localized, the source of noise is easier to find.

13.6.6 Miscellaneous Noise Sources

Chemicals and moisture present in the air of processing plants can cause corrosion at junctions of conductors, shields, conduit, and cable trays. Depending on the chemical products of corrosion, these junctions can be converted into DC voltage sources such as AC rectifiers or batteries. These DC voltage sources introduce errors in the transmitted signals. Corrosion of conduit or cable tray junctions reduces their electrical continuity to ground and increases the effect of current flow in producing normal and common mode noise.

Splices of conductors or shields should be avoided if at all possible. If junctions must be made, they should be made at terminal boards housed in gasketed junction boxes. Bags of desiccant in the junction box ensure a dry interior, reducing the corrosion rate. Both surfaces of conduit or cable trays should be clean and treated with a corrosion inhibitor prior to joining.

Another source of noise encountered in process control computer systems is due to the parallel connection of instrumentation. For economic reasons, a thermocouple or other transducer output might be connected to both an analog-input channel of the system and to a multipoint recorder in the control room. Depending on the nature of the analog-input subsystem and the recorder, this may cause an inaccurate reading in either the computer or the recorder. If the computer attempts a conversion at the same time that the multipoint recorder is connected to the point, the noise generated by the recorder may cause an error in the analog-to-digital conversion. In some cases, the action of the analog-input subsystem may cause an erroneous reading by the recorder. This might be the case, for example, when a flying capacitor multiplexer is used. As the capacitor is returned to the input side of the multiplexer circuit, a slight current surge recharges the capacitor. Depending on the characteristics of the recorder, this may result in a voltage spike which saturates the recorder amplifier. Similarly, chopper-stabilized amplifiers can inject a carrier signal into the instrumentation lines which can affect the operation of other equipment connected to the same lines. It is very difficult to predict such affects on other than an experimental basis. For this reason, most vendors recommend that separate instrumentation be used for the process control computer system inputs.

Other errors are also possible in both the data acquisition and control portions of the computer system. Many of these, such as thermal potentials, are discussed in Chap. 3 and Chap. 4. Although the discussion in these chapters is primarily concerned with sources of noise and errors internal to the analog-input subsystem, most of the discussion applies to the instrumentation system as well. The discussion of those chapters should be reviewed in the design and installation planning for the instrumentation system.

13.7 OUTPUT CONTROL SIGNALS

As discussed in previous chapters, the control signals generated by the computer for the actuation of control elements in the process can be either analog or digital. The analog signals are usually voltage signals in the range of 0–5 V or current signals in one of the standard ranges of 1–5 mA, 4–20 mA, or 10–50 mA. The digital signals may be either pulse duration signals or a fixed number of constant amplitude and duration pulses.

In the case of analog signals, the signal may be sustained by a digital-to-analog converter (DAC) on each channel. In essence, the DAC provides digital storage for the output control action. In other systems, a distributed analog-output system is provided in which a single DAC and multiplexer is used to store signals in analog memory circuits. The output of these circuits then drives the control element. The digital signals are most often used to actuate an electromechanical device such as a stepping motor. In this case, the control information is stored in the mechanical position of the device.

The installation planning and wiring considerations for analog-output signals are similar to those discussed in previous sections for analog-input signals. The control signals are high level, however, which usually means that noise and other problems have less effect on the system operation. In addition, the final control elements have limited bandwidth so the practices common to conventional analog control systems are usually applicable. Caution should still be exercised, however, to avoid ground loops and other sources of noise. Not only does this improve the performance of the analog-output system, but it may also prevent analog-input noise problems due to coupling between the input and output subsystems. A ground loop, for example, which is produced by an improperly grounded current-to-pneumatic transducer, may affect the operation of analog-input points located in the same vicinity.

Analog outputs used for graphical displays require installation practices similar to those used for high-level analog-input systems. In many cases, these are high-speed signals used for generating oscilloscope displays. Since the oscilloscope is a wide-bandwidth instrument, high-frequency noise can affect its operation. Appreciable noise can result in a fuzzy trace that reduces the clarity of displays. If the oscilloscope is used for alphameric display, the noise may cause small characters to be unreadable. Spikes and other transients on the analog-output signal can result in streaks or extended serifs on letters in the display. If severe enough, these extraneous traces can significantly reduce the contrast in the display and make it difficult to read under normal lighting conditions.

Pulse signals can also be affected by noise and the characteristics of the output signal wiring. If the receiving device is sensitive to narrow pulses or to amplitude variations, spikes or other noise superimposed on the pulse output can cause faulty operation. Owing to the nature of most ouput devices, this is not usually a serious problem. The control elements are limited in bandwidth and relatively insensitive to pulse variations. If long transmission lines are necessary, however, the characteristics of the lines can sometimes distort the output pulses to the point where the output device fails to respond properly. Where long transmission runs are anticipated, the specifications on the control equipment and the computer output should be carefully examined to determine the allowable degree of pulse distortion. In addition, the vendor specifications for output wiring should be followed. In special cases, the vendor can usually provide advice and, if necessary, equipment to minimize pulse distortion.

One problem that can be caused by output pulse signals is interference with analog-input signals. Compared with many analog-input signals, the output signals are relatively high power. As a result, they can couple into nearby low-level analog-input signal lines. In addition, they can cause pulse current to flow in the grounding system which can couple noise into the analog-input equipment.

For these reasons, the precautions discussed earlier for noise control in analog-input subsystems should be applied to both analog- and digital-output signals. This includes elimination of high-frequency transients in the output signals, shielding to prevent crosstalk, and proper grounding practices to reduce ground loops. Digital control signals should not be run in the same conduit or cable trays as analog-input and analog-output signals. Maintaining spatial separation between these signals is an effective control measure.

BIBLIOGRAPHY

Bean, K., "The Forgotten Costs of Installation," *Autom. Data Process.*, August 1962, pp. 34–35; September 1962, pp. 20–23.

Buchmann, A. S., "Noise Control in Low Level Data Systems," *Electromech. Des.*, September 1962, pp. 64–81.

Clifton, H. W., "Design Your Grounding System," *IEEE Trans. Aerosp.*, April 1964, pp. 589–596.

Considine, D. M., and S. D. Ross, (Eds.), *Handbook of Applied Instrumentation*, McGraw-Hill, New York, 1964.

Decker, R. O., "System Aspects of Control Computer Application," *Autom.*, September 1961, pp. 76–79.

Dorf, R. C., *Modern Control Systems*, Addison-Wesley, Palo Alto, Calif., 1967.

EID Staff, "Grounding Low-Level Instrumentation Systems—A Coherent Philosophy," *Electron. Instrum. Dig.*, January–February 1966, pp. 16–18.

Ficcke, R. F., "The Grounding of Structures," *IEEE Trans. Aerosp.*, August 1963, pp. 999–1007.

Fribance, A. E., *Industrial Instrumentation Fundamentals*, McGraw-Hill, New York, 1962.

Harris, R. A., "The Computer-Actuator Interface," *Control Eng.*, September 1966, pp. 110–113.

Hoffart, H. M., "The Grounding Concepts for Instrumentation Grounding as They Differ from Lighting and Power Fault Safety Grounding," *IEEE Natl. Conv. Rec.*, 1966, pp. 11 ff.

Hoffart, H. M., "Instrument Grounding Keeps Space Signals Clean," *Instrum. Technol.*, August 1967, pp. 41–44.

Howden, P. F., "Field-wiring Specifications for Low-Level Data Systems," *Electron. Eng.*, May 1968, pp. 249–253; June 1968, pp. 318–320.

Ida, E. S., "Reducing Electrical Interference," *Control Eng.*, February 1962, pp. 107–111.

Jackson, S. P., "Economics of Selecting Standby Power Sources," *ISA J.*, November 1966, pp. 35–39.

King, L. A., "Minimizing Process Computer Maintenance—Installation," *Instrum. Technol.*, January 1968, pp. 62–64.

Klipec, B., "Controlling Noise in Instrument Circuits," *Control Eng.*, March 1968, pp. 75–76.

Lawrence, J. A., "Wiring Recommendations for Computer Applications," General Electric Co., Phoenix, Ariz., n.d.

Loewe, R., "Instrumentation Enclosures for Hazardous Areas," *Instrumentation in the Chemical and Petroleum Industries—Vol. 4*, W. H. Matthews (Ed.), Plenum Press, New York 1968, pp. 13–18.

McAdam, W., and D. Vandeventer, "Solving Pickup Problems in Electronic Instrumentation," *ISA J.*, April 1960, pp. 48–52.

Magison, E. C., *Electrical Instruments in Hazardous Locations*, Plenum Press, New York, 1966.

Maher, J. R., "Cable Design and Practice for Computer Installations," *TAPPI*, May 1965, pp. 90A–93A.

Minor, E. O., "A New Way To Look at System Grounding," *IEEE Trans. Aerosp. and Electron. Syst.*, July 1966, pp. 754–761.

Morrison, R., "Shielding Signal Circuits," *ISA J.*, March 1966, pp. 58–59.

Morrison, R., *Shielding and Grounding in Instrumentation*, Wiley, New York, 1967.

Nalle, D., "Elimination of Noise in Low-Level Circuits," *ISA J.*, August 1965, pp. 59–68.

Ogle, F. T., "On Line Instrumentation and Control Systems," *Chem. Eng. Progr.*, October 1965, pp. 87–92.

Perry, V., D. Kirk, and J. Rolf, "Computer Control of a Fine Paper Machine," *TAPPI*, June 1965, pp. 52A–60A.

Phister, M., Jr., "Checklist for Large-Scale On-Line System Planning," *Control Eng.*, November 1967, pp. 87–89.

Rhodes, J. C., "The Computer-Instrument Interface," *Control Eng.*, September 1966, pp. 105–109.

Robinson, T. A., "Definition of Grounding Terms for Power, Electronics, and Aerospace," *IEEE Trans. Aerosp.*, August 1963, pp. 979–981.

Sladek, N. J., "Electromagnetic-Interference Control," *Electro-Technol.*, November 1966, pp. 85–94.

Soule, L. M., Jr., "Control Instrumentation/Computer Compatibility," *Instrum. Control Syst.*, May 1968, pp. 111–117.

Stewart, E. L., "Grounds, Grounds, and More Grounds," *Simulation*, August 1965, pp. 121–128.

Stewart, E. L., "Noise Reduction in Interconnect Lines," *Simulation*, September 1965, pp. 149–155.

Swartout, C. J., "The Design of a Compatible Electronic Control System for the Process Industries," *Proc. ISA*, 1962, pp. 32.3.62–32.2.71.

Togneri, M., "A New Approach to Instrument Wiring Design," *Proc. Texas A&M 22nd Ann. Symp. Instrum. for the Process Ind.,* January 1967, pp. 44–48.

Vogelman, J. H., "A Theoretical Analysis of Grounding," *IEEE Trans. Aerosp.,* April 1964, pp. 576–588.

von Loesecke, P., "Reducing Noise with Intelligent Cabling," *Instrum. Control Syst.,* August 1967, pp. 106–109.

von Loesecke, P., "More 'Noise' About Noise," *Proc. Texas A&M 22nd Ann. Symp. Instrumen. for the Process Ind.,* January 1967, pp. 49–56.

Weiss, M. D., "Getting the Control Systems Engineering Done," *Proc. ISA,* 1962, p. 6 ff.

Weiss, M. D., "Organizing for Systems Engineering," *Control Eng.,* May 1963, pp. 89–94.

Zoss, L. M., and B. C. Delahooke, *Industrial Process Control,* Delmar Publ., Albany, N.Y., 1961.

"Bibliography on Interference in Low Level Circuitry," *IEEE Trans. Ind. Electron. and Control Instrum.,* September 1964, pp. 50–54.

Grounding and Noise Reduction Practices for Instrumentation Systems, SDS Data Systems, Pamona, Calif., n.d.

"IBM 1800 Data Acquisition and Control System Installation Manual—Physical Planning," Form A26-5922, IBM Corp., San Jose, Calif., 1966.

"Manual on Installation of Refinery Instruments and Control Systems, API RP-550," American Petroleum Institute, New York, 1965.

"Selection Guide: Wire and Cable for Process Instrument Applications," Dekron Division, Samuel Moore & Co., Mantua, Ohio, n.d.

"Shielding, Grounding and Conduit Layout for Low Level Input Circuits," *Bulletin Form MB-2034-CSD,* Westinghouse Control Systems Div., n.d.

Chapter 14

Application Programming

GORDON W. MARKHAM

The task of the application programming team is to produce a set of instructions that direct a process control computer system to implement the control of some particular process. The programming systems described in Chap. 9 facilitate this task by providing programs and subroutines to handle general functions, such as compiling, managing the use of bulk storage, and the housekeeping involved in responding to interrupts. Using these general tools, the application team must produce routines to read signals from the process and convert them to engineering units, communicate the values to operating personnel, and compute and implement control actions. The team must also provide means for the process engineer to modify parameters and control actions.

Programming for process control is much like programming for any other application of computers, but there are some special problems. The problems are a result of four factors:

1. The economics sometimes dictate a small computer that must be programmed efficiently for many applications.

2. Many levels of interrupt are required to handle the execution priorities efficiently.

3. The state of the process control art is such that the detailed definition of the tasks to be accomplished changes as experience is accumulated; thus very flexible programs must be used.

4. Process control is usually a 24-hr operation. Heavy emphasis is placed on the requirement that the control system should operate continuously. As a result, it should be possible to perform minor program modifications on-line.

Any programming project goes through five phases: problem analysis, program design, coding, debugging, and documentation. In the management of the project, people must be assigned, schedules established, and progress controlled. Sections of this chapter discuss each of the phases of a process control programming project and the management of such an effort.

Until very recently there were no application programming systems. Every project team wrote a special set of programs for its own installation. The advent of better price/performance computers has enabled some of the higher speed to be traded for programming effort by using standard application programs. These standard packages contain programs to handle functions which are common to most process control projects, such as measuring, alarming, communication, and the common control algorithms. This chapter can be used either to guide the writing of application programs or to help in the selection of program packages.

14.1 PROBLEM ANALYSIS

The objective of the problem analysis phase of a programming project is a set of functional specifications. The specifications should describe what is to be done, but not *how* it is to be done. The responsibility for defining the tasks belongs to the process engineers, but the programmers must work closely with the engineers to ensure that all the required information is provided. The functional specification should cover the following:

1. Describe the operation to be performed.
2. Provide an estimate of the frequency at which the operation is required or, if the intervals are random, an estimate of the range of frequencies.
3. Give an estimate of the maximum allowable time between occurrence of the triggering condition (elapsed time or process status) and the implementation of control action.
4. State the range of parameters and whether the numbers are to be fixed for the whole process or whether they vary from measurement to measurement.
5. Describe the occurrence of exceptional conditions and action to be taken. This is especially true of any change between manual and automatic modes of control, start-up, and restart procedures.
6. Suggest messages to be typed or displayed and state whether these messages are to be interpreted by operators or engineers.
7. Suggest the parameters or procedures most likely to require on-line modification.

The functional specifications can be considered under four headings: process calculations, communication functions, maintenance routines, and executive routines. The term "executive" is sometimes used for the programming system described in Chap. 9; in the context of application programming, however, it describes the routines which initiate execution of the programs in the other three categories. Some of the questions that must be considered in specifying a process control program are discussed in this section.

14.1.1   Process Calculations

The process calculations acquire process data, convert them to engineering units, test for alarm conditions, and compute and implement control actions. The acquisition, conversion, and alarm checking functions are common to most installations. The control computations depend on the strategy to be implemented.

14.1.1.1   Acquiring Process Data. The quantities which characterize the state of a process, such as temperatures, pressures, and flows, are converted to electrical signals by transducers. The electrical signals are converted to digital values by the analog-input subsystem discussed in Chap. 3. The program to accomplish the task of reading these digital values employs one of three methods of controlling the analog-input subsystem:

1. Single point, in which the subsystem is commanded to read only one signal.
2. Sequential table, in which the subsystem is commanded to read many signals with consecutive multiplexer addresses.
3. Random table, in which the subsystem control is provided with a table of multiplexer addresses that are not necessarily consecutive.

The control may be initiated in one of three ways:

A. When another program determines a need for information and calls the analog-input subroutine.
B. At fixed intervals determined by an interval timer.
C. At system start-up, after which a continuous reading cycle is maintained with the first address following the last.

The specification of the data acquisition routines must include the control method and the initiation procedure. The parameters influencing the selections are the following: multiplexer plus ADC conversion time (sampling time); frequency and execution time of control calculations; total number of signals; and average conversion rate required.

The specifications must consider the timeliness of the readings. Control calculations performed repetitively must use "new" digital values, that is, values determined only a small fraction of the control cycle before the calculation is performed. Suppose, for example, that control calculations are performed every second for 100 signals, that the calculations require 4 msec, and that the sampling time is 10 msec. If one of the table control methods is used, the calculations are not synchronized with the acquisition of digital values. The nth calculation is performed at time $t = T + 0.004\,n$ on a digital value determined at time $t = T - 1 + 0.010\,n$, where T is the time at the start of a cycle. Thus the digital values are obtained from 400–994 msec before the calculation is begun. Such untimeliness can result in poor control. If the calculation frequency is reduced, or if a faster subsystem is used, a table method can be satisfactory. Suppose, for example, that the sampling time is 0.1 msec and a table scan is initiated twice a second, at the start of the interval and 250 msec later. The age of the digital values now ranges from about 0–249.6 msec and, therefore, the maximum age is about 25% of the cycle. An age up to 25% of a cycle has been used in practice without a noticeable degradation in control, although no theoretical investigation has been carried out.

In addition to minimizing the effects on the control, it is sometimes desirable to minimize the time between readings of different signals to obtain a "snapshot" of the process. For example, if the variation of a signal across a paper machine or a steel rolling mill is being investigated, the confusion between across-machine and along-machine variation is minimized by closely spacing the readings. The sequential table form of control is fastest since the selection of successive addresses can be performed by hardware if this feature is provided in the system.

There are three areas of concern, in addition to timeliness, that guide the specification of the data acquisition program: analog-input subsystem utilization, storage requirements, and program execution time. The analog-input subsystem utilization becomes important if the number of readings required per second approaches the capacity of the subsystem, or if it is desirable to avoid needless operation of the subsystem to reduce the maintenance requirements. The table methods can operate the subsystem closer to its maximum speed since the selection of the successive points requires no program execution. The avoidance of needless operation is determined by the method used to initiate reading. The continuous cycle method will probably obtain readings faster than they are used, and the interval timer method will obtain needless readings unless all the readings in a table are used every time they are obtained. If readings are obtained only when needed, there are no wasted operations.

Storage is required for data, the program which acquires the data, and the program which uses the data. The random table method of control usually uses more data storage than the other methods since it is often found necessary to store the multiplexer addresses both in the table and in another location. The execution times of the programs to acquire and use the data usually increase as the storage requirements increase.

Programs which use the process data are usually shorter if they can obtain data from a table in main storage since the programs are freed from the concern over the time between the request for a value and its availability. However, care must be taken to ensure that, on computer start-up, the using program does not execute until values have been read into the table. The data acquisition programs are also shorter for the table methods whether initiated periodically or linked in a continuous cycle.

The combinations of analog input and system control and initiation most frequently used in process control are the following:

1A. Single point when need is determined.
2B. Sequential table initiated periodically.
2C. Sequential table in a continuous cycle.
3B. Random table initiated periodically.
3C. Random table in a continuous cycle.

In Table 14-1 the five combinations are ranked for each of the considerations. The most desirable combination is denoted by a 1; thus, as far as timeliness of control is concerned, 1A is best, 2C and 3C next, with 2B and 3B last. The programmer should ensure that these characteristics are considered by the engineer when the specifications are written. If the combination of control and initiation cannot be selected when the specifications are written, then some indication of the relative importance of the considerations should be given.

14.1.1.2 Reasonableness and Limit Checks. Most specifications require that checks be applied to the acquired data; the checks typically consist of either comparing a value with a given number or comparing the rate of change of a value with a given number. There are two purposes for the checks, either to detect data not truly representing a

Table 14-1   Characteristics of Methods of Reading Process Signals

Characteristic	Technique	Ranking				
		1A	2B	2C	3B	3C
Reading program size		5	3	1	4	2
Reading program execution time		5	3	1	4	2
Data storage		1	2	3	4	5
Using program size		3	2	2	1	1
ADC utilization		1	2	3	2	3
Control timeliness		1	3	2	3	2
Snapshot timeliness		3	1	2	1	2

process condition or to detect abnormal process conditions. The erroneous data test is termed a reasonableness check; the process condition test is called a limit check.

An analog-to-digital converter reading can be unrepresentative if the analog-input subsystem is not operating correctly, if there is electrical noise on the signal lines, or if the transducer which should be measuring the process condition is faulty. If the subsystem is not operating correctly, there are usually violations of the reasonableness checks by readings from several input signals. Specifications usually include specific tests of the subsystem which are discussed in a later section. The following paragraphs discuss the effects of electrical noise and the common transducer failures.

The attenuation of the electrical noise by signal conditioning is described in Chap. 3. The signal conditioning filters, however, are ineffective when the noise consists of random spikes. These spikes must be eliminated either at the source or by the program. For example, relay coils and solenoids used in or near a process control system should be protected by diodes for DC operation or by zener diodes for AC operation. Failure to do so often results in high-frequency pulses of appreciable amplitude when the circuit to the coil is broken (the effect used in the ignition system of an automobile). In one installation a solenoid used to control an analyzer produced pulses of several volts at a frequency of 400 kHz. At such frequencies there is a great deal of radiation, so the pulses can appear on many signal lines and can be coupled around the filters. Other circuit and installation practices for minimizing this transient noise are discussed in Chap. 4 and Chap. 13.

Programs can be made less sensitive to this kind of noise by applying rate-of-change checks. If noise of this type is suspected the reading should be ignored. The digital filters discussed in the next section do not eliminate the spike. Instead they spread its effect over many cycles. For example, consider a process signal that is nominally around 25% of full scale and a filter that averages five successive readings. Table 14-2 shows the output of the filter in response to a spike of 50%, which is a frequently encountered value for some ADCs with binary weighting factors. The input and output values are shown in Fig. 14-1. It can be seen that errors on the order of 20% of the true value persist for an appreciable time.

The other common source of gross errors in the ADC reading is transducer failure. The two most common forms of transducers in process control are thermocouples for measuring temperatures and differential pressure-to-current converters for pressures, levels, and flows. Thermocouples fail in two ways; the wires either short or break. When shorted,

Table 14-2   Response of Averaging Filter to a Spike

Time Step	Filter Output (%)
0	25.00
1	30.00
2	30.00
3	30.00
4	30.00
5	30.00
6	25.00
7	25.00
8	25.00

the signal corresponds to the temperature at the short. This can be detected by the limit check since the temperature is usually very different from the process value. An open thermocouple is more troublesome and more common. If no hardware is installed to provide open-thermocouple detection, there is little the program can do. As a result of the charge stored in the signal conditioning filter, signals can remain in a valid

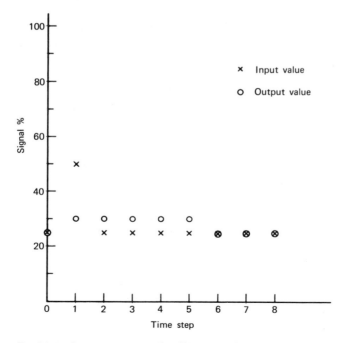

Fig. 14-1   Response of averaging filter to a spike.

range for many hours with only slight drifts up or down scale. The simplest hardware addition is a moderate sized resistor connected across the analog-input subsystem terminals. When the thermocouple opens, the filter capacitor discharges through this resistor. Thus an early warning is provided if the rate of discharge is faster than the process temperature can move. Eventually the indicated temperature is that of the computer terminals, and this is usually outside the process operating range. A disadvantage of this approach is that it reduces the effective input impedance of the analog-input subsystem. The resultant loading error must be considered in calibrating the system.

Current converters use one of the three output ranges 1–5 mA, 4–20 mA, or 10–50 mA. These ranges were chosen with a "live zero" (the 1-, 4-, or 10-mA value) to permit easy discrimination between a low process value and a transducer or power supply failure. Thus current transducer signals appreciably below the live zero can be considered unreasonable. The computer program can detect this condition by comparing the digital reading against a test value. To allow for some miscalibration, and to facilitate the use of the process control system in the calibration process, the test value should be slightly below the live zero value; 75% of the nominal value is often used. The test value for the live zero check is determined by the range of the output current from the transducer and the analog-input subsystem gain, expressed as computer digits per mA. Thus the test value is not a function of the process conditions.

Another kind of reasonableness check uses high and low test values which are specified for each process variable. The programming of this kind of check is very similar to the programming of the alarm checking described in the next paragraphs.

Most process control program specifications include the requirement that the program should check process data and alert the operating personnel if any abnormal process condition is detected, a function often called process monitoring. Although the basic function is simple enough, limit checking programs should be designed to avoid irritation to process operating people by responding to noise or by repeatedly sounding an alarm for values fluctuating around a limit.

The problem of fluctuating signals is avoided by adding deadbands to the alarm values. This is best illustrated by Fig. 14-2. Alarm action is taken at the point marked *AH*, but normal conditions are not considered to be reestablished until the process returns to the point marked *NH*. The difference between the values of the signals at the points *AH* and *NH* is termed the deadband. Similar comments apply to the points *AL* and *NL*. This technique is similar to that provided by process alarm relays since many of these relays have inherent deadbands resulting from the method of construction.

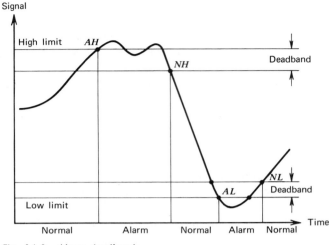

Fig. 14-2   Alarm deadbands.

To avoid alarming as a result of noise, it is usual to provide a 1- or 2-bit counter so that action is taken only after two, three, or four successive readings fall outside the normal range. Such repeated checks are especially necessary when the limit comparisons are made before digital filtering is applied. It is sometimes useful to establish two sets of limits. The inner limits might be used as warnings, the outer limits as test values for strong remedial action.

The standard actions taken on violating a limit are sounding an audible alarm and typing a message. It is frequently preferred that once the alarm has been sounded, no further sounding should take place until the alarm status for that variable is manually reset. The visual display in the form of alarm light or messages, however, continues to follow the state of the variable. The specifications of the alarm checking should state whether the alarm actions are the same for all process variables or whether different variables may have different actions. In the latter situation, the data area for the process variables must include the option selection parameters.

The digital filters described in the next paragraphs reduce fluctuations in a process value but at the same time they also reduce the speed of the response to a large signal change resulting from a transducer failure. Consequently, specifications usually state that reasonableness checks should be applied before digital filtering. Process monitoring, however, is usually applied after the filtering.

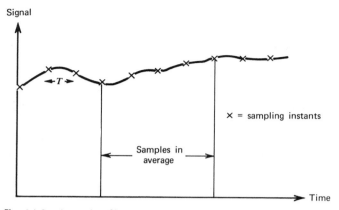

Fig. 14-3   Averaging filter—uniform sampling.

14.1.1.3   Digital Filtering. Once a reading has been determined to be reasonable, it is necessary to use digital filters to remove the effect of small disturbances which cannot be controlled, such as the waves on the top of a tank of liquid. The use of digital filters has received a great deal of attention in recent years. The reference by Rader and Gold (1967) is a general treatment of the available methods and contains a good list of reference material. In process control applications, the two filters most often specified are the averaging filter and the exponential smoothing filter.

There are two forms of the averaging filter, which are illustrated in Fig. 14-3 and Fig. 14-4. In both figures T is the time between calculations

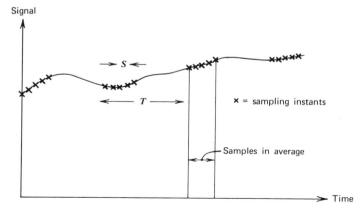

Fig. 14-4   Averaging filter—bunched sampling.

of the filter output. The filter represented by Fig. 14-3 uses a uniform sampling period and requires storage for n past values of the input. The filter represented by Fig. 14-4 accumulates n successive readings before computing the output and updates Y, the filter output, when all n readings have been obtained. This second filter requires only one storage word for the accumulator. Its frequency response is quite different from that of the first filter. Since the sampling interval is shorter, aliasing effects occur at higher frequencies. On the other hand, rejection of low-frequency noise is not as good since the sampling is bunched, rather than uniform. Figure 14-5 is a sketch of the frequency responses of the filters. The non-

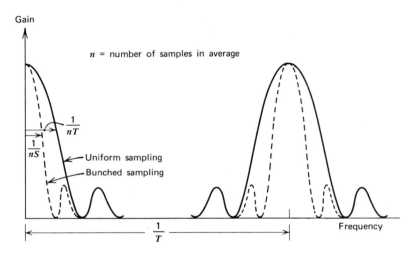

Fig. 14-5   Frequency response uniform and bunched sampling filters.

uniform sampling could lead to a more difficult signal-reading program, although the complication is not great. The biggest disadvantage of nonuniform sampling is in its inefficient utilization of the ADC.

The exponential filter is simple to program and provides less phase shift than the averaging filter. It is the sampled equivalent of a simple lag filter that is discussed as a single-section RC filter in Chap. 3. Past values of the input are exponentially weighted to provide the transfer equation

$$Y(i) = X(i - n)e^{-an} \tag{14-1}$$

where a and n are constants that determine the filter time constant. Equation (14-1) can be shown to be equivalent to the form

$$Y(i) = Y(i - 1) + K[X(i) - Y(i - 1)] \qquad (14\text{-}2)$$

The size of the exponential filter program is likely to be smaller than either of the averaging filter programs since there is no counting required. The execution time is also faster since there are just two adds and one multiply. These programming advantages, together with the smaller phase shift, account for the popularity of the exponential filter in process control applications.

14.1.1.4 Converting ADC Readings to Engineering Units. Once noise has been removed from the reading, the information must be converted to meaningful form. The digital readout (DRO) from the analog-to-digital converter is a linear representation of the voltage applied to the analog-input subsystem terminals. To present information to the process operating personnel in engineering units, sometimes termed physical units, some nonlinear conversion operation may be required. At the very least, linear scaling to reflect the range of the transducer is almost always required. The four common conversions are linear, square root, temperatures directly from thermocouples, and temperatures from thermocouple transmitter outputs.

Levels and pressures are usually measured by linear transducers. Thus it is only necessary to provide the linear scaling from DRO to engineering units. The constants necessary for the conversion can be specified either by stating the engineering value at both ends of the range or by specifying the value for zero DRO and the scale. The conversion formulas used are

$$Y = YMIN + (X - XMIN)\frac{YMAX - YMIN}{XMAX - XMIN} \qquad (14\text{-}3)$$

$$Y = AX + Y0 \qquad (14\text{-}4)$$

In Eq. (14-3), Y is the value in engineering units corresponding to the DRO value X, $YMAX$ and $YMIN$ are the engineering values at the ends of the range, with $XMIN$ and $XMAX$ the DRO corresponding to $YMIN$ and $YMAX$. In Eq. (14-4), $Y0$ is the engineering value for zero DRO, and A is the change in Y for a change of 1 digit in the DRO. Transducers are often specified by the first method, but the second approach is preferable since multiplication is faster than division. This linear scaling is often applied after a nonlinear conversion.

Flows are generally measured by differential pressure cells in which the differential pressure is proportional to the square of the flow. Flow conversion routines usually use a Newton-Raphson iteration technique. Chapter 1 of the first volume of the reference by Ralston and Wilf (1960) contains a description and flowchart for this technique. Iterations can be saved if the value from the previous sample period is used as a first approximation in the subsequent calculation. The routine should also recognize that small negative readings can be encountered, especially when the system is being used to calibrate a transducer. Although it may not be necessary to square root these negative values, the sign must be displayed to simplify the adjustment of the zero point. It is sometimes necessary to compensate a differential pressure transducer for changes in the density of the fluid, especially if an accurate material balance is being attempted. This density compensation involves measurements of the temperature and pressure of the fluid at the orifice. The correct formula for a particular application should be obtained from the process engineer.

Thermocouples provide a voltage which is a nonlinear function of both the process and the analog-input subsystem terminal temperature. A number of references (e.g., National Bureau of Standards Circular 561) tabulate the function for various types of thermocouple common in process control practice. The most common practice is to fit piecewise linear segments to the table values. The number of segments used depends on the required accuracy and the range of temperatures to be measured; thus the specifications must include this information. Four to seven segments are usually sufficient for the temperature ranges and accuracies required in most process applications. These segments must be chosen so that there is continuity at the boundaries. Otherwise cycling can occur if control is attempted near a discontinuity in a conversion equation. Occasionally higher order polynominals are used for compensation. In this case, execution time is reduced by programming, for example,

$$Y = A_3X^3 + A_2X^2 + A_1X + A_0$$

as

$$Y = B_3X[X + B_2(X + B_1)] + B_0$$

where

$$B_3 = A_3$$

$$B_2 = \frac{A_2}{A_3}$$

$$B_1 = \frac{A_1}{A_3}$$

$$B_0 = A_0$$

Grouping the terms in this fashion reduces the number of multiplications from six to three.

Whichever method is used, the DRO must first be corrected to reflect the effect of the temperature of the analog-input subsystem terminals.

A common technique is to maintain the terminals in an isothermal enclosure and to measure the isothermal temperature with a resistance thermometer. The formula for converting the DRO from the resistance thermometer to a temperature value is provided by the system vendor. The temperatures cannot merely be added, however; the terminal temperature must first be converted to an equivalent DRO and then added to the DRO from the thermocouple. The net DRO is used as the input to the nonlinear conversion.

A thermocouple transmitter is a device that compensates for the terminal temperature and converts the resulting voltage to one of the standard current ranges. The range of input millivoltages corresponding to the output current range is different for different transducers. Thus the conversion program does not provide terminal temperature compensation but it must rescale the DRO before using the polynomial or piecewise-linear algorithm.

In addition to conversion between the DRO and engineering units, it is often necessary to provide the inverse of the conversion. For example, quantities entered through display consoles in engineering units may have to be converted back to DRO units for limit checking. An inverse calculation routine can be provided for each nonlinear forward conversion routine. While this gives the fastest execution time, it uses more storage. Since the inversion is not performed frequently, it is usually preferable to save storage by using a standard algorithm which uses the forward conversion routines.

This standard algorithm is a binary search procedure, similar to that implemented in the successive approximation analog-to-digital converter described in Chap. 3. The flow diagram is given in Fig. 14-6. Initially the trial value of half-scale is converted from DRO units to engineering units using the forward conversion subroutine. The resulting engineering value is compared with the value to be inverted. If the latter is higher, the half-scale bit is left on and the next test value is three-quarters scale. Otherwise, the half-scale bit is turned off and the next test value is one-quarter scale. This procedure is continued until an equal comparison or the specified number of bits have been tested. This algorithm converges for any monotonically increasing function.

If the binary search inversion procedure is to be followed, it is preferable to divide the forward conversion into two parts. The first may be termed normalization, the second nonlinear conversion. Only the second portion is used in the inverse procedure. Normalization, in the case of a transducer signal, consists of the live zero check discussed earlier and a linear conversion which makes 0–100% of the process signal correspond to the range of positive numbers in the computer. Thermocouples are

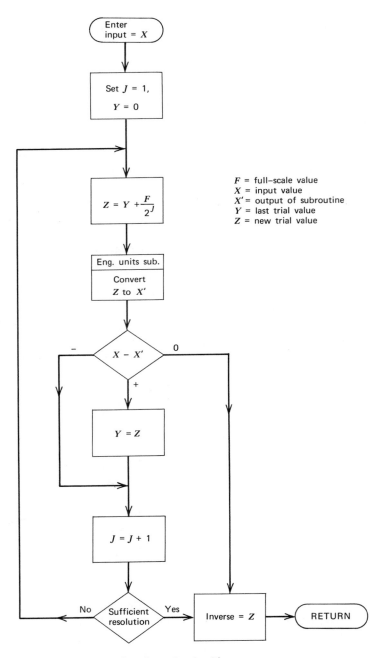

Fig. 14-6   Inverse engineering units algorithm.

normalized by calculating the cold junction compensation so that the normalized value is linearly related to the EMF the process temperature produces if the input terminals are at the reference temperature used in the nonlinear conversion.

14.1.2 Communication Functions

The effort required to specify the process calculations is often much less than that required to specify the communication functions. Unless it is easy for operating personnel to display process information and alter parameters, the computer control system will be switched to standby.

Establishing communication between the process under control and the operating personnel usually requires more programming effort than the control programs. Typically 4000–6000 machine language instructions are required. Good communication is essential for good operation, so the programs and the devices used to implement the man-machine control must receive a great deal of attention. The devices are discussed in Chap. 7 and the programs required depend very much on the hardware used to effect the communication. However, there are some general requirements which should be included in the specifications for the communication functions. These requirements relate to error sensitivity, limited access, entry checking, and recording changes.

14.1.2.1 Error Sensitivity. The major requirement is to ensure that the program is as insensitive to errors as possible since people do make mistakes. The programmer must continually remind himself that plant operators are not themselves programmers and that they often work under high stresses during plant emergencies.

The specification should require that any one erroneous entry or action does not cause the program to loop or branch to an invalid address. The entry or alteration of direct main or bulk memory addresses through consoles should be avoided unless the program can check the validity of the action. A better procedure is to enter or alter a number whose range can be checked. The actual hardware address is then computed from this number.

When card readers or typewriters are used as entry devices, a free format in which fields are separated by blanks or commas is highly desirable. Errors are very frequent when the fields are required to be in fixed positions. If fixed fields are used, the specification should include a check that no information appears outside the assigned areas since it is very easy to enter a number starting in an incorrect column and, therefore, be in error by a factor of 10.

14.1.2.2 Limited Access. Communication facilities are used by various groups of people such as engineers, instrument technicians, supervisors, foremen, and operators whose knowledge and authority vary. Thus a second requirement is the provision of controlled access to information. The hardware may contain a key switch which a program is to test before displaying, for example, tuning parameters. An alternative is to specify that some code number is to be entered before the tuning parameters are requested.

14.1.2.3 Entry Checking. In general it is necessary to provide high, low, and rate-of-change checks on numbers entered from a console. The permissible rate of change in the positive direction may differ from the permissible rate in the negative direction. For example, temperature might be restricted in its rate of increase, but need to be reduced rapidly in an emergency.

14.1.2.4 Recording Changes. Most specifications require that all changes that affect the process be recorded by a typewriter or printer. The recording should include the time, the type of change, and the new value. However, a change may have to be reversed soon after it is entered. Perhaps the wrong variable is addressed or the action may not affect the process as assumed. Hence it is very desirable that messages record previous, as well as new, values. Then it is easy to restore conditions to those existing before the incorrect change was executed.

It is generally better to avoid permanent recording of errors made by anyone using a communications console. If changes are not acceptable, some indication of this fact must be given to the user, together with the reason. However, public recording of errors leads to embarrassment and inhibits the use of the facility. The specifications should describe the possible errors and the actions to be taken.

14.1.2.5 Specific Functions. The above paragraphs describe some of the general requirements for the communication functions. The specifications must also describe each of the specific functions to be performed. For each function the specifications must provide such information as the following:

- Any mnemonic used to identify the function.
- The source of the information.
- The scaling and range of the information.
- The authority required to display and/or alter the data.
- The existence and source of any limits to be applied to entered numbers.
- Any program to be executed after data are entered.
- Any process status to be checked before displaying or entering data.

Although it may take a great deal of effort to write a good set of specifications for the general and specific communication functions, the time is well spent. A good communication system can help process operators cope with emergencies. A poor system might hinder operators so much that a minor problem deteriorates to a major difficulty and in one day loses more money than the system can save in a week.

14.1.3 Modification Functions

The communication programs allow operators or engineers to display and alter only a few values at any one time. More extensive modifications to the control strategy or the program itself are handled by the modification functions, also known as maintenance or utility functions.

The opening paragraphs of this section pointed out that specifications should attempt to forecast the parameters and procedures requiring on-line modification. The forecasts from the process control and communication specifications are the first inputs to the specifications of the modification functions. Some of the tasks that may be required and other inputs are discussed in the following paragraphs.

The process control specifications normally require that modification programs include the ability to load and save the process data tables. Tables may be loaded from cards, paper tape, magnetic tape, a disk, or a drum. The programs to save the tables reverse the loading procedure. If the specifications include these functions, it should be recognized that more than one version of the tables may exist, especially during the debugging phase and the start-up of the on-line control. Hence the specifications should include some way of labeling the version of the tables and a means to print the label of the tables in current use. The production of a new version may be the result of console operations, so it is also necessary to provide means to change a label before a table is saved.

In addition to modification of data tables, there is often a desire to permit on-line modifications to the control programs themselves. This desire, however, is tempered by the fear of the consequences of the errors which are to be expected. Some recent programs have found ways to implement on-line program changes using a higher level language. The reference by Ewing (1967) describes one such program, the Generalized Process Control Program, which was designed by the authors of the article and later made available as IBM's PROSPRO/1800. The requirement for on-line modification of strategies will probably be much more important in the future. If such a requirement is included in the specifications, there should be some means to negate a change so that the effects of errors can be minimized. If the modification program is on a lower interrupt level than the control, it cannot be used to stop the action; for example, if the error results in a program loop, the modification request would not be serviced. Hence under such circumstances some other procedure to disable control is required.

The process control specifications almost always call for modifications since one of the advantages of the computer control system is its flexibility, the facility to modify the control. On-line modification of the communication functions is not always specified, however. Required changes are less frequent so are often handled by off-line procedures. However, if the communication programs use tables to select options for functions, then modification of those tables could be implemented. If the tables are to be changed, then programs to load and save the tables may be required. The labeling required for the process data tables is not so necessary. The changes are infrequent and a new set of tables makes the previous set obsolete.

The on-line modification of the communication program itself is not often needed in normal operation, although such a facility might be useful during debugging. The more frequent requirement is the ability to disconnect and reconnect devices such as typewriters and card readers. Some programming systems include such a facility but if not, the application programs may need to provide it. Any such requirement should be explored thoroughly before the problem analysis phase is considered complete. If devices can be disconnected and reconnected, the specifications for programs that use the devices must include the testing of the status and the action to be taken if the device is unavailable.

14.1.4   Executive Routines

With the control, communication, and modification functions specified, the remaining step is to specify when the functions should be executed. The routines which handle these tasks are generally termed executive or supervisor routines. The specification of the executive routines should not be attempted before the specifications for the other programs are almost complete since additions or deletions of functions can significantly affect the executive. Programs can be initiated periodically, in response to an interrupt, or as a result of a program computation. Considerations for the specifications of these methods are discussed in the following paragraphs.

Most of the functions in the typical process control application are executed periodically; valve control calculations perhaps once per second, logging process conditions once per hour, for example. Timing routines are required to initiate these periodic functions. Once the functions themselves are specified, all the periodic requirements can be collected together. The specification of timing routines is not complex but two factors which should be carefully considered are alteration of the periods and phasing.

One should expect that the frequency of execution of any function might require alteration. Hence the parameter which defines the frequency should be easily accessible, not buried in the middle of a long section of program coding. Further, one should be wary of lumping functions together. Suppose programs A, B, C, and D should all be executed on a 5-min basis. If they are dependent (e.g., if B, C, and D must be executed immediately following A), then a single 5-min parameter could be used. However, if the programs are not dependent, one should expect that, sometime during the project, the frequency of one of the programs will require alteration. Either separate counters should be specified, or the ability to add and delete programs from the "5-min list" should be incorporated in the original design.

Periodic execution of programs requires two parameters, frequency and phase. Specifications should include not only how often a function should be performed but also at what instant relative to other functions or to the relative clock. For example, an hourly log is usually required to be *every hour, on the hour;* the latter phrase specifies the phase relative to the real-time clock. Suppose that there is a function which requires 0.1 min of calculation time, and should be performed every minute, and that there are five other independent functions to be performed every 5 min on the same interrupt level, each function requiring about 0.2 min of execution time. If all the programs are initiated on the same phase, the 1-min programs will be delayed. However, if the 5-min programs are assigned to separate minutes, the computation load will be balanced at 0.3 min in each 1-min cycle. In specifying executive routines for periodic functions, both frequency and phase should be considered.

A hardware interrupt can be looked upon as an exception condition detected by circuitry. Thus as far as an executive routine is concerned, interrupt and program initiation of functions are quite similar. The main consideration for the executive routines is the delay permissible between detection of the exception and the performance of the function. The two basic options are either to link directly to the response program or to place the response program name in some kind of queue. If queues are used, there must be some thought devoted to the size of the queue tables. The specifications should require such sizes to be readily modified.

Some of the specifications for executive routines may merely require the assignment of options in the programming system routines, others may require new programs. The functions assigned to the programming system vary from system to system. The features of the programming system are not crucial to the problem analysis phase of a project since analysis is concerned with what is to be accomplished, not how it should be performed. However, requirements are often flexible; hence the programmer should be able to advise the engineer whether a given function is easy or difficult to implement within the programming system. One should be careful that application program design effort is not squandered trying to achieve a difficult function that is nice, but not necessary.

14.2 APPLICATION PROGRAM DESIGN

There are decisions to be made in producing an application program that affect significant portions of the program. Program design is the definition of these "structural" decisions. The key to minimizing application program effort is to design for change. Strategies and procedures in process control are not at all rigid. Evolving experience and the modification of strategies to take advantage of the new control tool can lead to a significant amount of reprogramming. Applying the principles of good programming design can reduce the time for a modification from days to hours.

Once the process control problem has been analyzed by the engineers and the programmers, the programmers can proceed to the design phase. There are four major areas where decisions affect significant portions of the application program: the assignment of interrupt levels, the allocation of functions to subprograms, the location and format of process data, and the specification of program parameters and work areas. The factors to be considered in arriving at these decisions are discussed in the following sections.

14.2.1 Assignment of Priority Levels

The process control computer is attached to relatively slow devices such as typewriters, timers, and process equipment. Significant loss of computing power would occur if the program had to bear the burden of testing the devices for service requests. To overcome this problem, hardware typically is added to the interface between the device and the computer. The hardware tests for service requests and, finding one, signals the computer to interrupt its normal sequence of instructions to respond to the request. In most control computers there are many levels of interrupt arranged in a priority sequence such that a request at a particular priority interrupts programs of lower priority which are not serviced until higher level requests have been satisfied. The program design must assign the different functions to appropriate priority levels.

The adjective "priority" is often misunderstood; it correctly describes the sequence of the computer's response to interrupts. It is not necessarily related to the importance of the tasks. Assignment should usually be made on the basis of the frequency of interrupts and the time required to service the interrupts. High-frequency and low service time requests are assigned the higher priority levels to minimize the total waiting time of the program. It is rarely necessary to invoke queuing theory to optimize the allocation. The considerations are best explained through an example.

Suppose that the major justification for a control system accrues from the economic optimization of the plant. The program to compute the optimum operating conditions is run every hour and takes 10 min to execute. Another program logs process conditions on a typewriter every 10 min and takes 1 sec to collect the values. The typewriter, capable of writing 15 characters/sec, is serviced by an interrupt response routine which requires 0.5 msec to send a new character. Clearly, if the optimization, the most important task, is assigned the highest priority level, the logging program may wait 100% of its cycle before being executed. If the typewriter is on the lowest level, for example, interrupts are blocked for $10 + (6 \times \frac{1}{60}) = 10.1$ min/hr. Thus it operates at less than 83% of its capacity over the hour. However, if the priority assignments are reversed, the typewriter interrupts are serviced once every $\frac{1}{15}$ sec and the typewriter efficiency may be as high as

$$100 \times \frac{\frac{1}{15}}{\frac{1}{15} + 0.0005} = 99.25\%$$

Since the logging program operates every 10 min, its execution time is extended by at most

$$100 \times 15 \times 0.0005 = 0.75\%$$

owing to the interrupt servicing time. The optimization program operates every hour, and its execution time is extended by only

$$100 \times \frac{10 \times 15 \times 60 \times 0.0005 + 1}{10 \times 60} = 0.91\%$$

Thus assigning the logging task to a higher priority level represents a more balanced use of the resources and does not appreciably affect the more important optimization task.

Communication programs, however, sometimes violate this frequency rule. Although operator requests are infrequent, fairly high priority should be assigned. If the response is not fast, the operator becomes impatient and looks for alternate sources of information. The operator should feel his requests are considered important by the program or he will consider the process control system an inconvenience, rather than a tool to help him carry out his tasks. In general, the faster the device, the smaller the tolerable delay time; however, delays of less than 0.1 sec are indiscernible. If a typewriter is used as the communication device, delays of 1 or 2 sec are tolerable. If a graphic display device is used, then delays of the order of 0.2 sec are ideal and a 2 sec delay becomes satisfactory.

The frequency and service time rule is not applied when the response to a request must be given within a fixed period of time. The two most frequent examples of this type of situation are process hazard alarms and devices that retain information for a limited time, such as scanning analyzers. For this class of interrupts, the considerations are the maximum time between request and response, and the execution time of other programs. The fixed-time interrupts must be assigned to a higher priority level than any program which could run for longer than the required response time.

14.2.2 Allocating Functions to Subprograms

With the interrupt assignments complete, the designers can proceed to the allocation of the functions to subprograms. There are two reasons for dividing the program into sections or subprograms. The first is the application of the principle of "divide and conquer"; the complex functions are implemented by a collection of routines handling the component tasks. The second reason for the division is that some of the functions require similar sequences of steps applied, perhaps, to different data. Storage can often be saved by designing subroutines which collect all the necessary data and then perform the required steps. The considerations that affect the division of the program into subprograms are discussed in this section.

The term subroutine is applied to a sequence of instructions that may be entered from more than one location. An external subroutine is assembled or compiled as a single unit and depends on the program loader to provide the linkage address to the instructions which invoke the subroutine. An internal subroutine must be included with the statements which invoke it as part of the same assembly or compilation. This inclusion means that the symbol table of the invoking program is available to the subroutine. Thus an internal subroutine may be faster and shorter than an external subroutine since it can reference data directly.

Data can be provided to external subroutines in three ways: in machine registers, in a common work area, and by using a register to point to a string of values. The machine register method usually leads to shorter, faster routines but is not usable for FORTRAN programs since the registers are not accessible. Information in work areas ranks next in terms of speed. However, it has the disadvantage that any change in the allocation of the work area affects many programs. The third method is usable by either assembly or FORTRAN programs since compilers adopt a convention regarding the use of a register for subroutine data. Although speed is important, the choice of data transfer method is often influenced more by FORTRAN compatibility or reentrant requirements.

The need for reentrant subroutines is a subject to be carefully studied during program design. If the same function must be performed in different interrupt levels, one could provide a separate subroutine for each level. This is the simplest and the fastest solution but it requires more storage. If a single routine is provided for more than one interrupt level, however, a problem could arise if the subroutine is in execution when a higher level interrupt occurs that results in a request for the same subroutine. This could cause partial results to be destroyed. Interrupts must be prevented by the calling programs each time the subroutine is invoked or the subroutine must be reentrant. The former method is very undesirable since a basic programming principle is to do a job once, do it right, and have the machine handle the repetition. Reentrant subroutines are more difficult to write but the difficulties are concentrated.

The techniques to be used for writing reentrant subroutines depend heavily on the operating system. The problem that must be avoided is the alteration of a word in main or bulk storage by two instructions executed on different levels. If the operating system saves and restores a machine register in its interrupt handling routine, then that register can be used for arguments and partial results. To accommodate reentrant routines, some operating systems provide separate work areas for each interrupt level. The subroutine coding must then use a register to point to the work area and define all its changeable storage relative to the register. If the hardware or the operating system protects the return address during subroutine linkage, then the subroutine could inhibit interrupts until it has completed its use of changeable storage. This latter approach is simple but may prevent interrupts for an unacceptable period.

This glimpse of the techniques required should persuade the reader that the program designer should not use reentrant subroutines unless they are really necessary. However, one should not overlook the fact that they are sometimes highly desirable. The routines that reference bulk storage are sometimes overlooked. Reading or writing information to bulk storage involves the transfer of a record containing many data words. The fact that the different levels alter different words is not sufficient to avoid reentrant routines. During the update operation the whole record is written by all levels.

As functions are allocated to subprogams, the language to be used for coding the subprogram should be considered. The storage requirements and execution speed of subprograms are parameters that influence the design of the total program. The small main storage, speed, and bulk storage facilities of some process control computers may limit compilers to producing inefficient object code. Larger computers with more facilities can produce very efficient coding. If the inefficiencies can be tolerated, the high-level language should be used. The effort to code in FORTRAN, check, and document a subprogram is much less than is required to do the same task in an assembly language. The reference by Kipiniak and Quint (1968) is a comparison of the use of assembly and compiler languages for process control applications.

After reviewing the considerations of this section, the program designers should produce a set of specifications for each subprogram, stating the function, the data to be transferred, the language compatibility requirements, the reentrant requirements, the process data to be referenced, and the program parameters to be used.

14.2.3 Process Data

In many respects the process control application fits between commercial and scientific applications. The algorithms used are more complex than in commercial practice but do not usually require the careful numerical analysis of the typical scientific calculation. The quantity of process data is also intermediate, much greater than that used in most scientific applications but less than that used by the typical commercial program. There are, however, enough data so that very careful attention must be paid to the design of the data tables and the routines which access them. Here again, emphasis should be placed on the ease of change. Fields may be added, or deleted, and the optimum location of elements within a table may change as the control scheme evolves. It is certainly necessary to add or delete process variables from the tables without taking the system off line.

The decisions regarding the process data that must be made during the design phase are the form in which information is stored, the contents of the various tables containing the data, and the procedures to be used to access the data.

14.2.3.1 Data Form. Process data may be stored in tables located either in main storage or in bulk storage. If data usually reside in main storage, then, with present-day systems and strategies, the limiting factor in application program design is often the available main storage. If the data are stored on a disk or a drum, reducing the size of tables improves the transfer rate. Careful design can reduce table sizes by 30 or 40%. The costs of these reductions in data space are increases in size and execution time of the accessing programs. However, the penalties are usually worth the trade.

Every field in the data tables should be examined to see that its bits provide the maximum information content. The range of the field and the frequency of occurrence of values, specified during problem analysis, should be examined before choosing the form used to store information. The possibilities might be the following:

1. True floating point—two words.
2. Pseudo-floating point—one fixed point word with a small integer representing power of ten multiplier.
3. Fixed point number—sufficient bits to express largest value.
4. Index—sufficient bits to express the maximum number of choices.

Some examples show how these representations may be employed. First, suppose a rate-of-change check is to be applied to process measurements. The alarming value can be specified for each process variable in engineering units, in which case a two-word floating point value would be required. Alternately, the test value can be normalized by the range of the process measurement and the fraction of the range expressed as a fixed point number. However, reviewing the specifications might show that only the values 1.0, 0.1, 0.01, and 0.001 are necessary. Therefore, in this case only two bits are required in the data table to index the four possibilities.

As a second example, suppose it is required to specify the physical units of measurements on printed messages. Abbreviations such as MSCF/H for "thousands of standard cubic feet per hour" are commonly understood and can be used. If stored directly, six characters are required, and this uses 36 bits if a binary-coded decimal representation of the characters is used. However, in any one process it is unlikely that more than 10–20 different physical units are employed. Thus a 4- or 5-bit index into a table of units completely fulfills the requirements. This results in an 85% reduction in storage requirements for the field.

The choice between floating point and fixed point representation of a number is influenced by the skill of the programmers using the number and by the execution speed of the programs. Floating point is much easier to use but is much slower, especially if floating point hardware is not an installed machine feature.

In any design, the final result is a compromise. Selecting the form for process data is a compromise between the table space required and the speed and size of programs using the data. As program coding proceeds it will undoubtedly be necessary to modify the form of some of the data. In the design phase it is preferable to aim for tables that are as small as possible since they are usually easier to expand than to contract.

14.2.3.2 Data Tables. Having decided the forms to be used to represent the various process data fields, the next step is to consider the organization of the tables containing the data. The major questions to be decided are the number of tables and the table formats.

All the process data can be stored in a single table with one row of the table assigned to each process variable. This type of organization makes the on-line addition or deletion of process variables simpler since a row must merely be added or deleted. This form of organization is generally preferred if all the process data stays in main storage, as is typical in a direct digital control application, for example.

If process data are stored in bulk storage, the single table approach may require larger transfer times than a multiple table design. The routine operations in the process control program may use only a fraction of the data for a variable. For example, alarming and regulation routines may use normalized values; the engineering scale constants would then be required only by display, logging, and special calculation routines. Hence if separate tables are used for the high-use and low-use information, the average data transfer rate is lower.

Another basis for the division of process data into multiple tables is the operations performed. Suppose that a choice of computation sequence or frequency exists. An indicator in the process data table can be provided which is to be tested by an interpretive program executed routinely. Alternatively, separate data tables could be provided for each sequence or frequency. Then the table loading routine is the only program that needs to make the test. The multiple table method leads to faster running times at the expense of storage.

Each table can be considered as a two-dimensional array, with one subscript specifying the process variable and the other the data field. For each process variable, the number of words of data may be fixed or variable. If the number is variable, it must be specified directly or some kind of termination indicator provided. The indicator method can lead to longer execution times for routines which must search the table and is not usually used in process control.

Explicit or implicit specification of the number of process variables can be used. If a table has a fixed number of data words per variable, the explicit specification is preferable. The implicit method would waste almost one row of the table. If the number of data words per variable is not fixed, there is no particular advantage to either method. If multiple tables are used, organizing all tables analogously is likely to lead to fewer coding errors and may ease the coding of the table loading program.

14.2.3.3   Accessing Process Data.   With the form and content of the tables decided, the final design step is to specify the procedures to be used to access the data. The method used to identify a particular process variable and locate its table entries and the access to a particular entry must be specified.

Identification numbers may be random or sequential. Sequential numbering is used when an easy method of locating table entries is required. The identification number can be used to compute addresses in bulk or main storage. Access is rapid but storage is wasted when gaps appear in the sequence. This wasted space is often acceptable if the tables are in bulk storage. It may not be acceptable for main storage resident tables, however, since space is usually at a premium.

Random identification numbers are usually a coded translation of designations such as F123, T072, or P193, where F represents a flow, T a temperature, and P a pressure. The digits of such designations are often chosen in a memory-aiding manner. The process variables T267 and T367 might be corresponding temperatures in reactors 2 and 3, for example. If random identification numbers are used, each row of each table must contain the number or new tables of pointers must be specified. Hence this form of identification is not often used when multiple tables are employed. Table searching is more difficult with the random form since there is no direct way to locate an entry corresponding to a particular process variable.

A third possibility is to use the memory-aiding random numbers for communication between operators, engineers, and the computer, and to provide program conversion between the sequential and random numbers.

The remaining problem in the process data area is the access to particular fields. To aid the discussion, consider the access to a low alarm limit and to an index specifying the engineering units. Suppose that the alarm limit is stored in the right half of word 6 of some table and that the index is stored in bits 4–7 of word 5. In the examples below it is assumed that index register 1 points to word 0 and that the computer word contains 16 bits.

The limit could be reached by the following sequence of instructions:

LD   1   6   (Load the contents of the address formed by adding 6 to the contents of index register 1)

SLA   8   (Shift left to place field in most significant half of the accumulator)

It is almost inevitable that, as the project proceeds, it will be found necessary to change the location of the field and extensive programming changes will then be required. The word number and the bits assigned to the limit can be made symbolic using the following statements:

WDSIZ	EQU	16	(defines size of word)
ALRM	EQU	6	(limit is in word 6)
ALRML	EQU	8	(left-most bit is 8)
ALRMR	EQU	15	(right-most bit is 15)
	LD	1 ALRM	(load word 6)
	SRA	WDSIZ-1-ALRMR	(clear any information to right of field)
	SLA	WDSIZ-1-ALRMR + ALRML	(left justify)

For this particular example the Shift Right Accumulator (SRA) instruction is a waste of space and time since the amount of the shift is zero. In an actual case, however, it is required.

The engineering units index must be right justified in the accumulator so the corresponding set of statements is

UNIT	EQU	5	(word 5 contains value)
UNITL	EQU	4	(left-most bit is 4)
UNITR	EQU	7	(right-most bit is 7)
	LD	1 UNIT	(load word 5)
	SLA	UNITL	(clear information to left of field)
	SRA	WDSIZ-1-UNITR + UNITL	

This symbolic reference to fields costs a small amount of storage and execution time. However, rearrangement of the data fields is much easier. In an actual case using this method, a rearrangement was implemented in one day, whereas a previous program, designed for a similar application but using explicit values for word and bit numbers, required a week to effect the change.

14.2.4 Program Parameters

In addition to the process-related data, there are other parameters which are referenced by several subroutines. Parameters which specify the time base of the executive program, the device used for error messages, and the location of process data tables are three examples. During the design phase the access to this information should be specified; if individual programmers are allowed to choose their own methods, revision of the parameters is more difficult. In the following paragraphs some of the methods are discussed, using as an example the specification of the error printer as logical unit number 3.

Programs written in an assembly language can specify a parameter via an equate statement such as the following:

JERPT   EQU   3

This statement tells the assembler program to replace the symbol JERPT by the value 3 during the translation into machine language. In this approach the amount of main storage space and the execution time of the application program are not increased by the symbolic definition of the parameter. Some assembler programs allow one to define a set of symbols that can be incorporated in any program assembly. This is often termed a system symbol table. The use of such a facility allows the parameters to be defined once so the possibility of two programmers' using different values for JERPT is avoided. There are, however, two disadvantages to the equate method. First, assembling programs on a process control computer takes an appreciable time, so reassembly can be inconvenient. Second, and more important, on-line modification is difficult. The value is not explicit in the running program and may be used in several different places in one subroutine.

The value can be made explicit by defining a constant:

JERPT   DC   3

This statement causes the value 3 to be stored with the program. Thus it is not difficult to change the value using a patching program. If each subroutine uses its own constant, the possibility of confusion exists. This confusion can be eliminated by using one constant for all subroutines. There are four ways of specifying such a constant: It could be stored in a fixed address; it could be stored in an address which is relative to the start of a work area; it could be a dummy subroutine entry; or a subroutine could be used to provide the value.

Some program operating systems allow work areas to be defined at the beginning or the end of main storage. For example, the statements

ORG   192
JERPT   DC   3

tell the assembler to locate the constant at address 192. If this method is used, the subroutines should use a system symbol to reference the value such as:

AJERP   EQU   192

A system symbol should also be used if the constant is stored in a work area. A common technique used to locate such a constant is to use indexed instructions to load the constants; for example,

LD   3   AJERP

where index register 3 points to the start of the work area.

The third method of providing access to a parameter is to assign a dummy subroutine entry, such as

	ENT	JERPT	(Entry address)
JERPT	DC	3	(Value of constant)

The names of entry points are used by one subroutine to reference another, and the program loading routines supply the actual addresses. The dummy entry point could be used as follows:

LD   I   JJ + 1

.
.
.

JJ   CALL   JERPT

The program loader replaces the statement at JJ by two words, the second being the actual address of JERPT. The loading instruction specifies an indirect operation; it directs the computer to load the value whose *address* is located at JJ + 1.

Another sequence of statements uses a knowledge of the actual machine instruction representation:

CALL	JERPT	(Dummy call)
ORG	* − 2	(Set origin back to first word of call)
DC	/C400	(Machine language for load instruction)
ORG	* + 1	(Skip over the address)

These last two sets of instructions assume that the loader replaces the second word of the CALL statement by the actual address. Some loaders link subroutines through an intermediate table, and for these a more complex sequence of instructions may be required. The dummy entry point usually uses more time and space than the fixed area or work area methods. However, there may be limitations on the size of the areas.

Programs written in FORTRAN cannot use the symbolic method. They can, however, use the fixed or work area methods if the areas can be referenced using a COMMON statement. A subroutine could be used to provide the value of the constant to another program written either in FORTRAN or in assembly language.

Such a subroutine might contain the following statements:

	ENT	ERDVC	(Entry name)
ERDVC	DC	0	(Space for return address)
	LD	JERPT	(Load value)
	BSC I	ERDVC	(Branch to the instruction at the address contained in ERDVC)
JERPT	DC	3	(Constant)
	END		(End of subroutine)

An assembler language program could obtain the value by the statement

CALL   ERDVC   (Load device number into accumulator)

A FORTRAN program would consider the subroutine as a FUNCTION subprogram and could use statements such as

EXTERNAL   ERDVC

INERPT =   ERDVC

This last method is the slowest and takes more storage. Its advantage is that the information is more secure against program errors since it is not unusual for COMMON to be destroyed by a program error that computes the subscript of an array incorrectly.

Although the illustrations in this section were drawn from a specific machine, similar sets of statements can be defined for most process control computers. The multiplicity of methods underscores the need, in the design phase, to choose the way each of the key parameters is to be accessed. Careful attention to the decisions common to many subroutines before coding begins can reduce the need to revise coding and eases the checkout phase.

14.3   CODING AND DEBUGGING

Once the basic program design is complete, the coding of the individual routines can begin. Once coded, the routines must then be checked and errors, or "bugs," removed. There is always some overlap between the project phases since the output from one phase often suggests a modification to the previous phase. The overlap between the coding and debugging phases is especially pronounced. The following section discusses the activities in these two phases.

14.3.1   Coding

Two important thoughts to bear in mind during coding are that the program is to be used by someone other than the coder and that the program will probably need modification at some future date. Comments and remarks should be extensive.

Collecting parameters at the beginning or end of subroutines is also very desirable. The values used are more readily apparent in the documentation and program modification is simpler. Often when parameters are collected in a list it is found that different parameters use the same number. The temptation to reduce space by assigning both parameters the same address should be avoided. All too often one finds it necessary to change one parameter and not the other. If both are assigned the same address, it is impossible to make the change on-line. An area where frequent modification occurs is in the wording and format of messages and the assignment of the particular device on which a message is to be displayed or typed. If the application program is in FORTRAN, an entry message might be programmed as follows:

```
     DATA        JDVNO/3/
10 FORMAT        (I4,2X, 'OLD VALUE', F6.2,2X, 'NEW', F6.2)
     WRITE        (JDVNO, 10) JTIME,VALUO,VALUN
```

In this example the device, logical unit number 3, the number of spaces and so on, the format, and the arguments JTIME, VALUO, VALUN are specified separately. It is a very simple matter to change the device or to add more spaces. This same kind of principle can be followed in programs written in an assembly language. The number of spaces, carriage control functions, and wording can be produced from a specification list used by a subroutine which converts binary values to printer code. Then formats can be changed without affecting the program that computes the values.

All the messages that are to be typed or displayed should be collected together to review the possibilities of elimination of duplicate or redundant words. Messages can use a significant amount of storage. A very careful balance must be struck between the space required for a message and its clarity since short messages may be too brief for rapid comprehension. Vocabulary lists help to minimize storage since different messages may be constructed by juxtaposing words. For example, the word OUTPUT might be combined with the words HIGH or LOW to type an alarm message and with MAN or AUTO to type a change in control status. Lists can also be useful in reducing the space required to buffer a message. Only the position of the word in the list need be stored.

14.3.2 Utilities and Debugging Aids

Before beginning to code the project programs the list of utility and debugging aids should be reviewed. Programs in these categories can be found in vendor libraries or the library of past routines produced by the programming group. Any missing program should be coded and debugged first since the checkout of project programs proceeds much faster if good tools are available.

Utility programs should provide the facilities to list, reproduce, and resequence cards. Others should provide the ability to copy tapes or disk packs. All of these utility programs should run on the machine used for program checkout. Similar facilities located 15 min away or on a machine not controlled by the project manager are not so helpful.

A number of debugging aids have been found useful to a process control programmer. The simplest (and probably the most useful for debugging main storage resident routines) is a problem to print, on demand, the contents of a specified storage location without halting the program. It is most helpful if the location specification can include a relocation factor so that it corresponds to assembly or compilation listings.

Since a program may "hang up," another useful tool is a program which dumps selected storage locations onto a printer or typewriter. This routine should be completely self-contained since the hang-up may have resulted from destruction of a general output or code translation subroutine.

A more elaborate debugging aid is a "snapshot" program. A snapshot program replaces some of the instructions in the program under test by links to an analysis routine. The analysis routine compares the machine status with specified test conditions. If the conditions are satisfied, the machine status and selected portions of storage are printed. For example, suppose that a process control program computes a storage address incorrectly. A snapshot could be planted just after the address computation to print the storage locations of the information used to compute the address.

A program which bears some similarity to a snapshot is a trace routine. Some process control computers provide a switch setting that causes an interrupt after the execution of any instruction. This interrupt branches to a trace program which prints the status of all the machine registers. Since every instruction is interrupted, execution time is much slower than normal. However, since every step is traced the complete path through a program can be documented and studied at leisure.

Since assemblers and compilers on process control computers are often rather slow, a program which allows one to change, or patch, main or bulk storage can reduce the time wasted reassembling programs to correct minor errors.

Each of the debugging aids has its pros and cons and the choice is often a matter of taste. The essential steps are to make a choice and then to ensure that the debugging aids are available before getting very far into the project programs.

14.3.3   Debugging Project Programs

In Sec. 14.1 process control program tasks were analyzed in the following sequence: process calculations, communication functions, maintenance routines, and executive routines. The recommended debugging sequence is the converse of the analysis sequence.

The executive routines cause the execution of the other functions either on a periodic basis, as a result of switch setting, or in response to a process condition. The executive routines can be debugged using dummy routines for the other functions. These dummy routines should have the same name as the normal program but should merely print an identification number. Once the executive routines are operational, debugging of the other routines is easier since they can be initiated at will.

The objective of the communication programs is to allow operators and engineers to display and alter process data. Consequently, it is preferable to debug the maintenance programs first, so that they can be used to provide data for the checkout of the communication routines.

Once the process data can be loaded, displayed, and modified, the control programs can be tackled. It is very desirable to disable all audible alarms until a good fraction of the program has been checked out. The routines which measure process conditions are checked first. When the programmer feels that the routines are correct, the results should be checked with the instrument engineers. Cooperative effort is needed since discrepancies between computer values and instrument readings are often a result of a miscalibrated instrument. After the process conditions can be correctly measured, the output calculations can be checked, thus completing the debugging of the control functions.

During the checkout of the major functions, error conditions are always encountered so some of the error response routines are already checked. The final steps of the debugging phase should be to force as many errors as possible to complete the check of the error response routines.

14.4   DOCUMENTATION

During each of the first four project phases information is written down. After the checkout phase is complete, all this documentation must be collected, revised, and supplemented. The importance of the documentation cannot be overemphasized. The process control strategy is not static; as experience develops it is necessary to modify the program. These modifications may be implemented by people who were not members of the original project team and who must rely entirely on the written information. The three major classes of program documentation are functional description, operating instructions, and program description.

The problem analysis phase produces the first document in the functional description class. This is very detailed and is used by the programmers and process engineers. During the documentation phase it must be reviewed to ensure that all revisions have been incorporated.

An additional document should be written that describes the functions of the process control system in more general terms. This introductory document can be used to brief new project personnel and in the training of the operators.

Operating instructions for all the modification and communication functions must be prepared. These instructions should be complete and concise. However, brevity should not be obtained at the expense of clarity. For example, confusion can easily be generated if instructions jump from page to page. It is often better to repeat the description of steps rather than to refer from one operation to another. The operating instructions should be tested by someone unfamiliar with the program. By the time the instructions are written, the programmers know the operations too well to provide a valid test.

The program description documents are produced for the person who must modify the program, correct an undetected error, or revise a function. Typical documents are flowcharts, decision tables, narratives, and listings. Today there are many computer programs that print flowcharts. Coding the data for these programs does not take much more effort than is required to produce a neat, reproducible drawing. Modification is very much easier and the use of computer-drawn flowcharts is highly recommended. Decision tables have not been used much in process control projects. The technique is applicable but few process control programmers are familiar with the method (see Appendix B).

The program description narrative supports the flowcharts or decision tables. The description should not be written as though the charts and tables were nonexistent. The narrative should describe the linkage between programs and the names and values of significant parameters. An attempt should be made to forecast possible changes and to provide suggestions to assist the modification. A table that lists all the subroutines and the programs which call them is very useful but often overlooked.

Throughout the documentation phase the project personnel should remember that the process control system is designed to help the operation of the process. The documentation must be clear, concise, and complete. An efficient, sophisticated, error-free program is intellectual effort wasted if the process operators and engineers cannot understand the documentation well enough to operate the process control system.

14.5   PROJECT MANAGEMENT

The manager of any project must allocate resources to achieve the project objectives and then control the resources to ensure a satisfactory conclusion to the project. The magnitude of the effort must be estimated and people assigned to complete the effort in the required time. The progress of the project must be monitored and resources reallocated to ensure that schedules are met. The following sections offer some thoughts on each of these problems.

14.5.1   Estimating Total Programming Effort

The first task of the application programming project manager is to estimate the resources required to design, code, check, and document the program. Unfortunately this cannot be carried out in a precise fashion. The reference by Nelson (1967) explored data from 169 large projects undertaken for the U.S. Air Force in an attempt to derive estimating rules based on past experience. In all the categories the standard deviation of the man months per 1000 instructions exceeded the mean. Thus one cannot consider rules based on numeric information to be precise. However, they do provide a starting point. The best approach to estimation is to begin with quantitative rules, then to consider the factors which could modify the estimates, and finally to compare the current program with a similar effort that has already been completed.

A frequently used rule of thumb is that a man day produces six to ten documented and debugged instructions. This estimate, it should be noted, includes all five phases in programming: analysis, design, coding, debugging, and documentation. All too often programmers reject the formula because they equate it with just the coding and debugging phases. Clearly this rule requires an estimate of the number of instructions. Until the program has been analyzed, this estimate can be based only on experience with similar problems.

As the project progresses, the accomplishments can be used to refine the estimates. The reference by Delany (1966) quotes estimates of the proportions of the total manpower expended in the five phases; these

Table 14-3   Relative Effort in Programming Phases

Phase	Relative Effort	Remaining Effort
Analysis	20%	80%
Design	20%	60%
Code	25%	35%
Debug	25%	10%
Document	10%	0%

estimates are given in Table 14-3. As any phase is completed, the manpower expended can be used to forecast the effort remaining.

A second approach to estimating the effort required for each phase uses ratios between the measurable outputs. For example, about 15 pages of written documentation supports about 1000 source statements, and one block of a detailed flowchart results from six source statements. The functional description, produced in the analysis phase, and the narrative supporting the flowcharts each may range from one-third to one-half the written documentation of a process control program, the balance being devoted to operating instructions. Therefore, if a good functional specification is written during the analysis phase, the program size can be estimated on the basis of 150–200 source statements/page.

The estimates of the number of pages and flowcharts can also be used to calculate the documentation effort. One man day usually produces about four pages of flowchart narrative or ten pages of operating instructions. About 25 blocks of a hand-drawn detailed flowchart are produced per day. If machine-documented flowcharts are produced, the time to code and check must be added to this estimate. The coding time is very dependent on the particular program that is used to produce the charts.

Table 14-4 summarizes the estimates of one man day of work, and Table 14-5 summarizes the written documentation and flowcharts per

Table 14-4   Output per Man Day

Operation	Output
Complete project	6–10 instructions
Draw detailed flowchart	25 blocks
Write operating instructions	10 pages
Write flowchart narrative	4 pages
Code and debug	12–20 instructions

Table 14-5   Documentation per 1000 Instructions

Item	Quantity
Specifications	5–7 pages
Operating instructions	3–4 pages
Program description	5–7 pages
Flowchart	160 blocks

1000 source cards. Using such tables, the formulas in the reference by Nelson, or other rules of thumb, the first pass at the effort required for each of the phases can be produced.

The second pass should consider the factors that might cause the particular project to require modification of the rules. Many such factors could be considered; Nelson's work contains more than a hundred. For the process control application program the most important are the experience of the project personnel, the maturity of the programming system, the need to optimize time or storage, the size of the library of available subroutines, and the facilities provided to aid the programmers.

The experience required in the application programming project falls into two categories: familiarity with the application and familiarity with the hardware and the programming system. No general guidelines can be given for the time it takes to become familiar with an application. The project manager must base his estimates of this factor on his own experience and his knowledge of the individuals concerned.

The programming systems produced by computer vendors are general packages containing a number of options. The procedure of selecting the options and producing a running programming system is termed "system generation." The first attempt at system generation could take two days; after the sixth attempt the procedure might be completed in 3 hr. Programming systems adopt conventions to link to library subroutines, to provide diagnostic messages, and to provide debugging aids. The time required to debug a program operating under an unfamiliar system can be twice as long as the time required using an experienced programmer. Experience has shown that it takes about three months to become familiar with a programming system.

Testing of a large program such as an operating system is a difficult task. There is a very large number of combinations of conditions that might be encountered. The number is so large that a test group selects "the most significant subset" of the combinations. Therefore, one should expect that a new operating system will contain errors that were not detected by the testing procedures. As a system is used, new combinations of conditions occur which detect these errors. If a programmer is working with a new system, he can waste a lot of time looking for an error in his coding when the error is, in fact, in the operating system. Thus project planning should take into account the number of other users of the programming system and the time since it was first released.

The effort to write and debug a program can be significantly increased if its size or execution time must be minimized. Usually such a program is written and debugged once to check the logic and then rewritten to reduce the storage requirements or execution time. Thus doubling the estimates of the manpower required for optimized programs is a reasonable approach.

The estimates based on analysis of the problem and the forecast of the number of instructions assume that all the programming must be carried out by project personnel. Most vendors provide a large library of general routines which can be used to satisfy part of the project requirements. The availability of such routines should be used to reduce project effort as far as possible. The manager should be wary of the "not-invented-here" or NIH factor. There is a natural reluctance to learn how to use someone else's work. A general routine might be a little less efficient than a special routine written just for the particular project. However, programmers are often too optimistic in estimating the time required to rewrite, debug, and document a routine. Thus the manager must be cautious about accepting a recommendation against using an existing routine. There are times when such a step is necessary, but the burden of proof should be placed on the rewriter.

In estimating manpower requirements for a project, managers sometimes overlook the effect of the facilities available to the programmers. This can be quite important in a process control project where the computer may be installed at a location remote from other computer installations. The ability to duplicate, list, interpret, and sort cards can materially affect program coding and debugging effort. If a process control installation is remote from an off-line high-speed printer, for example, it may be desirable to add such a device to the process control computer during the debugging phase. The time required to type the written documentation can easily be halved if the typist is provided with a text-processing computer terminal or a typewriter with magnetic tape or card storage. Such facilities allow the typist and the programmer who edits the text to concentrate on the revisions, with the retyping of the unaltered text delegated to a machine.

After reviewing these factors of experience, programming system maturity, optimization requirements, program library, and available facilities, the estimates for the project phases can be revised. The factors also supply the basis for the manager's estimate of additional manpower margin that should be added to allow for errors in the estimates.

14.5.2 Project Manpower

With the estimates of total effort in hand, the manager must select the type of people and the number of people to complete the project. Application programming for process control should always be a team effort. The time between concept and start-up is usually too short for one man to be able to complete the task. More significant, perhaps, is the fact that it is the rare project that can afford to come to a halt if a man becomes ill or is assigned to a higher priority task. The most effective teams contain a mixture of engineers who can program and programmers who understand engineering. There is a need for understanding the process control requirements and a ready appreciation of the ease or difficulty of accomplishing those requirements when following any particular programming approach.

Present-day engineering curricula almost always include some computer courses. Hence most young engineers have a knowledge of programming. Older engineers who need to develop such knowledge can be sent to computer vendor schools, university short courses, or computer institutes. University short courses are also good vehicles to give programmers with a scientific or mathematical background an introduction to process control engineering.

The discussion in Sec. 14.5.1 emphasized the lack of precision in estimates of project manpower requirements. There are two ways to allow for this inaccuracy: start the project much earlier than the estimates would indicate or provide a pool of reserve manpower.

If the second approach is taken, the reserve manpower should be added to the project at the conclusion of the design phase. Once coding and debugging are under way, the project programmers begin to adopt conventions and establish the mnemonics for key parameters. These conventions and names are rarely written down until the documentation phase. Consequently, a person joining the project late in the coding and debugging phases must learn this information by asking questions. The time spent answering these questions could easily result in a net delay to the project. The reserve manpower can be initially assigned to programming tasks that are useful but not essential to the success of the project.

14.5.3 Project Control

Since the initial estimates of manpower requirements for a programming project are often imprecise, a project manager should not be severely criticized if additional effort is needed. He deserves criticism, however, if he does not discover the need until it is too late to meet the project schedule. Project control, the monitoring of progress, and the reallocation of resources are especially important when the estimates are imprecise. To borrow an analogy from control theory, feed-forward control is fine when a good model exists but feedback control is essential when the model is poor.

Many management texts provide good descriptions of aids to project control, such as PERT and bar charts. However, these aids are only as good as the numbers written on the charts or provided to the computer program. The numbers must be continually refined by monitoring the *measurable* output. The rules of thumb discussed in Sec. 14.5.1 should be reviewed as the project proceeds to determine, for example, whether the particular project is producing 6 or 10 instructions/man day.

Besides monitoring the measurable output, the project manager should also carefully control the sequence of the project activities. Tasks which involve other computers or the use of library routines should be tackled very early in the project. Such tasks often appear straightforward but, in practice, often require much more than the budgeted time. For example, if paper tape, magnetic tape, or disk packs are to be used to exchange information between machines, a trial run should be made as soon as possible. The character sets may be slightly different or there may be different header and trailer conventions.

The use of library subroutines can save a great deal of effort. However, people sometimes overlook the fine print and discover, too late, that the general routine does not perform the specific operation required for the project. Occasionally one finds a reproduction error which invalidates the particular copy of the library routine that is on hand. Such errors are rare but the low probability is no comfort if the project misses its deadline as a consequence.

Careful monitoring and tackling unfamiliar tasks early in the project can ensure that, as the project progresses, the estimates of time to completion become more precise. Meticulous planning and control by the manager helps the project team to produce an application program which turns the general-purpose hardware and software into a working process control system that operators and engineers can use to run the plant more efficiently.

REFERENCES

Delaney, W. A., "Predicting the Costs of Computer Programs," *Data Process. Mag.,* October 1966, pp. 32–34.

Ewing, R. W., G. L. Glahn, R. P. Larkins, and W. N. Zartman, "Generalized Process Control," *Chem. Eng. Progr.*, Vol. 63, No. 1, January 1967, pp. 104–110.

Kipiniak, W., and P. Quint, "Assembly vs. Compiler Languages," *Control Eng.*, February 1968, pp. 93–98.

Nelson, E. A. "Management Handbook for the Estimation of Computer Programming Costs" *U.S. Gov. Rep. AD 648750*, March 1967.

Rader, C., and B. Gold, "Digital Filter Design Techniques in the Frequency Domain," *Proc. IEEE*, Vol. 55, No. 2, February 1967, pp. 149–171.

Ralston, A., and H. Wilf, *Mathematical Methods for Digital Computers*, Vol. 1, 1960; Vol. 2, 1967, Wiley, New York.

BIBLIOGRAPHY

Clough, J., and A. Westerberg, "FORTRAN for On-Line Control," *Control Eng.*, March 1968, pp. 77–81.

Albers, D. M., "Developing Generalized Applications Packages," *Data Process. Mag.*, June 1968, pp. 24–27.

Brooks, G. M., "Interactions of Process Control Programming and Instrumentation," *Proc. Texas A&M 21st Symp. Instrum. for the Process Ind.*, January 1966, pp. 15–18.

Chorafas, D. N., *Control Systems Functions and Programming Approaches*, Academic Press, New York, 1965.

Davis, M. A., "The Efficient Use of a Digital Computer To Perform Sequential Logic for Process Control," *IEEE Gen. Meet. and Nucl. Radiat. Eff. Conf., Paper 63-974*, Toronto, June 1963.

Jirauch, D. H., "Software Design Techniques for Automatic Checkout," *IEEE Trans. Aerosp. and Electron. Syst.*, November 1967, pp. 934–940.

Kelly, V., "APEX: A Programming System for On-Line Process Control," *Data Process*, September 1967, pp. 232–237.

Lu, K., "Statistical Forecasting of Computer Program Testing: Concepts and Applications," in *Computer Impact on Engineering Management*, Instrument Society of America, Pittsburgh, Pa., 1968, pp. 60–67.

Mack, E. B., "Programming Real-Time Computer Systems," *Data Process. Mag.*, May–June 1965, pp. 143ff.

McIrvine, E. C., "Planning Software for a Manufacturing Line," *Control Eng.*, April 1968, pp. 100–103.

Markham, G. W., "Fill-in-the-Form Programming," *Control Eng.*, May 1968, pp. 87–91.

Neblet, J. B., "Stretching Process Computer Programmers," *ISA J.*, September 1966, pp. 26–27.

Rechter, R. "Undercover Signals are Unmasked," *Electron. Des.*, August 16, 1967, pp. 236–239.

Ryle, B. L., "An Engineering Approach to Software Configuration Management," *IEEE Trans. Aerosp. and Electron. Syst.*, November 1967, pp. 947–951.

Weiss, M. "Instrument Engineer's Guide to Digital Computer Control", *Instrum. Technol.*, September 1967, pp. 44–54.

Whitman, K., "Digital Computer Control: Organizing the Program for Control," *Chem. Eng.,* December 5, 1966, pp. 135–138.

Williams, T. J., "Software, More Standard Programs," *Control Eng.,* September 1966, pp. 126–130.

Williams, T. J., and S. J. Bailey, "Software: Critical Factor in Computer Usage," *Control Eng.,* October 1967, pp. 65–72.

Woodley, G., "Modelling and Programming for Direct Digital Control," *ISA J.,* March 1966, pp. 48–54.

Appendix A

Computer Fundamentals

J. PAUL HAMMER

The idea of a *computing machine* is not new. Many of the characteristics inherent in the modern high-speed digital computer were described several hundred years ago, and there were even attempts to build relatively complicated machines. Even further back in history, ancient civilizations devised mechanical devices to assist in performing mathematical calculations. It was not until World War II, however, that the need for powerful calculating machines and the advances in electronic technology combined to spark the development of the large-scale digital computer. In the last two decades, the digital computer has grown from a specialized scientific tool into an integral part of every business and industry.

It is used to balance our bank accounts, check our income tax and credit rating, control our traffic lights, control industrial processes, and to do many other tasks which affect our day-to-day living. This era will undoubtedly be called the "Computer Age" by future historians.

Despite the computer's involvement in every phase of our lives, its operation is understood by only a relatively small portion of our population. The mass media have popularized the computer as the "giant brain" or the "thinking machine." Whether a machine can "think" or has the attributes of a brain is a highly philosophical question, the answer to which greatly depends on the definition of "thinking" and intelligence. It is not necessary to resolve these arguments to concede that the modern digital computer is electronically and physically complex. This is obvious to anyone who views the thousands of circuits in the computer or who scans the rules that govern their detailed design. Despite this complexity, however, the computer is conceptually simple. It performs a relatively few simple operations that can be taught easily to a grade-school child who knows how to count and how to compare two one-digit numbers. Its computational power is derived from its ability to do these simple operations at extremely high speeds, often in less than 1 μsec.

It is certainly not necessary to be familiar with electrical engineering or its technology to understand the logical concept of digital computer systems which are constructed from electronic circuits. It is helpful to have an interest and ability in the solution of problems involving simple logical elements, however. This appendix is intended for persons with this problem solving ability and presents the conceptual basis for the operations performed by the modern computer.

A.1   LOGICAL ELEMENTS

The beginning of a problem normally consists of a list of definitions for terms and characters on which the problem solution is based. In this particular case, the terms and descriptions concern logical elements. Four of these elements are described and these are sufficient for the implementation of all the basic functions needed in a simple computer. (In reality, fewer elements are sufficient but using extra elements simplifies the discussion.) Each of the elements has at least one *input* and one, and only one, *output*. The function of the logical element is to generate an output based on the logical state(s) existing at the input. Each output and input can assume only one of two states, designated as **1** or **0**.

The names of the four logical elements are *AND*, *OR*, *EXCLUSIVE OR*, and *LATCH*. These are abbreviated as *A*, *O*, *XO*, and *L*, respectively.

The relationship between the input and output states of each of the logical elements is shown in Table A-1. The table indicates the conditions required to obtain a **1** output from the logical device. At any time when the output is not **1**, it is **0**.

Table A-1 shows that the logical element AND can have two or more inputs, each of which assumes a **1** or **0** value. In order to obtain a **1** value on the output of logical element *A*, all inputs must be in the **1** state; that is, input 1 *and* input 2 *and* input 3, and so on, must be in the **1** state.

Logical element OR also has two or more inputs, each of which assumes

Table A-1   Basic Logic Elements

Name	Symbol	Function
AND	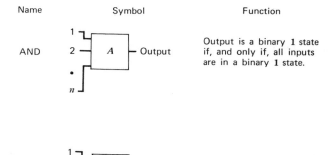	Output is a binary **1** state if, and only if, all inputs are in a binary **1** state.
OR	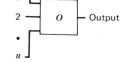	Output is a binary **1** state if one or more inputs is in a binary **1** state.
EXCLUSIVE OR	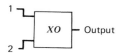	Output is a binary **1** state if one, and only one, input is in a binary **1** state.
LATCH	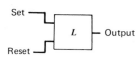	Output becomes a **1** following a **1** input to "set" and remains **1** even if the input is removed. A **1** input to "reset" changes the output to **0** where it remains until a **1** is appled to "set".

either the **1** or **0** state. In order to obtain a **1** output from logical element *O*, there must be at least one input that is in the **1** state; that is, input 1 *or* input 2 *or* input 3, and so on, must be in the **1** state.

The number of inputs for the EXCLUSIVE OR logical element is normally two. To obtain a **1** at the output of the logical element *XO*, one and only one input must be in the **1** state; that is, input 1 *or* input 2, but not both, must be **1**.

The output state of the LATCH is dependent on the sequence of events prior to the time the output state is examined. In this element, the two inputs are defined as "set" and "reset." If the output of the logical element *L* is **0**, a **1** applied to the set input changes the output to a **1**. If the output is **1**, it remains **1**. Applying a **1** to the reset input changes a **1** output to **0** but does not change the output state if it is already **0**. The major attribute of the LATCH is that it "stores" the set or reset information. That is, if the output is **0** and a set input is applied, the output changes to a **1** and remains in this state even if the set input state returns to **0**. The output is returned to the **0** state only when a **1** is applied to the reset input.

By interconnecting these logical elements with the output of one element serving as the input to one or more other elements, complex logical and arithmetic operations can be implemented. These operations are the basic functions required in the digital computer. In order to use them, however, the data on which the computer operates must be expressed in terms of the states **0** and **1**. In most modern computers, this is done by using the binary number system and a binary based code to designate decimal numbers, alphabetic information, and symbols.

A.2 BINARY NUMBERS AND ARITHMETIC

Logic elements with two states lend themselves quite naturally to the representation of quantities (numbers) as sequences of two symbols (numerals). For convenience, the two symbols are the numerals 0 and 1 which, by convention in this book, correspond to the two states **0** and **1**. By examining the structure of a decimal number, it is easy to see how a number can be represented as a binary number using only these two numerals.

If one were to take the decimal number 735 and indicate the significance of the position of each digit in that number, he would show the 5 multiplied by 10 to zero power, the 3 multiplied by 10 to the first power, and the 7 multiplied by 10 to the second power:

$$7 \times 10^2 + 3 \times 10^1 + 5 \times 10^0$$
$$700 \quad + \quad 30 \quad + \quad 5$$
$$735$$

In other words, the value of each digit in a decimal number is the product of the value of the digit and 10 raised to the power indicated by the position of the digit, starting from the right.

In the decimal number system, a number is represented using combinations of ten numerals 0, 1, 2, . . ., 9, and the positional significance is a power of 10. Since the logical elements can represent two numerals 0 and 1, a useful number system for representing quantities is the binary system which requires only two numerals and in which positional significance is a power of 2. That is, a binary number is interpreted by assigning positional significance based on 2 raised to the power indicated by the position of the digit in the number. For example, the value of the binary number 101101 is determined by summing 1 (the lowest order digit) times 2 to the zero power, 0 times 2 to the first power, 1 times 2 to the second power, 1 times 2 to the third power, 0 times 2 to the fourth power, and 1 times 2 to the fifth power. The sum in decimal notation is 45.

$$
\begin{array}{ccccccl}
1 & 0 & 1 & 1 & 0 & 1 & \text{Binary} \\
1 \times 2^5 + 0 \times 2^4 + 1 \times 2^3 + 1 \times 2^2 + 0 \times 2^1 + 1 \times 2^0 & & & & & & \text{Decimal} \\
32 \;+\; 0 \;+\; 8 \;+\; 4 \;+\; 0 \;+\; 1 & & & & & & \\
45 & & & & & &
\end{array}
$$

The procedure for converting any binary number to its decimal value is identical to the above description. Each digit in the binary number is multiplied by its positional significance and the products are summed. Example A-1 demonstrates how another binary number is converted to its decimal value.

One method of converting decimal numbers to binary numbers is successive division by powers of 2. Example A-2 demonstrates this method. The decimal number, in this case 625, is divided by a power of 2. Thus 512 goes into 625 one time with a remainder of 113. Next, the remainder

Example A-1   Binary to Decimal Conversion

1001110001	$1 \times 2^0 =$	1
100111000	$0 \times 2^1 =$	0
10011100	$0 \times 2^2 =$	0
1001110	$0 \times 2^3 =$	0
100111	$1 \times 2^4 =$	16
10011	$1 \times 2^5 =$	32
1001	$1 \times 2^6 =$	64
100	$0 \times 2^7 =$	0
10	$0 \times 2^8 =$	0
1	$1 \times 2^9 =$	512
		625

Example A-2   Decimal to Binary Conversion

$$625/512 = 1 + 113(\text{Rem.}) \qquad 1$$
$$113/256 = 0 + 113(\text{Rem.}) \qquad 10$$
$$113/128 = 0 + 113(\text{Rem.}) \qquad 100$$
$$113/64 \ = 1 + \ 49(\text{Rem.}) \qquad 1001$$
$$49/32 \ = 1 + \ 17(\text{Rem.}) \qquad 10011$$
$$17/16 \ = 1 + \ \ 1(\text{Rem.}) \qquad 100111$$
$$1/8 \ \ \ = 0 + \ \ 1(\text{Rem.}) \qquad 1001110$$
$$1/4 \ \ \ = 0 + \ \ 1(\text{Rem.}) \qquad 10011100$$
$$1/2 \ \ \ = 0 + \ \ 1(\text{Rem.}) \qquad 100111000$$
$$1/1 \ \ \ = 1 + \ \ 0(\text{Rem.}) \qquad 1001110001$$

is divided by the next lower power of 2, which is $2^8 = 256$. The 0 quotient is due to the fact that 256 is larger than 113. The result to this point is binary 1 followed by a binary 0. The division is carried on through all the powers of 2, down to 2 to the zero power which is unity. The binary number is the sequence of the 1 and 0 quotients resulting from the successive division. Therefore, the binary number equivalent of 625 is 1001110001.

This conversion demonstrates that the expression of numbers in binary notation is less efficient than decimal notation; the three-digit decimal number 625 is equivalent to a ten-digit binary number. This inefficiency would lead to considerable difficulty if binary numbers were commonly used for everyday activity. For example, a seven-digit phone number would require from 20–24 digits if expressed in binary form! However, the binary notation is ideally suited for use in digital computers because it permits numbers to be expressed in terms of two numerals. This allows the implementation of arithmetic operations using logic elements having only two states, and such elements are easily constructed using electronic circuits.

A.2.1   Binary Addition

The binary addition tables are greatly simplified since only two numerals are used in the binary number system. The addition rules are summarized in Table A-2. The number at the intersection of the row and

Table A-2   Binary Addition Table

		Addend	
		0	1
	0	0	1
Augend	1	1	10

column representing the numbers to be added is the sum. These sums are the same as those obtained in decimal addition except that the factors are expressed in binary notation and $1 + 1 = 10$.

The addition of two binary numbers is quite straightforward and is very similar to decimal addition. Example A-3 illustrates the addition

Example A-3   Binary Addition

Binary		Decimal
101101	Augend	45
1100101	Addend	101
11 11 1	Carry	
10010010	Sum	146

of the binary numbers 101101 and 1100101. Starting with the low order position, $1 + 1$ results in a sum of 0 with a carry into the next high order position. The second position then provides the sum of $0 + 0$ and the carry (1) from the low order position, the result being 1. The third order position again provides the sum of $1 + 1$, which is 0 plus a carry into the fourth order position of the number. The fourth order position provides the sum of $1 + 0$, which is 1, plus the carry from the next lower order which produces a zero in the sum and a carry into the fifth order position. Following this through, the sum is 10010010. The addition of the decimal equivalents of the binary numbers is also shown in the example.

A.2.2   Binary Subtraction

Binary subtraction is also similar to the decimal operation. The subtraction rules are easily derived from the addition table of Table A-2. The rules are the same as in decimal arithmetic only the notation is binary and $10 - 1 = 1$. As in decimal subtraction, a "borrow" operation is necessary if 1 is to be subtracted from 0. Example A-4 illustrates the

Example A-4   Binary Subtraction

Binary		Decimal
***	Borrow*	*
1100101	Minuend	101
101101	Subtrahend	45
111000	Difference	56

subtraction of two binary numbers. The binary number 101101 is subtracted from 1100101 which, in decimal notation, is the subtraction of 45 from 101. This example illustrates that the borrow operation is accomplished in the same manner in binary arithmetic as it is in decimal arithmetic.

In a computer, it is often easier to implement binary subtraction by using a form of subtraction known as complement addition. This allows the adder in the computer to be used for both addition and subtraction. In using complement addition for binary subtraction, the number to be subtracted, the subtrahend, is converted to its 2's complement which is then added to the other term, the minuend. The 2's complement is formed by changing each numeral 1 or 0 in the binary number to the opposite numeral and adding a 1 in the lowest order position. For example, the 2's complement of 101 is 010 + 001 = 011.

Example A-5 illustrates binary subtraction as accomplished using

Example A-5   Binary Subtraction Using Complement Addition

Subtrahend	0101101	
Two's complement	1010011	
	1100101	Minuend
	1010011	Subtrahend two's complement
	1 111	Carry
	10111000	Difference

complement addition. The minuend is 1100101 and the subtrahend is 0101101, which has a 2's complement of 1010011. The difference 10111000 is obtained by adding the minuend and the 2's complement of the subtrahend. The highest order numeral, a 1, is a carry beyond the highest order of either of the terms involved in the subtraction operation. This high order carry indicates that the answer is positive and in "true" (i.e., non-complemented) form. If the high order carry does not occur, the answer is negative and is in 2's complement form. In either case, the high order numeral is not part of the difference.

The use of complement addition for performing subtraction is not unique to the binary system. The same operation, known as 9's complement addition or "casting out 9's," is possible in decimal subtraction and was once commonly taught. In this method, each digit of the subtrahend is subtracted from 9 and a 1 is added in the lowest order position to form the 9's complement. The sum of the complement and the minuend then provides the difference. For example, the complement of 45 obtained in this manner is $954 + 1 = 955$. The sum of 101 and 955 is 1056, indicating that the difference of 101 and 45 is 56 and that the answer is positive.

A.2.3   Binary Multiplication

Binary multiplication also follows the form of the equivalent decimal operation. It is much simpler, of course, since the binary number system involves only the two numerals 0 and 1. Therefore, the four possible products are $1 \times 1 = 1$, $0 \times 1 = 0$, $1 \times 0 = 0$, and $0 \times 0 = 0$. Example A-6 shows the multiplication of two binary numbers. The product is

Example A-6   Binary Multiplication

1100101	Multiplicand	
101101	Multiplier	
1100101	0S	
00000000	1S	
110010100	2S	
1100101000	3S	Subproducts
00000000000	4S	
110010100000	5S	
1000111000001	Product	

formed as the sum of the subproducts which are listed on the lines 0S–5S. The rule for binary multiplication is particularly simple. When a 1 appears in the multiplier, the multiplicand is copied in the subproduct, shifted an appropriate number of places to the left. In the example, the top line designated 0S is a copy of the multiplicand shifted 0 places to the left; the third line, 2S, is a copy of the multiplicand shifted two places to the left, and so on. This simple algorithm makes the logical design of a binary multiplier very easy since the only operations involved are shifting and adding. The entire multiplication can be performed by a hardware adder with only minor modifications to provide for the shifting operation.

A.2.4   Binary Division

In decimal arithmetic, division can be accomplished by successive subtraction of the divisor from the dividend. The number of times the divisor can be subtracted from the dividend is the result of the division, the quotient. The alternate method, known as long division, involves selecting trial quotients digit by digit and subtracting partial products from the dividend. The same methods are applicable in binary division.

Example A-7　Binary Long Division

```
            11001
11001 | 1001110001
      −11001
        11100
       −11001
        11001
       −11001
        00000
```

Example A-7 demonstrates the long division method. It is identical in form to the calculation involved in decimal division. The example illustrates that the binary division operation consists of subtractions and right shifts of the divisor. Since the subtractions can be accomplished in a binary adder by using complement addition, the division operation can be implemented using a binary adder modified to provide the right shift function.

A.2.5　Information Coding

The arithmetic and conversion operations described in the previous sections are basic to the operation of a computer system. The operator of the system provides data to the system in decimal notation. The system converts these data to binary form and, when the results of the arithmetic calculations have been obtained, converts the binary result back to decimal notation for the convenience of the operator.

These operations require that the computer handle information in both decimal and binary form. However, the logic elements described earlier can assume only two states, **0** and **1**, which are used to represent the two numerals 0 and 1. By using a coded system, however, the two states **0** and **1** can be used for the representation of decimal, alphabetic, and graphical information.

Decimal coding can be illustrated by reference to a riddle which concerns a man who walked into a hotel and wanted to stay for 21 nights. Although he had no money with which to pay the bill, he did have a gold chain, each link of which was worth a night's lodging. He suggested to the hotel clerk that he submit the chain in payment for his bill. The hotel clerk declined his offer on the basis that he was not allowed to take prepayment for bills, nor could he allow the gentleman to stay in the hotel for 21 days before paying his bill. He further stated that under the circumstances he would like the gentleman to pay him each morning for the previous night's lodging.

The patron then offered to give the clerk one link of the chain each day in payment for the previous night's lodging. The clerk again declined the offer on the basis that 21 individual links would not be nearly as valuable as the complete chain. By the time he had the links put back together, the chain would have cost him a considerable amount of money. The patron then said to the clerk that he would cut only two links and yet assure the clerk that on each day during the 21-day period the clerk would have in his possession one link for each night that the patron had stayed at the hotel. The clerk accepted this offer and the chain was cut as shown in Fig. A-1.

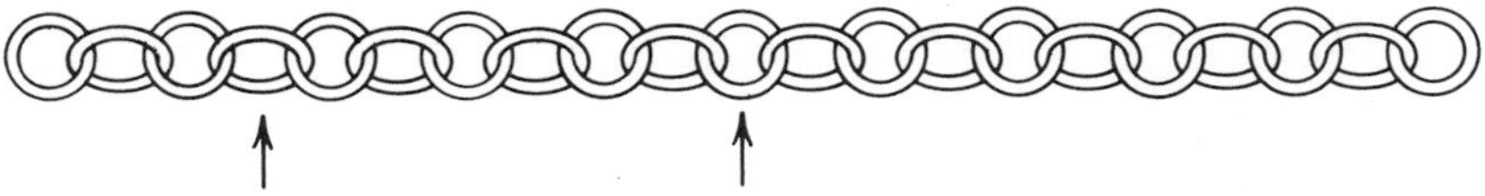

Night	1-bit	1-bit	3-bit	6-bit	10-bit
1	C	P	P	P	P
2	C	C	P	P	P
3	P	P	C	P	P
4	C	P	C	P	P
5	C	C	C	P	P
6	P	P	P	C	P
7	C	P	P	C	P
8	C	C	P	C	P
9	P	P	C	C	P
10	C	P	C	C	P
.	.	.	.	.	.
.	.	.	.	.	.
.	.	.	.	.	.
19	P	P	C	C	C
20	P	C	C	C	C
21	C	C	C	C	C

Fig. A-1   Gold chain riddle (C = clerk; P = patron).

If now the pieces of chain are designated as bits, there are 2 bits of chain with a weight equal to one link, 1 bit of chain with a weight equal to three links, 1 bit with a weight equal to six links, and 1 bit with a weight of ten links. The chart at the bottom of Fig. A-1 indicates how the 5 bits of chain were held by the two men throughout the 21-day period. After the first night's lodging, the clerk held one 1-bit piece of chain and the patron held a 1-bit, a 3-bit, a 6-bit, and a 10-bit piece.

After the second night's lodging the clerk held two 1-bit pieces of chain and the patron held the 3-bit, the 6-bit, and the 10-bit pieces. After the third night's lodging the clerk held the 3-bit piece of chain and the patron held the remaining pieces. Referring to the chart, the reader can see how the transaction was accomplished for the entire 21-day period.

To illustrate how this type of coding can be implemented with logic elements, suppose that four logic devices with outputs A, B, C, and D are assigned a bit value which is similar to the bit value designated for the pieces of chain. For example, the outputs of logic elements A and B are given a value of 1 bit each, the output of logic element C is assigned a 3-bit value, and the output D has a 6-bit value. Figure A-2 shows how

Decimal Digit	A 1-bit	B 1-bit	C 3-bit	D 6-bit	Logic Element Value
0	0	0	0	0	
1	1	0	0	0	
2	1	1	0	0	
3	0	0	1	0	
4	1	0	1	0	
5	1	1	1	0	
6	0	0	0	1	
7	1	0	0	1	
8	1	1	0	1	
9	0	0	1	1	

Fig. A-2   Information coding example.

the 1 or 0 state of these logic elements may be used to designate a decimal digit. To represent the decimal digit 0, all four elements are in a 0 state. To specify the digit 1, A is 1 and the remainder are 0. To designate the digit 2, A and B are 1 and the remainder are 0, and so forth, to the digit 9, which is designated by bits C and D being 1 and the remaining outputs being 0.

In computers, it is customary to code information that represents decimal digits and alphabetic characters in a manner similar to this example. Although coding is not standardized on an industry-wide basis at the present time, steps are now being taken toward standardization.

A.3   IMPLEMENTATION OF COMPUTER FUNCTIONS

The binary arithmetic and information coding concepts discussed in the previous section form the basis for digital computer operations. The use of weighted codes permits the representation of decimal values and similar codes are used for other forms of information such as alphabetic characters. Arithmetic operations implemented by means of logical elements provide the computational capability. Furthermore, computer functions implemented with the same elements provide the necessary controls for handling data in the computer. This includes the transfer of data between the various parts of the computer, the storage of data, and the decoding of data to determine specific storage locations or sequence of operations.

The implementation of these functions in the digital computer utilizes the logical elements introduced earlier. Each of these elements has inputs and an output that assume one of the two logical states 1 and 0. These two states are represented electrically by the magnitude of current flow or by a voltage level, the difference being strictly a matter of electrical implementation. Choices concerning the method of electrical implementation are based on cost, speed, reliability, and power requirements. Although these factors are important to the user of computing systems in terms of system performance, they are not fundamental to the understanding of computer operations.

A.3.1   Data Sensing and Transfer

A fundamental requirement in digital computers is the ability to sense the state at the output of a logical element and to transfer the information represented by that state to another part of the computer. These functions are easily implemented using the logical elements discussed earlier.

Figure A-3 illustrates one way of sensing and storing data. The switches labeled 1, 2, 4, 8, and 16 are closed and opened to represent numerical data in coded form. The switch labeled "sample" is closed after the data switches have been set to represent the value to be read into the computer. When closed, the output side of the sample switch assumes a binary 1 value. This is applied to one input of each of the AND circuits shown in the figure. Input number 2 of the AND logical element labeled A_1 is in a 0 state and, therefore, the output remains at a 0 value. This is also true of the AND circuits A_2, A_4, and A_{16}. However, the AND logical element labeled A_8 is connected to the output of a switch which is closed, so one of its inputs is connected to the binary 1 input. Therefore, both the inputs on this particular AND element are in a binary 1 state and the output assumes a value representing a logical 1. With a 1 on the output of the AND element A_8, there is a 1 on the "set" input of the latch element labeled L_8. Therefore, the latch is turned on and its output remains in the 1 state until reset.

The effect of this simple combination of logical elements is to sense the

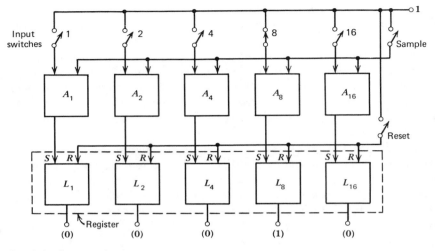

Fig. A-3   Data sensing.

status of the input switches when the sample switch is closed. Owing to the nature of the LATCH elements, the data represented by the position of the switches are "stored" in the LATCH elements even after the sample switch is opened. The set of latch circuits—which, taken collectively, is called a register—is reset to zero by applying a binary 1 to the reset inputs of each LATCH by momentarily closing the switch labeled "reset."

If the switches are now replaced by electrical contacts which are used for reading punched holes in paper tape or cards, or by contacts associated with keyboards or typewriters, the reader can visualize coded information being read into a data-processing system. The register shown in Fig. A-3 might then be referred to as an "input register."

The "sensing" switches in Fig. A-3 are labeled 1, 2, 4, 8, and 16 to indicate the binary bit weight. This decimal method of denoting weighted value is used in computing systems designed around the binary number system, in spite of the fact that the weights can be written in binary notation. This is done to facilitate conversation. The corresponding binary notation is 1, 10, 100, 1000, 10000. These binary numbers are said in the following manner: "one," "one-zero," "one-zero-zero," and so forth, rather than "one," "ten," "one hundred," and so forth. It is, in fact, extremely awkward to "speak binary." Therefore, systems designed around the binary number base have considerable decimal notation expressed in both written and verbal discussions of the computer to ease conversation and documentation.

The information previously represented by the state of the input switches of the register of Fig. A-3 is now stored in the latches L_1–L_{16}. Thus binary information can be stored in a register consisting of latch elements. Generally, however, registers of this type are used to provide storage only on a temporary basis. Other configurations of logical elements are designed to provide longer term storage in the computer. Computer storage is discussed in a subsequent section.

The hardware of Fig. A-4 is quite similar to that used to provide the

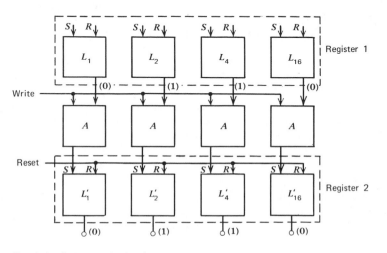

Fig. A-4   Data transfer hardware.

transfer of data between two registers. This is an important function since it allows data and control information to be routed into, and out of, special-purpose registers and other computer circuits such as adders. Figure A-4 illustrates an example of data transfer between two registers.

The data to be transferred are stored in register 1. Register 2 is assumed to be reset; that is, all latches are in the **0** binary state. When a write signal, represented by a binary **1** on the WRITE line, is applied, the outputs of the AND gates connected to the latches in register 1 which have a **1** output become **1**'s. This, in turn, sets the outputs of the corresponding latches in register 2. In Fig. A-4, the **1** outputs of latches L_1 and L_4 in register 1 are transferred to the corresponding latches in register **2**.

The data transfer hardware of Fig. A-4 requires that register 2 be reset before the data are transferred. Failure to reset the register results in a logical OR function in which the outputs of register 2 are 1's if, prior to the Write command, *either* corresponding latch is in a binary 1 state.

The requirement for resetting register 2 can be avoided with additional hardware. If the latch in register 2 is a 1 and the corresponding latch in register 1 is a 0, a reset pulse is applied to the latch in register 2 to reset it to 0. In all other cases, the latch in register 2 is actuated in the manner described above.

In either of these data transfer cases, the inputs to the latches of register 2 form what is called a data bus. In the case of the 4-bit register of Fig. A-4, the bus consists of four wires, one for each input in register 2. Figure A-5 illustrates how the registers, transfer gates (AND blocks), and data bus can be represented in a block diagram.

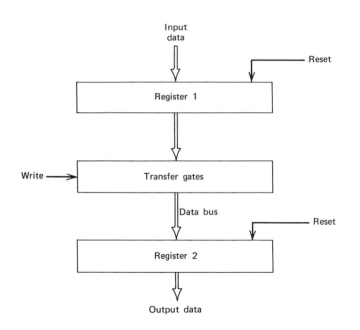

Fig. A-5   Data transfer—block diagram.

Figures A-3, A-4, and A-5 provide an indication of how switches, AND, and LATCH logic elements can be used for the sensing, storage, and transfer of information within an organization of logical elements. However, when a computer is mentioned, one normally thinks of a high-speed arithmetic device not an organization of logical elements which allows sensing and transfer of data without arithmetic manipulation. In the next section, an organization of logical elements which takes an input of two binary numbers and provides an output that represents the result of an arithmetic operation is discussed.

A.3.2   Arithmetic Operations

The previous discussion of binary arithmetic emphasized that subtraction, multiplication, and division of binary numbers can be performed by addition or a minor modification of addition. Thus the binary adder circuit can form the basis for implementing arithmetic operations in a digital computer.

Figure A-6 illustrates a 4-bit binary adder composed of the basic logical elements. This arrangement of elements develops the sum of two 4-bit factors A and B which are provided as inputs to the adder. (The **0** and **1** states in the figure refer to the specific example discussed below and should not be considered at this time.) The general structure of the adder is analyzed by starting at the least significant bit position in the upper right-hand corner of the figure.

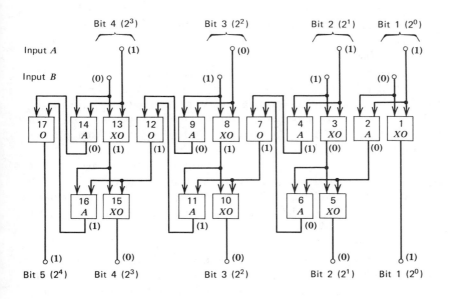

Output (sum)

```
1 0 1 1   A
0 1 1 0   B
1 1 1 0 0  Carry
1 0 0 0 1  Sum
```

Fig. A-6   Binary adder.

The input from the low order positions (2^0) of the two input numbers A and B are attached to an AND circuit and to an EXCLUSIVE OR circuit. Since the XO_1 circuit provides a 1 output when one, but not both, of the two inputs is 1, if either factor A or B has a 1 in the units position, the sum line at the output of the EXCLUSIVE OR circuit is in the 1 state. The AND circuit A_2 has a 1 output only if both inputs are 1. The output from the AND circuit represents the carry from the low order position to the next higher position. Neither of the two circuits involved in the low order positions of the adder has a 1 output if both inputs are 0.

To the left is the second order position (2^1). The second order bits of A and B are connected to an XO circuit and an AND circuit, as in the previous case. However, below these two circuits there is another stage where the partial sum of the two factors is added to the carry from the low order position. This stage consists of XO_5 and A_6. So again, the sum and the carry outputs are connected to an XO circuit as well as to an AND circuit. The output from this XO is the sum of the second order position of the input bits. The output from the AND circuit A_4 represents the carry from the second order position. This carry is ORed with the carry from the original two factors to provide the carry for the third order position. The third and fourth order positions repeat the form of the second order position.

The states representing the input factors A and B are usually the output states of other logical elements. They are most often the outputs of LATCH elements which form a register, as described previously. Similarly, the sum outputs of the adder are typically connected to the set input of the LATCH elements of another register which provides temporary storage of the sum.

The easiest way to understand the operation of the binary adder of Fig. A-6 is to trace through the specific example shown in the figure. Each of the logical elements is numbered to facilitate the discussion and the logical states at the inputs and outputs of each logical element for this specific example are indicated.

Starting in the upper right-hand corner of Fig. A-6, circuit 1 has a 1 and 0 as its inputs. Therefore, the output is a 1, which represents the sum of the low order digits. Circuit 2 has the same two input signals and, as it is an AND circuit, its output is 0, indicating that there is no carry. Circuit 3 has two 1 signals as inputs and, therefore, its output is 0. This output and the 0 output of circuit 2 are combined in circuit 5 to provide a 0 as the sum of the second order digits and the carry from the first order. The output of circuit 6 is also a 0, indicating that a second order carry into the third position is not required. The inputs to circuit 4, however, are both 1 so a carry is produced by this circuit. The outputs of circuit 6 and circuit 4 are ORed in circuit 7 to produce the actual carry into the next position of the adder.

Circuits 8 and 9 have a 1 and a 0 as input states representing the third order digits of the input numbers A and B. The output of circuit 8 is a 1 and the output of circuit 9 is a 0. In circuits 10 and 11, the partial sum represented by the output of circuit 8 and the carry from the second order position represented by the output from circuit 7 are combined. The output from circuit 10 is the full sum for the third order position of the answer. The output of circuit 11 is a 1 which, when combined with the 0 output of circuit 9, produces a 1 carry at the output of circuit 12.

At circuit 13, the fourth digit of the inputs from the input numbers A and B are a 1 and a 0. The 1 output from circuit 13 is combined with the carry from circuit 12 to produce a 0 as the fourth digit in the sum. These same inputs also produce a 1 output from circuit 16 which is combined with the 0 output of circuit 14 to produce a carry which is the high order 1 in the sum at the output of circuit 17.

The result, reading from the high order to the low order position, indicates a 1 followed by three 0's followed by a 1. Now if the reader looks at the example listed in Fig. A-6, he finds the problem worked out on paper and the result is 1001. Thus since 1 represents the binary numeral 1 and 0 represents 0, the binary adder performs the indicated addition.

This is a very straightforward adder design and the adder in any particular computer system differs depending upon the requirements placed upon the designer of that adder. However, for purposes of understanding the operation of a computer, the adder shown in Fig. A-6 is sufficient. If specific details are required about any particular computer system, the reader must seek the information from the supplier of the equipment. The adder in Fig. A-6 has the capability of adding two numbers, each having four binary digits. The adder can be expanded, however, by adding similar stages so that larger binary numbers can be added.

The binary adder is the basic logical unit required for arithmetic operations in digital computers. By adding appropriate control of the adder and of data flow to and from the adder, it can be used for multiplication, subtraction, and division. As shown in the section on binary arithmetic, multiplication can be accomplished by a series of additions and shifts of data. Similarly, subtraction can be accomplished by the use of complement addition and division can be done by shifting and subtracting. All these operations, however, require additional controls that are not included in the diagram of the adder shown in Fig. A-6.

Before describing the hardware necessary to accomplish binary multiplication, an additional data manipulation operation required in multi-plication is considered. Specifically, multiplication requires that the multiplicand be shifted to the left in generating partial products. This shift left operation can be implemented as a variation of the data transfer action described earlier. The data can be transferred from one register into a temporary holding register. The original register is then reset and the data are transferred back to the original register. In this second transfer operation, however, each bit is transferred to the position just to the left of its original position. Thus the shift left operation can be implemented as a three-step procedure.

An example of the hardware required to perform the three-step shift operation is shown in Fig. A-7. Assume that a 2-bit data word is stored

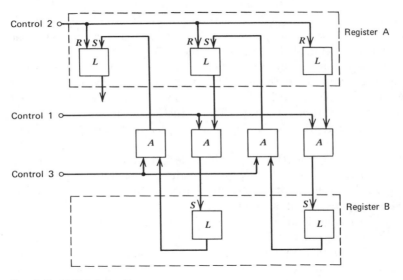

Fig. A-7 Shift left hardware.

in the right-most positions of register A. Actuating control 1 transfers this data into register B. The next control signal, control 2, resets regis-ter A to all 0's. The final step initiated by control 3 transfers the data from register B back into register A. Owing to the fact that each AND gate in the transfer gate hardware is connected from a bit in register B to the bit on the left of the corresponding bit in register A, the data word is shifted one place to the left.

As the data are shifted to the left, the right-most position of the data word is filled with a **0**. Similarly, the bit in the left-most position of the word is shifted out of the register and is "lost." For example, consider the 4-bit data word **1101** and assume that it is stored in a 4-bit register. As shift left operations are performed, the contents of the register are:

1101   Original data
1010   After first shift left
0100   After second shift left
1000   After third shift left
0000   After fourth shift left

The shift right operation is similar, as is the hardware required to implement it. In the hardware, the transfer gates which transfer the register from register B back to register A are connected to the bit to the right of the corresponding bit in register A. If the same data as in the above example were shifted to the right four times, the contents of the register would be

1101   Original data
0110   After first shift right
0011   After second shift right
0001   After third shift right
0000   After fourth shift right

In a manner analogous to the shift left operation, the left-hand positions of the register are filled with **0**'s as the data are shifted to the right and the right-most digits are shifted out of the register and lost.

Returning to the question of binary multiplication hardware, a multiplier can be implemented using registers, transfer gates, and shifting registers. As Example A-6 illustrates, the rules for binary multiplication are that the multiplicand is added into a partial sum if the multiplier bit is a 1; the multiplicand is then shifted one position to the left and the operation is repeated. Thus the hardware necessary to implement the multiplication operation must examine each bit of the multiplier to determine whether the multiplicand is to be added into the partial product sum.

Hardware which accomplishes the steps necessary to multiply two binary numbers is shown in Fig. A-8. The multiplication action requires as many cycles as there are bits in the multiplier. Each of the cycles, in this particular hardware implementation, consists of three steps initiated by control pulses. Thus control pulses 1, 2, and 3 control the first cycle, control pulses 4, 5, and 6 control the next cycle, and so forth. Referring to the figure it is assumed that the multiplier has been transferred into register B via transfer gates not explicitly shown. Similarly, the multi-

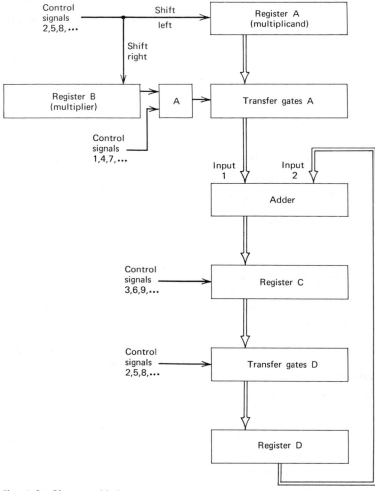

Fig. A-8   Binary multiplier.

plicand is transferred into register A. The AND gate connecting the right-most (least significant) bit position of register B controls the transfer of the shifted multiplicand into the adder by controlling an AND gate. If the least significant bit (LSB) of the multiplier is **0**, the transfer is inhibited and the input to the adder consists of all **0**'s.

As a specific example of the operation of this binary multiplier, consider the problem of multiplying 110 (the multiplicand) by 101 (the multiplier). As described above, the multiplicand and multiplier are transferred into registers B and A, respectively. Control signal 1 senses the LSB of the multiplier which, in this case, is a 1, so the multiplicand is transferred into the adder. The other input to the adder is 0 since register D is assumed to be reset before the multiplication operation is initiated. Thus the output from the adder is merely the multiplicand and this value is stored temporarily in register C. Control pulse 2 transfers the contents of register C into register D and also shifts the multiplicand in register A one position to the left and the multiplier in register B one position to the right. The next control signal, control 3, resets register C to all 0's. This completes the generation of the first subproduct in the multiplication procedure.

In the specific example being considered, the second multiplier bit is a 0, so the LSB in register B is now a 0. As a result, the transfer of the shifted multiplicand in register A into the adder is inhibited. Thus the output of the adder is still equal to the multiplicand 110 since this is input #2 to the adder and input #1 is 0. This result, stored in register C, is transferred into register D by the control signal 5 which also shifts the multiplicand and multiplier. The number in register A is now 1100 and the number in register B is 001. Since the LSB in register B is 1, the shifted multiplicand is added to the partial product stored in register D. Control 8 transfers this sum to register D and the multiplication operation is complete. The product is stored in register D and, in an actual machine, would be transferred to a storage location or to another register for subsequent processing.

This example and the description of the binary adder and multiplier have been brief and simplified. An actual computer utilizes logical configurations that are conceptually similar but, because of additional function, are considerably more complex. The description should be sufficient, however, to demonstrate how binary numbers consisting of the digits 1 and 0 and represented by the logical states 1 and 0 can be mathematically manipulated in a computer built of relatively simple logical elements.

Subtraction and division operations are implemented in a manner similar to that described above for multiplication and addition operations. As demonstrated in Sec. A.2, subtraction can be accomplished by complementary addition and division by subtraction of trial divisors generated by a shifting operation. It is, therefore, possible to implement these operations using combinations of registers, transfer gates, shifting registers, and the basic logic elements.

A.3.3 Decoding and Control

In the hardware implementation of basic arithmetic operations, sequences of control signals are used to initiate and control the transfer of information, the operation of the adder, and so on. The question that immediately arises is how the computer "knows" that it is supposed to perform a certain operation at any given time. As described in more detail in Appendix B, the detailed operations executed by the computer are specified by the programmer in the form of a sequence of instructions. These instructions specify each of the operations such as add, subtract, multiply, divide, transfer, and store. The instructions are stored in the storage of the computer. After the execution of each operation, the next instruction is transferred from storage into a register in the computer for interpretation. The interpretation, or decoding, of this instruction initiates the required actions to execute the operation specified by the instruction.

In terms of their representation in the storage and instruction register of the machine, the instructions have the same form as data; that is, they consist of a series of 1's and 0's which are represented as the binary states 1 and 0 in the hardware. The only difference between numerical data and an instruction is the manner in which they are manipulated and interpreted in the machine registers.

The machine instructions, in the simplest case, consist of two parts: the operation code, which specifies the operation to be performed, and an address, which specifies the storage address involved in the execution of the instruction. For example, the instruction set for a basic machine might include the binary words

Meaning	Instruction Word
Add	001XXXXX
Subtract	010XXXXX
Multiply	011XXXXX
Divide	100XXXXX
Load A	101XXXXX
Load B	110XXXXX

In each instruction, the first 3 bits are the operation code that identifies the operation. Thus, for example, the bits 110, when located in the first three positions of the instruction, are interpreted as the operation of transferring the data located at address XXXXX into register B. The symbols XXXXX in these instructions represent a combination of 0's and 1's which is the numerical address of the data in storage. Thus, for example, the sequence of instructions

    10100001   Load A with data in storage location 1
    00100010   Add data in storage location 2

could be interpreted as a program for adding the number in storage location 1 to the number in location 2. In some instructions, the operation is performed on numbers located in predefined registers so the address portion is left blank. For example, if the method used for multiplication requires that the multiplier and the multiplicand be loaded into registers A and B before the Multiply instruction is executed, the instruction for multiply would not have an address specified. Additional details concerning the representation of operation codes, addresses, and the generation of programs are discussed in Appendix B. For present purposes, it is only necessary to recognize that the operations executed by the computer are determined by a program of instructions and that the instructions have a binary representation similar to that described above.

The problem for the hardware designer is to design a combination of basic logic elements which decodes or interprets both the operation code and the address; that is, the hardware must generate a unique signal corresponding to each of the combinations of bits which represent an operation code and an address. The procedure for doing this is the same whether the sequence of bits is an operation code or an address. It is the position of the bits in the instruction that determines the significance of the binary number.

The hardware required to decode a 2-bit binary word is shown in Fig. A-9. Since there are four possible combinations of **0**'s and **1**'s in a 2-bit word (**00**, **01**, **10**, and **11**), this configuration of logic elements provides four unique output signals, one for each of the input combinations.

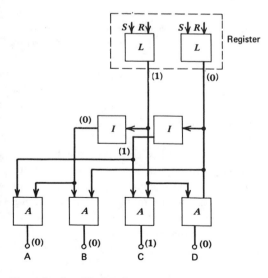

Fig. A-9   Two-bit decoder.

In Fig. A-9, the 2 input bits are stored in a register and one of the output lines A, B, C, or D is a 1 for each combination of input bits.

In order to simplify the logic configuration, a new logic element has been used in Fig. A-9. This is the INVERTER, shown as the block labeled *I*. The INVERTER is such that its output is always in the state opposite to that of its input. Thus if the input is a 1, the output is a 0, and, if the input is a 0, the output is a 1.

A specific decoding example is indicated in the figure. The number stored in the register is 10, as indicated by the binary states 1 and 0 at the LATCH outputs. The outputs at each of the other logic blocks are also indicated in parentheses. By checking the states which exist at the input to each of the AND elements, it is seen that the only one which has two 1's for inputs is C. Thus the only 1 output is C. As the reader can verify, the output states for the four possible input combinations in Fig. A-9 are

Input	Output ABCD
00	1 0 0 0
01	0 1 0 0
10	0 0 1 0
11	0 0 0 1

This example illustrates that the decoding network is the same whether the number stored in the register is an address or an operation code. The only difference is the manner in which the resulting output signal is used; in the case of an operation code, the signal initiates the operation specified by the operation code. If the number is an address, the output signal is used to control the transfer gates associated with the particular address. The example is simplified in that the decoding logic only decodes 2-bit words. In an actual computer, the same type of logic is expanded to provide decoding of up to 6 or 7 bits for the operation code and 10 or more bits for the address.

The decoding of the operation code provides the unique signal required to initiate the specified action. Some operations, however, require a sequence of control signals to complete the execution. For example, in the multiplier described in Sec. A.3.2, a sequence of three control signals is required for each bit in the multiplier. There are a number of ways in which such a sequence can be generated using logic elements and a pulse generator. For example, if four signals are required with each signal immediately following the previous one, the output of a binary counter could be decoded using the decoder of Fig. A-9. A binary counter is a combination of latches that counts the number of pulses applied at the counter input. In a 2-bit counter, the states of the two latches take on the values **00**, **01**, **10**, and **11** as the first four pulses are received. On the fifth pulse, the sequence starts again. Comparing this with the table describing the output of the decoder, it is seen that each of the outputs A, B, C, and D will be **1** in a sequential order.

The manner in which these pulses are used to control an operation requiring 3 pulses/cycle, such as in the case of the multiplier, is illustrated in Fig. A-10. The execute clock produces pulses on the three out-

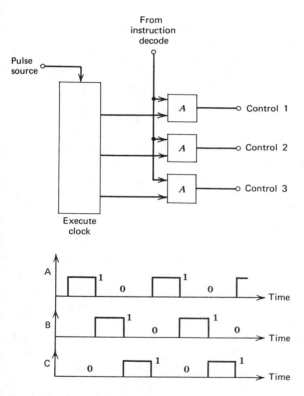

Fig. A-10   Control circuit.

put lines A, B, and C, as shown in the timing diagram. The execute clock is started at the beginning of the operation and continues to run until the execution is complete. The output pulses from the clock are distributed to the proper control lines by means of gates which are controlled by the output of the operation code decoder. Using the multiplier example, when the operation code for multiply is decoded, one side of each of the three AND gates is conditioned to the 1 state by the decoder output. As each of the signals A, B, and C switch to the 1 state, the appropriate control line is actuated.

The decoding and control hardware in an actual computer is very complex and, in fact, constitutes a major portion of the machine. The examples given here, however, are conceptually accurate and indicate the manner in which instructions are initiated and controlled.

A.3.4 Storage

As indicated in the previous section, computers execute only one instruction at a time. The execution time, however, is small, typically less than a few microseconds. In order to effectively utilize this high speed of execution, the computer must have access to the next instruction and to data in a similarly short period of time. The manual entry of instructions and data would limit the speed of execution to the speed of entry, and this is extremely slow compared with the electronic speeds at which the computer is capable of operating. To provide this rapid access to data and subsequent instructions, the instructions and data are stored in the computer system. By addressing the appropriate unit of storage, the next instruction or a specific piece of data can be transferred into the working registers of the computer at a speed consistent with the execution time of the machine.

There are several storage media available in the typical computer system. These differ primarily in speed and storage capacity. Since high-speed storage is relatively expensive, its capacity is usually limited to that necessary to store the information which must be available to the computer in a very short time. Less expensive storage devices having much greater capacity are provided for data that are not required immediately or that are required only occasionally.

The high-speed storage, often called main storage, typically has an access time of from a few microseconds to less than 1 μsec. In present generation machines, this storage is implemented using small ferrite cores, each of which can store 1 bit of information (see Fig. 1-3). A newer development is the monolithic memory or storage in which information is stored in registers consisting of latches similar to those discussed in previous sections. Although ferrite core storage is presently the most used main storage, monolithic storage is used as the basis of the following discussion because it is a natural extension of the circuits and configurations already presented.

In the main storage of the computer system, information can be stored in arrays of registers such as those illustrated in Fig. A-11. The figure

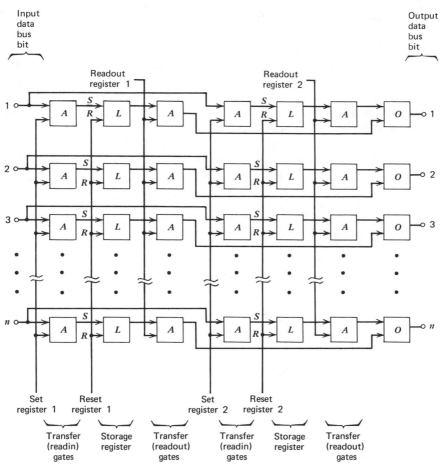

Fig. A-11   **Storage array.**

shows two vertical rows of latches with each row representing a register capable of storing one word of information. (Depending on the structure of the computer, the basic unit of storage may be other than a word, such as a byte or character.) Each horizontal row of latches represents a bit position in each word. To store information in a particular register, the register is reset and then set with the information existing on the input data bus. This operation is performed via transfer gates similar to those discussed previously. The particular address into which the data is transferred is controlled by the set line whose status is, in turn, determined as the output of an address decoder.

The readout operation is accomplished by actuating transfer gates which transfer information from a particular register onto the output data bus. A line in the output bus is connected to the corresponding bit in each register via an OR gate. As in the case of reading data into storage, the particular data selected are determined by the output of an address decoder. A single register is read at a time determined by the common output bus. In addition, hardware is generally included which prevents an attempt to read out a register during the same cycle that information is being stored in that register.

The storage array can be represented in abbreviated form as shown in Fig. A-12. The vertical lines in the matrix represent the set, reset, and

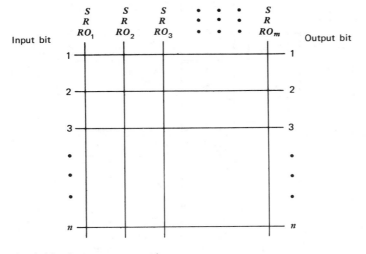

Fig. A-12 Storage array matrix.

readout lines for each of the m registers. The horizontal lines represent the input and output data bus for each of the n bits in the word. The data can be thought of as entering from the left and being read out to the right of the figure. This figure represents a two-dimensional storage array. If 32,000 words of storage are required, 32,000 registers must be provided. Associated with these 32,000 registers would be an equal number of set, reset, and readout lines. The decoding networks which control the state of these lines would be very extensive and, compared with other methods of arranging the array, quite expensive.

A storage arrangement that requires less decoding hardware, and the configuration that is used almost universally in today's computers, is the three-dimensional array. In this arrangement, shown in Fig. A-13, the two-dimensional arrays of the previous figure are combined to form

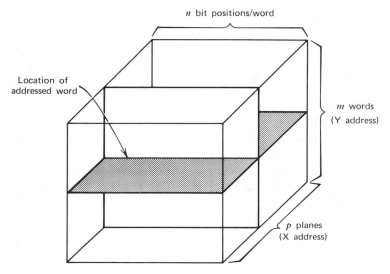

n bit positions/word

Location of
addressed word

m words
(Y address)

p planes
(X address)

Fig. A-13   Three-dimensional array.

a three-dimensional storage array. Each of the *p* planes in the array provides storage for *m* words of *n* bits each. To select a particular storage
word, a portion of the address is decoded to provide a signal which
actuates the particular plane; the remainder of the address is decoded
to provide the readout or readin signal for one of the *m* words in the
plane. For example, suppose that the address is given by the binary
word XXXYYY, where each of the *X*'s and *Y*'s is either a binary 1 or 0.
The 3 *X* bits are decoded to define which of eight planes is to be selected,
and the 3 *Y* bits are decoded to select one of the eight words in that
particular plane. This requires a total of 16 outputs from the address
decoding network compared with 64 outputs if the array is arranged in
two-dimensional form. The difference in decoding becomes even greater
as the size of the storage is increased. For example, in a 16,384-word
storage, a two-dimensional array requires 16,384 address lines compared
with only 192 when the storage is arranged as a three-dimensional array.

In order to accomplish the transfer of information into and out of
the three-dimensional storage array, the transfer gates shown in Fig.
A-11 must be modified to provide for the *X* and *Y* addressing lines. In
the two-dimensional array, the AND gates each have two inputs; specifically, the input transfer gates have one input from the data bus and
the other from the address decoder. In the three-dimensional array, the
AND gates have three inputs: the data bus input, the plane address line
corresponding to the *X* bits in the previous example, and the word

addressing line corresponding to the Y bits of the address. The X address line is connected to the AND gate associated with every latch in a particular vertical plane. Similarly, the Y address line is connected to all the AND gates in the particular horizontal plane.

As a particular example of addressing, consider an array consisting of eight planes of eight words, each of which is 16 bits long. If the address of the storage location is 101110, the last 3 bits are decoded to actuate the Y address line which is connected to each of the AND gates associated with the sixth word of each of the eight planes. Thus it is connected to a total of $8 \times 16 = 128$ gates. The 3 high order bits are decoded to provide a signal to an X address line which is connected to all the transfer gates in the fifth vertical plane, a total of 128 gates. The only latches in which both the addressing inputs to the transfer gates are actuated are those which lie at the intersection of the X address and the Y address. Thus this addressing method selects a single word out of the total array, as shown in Fig. A-13.

This method of addressing data in a three-dimensional array is usually called "half selection." The name results from the fact that the plane address and the word address each provides only half the conditions necessary to identify a particular word in storage. The same method of half selection is used to read out a storage location.

Until recently, main storage arrays have been implemented using small ferrite cores for storing each bit of information. The half selection addressing principle is employed in a manner essentially identical to that described above. The primary difference in the operation is that reading out the cores causes the register to be reset. In order to avoid losing the stored information, circuits are provided which re-store the information immediately following the readout operation. Thus the actual readout cycle consists of two phases, the first for reading the information and the second for re-storing it. The development of sophisticated integrated circuit techniques, however, is resulting in a low cost per circuit so that storage units consisting of latches or similar circuits are becoming practical. With a further decrease in costs, the integrated circuit storage arrays will undoubtedly gain in popularity and the operation of the three-dimensional array described above will be an accurate description of the hardware implementation as well as of the concept.

A.4   COMPUTER ORGANIZATION AND OPERATION

The previous sections describe some of the basic functions and hardware in the digital computer. There are, of course, additional functions implemented in an actual computer and the previous sections do not describe all the details involved in the physical hardware. The concepts and functions discussed, however, are sufficient to define a hypothetical computer which can be used to illustrate how these functions can be combined to form a simple computer capable of performing some simple sequences of operations.

The hypothetical computer which will be used for this purpose is shown in Fig. A-14. The storage unit with its X address decoder (X addr)

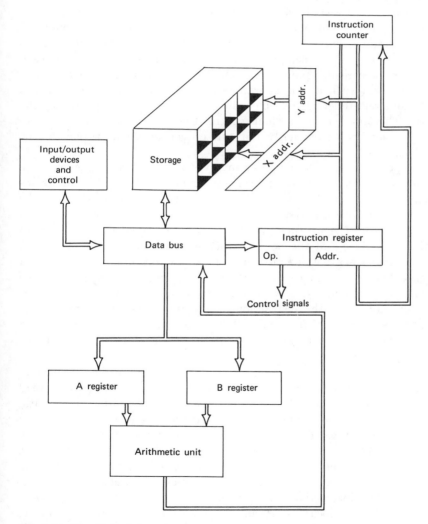

Fig. A-14   Hypothetical computer.

and Y address decoder (Y addr) could be implemented in hardware similar to that described in the previous section. In this computer, the storage is used to store instructions to be executed as well as the data, just as in an actual computer. The instruction register decodes the individual instructions to provide control signals and inputs to address decoding circuits. The A register and the B register are input registers for the arithmetic unit. This unit includes the circuits necessary for addition, subtraction, multiplication, division, and certain comparison operations. This hardware is a combination of some of the configurations discussed previously. The input/output control and device block represents the equipment and circuits necessary to read data into the computer and to read results out of the computer. Thus it includes such devices as card and paper tape reader/punches, typewriters, printers, and other peripheral equipment such as the operator's console. The data paths between the various registers and equipment in this computer are indicated by the arrows. The transfer gates which control this flow of data and instructions are not shown explicitly but can be considered as a part of the data bus.

The operation of this computer involves a basic two-step cycle. The instruction specified by the instruction counter is fetched from its location in storage and transferred into the instruction register via the data bus. The operation specified by the instruction is then executed. After execution, the instruction counter is incremented by one count to provide the address of the next instruction. The cycle is then repeated. In general, the instructions are stored in sequential order in storage and are executed in the same sequence. One exception, however, is when the sequence of instructions calls for a branching or skip operation. In this case, the number in the instruction counter must be altered to provide the address of the next instruction. This is accomplished by transferring the contents of the address portion of the Branch instruction into the instruction counter.

To illustrate the operation of this computer in more detail, consider the following problem: Values for the variables, X and Y, are to be read into the computer; the value of X is punched into the first, third, fifth, and so on, card of an input card deck, and the values of Y are punched into the even-numbered cards. If X is greater than 4.0, the expression $Z = (X + Y)/XY$ is to be evaluated and stored. If X is less than 4.0, the computer returns to the card reader and reads the next values of X and Y. A sequence of instructions which solves this problem is given in Fig. A-15. The instructions are assumed to be stored in locations 00–19 in storage. Storage locations 20 and 21 contain constants required in the execution of the program, and the values of Z are to be stored in locations 24, 25, 26, ...

Memory Location	Operator	Address	Remarks
00	RCD	22	Read card
01	RCD	23	Read card
02	LDA	22	
03	LDB	20	Form $X - 4$
04	SUB		
05	BRN	00	Go to location 00 if $X - 4 \leq 0$
06	LDB	23	
07	ADD		Form $X + Y$ and store in location 22
08	STO	22	
09	MPY		Form XY and store in location 23
10	STO	23	
11	LDB	23	
12	LDA	22	Form $(X + Y)/XY$ and store in location 24
13	DIV		
14	STO	24	
15	LDA	14	
16	LDB	21	Modify STO instruction so that next answer is stored in next storage location
17	ADD		
18	STO	14	
19	BR	00	Return to start of program
20		04	A constant
21		01	A constant
22			Temporary storage location
23			Temporary storage location
24			Location of first answer
25			Location of second answer

Fig. A-15 Sample program.

The sequence of instructions shown in Fig. A-15 represents a computer program. This and other types of programs are discussed in Appendix B. The program is a list of instructions, each of which consists of an operation code and an address (when one is required). In the figure, the computer operations are indicated with abbreviations, called mnemonics, and the addresses are given as decimal numbers. In the computer, of course, both the operation codes and the addresses are represented by binary numbers. The remarks column is provided in the figure merely as a guide, and this information is not stored in the computer.

With the program stored in the storage unit of the computer, execution of the program begins with the transfer of the instruction located at address 00 into the instruction register. This instruction requires that a card be read and that the data on the card be stored in location 22. The control signals resulting from the decoding of the operation code of this instruction cause this action to be executed. The instruction counter is incremented by 1 and the next instruction is transferred into the instruction register. This instruction reads the next card and stores the information in storage location 23.

The next sequence of instructions is used to determine whether the value of X stored in location 22 is greater than 4.0. Register A is first loaded with the value of X by the LDA instruction. Next, the constant 4.0 is transferred from storage location 20 into register B by the LDB instruction and the next instruction forms the difference $X - 4.0$. This subtraction is automatically performed by complementing the contents of register B and adding it to the contents of register A. Since the sign of the difference is indicated by the presence or lack of a carry in the high order position of the difference, the next instruction Branch on Negative (BRN) merely requires the testing of this bit to determine whether the carry is a 0 or a 1. If the difference is negative, the address portion of the BRN instruction is transferred into the instruction counter so that the next instruction performed is the instruction stored in location 00. If the difference is positive, the computer proceeds to the next instruction.

The next sequence of instructions evaluates the expression $Z = (X + Y)/XY$. Since the value of X is already stored in register A, it is only necessary to load register B with the value of Y which is stored in location 23. These numbers are summed and the result is stored in location 22. This destroys the value of X which was stored in location 22, but this is of no consequence since it is not required in further calculations. The numbers X and Y are still stored in registers A and B, so the next instruction specifies that they be multiplied. The product is stored in location 23. This destroys the value of Y but, as before, this does not matter. The next two steps form the required quotient and store it in location 24. This completes the required calculation but, before proceeding with the next set of data, the program must modify itself to ensure that the next answer is stored in the next storage location, location 25.

The modification of the storage location specified in the instruction located in storage location 14 is accomplished by the next four steps.

The word in location 14 is loaded into register A. A constant, equal to 1, is loaded into register B. These values are added and, since the address is located in the lower order positions of the instruction, the address portion of the instruction is changed from 24 to 25. The modified instruction is then re-stored in its original location, location 14. This modification of instructions is an important concept in digital computers. First, it indicates how the program can modify itself in the solution of a problem. Second, it emphasizes that the computer makes no differentiation between instructions and data, both are treated as binary numbers. In this case, an arithmetic operation was performed on an instruction by transferring the instruction into a register concerned with arithmetic operations. Thus the only basic difference between instructions and data in a digital computer is the manner in which they are manipulated.

With the completion of the address modification for the storage instruction, the next instruction to be interpreted is the unconditional Branch located in storage location 19. This causes the address portion of the instruction to be transferred to the instruction counter. Thus the next instruction to be executed is the one specified by the address in the BR instruction; in this case, the specified address is 00, so the program returns to the beginning and reads the next set of data from the next two cards.

In a real computer, some means of retrieving the computed results from computer storage and transforming them into human readable form is required. This is not included in the sample program. In essence, what should be added to the program are instructions which transfer the contents of locations 25, 26, 27, . . . to the I/O control unit which, in turn, would control an output printer.

Although greatly simplified, this example illustrates the step-by-step manner in which a digital computer operates. The real computer is obviously much more complicated than this, often involving more than 100 operation codes and the control of many different peripheral devices. Nevertheless, the hypothetical computer discussed in this section and the hardware discussed in previous sections are conceptually accurate. This discussion should provide the reader who is unfamiliar with the operation of a digital computer with some idea of how such a machine operates.

BIBLIOGRAPHY

Davis, G. B., *An Introduction to Electronic Computers,* McGraw-Hill, New York, 1965.

Harrison, T. J., "Hardware: A Matter of Logic, Memory, and Timing," *Control Eng.*, November 1967, pp. 74–79.

Hodge, B., and J. P. Mantey, "Computer Refresher," *Chem. Eng.* (a series of monthly articles starting September 25, 1967).

Litton Industries, Inc., Data Systems Division Tech., Training Group, *Digital Computer Fundamentals*, Prentice-Hall, Englewood Cliffs, N.J., 1965.

Parton, K. C., *The Digital Computer*, Pergamon Press, New York, 1964.

Weinstein, S. M., and A. Keim, *Fundamentals of Digital Computers*, Holt, Rinehart and Winston, New York, 1965.

Appendix B

Programming Fundamentals

ROGER M. SIMONS

In the parlance of the computer industry, the previous appendix is concerned with digital computer *hardware*. The term is appropriate since the electronic devices used to fabricate the computer are real, physical things; in a sense, they are "hard."

As described in the previous appendix, the hardware consists of devices assembled into circuits, each of which is designed to perform a simple logical function. For example, several transistors and resistors are combined to form a circuit which performs a logical AND function. The output of this circuit is in a certain electronic state, represented by a voltage or current level, if, and only if, all of the inputs are in the same state. The electronic circuits, each representing a simple logical function, are combined to perform more complex logical and arithmetic operations such as adding binary numbers, comparing two binary numbers, and providing a unique output corresponding to a pattern of input states.

This, in a grossly simplified sense, is the hardware of a digital computer. The large-scale digital computer contains tens of thousands of these circuits composed of hundreds of thousands of devices. These circuits, each of which may represent an area only several thousandths of an inch square on a chip of silicon, are capable of performing their simple logical function in as little as a few nanoseconds (10^{-9} sec) to provide a machine that can perform relatively complex tasks, such as multiplication, millions of times in a single second.

This logical ability and the circuits that provide it, however, are only part of the story of the digital computer. In order to utilize the computer in the solution of problems, the basic operations that the computer is capable of performing must be organized into a specific sequence of operations. Those operations are defined by a sequence of instructions, called a *program*. The programs used in a computer are referred to as the computer *software*. As in the case of hardware, the term is descriptive in the sense that the program is what shapes the logical power of the computer into useful form, in much the same way that soft clay is shaped into a meaningful physical form by an artist.

The computer responds to the instructions in the program in a fixed and unvarying manner determined by the hardware. The computer does nothing more, and nothing less, than the program instructs it to do. Its effective use, therefore, depends on a precise definition of the operation sequence required to solve a particular problem.

The preparation of a program can, in a simplified way, be exemplified by the instructions given a small child to perform a simple task. The child possesses simple motor abilities analogous to the simple functional capabilities of the computer. The analogy to the computer program is the parent who, in a step-by-step sequence, tells the child what to do. For example, in instructing him to brush his teeth, the parent tells him to open the toothpaste, to wet the toothbrush, to apply the paste, and to manipulate the brush in a certain manner. If not told otherwise, the child continues to brush the same spot indefinitely. Failing to include any instruction, no matter how simple, often results in less than the desired result.

The analogy with programming a computer is quite exact in that the computer must be instructed in minute detail if the desired result is to be obtained. The analogy can be carried one step further by noting that as the child matures, the simple statement, "brush your teeth!" causes (hopefully) the child to perform the sequence of steps necessary to obtain the desired result. As discussed later, higher level languages exist which allow the computer user to instruct the computer in similarly broad terms. A single statement in the higher level language causes the computer to perform a sequence of steps which produce the desired result.

Programming, therefore, is the art of instructing a computer to do a job. That programming is still somewhat of an art, and not a science, is a result of the newness of the field and the fact that there is no unified body of knowledge on which it is based. Nevertheless, it is possible to describe the programming process and to exemplify it as it is applied to specific problems.

B.1   ELEMENTS OF PROBLEM SOLUTIONS

The steps in solving a problem on a digital computer can be categorized as analysis, programming, coding, and checkout. Analysis is the most highly skilled step in using a computer. In an engineering or scientific computer application, it is the analyst who translates the problem statement into terms that are amenable to computer solution. This may involve approximating a continuous process by a discrete one, approximating an infinite range by a finite one, or accounting for the effects of truncation and round-off errors that affect some types of numerical computation. The choice of numerical methods and algorithms is also the responsibility of the analyst.

As a simple example, consider the problem of finding the roots of the algebraic equation

$$ax^2 + bx + c = 0$$

A straightforward, but naive approach, is to compute one of the roots using the formula

$$x_1 = \frac{-b + \text{SQRT}(b^2 - 4ac)}{2a}$$

Three potential pitfalls are inherent in this approach:

1. If $a = 0$, the division by zero causes the computer to halt, produce erratic results, branch to an error routine, or terminate execution of the entire job, depending on the characteristics of the computer and its programming system. The possibility that $a = 0$ must be anticipated to avoid division by zero.

2. If $4ac \ll b^2$, round-off in the subtraction under the radical can produce the root $x_1 = 0$. However, by expanding the square root using the binomial expansion, the root is approximated by

$$x_1 = -\frac{c}{b}\left[1 + \mathbf{O}\left(\frac{ac}{b^2}\right)\right]$$

where $\mathbf{O}(-)$ is the order-of-magnitude function. Thus the loss of precision caused by round-off is avoided in this special case by using this alternate formula for the root.

3. If $b^2 - 4ac < 0$, the roots are complex. However, an attempt to compute the square root of a negative number using a real number square root routine causes the program to branch to an error routine or, in some cases, the square root of the absolute value of the quantity under the radical is computed. In either case, the desired result is not obtained.

This example illustrates that analysis requires careful attention to all situations that might arise during a computation. The failure to anticipate problems caused by certain combinations of input data can result in a program malfunction or in an apparently valid answer which is, in reality, incorrect. It is for these reasons that analysis is one of the most demanding and important aspects of programming.

After analysis, the next step in preparing a problem for a computer is that of programming. As used here, the term "programming" is only one of the four steps in the process of using a computer. The programmer decides how to use the computing resources at his disposal. He chooses the appropriate computer if more than one is available, chooses the programming language, allocates internal storage to data and programs, and decides how to use the auxiliary storage and input/output devices.

An important decision for the programmer is the trade-off between conserving storage space or conserving computing time. For example, if storage is at a premium, the programmer may decide to pack several pieces of data into a single machine word, even though packing and unpacking require additional time. Similarly, arrays of data or programs may be placed on an auxiliary storage device, such as tape, disk, or drum, and brought into main storage only when needed.

To minimize the computer time required to solve the problem, on the other hand, the programmer may use extensive tables to minimize computation. For example, if a function $f(x)$ is evaluated using the Taylor's series expansion about $x = 0$, many terms may be required in order to obtain the desired accuracy. An alternative is to store values of $f(x)$ and its first few derivatives for many values of x in the range of interest. Then for a particular value of x, the tabular entry closest to x is found and the Taylor's series expansion is made about this tabular value. By judicious use of this technique, two or three terms of the expansion are often sufficient. Programming decisions of this type can affect the quality of the resulting program markedly.

After making these overall decisions, the programmer describes the logic of the program by means of a flowchart or decision table. These techniques, described in detail in Sec. B-2, are graphical techniques which assist the programmer in organizing the problem solution.

The next step, coding, is the process of preparing the instructions which cause the computer to perform the detailed steps necessary to carry out the computation and the logic as specified in the flowchart or decision tables. As noted earlier, a computer must be instructed in great detail just how to carry out a calculation. Indeed, even to find the sum of two numbers in storage requires two instructions, one to load the first addend into the accumulator register from storage, and a second to add the other addend into the accumulator register. A third instruction is required if the result is to be re-stored.

If coding is done at the machine-language level, the coder must be familiar with the functional operation of the computer. As described later, he must know the formats of words used for instructions and data, as well as the use of all special registers. Figure B-1 shows a coding form with instructions of this type written on it. Using a high-level language like FORTRAN, to be discussed later, the coder need not be concerned with the details of the computer and is free to concentrate on the problem being solved. A FORTRAN coding form and several statements are shown in Fig. B-2. In general, coding is the process of transforming the program logic and computation into an appropriate computer language.

After coding, the program must be subjected to checkout. This phase of programming, often called *debugging*, is an iterative procedure in which the program is thoroughly tested to detect errors in coding and logic. A thorough checkout includes test cases involving all types of expected input data, as well as a variety of erroneous input data to ensure that error conditions are identified and properly handled. Further, test cases should be chosen to cause all parts of the program to be

IBM

IBM 1800 Assembler
Coding Form

Form X26-5908-1
Printed in U.S.A.

Program __CHECKSUM (FRAGMENT)__   Date __11/5/68__

Programmed by __K. UTLEY__   Page No. __1__ of __10__

Label	Operation	F	T	Operands & Remarks	Identification
*					
* CALCULATE NEW CHECKSUM					C.KSM
*	LDX	I	1	ADROT X,E,1,=ADDR CARD TO PUNCH	
	LDX		2	53 CALCULATE NEW CHECKSUM	
	LD	L		PACKD+53 START WITH LAST COL	
	BSC			C TURN CARRY OFF	
	NOP				
CKSUM	A	L	2	PACKD,-1 ADD NEW-TO-LAST COL	
	BSC			C	
	A			D0001 ADD +1 IF CARRY	
	MDX		2	-1 Q. DONE 55	
	MDX			CKSUM	
	EOR			H.FFFF GET 1'S COMPLEMENT	
	A			D0001 GET 2'S COMPLEMENT	
	BSI			STORE STORE NEW CKSUM	
*					
* PUNCH COMPRESSED WITH NEW CKSUM (+ SEQ NO.)					
*					
	LIBF			CARD1 PUNCH 1 CARD	
	DC			/2000	
PUNCH	DC			*-*	
	DC			CDERR	
	MDX	L		COUNT,-1 DECR NO. CARDS TO PUNCH	
	NOP			IGNORE SKIP IF GOES TO 0	
	LDX	L	1	ADROT	
UPDTE	LD		1	0 INCREMENT I/O ADDR FOR	

Fig. B-1   Assembly language coding form.

executed. Applied to the previous example of finding the roots of the quadratic equation, the following special cases must be checked:

$$b^2 - 4ac > 0$$
$$b^2 - 4ac = 0$$
$$b^2 - 4ac < 0$$
$$b^2 \gg 4ac$$
$$a = 0, b \neq 0$$
$$a = b = 0$$

This particular example is misleading in its simplicity, as all possibilities can be considered and checked out. In a large program involving thousands of instructions, this is not always possible. Even though all the obvious special cases are checked, it is not uncommon in large programs to discover "bugs" after the program has been in use for months or years. It is this experience that leads to the observation that large programs are never completely debugged.

The foregoing discussion divides the preparation for using a computer

IBM

Form X28-7327-4
Printed in U.S.A.

FORTRAN CODING FORM

			Punching Instructions								Page / of /

| Program | FIND RANGE | | Graphic | | | | | | Card Form # | * | Identification |
| Programmer | R. SIMONS | Date | Punch | | | | | | | | 73   80 |

C FOR COMMENT

STATEMENT NUMBER		FORTRAN STATEMENT
1 5	6 7 10 15 20 25 30 / 35 40 45 50 55 60 65 70 72	

```
C      PROGRAM TO FIND RANGE OF NUMBERS PUNCHED
C      IN COLUMNS 1-8, OF INPUT CARD DECK

       READ(5, 1) XMIN
    1  FORMAT (F8.0)
       XMAX = XMIN
    2  READ (5, 1, 6) X
       IF(X - XMIN) 3, 2, 4
    3  XMIN = X
       GO TO 2
    4  IF (X - XMAX) 2, 2, 5
    5  XMAX = X
       GO TO 2
    6  WRITE(6, 7) XMIN, XMAX
    7  FORMAT (' RANGE IS: ', F8.0 ' TO ', F8.0)
       RETURN
       END
```

* A standard card form, IBM electro 888157, is available for punching source statements from this form.

Fig. B-2 FORTRAN coding form.

into four distinct phases: analysis, programming, coding, and checkout. Actually these phases are not distinct; they blend into each other. In fact, one or two individuals are typically responsible for all phases. These four phases, taken as a whole, are commonly referred to as programming. In the remainder of this chapter, the word *programming* is used in this larger context.

B.2 PROBLEM ANALYSIS

To solve a problem on a computer requires that the problem and its method of solution be stated precisely. This, in turn, requires a careful analysis of the problem to verify that the problem is amenable to computer solution, to identify conditions that can result in program malfunctions or errors, and to organize the problem solution in preparation for writing the actual program code. The insight gained by preparing a problem for solution on a computer is sometimes more valuable than the actual solution.

Programming generally proceeds from overall organization to the specific, and this is the order in which we will discuss problem analysis as it relates to programming.

B.2.1   Problem Identification

Problems that are best suited to solution by computers are characterized by repetition, complexity, time limiting, or a combination of these characteristics. By way of contrast, to solve one quadratic equation involves none of the foregoing characteristics and, therefore, a digital computer is not required or recommended for its solution. However, in the solution of hundreds of quadratic equations, a computer can be used to advantage. Similarly, financial institutions use computers to process mortgage loan account payments, a relatively simple calculation but one which is performed thousands of times each month. In both these examples of repetition, we speak of moving data through the computer or external repetition. In the first case, successive sets of coefficients of quadratic equations are read, each equation is solved, and then the next set of coefficients is read into the computer. Similarly, each set of data pertaining to a mortgage loan account is read into the computer, processed, and the same procedure is repeated with the next set of data. Moving data through the computer is typical of many data reduction and business applications of computers.

A second type of iteration involves moving the program over a fixed set of data, or internal repetition. This type of repetition is exemplified by a matrix inversion program in which the program scans the elements successively, row after row. This type of iteration, on a fixed amount of data, is typical of many engineering and scientific applications.

Complexity is the second characteristic of problems that are suitable for computer solution. An example of a complex problem is the analysis of a move in the game of chess. Note, however, that this complex problem also contains elements of internal repetition as many tentative moves and countermoves are evaluated.

Finally, time-limited problems are also appropriate for computer solution. Examples of time-limited problems are weather prediction where the prediction must be made before the weather develops, airline ticket reservations where the transaction must be completed before the customer becomes impatient, and process control where the computations must be completed before the data on which they are based become invalid and also before the process becomes dangerous. Again, it is noted that time-limited problems also have elements of repetition and, possibly, complexity.

B.2.2 Algorithms and Heuristics

After the specific problem is identified, it is necessary to decide on a solution procedure. Depending on the nature of the problem, this may require a knowledge of topics such as numerical methods, linguistics, circuit analysis, or petroleum refining. In general, the transition from problem analysis to problem solution depends on a knowledge of the problem area and techniques of solving problems in that discipline. Algorithms and heuristics are two general classes of problem solving procedures that will be discussed. The distinction between algorithms and heuristics is not always clear cut; indeed, as one's knowledge of a procedure increases, what was originally regarded as a heuristic may later appear to be an algorithm.

An *algorithm* is a well defined procedure, or set of rules, for solving a problem in a finite number of steps. To find a root of the equation $f(x) = 0$, for example, Newton's Method can be used. In this algorithmic computation, successive iterates to a solution are expressed by the formula $x_{i+1} = x_i - f(x_i)/f'(x_i)$, where $f'(x)$ is the first derivative $df(x)/dx$. Once an appropriate initial iterate x_1 is chosen, Newton's Method completely specifies the computation process. Subject to certain known limitations on x_1, this algorithm converges to a solution in a finite number of steps.

Algorithms can be stated in various ways—as a written procedure, as a flowchart or decision table, or as a computer program. Mathematical algorithms have been widely published in the mathematical and computing literature. A noteworthy collection of mathematical algorithms is published as a regular feature in the *Communications of the ACM*.

A *heuristic*, on the other hand, uses exploratory methods of problem solving in which solutions are discovered by evaluation of the progress made toward the final result. Also, a rule of thumb, for which a mathematical proof is lacking, is a heuristic. For example, experienced chess players attempt to control the center of the board. Although no proof exists for substantiating the value of this control, experience shows that this heuristic is helpful in playing chess. Heuristic methods are often used in artificial intelligence applications such as pattern recognition, game playing, theorem proving, and some optimization problems.

B.2.3 Program Organization

Whatever the original definition of a problem, it is very likely to change before the program becomes operational or, in any case, to be modified by subsequent variations of the problem. Therefore, a program should be as general as possible; it should anticipate possible modifications and should be organized such that modifications can be implemented easily. For example, any quantity that conceivably could change should be a parameter rather than a constant in the program. Parameters should be read in as data rather than imbedded in the program itself.

Program modifications are simplified if the program is segmented into a main program and several subprograms since a change in the problem definition often affects only a portion of the program. The main program contains the overall solution logic and consists of a sequence of *calls* or invocations of the subprograms, each of which provides a well-defined calculation or operation. Thus, for example, the program might be segmented into subprograms which perform a specific calculation such as a function evaluation or a matrix inversion, provide tests necessary to determine validity of data or results, and control input and output functions such as card reading or punching.

This modularity of subprograms has several advantages:

1. Each subprogram is written and checked independently. Errors can often be detected without becoming involved in the overall program logic.

2. The subprograms may be generally useful and so can be included in a library of subprograms available to other programmers, as discussed in Sec. B.2.6. This reduces duplication of programming effort.

3. Changes in the computation or logic are often isolated to a single subprogram, thereby simplifying program changes.

4. Documentation and understanding of a program are often simplified if the program is modular.

As a part of organizing the overall program, the programmer must specify the layout of all input and output data. For example, if parameter values are to be read into the computer on punched cards, data fields for each piece of data must be precisely defined, as shown in Fig. B-3. When using the higher level languages discussed in subsequent sections, the representation of data as floating or fixed point numerals or as alphabetic characters must also be defined. Similarly, the layout of output reports must be specified, including the report title, the column headings, the representation of numerical data as fixed or floating point, and their precision. The exact manner in which input and output data formats are specified depends on the input/output medium and the programming language being used. In any case, however, these definitions are an important part of the program organization.

Utilization of computer resources is also an important part of overall

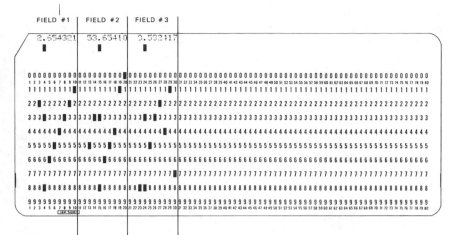

Fig. B-3 Data fields on punched card.

program organization. Many computers have a hierarchy of storage devices available, each of which has its own characteristics of capacity, accessibility, and data transfer rate. An important programming decision involves the allocation of data and programs to these various storage devices. For example, if one applies relaxation techniques to the solution of a partial differential equation, the values of variables at grid points are scanned essentially sequentially. Thus magnetic tape is an acceptable storage medium for the variables since data are stored sequentially. On the other hand, if a large amount of data is to be accessed at random, as in an on-line terminal application, a sequential access device is a poor storage medium because of the large amount of data that must be passed over to reach the desired data. A better choice in this instance is a disk or drum which provides more direct accessibility, or main storage which provides true random access.

Just as a hierarchy of storage devices may be available, so may there be several input/output devices available. The traditional method of entering data manually is by punching machine readable cards or paper tape. Although punched cards can be read directly into a computer, they are usually transcribed to magnetic tape, disk, or drum independent of the main computing process because of the relatively slow speed of mechanical card readers when compared with the much higher speed of the computing circuits. By using off-line transcription, input data from punched cards are read into the computer from the faster medium when the data are required. The use of typewriter and graphic terminals is now beginning to supplant punched cards for original data entry. This results from the development of terminal-based computing systems and the fact that the computer can assist in the data entry operation by providing editing, checking, and correcting capabilities.

In addition to data entry, a programmer must also be able to handle intermediate data generated by the computer. If extensive, intermediate data must be stored on an external device such as a magnetic tape, disk, or drum. As stated previously, the choice of intermediate storage device depends on the quantity of data, the access method—sequential or random—and the speed and frequency with which the data must be accessed.

Because of their speed, computers can generate large quantities of output data which can be printed very rapidly. Indeed, computers have been called "high-speed printing presses." A programmer should pay particular attention to deciding what types and amount of output to produce. If he prints too little, he runs the risk of wasting a computer run because some significant data are missing. On the other hand, if he prints too much, he runs the risk of becoming inundated with paper, only a small percentage of which is significant. In each case, it is important to know for whom the printed results are intended and how one plans to use them. In this way, the amount of detail required in the printed results can be determined.

B.2.4 Flowcharts

A *flowchart* is a graphical representation in which the major steps of the program are represented by distinctively shaped boxes connected by directed line segments showing flow of control. During the organization of the overall program, the programmer normally prepares a flowchart showing the logical flow of the program. Flowcharts can be prepared to any level of detail desired, and a large program usually is represented by at least two levels of detail—an overall coarse flowchart, each of whose parts is further refined by a more detailed second-level flowchart.

As a simple example, suppose we want to find the range of a set of 100 numbers, each of which is punched on a separate card. If the plan is to read all 100 numbers into memory, compute the range, and print the results, the first-level flowchart illustrated in Fig. B-4 shows the overall program organization. The different shapes used for each of the boxes are flow-charting conventions that specify various functions. A widely accepted symbol convention is illustrated by the template in Fig. B-5.

The processing step "compute range" in Fig. B-4 merits further elaboration, as given in the second-level flowchart of Fig. B-6. We assume that

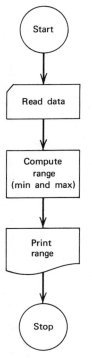

Fig. B-4   Example first-level flowchart.

the step labeled "read data" in Fig. B-4 has read 100 numbers into an array labeled X in Fig. B-6, which has elements X_i.

The left arrow $\leftarrow$ used in the flow diagram indicates replacement. Thus $i \leftarrow i + 1$ means that the value of i is to be replaced by the value obtained by adding 1 to the current value of i. The colon stands for a relational operator such as $\leq$, $<$, $=$, $\neq$, $>$, or $\geq$. In general, the possible relational operators that the colon stands for are used to identify the paths leaving the decision box. For example, at the decision symbol labeled $i:100$, if $i = 100$, the program proceeds to the termination point labeled END. If $i \neq 100$, the program proceeds to the program modification box labeled $i \leftarrow i + 1$.

Programmers differ in their use of flowcharts. Some programmers flow chart a program in great detail before starting to code. If this is done, coding is a relatively easy task. Other programmers, particularly those using higher level languages, either do not flow chart at all or do so only in the form of overall first-level charts. Whether the flowchart is drawn before coding or after checkout, it represents an excellent means of program documentation.

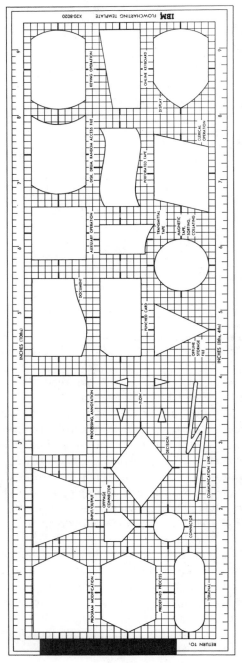

Fig. B-5   Flowchart template. From IBM Corporation.

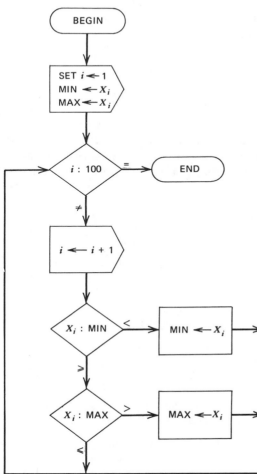

Fig. B-6   Example second-level flowchart.

B.2.5   Decision Tables

Another method for specifying the logic and organization of a program is the use of decision tables. Unlike a flowchart which gives a computer-dependent, step-by-step method of calculation, a decision table provides a problem-oriented statement of the program logic.

Figure B-7 illustrates the four parts of a decision table separated by double horizontal and double vertical lines. The *condition stub* lists all the conditions that determine which of the actions listed in the *action*

Condition stub	Condition entry
Action stub	Action entry

Fig. B-7   Decision table.

stub are to be taken. The *condition entry* specifies various combinations of conditions that are possible, and the *action entry* specifies the actions to be taken for each combination of conditions in the condition entry. Thus, in the example problem of finding the range of a set of numbers, the decision table, shown in Fig. B-8, lists six conditions in the condition stub,

$$i \le 100 \qquad\qquad i > 100$$
$$X_i < \text{MIN} \qquad\qquad X_i \ge \text{MIN}$$
$$X_i > \text{MAX} \qquad\qquad X_i \le \text{MAX}$$

and five actions in the action stub,

$$\text{MIN} \leftarrow X_i \qquad \text{(replace MIN by } X_i)$$
$$\text{MAX} \leftarrow X_i \qquad \text{(replace MAX by} X_i)$$
$$i \leftarrow i + 1 \qquad \text{(replace } i \text{ by } i + 1)$$
$$\text{GO TO TABLE 1}$$
$$\text{GO TO END}$$

Figure B-8 is an example of a *limited entry form* table in which the condition entries are one of three characters,

> Y, meaning yes, this condition is true
> N, meaning no, this condition is false
> —, meaning this condition is irrelevant

The action entries are one of two characters,

> X, meaning execute this action
> —, meaning this action is irrelevant

The entry portions of Fig. B-8 are divided into four vertical columns called *rules*. The rules are interpreted by reading the columns from top to bottom. Thus, for example, rule 1 states that, if $i \le 100$ and $X_i < \text{MIN}$, then MIN is replaced by X_i, i is replaced by $i + 1$, and the program repeats table 1 again. Rule 4 states that if $i > 100$, regardless of the value of X_i, control of the program goes to another decision table labeled END. In the example, the table END represents the subprogram labeled "print range" in the first-level flowchart of Fig. B-4.

TABLE 1	Rule 1	Rule 2	Rule 3	Rule 4
$i \leq 100$	Y	Y	Y	N
$i > 100$	—	—	—	Y
$X_i < \text{MIN}$	Y	—	N	—
$X_i \geq \text{MIN}$	—	—	Y	—
$X_i > \text{MAX}$	—	Y	N	—
$X_i \leq \text{MAX}$	—	—	Y	—
$\text{MIN} \leftarrow X_i$	X	—	—	—
$\text{MAX} \leftarrow X_i$	—	X	—	—
$i \leftarrow i + 1$	X	X	X	—
GO TO TABLE 1	X	X	X	—
GO TO END	—	—	—	X

Fig. B-8   Limited entry decision table.

Note that the first two conditions in the condition stub ($i \leq 100$ and $i > 100$) are mutually exclusive, as are the other two pairs of conditions and the last two actions which control the program execution sequence.

The same logic specified in Table 1 can be specified as shown in Table 2 of Fig. B-9. Here either part of the condition and/or part of the action

TABLE 2	Rule 1	Rule 2	Rule 3	Rule 4
$i \leq 100$	Y	Y	Y	N
X_i	< MIN	> MAX	≥ MIN & ≤ MAX	—
$\text{MIN} \leftarrow X_i$	X	—	—	—
$\text{MAX} \leftarrow X_i$	—	X	—	—
$i \leftarrow i + 1$	X	X	X	—
GO TO	TABLE 2	TABLE 2	TABLE 2	END

Fig. B-9   Mixed entry decision table.

extends into the entry portion of the table. If all the conditions and actions are extended into the entry portion of the table, it is called an *extended entry form;* if some rows of the table are of the limited entry form and other rows of the extended entry form, as in Fig. B-9, the table is called a *mixed entry form.*

The order of executing the actions in the action stub is significant. Thus in rule 1 of table 2, MIN must be replaced by X_i before incrementing i, and the sequencing operation GO TO TABLE 2 must be executed last. By convention, the actions are executed in the order they are listed in the action stub.

A complete program is usually represented by several decision tables just as it may be represented by several flowcharts. A flowchart specifies the logic of a program sequentially, whereas a decision table specifies the logic in parallel. Thus the programmer working from a decision table must consider the frequency of executing each rule in order to optimize the efficiency of the resulting program.

Decision tables offer certain advantages over flowcharts:

- The logic of a program is displayed compactly.
- Logical inconsistencies are easy to detect.
- Incomplete specifications are easy to detect.
- Changes to the logic are easy to implement and the consequence of a change in logic can readily be identified.
- Documentation of program logic, as distinguished from procedure, is aided by tables.

B.2.6 Program Libraries

There are many computations that are common to a wide class of programs. Examples of common computations include evaluation of elementary functions, solution of polynomial equations, solution of initial value problems, matrix computations, pseudo-random number generation, data reduction, and statistical measures. In order that commonly used program fragments such as these can be used by many programmers, they are written in a form called a subroutine and organized into libraries. A *subroutine* is a program to which control is passed along with arguments. After completing its calculation, the subroutine passes control back to the calling program and returns calculated results. In the FORTRAN program discussed in Sec. B.3.2, the statement

$$\text{CALL INVERT (A, 30)}$$

is used to pass control to a subroutine named INVERT, along with two arguments, an array A and the constant 30 which is the order of the matrix A. After performing the matrix inversion, the subroutine passes control back to the next executable statement in the FORTRAN program. The matrix inversion subroutine INVERT can be thought of as a "function box" which can be used in any FORTRAN program.

Such commonly used subroutines are organized into libraries of programs so that they are readily accessible to many programmers. One of the first such program libraries, and one of the most extensive, was started in 1955 by an organization of IBM 704 users called SHARE. This organization is still active in programming methodology, but its scope has increased with time to include programs for later IBM computers including the 709, 7090/94, and, most recently, the larger models of Systems 360 and 370. Similar user organizations exist to support other computing systems including those used in control applications.

A somewhat different approach is taken by the *Communications of the ACM* (*CACM*) in its Algorithms Department. Here, mathematical algorithms are published, usually in the higher level language ALGOL, without reference to any particular computing system. In addition to publishing contributed algorithms, the *CACM* publishes certifications of the validity of algorithms as actually implemented for a particular computer. These algorithms represent an important collection of mathematical methods in a form suitable for inclusion (possibly translated to another programming language) in many program libraries.

Program libraries represent an important asset to a computer installation. By judicious use of subroutines from a library, a programmer often reduces substantially the time required to complete a required program.

B.3   PROGRAMMING LANGUAGES

As noted in Sec. B.1, a computer program is the step-by-step procedure that instructs the computer to carry out a desired computation. A programmer must express the program in a form meaningful both to the programmer and to the computer. This requires a communication medium, that is, a language.

A programming language consists of symbols, syntax, and a program structure. Specific programming languages are discussed in this section and examples of symbols, syntax, and structures for each language are illustrated.

Machine language is the sequence of binary digits or alphanumeric characters that is meaningful to the electronic circuits of the computer. In a typical computer, instructions are composed of elements called *words,* where a word consists of a fixed number of binary digits having a defined structure. For example, the structure of a 16 binary digit instruction might be

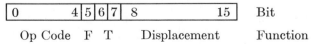

| 0 | 4 | 5 | 6 | 7 | 8 | 15 | Bit |

Op Code   F   T   Displacement   Function

This particular instruction has four parts. The first 5 bits (binary digits) define the operation code, that is, the operation to be performed by the computer. In this computer, there are 32 (2^5) possible operation codes. Examples of operations are add, subtract, multiply, store a number in main storage, read a card, shift the bits in a register left, etc. The next bit of the instruction (F) indicates the instruction format: in this case, F = 0 indicates a one-word instruction as pictured above; F = 1 indicates a two-word instruction having 32 bits. Bits 6 and 7, denoted by T, specify which of four registers participate in the addressing of data. The last 8 bits, known as the displacement, specify with T the specific operand of the instruction.

For example, the instruction

$$10010 \ 0 \ 11 \ 00001000$$

has an operation code 10010 which means subtract, a format bit of 0 meaning a single-word instruction, a tag T of 11 indicating that register 3 participates in the addressing, and a displacement of 00001000, the binary equivalent of decimal 8.

Suppose that the contents of register 3 is 3972. The effective address for this instruction is obtained by the computer by adding the contents of register 3 to the displacement or

$$3972 + 8 = 3980$$

Thus this example instruction causes the computer to subtract from the accumulator the contents of storage location 3980.

From the above example, it is clear that machine language is so cumbersome and unwieldy that it is impractical to program directly at this level. Fortunately, it is not necessary to do so. Higher level languages are available that permit a programmer to communicate with a computer in terms that are more meaningful than is machine language. These higher level programming languages are not meaningful to the computer directly, but rather to another computer program whose function is to translate the language used by the programmer into machine language. The language used by the programmer is called *source language,* and the program written in source language is called the *source program.* The program which is the result of translating the source program into machine language is called the *object program.* This section discusses four classes of programming languages with examples of each.

B.3.1 Assemblers

An assembler consists of two parts. The first is an *assembly language* (the source language) used by the programmer to code a program. The second is the *assembly program,* or processor, that translates the programmer's code expressed in the assembly language into machine code.

For most instructions, the assembly program produces one machine language instruction for each assembly language instruction; that is, it translates on a one-for-one basis. In translating from assembly language to machine language, the assembly program performs several additional functions. It assigns storage locations based on symbolic addresses specified by the programmer, and it assembles the numerical operation code based on a symbolic or mnemonic operation code specified by the programmer. Most assembly programs provide additional aids to programmers by checking for syntactical errors, undefined or multiply defined symbols, invalid operation codes, etc. Many assembly programs also provide an index of programmer-defined symbols and a cross reference showing where each symbol is used. These aids are useful in debugging a program.

Using an assembly language, the programmer expresses the various parts of a computer instruction symbolically. For example, symbolic operation codes of a typical computer in an assembly language are given in Fig. B-10 along with their machine equivalents and their mean-

Assembly language	Machine language	Meaning
LD	11000	Load accumulator from memory
STO	00100	Store accumulator in memory
A	10000	Add to accumulator from memory
S	10010	Subtract from accumulator from memory
M	10011	Multiply accumulator from memory
D	10100	Divide accumulator from memory
CMP	10110	Compare accumulator with memory
BSI	01000	Branch and store instruction counter
BSC	01001	Branch or skip conditionally
BOSC	01001	Branch out or skip conditionally
MDX	01110	Modify index and skip

Fig. B-10 Operation codes of a typical computer.

ings. Notice that the symbolic operation codes in this assembly language are suggestive of their meanings. In general, assembly language operation codes are chosen to have this mnemonic quality. For this reason they are often called *mnemonic operation codes* or *mnemonics.*

Operands of an instruction often refer to storage locations and an assembly lanaguage normally provides for symbolic reference to storage. Thus, instead of referring to numerical addresses, the programmer chooses symbols having mnemonic value and the assembly program replaces these symbolic references with numerical addresses. For example, to load the accumulator with the contents of storage location 2062, one writes in assembly language

LD   SUM

where SUM is the mnemonic symbol assigned to location 2062. The assembler, on encountering this instruction, performs table look-ups to find the machine code equivalents of the operation code LD and the symbol SUM. It then generates the following machine language instruction:

11000 0 10 00001110

Of course, the spaces in the above sequence of bits have no significance other than to set off the four parts of this instruction for illustrative purposes. The assembly program interprets the symbolic operation code LD and assembles the appropriate numerical operation code 11000 and the appropriate flag bit of 0, indicating a single-word instruction. The numerical address of the operand SUM, 2062, is too large to fit in the 8-bit displacement part of the instruction. Accordingly, register number 2, specified by the tag of 10, was previously loaded with a number that permits addressing SUM and other quantities of interest to the programmer. In this example, we suppose that register 2 contains 2048. Accordingly, the effective address of this instruction is

contents of register 2 + displacement = 2048 + 14 = 2062

From this example it is clear that the assembler relieves the coder of a great deal of routine clerical work in writing a program.

In addition to the operation code and operand, an assembly language instruction optionally has a label. The label, if it appears, is written to the left of the instruction and is used by other instructions to reference the labeled instruction. If we label the above symbolic instruction with the label XX, it would be written

XX   LD   SUM

Symbols are defined by their appearance in the label field of the program.

On encountering this instruction, the assembler assigns to the label XX the location of this instruction. This enables one to refer to this instruction as in the following branch instruction:

BSC   XX

This instruction causes the program to branch to the instruction labeled XX.

Actually, the value assigned to XX is the relative location of the instruction labeled XX with respect to the beginning of the program. The distinction between absolute address and relative address is significant at the time the program is loaded into storage. Since many programs are ordinarily in main storage at the same time, each program must be displaced or relocated from the beginning of storage in such a way that no two programs overlap. Thus the *relative address* is independent of the location of the program since it is relative to the origin of the program. The *absolute address* is the actual location of the instruction in storage. Thus the absolute address is the relative address plus the absolute address of the origin of the program.

For example, the programs A, B, and C represented in Fig. B-11 begin

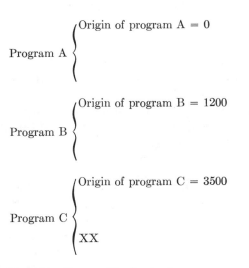

Program A {
Origin of program A = 0

Program B {
Origin of program B = 1200

Program C {
Origin of program C = 3500

XX

Fig. B-11   Storage utilization for three programs.

at locations 0, 1200, and 3500, respectively. If the relative address assigned to the label XX in Program C is 200, then its absolute address is $3700 = 3500 + 200$. If programs B and C are loaded in the reverse order, the absolute address of XX would be $1400 = 1200 + 200$. The absolute locations into which a program is loaded into the computer are determined by a *loader* program. The loader also computes absolute addresses from the program origin and relative addresses.

In addition to machine instructions, assembly languages normally provide several pseudo-instructions for the convenience of the programmer. A *pseudo-instruction* is an instruction to the assembler that does not necessarily result in object code. The following table lists some typical pseudo-instructions:

Mnemonic	Meaning
DC	Define constant
BSS	Block started by symbol
EQU	Equate symbol
ORG	Define origin

The DC instruction is the only one of these four instructions that generates object code. It provides a convenient means to introduce constants of various formats into the program. The instruction BSS reserves a block of storage whose length is specified as the operand. EQU is used to define a symbol arbitrarily. For example,

$$\text{A} \qquad \text{EQU} \qquad 4*\text{N-2}$$

defines A to be the value 4N-2 where the symbol N must be defined by appearing as a label in a previous statement. The instruction ORG is used to specify the address at which the following instructions are to be assembled. Thus the programmer can control the positioning of various sections of code. Assembly languages have many pseudo instructions and the preceding examples are typical of their function.

As an example of the use of an assembly language, the following program fragment determines which of two numbers, X or Y, is the larger and stores it in a location designated by MAX. More precisely, this program determines which of the two numbers stored in locations X and Y is the larger. The confusion between a storage location and the contents of that location emphasizes the value of referring to storage locations with mnemonic labels.

MAX	BSS	1	RESERVE 1 WORD FOR MAX
START	LD	X	BEGIN EXECUTION, LOAD X INTO ACCUMULATOR
	S	Y	X − Y
	BSC	XHI, +	BRANCH TO XHI IF DIFFERENCE IS PLUS
	LD	Y	OTHERWISE LOAD Y IN ACCUMULATOR
	BSC	XHI + 1	UNCONDITIONAL BRANCH
XHI	LD	X	
	STO	MAX	STORE MAXIMUM

This program fragment illustrates relative addressing in that XHI + 1 is used to address the first instruction following that labeled XHI. It also illustrates the use of the remarks field to the right of the instruction. Copious use of remarks is good programming practice and contributes to good program documentation. Two forms of the BSC instruction are used in the example program. The first is a conditional branch; the branch occurs only if the accumulator is positive. If the accumulator is not positive when this conditional branch instruction is executed, control passes to the next instruction in sequence. The second use of BSC is an unconditional branch; that is, the branching operation is executed independent of what is in the accumulator.

In summary, an assembly language provides a shorthand notation using mnemonic symbols and is much simpler to use than machine language. The assembly program typically provides a one-for-one translation of symbolic assembly language instructions into machine language instructions. Although assemblers reduce the coding effort considerably, use of assembly language is still tedious and error prone. Accordingly, a simpler method of coding is frequently more desirable, and this is discussed in the next section.

B.3.2   Compilers

From the previous discussion of assemblers, the reader may observe that a great deal of attention must be paid to details unrelated to the problem he is solving. He must deal with some 50–200 operation codes, as many as several thousand symbolic addresses, and many other machine level details. By using a language based on a compiler, a programmer relieves himself of many detailed considerations concerning the computer and computing process, thereby concentrating on the problem to be solved. Like assemblers, compilers consist of two parts. First there is a language (source language) used by the programmer to code a program. A program statement in the FORTRAN language, a compiler language oriented toward scientific computational needs, might appear as follows:

$$X = (-B + SQRT(B**2 - 4.*A*C))/2.*A$$

This statement resembles the language of mathematical notation so closely that its meaning is clear without any explanation other than to point out that the asterisk * denotes multiplication, SQRT denotes the square root function, and the double asterisk ** indicates exponentiation.

In addition to a language, a compiler has a processor, that is, a computer program that accepts source language statements as inputs and produces machine language object code as output.

Compiler languages are further removed from the details of the computer operation; indeed, a programmer using a compiler language need know nothing about the instruction repertoire of the computer for which he is programming. In fact, many compilers have processors for many different computers. The significance of this is that a program written in a compiler language can be processed on any computer for which an appropriate processor exists. In other words, compiler languages tend to be machine independent and are often called *higher level languages.*

Three popular compiler languages are FORTRAN, COBOL, and PL/I. FORTRAN is oriented to scientific and engineering problems and, as shown by the above example, uses statements very similar to ordinary algebra. It is the first compiler to win widespread acceptance and has evolved through many versions. COBOL is oriented to business or data-processing problems and is widely used in that area of computer application. PL/I, a relatively new higher level language, combines the best features of FORTRAN and COBOL, as well as provides significant new language features. PL/I is a rich language that gives promise of supplanting both FORTRAN and COBOL over the next several years. The remainder of this section gives examples of the power and ease of using higher level languages by means of examples in FORTRAN.

The following FORTRAN program fragment computes the sum of 100 numbers.

$$\text{TOTAL} = 0.$$
$$\text{DO } 17 \text{ I} = 1, 100$$
$$17 \text{ TOTAL} = \text{TOTAL} + \text{A(I)}$$

The 100 numbers whose sum is to be computed are assumed to be in the array A whose elements, denoted by subscripts, are as follows: A(1), A(2), . . . , A(100). In this example, TOTAL, I, and A are variables; the numbers 0., 1, and 100 are constants, and 17 is a statement number. The equal sign (=) does not denote equality; rather, it means replace the value of the variable on the left side with the value of the expression on the right side. Thus the first statement in the above example initializes the variable TOTAL by replacing whatever value it had previously with the constant 0.

The second statement is called a "DO statement." It provides a convenient way to specify iterations in which one or more statements are executed repetitively while a variable changes in a prescribed manner. In the example, the DO statement specifies that the single statement following it, numbered 17, should be executed 100 times, first with I = 1, then with I = 2, and so forth until I = 100. Each time statement 17 is executed, the effect is to replace the value of TOTAL by the previous value of TOTAL plus the next element in the A array. Thus after execution of this program fragment, the value of TOTAL is the sum of the elements of the A array

$$A(1) + A(2) + A(3) + \cdots + A(100)$$

This program fragment uses the arithmetic and DO statements of the FORTRAN language. The next example illustrates other features of the language. Suppose we are given a matrix A of dimension 30×30, and a sequence of vectors X1, X2, . . . each of dimension 30. It is required to find the sequence of vectors

$$A^{-1} \cdot X1, \qquad A^{-1} \cdot X2, \ldots$$

and the length of each of these resulting vectors. The length of the vector is defined as the square root of the sum of the squares of the elements in the vector. Our method of solution is to read into the computer the 900 elements of A, and use a standard matrix inversion library routine to compute the matrix inverse A^{-1}. We then read an X vector, compute the matrix by vector product $A^{-1} \cdot X$, compute the length of this vector, and print out the results. This process is illustrated in the first-level flowchart shown in Fig. B-12.

Although it is not the purpose of this chapter to teach FORTRAN programming, an explanation of the FORTRAN program which implements this flowchart will enhance the reader's understanding of the power of a higher level language.

The example program, shown in Fig. B-13, demonstrates several features of the FORTRAN language. First, one observes that commentary can be introduced to help document what the program is to do. In Fig. B-13, the first three lines are comments, as denoted by a C in column 1. (Generally, input to FORTRAN consists of a sequence of 80-column cards, in which the first five columns are reserved for a statement number, column 6 is a continuation column, and columns 7–72 are used for the FORTRAN statement. Columns 73–80 can be used for any purpose, but usually they contain a program identification and sequence number.) In the sample program each statement is numbered sequentially from 1–18. There is no requirement to number each statement; only those statements that are referred to by other statements (statements 3, 6, 12, 13, and 16 in the example) must be numbered. The remaining statements are numbered in this example simply to aid in explanation of the program. The use of the continuation feature is illustrated by statement 16. The fact that statement 16 continues onto the next card is denoted by the 1 in column 6 of the continuation card.

Statement 1, a DIMENSION statement, is used to declare that the variables A, X and P are arrays. Specifically, A is declared to be a two-

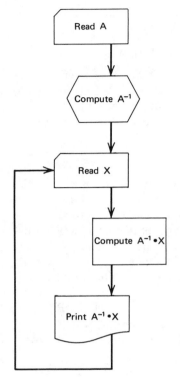

Fig. B-12 Flowchart for example program.

dimensional array with dimensions 30×30; X and P are one-dimensional arrays or vectors, each of dimension 30.

Statement 2 is a READ statement which causes the elements of the array A to be read from input device 5, according to the format specified by statement 3. In this example, input device 5 is a numerical designation for a magnetic tape unit. READ statements can be more selective in reading only certain elements of the array A, but in the case of statement 2, where only the name of the array is specified, it is implied that all 900 elements of A are to be read. The FORMAT statement 3 specifies that eight elements of the array are to be read per card image, with each element occupying ten columns. Thus A(1, 1) is punched in columns 1–10 of the first card, A(1, 2) is punched in columns 11–20 of the first card, and so on through A(1, 8). The next element, A(1, 9), is punched in the first ten columns of the second card, and so on through the entire array A. If a decimal point is not punched in the card, the format statement implies that five digits are to be considered to the right of the decimal point.

```
C  PROGRAM TO READ MATRIX A OF ORDER 30, INVERT A, READ
C  SEQUENCE OF VECTORS X OF ORDER 30. FOR EACH X, COMPUTE
C  AND PRINT A INVERSE TIMES X AND ITS LENGTH.
   1 DIMENSION A(30, 30), X(30), P(30)
   2 READ (5, 3) A
   3 FORMAT (8F10.5)
   4 CALL INVERT (A, 30)
   5 N = 0
   6 READ (5, 3) X
   7 N = N + 1
   8 PNORM = 0.0
   9 DO 13 I = 1, 30
  10 P(I) = 0.0
  11 DO 12 J = 1, 30
  12 P(I) = P(I) + A(I, J)*X(J)
  13 PNORM = PNORM + P(I)**2
  14 PNORM = PNORM**(0.5)
  15 WRITE (6, 16) N, PNORM, ((I, P(I), I = L, 30, 10), L = 1, 10)
  16 FORMAT ('1', 22X, 'CASE', I4, 10X, 'LENGTH =', E12.5//
   1           3(I10, E15.5))
  17 GO TO 6
  18 END
```

Fig. B-13   FORTRAN program example.

A more practical program for reading the elements of the matrix A would check to make certain that all 900 elements of the array A are actually present and, in fact, would identify each card image with a number which could be checked by the program to make certain that the data are in correct order. Alternatively, if one expected the matrix A to be sparse, it would be convenient for the program initially to set all elements of A to 0 and then read in only nonzero elements. In this case, it would be essential to identify the elements by their row and column position. For the sake of simplicity, these refinements are omitted from the example.

Statements 1 and 3 are examples of nonexecutable statements. The first statement executed in this program is statement 2, the READ statement, and the next statement executed is statement 4. Statement 4 is an example of a CALL to a subroutine. The subroutine is named INVERT, and it has two arguments. The first is the name of the matrix whose inverse is desired, and the second is the order of the matrix. The subroutine INVERT is a general matrix inversion program for arbitrary size matrices, not just matrices of order 30, hence the requirement to specify the order of the particular matrix being inverted in this example.

The matrix inversion subroutine INVERT inverts the matrix A and stores the inverse in place of the original matrix A. In other words, the original matrix A is destroyed. If it were important to retain the original matrix A, either the main program could copy A onto another matrix, or a matrix inversion routine that stored the inverse as a separate matrix could be used.

Statement 5 sets a variable N to 0. This variable is used to count the successive vectors X on which the program operates. Statement 6 is another READ statement, again specifying input device 5 and the same format statement 3, as used previously. This time, however, the elements to be read are the elements of the vector X. Since the vector X is read according to the same format as the matrix A, four cards are required for each vector X. The first three cards contain the first 24 elements of X, and the remaining elements are read from the fourth card.

Statement 7 increments the counter N by one, indicating that one vector X has been read. The length of the vector A·X is denoted by PNORM, and statement 8 initializes this variable by setting it to 0.

Statement 9 is a DO statement whose range is statements 10, 11, 12 and 13. In other words, this DO statement causes statements 10–13 to be executed successively 30 times, first with $I = 1$, next with $I = 2$, and so on through $I = 30$. The purpose of the calculation performed in this DO loop is to compute the matrix by vector product A·X, where A now is the inverse of the original matrix A, and also to compute the length of this resultant vector.

Statement 10 initializes the resultant product vector element $P(I)$ to 0. Statements 11 and 12 are an inner DO loop. Statement 11 is a DO whose range is the single statement 12 under control of the index variable J. This inner loop computes the element $P(I)$ by adding 30 terms, each of which is the product of the Jth element of the Ith row of A times the Jth element of X. This example illustrates that DO statements can be *nested*. In this case, statement 12 is executed 900 times for each vector X that is read: 30 times for each execution of the inner DO on J, which in turn is executed 30 times for each execution of the outer DO on I. After the Ith element of P is evaluated, control passes to statement 13 which adds the square of the Ith element of P to PNORM.

Statement 13 is the last statement in the range of the outer DO statement. Thus if I is less than 30, the next statement that is executed is statement 10 with the index I incremented by 1. After statement 13 has been executed with $I = 30$, all 30 elements of the product vector P have been calculated, as well as the square of the length of the product vector, whose name is PNORM. Statement 14 evaluates the square root of this number and stores it in PNORM. Therefore, after statement 14 has been executed, PNORM contains the length of the product vector. This completes the required calculations.

Statement 15 writes out the desired results on output device 6 which, for the purpose of this example, is a printer. Figure B-14 shows the output produced by the WRITE statement 15. It shows the vector number N and the length PNORM on the first line, followed by ten lines, each with six entries. The first column of data contains the indices 1–10,

CASE 27			LENGTH = 0.23295E 03		
1	0.14286E 01	11	0.14281E 02	21	0.79363E 02
2	0.10704E 02	12	0.20166E 02	22	0.93137E 02
3	0.73385E 02	13	0.33791E 02	23	0.72727E 02
4	0.27297E 02	14	0.96602E 01	24	0.46262E 02
5	0.26097E 02	15	0.67093E 02	25	0.12305E 01
6	0.57466E 01	16	0.78633E 01	26	0.80703E 01
7	0.65616E 02	17	0.96558E 01	27	0.26146E 02
8	0.27766E 02	18	0.66668E 02	28	0.18502E 02
9	0.55150E 02	19	0.23898E 02	29	0.12066E 02
10	0.79219E 01	20	0.56176E 02	30	0.28461E 02

Fig. B-14   Result of example program.

and column 2 contains the corresponding elements of the product vector, $P(1), \ldots, P(10)$. The second pair of columns shows the indices 11–20 and the corresponding elements of the product vector, etc.

This output form is the result of the WRITE and FORMAT statements 15 and 16. The WRITE statement 15 specifies that the first two elements to be written out are N and PNORM. Next, the 30 indices and 30 vector elements are specified in a manner analogous to a nested DO loop, where the inner DO loop is on the index I and the outer DO loop is on the index L. One can think of the index L, which runs from 1–10 as specifying line numbers 1–10, and the index I as representing the actual element being written, since it runs from L–30 in steps of ten. The index I does not necessarily take on the value 30; what is meant by the indicated notation is that I takes on the values L, L + 10, and L + 20, but does not exceed 30.

Format statement 16 specifies that the first character to be written is a "1." The 1 in the FORMAT specification is not printed, but rather causes the printer to skip to a new page before it begins printing data. It is called a carriage control character. The next FORMAT item "22X" results in 22 blank character positions and is used to center the information on the first line. The next item "case" is called a literal, and results in the 4 characters CASE being printed in print positions 23–26 of line 1.

The next format item, I4, indicates that the first item from the WRITE statement, namely N, is written as an integer in the next four print positions; thus when printing out the results for the 27th vector X the printer prints "CASE 27" at the beginning of a new page. The 10X in the format statement indicates that the next ten print positions are to be blank. Next, the format statement specifies the eight characters "LENGTH $=$" followed by the number PNORM in a format E12.5. This format calls for a floating point number which occupies twelve print positions with five digits printed to the right of the decimal point. A slash in the format statement indicates the end of a line, so, in our example, the printer skips one line between the heading information and data.

The remainder of the format statement, which happens to be on a continuation card, specifies that all of the remaining lines are to have the same format with each such line consisting of three pairs of fields. The first field of each pair, used for the index I, occupies ten print positions. The other field, for the element P(I), occupies the next 15 print positions and is to be printed in floating point format with five digits to the right of the decimal point.

The next executable statement is 17. This statement is an unconditional transfer of control to statement 6. Thus after completing the processing for a vector X and printing the results, control passes back to statement 6 which reads in a new vector X and the process repeats until no further input data on input device 5 exist, at which time the program terminates.

Statement 18, the END statement, is simply a signal to the FORTRAN processor that it has reached the end of the source program. It is not an executable statement.

From this sample FORTRAN program, it is clear that numerical computations, data entry, and printing of results are programmed simply by the use of this high-level language. Of course, FORTRAN has other statement types for handling complex numbers and logical expressions, comparing variables, controlling other types of input-output devices, and many other facilities. Further details of the FORTRAN language and other high-level languages are found in the references and bibliography.

B.3.3 Interpreters

Some computers do not have *floating point* hardware. Consequently, it is necessary to program floating point arithmetic using the available fixed point instructions. The resulting programmed floating point capability sometimes takes the form of an *interpreter* that allows the programmer to code in a pseudo-language that is interpreted according to a prescribed set of rules. The interpretive language need bear no relation to machine instructions. Often, however, the interpretive language is very similar to machine language, except that the floating point arithmetic capability is added. Interpreters are not restricted to providing floating point arithmetic; indeed, many modern computers have floating point hardware built in. Interpreters generally can provide any desired capability such as matrix manipulations, solution of initial value problems, complex arithmetic, and string manipulation. These capabilities can also, of course, be programmed using a compiler such as FORTRAN.

The essential distinction between compilers and interpreters is that compilers have a processor which translates the source program into machine language, but interpreters do not. Consequently, once a compiler source program is translated into machine language, the resulting object program can be executed as often as required without further translation. A program written in an interpretive language, on the other hand, is interpreted each time it is executed. Thus interpretive programs execute slower than a corresponding program written in a compiler language. The less efficient execution is sometimes counterbalanced by the ease and speed of programming of the interpretive language.

An example of an interpreter is a program to interpret FORTRAN. Although FORTRAN is normally implemented as a compiler, an interpreter would skip the translation into machine language by the FORTRAN processor and instead interpret the FORTRAN statements during execution.

A second example of an interpreter is a computer simulator. Suppose one wishes to program in the language of machine A, but only machine B is available. An interpreter can be written for machine B that interprets machine A instructions and simulates their performance on machine B.

B.3.4   Problem-Oriented Languages

When the need for a computer program arises, it may not be necessary to program it from the ground up. There is a growing trend to use general-purpose application programs with user-supplied input data. General purpose application programs are available for many problems, and many of them provide versatile input, output, and computational capabilities.

Basically, application programs require only input data and control information to be specified by the user. For the more elaborate application programs, the form of the input can actually be called a language, hence the name *problem-oriented language*. In this section, two application programs are described—one for solving linear programming problems, and the other, a systems simulator.

In linear programming, or mathematical programming as it is sometimes called, the problem to be solved can be stated as follows. Given a matrix A and vectors B and C, find a vector X satisfying the linear inequalities

$$AX \leq B$$

which maximizes the linear form CX. Problems in resource allocation and blending of ingredients can often be formulated as a linear programming problem. Because linear programming is a widespread computer application, many compunter programs have been written to solve this and related problems.

Typically, the user specifies the constraint equations by means of A and B, and the linear form to be maximized by means of C. Rather than working with the abstract matrices and vectors A, B, and C, the user provides these inputs in a simple, problem-oriented context. The left-hand side of each scalar inequality, represented by a row of A, is given a name that is meaningful in the context of the problem being solved. The difference between the right-and left-hand sides of a scalar inequality is called slack. To assist in interpreting results, each slack quantity is associated with the name of its corresponding scalar inequality. Similarly, the elements of the C vector are identified by meaningful names.

The linear programming problem is usually solved by an iterative procedure and the progress toward solution can be optionally printed out after each iteration. If the solution is nonunique, alternative solutions can be requested by the user. Finally, the user can request the perturbations in A, B, or C required to change the solution. With such flexibility provided by a general purpose linear programming program, it is clear that available programs of this type should be carefully evaluated before attempting to write a new program for linear programming.

General purpose simulation system (GPSS) is an IBM-developed application program that is useful for simulating discrete particle flow (see References). Suppose one wishes to simulate the flow of automobiles through a grid of streets subject to control of traffic lights and taking account of pedestrian traffic, interference of left-hand turns, and varying driving patterns of motorists. This problem is typical of discrete particle flow problems where, in this case, the discrete particles are the automobiles and pedestrians. Although one could use a programming language like FORTRAN or an assembly language to simulate the traffic flow model, a problem-oriented language like GPSS is generally preferable since many of the required details are provided automatically.

As an example, the mean and range can be specified for the inter-arrival time of particles entering the system. More generally, any frequency distribution of interarrival times can be specified. In the language of GPSS, the statement GENERATE specifies that particles are created according to interarrival time specified by arguments of this statement. In addition, several parameters associated with the particle can also be specified at the time the particle is generated.

Other entities in GPSS are facilities, stores, and queues. A facility has the property that when it is in use, any subsequent demands for the same facility cannot be satisfied until the facility is released by the particle that "seized" it originally. In the traffic model, a crosswalk is a facility which may be seized by a pedestrian, thus requiring any automobile traffic to form a queue until the pedestrian releases the crosswalk to vehicular traffic. In GPSS, the statements SEIZE and RELEASE control the use of facilities.

A store is used to collect particles up to a specified capacity. A parking lot is represented by a store in our traffic model. Cars can freely enter the parking lot so long as space remains available, after which no further cars can enter until cars leave the lot. Stores are specified and controlled in GPSS with the statements CAPACITY, ENTER, and LEAVE.

Statistics gathering capability is built into GPSS. The collected statistics are available on request during the course of the simulation in tabular form and histograms. Thus the user can observe the characteristics of the system, alter the model, and simulate the altered model to determine the effect of the change.

This brief description of GPSS only alludes to the total capability of this general purpose application program to simulate discrete particle flow. The discussion illustrates, however, that a problem-oriented language like GPSS often represents a more appropriate language than FORTRAN, for example, for certain classes of problems.

B.4   OPERATING ENVIRONMENTS

In the early days of computers, each program was run as a separate entity, each requiring nonproductive time for setup and make ready. In this environment, it was commonplace for both the computer operator and programmer to intervene in the operation of the program. Indeed, a common method of program debugging consisted of running the program until a particular condition occurred and then manually inspecting the contents of registers in an attempt to debug the program while it was in the machine. This manner of debugging, while possibly utilizing programmer time effiiciently, wastes large amounts of computer time, thus decreasing computer throughput. As the speed of computers has increased, the significance of this waste time has increased correspondingly.

The emphasis in early computers centered on the central processing unit. Main storage was extremely small by today's standards, and input/output devices were relatively simple and inflexible. By contrast, it is not uncommon in modern computing systems to have 8×10^6 bits of main storage. This compares with typical early computers whose main storage contained on the order of 0.07×10^6 bits. In addition to main storage, modern computers typically have on-line direct access auxiliary storage ranging from 8×10^6 to $10,000 \times 10^6$ bits. Similarly, input/output devices have increased in complexity from single card readers and printers to include such devices as graphic consoles, multiple typewriter terminals, and a variety of sensing devices used for process control.

As computer hardware has become faster, larger, and more versatile, there has developed a corresponding need for more sophisticated software, that is, programming. It is no longer possible for an individual user to approach the computer with a deck of cards and run his program, every instruction of which he has written himself. Nowadays, computer users rely on extensive programming support to take advantage of the advances made in computing systems. In addition to application programs like GPSS and library subroutines like matrix inverters, this programming support includes other types of programs called systems programs. Systems programs include assemblers, compilers, loaders, and the operating systems discussed in the next section. The throughput of a computer installation now depends more on the systems programming support than it does on the cleverness of individual applications programmers in optimizing some aspect of their programs, such as overlapping input/output operations with computing. In this section, three aspects of computer usage that transcend individual application programs are discussed. These are operating systems, multiprogramming, and multiprocessing.

B.4.1 Operating Systems

Operating systems were developed to increase computer throughput and to provide additional programmer flexibility. An *operating system* is a collection of computer programs which provide automatic sequencing of jobs through the computer. It minimizes operator intervention by queuing jobs according to a predefined algorithm and notifying the operator when, and only when, his assistance is required, for example, to mount a tape reel. In addition to sequencing jobs, many operating systems manage the flow of data to and from auxiliary storage, manage peripheral operations such as card reading and printing, and perform job accounting. Operating systems are virtually indispensable in the operation of a computing facility. Other terms such as monitor or executive program are sometimes used interchangeably with operating system. Operating systems are normally provided by the computer manufacturer.

B.4.2 Multiprogramming

Multiprogramming is the ability to have two or more programs operating concurrently. It is an attempt to utilize all the resources of a computer all the time. Suppose three programs, A, B, and C, are loaded in main storage and are ready to run; that is, all resources such as tapes, storage space, and peripherals are available for all three programs. For definiteness, let program A have highest priority and program C lowest priority. The three programs may be parts of the same job but more generally they are parts of unrelated jobs such as a FORTRAN compilation, a linear programming program, and a sort of input records for a mortgage loan accounting program. Initially, the operating system gives control to the program having highest priority, program A in our example. Program A retains control either until it terminates or until it must wait, for example, for a record to be read from a disk file. When program A must wait, it relinquishes control to the operating system which in turn passes control to program B, which has next lower priority. Program B runs either until it terminates, until it must wait, or until program A is again ready to run. If program B relinquishes control before program A is ready to run, then control passes to program C. In this manner, control is always given to the program of highest priority which is ready to run. With several programs running in a multiprogrammed environment, the computing system is kept busier with a corresponding increase of throughput as compared with a single-program environment.

To realize the maximum advantage of multiprogramming, programs having different types of requirements should be competing for computer resources. An example of this is a compute limited job and an input/output limited job. If the input/output job has higher priority, then it always has control when it has something to do, but during nonproductive periods when it is waiting for an input/output operation to be completed, the computing system is still productive with the compute limited job.

A control system is a particularly good example of an environment in which multiprogramming is used to advantage. Here the computing system is dedicated to controlling an on-line process, often 24 hours a day, and real-time response is a prime requirement. Since the computing demands of a control system are often modest, multiprogramming offers the opportunity to service other computing requirements concurrently without appreciably affecting the capability of the system to respond to the process control needs on a real-time basis.

Teleprocessing and terminal oriented applications are other areas where multiprogramming is effectively used. In these cases, a background or batch stream of jobs is processed concurrently with the on-line application in order to utilize available computing capacity.

Multiprogramming is made possible by capabilities of both the hardware and the software. The hardware must provide for asynchronous input/output and computing. It must also provide a program interrupt, or trap, which functions as follows. When an input/output operation terminates, detects an error, or in general encounters a status that requires intervention by the computer program, the normal sequence of program instruction is interrupted and control passes to an appropriate program for handling the input/output status. In this way, control of the computer can pass back and forth between programs that handle input/output and programs that perform computing.

B.4.3 Multiprocessing

The cooperation of two or more computers (processors) in an interconnected system is known as *multiprocessing*. In some cases, only the central processing unit is duplexed with both CPUs sharing the same main and auxiliary storage. In others, two or more complete computing systems are interconnected. The two principal reasons for multiprocessing are increased throughput and increased reliability.

Increased throughput results from multiprocessing since more than one CPU is available to process multiple jobs simultaneously. Note that multiprocessing is not equivalent to multiprogramming; the former implies multiple processors, the latter multiple programs. Of course, multiprocessing may be combined with multiprogramming in the same system.

Increased system reliability is the most common reason for multiprocessing. In the case of a system having two CPUs, when one CPU is being repaired, the other is available for productive work. The entire system is inoperative, so far as the user is concerned, only when both CPUs are inoperative at the same time. In nonmultiprocessing systems, the failure of the CPU or other critical components, such as the only channel connected to auxiliary storage, results in the entire system becoming inoperative. In other words, it is either "go" or "no-go."

In applications where high reliability is essential, such as on-line process control, airline ticket reservation, or airplane traffic control, critical functions are designed in such a way that the failure of such a component only degrades system performance instead of disabling the entire system. This ability to continue partial operation when a component of a system fails is called *graceful degradation* or *fail soft*. In the case of a process control application, the failure of a CPU in a multiprocessing system might terminate background batch operations; failure of a duplexed channel might restrict the data sampling from sensors to a fraction of the normal rate, and so forth. Despite this, however, the design of a process control application with graceful degradation ensures that the most essential applications continue, while less essential ones are deferred as system components fail.

B.5 PROGRAMMING MILIEU

Considerable folklore exists concerning the personal traits, abilities, and interests of computer programmers. It is commonly asserted that programmers like to solve puzzles, play bridge and chess, are somewhat eccentric in their personal attire, and are highly educated and geniuses in mathematics. Actually, programmers are a heterogeneous group of people who vary widely in these characteristics. In this section, the selection, training, and professional development of this heterogeneous group are briefly considered.

B.5.1 Programmer Selection

In the early days of computing, programmers were frequently selected on the basis of a mathematics background. As the diverse functions performed by programmers became better understood, it became clear that mathematics was not required for many programming tasks. Indeed, business applications of computers require programmers who are familiar with their company's business operations such as inventory control, payroll, accounts receivable, and so forth. Accordingly, some firms have been successful in selecting programmer trainees from their own ranks and teaching them programming, rather than hiring experienced programmers and teaching them about business. Similarly, systems programmers—those who write assemblers, compilers, and operating systems —require special competence in more advanced programming techniques but rarely do they require mathematics. Scientific application programmers, on the other hand, should have a good background in applied mathematics, including numerical analysis, as well as having some familiarity with an applied discipline such as engineering, physics, or chemistry.

Many tests have been used in an attempt to select promising candidates for programming jobs. Included are tests of intelligence, aptitude, personality, and interest. Mayer and Stalnaker (1968) summarize the current state of knowledge of computer personnel research as follows: "The instruments for predicting training grades of a computer programming course are fair-to-good; however, prediction with any instrument for on-the-job performance is still very questionable." Thus there is no simple method for selecting computer programmers.

B.5.2 Programmer Training

Computer manufacturers commonly provide computer programming courses for their customer personnel. There are also trade school courses available. For simple types of programming, with emphasis on coding, these short courses may suffice, particularly when supplemented by on-the-job training.

Formal academic education in the broader aspects of computing is becoming more widely available. Universities typically have one or more computer installations serving the academic community, and many offer advanced degrees in computer science, information processing, or a discipline with a similar name. The educational content of university courses in computing far transcends programming (analysis, coding, etc.). In fact, there is considerable discussion currently in the computing community concerning the appropriate content for university level education in the computing field (Finerman, 1968).

In addition to the specialized education offered by departments of computer science, almost all engineering and science departments require some exposure to computer applications and programming. There is even a trend in some schools to offer survey courses in computing to liberal arts majors in recognition of its growing importance to society.

B.5.3 Professional Organizations

The Association for Computing Machinery is the principal professional organization dedicated to the development of information processing as a discipline and to the responsible use of computers in an increasing diversity of applications. The ACM, with 12,000 members and 98 chapters, is relevant to mathematicians, engineers, physical scientists, business system specialists, analysts, and social scientists interested in computing and data processing. The association publishes *Communications of the ACM* (monthly), *Journal of the ACM* (quarterly), *Computing Reviews* (bimonthly), and *Computing Surveys* (quarterly).

The Institute of Electrical and Electronics Engineers (IEEE), with 145,000 members and 250 sections, is of interest to engineers and scientists in electrical engineering and allied fields. The Computer Group of the IEEE is an association of IEEE members with professional interest in the field of computers. It publishes *Transactions of the IEEE Group on Computers* (monthly).

The Data Processing Management Association, with 20,500 members and 213 chapters, is of interest to managers and supervisors of data-processing installations, including certified public accountants, systems and procedures personnel, programmers, and machinery and equipment manufacturing representatives. This trade association publishes the *Journal of Data Management* (monthly) and *DPMA* (monthly).

REFERENCES

Finerman, Aaron, *University Education in Computing Science,* Academic Press, New York, 1968.

Mayer, David, B., and Ashford W. Stalnaker, *Selection and Evaluation of Computer Personnel—The Research History of SIG/CPR, Proc. 23rd Natl. Conf. Assoc. Comput. Mach.,* 1968.

"IBM General Purpose Simulation System User's Manuals," Forms H20-0304 and H20-0326, IBM Corp., White Plains, 1968.

BIBLIOGRAPHY

Cole, R. W., *Introduction to Computing,* McGraw-Hill, New York, 1969.

Collins, J. S., and M. Almond, *Principles of ALGOL Programming,* Harrap, London, 1966.

Gear, C. W., *Computer Organization and Programming,* McGraw-Hill, New York, 1969.

Harrison, T. J., "How Hardware Responds to Software," *Control Eng.,* December 1967, pp. 65–70.

Hastings, Cecil, Jr., *Approximations for Digital Computers,* Princeton Univ. Press, Princeton, N.J., 1955.

Iverson, Kenneth E., *A Programming Language,* Wiley, New York, 1962.

Leeds, Herbert D., and Gerald M. Weinberg, *Computer Programming Fundamentals,* McGraw-Hill, New York, 1961.

Louden, R. K., *Programming the IBM 1130 and 1800,* Prentice-Hall, Englewood Cliffs, N.J., 1967.

Montalbano, "Tables, Flow Charts, and Program Logic," *IBM Systems Journal,* Vol. 1, September 1962, pp. 51–64.

Organick, Elliot I., *A FORTRAN Primer,* Addison-Wesley, Reading, Mass., 1963.

Ralston, Anthony, and Herbert S. Wilf, *Mathematical Methods for Digital Computers,* Wiley, New York, 1963.

Rice, J. K., and J. R. Rice, *Introduction to Computer Science,* Holt, Rinehart and Winston, New York, 1969.

Rosen, S., *Programming Systems and Languages,* McGraw-Hill, New York, 1967.

Salzman, R. M., "Trends in the Programming Profession," *Data Process.*, November 1968, pp. 20–22, 24, 26.

Sherman, Philip M., *Programming and Coding Digital Computers*, Wiley, New York, 1963.

Sippl, Charles J., *Computer Dictionary and Handbook*, Howard W. Sams, Indianapolis, Ind., 1966.

Swallow, K. P., and W. T. Price, *Elements of Computer Programming*, Holt, Rinehart and Winston, New York, 1965.

Weinberg, Gerald M., *PL/I Programming Primer*, McGraw-Hill, New York, 1961.

"IBM System/360 PL/I Reference Manual," Form C28-8201, IBM Corp., White Plains, N.Y., March, 1968.

"IBM System/360 Operating System Concepts and Facilities," Form C28-6535, IBM Corp., White Plains, N.Y., October, 1968.

"A Programmer's Introduction to the IBM System/360 Architecture, Instructions, and Assembler Language," Form C20-1646, IBM Corp., White Plains, N.Y., September 1968.

"IBM 1800 Functional Characteristics," Form A26-5918, IBM Corp., White Plains, N.Y., June 1968.

"IBM 1800 Assembler Language," Form C26-5882, IBM Corp., White Plains, N.Y. 1968.

"Decision Tables—A Systems Analysis and Documentation Technique," Form F20-8102, IBM Corp., White Plains, N.Y., 1962.

"IBM System/360 Scientific Subroutine Package, Version III, Programmer's Manual," Form H20-0205, IBM Corp., White Plains, N.Y., March 1968.

Appendix C

Application Bibliography

THOMAS J. HARRISON

The different applications in which industrial control computers are used number in the hundreds. If the applications in which data acquisition computers are used are added to this number, there are probably thousands of different applications. Some applications are relatively simple, involving no more than the collection of analog and digital data for subsequent analysis on another computer; others are extremely complex, requiring the collection of data, the calculation of control actions according to a control algorithm, and the implementation of the control action, all on a real-time basis. The computer systems utilized in these applications also range from basic configurations to very large systems involving thousands of input data points, many control outputs, and extensive peripheral equipment such as display consoles and communication equipment.

The hundreds of different applications and the thousands of process control computers are a product of the 1960s. Prior to 1960, only a few process control computers were installed and most of these were considered to be experimental in nature. This is no longer true and computer systems are often considered a necessary part of many new plant designs. In addition to installations in application areas such as chemical, petroleum, and paper, in which the use of such systems is well established, new applications are constantly being reported. Examples of newer applications include automobile engine diagnostic systems, traffic control, and the control of complex discrete manufacturing processes.

As discussed in previous chapters, the process control computer is characterized by the range and versatility of input/output equipment which is used to interface with analog and digital sensor and control equipment. These same functions are also required in applications not generally considered as process control. For example, the control of a space probe utilizes computers with many of the same functional capabilities which characterize the process control computer. Similarly, medical research teams are using sensor-based computer systems in exploring new approaches to intensive patient care. In essence, process control computer applications are only a subset of a total spectrum of computer applications which involve real-time sensor-based data acquisition.

The process control subset of applications is quite large and, in terms of the number of computers installed for data acquisition and/or control, it probably represents the major portion of the total spectrum of applications. This large number of installations and the large variety of applications makes it difficult to provide a comprehensive discussion of process control computer applications. Rather than attempt such a discussion, this chapter provides an extensive bibliography of articles which have appeared in the technical and trade press. By this means, the reader interested in computer process control can find a number of articles in his particular field of interest.

Although bibliographies necessarily become outdated during the life of any book, they still provide a valuable function in that they indicate which journals and magazines typically publish material concerning computer control. With this information, the user can easily update the bibliography.

C.1   BIBLIOGRAPHY ORGANIZATION

The bibliography in this appendix is divided into seven primary categories. Five of these relate to specific application areas such as paper, rubber, and textiles. The articles in these categories generally describe some aspect of computer control as it is being practiced in the commercial industry. The sixth category includes articles of general interest and articles which describe applications not included in the previous sections. Included in this category are applications in areas such as aerospace, traffic control, test and inspection, and food processing. Many of these categories are important applications of control computers but they do not fall into the category of process control in its classical sense. The last category of articles covers research applications. These applications may include research in one of the application areas in the other categories, but the work is not commercially oriented.

Each category is considered in a separate section of this appendix. The section is introduced with a brief statement of the scope of the category. Following this, the entries are arranged alphabetically by author. The citations are in a standard bibliographic form.

In general, articles published before 1963 have not been included in the bibliography. Those few that are included are felt to be of significant interest for one of several reasons. Every attempt has been made to verify the accuracy of the citations. In a bibliography of this length, however, some errors are almost inevitable. Whenever possible, the entries have been verified through the use of multiple index sources and, in most cases, the author of this appendix has reviewed the actual article.

C.2   PETROLEUM AND CHEMICAL APPLICATIONS

Petroleum and chemical applications of process control computers are among the most numerous and sophisticated to date. Articles included in the following bibliography include discussions of process control computer use in typical petroleum and chemical processes such as catalytic crackers, distillation columns, production of chemicals and artificial fibers, and the control of analytical instrumentation. Also included are articles dealing with related topics such as pipeline control and pilot plant instrumentation and control.

Ackerman, C. D., G. P. Huling, and E. M. Palmer, "A Computerized On-Line Pilot-Plant Data System," *Proc. 23rd ISA Conf., Pap. 68-860,* 1968.

Adams, C. F., "Closed Loop Control of an 85,000 Barrels per Day Crude Oil Unit," *Oil Gas J.,* May 21, 1962, pp. 114–118.

Amrehn, H., "Computers Direct Batching in 80-Reactor PVC Plant," *Instrumen. Technol.,* November 1967, pp. 45–48.

Andrews, A. J., "Phillips Mobile Plant Testing Laboratory," *Proc. Texas A&M Ann. Symp. Instrum. for the Process Ind.,* January 1962, pp. 35–41.

Andrews, A. J., G. D. Stacy, and P. U. Webb, "A Control Engineer Looks at the Impact of Plant Computers," *Proc. Texas A&M Ann. Symp. Instrum. for the Process Ind.*, January 1965, pp. 15–22.

Bacher, S., and A. Kaufman, "Computer-Controlled Batch Chemical Reactions," *Ind. & Eng. Chem.*, January 1970, pp. 53–61.

Baker, W. J., and J. C. Weber, "Direct Digital Control of Batch Processes Pays Off," *Chem. Eng.*, December 15, 1969, pp. 121–128.

Bernard, J. W., and J. W. Wujkowski, "DDC Experience in a Chemical Process," *ISA Conf. Prepr.*, January 3, 1965; see also "Direct Digital Control Experiences in a Chemical Process," *ISA J.*, December 1965. pp. 43–47.

Carter, R. J., "Control of Continuous Distillation Columns," *Control,* January 1965, pp. 21–25.

Clark, G. L., "Time-Shared Program Control of Two Product Pipelines," *IEEE Trans. Ind. Gen. Appl.*, March 1968, pp. 167–170.

Cline, R. P., "A View of Refinery Computer Control," *ISA Conf. Prepr.*, 1965.

Colville, A. R., "Computer Control of Fluid Catalytic Cracking," *Proc. Texas A&M 18th Ann. Symp. Instrum. for the Process Ind.*, January 1963, pp. 91–96.

Crowther, R. H., J. E. Pitrak, and E. N. Ply, "Computer Control at American Oil, Application to the Process Unit," *Chem. Eng. Progr.*, June 1961, pp. 39–44.

Davidson, B., and M. J. Shah, "Simulation of the Catalytic Cracking Process for Styrene Production, *IBM J. Res. and Dev.*, November 1965, pp. 388–399.

Douloff, A., "Gas Pipeline's Performance Is Improved with Aid of Computer," *Oil Gas J.*, May 9, 1966, pp. 102–117.

Bleakley, W. B., "Streamlined Production Is Payoff of Computer Control," *Oil Gas J.*, April 12, 1965, pp. 90–94.

Block, W. E., and J. A. Bodine, "Computer-Controlled Petroleum Production Systems," *IEEE Trans. Ind. Gen. Appl.*, July 1969, pp. 403–410.

Bodine, J. A., "A Systems Approach to Centralized Control of Refinery Off-Plot Facilities," *IEEE Trans. Ind. Gen. Appl.*, March–April 1965, pp. 107–118.

Bozeman, H. C., "Closed Loop Computer Control Is Used on an Ammonia Plant," *Oil Gas J.*, November 14, 1960, pp. 148–150.

Brandon, D. B., "Digital Computer Control and Gasoline Blending," *Instrum. and Control Syst.*, December 1963, pp. 123–124.

Budzilovich, P. N., "Central Refinery Computer Control," *Control Eng.*, January 1969, pp. 103–107.

Buell, K., "Automating Plant Operations," *Pet. Eng.*, March 1970, pp. 8–12.

Butler, P. A., "On-Line Petrochemical Plant Control," *Chem. & Process Eng.*, July 1970, pp. 88–90.

Farrar, G. L., "Computer Control in the Oil Industry," *Oil Gas J.*, October 25, 1965, pp. 91–111.

Flukinger, T. E., and G. R. Smith, "Liquid Pipeline Scheduling and Control with an On-Line Multiprogramming Process Computer," *IEEE Trans. Ind. Gen. Appl.*, July 1969, pp. 389–402.

Forgotson, J. M., Jr., and C. F. Iglehart, "Current Uses of Computers by Exploration Geologists," *Am. Assoc. Pet. Geol. Bull.*, July 1967, pp. 1202–1224.

Fraade, D. J., "A Review of Closed Loop Control of an 85,000 Barrels per Day Crude Unit," *Digital Computer Applications to Process Control*, W. E. Miller (Ed.), Plenum Press, New York, 1965, pp. 1–24.

Franklin, J., "Digital Control of In Line Blending," *Ind. Electron.*, December 1965, pp. 568–571.

Friedli, F. L., "Direct Digital Control of Pipeline Compressors," *Instrum. & Control Syst.*, September 1968, pp. 122–124.

Gaines, N. W., *et al.*, "Union Carbide Integrates Multi-Computer Process Control," *Instrum. Technol.*, March 1967, pp. 49–51.

Gandsey, L. J., "Two Way Communication Between Operator and Computers," *Chem. Eng. Progr.*, October 1965, pp. 93–98.

Haalman, A., K. Hoogendorn, and V. M. J. Evers, "In-Line Computer Control of Ethylene Production," *Proc. IFAC/IFIP Conf. Comput. Control*, Toronto, Canada, 1968.

Harders, H., G. Heller, and P. R. Laurer, "Experience Gained in the Optimization of a Petrochemical Synthesis Using a Process Control Computer," *Chem. Ing. Tech.*, June 1963, pp. 405–410.

Harders, H., *et al.*, "Computer Control of the Oxo-Synthesis Process," *Digital Computer Applications to Process Control*, W. E. Miller (Ed.), Plenum Press, New York, 1965, pp. 89–118.

Howard, E. W., and D. W. Pierce, "Offshore Gas System Put On Computer Controls," *Pet. Eng.*, June 1967, pp. 50–52.

Heimstead, J. E., and L. J. Nelson, "Systems Approach to Automation of Oil and Gas Production; Conventional or Visionary," *IEEE Trans. Ind. Gen. Appli.*, January 1967, pp. 31–39.

Hix, A. H., "Status of Process Control Computers in the Chemical Industry," *Proc. IEEE*, January 1970, pp. 4–10.

Holt, A. B., "Improved Fractionation Thru Computer Control," *Instrum. Technol.*, January 1971, pp. 43–48.

Inkofer, W. A., "The Transportation and Distribution of Anhydrous Ammonia by Pipeline," *1969 ISA Ann. Conf., Paper 682*, October 1969.

Imbert, J., "Remote Control and Supervision System for an Oil Pipeline and Pumping Station," *Ann. de Radioelectr.*, October 1965, pp. 344–355.

Karp, H. R., "Direct Digital Control Up to the Hilt," *ISA J.*, February 1966, pp. 13, 15.

Killick, R. D., "Computer Control of Ammonia Production," *Chem. and Process Eng.*, July 1962, pp. 324–329.

Lane, J. W., "Digital Computer Control of a Catalytic Reforming Unit," *Trans. ISA*, July 1962, pp. 291–297.

Lasher, R. J., "Computer Controls Blending of Motor Gasoline," *Oil Gas J.*, December 4, 1967, pp. 78–82.

Lauher, V. A., "Intercomputer Talk at Chocolate Bayou," *Proc. JACC*, 1964, pp. 131–137.

Lauher, V. A., R. Middleton, and T. J. Williams, "The Coordinated Multi-Computer System for Monsanto's Chocolate Bayou Plant," *Digital Computer*

Applications to Process Control, W. E. Miller (Ed.), Plenum Press, New York, 1965, pp. 25–54.

Lee, E. S., and E. H. Gray, "Optimizing Complex Chemical Plants by Mathematical Modeling Techniques," *Chem. Eng.,* August 28, 1967, pp. 131–138.

Lemay, L. P., "On Line Computer Control of a Chemical Plant," *Proc. 3rd Conf. Comput. and Data Process. Soc. Can.,* June 1962, pp. 258–276.

Lubyen, W. L., "Feedforward Control of Distillation Columns," *Chem. Eng.,* August 1965, pp. 74–78.

Lumb, A. C., "Computer Runs Amine Plant and Reduces Analyzer Data," *Instrum. Technol.,* April 1969, pp. 56–60.

Maddock, J., "Check Furnace Performance by Computer," *Hydrocarbon Process,* June 1967, pp. 161–168.

Mason, R., "Process Control—Easier via Computers," *Can. Chem. Process.,* September 1965, pp. 59–64.

Merley, R. A., and C. M. Cundall, "The Ferranti System and Experience with Direct Digital Control," *Process Control and Autom.,* July 1965, pp. 289–295.

Middleton, J. R.. "Chocolate Bayou Instrument and Computer System," *Proc. Texas A&M 18th Ann. Symp. Instrum. for the Process Ind.,* January 1963, pp. 83–90.

Morello, V. S., "Application of a Digital Computer to Control of Styrene Cracking," *Proc. Texas A&M 19th Ann. Symp. Instrum. for the Process Ind.,* January 1964, pp. 11–15.

Nofrey, C. E., "North African Pipeline Is Remotely Controlled," *Oil Gas J.,* April 12, 1965, pp. 90–94.

O'Donnell, J., "Crude Products Line Complex Controlled from Single Center," *Oil Gas J.,* July 12, 1965, pp. 82–83.

Ogle, F. T., "On Line Instrumentation and Control Systems," *Chem. Eng. Progr.,* October 1965, pp. 87–92.

Rijnsdorp, J. E., "Chemical Process Systems and Automatic Control," *Chem. Eng. Progr.,* July 1967, pp. 97–116.

Routley, J. H., "Application of an On-Line Computer to an Acetic-Acid Process," *Proc. IEE,* September 1969, pp. 1585–1588.

Sawyer, J. R., "Computer Directs Black Lake Complex," *Pet. Eng.,* June 1967, pp. 63–64+.

Schuyter, H., "Lower Costs Improve Trend to On-Line Digital Computer Control," *Hydrocarbon Process,* April 1967, pp. 155–157.

Shah, M. J., "Computer Control of Ethylene Production," *Ind. and Eng. Chem.,* May 1967, pp. 70–85.

Sharp, D. A., "Computer Applications in the Petroleum and Mineral Industries," *Can. Min. and Met. Bull.,* September 1967, pp. 1046–1050.

Shinsky, F. G., "Feedforward Control for Distillation: Why and How," *Proc. Texas A&M 20th Ann. Symp. for the Process Ind.,* January 1965, pp. 59–65.

Skalnik, C. R., "Centralization + Modernization + Computer = Profits," *Proc. Texas A&M 21st Symp. for the Process Ind.,* January 1966, pp. 9–14.

Stout, T. M. "Computer Control of Butane Isomerization," *ISA J.,* September 1959, pp. 98–103.

Tau, C., "Digital Computer Mirrors a Real Chemical Plant," *Chem. Eng.*, September 27, 1965, pp. 84–86.

Wadhwani, N. M., "Fractionation Column Control," *IMDA J.*, March 1965, pp. 20–23.

Washimi, K., and M. Asakura, "Computer Control of a Vinyl Chloride Plant Provides Optimization," *Chem. Eng.*, October 24, 1966, pp. 133–138; November 21, 1966, pp. 121–126.

Watson, D., "Computer Control of Grinding Mill," *Chem. & Process Eng.*, January 1970, pp. 45–48.

Youle, P. V., and L. A. Duncanson, "On-Line Control of Olefine Plant," *Chem. & Process Eng.*, May 1970, pp. 49–52.

Young, F. S., Jr., "Computerized Drilling Control," *J. Pet. Technol.*, April 1969, pp. 843–896.

"Computer Control of Refinery Operations: A Status Report," *Instrum. Technol.*, June 1968, pp. 12 ff.

"Computer Process Control Forging Ahead at Accelerated Pace," *Oil Gas J.*, November 21, 1966, pp. 178–182.

"Computer Simplifies Control of Candu," *Can. Chem. Process.*, May 1964, pp. 65–67.

"Computer Supervises Oil-Well Drilling," *Control Eng.*, February 1968, p. 54.

"Computer Takes Control of Japanese Urea Unit," *Chem. and Eng. News*, February 13, 1967, pp. 26–28.

"Direct Digital Control at Oil Refinery," *Chem. Process.*, March 1966, pp. 44–45.

"Direct Digital Control Putting the Computer in Charge," *Can. Chem. Process.*, September 1965, pp. 65–70.

"ESSO-Fawley To Install Multiple Computer for DDC," *Instrum. Pract.*, February 1966, pp. 154–155.

"Joint Computer Optimization on Acetic Acid Plant," *Instrum. Pract.*, October 1965, pp. 929–931.

"Mobile Computerizes at Pegasus," *Pet. Eng. Int.*, June 1967, pp. 53–55.

"More Computer Control—IBM and Dupont Study a Tricky New Process," *Control Eng.*, February 1961, pp. 175–176.

"Optimizing Process Profitability by DDC," *Instrum. Pract.*, December 1964, pp. 1260–1262.

C.3   METAL INDUSTRY APPLICATIONS

In recent years the number of process control computers installed in the metal industries has increased significantly. Applications include all phases of processing ranging from the reduction of ore through finishing. Substantial economic justification has been reported in a number of different applications with improved quality control being an often cited factor. In the bibliographic entries listed in this section, applications include blast furnace control, control of other reduction processes such as the basic oxygen furnace, rolling and forming processes, finishing such as annealing and plating, and auxiliary applications such as the control of utility and fuel resources.

Auricoste, J., and P. Westercamp, "Computer Control of an Oxygen Steelmaking Process," *Digital Computer Applications to Process Control,* W. E. Miller (Ed.), Plenum Press, New York, 1965, pp. 143–166.

Barnes, F. W., and E. J. Denner, "Armco-Middletown's 86-In. Continuous Hot Strip Mill," *Iron & Steel Eng.,* January 1970, pp. 59–71.

Baulk, R. H., R. J. Jakeways, and K. C. Padley, "Experiences with an On-Line Process Control Computer in a Steel Works," *Digital Computer Applications to Process Control,* W. E. Miller (Ed.), Plenum Press, New York, 1965, pp. 207–232.

Beswick, A. A., et al., "Blast Furnace Control by Computer," *J. Iron Steel Inst.,* June 1969, pp. 743–750.

Binning, J. E., and D. R. Berg, "Computer Control of Kaldo Oxygen Steelmaking," *J. Met.,* July 1965, p. 725.

Blum, B., et al., "Closed-Loop Computer Control of Basic Oxygen Steelmaking," *Iron & Steel Eng.,* June 1967, pp. 111–119.

Borrebach, E. J., "Digital Computer Control for the Basic Oxygen Steelmaking Process," *Westinghouse Eng.,* March 1963, pp. 40–43.

Brower, A. S., "Applying Digital Control Computers to Metals Industry Processes," *Proc. ISA,* January 2, 1964.

Brower, A. S., "Digital Control Computers for the Metals Industry," *ISA J.,* February, 1965, pp. 51–63.

Brower, A. S., "The Spread of Computer Control in Steel Mills," *IEEE Trans. Ind. Electron. and Control Instrum.,* April 1966, pp. 16–23.

Cartwright, W. F., and G. W. Thomas, "The Integration of Production Planning, Electronic Data Processing and Process Control," *J. Iron Steel Inst.,* 198(3), pp. 250–256.

Cesselin, P., et al., "Closed Loop Computer Control of the OLP Process," *J. Met.,* July 1965, pp. 722–724.

Darnell, R. J., "Computer Control on Inland's 80-in. Hot Strip Mill," *Iron & Steel Eng.,* April 1968, pp. 128–134.

Dickinson, L. I., "Computers in Utilities and Fuel Dispatching," *Iron & Steel Eng.,* October 1962, pp. 89–95.

Drewry, H. S., and W. R. Edens, "Minicomputer Grades Steel Strip On-Line," *Instrum. Technol.,* January 1971, pp. 49–53.

Fapiano, D. J., "Technical and Economic Consideration in Plate Mill Process Control," *Iron & Steel Eng.,* October 1966, pp. 101–112.

Fear, D., "Computer Control Applied to a Plate Mill," *Eng.,* August 20, 1965, pp. 291–293.

Fuchigami, T., "Computer Schedules Ingot, Slab, and Plate Production," *Control Eng.,* September 1966, pp. 138–139.

Gloven, D. O., "Computer Control for Electric Furnace Steelmaking," *J. Met.*, December 1964, pp. 963–966.

Glow, F. W., and S. J. Bailey, "On-Line Sampling Saves Valuable Ore," *Control Eng.*, January 1969, pp. 117–119.

Greaves, M. J., and J. C. Ponstingle, "Automation Today—Charging Blast Furnaces," *Autom.*, 12(5), 76(1965).

Greene, A. M., "Steel in a World of Computers," *Iron Age*, February 26, 1970, pp. 67-71.

Hittinger, N. J., "Control and Automation of Hot Strip Mills: Criteria for Computers," *Iron & Steel Eng.*, September 1969, pp. 74–75.

Jones, J. T., "Computer Trio Runs the Works at Big British Mill," *Electron.*, January 25, 1965, pp. 80–89.

Jones, J. T., and N. J. Williams, "Computer Control of Steelworks Production," *Proc. IEE*, June 1964, pp. 1183–1192.

Keenon, D. L., N. R. Carlson, and L. F. Martz, "Dynamic Control of Basic Oxygen Steel Process, *Instrum. and Control Syst.*, May 1967, pp. 139–144.

Kelly, G. H., "Integrated Computer Control of Steelworks Production," *Engl. Elec. J.*, August 1964, pp. 22–40.

Kirkland, R. W., "Process Computers—Their Place in the Steel Industry," *Iron & Steel Eng.*, February 1965, pp. 115–124.

Kung, E. Y., "Analysis and Control of a Continuous Annealing Line," *IEEE Gen. Meet. and Nucl. Eff. Conf.*, Toronto, June 1963, Pap. 63-911.

Lasiewicz, T. W., "Automatic Data Handling, Scheduling and Control of the Batch Anneal," *Iron & Steel Eng.*, September 1965, pp. 191–194.

Lodics, E. S., "A Computer Runs a Hot Strip Rolling Mill," *Control Eng.*, April 1969, pp. 89-93.

Lopresti, P. V., and T. N. Patton, "An Approach to Minimum Cost Steel Rolling," *Proc. IEEE*, January 1970, pp. 23–30.

MacQueen, A. J. F., "Finding a Practical Method for Calculating Roll Force in Wide Reversing Cold Mills," *Iron & Steel Eng.*, June 1967, pp. 95–110.

Massey, R. G., "Computers in New Steelworks," *Comput. J.*, January 1963, pp. 271–275.

McGill, J. D., Jr., "Computer Controls Pipe Foundry," *Foundry*, December 1968, pp. 44–49.

Meridith, R., J. T. Fisher, and R. W. Kirkland, "Computer Control of Hot Strip Mills at Spencer Works," *Iron & Steel Eng.*, December 1965, pp. 73–81.

Morgan, H. D., and R. W. Kirkland, "The Application of Automatic Gauge Control to an Existing Hot Strip Mill," *Digital Computer Applications to Process Control*, W. E. Miller (Ed.), Plenum Press, New York, 1965, pp. 167–206.

Ohnari, M., Y. Morooka, and Y. Yamamoto, "Experience in Installing a Computer Control System in a Hot Strip Mill," *Proc. IEEE*, January 1970, pp. 30–37.

Pellinat, W. K., "Computer Control of Scheduling, Melting and Continuous Casting in the Melt Shop," *Iron & Steel Eng.*, August 1968, pp. 121–133, 134–136.

Pigott, P. M., and M. Graf, "Automatically Controlled Billet-Handling Crane," *Control*, April 1965, pp. 187 ff.

Putnam, R. E. J., "New Developments in Process Control Using Computers," *Min. Congr. J.*, December 1967, pp. 34–38.

Price, F. C., "The New Technology of Iron and Steel," *Chem. Eng.*, September 14, 1964, pp. 179–194.

Ray, D. J., "Present and Future Trends in Hot Strip Mill Computer Control," *J. Iron & Steel Inst.*, June 1969, pp. 907–915.

Robins, N. A., "Computer Control of Inland's No. 5 Blast Furnace," *Iron & Steel Eng.*, December 1967, pp. 129–135.

Roth, J. F., "Collection of Analogue Plant Data for On-Line Computer Processing," *Ind. Electr.*, March 1963, pp. 326–329.

Roth, J. F., "Computer Control Hierarchy in Steel Plant," *Iron & Steel*, March 4, 1964, pp. 93–98.

Samuel, G. H., "On-Line Computer Boosts Steel Mill Output," *Control Eng.*, June 1967, pp. 80–84.

Schuerger, T. R., and F. Slamar, "Control in the Iron and Steel Industry," *Datamat.*, February 1966, pp. 28–32.

Shibuya, J., "Computer Control of Blast Furnace," *Iron & Steel Eng.*, February 1964, pp. 81–86.

Slamar, F., "Status Report on Computers in Steel Making," *Autom.*, June 1963, pp. 52–56.

Smith, A. W., and L. P. Gripp, "On Line Computer Control for a Reversing Plate Mill," *Iron & Steel Eng.*, August 1962, pp. 77–82.

Smith, H. W., and C. L. Lewis, "Computer Control Experiments at Lake Dufault," *Can. Min. and Met. Bull.*, February 1969, pp. 109–115.

Smith, H., Jr., "Control and Automation of Hot Strip Mills: Most Beneficial Areas for Computer Control," *Iron & Steel Eng.*, September 1969, pp. 72–74.

Teschner, H. G., "Controlling Data Flow in a Steel Mill," *Control Eng.*, April 1968, pp. 76–80.

Thompson, P. C., "Hot Strip Mill Control," *Eng. J.*, August 1967, pp. 13–17.

Tippett, J., A. Whitwell, and L. H. Fielder, "Controlling Megawatts in Steel-making," *Control Eng.*, June 1965, pp. 68–70.

Tocher, T. D., and M. Splaine, "Computer Control of the Bessemer Process," *J. Iron Steel Inst.*, February 1966, pp. 81–86.

Trace, J. J., and M. E. Murphy, "Granite City Steel's Computerized 80-in. Hot Strip Mill," *Iron & Steel Eng.*, March 1970, pp. 51–62.

Van Langen, J. M., *et al.*, "Burden Preparation on Computer Control of the Blast Furnace," *J. Met.*, March 1968, pp. 68–74.

Veeder, N. P., "Hot Strip Mill Start-Up—Then and Now," *Iron & Steel Eng.*, February 1969, pp. 77–79.

Von Loesecke, P. H., and J. W. Peirce, "Making Basic Oxygen Furnace Analog Signals Compatible to Digital Computer Accuracy," *Iron & Steel Eng.*, August 1965, pp. 97–101.

Walter, L., "Computer Control of Continuous Tin Plating Annealing Line," *Sheet Met. J.*, September 1962, pp. 635–642.

Wisdon, K., "Power to Furnace Under Computer Control," *Data and Control*, July 1963, pp. 23–24.

Wyrod, M. W., "Controlling Oxygen Steelmaking," *ISA J.*, January 1963, pp. 61–64.

"Computer-Controlled Continuous Tube Mill: Thyssen Röhrenwerke," *Eng.*, January 6, 1967, pp. 41–42.

"Computer Cuts Energy Consumption at Bethlehem," *Control Eng.*, April 1967, pp. 20–21.

"Digital Power Control Computer Being Installed by Northwestern Steel and Wire," *Iron & Steel Eng.*, November 1967, pp. 153–154.

"Discussion on Computers in the Iron and Steel Industry, Computers for Process Control," *J. Iron Steel Inst.*, February 1962, pp. 126–131.

"New Programmed Automatic Plating Plant," *Process Control and Autom.*, April 1966, pp. 48–50.

"Two Years Progress at Spencer Works," *Control*, February 1965, pp. 83–84.

C.4   POWER AND UTILITY APPLICATIONS

The economic dispatch of electric power is one of the most often cited applications of control computers in the utility industry. This particular application often involves complex optimization procedures which utilize some of the most sophisticated mathematics in use in the control industries. The control problem is complicated by the nature of the large interconnected power distribution systems which are common in the United States. National attention was directed toward this particular control problem in recent years as a result of the Northeast blackout and other major system faults.

In addition to the electric utility dispatch application, however, control computers are utilized in other phases of the electric utility industry. Applications include optimization of fuel resources, control of boilers and turbines, and sequence control for start-up and shutdown of generating and distribution equipment.

Control computers are also used in other utility and power applications such as those found in the gas industry. They are used for control of pipelines, compressor stations, and distribution facilities. In addition, they are used to forecast pipeline loads, including factors such as weather conditions and the distributed load along the length of the pipeline.

Battuello, F. L., "Digital Computer Sparks Major Changes in Steam Stations," *Power Eng.*, July 1965, pp. 37–40.

Beechey, M. A., and R. F. E. Crump, "Instrumentation and Control of Modern Boiler-Turbine Units in CEGB Power Stations," *Control* (a monthly series of articles from July, 1965, through April, 1966).

Bevilacqua, F., and E. A. Yuile, "Applications of Digital Computer Systems in Nuclear Power Plants," *IEEE Trans. Nucl. Sci.*, February 1969, pp. 196–206.

Bishop, B. W., "Modern Supervisory Control Systems Make Increasing Use of the Computer," *Power Eng.*, January 1967, pp. 46–49.

Bjorklund, G. J., "Operating Experience with Computer Controlled Generating Stations," *Combust.*, May 1967, pp. 12–15.

Blake, N. B., and J. M. Reeve, "The Overseer System," *Control,* June 1965, pp. 310–313.

Bolton, R. E., "Computer Analysis of Alarms at Oldbury Nuclear Power Station," *Process Control and Autom.*, November 1965, pp. 463–465.

Carpentier, J., "Operation Optimization in an Electric Power System," *Digital Computer Applications to Process Control,* W. E. Miller (Ed.), Plenum Press, New York, 1965, pp. 299–318.

Cohn, N., "The Automatic Control of Electric Power in the United States," *IEEE Spectrum,* November 1965, pp. 67–77.

Cohn, N., "Computers for Electric Utility Industry," *Electron.*, April 19, 1965, pp. 124 ff.

Cohn, N., *et al.*, "On-Line Computer Applications in the Electric Power Industry," *Proc. IEEE,* January 1970, pp. 78–87.

Couch, C. S., G. M. Hugh, and D. J. Newton, "Combined Use of a Very Large Telemetry Network and a Process Control Computer by a Natural Gas Pipeline," *Proc. Natl. Telem. Conf.,* May 1966.

Couch, C. S., *et al.*, "Telemetry and Process Control Application in Natural Gas Dispatching," *IEEE Trans. Comm. Tech.,* February 1967, pp. 96–102.

Crowther, R. L., and L. K. Holland, "Nuclear Applications of On-Line Process Computers," *Am. Nucl. News,* May 1968.

Derks, R. A., and E. H. Preston, "Dispatch Computer System Reduces Peaking Capacity Requirements," *Power Eng.*, September 1965, pp. 58–60.

Drewry, H. S., "Computer Controlled Power Station," *Proc. 15th Conf. Instrum. Iron and Steel Ind.*, 1965.

Drewry, H. S., and J. R. Howard, "Taking the Giant Step in Power Plant Automation," *ISA J.,* July 1964, pp. 53–60.

Drewry, H. S., and J. R. Howard, "For Successful Computing Control—Start Simple," *Control Eng.,* August 1965, pp. 76–82.

Dukelow, S. G., "The Revolution in Boiler and Power Plant Control," *Instrum. Control Syst.,* December 1965, pp. 83ff.

Dzhigit, G. A., "Automatic Boiler Start-Up System with Nearly Optimum Response," *Autom. Remote Control,* 26(11), 1911(1965).

Ender, R. C., "System Automation Promises Savings," *Electr. World,* January 1966, pp. 64–65, 126.

Evans, R. K., "Computer Control, Power Plants Join Industry Wide Advance," *Power,* September 1964, pp. 189–194.

Fiedler, H. J., and L. K. Kirchmayer, "Automation Developments in the Control of Interconnected Electric Utility Systems," *Proc. IEEE,* May 1969, pp. 744–750.

Friedlander, G. D., "Computer Controlled Power Systems, Part I—Boiler Turbine Unit Controls," April 1965, pp. 60–92; "Part II—Area Control and Load Dispatch," May 1965, pp. 72–91; *IEEE Spectrum.*

Golze, A. R., "Aqueduct Control System Saves $100 Million," *Power Eng.*, October 1969, pp. 30–35.

Hatton, C., "Controlling a Gas Compressor Station Electronically," *Instrum. and Control Syst.*, May 1967, pp. 145–148.

Howard, J. R., "Experience in DDC Turbine Start-Up," *ISA J.*, July 1966, pp. 61–65.

Humpage, W. D., and T. N. Saha, "Digital-Computer Methods in Dynamic-Response Analyses of Turbogenerator Units," *Proc. IEE,* August 1967, pp. 1115–1130.

Jervis, M. W., and P. R. Maddock, "Electric Power Start Up and Control," *Digital Computer Applications to Process Control,* W. E. Miller (Ed.), Plenum Press, New York, 1965, pp. 251–298.

Jonon, P., "The Use of Digital Computers in French Thermal and Nuclear Power Stations," *Digital Computer Applications to Process Control,* W. E. Miller (Ed.), Plenum Press, New York, 1965, pp. 343–374.

Jonon, P., "Use of Digital Computers in French Conventional and Nuclear Power Stations," *Automatisme,* January 1965, pp. 14–25.

Kruse, H., and R. Quack, "Computer Control of a Power Plant Producing Both Electricity and Heat," *Digital Computer Applications to Process Control,* W. E. Miller (Ed.), Plenum Press, New York, 1965, pp. 319–342.

Kunz, J. M., "Automation of the Alberta-California Natural Gas Pipeline," *ISA Conf. Prepr.,* January 1, 1965, pp. 7 ff.

Lindgren, R. W., "Pipeline Telemetering System," *Instrum. and Control Syst.,* June 1966, pp. 107–111.

Long, F. V., "The Latest Thing in the North Sea—Computer Control from Gas Well to Customer," *1969 ISA Ann. Conf., Pap. 683,* October 1969.

McDaniel, G. H., and C. P. Zimmerman, "Hybrid System Upgrades Dispatching, Speeds Information Processing," *Electron. World,* August 1965, pp. 56–58.

Mary, A. J., F. E. Hevert, and R. E. Smith, "Digital to Analog Computers Optimize Dispatching," *Electr. World,* April 5, 1965, pp. 62–63.

Maurin, L. V., and J. A. Calvo, "Computing Control at Little Gypsy Generating Station," *ISA J.,* May 1966, pp. 34–38.

Meykar, O. A., "Prevention of Power Failures in Major Networks," *IEEE Trans. Aerosp. and Electron. Syst.,* March 1970, pp. 119–128.

Miller, G. S., "Computer Control of Conventional Power Generation," *1969 ISA Ann. Conf., Paper 676,* October, 1969.

Nelson, G. B., "Multiplex Transmission of Digital and Analog Information," *Instrum. and Control Syst.,* October 1965, pp. 111–114.

Quack, R., "Integrated Electricity System Operation," *Digital Computer Applications to Process Control,* W. E. Miller (Ed.), Plenum Press, New York, 1965, pp. 383–408.

Quazza, G., "On Line Digital Computers in Power Stations in Italy," *Digital Computer Applications to Process Control,* W. E. Miller (Ed.), Plenum Press, New York, 1965, pp. 409–426.

Ringlee, R. J., "Sensitivity Methods for Economic Dispatch of Hydroelectric Plants," *IEEE Trans. Autom. Control,* July 1965, pp. 315 ff.

Roberts, J. A., "Application of Digital Control to Power Generation in the United Kingdom," *Digital Computer Applications to Process Control,* W. E. Miller (Ed.), Plenum Press, New York, 1965, pp. 427 ff.

Ruskin, V. W., "Computer Simulation Models for Gas Companies and Systems," *Gas,* February 1968, pp. 60 ff.

Ryan, F., "Advanced Energy Control for Houston Lighting and Power," *Control Eng.,* April 1967, p. 21.

Shackcloth, B., V. R. Robson, and R. S. Hackett, "Town Gas Produced Under Computer Control," *Control Eng.,* November 1969, pp. 112–117.

Skotzki, B. G. A., "Automatic Computer Control, A Time of Trial," *Power,* September 1963, pp. 191–198.

Tuppek, F., "An IBM 1710 Process Computer Controls a Gas Pipeline Network," *Digital Computer Applications to Process Control,* W. E. Miller (Ed.), Plenum Press, New York, 1965, pp. 465–484.

Turbeville, J. F., "Automation Potential for Gas Operations," *ISA Conf. Prepr.,* 1965.

Turbeville, J. F., "Computer System Upgrades Gas Dispatching," *Gas,* January 1968, pp. 37–40.

Willson, G. C., "Remote Supervisory Control and Telemetering," *Instrum. and Control Syst.,* December 1967, pp. 115–117.

"Automatic System Control—Progress with Bristol Scheme," *Electr. Rev. (GB),* January 1966, pp. 123–124.

"Blue Collar Computers Utilities Big Users," *Electron. World,* September 27, 1965, pp. 31–34.

"CEGB Automatic Control System Experiment," *Control,* March 1966, pp. 117–119.

"Computer Control of Power Station," *Eng.,* October 2, 1964, pp. 535–536.

"A Computer To Cope with Catastrophe," *Eng.,* October 9, 1964, pp. 450–461.

"Multi-Computer System for Central Power Stations," *Mech. Eng.,* December 1964, pp. 48–50.

C.5 CEMENT AND GLASS INDUSTRY APPLICATIONS

The first application of a control computer is generally cited as the installation of a computer in a cement plant at Riverside, California, in 1956. Since then, the application of control computers in this industry has continued to increase. Computers are used in all phases of the operation from quarry operation and material blending through kiln control and clinker composition studies. Significant economic returns have been reported.

The application of control computers to the glass industry is somewhat more recent. The systems are used for furnace control and inspection in a number of different glass production processes. They are also used for the control of processes which produce glass fibers.

Aiken, W. S., "Don't Measure—Compute," *ISA J.,* January 1963, pp. 59–60.

Bay, T., *et al.,* "Dynamic Control of the Cement Process with a Digital Computer System," *IEEE Trans. Ind. and Gen. Appl.,* May 1968, pp. 294–303.

Beattie, J. R., "Adapting Automatic Inspection to a Glass Line," *ISA J.,* February 1966, pp. 67–70.

Bedworth, D. D., and J. R. Gaillace, "Instrumenting Cement Plants for Digital Computer Control," *ISA J.,* November 1963, pp. 47–54.

Bernt, J. O., "Automatic Control of Feed Rates to Grinding Mills in the Cement Industry," *Proc. 19th Ann. ISA Conf.,* October 1964.

Butler, K. J., and R. A. Merryweather, "Direct Digital Control of a Pilkington Fibre Glass Furnace," *Proc. 23rd Ann. ISA Conf., Pap. 68-842,* October 1968.

Egger, H., "Design and Conception of Integrated Automated Cement Plants," *1969 Proc. IEEE Cem. Ind. Tech. Conf.,* Toronto, Canada, May 1969.

Gleeman, J. P., *et al.,* "Riverside Puts Kiln Under Closed Loop Computer Control," *Pit & Quarry,* May 1962, pp. 152–166.

Greer, M. J., "Quarry Process Control and Optimization Using a Digital Computer," *Pit & Quarry,* October 1969, pp. 112–115 ff; November 1969, pp. 137–141.

Griem, P. D., "Direct Digital Control of a Glass Furnace," *ISA Conf. Prepr.,* October 1965.

Hay, I., "Computer Control of Cement Kiln Feed Blending," *Instrum. and Control Syst.,* July 1964, pp. 134–138.

Heaton, N. M., "Computer Control System for Cement Manufacture," *Electron.,* December 1968, pp. 492–496.

Kaiser, V. A., and J. W. Lane, "Optimizing Cement Plant Operations with a Centralized Computer Control System," *Proc. 1963 IEEE Cem. Ind. Tech. Conf.,* Cleveland, Ohio, April 1963.

Kaiser, V. A., "Computer Control in the Cement Industry," *Proc. IEEE,* January 1970, pp. 70–77.

Lebel, F., *et al.,* "Computer Direction of Quarry Operations," *Pit & Quarry,* March 1967, pp. 106–108 ff.

Lee, W. G., "Cement Manufacture Under Computer Control," *Lime and Gravel,* March 1963, pp. 79–84.

Levine, S., "Sophisticated Sampling Systems Optimize Computer Operations at Allentown Portland Cement," *Rock Prod.,* April 1967, pp. 67–72.

Mehta, P. K., and M. J. Shah, "A Rapid Method for Determining Compound Composition of Cement Clinker: Application to Closed Loop Kiln Control," *IBM J.,* January 1967, pp. 106–114.

Muldoon, J. F., "Computer Takes Guesswork Out of Cement Blending," *Rock Prod.,* December 1966, pp. 78–80 ff.

Ostberg, W., "Computer Ups Cement Production 5 Percent," *Control Eng.,* September 1966, pp. 136–138.

Patel, N. R., "A DDC Software Package for a Float Glass Process," *Control Eng.,* July 1968, pp. 68–71.

Patel, N. R., "Computer Control of Float Glass at Ford," *IEEE Trans. Ind. & Gen. Appl.,* July 1970, pp. 375–377.

Phillips, R. A., "Automation of a Portland Cement Plant Using a Digital Control Computer," *Proc. 2nd IFAC*, 1963, pp. 347–357.

Raney, A. K., "Computers and Their Application in Glass Manufacturing," *Glass Ind.*, January 1967, pp. 29–32; February 1967, pp. 82–87; March 1967, pp. 137–139.

Romig, J. R., W. R. Morton, and R. A. Phillips, "Making Cement with a Computer System," *Pit & Quarry*, December 1964, pp. 62–67.

Romig, J. R., and W. R. Morton, "Application of a Digital Computer to the Cement Making Process," *Proc. 1st Int. Congr. IFAC-IFIP* (Stockholm, 1964), Plenum Press, New York, 1965, pp. 484–505.

Utley, H. F., "Electronic Computer Controls Cement Production," *Pit & Quarry*, March 1960, pp. 86–91.

Watson, I. M., *et al.*, "Superstaticlogilisticsensoraliocious or Automation in Cement Plants," *IEEE Trans. Ind. and Gen. Appl.*, January 1968, pp. 11–29.

Weeks, L. W., "Control of Kiln Operations with a Digital Computer," *Min. Congr. J.*, June 1964.

Yakas, J. T., "Physical Planning of a Glass Plant Process Control Computer Installation," *IEEE Trans. Ind. and Gen. Appl.*, May 1970, pp. 205–215.

"On Line Control in the Cement Industry," *Chem. Process.*, August 1965, pp. 36–38.

"Scoreboard of Process Control Computers for the Cement Industry," *1969 IEEE Cem. Ind. Tech. Conf.*, Toronto, Canada, May 1969.

"Universal Atlas' New Hannibal, Mo., Plant," *Pit & Quarry*, July 1967, pp. 154–157.

C.6 PAPER, RUBBER, AND TEXTILE INDUSTRY APPLICATIONS

Of these three industries, paper accounts for the majority of the installations. Despite the fact that the paper process is difficult to control because of its complex chemistry and mechanical characteristics, computers are being applied to all phases of the industry. Applications include water treatment, digester and jordan control, steam control, and bleach plant operations. In addition, the control computer is being used to control both the wet and dry ends of Fourdriner and cylinder machines to produce a wide range of products from linerboard to fine paper stock. In addition to control applications, the computer is also used to provide optimum trimming operations of the finished product.

Applications in the rubber industry are limited but include control of synthetic rubber production and the control of manufacturing processes such as tire making. The textile industry utilizes computer for loom control, optimum scheduling of production among different machines, and the control of dying and printing operations.

Bairstow, J. N., "Goal: Five Paper Machines Under Computer Control," *Control Eng.*, January 1969, pp. 120–123.

Bettinger, D., et al., "Computer Speeds Sponge Process," *Rubber World,* September 1968, pp. 43–47.

Bibbs, F. C., "Data Processing in the Corrugated Box Plant," *TAPPI,* March 1965, pp. 105A–107A.

Birkett, R. B., et al., "Process Control for the 1970 Mill," *TAPPI,* October 1968, pp. 83A–89A.

Bowers, R. T., "Mill Computer System Installation Requirements," *TAPPI,* October 1965, pp. 57A–59A.

Brewster, D. B., and A. K. Bjerring, "Computer Control in Pulp and Paper 1961–1969," *Proc. IEEE,* January 1970, pp. 49–69.

Bujak, P. M., et al., "Computerized Tire Fabric Processing," *Rubber Age,* July 1970, pp. 55–59.

Burkhardt, R. C., "Draft Pulp Bleaching Under Computer Control—When Vendor and User Collaborate," *1969 ISA Ann. Conf., Pap. 689,* October 1969.

Cauthen, J. C., Jr., "Computer Control of a Bleach Plant Operation; Progress Report," *Am. Pap. Ind.,* June 1967, pp. 36 ff.

Church, D. F., "Installation of a Control Computer on a Crown Zellerbach Machine," *Pap. Trade J.,* January 31, 1966, pp. 41–44.

Coleman, H. C., and J. H. Smith, "Computer Calibration and Operation of a Karnyr Digestor," *Pap. Ind.,* December 1964, pp. 846–849.

Couture, H. D., Jr., "A Paper Industry Application of Systems Engineering and Direct Digital Control," *1969 ISA Ann. Conf., Pap. 691,* October 1969.

Cross, B., "Control Computers in Papermaking, Results Begin To Show at Potlatch," *Control Eng.,* May 1963, pp. 24–26.

Dahlin, E. B., "Interactive Control of Paper Machines," *Control Eng.,* January 1970, pp. 76–81.

Dahlin, E. B., and R. N. Linebarger, "Digital Simulation Applied to Paper Machine Dryer Studies," *6th ISA Int. Pulp and Pap. Instrum. Symp.,* Green Bay, Wis., May 4–8, 1965.

Davidson, H. R., "Advances in Instrumentation for Shade Matching; Use of Computers," *Am Dyestuff Rep.,* June 19, 1967, pp. 443–445.

Diaz, R. P., "Computer Process Control at International Paper Company," *Pap. Trade J.,* March 7, 1966, pp. 56–57.

Duffin, C. H., "The Trimming of Paper Machines by Computer," *Pulp and Pap. Mag. of Can.,* November 1965, pp. 551–552.

Dyck, A. W. J., "Mead's No. 5 Kingsport Machine Designed for Computer Control," *Am. Pap. Ind.,* November 1966, pp. 26–37.

Dyck, A. W. J., "Process Control at Catawba Newsprint: A Complete Success," *Am. Pap. Ind.,* March 1970, pp. 18–19.

Erard, D., "Control of Pulp and Paper Manufacturing by Means of Computers," *Papeterie,* July 1965, pp. 926–932.

Farrell, R. J., "The Status of Process Control Computers in the Paper Industry," *51st Ann. Mtg. TAPPI,* New York, February 24, 1966.

Fish, K. L., "Computer Application in Steam and Power Reporting," *TAPPI,* December 1965, pp. 87A–89A.

Fritz, J. C., "Determination of Filtration Resistance on a Paper Machine with a Digital Computer Control System," *TAPPI,* March 1965, pp. 99A–102A.

Guillory, A., "Progress Report on Computer Control of EasTex's Bleach Plant," *Pap. Trade J.,* January 31, 1966, pp. 35–36.

Haas, L., "On Line Computer in UK," *Pulp and Pap. Int.,* November 1965, pp. 54–58.

Hiestand, J. H., "Computer Control of Parallel Refiners," *TAPPI,* May 1970, pp. 879–882.

Hodges, G., "Great Southern—New Kraft Linerboard Mill Designed for Computer Control," *Pulp and Pap.,* April 20, 1964, pp. 17–26.

Hollander, A., "Process Computer—Why and How," *TAPPI,* March 1963, pp. 181A–183A.

Hyvarinen, L., "Experiences with an On Line Computer in Papermaking," *Digital Computer Applications to Process Control,* W. E. Miller (Ed.), Plenum Press, New York, 1965, pp. 465–484.

Inserra, P., "Mead Takes Giant Step in Computer Control," *Pulp and Pap.,* May 10, 1965, pp. 30–32.

Kirk, D. E., "An Overview of Computer Control in the Pulp and Paper Industry," *Pap. Trade J.,* January 31, 1966, pp. 34 ff.

Kott, G. L., "Process Computer Control of a Bleach Plant," *TAPPI,* July 1967, pp. 41A–42A.

Kramer, W., "Computer Directs Process Control at Textile Plant," *Control Eng.,* October 1966, pp. 20–21.

Kuhn, H., "Initiating Computer Control in Paper Making," *Res./Dev.,* January 1963, pp. 24–27.

Lavcock, G. H., "Application of Digital Control Computer to the Pulp and Paper Industry," *Process Control and Autom.,* May 1963, pp. 189–194.

Maloney, J. D., "Status of Paper Machine Process Control Computers at Mead Corporation," *Pap. Trade J.,* January 31, 1966, pp. 37–40.

Maloney, J. D., "Some Experiences with Process Control Computers," *Proc. 6th Int. Pulp and Pap. Symp.,* 1965.

McMahon, T. K., "Control Model Formulation for the Pulp and Paper Industry," *TAPPI,* February 1964, pp. 84–88.

Mardon, J., J. E. Barrett, and W. H. Mehaffey, "A System Approach to Paper-machine Instrumentation," *APPITA* (Australia), May 1965, pp. 17–40.

Marton, F. D., "Process Control Systems," *Instrum. and Control Syst.,* April 1965, pp. 131–132.

Mason, R. E. A., "Can Computer Control Be Applied to Board Machines," *Pulp and Pap. Mag. of Can.,* November 1965, pp. 545–550.

Mehaffey, W. H., and J. Mardon, "System Engineering Leads Way to Computer Control of Machines," *Pap. Trade J.,* January 24, 1966, pp. 37–44, 46.

Mulligan, B., "Kelly-Springfield Uses On-Line Computer to Control, Schedule Tire Output," *Rubber World,* December 1967, pp. 41–46.

Ochtman, E., and P. R. Sullivan, "DDC Software for a Paper Machine," *Control Eng.,* November 1968, pp. 90–95.

Oughton, K. D., "Digital Computer Control Paper Machine," *Ind. Electron.,* August 1965, pp. 358–362.

Perrin, L., "Use of Electronic Computers and Automatic Accounting Systems in the Manufacture of Corrugated Cardboard," *Int. Paperboard Ind.,* November 1964, pp. 12–14, 16–17.

Perry, V., D. Kirk, and J. Rolf, "Computer Control of a Fine Paper Machine," *TAPPI,* June 1965, pp. 52A–60A.

Perry, V., and D. Wilder, "Paper Production Up 7 Percent—Thanks to Digital Computer," *Chem. Process.,* September 1965, pp. 46–47, 49.

Rackley, N. G., *et al.,* "Application of Computers in the Paper Industry," *Pap. Technol.,* August 1965, pp. 285–348.

Savas, E. S., "Computer Control in the Paper Mill," *TAPPI,* "Part I: A Management Operating System for the Paper Mill," February 1964, pp. 164A–168A; Part II: Paper Machine Monitoring," March 1964, pp. 149A–152A; "Part IIIA: Operator Guide Control of the Paper Machine," April 1964, pp. 181A–188A; "Part IIIB: Operator Guide Control of the Paper Machine," May 1964, pp. 127A–133A.

Scheffler, B., "The Use of Electronic Computers in the Paper Industry," *Zellst. und Pap.,* July 1964, pp. 1ϟ   ̒97 (in English).

Sidebottom, A. W., "How the Wolvercote Paper Machine Computer Installation Operates," *Pap. Trade J.,* August 16, 1965, pp. 32–33.

Sidebottom, A. W., "Computer Control of Paper-Making Plant," *Proc. IEE,* October 1969, pp. 1755–1758.

Smith, E. J., "A Computerized Pulp and Paper Mill Instrumentation and Control System," *IEEE Trans. Ind. Electron. and Control Instrum.,* April 1966, pp. 10–15.

Smith, W. E., "Mill Operations—2. Plant Engineering for Computer Control," *Pulp and Pap.,* August 5, 1963, pp. 40–44.

Surgen, D. H., "Computers in the Cutting Room: A Perfect Fit," *Control Eng.,* October 1969, pp. 115–118.

Vanderslice, C., "Firestone Applies Computer Control to Its SBR Plant," *Oil and Gas J.,* August 9, 1965, pp. 85–87.

Wardlaw, D. J., "How To Automate a Coating Kitchen," *Pulp and Pap. Int.,* January 1966, pp. 52–53.

Wickstrom, W. A., and M. Horner, "Closed-Loop Color Control for Printing Papers," *TAPPI,* May 1970, pp. 784–791.

Wise, C. T., E. R. Tims, and J. F. Steedley, "Automated Process Water Treatment," *TAPPI,* September 1965, pp. 129A–131A.

Wurmser, D., "An Automation Tool in the Paper Industry—The IBM 1800 Computer," *Papeterie,* May 1965, pp. 552–562.

"AATT Hears How Computers Can Aid in Color Matching," *Am. Dyestuff Rep.,* March 25, 1968, pp. 217–218.

"Closed Loop On Line Computer in Paper Industry," *Instrum. Pract.,* April 1964, pp. 395–396.

"Computer Control—Country by Country Reports," *Pulp and Pap. Int.,* November 1965, pp. 59–64.

"Computer Control Is at Hand in Pulping," *Can. Chem. Process,* November 1967, pp. 82–83 ff.

"Computer Control of Continuous Mill in Alabama," *Eng.,* January 17, 1964, pp. 154–157.

"Computer Control of Paper Making," *Process Control and Autom.,* May 1965, pp. 213–215.

"Computer Directed Control System," *Mod. Text. Mag.,* April 1967, pp. 37–38ff.

"Computerized Textile Dyeing," *Instrum. and Control Syst.,* February 1967, pp. 139–140.

"Mead's Computer Control System on Paper Machine Working Well," *South. Pulp and Pap. Mfr.,* July 10, 1965, pp. 76–82.

"Papermaking No Longer an Art," *Chem. and Eng. News,* May 24, 1965, pp. 58–60.

"Predictive Quality Control," *Meas. and Control,* January 1964, pp. 20–25.

"Wolvercote Computer Installation a Major Breakthrough," *World Pap. Trade Rev.,* April 8, 1965, pp. 103 ff.

C.7   GENERAL APPLICATIONS

The references in this category demonstrate the wide spectrum of computer control usage. They also illustrate the difficulty in defining the bounds on computer process control. Many of the applications in this category do not fall within the bounds of the classical process control definition, as is the case with most of the applications cited in previous sections. Nevertheless, applications such as antenna control exhibit the same characteristics and have many of the same requirements as the control of continuous industrial processes. Because of the wide spectrum of these applications, the bibliographic entries of this chapter merely provide an overview of some of the areas in which control computers are being utilized.

Applications include the control of discrete processes and phenomena such as machine shop control, traffic control in metropolitan areas, and control of railroad yards and systems. Engine testing and medical applications such as intensive care of patients are examples of continuous processes which closely resemble some industrial processes. Medical applications, in particular, are receiving considerable attention in the technical literature and significant applications of computers in this area are being reported.

The food industry is another example of process control that is often not considered; yet, it has many of the problems and characteristics of the more traditional application problems. Computers are being used to control the blending of ingredients, the baking and cooling processes, and the packaging operation. Specific examples include the control of beer brewing, potato chip machines, and coffee roasting ovens.

Abegglen, P. C., W. R. Faris, and W. J. Hankley, "Design of a Real-Time Central Data Acquisition and Analysis System," *Proc. IEEE,* January 1970, pp. 38–48.

Ainslie, T. C., and J. J. Steranko, "Computer Controlled Manufacturing Line," *Autom.,* January 1967, pp. 66–74.

Aronson, R. L., "Machine Monitoring by Process Computer Pays Off in Appliance Manufacture," *Control Eng.,* February 1971, pp. 68–69.

Bailey, S. J., "Materials Handling Control," *Control Eng.,* September 1968, pp. 81–89.

Bailey, S. J., "On-Line Computer Users Polled," *Control Eng.,* January 1969, pp. 86–94.

Bairstow, J. N., "What's Wrong with Urban Traffic Control," *Control Eng.,* February 1969, pp. 78–84.

Barnett, G. O., and J. H. Hughes, "On the Horizon—Better Hospital Care Through Computer Time Sharing," *Electron.,* January 24, 1966, pp. 93–98.

Barnes, L. B., "A Computer Tests and Adjusts Automobile Carburetors," *Control Eng.,* January 1969, pp. 95–99.

Birrell, K. E., "Computers Are Being Applied to Dimensional Measurement in Production of Discrete Parts," *J. SAE,* November 1969, pp. 32–38.

Bobroff, D. A., "Computer Controlled Testing," *Autom.,* March 1970, pp. 118–120.

Bogod, J. L., "Computer Application in Hospitals," *Ind. Electron.,* March 1967, pp. 98–103.

Bollinger, J. G., "Will Computers Help Dynamics Match Control," *Iron Age,* September 28, 1967, pp. 97–98.

Briggs, P. P., "Computer-Controlled Chromatographs," *Control Eng.,* September 1967, pp. 75–80.

Brodovicz, B. A., *et al.,* "Pennsylvania's Computerized Air Monitoring System." *Air Poll. Control Assoc. J.,* July 1969, pp. 484–489.

Clancy, J. J., and M. S. Fineberg, "Hybrid Computing—A User's View," *Simulation,* August 1965, pp. 104–112.

Cole, M. H., L. E. Wise, and J. Henderson, "Real Time Engine Test System," *Instrum. and Control Syst.,* June 1968, pp. 95–99.

Cox, G. E., and D. H. Ralston, "The Application of Computers to Shop Control," *Prod. Eng.,* January 1967.

Currie, R., "First Computer-Controlled Injection Machine," *Mod. Plast.,* October 1968, pp. 122–124 ff.

Davis, J. F., "Computers in Medicine," *Int. Sci. and Tech.,* December 1966, pp. 40–48.

Dexter, C. V., and M. L. Larsen, "Centralized Large-Scale Clinical Testing in a Commercial Environment," *Proc. IEEE,* November 1969, pp. 1988–1995.

DiCicco, J. J., "Computer Organized Data Flow Smooths Chrysler's Assembly Operations," *Control Eng.,* September 1966, pp. 139–141.

Elbrond, J., "Optimized Rail Transport in the Kiruna Mine of LKAB," *Digital Computer Applications to Process Control,* W. E. Miller (Ed.), Plenum Press, New York, 1965, pp. 543–556.

Feinberg, B., "Computer Boosts Transfer-Line Efficiency," *Tool & Manuf. Eng.,* October 1969, pp. 14–16.

Felton, H. R., H. A. Hancock, and J. L. Knupp, Jr., "Integrated Computer/Gas Chromatograph System," *Instrum. and Control Syst.*, August 1967, pp. 83–85.

Fichtenbaum, M., "Computer-Controlled Testing Can Be Fast and Reliable and Economical Without Extensive Operator Training," *Electron.*, January 19, 1970, pp. 82–86.

Flanagan, T. P., "Bakers Prove Receptive to Control Engineering Ideas," *Control,* April 1966, pp. 173–175.

Fleet, J. J., "Traffic Automation, Part I," *Data and Control,* January 1966, pp. 20–23; Part II: February 1966, pp. 24–27.

Fraade, D. J., and R. S. Davis, "Centralized Process and Chromatograph Control Using Digital Computers," *Instrumentation in the Chemical and Petroleum Industries—Vol. 4,* W. H. Matthews (Ed.), Plenum Press, New York, 1968, pp. 53–60.

Friedlander, G. D., "Computer-Controlled Vehicular Traffic," *IEEE Spectrum,* February 1969, pp. 30–43.

Greene, P., "Digital Control of a Nuclear Reactor," *Instrum. and Control Syst.,* June 1965, pp. 85 ff.

Gomolak, L. S., "Take Two Tons of Flour," *Electron.,* April 20, 1964, pp. 82–89.

Heart, F. E., P. D. Smith, and A. A. Mathiasen, "The Haystack Computer Control System," *Tech. Rep.,* MIT Lincoln Lab., 1965.

Heumann, G. W., and W. Kramer, "Digital Computers Rev Up Traffic Control," *Control Eng.,* August 1967, pp. 95–100.

Hilado, C. J., "Computer Applications in Material Evaluation Programs," *Ind. & Eng. Prod. Res. & Dev.,* December 1966, pp. 301–305.

Hodges, J. D., Jr., and D. W. Whitehead, "Automatic Traffic Signal Control Systems—The Metropolitan Toronto Experience," *Proc. SJCC,* 1969, pp. 529–535.

Holland, T., "London's Computerized Traffic," *Eng.,* March 7, 1969, pp. 374–375.

Hope, G. S., and S. J. Jura, "Detecting Transmission Line Abnormalities by Computer," *Instrum. Technol.,* February 1971, pp. 39–45.

Horgan, J. D., "Computer Applications in Medicine," *Proc. 23rd ISA Conf., Pap. 68-979,* 1968.

Hovey, R. K., "Integrated Large Scale Analog to Digital Control Systems," *Proc. ISA,* 1963, pp. 63.1–63.12.

Humphreys, K. K., *et al.,* "Computers for Coal," *Coal Age,* June 1969, pp. 87–88 ff.; September 1969, pp. 88–91 ff.

Hutchins, D. L., "A General Laboratory Computing Facility," *1969 ISA Ann. Conf., Pap. 506,* October 1969.

Jagt, D. J., "Your Computer as an Instrument Maintenance Tool," *Instrum. Technol.,* July 1968, pp. 56–59.

John Diebold & Assoc., "Numerical Control—The Changing Market," *Data and Control,* July 1965, pp. 26–28.

Jurgen, R. K., "Minicomputer Applications in the Seventies," *IEEE Spectrum,* August 1970, pp. 37–52.

Keenan, T. A., "Computer Controls Stacker Cranes," *Autom.,* January 1967, p. 40.

Khol, R., "Controlling Machines by Computer," *Mach. Des.,* July 23, 1970, pp. 118–125.

Kohnert, D., and A. Musiol, "Traffic Signal Control by Computer," *Control Eng.*, June 1969, pp. 115–120.

Kramer, W., "Control is TOPS at Southern Pacific," *Control Eng.*, November 1966, pp. 91–93.

Kruse, W., "Computerized Weighing System," *Chem. & Process Eng.*, September 1969, pp. 103–104 ff.

Landoll, J. R., and C. A. Caceres, "Automation of Data Acquisition in Patient Testing," *Proc. IEEE*, November 1969, pp. 2017–2035.

Lapidus, G., "A Look at Minicomputer Applications," *Control Eng.*, November 1969, pp. 82–91.

Laspe, C., "Look at Computer Control in the Process Industries," *IEEE Trans. Ind. and Gen. Appl.*, January 1967, pp. 11–14.

Lauer, G., *et al.*, "Electrochemical Data Acquisition and Analysis System Based on a Digital Computer," *Anal. Chem.*, June 1967, pp. 765–769.

Lauria, J., and J. F. Polizzotto, "Computer Controls LSI Mask Making," *Control Eng.*, January 1969, pp. 100–102.

Layer, W. J., "Computer Applications for Parameter Control and Data Processing in a Metrology Laboratory," *IEEE Trans. Instrum. and Meas.*, December 1969, pp. 294–299.

Lees, L. H., and F. G. Willetts, "An Advanced Data Handling System for Environmental Testing of Satellites," *IEEE Trans. Ind. Electron. and Control Instrum.*, August 1970, pp. 384–391.

Lippitt, D. L., and D. A. Swann, "Industrial Process Control by Digital Computer," *ISA Trans.*, October 1965, pp. 319ff.

Malim, T. H., "Process Control Computers: The Controversy over Control," *Iron Age*, March 2, 1967, pp. 61–68.

Mann, P. J., "Computerized Control: A Progress Report," *Mater. Handl. Eng.*, December 1968, pp. 70–72.

McCrea, A. F., "Features of a Computerized Building Automation," *1969 ISA Ann. Conf., Pap. 508*, October 1969.

Mears, F. C., "Success Is Word for Computer-Controlled Chromatograph Unit," *Oil Gas J.*, December 18, 1967, pp. 90–93.

Michael, P. C., "Small Computer Speeds Circuit Testing," *Electron Eng.*, December 1970, pp. 56–59.

Miller, W. E., *Digital Computer Applications to Process Control*, Plenum Press, New York, 1965.

Mori, M., and S. Yamashita, "Sequence Control System Runs Fermentation Batches," *Control Eng.*, July 1967, pp. 66–70.

Mouly, R. J., "Integrated Process Control," *Instrumentation in the Chemical and Petroleum Industries—Vol. 4*, W. H. Matthews (Ed.), Plenum Press, New York, 1968, pp. 31–36.

Murray, R. W., "Computers Bid for On-Line Inspection," *Iron Age*, October 26, 1967, pp. 79–81.

Noonan, R. P., "Computer Control of Materials Handling Systems," *Proc. Texas A&M 20th Ann. Symp. Instrum. for the Process Ind.*, January 1965, pp. 23–31.

Noonan, R. P., "Testing and Inspecting with Computers," *Autom.*, August 1970, pp. 27–31.

Ohlsen, G. T., "Digital Control of Machine Tools," *Electron. Equip. News,* December 1965, pp. 30–36.

Opladen, H. B., "Computer-Optimized Fire Reduces Air Pollution," *Instrumen. Technol.* August 1968, pp. 63–66.

Papaioannou, A. C., and G. P. Trearchis, "Control Food Processes with Digital Computer?" *Food Eng.,* November 1967, pp. 67–71.

Peters, R. A. P., and F. Thumser. "On-Line Production Tracking Aids Integrated Management System," *Control Eng.,* January 1970, pp. 89–91.

Puncochar, G. A., "The First Computer Directed Food Plant," *Proc. ISA,* 1963, p. 7 ff.

Rader, L. T., "Future of Computers in Manufacturing," *Autom.,* June 1970, pp. 50–55.

Ralston, E. L., "On-Line Computerized Control," *Mater. Handl. Eng.,* December 1968, pp. 73–75.

Renouard, C. A., "A Manufacturing Information and Control System Design," *Control Eng.,* September 1970, pp. 78–81.

Rosenbaum, M., "Implementing Computer Control at the Baytown Fuels Control Center," *1969 ISA Ann. Conf., Pap. 544,* October 1969.

Rout, F. P., "Control Trends in the Brewing and Allied Industries," *Instrum. Pract.,* April 1964, pp. 335–339.

Ryan, W. J., "Tailoring Chromatographs to a Monitoring Computer," *Instrum. Technol.,* November 1969, pp. 57–62.

Sanders, R. P., "New Data-Gathering and Control System," *IEEE Trans. Comm. Tech.,* June 1967, pp. 444–449.

Savas, E. S., *Computer Control of Industrial Processes,* McGraw-Hill, New York, 1965.

Schaffer, G., "NC and the Computer," *Am. Mach.,* March 10, 1969, pp. 145–153.

Schmitz, W., "Some Main Problems Involved in the Use of Computers for the Automation of Railway Operating Procedures," *Digital Computer Applications to Process Control,* W. E. Miller, (Ed.), Plenum Press, New York, 1965, pp. 517–542.

Schumann, P. A., "The Computer in Environmental Control Systems, *1969 ISA Ann. Conf., Pap. 509,* October 1969.

Shergalis, L. D., "Data Acquisition System Expects the Unexpected," *Electron.,* June 1, 1964, pp. 57–62.

Smith, J. K., "Chrysler Instantaneous Quality Reporting," *Ind. Qual. Control,* January 1967, pp. 325–328.

Smith, K. F., and K. D. Smith, "Computer Controlled Metal Thick Oxide Silicon Large Scale Integrated Circuits Tester," *Comput. Des.,* September 1968, pp. 51–57.

Stark, D. E., and R. A. Sylvester, "A Multi-Station Remote Control and Data Handling System with Flexibility," *Proc. ISA,* 1963, pp. 6 ff.

Stevens, A. H., "Computer Controls Freezer Storage," *Food Eng.,* January 1969, pp. 82–84.

Thompson, A., "Operating Experience with Direct Digital Control," *Proc. 1st*

Int. Conf. IFAC-IFIP (Stockholm, 1964), Plenum Press, New York, 1965; also *Chem. Eng.,* May 1965, pp. 96–101.

Van Dyke, R. P., "Computerized Operation of Distribution Systems," *J. Am. Water Works Assoc.,* April 1970, pp. 234–240.

Wherry, T. C., *et al.,* "Process Controls and Computers, Direct Digital Control," *Chem. Eng. Progr.,* April 1968, pp. 33–38.

Whitehead, D. W., "Toronto System: Intersection Evaluation and Control," *Traffic Eng.,* September 1969, pp. 28–30.

Whitman, K., "Direct Digital Control," *Chem. Eng.,* December 5, 1966, pp. 135–138; January 2, 1967, pp. 81–86.

Wick, C. H., "Economical Direct Computer Control for Machine Tools," *Mach.,* March 1970, pp. 53–56.

Wick, C. H., "Computer Control Speeds Automobile Parts Machining," *Mach.,* March 1970, pp. 69–71.

Williams, T. J., "Development of Computer Control in the Continuous Process Industries," *Proc. IEEE,* December 1966, pp. 1751–1756.

Williams, T. J., "Computers and Process Control: Annual Review," *Ind. & Eng. Chem.,* December 1966, pp. 55–70.

Williams, T. J., "Computers and Process Control: Annual Review," *Ind. & Eng. Chem.,* December 1967, pp. 53–68.

Williams, T. J., "Computers and Process Control: Annual Review," *Ind. & Eng. Chem.,* January 1969, pp. 76–89.

Williams, T. J., "The Coming Years . . . The Era of Computing Control," *Instrum. Technol.,* January 1970, pp. 57–63.

Williams, T. J., "Computers and Process Control: Annual Review," *Ind. & Eng. Chem.,* February 1970, pp. 28–40.

Williamson, D. T. N., "Integrated Manufacturing Control," *Control Eng.,* September 1967, pp. 66–74.

Wulfmeyer, H., "Centralized Control in the Coffee Roasting Process," *Regelungstech. Prax.,* January 1966, pp. 9–11.

"Computers in Food Technology," *Food Technol.,* June 1967, pp. 816ff.; January 1968, pp. 44–47.

"Computers Start to Study Medicine," *Bus. Week,* July 9, 1966, pp. 142–148.

"Computer to Control West London Traffic," *Process Control and Autom.,* January 1966, pp. 48–50.

"In Three Seconds, Complete Shop Data; Martin Marietta Corp.," *Mod. Mater. Handl.,* September 1968, pp. 50–55.

"Over 800 Areas of Application of Computers," *Comput. and Autom.,* June 1965, p. 83.

C.8 RESEARCH APPLICATIONS

The applications grouped under the category of research exhibit the same diversity of subject as those considered in the previous section. Many of them deal with subjects related to the specific application area where computer control is well established; others are concerned with topics that are the precursors of future industry applications. In this section, a small sample of the papers dealing with research is presented. The list is not exhaustive since the range of applications and the boundaries of the category are difficult to define. Nevertheless, they represent important contributions to the application of control computers and indicate the direction of current utilization of the real-time, sensor-based computer system.

Almasy, G., "The Use of Computers for Research in the Petroleum Industry," *Int. Chem. Eng.*, October 1965, pp. 590ff.

Brown, R. M., *et al.*, "The SLAC High-Energy Spectrometer Data Acquisition and Analysis System," *Proc. IEEE*, December 1966, pp. 1730–1751.

Brown, D. W., and D. S. Groome, "On-Line Computer for Nuclear Medicine," *Electro-Technol.* September 1969, pp. 53–54.

Bullis, W. M., *et al.*, "Time Shared Computers in Experimental Environments," *Instrum. and Control Syst.*, October 1970, pp. 105–110.

Cowley, P. E. A., "Pilot Plant Puts DDC to the Test," *Control Eng.*, February 1966, pp. 53–56.

Denes, P. B., and M. V. Mathews, "Laboratory Computers; Their Capabilities and How To Make Them Work for You," *Proc. IEEE*, April 1970, pp. 520–530.

Frederick, D. E., and T. Marshall, "A Computer Based Multiparameter Analyzer," *IEEE Trans. Nucl. Sci.*, February 1966, pp. 144–150.

Gardiner, T., and J. Levin, "The LASL Tandem Accelerator Facility Data System," *IEEE Trans. Nucl. Sci.*, February 1966, pp. 151–157.

Gingell, C. E. L., *et al.*, "A Digital Control System for Nuclear Physics Laboratory," *IEEE Trans. Nucl. Sci.*, February 1969, pp. 165–170.

Gladney, H. M., B. F. Dowden, and J. D. Swalen, "Computer-Assisted Gas-Liquid Chromatography," *Anal. Chem.*, June 1969, pp. 883–888.

Gossling, T. H., "Digital Computer Control in Radio Astronomy," *New Sci.*, December 26, 1963, pp. 778–779.

Greene, P. J., "Direct Digital Control of a Nuclear Reactor," *Instrum. and Control Syst.*, June 1965, pp. 85–88.

Kaplow, R., and J. Posen, "On-Line Computer Analysis and Control of Experiments," *J. Appl. Phys.*, November 1969, pp. 4948–4953.

Katona, P. G., "Automated Chemistry Laboratory; Application of a Novel Time-Shared Computer System," *Proc. IEEE*, November 1969, pp. 2000–2006.5.

Kracht, U., "Data Logging on a Proton Linear Accelerator and a Description of the Analog to Digital Converter Used," *Nucl. Instrum. and Methods*, December 1965, pp. 231–239.

Lewis, L. G., "Computer Control of High Energy Accelerators," *IEEE Trans. Nucl. Sci.*, June 1965, pp. 1–8.

Lewis, J. G., L. K. Holland, and R. L. Holladay, "On Line Computer at Big Rock Point," *Nucleon.*, November 1964, pp. 46–47.

Lindenbaum, S. J., "The On Line Computer Counter and Digitalized Spark Chamber Technique," *Phys. Today,* April 1965, pp. 19–24, 26–28.

Ramaley, L., and G. S. Wilson, "General Purpose System for Computer Data Acquisition and Control," *Anal. Chem.,* May 1970, pp. 606–611.

Robinson, L., "Applications of Small Computers in Nuclear Spectroscopy," *IEEE Trans. Nucl. Sci.,* February 1966, pp. 161–166.

Stuckert, P. E., "Computer Augmented Oscilloscope System," *IEEE Trans. Instrum. and Meas.,* December 1969, pp. 299–306.

Williams, T. J., "University Research Results for Industrial Process Control," *Instrum. Technol.,* November 1967, pp. 49–56.

"Development of Techniques for the Automatic Control of Experiments in a Psychology Laboratory," *U.S. Gov. Rep. AD 614793,* February 1965.

"On Line Computers in Nuclear Research," *Nucleon.,* January 1967, pp. 34–40; February, pp. 45–51; May, pp. 64–71.

INDEX

1035